Lecture Notes in Artificial Intelligence 1856

Subseries of Lecture Notes in Computer Science
Edited by J. G. Carbonell and J. Siekmann

Lecture Notes in Computer Science
Edited by G. Goos, J. Hartmanis and J. van Leeuwen

Springer
*Berlin
Heidelberg
New York
Barcelona
Hong Kong
London
Milan
Paris
Singapore
Tokyo*

Manuela Veloso   Enrico Pagello
Hiroaki Kitano (Eds.)

RoboCup-99: Robot Soccer World Cup III

 Springer

Series Editors

Jaime G. Carbonell, Carnegie Mellon University, Pittsburgh, PA, USA
Jörg Siekmann, University of Saarland, Saabrücken, Germany

Volume Editors

Manuela Veloso
Carnegie Mellon University, School of Computer Science
Computer Science Department, Pittsburgh, PA 15213-3890, USA
E-mail: veloso@cs.cmu.edu

Enrico Pagello
The University of Padua, Department of Electronics and Informatics (DEI)
Via Gradenigo 6/a, 35131 Padova, Italy
E-mail: epv@dei.unipd.it

Hiroaki Kitano
Sony Computer Science Laboratories, Inc.
3-14-13 Higashi-Gotanda, Shinagawa, Tokyo 141-0022, Japan
E-mail: kitano@csl.sony.co.jp

Cataloging-in-Publication Data applied for

Die Deutsche Bibliothek - CIP-Einheitsaufnahme

RoboCup <3, 1999, Stockholm>:
Robot Soccer World Cup III / RoboCup-99. Manuela Veloso ... (ed.). -
Berlin ; Heidelberg ; New York ; Barcelona ; Hong Kong ; London ;
Milan ; Paris ; Singapore ; Tokyo : Springer, 2000
 (Lecture notes in computer science ; Vol. 1856 : Lecture notes in
 artificial intelligence)
 ISBN 3-540-41043-0

CR Subject Classification (1998): I.2, C.2.4, D.2.7, H.5, I.5.4, I.6, J.4

ISBN 3-540-41043-0 Springer-Verlag Berlin Heidelberg New York

This work is subject to copyright. All rights are reserved, whether the whole or part of the material is
concerned, specifically the rights of translation, reprinting, re-use of illustrations, recitation, broadcasting,
reproduction on microfilms or in any other way, and storage in data banks. Duplication of this publication
or parts thereof is permitted only under the provisions of the German Copyright Law of September 9, 1965,
in its current version, and permission for use must always be obtained from Springer-Verlag. Violations are
liable for prosecution under the German Copyright Law.

Springer-Verlag Berlin Heidelberg New York
a member of BertelsmannSpringer Science+Business Media GmbH
© Springer-Verlag Berlin Heidelberg 2000
Printed in Germany

Typesetting: Camera-ready by author
Printed on acid-free paper   SPIN: 10722175   06/3142   5 4 3 2 1 0

Preface

RoboCup-99, the Third Robot World Cup Soccer Games and Conferences, was held July 27 to August 6, 1999 in Stockholm, Sweden. RoboCup-99 included a technical workshop and competitions in four different leagues, namely the simulation league, the small-size (F180) robot league, the middle-size (F2000) robot league, and, for the first time officially, the Sony legged robot league. RoboCup-99 also included demonstrations of RoboCup, Jr., especially designed for children's participation.

This book consists of four parts: (i) overview; (ii) the research papers of the champion teams and of the scientific award winners; (iii) the technical papers presented at the RoboCup-99 workshop; and (iv) the team descriptions of a large number of the competing teams.

The first chapter contributes a comprehensive overview of RoboCup-99, mainly introducing the leagues and the research challenges presented to the competition participants. It also includes the results of all of the games played in the RoboCup-99 competitions.

The RoboCup-99 workshop was a very successful event with more than 120 participants. It is indeed a great technical challenge to completely develop a team of agents, software or robots, to actually carry out a soccer game. At the RoboCup-99 workshop, and as presented in this book, the research papers addressed a rich spectrum of topics, including: machine vision, multiagent system architectures, real-time perception, sensing and reasoning, small-sized robotics, autonomous localization, team coordination, control learning, opponent modeling, adaptation to the opponent, robotics education, graphics, simulation, and visualization. The scientific challenge award was given to work on the automated statistical analysis of games.

The competing teams contributed to this book technical research papers presented at the workshop and, or, focused descriptions of their teams. These papers are of great value to the other competitors and to general observers. The RoboCup-99 final competitions were attended by a very large crowd of interested, curious, and enthusiastic researchers, both RoboCup-99 and IJCAI-99 participants. We expect this book to be of particular interest to all who participated and attended RoboCup-99.

The next steps for RoboCup are the first RoboCup European Championship in Amsterdam in May 2000, and the International RoboCup-2000 in Melbourne, Australia, in August 2000. We are strongly interested in the transfer of issues and solutions from robotic soccer to other tasks. Hence, we are explicitly researching in other multiagent domains, such as robotics and education, search and rescue environments, and general entertainment robotics. The future RoboCup events will continue our on-going effort to be a forum for a rich spectrum of demonstrations and innovative contributions to challenging scientific problems.

June 2000   Manuela Veloso, Enrico Pagello, and Hiroaki Kitano

RoboCup-99 Organization and Support

General Chair:
Silvia Coradeschi, Linköping University, Sweden

Vice Chairs:
Magnus Boman, Stockholm University, Sweden
Lars Karlsson, Linköping University, Sweden

Organizing Committee:
Minoru Asada, Osaka University, Japan
Dominique Duhaut, Laboratoire de Robotique de Paris, France
Wei-Min Shen, University of Southern California, U.S.A.

Simulation League Committee:
Peter Stone (chair), Carnegie Mellon University, U.S.A.
Hans-Dieter Burkhard, Humboldt University of Berlin, Germany
Itsiku Noda, ETL, Japan
Paul Scerri, Linköping University, Sweden

Small-Size Robot League (F-180) Committee:
Tucker Balch (chair), Carnegie Mellon University, U.S.A.
Paulo Costa, Faculdade de Engenharia da Universidade do Porto, Portugal
Yuki Nakagawa, Tokyo Institute of Technology, Japan
Christian Balkenius, Lund University Cognitive Science, Sweden

Middle-Size Robot League (F-2000) Committee:
Gerhard Kraetzschmar (chair), University of Ulm, Germany
Andrew Jennings, Royal Melbourne Institute of Technology, Australia
Sho'ji Suzuki, Osaka University, Japan
Christian Balkenius, Lund University Cognitive Science, Sweden

Sony Legged Robot League Committee:
Masahiro Fujita (chair), Sony Inc., Japan
Minoru Asada, Osaka University, Japan
Dominique Duhaut, Laboratoire de Robotique de Paris, France
Manuela Veloso, Carnegie Mellon University, U.S.A.

Workshop Co-chairs:
Manuela Veloso, Carnegie Mellon University, U.S.A.
Enrico Pagello, The University of Padua, Italy
Hiroaki Kitano, Sony CSL, Japan

Supported by:
Linköping Univ., Stockholm Univ., The Royal Institute of Technology
Sun Microsystems, Inc. (official and exclusive computer system sponsor)
The Foundation for Knowledge and Competence Development
The Swedish Council for Planning and Coordination of Research
The Swedish Foundation for Strategic Research
NUTEK, WITAS, First Hotel

RoboCup World Wide Sponsors:
Sony Corporation

Table of Contents

Overview of RoboCup-99 1
  Manuela Veloso, Hiroaki Kitano, Enrico Pagello, Gerhard Kraetzschmar, Peter Stone, Tucker Balch, Minoru Asada, Silvia Coradeschi, Lars Karlsson, and Masahiro Fujita

Champion Teams

The CMUnited-99 Champion Simulator Team 35
  Peter Stone, Patrick Riley, and Manuela Veloso

Big Red: The Cornell Small League Robot Soccer Team 49
  Raffaello D'Andrea, Jin-Woo Lee, Andrew Hoffman, Aris Samad-Yahaja, Lars B. Cremean, and Thomas Karpati

Middle Sized Soccer Robots: ARVAND 61
  M. Jamzad, A. Foroughnassiraei, E. Chiniforooshan, R. Ghorbani, M. Kazemi, H. Chitsaz, F. Mobasser, and S. B. Sadjad

Vision Based Behavior Strategy to Play Soccer with Legged Robots 74
  Vincent Hugel, Patrick Bonnin, Ludovic Raulet, Pierre Blazevic, and Dominique Duhaut

Scientific Challenge Award Papers

Automated Assistants to Aid Humans in Understanding Team Behaviors .. 85
  Taylor Raines, Milind Tambe, and Stacy Marsella

LogMonitor: From Player's Action Analysis to Collaboration Analysis and Advice on Formation 103
  Tomoichi Takahashi

A Statistical Perspective on the RoboCup Simulator League: Progress and Prospects 114
  Kumiko Tanaka-Ishii, Ian Frank, Itsuki Noda, and Hitoshi Matsubara

Technical Papers

Real-Time Color Detection System Using Custom LSI for High-Speed Machine Vision 128
  Junichi Akita

A Segmentation System for Soccer Robot Based on Neural Networks 136
  Carmelo Amoroso, Antonio Chella, Vito Morreale, and Pietro Storniolo

Practical Camera and Colour Calibration for Large Rooms 148
  Jacky Baltes

Path-Tracking Control of Non-holonomic Car-Like Robot with Reinforcement Learning 162
  Jacky Baltes and Yuming Lin

Fast Image Segmentation, Object Recognition and Localization in a
RoboCup Scenario 174
  Thorsten Bandlow, Michael Klupsch, Robert Hanek, and Thorsten Schmitt

Using Hierarchical Dynamical Systems to Control Reactive Behavior 186
  *Sven Behnke, Bernhard Frötschl, Raúl Rojas, Peter Ackers,
  Wolf Lindstrot, Manuel de Melo, Andreas Schebesch, Mark Simon,
  Martin Sprengel, and Oliver Tenchio*

Heterogeneity and On-Board Control in the Small Robots League 196
  Andreas Birk and Holger Kenn

The Body, the Mind or the Eye, First? 210
  Andrea Bonarini

Motion Control in Dynamic Multi-Robot Environments 222
  Michael Bowling and Manuela Veloso

Behavior Engineering with "Dual Dynamics" Models and Design Tools ... 231
  *Ansgar Bredenfeld, Thomas Christaller, Wolf Göhring, Horst Günther,
  Herbert Jaeger, Hans-Ulrich Kobialka, Paul-Gerard Plöger, Peter Schöll,
  Andrea Siegberg, Arend Streit, Christian Verbeek, and Jörg Wilberg*

Techniques for Obtaining Robust, Real-Time, Colour-Based Vision
for Robotics 243
  James Brusey and Lin Padgham

Design Issues for a Robocup Goalkeeper 254
  Riccardo Cassinis and Alessandro Rizzi

Layered Reactive Planning in the IALP Team 263
  Antonio Cisternino and Maria Simi

From a Concurrent Architecture to a Concurrent Autonomous Agents
Architecture 274
  Augusto Loureiro da Costa and Guilherme Bittencourt

Tracking and Identifying in Real Time the Robots of a F-180 Team 286
  *Paulo Costa, Paulo Marques, António Moreira, Armando Sousa,
  and Pedro Costa*

VQQL. Applying Vector Quantization to Reinforcement Learning 292
  Fernando Fernández and Daniel Borrajo

Fast, Accurate, and Robust Self-Localization in the RoboCup
Environment 304
  Jens-Steffen Gutmann, Thilo Weigel, and Bernhard Nebel

Self-Localization in the RoboCup Environment 318
  Luca Iocchi and Daniele Nardi

Virtual RoboCup: Real-Time 3D Visualization of 2D Soccer Games 331
  Bernhard Jung, Markus Oesker, and Heiko Hecht

The RoboCup-98 Teamwork Evaluation Session: A Preliminary Report ... 345
  Gal A. Kaminka

Towards a Distributed Multi-Agent System for a Robotic Soccer Team ... 357
  Nadir Ould Khessal

A Multi-threaded Approach to Simulated Soccer Agents for the
RoboCup Competition 366
  Kostas Kostiadis and Huosheng Hu

A Functional Architecture for a Team of Fully Autonomous
Cooperative Robots 378
  Pedro Lima, Rodrigo Ventura, Pedro Aparício, and Luís Custódio

Extension of the Behaviour Oriented Commands (BOC) Model
for the Design of a Team of Soccer Players Robots 390
  C. Moreno, A. Suárez, E. González, Y. Amirat, and H. Loaiza

Modular Simulator: A Draft of New Simulator for RoboCup 400
  Itsuki Noda

Programming Real Time Distributed Multiple Robotic Systems 412
  Maurizio Piaggio, Antonio Sgorbissa, and Renato Zaccaria

The Attempto RoboCup Robot Team 424
  *Michael Plagge, Richard Günther, Jörn Ihlenburg, Dirk Jung,
  and Andreas Zell*

Rogi Team Real: Dynamical Physical Agents 434
  *Josep Lluís de la Rosa, Bianca Innocenti, Israel Muñoz, Albert Figueras,
  Josep Antoni Ramon, and Miquel Montaner*

Learning to Behave by Environment Reinforcement 439
  Leonardo A. Scardua, Anna H. Reali Costa, and Jose Jaime da Cruz

End User Specification of RoboCup Teams 450
  Paul Scerri and Johan Ydrén

Purposeful Behavior in Robot Soccer Team Play 460
  Wei-Min Shen, Rogelio Adobbati, Jay Modi, and Behnam Salemi

Autonomous Information Indication System 469
  Atsushi Shinjoh and Shigeki Yoshida

Spatial Agents Implemented in a Logical Expressible Language 481
  Frieder Stolzenburg, Oliver Obst, Jan Murray, and Björn Bremer

Layered Learning and Flexible Teamwork in RoboCup Simulation Agents 495
  Peter Stone and Manuela Veloso

A Method for Localization by Integration of Imprecise Vision and
a Field Model 509
    *Kazunori Terada, Kouji Mochizuki, Atsushi Ueno, Hideaki Takeda,
Toyoaki Nishida, Takayuki Nakamura, Akihiro Ebina,
and Hiromitsu Fujiwara*

Multiple Reward Criterion for Cooperative Behavior Acquisition
in a Multiagent Environment 519
    Eiji Uchibe and Minoru Asada

BDI Design Principles and Cooperative Implementation in RoboCup 531
    *Jan Wendler, Markus Hannebauer, Hans-Dieter Burkhard,
Helmut Myritz, Gerd Sander, and Thomas Meinert*

Team Descriptions
Simulation League

AT Humboldt in RoboCup-99 542
    *Hans-Dieter Burkhard, Jan Wendler, Thomas Meinert, Helmut Myritz,
and Gerd Sander*

Cyberoos'99: Tactical Agents in the RoboCup Simulation League 546
    Mikhail Prokopenko, Marc Butler, Wai Yat Wong, and Thomas Howard

11Monkeys Description 550
    Shuhei Kinoshita and Yoshikazu Yamamoto

Team Erika 554
    Takeshi Matsumura

Essex Wizards'99 Team Description 558
    H. Hu, K. Kostiadis, M. Hunter, and M. Seabrook

FCFoo99 563
    Fredrik Heintz

Footux Team Description: A Hybrid Recursive Based Agent Architecture . 567
    François Girault and Serge Stinckwich

Gongeroos'99 572
    Chee Fon Chang, Aditya Ghose, Justin Lipman, and Peter Harvey

Headless Chickens III 576
    *Paul Scerri, Johan Ydrén, Tobias Wiren, Mikael Lönneberg,
and Pelle Nilsson*

IALP 580
    Antonio Cisternino and Maria Simi

Kappa-II 584
    Itsuki Noda

Karlsruhe Brainstormers - Design Principles 588
  M. Riedmiller, S. Buck, A. Merke, R. Ehrmann, O. Thate, S. Dilger,
  A. Sinner, A. Hofmann, and L. Frommberger

Kasugabito III 592
  Tomoichi Takahashi

RoboCup-99 Simulation League: Team KU-Sakura2 596
  Harukazu Igarashi, Shougo Kosue, and Takashi Sakurai

The magmaFreiburg Soccer Team 600
  Klaus Dorer

Mainz Rolling Brains 604
  Daniel Polani and Thomas Uthmann

NITStones-99 608
  Kouichi Nakagawa, Noriaki Asai, Nobuhiro Ito, Xiaoyong Du,
  and Naohiro Ishii

Oulu 99 611
  Jarkko Kemppainen, Jouko Kylmäoja, Janne Räsänen,
  and Ville Voutilainen

Pardis 614
  Shahriar Pourazin

PaSo-Team'99 618
  Carlo Ferrari, Francesco Garelli, and Enrico Pagello

PSI Team 623
  Alexander N. Kozhushkin

RoboLog Koblenz 628
  Jan Murray, Oliver Obst, and Frieder Stolzenburg

Rational Agents by Reviewing Techniques 632
  Josep Lluís de la Rosa, Bianca Innocenti, Israel Muñoz,
  and Miquel Montaner

The Ulm Sparrows 99 638
  Stefan Sablatnög, Stefan Enderle, Mark Dettinger, Thomas Boß,
  Mohammed Livani, Michael Dietz, Jan Giebel, Urban Meis, Heiko Folkerts,
  Alexander Neubeck, Peter Schaeffer, Marcus Ritter, Hans Braxmeier,
  Dominik Maschke, Gerhard Kraetzschmar, Jörg Kaiser, and Günther Palm

UBU Team 642
  Johan Kummeneje, David Lybäck, Håkan Younes, and Magnus Boman

YowAI 646
  Takashi Suzuki

Zeng99: RoboCup Simulation Team with Hierarchical Fuzzy Intelligent
Control and Cooperative Development 649
*Junji Nishino, Tomomi Kawarabayashi, Takuya Morishita,
Takenori Kubo, Hiroki Shimora, Hironori Aoyagi, Kyoichi Hiroshima,
and Hisakazu Ogura*

Small-Size Robot (F180) League

All Botz 653
Jacky Baltes, Nicholas Hildreth, and David Maplesden

Big Red: The Cornell Small League Robot Soccer Team 657
*Raffaello D'Andrea, Jin-Woo Lee, Andrew Hoffman,
Aris Samad-Yahaja, Lars B. Cremean, and Thomas Karpati*

CMUnited-99: Small-Size Robot Team 661
Manuela Veloso, Michael Bowling, and Sorin Achim

5dpo Team Description 663
*Paulo Costa, António Moreira, Armando Sousa, Paulo Marques,
Pedro Costa, and Aníbal Matos*

FU-Fighters Team Description 667
*Sven Behnke, Bernhard Frötschl, Raúl Rojas, Peter Ackers,
Wolf Lindstrot, Manuel de Melo, Andreas Schebesch, Mark Simon,
Martin Sprengel, and Oliver Tenchio*

Linked99 671
*Junichi Akita, Jun Sese, Toshihide Saka, Masahiro Aono,
Tomomi Kawarabayashi, and Junji Nishino*

OWARI-BITO 675
*Tadashi Naruse, Tomoichi Takahashi, Kazuhito Murakami,
Yasunori Nagasaka, Katsutoshi Ishiwata, Masahiro Nagami,
and Yasuo Mori*

Rogi 2 Team Description 679
*Josep Lluís de la Rosa, Rafel García, Bianca Innocenti,
Israel Muñoz, Albert Figueras, and Josep Antoni Ramon*

Temasek Polytechnic RoboCup Team-TPOTs 683
Nadir Ould Khessal

The VUB AI-lab RoboCup'99 Small League Team 687
Andreas Birk, Thomas Walle, Tony Belpaeme, and Holger Kenn

Middle-Size Robot (F2000) League

Agilo RoboCuppers: RoboCup Team Description 691
Thorsten Bandlow, Robert Hanek, Michael Klupsch, and Thorsten Schmitt

ART99 - Azzurra Robot Team 695
  *Daniele Nardi, Giovanni Adorni, Andrea Bonarini, Antonio Chella,
  Giorgio Clemente, Enrico Pagello, and Maurizio Piaggio*

CoPS-Team Description 699
  *N. Oswald, M. Becht, T. Buchheim, G. Hetzel, G. Kindermann,
  R. Lafrenz, P. Levi, M. Muscholl, M. Schanz, and M. Schulé*

CS Freiburg'99 703
  B. Nebel, J.-S. Gutmann, and W. Hatzack

DREAMTEAM 99: Team Description Paper 707
  *Wei-Min Shen, Jafar Adibi, Rogelio Adobbati, Jay Modi, Hadi Moradi,
  Behnam Salemi, and Sheila Tejada*

Description of the GMD RoboCup-99 Team 711
  *Ansgar Bredenfeld, Wolf Göhring, Horst Günther, Herbert Jaeger,
  Hans-Ulrich Kobialka, Paul-Gerhard Plöger, Peter Schöll,
  Andrea Siegberg, Arend Streit, Christian Verbeek, and Jörg Wilberg*

ISocRob - Intelligent Society of Robots 715
  *Rodrigo Ventura, Pedro Aparício, Carlos Marques, Pedro Lima,
  and Luís Custódio*

KIRC: Kyutech Intelligent Robot Club 719
  *Takeshi Ohashi, Masato Fukuda, Shuichi Enokida, Takaichi Yoshida,
  and Toshiaki Ejima*

The Concept of Matto 723
  Kosei Demura, Kenji Miwa, Hiroki Igarashi, and Daitoshi Ishihara

The RoboCup-NAIST 727
  T. Nakamura, K. Terada, H. Takeda, A. Ebina, and H. Fujiwara

Robot Football Team from Minho University 731
  Carlos Machado, Ilídio Costa, Sérgio Sampaio, and Fernando Ribeiro

Real MagiCol 99: Team Description 735
  C. Moreno, A. Suárez, Y. Amirat, E. González, and H. Loaiza

RMIT Raiders 741
  *James Brusey, Andrew Jennings, Mark Makies, Chris Keen,
  Anthony Kendall, Lin Padgham, and Dhirendra Singh*

Design and Construction of a Soccer Player Robot ARVAND 745
  *M. Jamzad, A. Foroughnassiraei, E. Chiniforooshan, R. Ghorbani,
  M. Kazemi, H. Chitsaz, F. Mobasser, and S. B. Sadjad*

The Team Description of Osaka University "Trackies-99" 750
  *Sho'ji Suzuki, Tatsunori Kato, Hiroshi Ishizuka, Hiroyoshi Kawanishi,
  Takashi Tamura, Masakazu Yanase, Yasutake Takahashi, Eiji Uchibe,
  and Minoru Asada*

5dpo-2000 Team Description 754
    Paulo Costa, António Moreira, Armando Sousa, Paulo Marques,
    Pedro Costa, and Aníbal Matos

Sony Legged Robot League

Team ARAIBO 758
    Yuichi Kobayashi and Hideo Yuasa

BabyTigers-99: Osaka Legged Robot Team 762
    Noriaki Mitsunaga and Minoru Asada

CM-Trio-99 766
    Manuela Veloso, Scott Lenser, Elly Winner, and James Bruce

Humboldt Hereos in RoboCup-99 770
    Hans-Dieter Burkhard, Matthias Werner, Michael Ritzschke,
    Frank Winkler, Jan Wendler, Andrej Georgi, Uwe Düffert,
    and Helmut Myritz

McGill RedDogs 774
    Richard Unger

Team Sweden 784
    M. Boman, K. LeBlanc, C. Guttmann, and A. Saffiotti

UNSW United 788
    Mike Lawther and John Dalgliesh

UPennalizers: The University of Pennsylvania RoboCup Legged
Soccer Team 792
    James P. Ostrowski

Author Index 799

Overview of RoboCup-99

Manuela Veloso[1], Hiroaki Kitano[2], Enrico Pagello[3],
Gerhard Kraetzschmar[4], Peter Stone[5], Tucker Balch[1],
Minoru Asada[6], Silvia Coradeschi[7], Lars Karlsson[7], and Masahiro Fujita[8]

[1] School of Computer Science, Carnegie Mellon University, Pittsburgh, USA
[2] Sony Computer Science Laboratories, Inc., Tokyo, Japan
[3] Dept. of Electronics and Informatics, The University of Padova, Italy
[4] Neural Information Processing, University of Ulm, Ulm, Germany
[5] AT&T Labs — Research, 180 Park Ave., Florham Park, USA
[6] Adaptive Machine Systems, Osaka University, Osaka, Japan
[7] Örebro University, Örebro, Sweden
[8] Sony Corp., Tokyo, Japan

Abstract. RoboCup-99, the third Robot World Cup Soccer Games and Conferences, was held in conjunction with IJCAI-99 in Stockholm. RoboCup has now clearly demonstrated that it provides a remarkable framework for advanced research in Robotics and Artificial Intelligence. The yearly RoboCup event has included a technical workshop and competitions in different leagues. This chapter presents a comprehensive overview of RoboCup-99 and the scientific and engineering challenges presented to the participating researchers. There were four RoboCup-99 competitions: the simulation league, the small-size robot league, the middle-size robot league, and, for the first time officially, the Sony legged robot league. The champion teams were CMUnited-99 (Carnegie Mellon University, USA) for the simulation league, Sharif CE (Sharif University of Technology, Iran) for the middle-size league, Big Red (Cornell University, USA) for the small-size league, and "Les 3 Mousquetaires" (Laboratoire de Robotique de Paris, France) for the Sony legged robot league. The Scientific Challenge Award was given to three papers on innovative research for the automated statistical analysis of the games, from the University of Southern California (ISI/USC), USA, the Electrotechnical Laboratory (ETL), Japan, and Chubu University, Japan. There will be the first RoboCup European Championship in Amsterdam in May 2000, and the International RoboCup-2000 will take place in Melbourne, Australia, in August 2000.

1 Introduction

The RoboCup Initiative, the Robot World Cup Soccer Games and Conferences, provides a large spectrum of research and development issues in Artificial Intelligence (AI) and Robotics. In particular, it remarkably provides a common task, namely robotic soccer, for the investigation and evaluation of different approaches, theories, algorithms, and architectures for multiagent software and robotic systems.

RoboCup-99, held in Stockholm, followed the successful RoboCup-97 in Nagoya [6] and RoboCup-98 in Paris [3]. The RoboCup-99 event included a technical workshop, robotic soccer competitions in four different leagues, and a variety of demonstrations.

This chapter introduces the different leagues in detail, summarizes the challenging research problems underlying each league, and overviews RoboCup-99. It further includes several appendices with the results of all the games in the four competition leagues. It shows the results of the preliminary round-robin phases and the results from the elimination rounds. The book contains the technical contributions on how each of the multiple teams concretely addressed these research challenges at the RoboCup-99 competitions. The chapter includes a brief discussions of the lessons learned and future directions. A new RoboCup search and rescue task is under development and will be part of the next RoboCup events.

The RoboCup events are held every year. RoboCup has been held in conjunction with international technical conferences. It has been attended by the research community and by the general public. RoboCup-97 and RoboCup-99 were held at the biannual International Joint Conference on Artificial Intelligence (IJCAI). RoboCup-98 was held with the International Conference on Multiagent Systems (ICMAS) in Paris. RoboCup-98, in particular, attracted a large audience, as it took place mostly at the same time as the human World Cup.

RoboCup-99, the Third Robot World Cup Soccer Games and Conferences, was held on July 27th through August 4th, 1999 in Stockholm. It was organized by Linköping University with the cooperation of Stockholm University, and it was sponsored by Sony Corporation, Sun Microsystems, Futurniture, First Hotel, The Foundation for Knowledge and Competence Development, The Swedish Council for Planning and Coordination of Research, The Swedish Foundation for Strategic Research, NUTEK, and WITAS.

The purpose of RoboCup is to provide a common task for evaluation of different algorithms and their performance, theories, and robot architectures [8]. In addition, as soccer, as a game, is quite accessible to both experts and non-experts, RoboCup has also shown to provide an interesting popular demonstration of research in AI and Robotics.

RoboCup-99 had four different leagues, each one with its specific architectural constraints and challenges, but sharing the goal of developing teams of autonomous agents with action, perception, and cognition. RoboCup-99 also included the RoboCup Jr. event targeted at allowing children to experiment with automated robotic systems.

The Scientific Challenge Award is given each year to people or groups that have made significant scientific contributions to RoboCup. At RoboCup-99, the Scientific Challenge Award was given to three papers on innovative research for the automated statistical analysis of the games, from the University of Southern California (ISI/USC), USA, the Electrotechnical Laboratory (ETL), Japan, and Chubu University, Japan.

2 Simulation League

The simulation league continues to be the most popular part of the RoboCup leagues, with 37 teams participating in RoboCup-99, which is a slight increase over the number of participants at RoboCup-98. In this section, we briefly describe the RoboCup simulator; we present the major research challenges and some of the ways in which they have been addressed in the passed; and we summarize the 1999 competition.

2.1 The RoboCup Simulator

The RoboCup-99 simulation competition was held using the RoboCup soccer server [11], which has been used as the basis for previous successful international competitions and research challenges [8]. The soccer server is a complex and realistic domain, embracing as many real-world complexities as possible. It models a hypothetical robotic system, merging characteristics from different existing and planned systems as well as from human soccer players. The server's sensor and actuator noise models are motivated by typical robotic systems, while many other characteristics, such as limited stamina and vision, are motivated by human parameters.

The simulator includes a visualization tool, pictured in Figure 1. Each player is represented as a two-halved circle. The light side is the side towards which the player is facing. In Figure 1, all of the 22 players are facing the ball, which is in the middle of the field. The black bars on the left and right sides of the field are the goals.

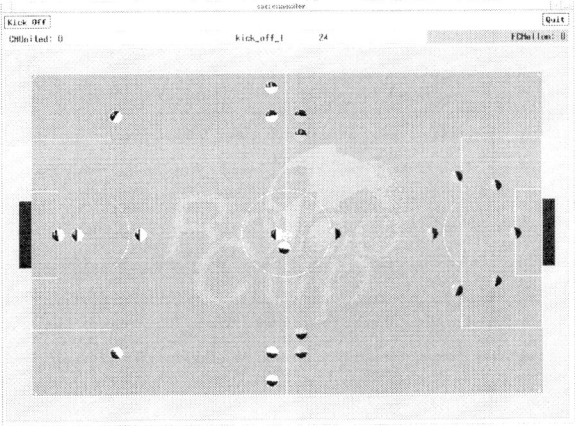

Fig. 1. The soccer server display.

The simulator also includes a *referee*, which enforces the rules of the game. It indicates changes in play mode, such as when the ball goes out of bounds, when a goal is scored, or when the game ends. It also enforces the *offsides* rule. Like in real soccer, a player is offsides if it is in the opponent's half of the field and closer to the opponent's goal line (the line along which the goal is located) than all or all but one of the opponent players when the ball is passed to it. The crucial moment for an offsides call is when the ball is kicked, not when it is received: a player can be behind all of the opponent defenders when it receives a pass, but not when a teammate kicks the ball towards it.[1] The offsides rule, which typically plays an important role in shaping soccer strategies, is not enforced in any of the other RoboCup leagues.

The simulator, acting as a server, provides a domain and supports users who wish to build their own *agents* (also referred to as *clients* or *players*). Client programs connect to the server via UDP sockets, each controlling a single player. The soccer server simulates the movements of all of the objects in the world, while each client acts as the brain of one player, sending *movement commands* to the server. The server causes the player being controlled by the client to execute the movement commands and sends sensory information from that player's perspective back to the client.

When a game is to be played, two teams of 11 independently controlled clients connect to the server. Thus, it is a fully distributed, multiagent domain with both teammates and adversaries. The simulation league is the only RoboCup league that uses teams of 11 players as in real soccer.

The sensory information sent from the server to each client provides only a partial world view at any given moment. Each player can only "see" objects within a limited angle of the direction it is facing, and both the accuracy and description-detail of seen objects degrades with distance. In particular, sensory information is partial and noisy. Both agent action and object movement are noisy as well.

Another of the real-world complexities embraced by the soccer server is asynchronous sensing and acting. Whereas most AI simulators use synchronous sensing and acting: an agent senses the world, acts, senses the result, acts again, and so on. In this paradigm, sensations trigger actions. On the other hand, both people and complex robotic systems have independent sensing and acting rates. Sensory information arrives via different sensors at different rates, often unpredictably (e.g. sound). Meanwhile, multiple actions may be possible in between sensations or multiple sensations may arrive between action opportunities.

The soccer server communication paradigm models a crowded, low-bandwidth environment. All 22 agents use a single, unreliable communication channel. When an agent "speaks," nearby agents on both teams can hear the message. Agents have a limited communication range and a limited communication capacity, both in terms of message length and frequency.

Another limited resource of the agents is stamina. The more the agents run, the more tired they get, so that future running is less effective. Stamina has both a renewable component, that replenishes if the agents stands still, and an unrenewable component that can degrade over the course of the game.

---

[1] The soccer server operationalizes the offsides rule making it an objective call.

Finally, soccer server agents, like their robotic counterparts, must act in real time. The simulator uses a discrete action model, collecting player actions over the course of a 100 msec cycle, but only executes them and updates the world at the end of the cycle. If a client sends more than one movement command in a simulator cycle, the server chooses one randomly for execution. Thus, it is in each client's interest to try to send at most one movement command each simulator cycle. On the other hand, if a client sends no movement commands during a simulator cycle, it loses the opportunity to act during that cycle, which can be a significant disadvantage in a real-time adversarial domain: while the agent remains idle, opponents may gain an advantage.

In summary, the RoboCup soccer server is a fully *distributed, multiagent* domain with both *teammates* and *adversaries*. There is *hidden state*, meaning that each agent has only a partial world view at any given moment. The agents also have *noisy sensors and actuators*, meaning that they do not perceive the world exactly as it is, nor can they affect the world exactly as intended. In addition, the perception and action cycles are *asynchronous*, prohibiting the traditional AI paradigm of using perceptual input to trigger actions. *Communication* opportunities are limited; the agents have *limited stamina* and the agents must make their decisions in *real-time*. These italicized domain characteristics combine to make the RoboCup soccer server a realistic and challenging domain.

2.2 Research Challenges

Research directions in the RoboCup simulation league are quite varied, as is evident from the articles in this book that are based on simulation research. This section presents a small sample of these directions.

The RoboCup synthetic agent challenge [8] identifies three major simulation-based challenges as being:

1. machine learning in a multiagent, collaborative and adversarial environment,
2. multiagent architectures, enabling real-time multiagent planning and decision-making, in service of teamwork, and
3. opponent modeling

Much past research has been devoted to these topics as reflected in this and the past RoboCup books [6, 3].

Several other challenges are suggested by the characteristics of the soccer server presented above. For example, asynchronous sensing and acting, especially when the sensing can happen at unpredictable intervals, is a very challenging paradigm for agents to handle. Agents must balance the need to act regularly and as quickly as possible with the need to gather information about the environment. For example, the runner-up of the 1999 competition, magmaFreiburg, used an action-selection method based on extended behavior networks that generated decisions very quickly. This method was used primarily for times when an agent was in possession of the ball.

Some other research areas related to agent-development in the simulation league include:

- communication in single-channel, low-bandwidth communication environments,
- social conventions, or coordination without communication,
- distributed sensing, and
- resource management.

It is interesting to note that different techniques are generally used for agent control when the agents are not in possession of the ball. Many teams use the concept of flexible formations in which agents adjust their positions based on the ball's location (e.g., [13]). Some research is focussed on using machine learning or linear programming techniques to allow agents to adapt their positioning based on the locations of the opponent players during the course of a game (e.g., [1]).

In addition to soccer-playing agent development, the soccer server has been used as a substrate for 3-dimensional visualization, real-time natural language commentary, and education research.

Figure 1 shows the 2-dimensional visualization tool that is included in the soccer server software. SPACE [12] converts the 2-dimensional image into a 3-dimensional image, changing camera angle and rendering images in real time.

Another research challenge being addressed within the soccer server is producing natural language commentary of games as they proceed. Researchers aim to provide both low-level descriptions of the action, for example announcing which team is in possession of the ball, and high-level analysis of the play, for example commenting on the team strategies being used by the different teams. Commentator systems for the soccer server include ROCCO [2], MIKE [10], and Byrne [4].

Robotic soccer has also been used as the basis for education research. A survey of RoboCup-97 participants indicates that the majority of participants were students motivated principally by the research opportunities provided by the domain [14]. There has also been an undergraduate AI programming course based on teaching students to create robotic soccer-playing agents in the soccer server [5].

2.3 The RoboCup-99 Tournament

As with RoboCup-97 and RoboCup-98, teams were divided into leagues. In the preliminary round, teams played within leagues in a round-robin fashion, and that was followed by a double-elimination round (where a team has to lose twice to be eliminated) to determine the first three teams. Many of the games were extremely exciting, leading up to the final—watched by several hundred people—in which CMUNITED-99 defeated MAGMA FREIBURG by a score of 4–0.

With respect to the competition entrants themselves, there is concrete evidence that the overall level improved significantly over the previous year. The defending champion team, the CMUNITED-98 simulator team was entered in the competition. Its code was left unaltered from that used at RoboCup-98 except for minor changes necessary to update from version 4 to version 5 of the soccer simulator. In 1998, this team won all of its matches and suffered no goals against. However, this year, after advancing to the elimination round, it won only one game before being eliminated.

An interesting improvement to the soccer simulator in 1999 was the addition of an on-line coach. Each team was permitted to use a single agent with an overhead view of the field that could communicate with all teammates whenever play was stopped (i.e. the ball was out of bounds). At least one team took advantage of this feature to have the coach give advice to the team regarding the overall formation of the team, which could range from offensive to defensive, and "narrow" (concentrated near the middle of the field) to wide.

Building on the success of the 1999 tournament, the RoboCup-2000 simulator tournament has even more entrants and promises to be another exciting event spawning new research approaches and successes.

3 F-180: Small-Size Robot League

The F-180, or "small-size" RoboCup league, features up to five robots on each team in matches on a field the size of a ping-pong table. Each robot can extend up to $18cm$ along any diagonal and occupy up to $180cm^2$ of the pitch. Color markers on the field, the robots and the ball help computerized vision systems locate important objects in the game. The robots are often controlled remotely by a separate computer that processes an an image of the field provided by an overhead camera. A couple of teams, and probably more in the future, included on-board vision.

In this section we will review the characteristics of the F-180 league, the research challenges facing teams competing in the league, and recent research contributions by some of the competing teams.

3.1 Characteristics of the F-180 League

The playing surface consists of a green ping-pong table enclosed by white walls. One goal area is painted yellow, while the other is painted blue – these colors help robots with onboard vision find the goals. In 2000, the league is moving to a carpeted surface of the same dimensions.

To help competitors locate their opponents, each robot carries a single colored ping-pong ball provided by the RoboCup organization. The marker is located at the geometric center of the robot as viewed from above. One team is fitted with yellow markers, while the other is equipped with blue ones. At RoboCup-99, the team carrying yellow markers attacks the blue goal and blue team attacks the yellow goal. In addition, the robots may be colored with additional markers to help computer controllers locate and orient them.

3.2 Research and Engineering Challenges

The core issues faced by RoboCup F-180 researchers include the construction of the robots, development of individual robot skills, reliability in dynamic, uncertain and adversarial environments, and importantly, cooperative team coordination. In the F-180 league these capabilities depend significantly on underlying engineering like reliable real-time vision and high-performance feedback control of small robots.

Competitors in the F-180 league must address most of the challenges faced by teams engaged in the simulator competition (e.g., cooperation, localization, strategy and tactics). Additionally, however, vision systems for tracking the robots must be developed and hardware to execute the control commands (the robots themselves) must be built. In terms of the autonomy required of robots, the F-180 league lies somewhere between the F-2000 league and simulation. The difficulties of locating the ball, other robots, and opponents is reduced in comparison with the F-2000 league because an overhead camera is allowed. However, the technical challenges of real-time visual tracking, feedback control and team play remain.

The visual tracking problem for the small-size league can actually be seen as more difficult in some ways than for robots in the middle-size league. The computer responsible for processing images from an overhead camera must be able to simultaneously estimate the locations and velocities of 10 robots and the ball. In some cases these robots move as fast as $2m/s$, while the ball has been recorded at speeds of $6m/s$. This vision task is being addressed using a wide range of technologies including: specialized Digital Signal Processing (DSP) hardware, commercial color tracking systems, and fast PCs equipped with commodity color capture hardware but programmed with highly-optimized image processing software. Pioneered by the CMUnited-97 and CMUnited-98 vision processing algorithms, several teams currently predict the future trajectory of the ball, and use this prediction to intersect the ball. At RoboCup-99, most of the top teams, in particular the three small robots of the RobotIS team, impressively intersected and controlled rather fast moving balls.

In addition to addressing vision and position control issues, robots must be able to manipulate the ball. Skills such as dribbling, passing and shooting are critical to successful play. It is also important for robots to be able to remove the ball from along the walls. Many robots are also equipped with devices for kicking the ball. Determining when to activate the kicker can be a tricky tactical decision.

RoboCup-99 saw a substantial increase in the mechanical capabilities of robots. In past years, a majority of the teams mainly focused with vision processing, obstacle avoidance, and cooperative behaviors. This year several teams seemed to have focused on player skills.

One of the most interesting developments concerned ball kicking technologies. At RoboCup-98, only a few teams had their own kicking devices, in particular the winning CMUnited-98 team. However the devices used in 1998 did not seem to be significantly effective. At RoboCup-99 nearly half of the participating teams utilized some sort of kicking device. One team (the FU-FIGHTERS from Berlin) was remarkably able to propel the ball so fast that observers could barely track it (see Figure 2).

Fig. 2. The FuFighters robots with their kicking device.

Another interesting development was a new spinning technique for removing stuck balls from along the wall or in corners. The RoboCup-99 winning Big Red team demonstrated this early in the tournament and several others were able to also adopt the spinning behavior.

3.3 The RoboCup-99 Tournament

Participation in the Small-Size league RoboCup soccer continues to grow at a remarkable pace. Competitions in 1997 and 1998 included five and eleven competitors respectively. In anticipation of even more participants in 1999, the league instituted qualification rules to limit the field to a manageable number and to ensure groups did not travel to Stockholm with no reasonable hope of competing. In order to qualify, each team had to submit a video tape by April demonstrating at least one robot able to move the ball across the field and score (this may sound easy, but it is in fact a very challenging problem). Eighteen teams from around the world qualified for the third annual competition. The group included teams from Australia, Belgium, France, Germany, Japan, Korea, New Zealand, Portugal, Singapore, Spain, and the USA.

For the round-robin phase, the 18 teams were split into four groups of four or five teams each. In an effort to ensure equally competitive divisions each group included one of the top four finishers from RoboCup-98 and one or two new competitors. Also, no two teams from the same country were placed in the same group. During the round-robin phase, each team in each group played each of the other teams in its group. Group standings were determined by awarding three points to a team for each game it won and one point for each tie. The top two teams from each division progressed to the single elimination tournament.

Because of the large number of teams, four separate fields were required for the round-robin. Scheduling the round-robin tournament was challenging because teams sometimes ran into technical problems and asked for delays. The task was complicated by the fact that many teams used the same frequencies for controlling their robots, and therefore could not play at the same time. Games were played twelve to thirteen hours a day for two days; there were almost always two games running concurrently.

At the end of the round-robin, the top two teams from each group (eight teams in all) took a day to move to the central conference location for the finals. This move was a bit more difficult than had been anticipated. Fortunately, all of the teams were able to adapt to the new lighting conditions and slightly cramped environment. The new location boosted attendance and crowd participation significantly. The quarter finals, semi-finals and final match conducted over the next two days in standing-room-only conditions. The final game resulted in the runner-up FuFighters and the winning BigRed team (see Figure 3).

Fig. 3. The FuFighters runner-up team and the BigRed champion team.

3.4 Evolution of the Rules

Rules for F-180 league robotic soccer continue to evolve. Of course the long term vision for RoboCup is participation in the real human World Cup, so our robots must eventually be capable of play according to FIFA (the World Cup rules-making body) regulations. For now, however, we adjust FIFA's rules to accommodate our robots. Examples of RoboCup adjustments to the rules include special markings to help with vision issues and walls around the pitch to keep the ball from departing the playing surface.

One detail of the rules is particularly interesting from a philosophical point of view. In real soccer, yellow cards are assigned to individual players who commit serious fouls; this approach is also used at RoboCup. In both FIFA and RoboCup soccer, when an individual receives two yellow cards, he/she/it is ejected from the game and cannot be replaced (reducing the number of players on the field). When a star human player receives a yellow card, the team's coach is faced with an important decision: should the star player be kept in the game and bear the risk that of receiving another yellow card, or should the star be replaced with a substitute? The situation is completely different for robot teams. Competitors often have a number of identical "spare" robots that can be immediately substituted for a penalized player — several teams followed this strategy.

This kind of substitution was perfectly legal, but seems to violate the spirit of the rule which is intended to punish the "offender." But which is the offender, the robot hardware or the software? Should the physical hardware be tagged with the yellow card, or should it apply to the software controlling it? This issue has been addressed in 2000 by changing the manner in which yellow cards are tracked: now they are tracked against the team as a whole. Every time two yellow cards are assigned, one player must be removed from the field.

3.5 Lessons Learned and Current State of the League

Probably the most frequent difficulty faced by teams in the F-180 league concerns fast vision processing. Even though many teams' vision systems work perfectly in the lab, after being re-located half-way around the world it is often a great challenge to re-calibrate them in a new environment. Problems are caused by the specific height of the camera, the variable intensity of field lighting, and the spectrum of illumination provided by the lights. Still another source of vision problems concerned the colored markers worn by opponent teams. These difficulties highlight the importance of robust vision for robots – this is a substantial challenge in nearly all domains of robotics research.

Another important lesson from successful teams is that as much effort must be applied to software development as is devoted to hardware design. It is common to see beautiful hardware designs with poor or very slow control algorithms. The winners and other top-ranked teams in previous RoboCups and also at RoboCup-99 clearly balanced their development effort between hardware and software.

The league is in great shape. It continues to draw more researchers each year. We expect about 20 teams to compete in RoboCup-2000 at Melbourne. The rules continue to be dynamic, and to reflect the research interests and directions of the participants. Two significant changes for the future include a shift to a more realistic carpet surface, and a switch to angled walls that allow the ball to leave the field more easily. In the future we hope to remove walls altogether. RoboCup in the 2000s promises to be even more exciting than in the last millennium!

4 F-2000: Middle-Size Robot League

The RoboCup F-2000 League, commonly known as middle-size robot league, poses a unique combination of research problems, which has drawn the attention of well over 30 research groups world-wide. In this section, we briefly describe the fundamental characteristics of the league and discuss its major differences to other leagues. Then we present some typical research problems and the solutions developed by F-2000 teams. This overview is selective, while this book provides a rather complete survey of the research performed in F-2000 league. In particular, the references to the technical contributions from the different teams can be found as chapters in this book. We conclude with a summary of the RoboCup-99 middle-size league tournament in Stockholm and a few observations on team performance and the current state of the league.

4.1 Characteristics of F-2000 League

Two major factors influence the design of teams and robotic soccer players for middle-size robot league: (a) the playing environment, in particular, the field, and (b) constraints imposed on robot design.

The playing environment is carefully designed such that the perceptual and motory problems to be solved are reasonably simple, but still challenging enough to ignite interesting and serious research efforts. The field size is variable within certain bounds; in Stockholm, the field size was $9m \times 5m$. The goals do not have nets, but colored walls in the back and on the sides (yellow/blue). The field is surrounded by white walls (50cm height) that carry a few extra markings (squared black markers of 10cm size plus black-and-white logos of sponsors in large letters). A special corner design is used and marked with two green lines. The goal lines, goal area, center line and center circle are all marked with white lines. The ball is dark orange. Illumination of the field is constrained to be within 500 and 1500 lux. Matches are played with teams of four robots, including the goalie.

The robots must have a black body and carry color tags for team identification (light blue/magenta). Quite elaborate constraints exist for robot size, weight, and shape; roughly, a robot body may have about 50cm diameter and be up to 80cm high, must weigh less than 80kg, and feature no concavities large enough to take up more than one-third of the ball diameter. The robots must carry all sensors and actuators on-board; no global sensing system is allowed. Wireless communication is permitted both between robots and between robots and outside computers.

4.2 Research and Engineering Challenges

A general survey of RoboCup research issues can be found in [6, 3]. An interesting perspective on middle-size league research issues results when compared with the simulation league. Ultimately, all major research issues in simulation league, like coordinated team play, opponent modeling, game strategy and tactics, in-game adaptation to opponent tactics, etc., have to be solved in middle-size league as well. However, while agent design in simulation league can build upon a set of reasonably reliable perception and action commands (e.g., it is possible to precisely and deterministically determine a player's position and orientation on the field with little computational effort), it is a non-trivial task to achieve this level of player capability with real robots. Thus, building a F-2000 robot team starts with a combined engineering and research challenge: choosing or designing appropriate robots.

A basic decision with far-reaching consequences is the solution selected to achieve mobility. A wide spectrum of alternative solutions developed in classical robotics is available; for example, omnidirectional drive systems allow to design very agile robots, but can be quite complex to control. Differential drive systems are much easier to control, but often require more complex movement maneuvers during play. Another mechanical engineering problem is the design of mechanisms for handling the ball. Because ready-to-use solutions were hardly available, this problem has led to a wide range of different approaches and interesting new designs.

Although a number of small, commercial robot platforms have become available over the past few years (e.g., the Pioneer series by Activmedia and the Scout robots by Nomadics, both of which use differential drive systems), almost none of them can be considered a complete robotic soccer player. Typical items teams found necessary to add include mechanical kicking devices, additional sensors like bumpers, compasses, laser scanners, unidirectional and omnidirectional cameras, and additional computing power (notebook computers, embedded computers, DSP boards). Because off-the-shelf soccer robots are not available, a substantial number of teams decided to build their own robots. In either case, the time and effort needed to design and construct (or acquire) all necessary components and to *integrate* them into a reliably working soccer robot is often grossly under-estimated and under-valued. At the very least, this kind of system's integration work is a great educational experience for students.

Once a functional physical robot is available, a number of basic perceptual and behavioral problems must be solved. Perception and action commands are needed for the following functions:

- Detection and tracking of the ball.
- Detection and tracking of the goals, corners, lines, and other landmark features of the field.
- Detection and recognition of teammates and opponents.
- Kicking and passing the ball.
- Dribbling the ball.
- Goalie.

All this basic functionality, which has proven to be very hard to obtain in a robust and reliable manner, is already available for designers of soccer players in the simulation league. Also, this functionality is achieved in the small-size robot league through an outside camera mounted above the field. The images to be processed in the small-size league are characterized by comparatively stable lighting conditions (lighting comes from the same direction as the camera) and little optical flow. The goals and the walls surrounding the field remain stable, and only the position of the ball and the players change. Vision processing in the F-180 league is however still very challenging due to the typically very high speed of both robots and the ball. The small-size researchers have developed fast vision processing routines (up to 60 Hz frame rate) to detect and track in real time up to the eleven fast moving objects on the field.

The situation is completely different in the F-2000 league, where the camera is near the floor (lighting direction is almost perpendicular to camera direction) and the lighting situation is far less stable. As a consequence, for example cameras see the ball as an ensemble of three differently colored regions: a red portion in the middle, a white portion at the top (reflection of lighting for the field), and a black portion at the bottom (the shadowed lower part of the ball and its own shadow). Also, the camera is actively moved through the environment, resulting in images where everything constantly changes, especially the distances to objects and landmark features to be recognized. Usually, only a small portion of the environment is visible at any given time; 30 to 120 degree visual angle for unidirectional cameras and a visual field of about 2m around the robot for omnidirectional cameras are typical. Detecting and tracking the relevant objects and landmark features requires robust and reliable techniques for color-based and texture-based image segmentation, line detection, and the combination of color, shape, and texture feature for object recognition and tracking. In summary, the middle-size research platform offers a very challenging setting regarding the perceptual situation of robotic soccer players.

On the behavioral side, hand-crafting robust and reliable action commands for kicking or passing the ball into a certain direction (and possibly, with varying strength) as well as moving with ball such that the robot maintains control over the ball (necessary for dribbles) often require substantial programming and tuning effort. An interesting research challenge is to develop tools for programming and debugging such behaviors modules and to apply learning techniques to this problem.

Due to these constraints, constructing the basic functionalities listed above proves to be a very hard problem. The quality of the solutions achieved for these problems usually directly influences the performance level for the next level of functionality, which includes

- world modeling,
- self-localization,
- obstacle detection and avoidance of or recovery from collisions, and
- behavior engineering, especially behaviors for finding the ball, dribbling the ball, passing to teammates, shooting a goal, performing a penalty kick, etc.

World modeling and self-localization in RoboCup are interesting because the environment is highly dynamic, currently containing nine almost constantly moving objects. Several other state-of-the-art mobile robot applications, where robots are also in highly dynamic environments like museums, treat all dynamic objects as obstacles which have no direct relevance to the task at hand. On the contrary, soccer robots cannot take this simplified view. The development of probabilistic representations for highly dynamic environments, like robotic soccer, is a challenging and still open research problem. Accordingly, adapting existing techniques for self-localization to work with such representations is required. For teams following an approach with some kind of central team coordinator (often by an outside computer) the integration of partial (and possibly inconsistent) world models provided by individual players is another research topic, which is now also investigated outside of the RoboCup community.

Obstacle detection and avoidance or recovery from collisions is a difficult problem in RoboCup, because of the intricate rulings on charging and foul play. Although soccer is a sport where physical contact is not always avoidable, there is mutual understanding in the community that pure robot strengths should not be a "winning factor"; charging fouls have drastic consequences, up to exclusion from the tournament. On the other hand, robots being overly cautious to avoid physical contact may give way to their opponents too easily. Thus, we encounter a very difficult situation assessment and classification problem.

The effort of hand-crafting more complex behaviors, like dribbling the ball or performing a penalty kick, is even higher than those mentioned before. Thus, there is a large need for behavior engineering tools, and for techniques applying learning and on-line adaptation to the behavior engineering and action selection problems.

Above the level described so far, the research challenges are quite similar to those in the simulation league. Some research groups are actually active in more than one league (e.g, CMU, Ulm, Italy, Portugal) and hope to apply results regarding strategic and tactical play from their simulation team to the robot team in the near future. An interesting design issue is that while each robot of most teams showed the same technical characteristics of its playmates, or at maximum differentiated with respect to the goalkeeper, the ART team was forced to put playing together all kinds of robots. The necessity of forming a team with robots, having different mechanics, different hardware, different software architectures, and different sensors, led to the development of a specific ability of organizing a heterogeneous multi-robot system where it would be possible to replace, at any time during the game, a specific robot with a different one. This ability was achieved by the dynamic assignment of different roles through the evaluation of some utility functions (team Italy).

For the sake of completeness, it should be noted that, in the F-2000 league, modeling player stamina is usually not investigated, while it is a considerable problem for many simulation teams. Also, F-2000 players may use considerable bandwidth in communication both between themselves and with outside computers, which allows teams to apply more centralized team architectures (team Osaka). Constraints on communication are stricter in simulation, and all soccer agents must have a high degree of autonomy, while in F-2000 only few teams follow this idea.

4.3 Examples of Engineering and Research Results

Substantial effort has been spent in most teams on actually designing robotic soccer players. The Australian team RMIT Raiders won the RoboCup-97 technical innovation award for their omnidirectional drive system design that is modeled after a computer mouse. The Japanese team Uttori United, a joint effort by three research labs, developed another omnidirectional drive design in 97 and 98 that uses four so-called Swedish wheels (wheels with free rollers at the rim) arranged in a rectangular setup. Such drive designs are usually quite complex to control, but the Uttori design applies three actuators and an elaborate transmission mechanism to decouple the various degrees of freedom (DoFs): each actuator contributes only to its corresponding DoF. In RoboCup-99, the team from Sharif University of Technology (Iran) presented robots with 4 DoFs mobile base consisting of two independently steered and actuated wheels plus additional castor wheels. This design provided excellent mobility and speed that contributed much to the overall success of the team in RoboCup-99.

A wide range of different kicking devices have been developed. Several teams use electrically activated pneumatic cylinders as actuators for kicking in order to get sufficient kicking power (e.g., teams Italy, Matto, and Ulm). In particular, Bart and Homer (the two robots designed at the University of Padua that played with the ART team), were equipped with a flexible directional kicker that had a left and right side able to slide one to each other, in order to acquire high flexibility and accuracy. Some teams built complete robots from scratch using standard industrial equipment as components where possible (e.g., teams GMD, Italy, Matto, and Ulm). An interesting method for easy integration of microcontroller-driven sensing and actuating devices is based on the CAN bus that allows to connect up to 64 devices on a single 1 Mbit bus (team Ulm).

Aside of the mechanical design and engineering questions, the research efforts of teams in the middle-size robot league clearly indicate several focal points: vision, localization, and behavior engineering.

In the vision area, methods for fast color image segmentation have been developed and continue under research. Several teams use omnidirectional cameras and develop methods for processing the images, in particular for self-localization and object recognition (e.g., teams Italy, Tübingen, Osaka, Matto, and Portugal). The Italian team developed special mirror designs in order to extend the field of view in general and to combine a view of the local surrounding with a more global view of localization-relevant parts of the environment (walls and goals).

A scan-matching approach to self-localization has been the key to the success of the CS Freiburg team in RoboCup-98. Although the use of laser scanners on every robot means significantly more weight and power consumption, having a specialized sensing system for localization tends to make camera control and vision processing simpler. In fact, visually tracking the ball and opponents and trying to find localization-relevant visual features of the environment (corners, walls, goals) at the same time (or interleaved) often causes conflicts in view direction.

Another interesting direction in localization investigates approaches to vision-based self-localization. Iocchi et al. use a Hough transform for localization purposes in the team from Italy. Ritter et al. extended Monte-Carlo-Localization, a model-based method developed mainly for use with laser range finders, to work with features extracted from camera images. Other effective extensions of MCL were achieved within by Carnegie Mellon within the Sony legged robot league. Compared to standard MCL, vision-based features in RoboCup are very few and can be detected far less frequently and reliable than laser scan points, but recent results prove that the extended MCL framework is functional even under these restrictive constraints. The Agilo RoboCuppers from Munich and the team Attempto from Tübingen both use model-matching methods for localization.

Perceiving the relevant objects and knowing where the robot is are important prerequisites for generating successful soccer playing behaviors. The team from GMD uses a behavior-based approach, called dual dynamics, and presented tools for designing behaviors for a single player without giving particular attention to cooperative play. The Italian team also developed tools for designing behaviors effectively.

A couple of teams already started to seriously investigate methods for generating cooperative playing skills. In particular, Bart and Homer, from the University of Padua, achieved the cooperative ability of "exchanging the ball" between two players (a kind of action less complex than "passing a ball") through the implementation of efficient collision avoidance algorithms activated in the framework of the dynamic role assignment used by ART. The team from Osaka University has already significant experience in methods for generating cooperative playing skills and in applying reinforcement learning techniques to this problem.

4.4 The RoboCup-99 Tournament

The middle-size (F-2000) RoboCup-99 tournament went very smoothly. Twenty teams participated and played a total of 62 games, giving all teams ample opportunity to gain practical playing experience. The new rule structure for the middle-size robot league, which is based upon the official FIFA rules, proved to be quite successful and helped to focus on real research issues instead of rule discussions.

Just as in real soccer, the games were very exciting and unpredictable (see Figure 4). Several teams, which performed well in the past and have already won a cup, suffered unexpected losses, often against strong newcomers like SHARIF-CE (Sharif University of Technology, Iran), ALPHA++ (Ngee Ann Polytechnic, Singapore), and WISELY (Singapore Polytechnic, Singapore), and did not survive the preliminary rounds. The 20 participants were distributed into three groups, which came up with eight finalist teams.

The most struggled game was one of the semi-finals, when the Italian Team (ART) won over the then-undefeated champion of RoboCup-98, CS FREIBURG, in a match that required one penalty kick round, and two technical challenge rounds to come up with a decision. This game showed the real achievement of

Fig. 4. A view of a middle-size game.

the third RoboCup Physical Agent Challenge (the explicit passing of the ball between two players), when a German robot, which was controlling the ball near the opposite goal, waited for its playmate to reach a good position and then passed the ball to it. However, high-level reasoning capabilities of CS FREIBURG robots in general were not sufficient by themselves to defeat ART robots. Based on a set of reactive behaviors, especially when Bart and Homer played together, the ART team was able to generate emergent cooperative abilities.

In a very exciting final game, a crowd of several hundred spectators watched how the team from Sharif University, SHARIF CE, defeated the ART team by 3:1. Also this game showed the achievement of a difficult challenge, when an Iranian robot was able to perform a perfect dribbling and scored a goal in a few seconds from the start of the game.

4.5 Lessons Learned and Current State of the League

One thing one can learn from the tournament is that hardware alone does not buy success. Several teams, in spite of a sophisticated robot design with advanced tools like both directional and omni-directional cameras, cognachrome vision systems, laser scanning, expensive on-board laptops, did not always perform better than less high-tech teams. This fact proves that complex hardware requires substantial time to develop adequate software that can actually exploit the hardware features. On the other hand, hardware innovations can also be the foundation for success. The 1999 champion, SHARIF CE, benefited substantially from the agility of their robots, which arose from a combination of clever drive design and speed. Overall, systems that manage to exhibit relatively few behaviors, in a very robust and reliable manner, seemed to be more successful than more complex, but less reliable, systems.

Overall, the league is in good shape. Worldwide, well over 30 teams are working on building and improving a middle-size robot team. Provided that the rules and the playing field remain reasonably stable for the near future, we expect significantly enhanced vision capabilities, much improved ball control, smoother individual behaviors, and increasingly more cooperative playing behaviors. It will be a lot of fun to watch the RoboCup-2000 and RoboCup-2001 tournaments!

5 Sony Legged Robot League

Sony Legged Robot League is a new official RoboCup league since RoboCup-99. Four-legged autonomous robots compete in three-on-three soccer matches. The robot platform used is almost the same as the Sony AIBO entertainment robot that was introduced into a general consumer's market last July. 5000 sets were immediately sold, namely 3000 in Japan in 20 minutes, and 2000 in the US in four days.

The robot platform used in this league is modified from the commercial product version so that the RoboCup participating teams can develop their own programs to control the robots. Since hardware modifications to the robots are not allowed, the games are decided by who has developed the best software. The robots are equipped with a CCD vision camera.

The playing field is carpeted and slightly wider and larger than the small-size field. In order for a robot to be able to localize itself, the important game items, namely the ball and the goals, are painted in different colors. The ball is orange and the goals are yellow and blue. The field also includes six colored distinguishable landmark poles at the corners and the middle of the field. Robots can use the colored landmarks to localize themselves in the field.

Each team has three players, and a game consists of two 10min halves with a 10min break. If the game is a draw at the end of the 20 minutes, penalty kicks are carried out. There is a penalty area where only one robot can defend the goal. A referee can pick up and replace robots to other locations, if multiple robots are entangled usually while competing for the ball.

5.1 Research Issues

Vision: Since the robots easily lose sight of the ball due to occlusion by other robots and due to the limited visual angle of their camera, they need to effectively search for the ball. Research teams need to develop image processing algorithms combined with object recognition and search.

Navigation: Most teams used the four-legged walking programs provided by Sony due to the limited time available for development of new walking algorithms and/or because they preferred to focus on the tactics of the game. A few teams developed their own walking programs, for instance LRP developed a stable and robust walking program, and Osaka developed a trot walking to increase the speed of walking. The former could show the good performance during matches while the latter did not seem consistently robust although the speed itself was better than the former.

Playing Skills: Since the motor torque at each joint of the leg is not very powerful, kicking is not actually very effective, and pushing showed to be sufficient. Most teams used simple pushing as their shooting behavior. One exception was performed by the team ARAIBO (U. of Tokyo), as they used a heading shoot having the robots fall forward and performing interestingly.

Localization: Self-localization is one of the most important issues as the robots need to know where they are on the field. The teams attempted to use the colored landmarks for localization based on triangulation. CM-Trio-99 (Carnegie Mellon Univ.) introduced a new algorithm based on probabilistic sampling that allowed the robots to effectively process poorly modeled robot movement and unexpected errors, such as the change of robot location by the referees. The team from Osaka used the landmarks for task accomplishment. Based on an information criterium, the robot decides if more observation is necessary to determine the optimal action. CM-Trio-99 also introduced multi-fidelity behaviors that degrade and upgrade gracefully with different localization knowledge. The winning team from LRP, France mainly used the goals for localization moving fast and successfully towards the offensive goal.

Teamwork: Each player is marked with the team color, namely red and dark blue. So far, it has shown to be rather difficult to reliably detect the other robots based on their team colored patches. Most teams achieved basic teamwork through the assignment of roles, as one goalie robot and two attackers. The Osaka played with no specific goalie. The runner-up team from UNSW and CM-Trio-99 (3rd place) demonstrated interesting goalie behaviors. The attackers from LRP were quite effective at moving towards the goal. Teams have not yet achieved more sophisticated cooperative behaviors, such as passing.

5.2 The RoboCup-99 Tournament

An initial competition, as a demonstration, was held at RoboCup-98 in Paris. The prototype "AIBO" robots were used by three teams, Osaka University BABYTIGERS (Japan), Carnegie Mellon University CM-TRIO-98 (USA), and Laboratoire de Robotique de Paris (LRP) LES TITIS PARISIENS (France). CM-Trio-98 was the winner of this RoboCup-98 competition.

At RoboCup-99, in addition to the three seeded teams from RoboCup-98, namely BabyTigers-99, CM-Trio-99, and "Les 3 Mousquetaires" from LRP, there were six new teams: from Sweden (Örebro, Stockholm, Ronneby and others), Humboldt University (Germany), University of Tokyo (Japan), University of New South Wales (Australia), University of Pennsylvania (USA), and McGill University (Canada).

The teams were divided into three groups, each of which included one seeded team. After the round-robin phase, all three seeded teams advanced to the elimination phases. Although the seeded teams had to develop new algorithms and implementations for the new AIBO robots, they were probably still in advantage, as the six new entering teams had only two months to develop their teams. The wild card for the fourth participant in the semi-finals was decided by the RoboCup Challenge. Each team had one try to have a single robot on the field score a goal in three different unknown situations. Although none of the teams successfully performed and scored, the UNSW secured the wild card spot with their steady performance (the shortest distance between the ball and the opponent goal).

In the first semi-final, Osaka gained one goal in the first half but LRP quickly recovered and scored two goals. Osaka attacked LRP's goal many times in the second half, but their attacks were blocked by the LRP's robust defense. Since Osaka had no goalie, LRP gained two more goals in the second half. LRP's walking and image processing seem to be very robust.

The second semi-final between UNSW and CM-Trio-99 was an interesting game. The two teams had already played in the round-robin phase and CM-Trio-99 had won. However, in this game, the CM-Trio-99 team encountered some unexpected problems and lost 2-1 to the UNSW team. UNSW scored two goals in the first half, the first goal by squeezing the ball into the goal behind CM-Trio-99's goalie. In the second half CM-Trio-99, partially recovered, could not score more than one goal, as the UNSW goalie was notably strong.

The final between UNSW and LRP started at 1:30pm on August 4, 1999, being observed by a large crowd. The first half ended with UNSW scoring a goal into its own goal. In the second half, many attacks by LRP showed their superiority and gained three goals in spite of the nice defense by the UNSW goalie. UNSW still gained one goal through a quick attack just after the kickoff. LRP won the championship. We will have twelve teams in RoboCup-2000 in Melbourne.

Fig. 5. Two Sony dog robots in a game.

6 RoboCup Jr. Exhibition

RoboCup Jr. is an initiative where children can get hands-on experience with advanced robotic topics. It is an educational project aiming at providing an environment for children to learn general science and technology through robotics [7, 9]. Due to the long range goal of RoboCup aiming at having a robotic team play a real human team in 2050, it is essential that younger generations get involved in science and technology in general, and in RoboCup activities in particular.

RoboCup Jr. is designed to enhance education using the excitement of the soccer game and the sense of the technical complexity of the real world by actually programming physical entities, instead of virtual creatures.

Unlike in other RoboCup soccer leagues that are designed for top-level research institutions, RoboCup Jr. has flexible and easy to start setup with using several robot platforms, such as LEGO Mindstorms, and easy-to-program environments. Children are expected to have a hands-on experience actually building and program robots, playing games, and learning general technical lessons from their experiences.

While RoboCup Jr. is in its infancy, we are planning to enlarge this activity to have a wide variety of educational programs in a very systematic manner with solid support from education and developmental psychology research.

6.1 Characteristics of RoboCup Jr.

For children to easily get involved in RoboCup Jr., several commercial robot platforms and robot kits are used with specific size and configurations.

Leagues consist of competition division, such as RoboCup Jr. Soccer League, and collaborative division, such as RoboCup Jr. Performers that involves parade and dancing. Even within the RoboCup Jr. Soccer League, several levels are planned to be provided depending upon the educational needs.

The easiest level focused on assembly of robot kits to learn how to construct robots. Similarly, the focus can be placed on how to program the robots whereas the robot themselves are already provided. The second level consists of building and program a simple robot and play a game or do parade. Both aspects of craftsmanship and programming are required. Higher levels may be arranged for more advanced children and for undergraduates that are yet to get involved in the research-oriented RoboCup leagues. The numbers of robots used may vary. A simple game can be just a one on one robot game.

The standard set up for the entry-level league is to have a field of size $120cm$ by $90cm$ with $5cm$ walls around. The field has gray scale from one end to the other so that a simple sensor can detect the approximate location of robots in the field along one axis. Pre-assembled LEGO Mindstorms or equivalent are provided and children program the robots. A programming environment is available so that children can program and play a game in a short time, approximately within one hour.

6.2 Research Challenges

Many research challenges arise within RoboCup Jr. The first issue is to define a comprehensive system for robotics education, and science and technology education through robotics. While there are major on-going efforts in Information Technology education, there are only a few efforts made on education with robotics as the central theme. We believe that the use of robotics greatly enhances technical education due to the sense of reality involved in using real physical objects that can be programmed.

Secondly, the development of appropriate infrastructures, such as robot kits, programming environments, and educational materials, is a major challenge. Since RoboCup Jr. aims at wide-spread activities of world-wide scope, a rigid and well-designed environment is essential. This is also an important challenge from the aspect of human-computer interaction.

While RoboCup Jr. mainly targets education for children, it can be applied for science and technology literacy for the general public with non-scientific or technological background.

6.3 RoboCup Jr. at RoboCup-99

In RoboCup-99, Stockholm, a small experimental exhibition was carried out using one field to play with two PCs for programming. Children showed up on the day and programmed robots to play one-vs-one soccer using pre-assembled LEGO Mindstorms. Over 60 children participated over 2 days of exhibitions and a large number of games were played, as well as informal tournaments. Children were very involved and participated actively (see Figure 6).

Fig. 6. Children participating at RoboCup Jr.

RoboCup-99 included three main demonstrations of RoboCup Jr. An Israeli team showed the use of robot soccer in high school education with a penalty shooting robot developed by high school students (age approximately 16-17). Bandai showed a remotely-controlled soccer game with two robots, each with a holding and a shooting device, on each team. The LEGO Lab, as described above, developed by the University of Aarhus, arranged open sessions for children (ages 7-14) to develop their own robot soccer players and to participate in a tournament.

A few additional interesting observations can be made. First, it was confirmed that 7-9 years old children can program robots, such as the Lego Mindstorms, within 30 minutes when the appropriate environment is provided and a task is well defined. It is less feasible to expect children to be able to write code in some higher-level language. Instead there was clear evidence that an easy visual programming environment showed to be of great help.

Second, it was noticed that different children approach the robot programming task in different ways. Some immediately program the robots and others cautiously delay their programming until they have observed the essence of the games. After RoboCup-99, Henrik Lund and one of his colleagues carried out RoboCup Jr. with two robots per team at the Mindfest at the MIT Media lab., and observed other emotional reactions of children.

RoboCup Jr. events are planned for the RoboCup European Championship in Amsterdam, in May 2000 and for RoboCup-2000, in Melbourne, in August.

7 Conclusion

RoboCup is growing and expanding in many respects. The number of participants is increasing, and so is the complexity of the organization. A new league with the Sony legged robots was officially introduced this year. The performance of all the teams in the different leagues is clearly increasing. The research contributions are getting increasingly well identified and reported. RoboCup-99 attracted the interest of many researchers, of the general public, and of the media. RoboCup-99 continues to pursue its core goal as a research environment, stimulating and generating novel approaches in artificial intelligence and robotics. The RoboCup environment provides a remarkable concrete platform for researchers interested in handling the complexities of real-world problems.

In 2000, there will be the first RoboCup European Championship in Amsterdam, in May, with a workshop, RoboCup Jr,. and competitions in the simulation, small-size, and middle-size leagues. The annual international RoboCup-2000 will be held in Melbourne, Australia, in August, in connection with the Sixth Pacific Rim International Conference on Artificial Intelligence (PRICAI-2000). RoboCup-2000 will include a technical workshop, competitions in all of the four leagues (simulation, small-size and middle-size robots, and Sony legged robots), RoboCup Jr., and two new demonstrations towards a RoboCup Humanoid league and a RoboCup Search and Rescue league.

Appendix A: Round-Robin Simulation League

This appendix includes all the results from the preliminary games in the simulation leagues. The tables show the numbers of goals, the number of wins, losses, and draws (W/L/D), and the rank of each team within each group. A win was counted against a team that forfeited a game.

Group A

- A1: CMUnited-99
- A2: The Ulm Sparrows
- A3: Headless Chickens III
- A4: Zeng99
- A5: Dash (forfeited the games)

	A1	A2	A3	A4	W/L/D	Rank
A1	-	29-0	17-0	11-0	4/0/0	1
A2	0-29	-	0-12	1-0	2/2/0	3
A3	0-17	12-0	-	3-0	3/1/0	2
A4	0-11	0-1	0-3	-	1/3/0	4

Group B

- B1: 11 Monkeys
- B2: Sibiu Team
- B3: FCFoo
- B4: Pardis
- B5: Gongeroos

	B1	B2	B3	B4	B5	W/L/D	Rank
B1	-	17-0	18-0	2-0	4-0	4/0/0	1
B2	0-17	-	0-5	6-1	0-3	1/3/0	4
B3	0-18	5-0	-	2-0	2-4	2/2/0	3
B4	0-2	1-6	0-2	-	0-9	0/4/0	5
B5	0-4	3-0	4-2	9-0	-	3/1/0	2

Group C

- C1: Mainz Rolling Brains
- C2: IALP
- C3: RoboCup Koblenz
- C4: Erika
- C5: Polytech100

	C1	C2	C3	C4	C5	W/L/D	Rank
C1	-	17-0	25-0	22-0	16-0	4/0/0	1
C2	0-17	-	1-1	0-8	0-2	0/3/1	5
C3	0-25	1-1	-	0-0	0-0	0/1/3	4
C4	0-22	8-0	0-0	-	0-0	1/1/2	2
C5	0-16	2-0	0-0	0-0	-	1/1/2	3

Group D

- D1: Gemini
- D2: Footux-99
- D3: UBU
- D4: KU-Sakura-2

	D1	D2	D3	D4	W/L/D	Rank
D1	-	12-0	14-0	0-0	2/0/1	1
D2	0-12	-	5-0	0-11	1/2/0	3
D3	0-14	0-5	-	0-9	0/3/0	4
D4	0-0	11-0	9-0	-	2/0/1	2

Group E

- E1: ATHumboldt-99
- E2: Paso-Team
- E3: AIACS
- E4: NITStones
- E5: magmaFreiburg

	E1	E2	E3	E4	E5	W/L/D	Rank
E1	-	16-0	18-0	9-0	0-9	3/1/0	2
E2	0-16	-	0-9	0-12	0-11	0/4/0	5
E3	0-18	9-0	-	1-6	0-19	1/3/0	4
E4	0-9	12-0	6-1	-	0-20	2/2/0	3
E5	9-0	11-0	19-0	20-0	-	4/0/0	1

Group F

- F1: CMUnited-98
- F2: Essex Wizards
- F3: Karlsruhe Brainstorms
- F4: Kasuga-bito III
- F5: Cyberoos

	F1	F2	F3	F4	F5	W/L/D	Rank
F1	-	0-0	1-0	19-0	1-0	3/0/1	2
F2	0-0	-	1-0	20-0	3-0	3/0/1	1
F3	0-1	0-1	-	32-0	6-2	2/2/0	3
F4	0-19	0-20	0-32	-	2-0	1/3/0	4
F5	0-1	0-3	2-6	0-2	-	0/4/0	5

Group G

- G1: UvA-Team
- G2: KULRoT
- G3: Smackers-99
- G4: YowAI

	G1	G2	G3	G4	W/L/D	Rank
G1	-	13-0	3-2	0-2	2/1/0	2
G2	0-13	-	0-11	0-20	0/3/0	4
G3	2-3	11-0	-	0-21	1/2/0	3
G4	2-0	20-0	21-0	-	3/0/0	1

Group H

- H1: Oulu-99
- H2: PSI Team
- H3: Rogi2 Soft Team
- H4: SP-United
- H5: Kappa-II

	H1	H2	H3	H4	H5	W/L/D	Rank
H1	-	1-0	5-1	1-0	0-11	3/1/0	2
H2	0-1	-	0-5	0-1	0-2	0/4/0	5
H3	1-5	5-0	-	0-0	0-13	1/2/1	4
H4	0-1	1-0	0-0	-	0-5	1/2/1	3
H5	11-0	2-0	13-0	5-0	-	4/0/0	1

Appendix B: Round-Robin Small-Size League

This appendix includes all the results from the preliminary games in the small-size league. The tables show the numbers of goals, the number of wins, losses, and draws (W/L/D), and the rank of each team within each group. A win was counted against a team that forfeited a game.

Group A

- A1: AllBotz
- A2: Big Red
- A3: RobotIS
- A4: 5dpo

	A1	A2	A3	A4	W/L/D	Rank
A1	-	0-33	0-35	3-1	1/2/0	3
A2	33-0	-	2-1	8-0	3/0/0	1
A3	35-0	1-2	-	4-0	2/1/0	2
A4	1-3	0-8	0-4	-	0/3/0	4

Group B

- B1: Rogi2
- B2: CMUnited-99
- B3: Linked99
- B4: VUB (forfeited the games)
- B5: TPOTS

	B1	B2	B3	B5	W/L/D	Rank
B1	-	2-12	5-0	6-4	3/1/0	2
B2	12-2	-	10-0	4-0	4/0/0	1
B3	0-5	0-10	-	0-9	1/3/0	4
B5	4-6	0-4	9-0	-	2/2/0	3

Group C

- C1: Dynamo (forfeited the games after 4 goals suffered)
- C2: RoboRoos-99
- C3: SingPoly
- C4: Crimson
- C5: Owaribitos

	C1	C2	C3	C4	C5	W/L/D	Rank
C1	-	0-4	0-4	0-4	0-4	0/4/0	5
C2	4-0	-	17-1	10-0	29-0	4/0/0	1
C3	4-0	1-17	-	5-2	10-1	3/1/0	2
C4	4-0	0-10	2-5	-	8-3	2/2/0	3
C5	4-0	0-29	1-10	3-8	-	1/3/0	4

Group D

- D1: J-Star-99
- D2: FU-Fighters
- D3: LuckyStar
- D4: Microb3 (forfeited the games after 4 goals suffered)

	D1	D2	D3	D4	W/L/D	Rank
D1	-	1-10	0-27	4-0	1/2/0	3
D2	10-1	-	2-1	4-0	3/0/0	1
D3	27-0	1-2	-	4-0	2/1/0	2
D4	0-4	0-4	0-4	-	0/3/0	4

Appendix C: Round-Robin Middle-Size League

This appendix includes all the results from the preliminary games in the middle-size league. The tables show the numbers of goals, the number of wins, losses, and draws (W/L/D), and the rank of each team within each group.

Group A

- A1: CS Freiburg
- A2: ISocRob
- A3: Raiders
- A4: GMD Robots
- A5: KIRC
- A6: Wisely

	A1	A2	A3	A4	A5	A6	W/L/D	Rank
A1	-	5-0	6-0	6-0	6-0	4-0	5/0/0	1
A2	0-5	-	2-1	1-0	1-2	1-1	2/2/1	3
A3	0-6	1-2	-	1-0	0-1	0-3	1/4/0	5
A4	0-6	0-1	0-1	-	5-0	0-4	1/4/0	5
A5	0-6	2-1	1-0	0-5	-	1-1	2/2/1	3
A6	0-4	1-1	3-0	4-0	1-1	-	2/1/2	2

Group B

- B1: Attempto!
- B2: NAIST
- B3: The Ulm Sparrows
- B4: RealMagiCol
- B5: 5dpo-2000
- B6: Matto
- B7: Alpha++

	B1	B2	B3	B4	B5	B6	B7	W/L/D	Rank
B1	-	0-1	0-0	2-0	2-0	3-1	0-1	3/2/1	4
B2	1-0	-	0-3	2-0	3-0	1-0	0-5	4/2/0	3
B3	0-0	3-0	-	2-0	3-0	2-1	0-1	4/1/1	2
B4	0-2	0-2	0-2	-	0-2	0-2	0-2	0/6/0	7
B5	0-2	0-3	0-3	2-0	-	0-1	0-2	1/5/0	6
B6	1-3	0-1	1-2	2-0	1-0	-	0-5	2/4/0	5
B7	1-0	5-0	1-0	2-0	2-0	5-0	-	6/0/0	1

Group C

- C1: Trackies-99
- C2: Agilo
- C3: ART-99
- C4: USC Dream Team-99
- C5: Patriarcas
- C6: Sharif CE
- C7: CoPS Stuttgart

	C1	C2	C3	C4	C5	C6	C7	W/L/D	Rank
C1	-	1-0	1-2	2-1	2-0	1-2	1-3	2/4/0	5
C2	0-1	-	0-0	1-0	3-0	2-1	4-1	4/1/1	1
C3	2-1	0-0	-	1-0	2-0	1-0	0-1	4/1/1	2
C4	1-2	0-1	0-1	-	2-0	0-10	0-3	1/5/0	6
C5	0-2	0-3	0-2	0-2	-	0-4	0-5	0/6/0	7
C6	2-1	1-2	0-1	10-0	4-0	-	2-1	4/2/0	3
C7	3-1	1-4	1-0	3-0	5-0	1-2	-	4/2/0	4

Wild Card Match

	ISocRob	NAIST	SharifCE	W/L/D	Rank
ISocRob	-	1-2	0-3	0/2/0	3
NAIST	2-1	-	0-1	1/1/0	2
SharifCE	3-0	1-0	-	2/0/0	1

Appendix D: Round-Robin Sony Legged Robot League

This appendix includes all the results from the preliminary games in the middle-size league. The tables show the numbers of goals, the number of wins, losses, and draws (W/L/D), and the rank of each team within each group.

Group A

- A1: CM-Trio99 (Carnegie Mellon Univ., U.S.A.)
- A2: Humboldt Hereos (Humboldt Univ., Germany)
- A3: UNSW United (Univ. of New South Wales, Australia)

	A1	A2	A3	W/L/D	Rank
A1	-	2-1	3-0	2/0/0	1
A2	1-2	-	1-2	0/2/0	3
A3	0-3	2-1	-	1/1/0	2

Group B

- B1: BabyTigers-99 (Osaka Univ., Japan)
- B2: UPennalizers (Univ. of Pennsylvania, U.S.A.)
- B3: Team Sweden (Örebro Univ. and Stockholm Univ., Sweden)

	B1	B2	B3	W/L/D	Rank
B1	-	2-0	1-0	2/0/0	1
B2	0-2	-	2-0	1/1/0	2
B3	0-1	0-2	-	0/2/0	3

Group C

- C1: Les 3 Mousquetaires (LRP, France)
- C2: McGill RedDogs (McGill Univ., Canada)
- C3: Araibo (Univ. of Tokyo, Japan)

	C1	C2	C3	W/L/D	Rank
C1	-	3-1	2-0	2/0/0	1
C2	1-3	-	0-1	0/2/0	3
C3	0-2	1-0	-	1/1/0	2

Appendix F: Finals

This appendix includes the final results from the eliminatory games in the four leagues. The simulation league had a double-elimination tournament, while the other leagues had a single-elimination tournament.

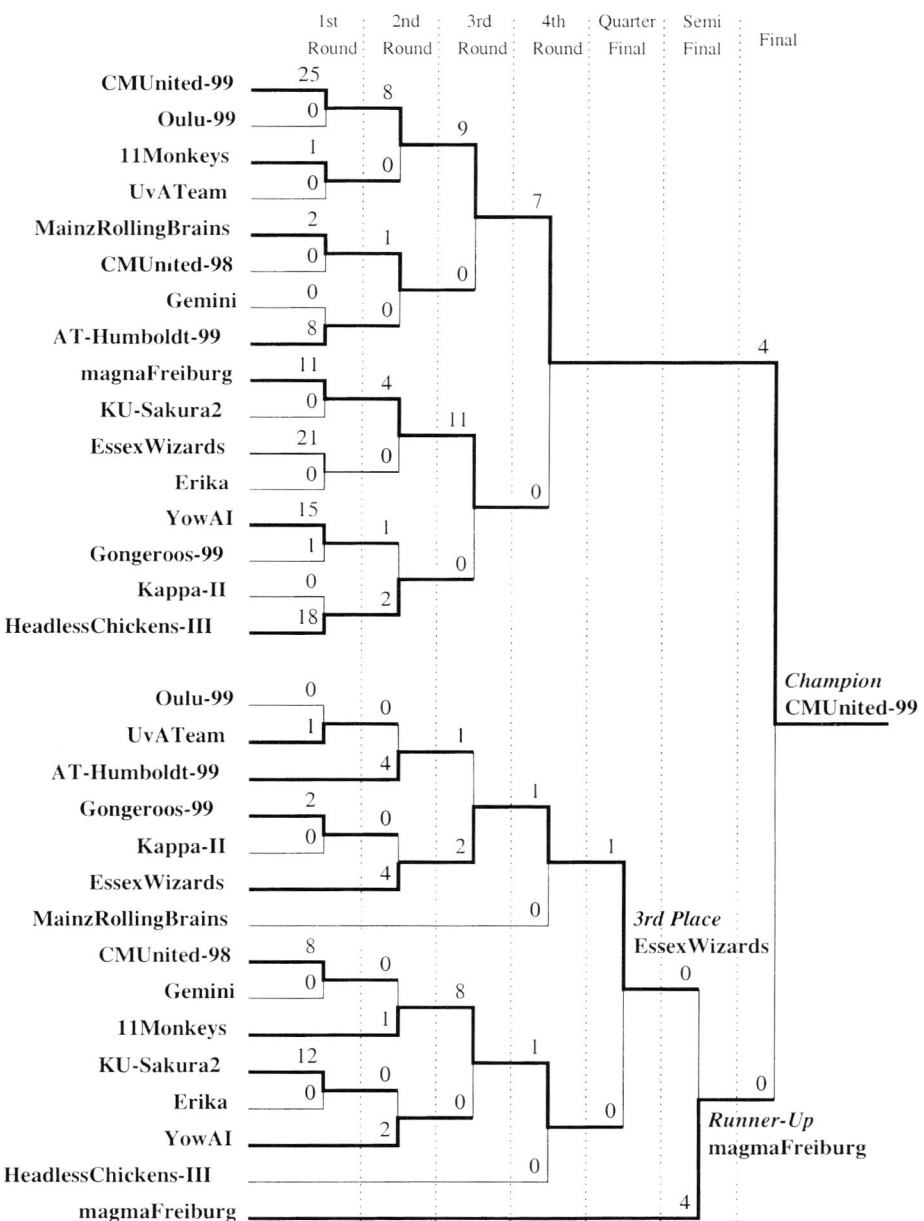

Fig. 7. Double-Elimination Tournament: Simulation League

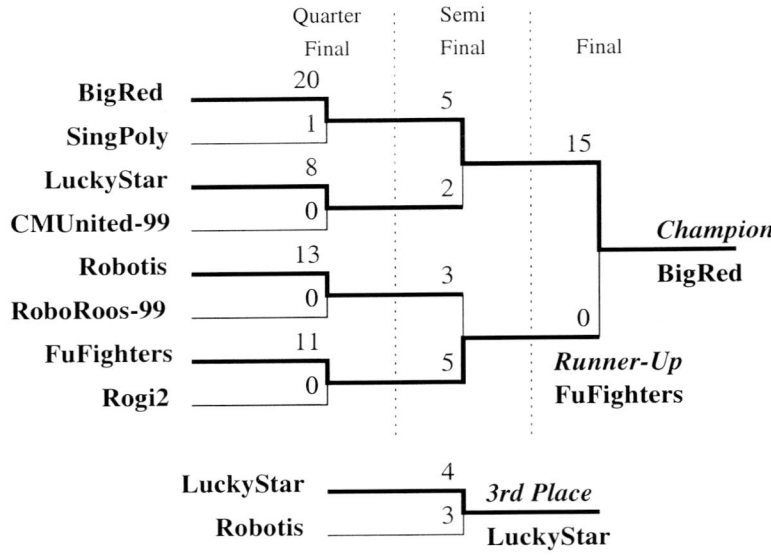

Fig. 8. Single-Elimination Tournament: F-180 Small-Size Robot League

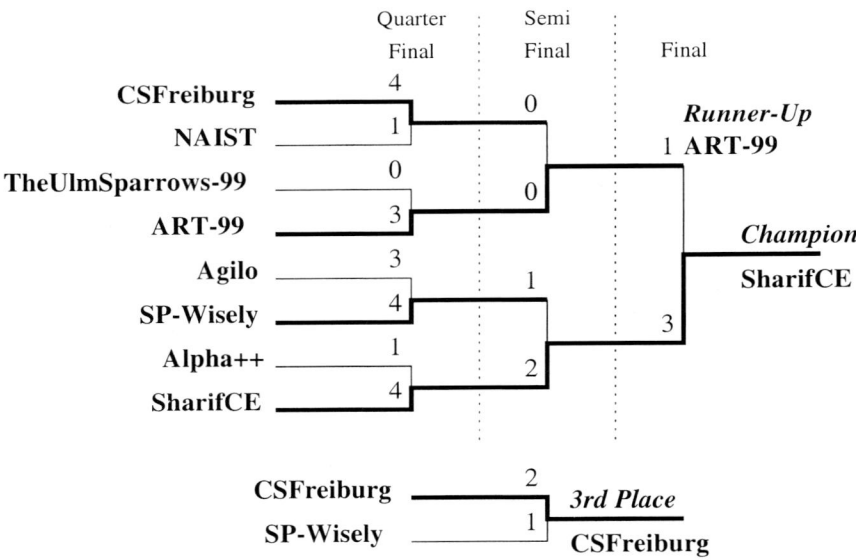

Fig. 9. Single-Elimination Tournament: F-2000 Middle-Size Robot League

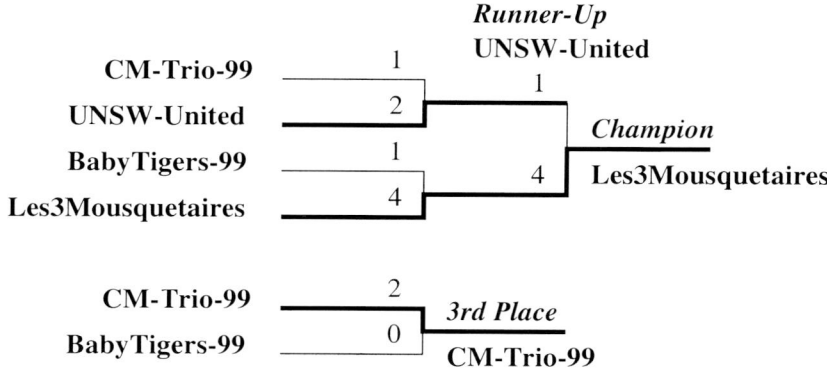

Fig. 10. Single-Elimination Tournament: Sony Legged Robot League

References

1. T. Andou. Refinement of soccer agents' positions using reinforcement learning. In Hiroaki Kitano, editor, *RoboCup-97: Robot Soccer World Cup I*, pages 373–388. Springer Verlag, Berlin, 1998.
2. E. Andre, G. Herzog, and T. Rist. Generating multimedia presentations for RoboCup soccer games. In Hiroaki Kitano, editor, *RoboCup-97: Robot Soccer World Cup I*, pages 200–215. Springer Verlag, Berlin, 1998.
3. M. Asada and H. Kitano, editors. *RoboCup-98: Robot Soccer World Cup II*. Springer, Lecture Notes in Artificial Intelligence, 1999.
4. K. Binsted. Character design for soccer commentary. In Minoru Asada and Hiroaki Kitano, editors, *RoboCup-98: Robot Soccer World Cup II*. Springer Verlag, Berlin, 1999.
5. S. Coradeschi and J. Malec. How to make a challenging AI course enjoyable using the RoboCup soccer simulation system. In Minoru Asada and Hiroaki Kitano, editors, *RoboCup-98: Robot Soccer World Cup II*. Springer Verlag, Berlin, 1999.
6. H. Kitano, editor. *RoboCup-97: Robot Soccer World Cup I*. Springer, Lecture Notes in Artificial Intelligence, 1998.
7. H. Kitano, S. Suzuki, and J. Akita. RoboCup Jr.: RoboCup for Entertainment. In *Proceedings of IEEE International Conference on Systems, Man, and Cybernetics 1999 (SMC-99)*, Tokyo, 1999.
8. H. Kitano, M. Tambe, P. Stone, M. Veloso, S. Coradeschi, E. Osawa, H. Matsubara, I. Noda, and M. Asada. The RoboCup synthetic agent challenge 97. In *Proceedings of the Fifteenth International Joint Conference on Artificial Intelligence*, 1997.
9. H. Lund and S. Pagliarini. RoboCup Jr. with Lego Mindstorms. In *Proceedings of IEEE International Conference on Systems, Man, and Cybernetics 1999 (SMC-99)*, Tokyo, 1999.
10. H. Matsubara, I. Frank, K. Tanaka-Ishii, I. Noda, H. Nakashima, and K. Hasida. Automatic soccer commentary and RoboCup. In Minoru Asada and Hiroaki Kitano, editors, *RoboCup-98: Robot Soccer World Cup II*. Springer Verlag, Berlin, 1999.
11. I. Noda, H. Matsubara, K. Hiraki, and I. Frank. Soccer server: A tool for research on multiagent systems. *Applied Artificial Intelligence*, 12:233–250, 1998.

12. A. Shinjoh. RoboCup-3D: The construction of intelligent navigation system. In Hiroaki Kitano, editor, *RoboCup-97: Robot Soccer World Cup I*, pages 188–199. Springer Verlag, Berlin, 1998.
13. P. Stone and M. Veloso. Task decomposition, dynamic role assignment, and low-bandwidth communication for real-time strategic teamwork. *Artificial Intelligence*, 110(2):241–273, June 1999.
14. I. Verner. RoboCup: A challenging environment for engineering education. In Minoru Asada and Hiroaki Kitano, editors, *RoboCup-98: Robot Soccer World Cup II*. Springer Verlag, Berlin, 1999.

The CMUnited-99 Champion Simulator Team

Peter Stone[1], Patrick Riley[2], and Manuela Veloso[2]

[1]AT&T Labs — Research
180 Park Ave., room A273
Florham Park, NJ 07932
pstone@research.att.com
http://www.research.att.com/~pstone

[2]Computer Science Department
Carnegie Mellon University
Pittsburgh, PA 15213
{pfr,veloso}@cs.cmu.edu
http://www.cs.cmu.edu/{~pfr,~mmv}

Abstract. The CMUnited-99 simulator team became the 1999 RoboCup simulator league champion by winning all 8 of its games, outscoring opponents by a combined score of 110–0. CMUnited-99 builds upon the successful CMUnited-98 implementation, but also improves upon it in many ways. This paper gives a detailed presentation of CMUnited-99's improvements over CMUnited-98.

1 Introduction

The CMUnited robotic soccer project is an ongoing effort concerned with the creation of collaborative and adversarial intelligent agents operating in real-time, dynamic environments. CMUnited teams have been active and successful participants in the international RoboCup (robot soccer world cup) competitions [1, 2, 15]. In particular, the CMUnited-97 simulator team made it to the semi-finals of the first RoboCup competition in Nagoya, Japan [9], the CMUnited-98 simulator team won the second RoboCup competition in Paris, France [13], and the latest CMUnited-99 simulator team won the third RoboCup competition in Stockholm, Sweden [1].

The CMUnited-99 simulator team is modeled closely after its two predecessors. Like CMUnited-97 and CMUnited-98, it uses layered learning [12] and a flexible team structure [11]. In addition, many of the CMUnited-99 agent skills, such as goaltending, dribbling, kicking, and defending, are closely based upon the CMUnited-98 agent skills. However, CMUnited-99 improves upon CMUnited-98 in many ways. This paper focuses on the research innovations that contribute to CMUnited-99's improvements.

Coupled with the publicly-available CMUnited-99 source code [8], this article is designed to help researchers involved in the RoboCup software challenge [3] build upon our success. Throughout the article, we assume that the reader is familiar with the RoboCup simulator, or "soccer server" [5]. A detailed overview of the soccer server, including agent perception and actuator capabilities, is given in [7].

Section 2 describes the improvements in CMUnited-99's low-level skills, including the introduction of teammate and opponent modeling capabilities. Sec-

---

[1] The CMUnited small-robot team is also a two-time RoboCup champion [14, 16].

tion 3 presents the improvements in CMUnited-99's ball handling decision. Section 4 focuses on the process by which the low-level skills were improved. Section 5 introduces the concept of layered extrospection, a key advance in our development methodology. Section 6 summarizes CMUnited-99's successful performance at RoboCup-99 and concludes.

2 Agent Skills

CMUnited-99's basic skills are built mostly on CMUnited-98's skills. This section focusses on CMUnited-99's improvements in low-level skills.

2.1 Ball Velocity Estimation

One of the most important part of good ball handling skills is an accurate estimation of the ball's velocity. When a player is facing the ball, an estimate of the ball's velocity is "visible" via the player's sensory perceptions. However, in both CMUnited-98 and CMUnited-99, when an agent is handling the ball, it uses *position based velocity estimation*. That is, if the agent observes the ball on two successive cycles, it knows the actual path which the ball traveled, and therefore its current velocity.

While this is intuitively a fairly simple idea, there are several complications. First, each agent needs to keep track of the kicks it performed in order to accurately estimate the ball velocity. This is for cases where the agent is not receiving sensations every cycle. The server gives information about kicks that it received, and it is important to note when requested kicks are not executed by the server.

The second complication is somewhat of an artifact of our world model. Our agents store the current position of all objects in global coordinates by converting the objects' sensed relative positions to global coordinates based on the agent's estimated current position. When an agent gets new visual information, it re-estimates its current position. Both of the estimates are quite noisy since they are based on usually distant flags. This means that the ball's old position and new position are in essentially different coordinate frames. In Figure 1, objects in the old coordinate frame are represented in dark grey and objects in the new coordinate frame are lighter. As shown, our agents can calculate the disparity between the coordinate frames by taking the difference of the player's predicted position at time t (judged in the coordinate frame from time $t-1$) and the player's observed position at time t. The ball's position at time $t-1$ can then be moved to the new coordinate frame. The ball velocity is then simply the difference between it's position at time t and position at time $t-1$. This gives a good estimate of the ball's velocity because the only error left is the error in the reported ball positions. When the ball is close to the player, this error is quite small.

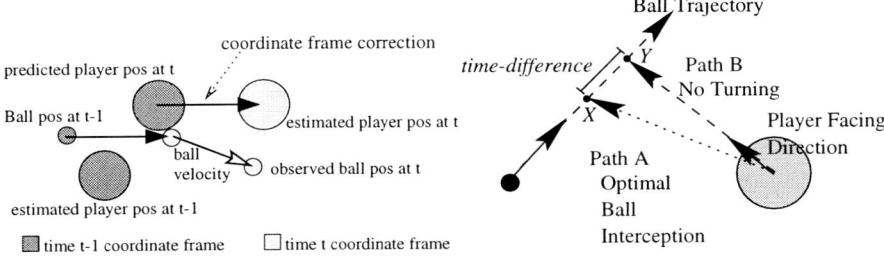

Fig. 1: Correcting for position error in ball velocity estimation.

Fig. 2: Deciding whether to turn in ball interception.

2.2 Ball Interception

The basic structure of our ball interception strategy is the same as in CMUnited-98. We successively simulate the ball's positions on future cycles and then determine if the agent can reach a spot that the ball will occupy before the ball does.

An important change to this scheme is an emphasis on dashing instead of turning. In order to execute less turns while in pursuit of the ball, the agents now first calculate how long it would take to intercept the ball with no turning at all. This is just a simple ray-ray intersection as shown in Figure 2, path B. *time-difference* is the distance between the calculated optimal (point X) and the path with no turning (point Y), judged by how long it would take the ball to get from X to Y. If *time-difference* is below a threshold of a few cycles, then the agent will proceed along path B instead of path A. Proceeding along path B will always result in a dash instead of a turn.

2.3 Using Models of Opponents and Teammates

Deciding when to shoot at the goal has a huge impact on the performance of a team. CMUnited-98 made this decision based on several very inaccurate heuristics. CMUnited-99 makes this decision in a more principled way by using a model of an "optimal" goalie. That is, we use a model of a goalie that reacts instantaneously to a kick, moves to exactly the right position to stop the ball, and catches with perfect accuracy.

When deciding whether to shoot, the agent first identifies its best shot target. It generally considers two spots, just inside of the two sides of the goal. The agent then considers the lines from the ball to each of these possible shot targets. *shot-target* is the position whose line is further from the goalie's current position.

The agent then predicts, given a shot at *shot-target*, the ball's position and goalie's reaction using the optimal goalie model. We use the following predicates:

blocking-point The point on the ball's path for which an optimal goalie heads.
ball-to-goalie-cycles The number of cycles for the ball to get to the *blocking-point*
goalie-to-ball-cycles The number of cycles for the goalie to get to the *blocking-point*

shot-margin $=$ *ball-to-goalie-cycles$-$goalie-to-ball-cycles*

better-shot(k) Whether teammate k has a better shot than the agent with the ball, as judged by *shot-margin*

The value *shot-margin* is a measure of the quality of the shot. The smaller the value of *shot-margin*, the more difficult it will be for the goalie to stop the shot. For example, for a long shot, the ball may reach the *blocking-point* in 20 cycles (*ball-to-goalie-cycles*$=$ 20), while the goalie can get there in 5 cycles (*goalie-to-ball-cycles*$=$ 5) This gives a *shot-margin* of 15. This is a much worse shot than if it takes the ball only 12 cycles (*ball-to-goalie-cycles*$=$ 12) and the goalie 10 cycles to reach the *blocking-point* (*goalie-to-ball-cycles*$=$ 10). The latter shot has a *shot-margin* of only 2. Further, if *shot-margin*$<$ 0, then the "optimal" goalie could not reach the ball in time, and the shot should succeed.

Using a model of opponent behavior gives us a more reliable and adaptive way of making the shooting decision. We can also use it to make better passing decisions. When near the goal, the agent may often be faced with the decision about whether to pass or shoot the ball. The agent with the ball simulates the situation where its teammate is controlling the ball, using the goalie model to determine how good of a shot the teammate has. If the teammate has a much better shot, then the predicate *better-shot*(k) will be true. This will tend to make the agent pass the ball, as described in Section 3.

Note that this analysis of shooting ignores the presence of defenders. Just because the goalie can not stop the shot (as judged by the optimal goalie model) does not mean that a nearby defender can not run in to kick the ball away.

2.4 Breakaway

An important idea in many team ball sports like soccer is the idea of a "breakaway." Intuitively, this is when some number of offensive players get the ball and themselves past the defenders, leaving only perhaps a goalie preventing them from scoring. After looking at logfiles from previous competitions, we saw many opportunities for breakaways which were not taken advantage of.

The first question which has to be answered is "What exactly is a breakaway?" This is built upon several predicates (note that we can naturally reflect these to the other side of the field):

controlling-teammate Which teammate (if any) is currently controlling the ball. "Control" is judged by whether the ball is within the kickable area of a player.

controlling-opponent Which opponent (if any) is currently controlling the ball

opponents-in-breakaway-cone The breakaway cone is shown in Figure 3. The cone has its vertex at the player with the ball and extends to the opponents goal posts.

our-breakaway $= $ (*controlling-teammate*$\neq$ None) $\wedge$ (*controlling-opponent*$=$None) $\wedge$ (*opponents-in-breakaway-cone*$\leq$1)

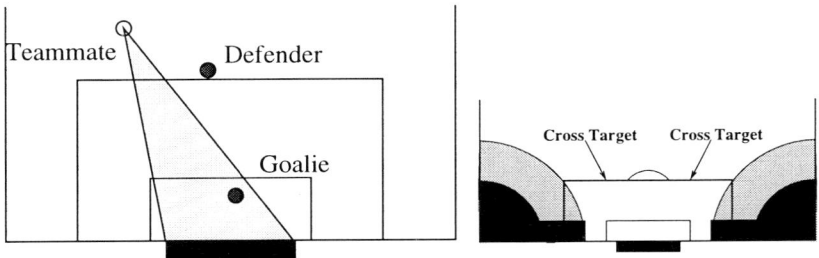

Fig. 3: The Breakaway Cone

Fig. 4: Important regions for the "cross" behavior

The first new skill we use in breakaways is a generalization of dribbling called "kick and run." When executing the normal dribbling skill, the agent aims to do one kick and then one dash and have the ball end up in its kickable area again. However, this causes a player dribbling the ball to move only half as quickly as a player without the ball since half of its action opportunities are spent kicking rather than moving. Therefore, defenders are able to easily catch up to dribbling players. For kick and run, the agents aim for one kick and n dashes before being in control of the ball again. In effect, they kick the ball harder to allow them to spend more of their time running.

We use the optimal model described in Section 2.3 to help make the decision about when to shoot. During a breakaway, the agent shoots when either one of the following is true:

- *shot-margin* (defined in Section 2.3) gets below a certain threshold (1 cycle in CMUnited-99)
- The time that it would take for the goalie to proceed directly to the ball and steal it gets below a certain threshold (6 cycles in CMUnited-99).

This skill was extremely effective in the competition, with the vast majority of our goals being scored using the specialized breakaway code.

3 Ball Handling Decision

One crucial improvement in CMUnited-99 is the agents' decision-making process when in control of the ball. The decisions made at these times are the most crucial in the robotic soccer domain. Which an agent chooses affects the future options of teammates and opponents.

The agent uses a complex heuristic decision mechanism, incorporating a machine learning module, to choose its action. The most significant changes from CMUnited-98 are that the agents use special-purpose code for breakaways (see Section 2.4); that the pass-evaluation decision tree [10] has been retrained during practice games to capture the agents' improved ball-interception ability (see Section 2.2; that the agents can cross the ball (see below); and that the agents consider whether there is a teammate in a better position than they are to shoot the ball (see Section 2.3).

In presenting the agent decision-making process, we make use of the predicates defined in Section 2 as well as the following:

distance-to-their-goal The distance to the opponent's goal
distance-to-our-goal The distance to own goal
opponents-in-front-of-goal The number of opponents (including the goalie) in the breakaway cone shown in Figure 3
closest-opponent The distance to the closest opponent
closer-to-goal(k) Whether teammate k is closer to the opponent's goal than the agent with the ball
can-shoot Whether *distance-to-their-goal* < 25 and (*opponents-in-front-of-goal* ≤ 1 and *shot-margin* ≤ 6.
can-shoot(k) Same as above but from teammate k's position
congestion = $\sum_{opponents} \frac{1}{(distance-to-opponent)^2}$)
congestion(k) Same as above but from teammate k's position
can-dribble-to(x) No defender is nearby or in a cone extending towards the point x

Following is a rough sketch of the decision-making process without all of the parametric details. In all cases, passes are only made to teammates that the decision tree predicts will be able to successfully receive the pass (called *potential receivers* or PR below). If there is more than one potential receiver satisfying the given criteria, then the one predicted with the highest confidence to receive the pass is chosen.

- If $\exists r \in$ PR s.t. *better-shot*(r): pass to r.
- If *our-breakaway*: execute special-purpose breakaway code (see Section 2.4).
- If (*distance-to-their-goal* < 17 and *opponents-in-front-of-goal* ≤ 1) or *shot-margin* ≤ 3: shoot on goal.
- At the other extreme, if *distance-to-our-goal* < 25 or *closest-opponent* < 10: clear the ball (kick it towards a sidelines at midfield and not towards an opponent [13]).
- If $\exists r \in$ PR s.t. *closer-to-goal*(r) and *can-shoot*(r) and *congestion*(r) $\leq$ *congestion*: pass to r.
- If *can-dribble-to*(opponent's goal): dribble towards the goal.
- If $\exists r \in$ PR s.t. *closer-to-goal*(r) and *congestion*(r) $\leq$ *congestion* (even if unable to shoot): pass to r.
- If close to a corner of the field (within a grey or black area in Figure 4) then *cross* the ball as follows.
  - if very near the base line or the corner (in the black area): kick the ball across the field (to "cross target"), even if no teammate is present ("cross it").
  - If able to dribble towards the baseline: dribble towards the baseline (for a later cross).
  - If able to dribble towards the corner: dribble towards the corner.
  - Otherwise, cross it.

Even though the cross doesn't depend on a teammate being present to receive the ball, we observed many goals scored shortly after crosses due to teammates being able to catch up to the ball and shoot on goal.
- If *can-shoot*: shoot.
- *can-dribble-to*(one of the corner flags): dribble towards the corner flag.
- If approaching the line of the last opponent defender (the offsides line): send the ball (clear) past the defender.
- If $\exists r \in$ PR s.t. *closer-to-goal*(r) or *congestion*(r) $\leq$ *congestion*: pass to r.
- no opponent is nearby: hold the ball (i.e. essentially do nothing and wait for one of the above conditions to fire).
- If $\exists r \in$ PR s.t. no opponent is within 10 or r: pass to r.
- Otherwise: Kick the ball away (clear).

Notice that such a ball-handling strategy can potentially lead to players passing the ball backwards, or away from the opponent's goal. Indeed, we observed such passes several times during the course of games. However, the forward passes and shots are further up in the ball-handling decision, and therefore will generally get executed more often.

4 Off-line Training

For the various agent skills described in Section 2 and in [13], there are many parameters affecting the details of the skill execution. For example, in the ball skill of dribbling, there are parameters which affect how quickly the agent dashes, how far ahead it aims the ball, and how opponents affect the location of the ball during dribbling.

The settings for these parameters usually involve a tradeoff, such as speed versus safety, or power versus accuracy. It is important to gain an understanding of what exactly those tradeoffs are before "correct" parameter settings can be made.

We created a trainer client that connects to the server as an omniscient off-line coach client (this is separate from the on-line coach). The trainer is responsible for three things:

1. **Repeatedly setting up a particular training scenario.** In the dribbling skill, for example, the trainer would repeatedly put a single agent and the ball at a particular spot. The agent would then try to dribble the ball to a fixed target point.
2. **Recording the performance of the agent on the task.** Here we use a hand-coded performance metrics, generally with very simple intuitive ideas. In the kicking skill, for example, we record how quickly the ball is moving, how accurate the kicking direction is, and how long it took to kick the ball.
3. **Iterating through different parameter settings.** Using the server's communication mechanism, the trainer can instruct the client on which parameter settings to use. The trainer records the performance of the agent for each set of parameter values.

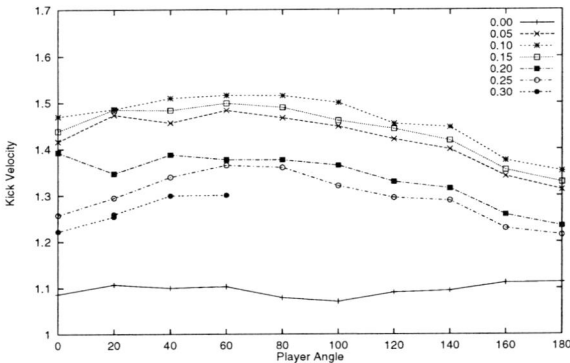

Fig. 5: An example of training data

Once the scenario is set up, the system runs autonomously. Since most skills only involve one or two clients, we could afford to have the trainer iterate over many possible parameter values, taking several hours or days.

Once the trainer has gathered the data, we would depict the results graphically and decide which parameters to use. An example for the hard kicking skill is shown in Figure 5. The two parameters varied for the test shown are the angle the agent is facing relative to the kicking angle (the x-axis), and the buffer around the player out of which the agent tries to keep the ball (the different lines).

Sometimes, the "optimal" parameter selection was fairly clear. For example, in Figure 5, we are trying to maximize the kick velocity. Therefore, we would select a player angle of approximately 60 degrees and a buffer of 0.10. Other times, the data looked much noisier. In those cases we could narrow our search down somewhat and get more data over the relevant parts of the parameter space.

We were sometimes limited by processing power in the breadth or resolution of the parameter space that we could examine. A more adaptive searching strategy, such as might be given by various learning techniques like genetic programming [4], would be a useful addition.

5 Layered Disclosure

A perennial challenge in creating and using complex autonomous agents is following their choice of actions as the world changes dynamically, and understanding why they act as they do. In complex scenarios, even the human computer-agent developer is often unable to identify what exactly caused an agent to act as it did in a given situation. Adding support for human developers and observes to better follow and understand the actions of autonomous agents can be a significant enhancement to processes of development and use of agents.

To this end, we introduce the concept of *layered disclosure* by which autonomous agents include in their architecture the foundations necessary to allow a person to probe into the specific reasons for an agent's action. This probing may be done at any level of detail, and either retroactively or while the agent is acting.

A key component of layered disclosure is that the relevant agent information is organized in *layers*. In general, there is far too much information available to display all of it at all times. The imposed hierarchy allows the user to select at which level of detail he or she would like to probe into the agent in question.

When an agent does something unexpected or undesirable, it is particularly useful to be able to isolate precisely *why* it took such an action. Using layered disclosure, a developer can probe inside the agent at any level of detail to determine precisely what needs to be altered in order to attain the desired agent behavior.

Layered disclosure was a significant part of the development of CMUnited-99, and led to many of the improvements in the team over CMUnited-98. Our development of layered disclosure was inspired in part by our own inability to trace the reasons behind the actions of CMUnited-98. For example, whenever a player kicks the ball towards its own goal, we would wonder whether the agent was mistaken about its own location in the world, whether it was mistaken about the ball's or other agents' locations, or if it "meant" to kick the ball where it did, and why. Due to the dynamic, uncertain nature of the environment, it is usually impossible to recreate the situation exactly in order to retroactively figure out what happened.

Our layered disclosure implementation is publicly available[8]. It can easily be adapted for use with other RoboCup simulator teams.

During the course of a game, our agents store detailed records of selected information in their perceived world states, their determination of their short-term goals, and their selections of which actions will achieve these goals, along with any relevant intermediate decisions that lead to their action selections.

After the game is over, the logfile can be replayed using the standard "log-player" program which comes with the soccer server. Our disclosure module, implemented as an extension to this logplayer, makes it possible to inspect the details of an individual player's decision-making process at any point.

In the remainder of this section we provide two examples illustrating the usefulness of layered disclosure.

5.1 Discovering Agent Beliefs

When observing an agent team performing, it is tempting, especially for a person familiar with the agents' architectures, to infer high level beliefs and intentions from the observed actions. Sometimes, this can be helpful to describe the events in the world, but misinterpretation is a significant danger.

Consider the example in Figure 7. Here, two defenders seem to pass the ball back and forth while quite close to their own goal. In general, this sort of passing back and forth in a short time span is undesirable, and it is exceptionally dangerous near the agents' own goal. Using the layered disclosure tool, we get the information displayed in Figure 6.

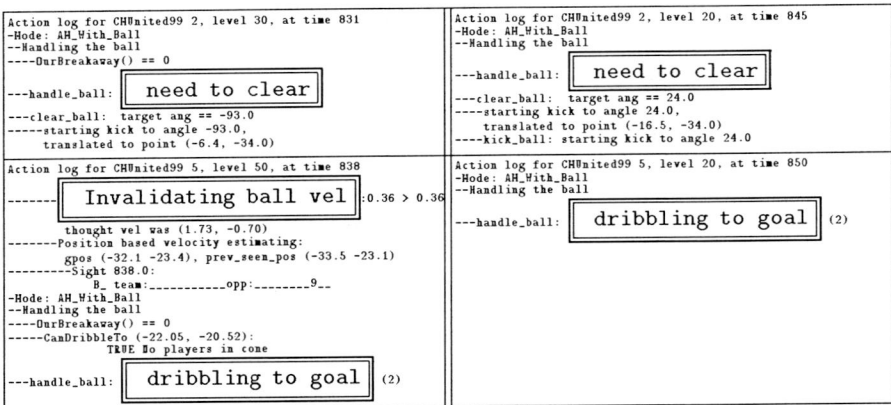

Fig. 6: Layered disclosure for the passing example (the boxes have been added for emphasis).

First, we see that in both cases that player number 2 was in control of the ball (time 831 and 845), it was trying to clear it (just kick it away from the goal), not pass to player number 5. Given the proximity of the goal and opponents, clearing is a reasonable behavior here. If a teammate happens to intercept a clear, then our team is still in control of the ball. Therefore, we conclude that this agent's behavior matches what we want and expect.

Next, we can see that player number 5 was trying to dribble towards the opponent's goal in both cases that he controlled the ball (time 838 and 850). There are no opponents immediately around him, and the path on the way to the goal is clear. This agent's intention is certainly reasonable.

However, at time 838, player number 5 does not perform as it intended. Rather than dribbling forward with the ball, it kicked the ball backwards. This points to some problem with the dribbling behavior. As we go down in the layers, we see that the agent invalidated the ball's velocity. This means that it thought the ball's observed position was so far off of its predicted position that the agent's estimate for the ball's velocity could not possibly be right. The agent then computed a new estimate for the ball's velocity based on its past and current positions (see Section 2.1).

Given this estimation of the ball's velocity (which is crucial for accurate ball handling), we are led to look further into how this velocity is estimated. Also, we can compare the estimate of the velocity to the recorded world state. In the end, we find that the ball collided with the player. Therefore, it was invalid to estimate the ball's velocity based on position. In fact, this led us to more careful application of this velocity estimation technique.

In this case, inferring the intentions of the players was extremely challenging given their behaviors. Without layered disclosure, the natural place to look to correct this undesirable behavior would have been in the passing decisions of the players. It would have been difficult or impossible to determine that the problem was with the estimation of the ball's velocity.

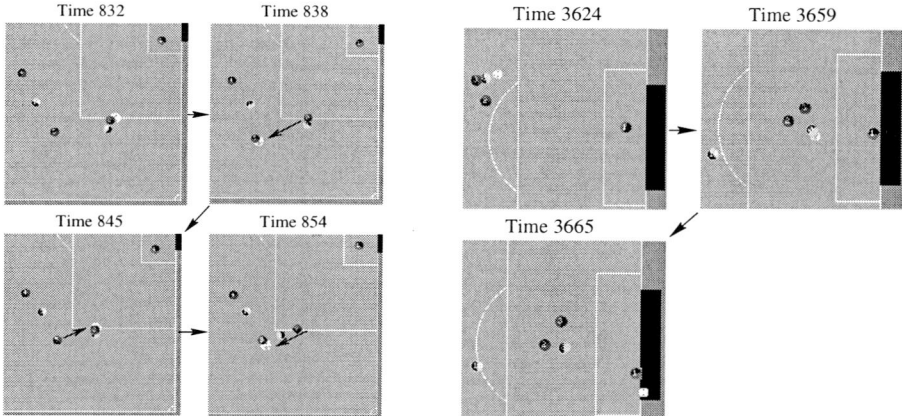

Fig. 7: Undesired passing behavior   **Fig. 8**: Poorly performing defenders

5.2 The Use of Layers

The fact that the agents' recordings are layered is quite important. One important effect is that the layers allow the observer to look at just higher levels, then explore each case more deeply as required.

Consider the example depicted in Figure 8. Here, two defenders are unable to catch up and stop one offensive player with the ball, even though the defenders were in a good position to begin with.

Since this is a scenario that unfolds over many time steps, we need to be able to understand what happens over that time sequence. The first pass at this is to just look at the highest level decision. The first decision our agents make is in which "action mode" they are [13]. This decision is based on which team is controlling the ball, current location, role in the team structure, etc. Usually the fastest player to the ball is in one mode and the second fastest in another.

By using the layered disclosure tool to look at just the highest level of output, two facts can be learned: the defenders switch modes (between fastest and second fastest) frequently, and are often unsure about which team is controlling the ball. The different modes tell the agent to go to different spots on the field. By switching back and forth, the agents will waste a great deal of time turning to face the direction they want to go instead of actually going. Therefore, the agents do not catch up.

Further, The decision about what mode to go into is sometimes affected by which team the agent believes is controlling the ball. Realizing that this value is often unknown should lead to changes in the way that value is determined, or changes in the manner in which it is used.

In this case, making use of layered disclosure to examine just the high-level reasoning decisions of a pair of agents allows us to focus on a problem that would have otherwise been easily overlooked.

Opponent	Affiliation	Score (CMU–Opp.)
Kasuga-Bitos III	Chubu University, Japan	19 – 0
Karlsruhe Brainstormers	University of Karlsruhe, Germany	1 – 0
Cyberoos	CSIRO, Australia	1 – 0
Essex Wizards	University of Essex, UK	0 – 0
Mainz Rolling Brains	University of Mainz, Germany	0 – 2
Gemini	Tokyo Institute of Technology, Japan	8 – 0
11 Monkeys	Keio University, Japan	0 – 1
TOTAL		29 – 3

Table 1: The scores of CMUnited-98's games in the simulator league of RoboCup-99. CMUnited-98 won 3, lost 2, and tied 1 game.

Note 1. The last game was lost by one goal in overtime.

We envision that layered disclosure will continue to be useful in the RoboCup simulator and in other agent development projects, particularly those with complex agents acting in complex, dynamic environments.

6 Results and Conclusion

The third international RoboCup championship, RoboCup-99, was held on July 28–August 4, 1999 in Stockholm, Sweden in conjunction with the IJCAI-99 conference [15]. As the defending champion team, the CMUnited-98 simulator team was entered in the competition. Its code was left unaltered from that used at RoboCup-98 except for minor changes necessary to update to version 5 of the soccer server. Server parameter changes that reduced player size, speed, and kickable area required adjustments in the CMUnited-98 code. However CMUnited-98 did not take advantage of additions to the players' capabilities such as the ability to look in a direction other than straight ahead (simulation of a neck).

The CMUnited-98 team became publicly available soon after RoboCup-98 so that other people could build upon our research. Thus, we expected there to be several teams at RoboCup-99 that could beat CMUnited-98, and indeed there were. Nonetheless, CMUnited-98 performed respectably, winning 3 games, losing 2, and tying 1 and outscoring its opponents by a combined score of 29–3. Table 1 presents the details of CMUnited-98's matches.

Meanwhile, the CMUnited-99 team was even more successful at the RoboCup-99 competition than was its predecessor at RoboCup-98. It won all 8 of its games by a combined score of 110–0, finishing 1st in a field of 37 teams. Table 2 shows CMUnited-99's game results.

Qualitatively, there were other significant differences between CMUnited-98's and CMUnited-99's performances. In RoboCup-98, several of CMUnited-98's matches were quite close, with many offensive and defensive sequences for both teams. CMUnited-98's goalie performed quite well, stopping many shots. In RoboCup-99, CMUnited-99's goalie only had to touch the ball three times over all 8 games. Only two teams (Zeng99 and Mainz Rolling Brains) were able to

Opponent	Affiliation	Score (CMU–Opp.)
Ulm Sparrows	University of Ulm, Germany	29 – 0
Zeng99	Fukui University, Japan	11 – 0
Headless Chickens III	Linköping University, Sweden	17 – 0
Oulu99	University of Oulu, Finland	25 – 0
11 Monkeys	Keio University, Japan	8 – 0
Mainz Rolling Brains	University of Mainz, Germany	9 – 0
Magma Freiburg	Freiburg University, Germany	7 – 0
Magma Freiburg	Freiburg University, Germany	4 – 0
TOTAL		110 – 0

Table 2: The scores of CMUnited-99's games in the simulator league of RoboCup-99. CMUnited-99 won all 8 games, finishing in 1st place out of 37 teams.

create enough of an offense in order to get shots on our goal. Improvements in ball velocity estimation (Section 2.1), ball interception (Section 2.2), and a myriad of small improvements made possible by layered extrospection (Section 5) greatly improved CMUnited-99's midfield play over CMUnited-98.

Another qualitative accomplishment of CMUnited-99 was how closely its actions matched our ideas of what should be done. When watching games progress, we would often just be starting to say "Pass the ball!" or "Shoot it" when the agents would do exactly that. While this is certainly not a solid criterion on which to judge a team in general, it is a testament to our development techniques that we were able to refine behaviors in such a complex domain to match our high level expectations.

There are certainly many improvements to be made. For example, in CMUnited-99's game against Zeng99, our breakaway behavior (Section 2.4) was much less effective in general. This was because the Zeng99 team put an extra defender behind the goalie. CMUnited-99's agents assumed the defender closest to the goal was the goalie. Therefore, the agents applied the goalie model to that defender instead of to the real goalie. This allowed the real goalie to stop many shots which our agents did not anticipate could be stopped. Creating models of other opponents and using them more intelligently could improve this behavior.

Further, adapting models to opponents during play, as well as changing team strategy is a promising future direction. We have done some experimentation with approaches to quick adaptation in complex domains like robotic soccer[6]. Other researchers associated with RoboCup are also looking in this direction, especially with the newly introduced coach agent.

Various software from the team is available [8]. The binaries for the player and coach agents are available. Full source code for the coach agent, the trainer agent, and the layered extrospection tool are also available. Further, skeleton source code for the player agents, including the low level skills, is also available.

References

1. Minoru Asada and Hiroaki Kitano, editors. *RoboCup-98: Robot Soccer World Cup II*. Springer Verlag, Berlin, 1999.
2. Hiroaki Kitano, editor. *RoboCup-97: Robot Soccer World Cup I*. Springer Verlag, Berlin, 1998.
3. Hiroaki Kitano, Milind Tambe, Peter Stone, Manuela Veloso, Silvia Coradeschi, Eiichi Osawa, Hitoshi Matsubara, Itsuki Noda, and Minoru Asada. The RoboCup synthetic agent challenge 97. In *Proceedings of the Fifteenth International Joint Conference on Artificial Intelligence*, pages 24–29, San Francisco, CA, 1997. Morgan Kaufmann.
4. John R. Koza. *Genetic Programming*. MIT Press, 1992.
5. Itsuki Noda, Hitoshi Matsubara, Kazuo Hiraki, and Ian Frank. Soccer server: A tool for research on multiagent systems. *Applied Artificial Intelligence*, 12:233–250, 1998.
6. Patrick Riley and Manuela Veloso. Towards behavior classification: A case study in robotic soccer. In *Proceeding of the Seventeenth National Conference on Artificial Intelligence*, Menlo Park, CA, 2000. AAAI Press. (to appear).
7. Peter Stone. *Layered Learning in Multi-Agent Systems*. PhD thesis, Computer Science Department, Carnegie Mellon University, Pittsburgh, PA, December 1998. Available as technical report CMU-CS-98-187.
8. Peter Stone, Patrick Riley, and Manuela Veloso. CMUnited-99 source code, 1999. Accessible from http://www.cs.cmu.edu/~pstone/RoboCup/CMUnited99-sim.html.
9. Peter Stone and Manuela Veloso. The CMUnited-97 simulator team. In Hiroaki Kitano, editor, *RoboCup-97: Robot Soccer World Cup I*. Springer Verlag, Berlin, 1998.
10. Peter Stone and Manuela Veloso. A layered approach to learning client behaviors in the RoboCup soccer server. *Applied Artificial Intelligence*, 12:165–188, 1998.
11. Peter Stone and Manuela Veloso. Task decomposition, dynamic role assignment, and low-bandwidth communication for real-time strategic teamwork. *Artificial Intelligence*, 110(2):241–273, June 1999.
12. Peter Stone and Manuela Veloso. Layered learning and flexible teamwork in robocup simulation agents. In Manuela Veloso, Enrico Pagello, and Hiroaki Kitano, editors, *RoboCup-99: Robot Soccer World Cup III*, Berlin, 2000. Springer Verlag.
13. Peter Stone, Manuela Veloso, and Patrick Riley. The CMUnited-98 champion simulator team. In Minoru Asada and Hiroaki Kitano, editors, *RoboCup-98: Robot Soccer World Cup II*. Springer Verlag, Berlin, 1999.
14. Manuela Veloso, Michael Bowling, Sorin Achim, Kwun Han, and Peter Stone. The CMUnited-98 champion small robot team. In Minoru Asada and Hiroaki Kitano, editors, *RoboCup-98: Robot Soccer World Cup II*. Springer Verlag, Berlin, 1999.
15. Manuela Veloso, Enrico Pagello, and Hiroaki Kitano, editors. *RoboCup-99: Robot Soccer World Cup III*. Springer Verlag, Berlin, 2000. To appear.
16. Manuela Veloso, Peter Stone, Kwun Han, and Sorin Achim. The CMUnited-97 small-robot team. In Hiroaki Kitano, editor, *RoboCup-97: Robot Soccer World Cup I*, pages 242–256. Springer Verlag, Berlin, 1998.

Big Red: The Cornell Small League Robot Soccer Team

Raffaello D'Andrea, Jin-Woo Lee, Andrew Hoffman, Aris Samad-Yahaya, Lars B. Cremean, and Thomas Karpati

Mech. & Aero. Engr., Cornell University, Ithaca, NY 14853, USA
{rd28, jl206, aeh5, as103, lbc4, tck4}@cornell.edu
http://www.mae.cornell.edu/robocup

Abstract. In this paper we describe Big Red, the Cornell University Robot Soccer team. The success of our team at the 1999 competition can be mainly attributed to three points: 1) An integrated design approach; students from mechanical engineering, electrical engineering, operations research, and computer science were involved in the project, and a rigorous and systematic design process was utilized. 2) A thorough understanding of the system dynamics, and ensuing control. 3) A high fidelity simulation environment that allowed us to quickly explore artificial intelligence and control strategies well in advance of working prototypes.

1 Introduction

In this paper we describe Big Red, the Cornell University Robot Soccer team. The success of our team at the 1999 competition can be mainly attributed to three points:

1. An integrated design approach; students from mechanical engineering, electrical engineering, operations research, and computer science were involved in the project, and a rigorous and systematic design process[6] was utilized.
2. A thorough understanding of the system dynamics, and ensuing control.
3. A high fidelity simulation environment that allowed us to quickly explore AI and control strategies well in advance of working prototypes.

The paper is organized as follows. In Section 3, we describe the electrical and mechanical aspects of the project, followed by a description of the global vision system in Section 4. The team skills are described in Section 5, followed by the artificial intelligence and strategy in Section 6. We include some of the other features of our team in Section 7.

2 Team Development

Team Leader: Raffaello D'Andrea[Assistant Professor]

Team Members:
Dr. Jin-Woo Lee[Visiting Lecturer]
Andrew Hoffman[Master of Engineering student]
Aris Samad-Yahaya[Master of Engineering student]
Lars B. Cremean[Undergraduate student]
Thomas Karpati[Master of Engineering student]
Affiliation: Cornell University, U.S.A
Web page http://www.mae.cornell.edu/RoboCup

3 Electro-mechanical System

3.1 Mechanical Design

The Cornell University team consists of two mechanical designs, one for the field players and the second for goalkeeper. All of the robots have a unidirectional kicking mechanism powered by one solenoid (two for the goalkeeper).

The robots have a design mass of 1.5 kg, a maximum linear acceleration of 5.1 m/s^2, and a maximum linear velocity of 2.5m/s. The goalkeeper has a different design from the field players. It is equipped with a holding and kicking mechanism that can catch a front shot on goal, hold it for an indefinite amount of time, and make a pass. All of the designs were performed using ProE[10].

Listed below are the main characteristics of our robots:

Characteristic	Goal Keeper	Field Player
Weight	1.78 kg	1.65 kg
Max. Acceleration	5.90 m/s^2	5.10 m/s^2
Max. Velocity	1.68 m/s	2.53 m/s
Max. Kicking Speed	4.18m/s	2.6 m/s
Operating time	30 min per battery pack	
Special function	Ball Holding mechanism	

3.2 On-Board Electronics

The main function of the on-board electronics is to receive left and right wheel velocity signals via wireless communication and to implement local feedback control to ensure that the wheel velocity were regulated about the desired values. Considering the speed, memory space, I/O capability, and the extension flexibility, 16bit 50MHz microcontrollers are used.

In order to get a precise kick, a ball detecting system separate from the global vision system is implemented. An infrared system is used to detect the ball. It informs the microcontroller when the ball has come into contact with the front of the robot. When the global vision system makes the observation that the robot is in a position to kick, a command is sent to inform the robot to kick the ball if and only if the infrared system detects the ball. All the capture and layout for the on-board electronics were performed in-house using OrCad[11].

3.3 Communication

After careful considerations and trade-off analysis, the wireless communication was limited to one-way transmission from the global AI workstation to each robot. The main justification for the decision is the lack of on-board local sensing information. The one-way transmission saves the communication time, as compared to the two-way communication, and simplifies the AI strategy and the on-board firmware. The wireless communication system takes 12.5ms to transmit the information from the AI workstation to the robots.

With the experimentally verified assumption that the robots do not drift far from the desired position between frames, the need for local sensing and correction is minimal. Based on the current design, the robots can drift 5cm at maximum speed. For debugging purposes, each robot has the capability to transmit data back to the AI workstation.

4 Visual Tracking Algorithm

A dedicated global vision system identifies the ball and robot locations as well as the orientation of our robots. In order to determine the identity of each robot and their orientation, blob analysis[5] is used as a basic algorithm. The vision system perceives the current state of the game and communicates this state to the AI workstation allowing decisions to be made in real-time in response to the current game play. The end result of the vision system is the reliable real-time perception of the position and velocity of the ball and the players, and also the orientation and the identity of the Cornell players. The vision system captures frames at a resolution of 320x240 and a rate of 35 Hz.

4.1 Interest Determination

Color segmentation of a frame often results in spurious blobs that do not correspond to the ball, or the robots. These points can be resultant from areas outside of the field, highlights from the lighting, deep shadows, the goals, and aliasing. Computation time of features of these blobs can result in a significant slowdown in system performance. To eliminate this slowdown and to produce a clean image of only the objects of interest, a single frame is captured and saved previous to the beginning of game play. This frame consists of the empty field only, without robots or the ball. This frame is later subtracted from the currently captured frame producing a difference image. Regions of this image where the disparity is high are postulated as areas where objects of interest lie.

4.2 Color Segmentation and Feature Extraction

Once areas of interest are determined, color segmentation is performed on the image. The segmentation is done by independently thresholding each of the Red, Green, and Blue color channels and performing a logical AND on the results of color thresholding. This logical operation extracts a sub-cube from the RGB color space. All objects are then classified based on the sub-cube that the corresponding blob falls into. This approach is well suited for the colors that are determined by the RoboCup Federation. The ideal pure colors that are defined: Orange (which is mainly Red), Green, Blue, Yellow, White, and Black, can be found on the corners of the RGB color cube. Two remaining colors exist and are used for our purposes, Cyan for initial orientation information and Magenta for robot identification. Since these eight colors are so separated in RGB color space, no color space conversion is performed on the input image, which is computationally costly.

Once the color segmentation is completed, blob features are computed including position, size, and perimeter length. These features are then used to filter any salt and pepper noise that may have been the result of incorrect color thresholding.

4.3 Tracking

Further rejection of false object classification is performed in the tracking stage. During tracking each orientation and identification blob is attempted to be registered to the blobs that correspond to the appropriate Cornell team color. All blobs that are not located within an appropriate physically realizable distance from the team color blob are thrown away. This step will reject any colors that are of interest, but are found on the opponent robots, for example. The team marker blobs which have an orientation marker and one or more identification markers registered with them are considered initially for identification and localization. The other team markers are afterward considered if there are any robots that are not found in the first set.

From these markers, and initial orientation is computed and the positions of the identification marker with respect to this orientation marker. The identification markers can be located in three of the four corners of the robot cover. The cover is divided into four quadrants, and each marker is classified by the angle that is produced relative to the orientation marker. The pattern of the identification markers can then be determined and the robot identified invariant to the robot orientation on the field. Once the robot pattern is identified, the final orientation is computed using all of markers on the cover.

Robot identification is also simplified by the physical constraints imposed by the maximum velocity and acceleration attainable. Each candidate position for a specific robot is compared to a predicted location and velocity determined by the two previous locations. If the position is outside the physically realizable radius of this predicted position, the candidate is rejected. Among several candidate positions that are located within this attainable radius, the necessary velocity to reach that position is computed and the position that most closely matches the predicted position and velocity is chosen as the true position of the robot.

4.4 Filtering

Although, the resolution of the camera is quite high and the blob analysis can compute the center of gravity of a blob to sub-pixel accuracy, these centers of gravity contain noise that may be modeled as white Gaussian noise in the system. While position calculations are fairly accurate, the orientation calculation suffers from this noise. The spatial proximity of the orientation marker and the team marker on the cover of the robot results in orientation errors of up to 10 degrees. To compensate for these errors all of the markers on top of the robot are used for the orientation. The true positions of the markers are known, thus the optimal rotation of the robot can be computed by using a least squares fit of the perceived marker locations and the actual locations on the cover. This fit becomes more precise as the number of markers on the cover of the robot increases.

5 Skills

The sophistication of the trajectory control algorithms described below together with very tight PID velocity control[4] enable our robots to get to a desired final state (of position, orientation and wheel velocities) in a fast manner.

This, combined with a prediction algorithm[3] for the ball, makes for effective real-time interception for both a stopped ball and a moving ball. The limitation on robot speed of maneuverability comes primarily from a system latency, described later in this paper.

Ball control is achieved with a front surface that is slightly recessed from the front corner bumpers, nominally allowing a player to change the direction of the ball's motion. An energy absorbing contact surface affords greater control. Dribbling is not a dominant skill. Passing is accomplished in an emergent manner, as a result of clever positioning of players that are not assigned to the ball.

Kicking is accomplished by a unidirectional solenoid with a front plate attached to its shaft. Robots will only kick in potential goal-scoring situations, and the timing of the kick is done with the use of an infrared sensor circuit that detects when the ball is directly in front of the robot. Typical kicks impart an additional 1 m/s to the ball.

The goalkeeper design is independent of the field player design, and thus the goalkeeper exhibits significantly different skills. The goalkeeper is equipped with a holding and kicking mechanism that can catch a front shot on goal, hold it for an indefinite amount of time required to find a clear pass to a teammate, and make this pass.

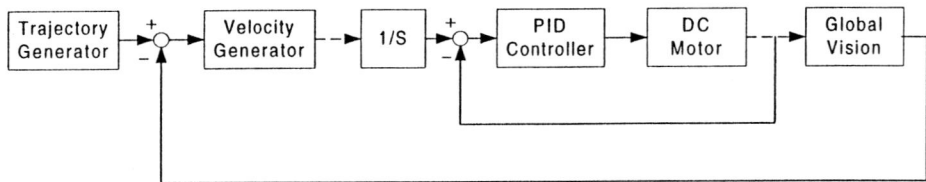

Fig. 1. Schematic diagram for Robot Control

6 Artificial Intelligence

6.1 Role Based Strategy

The artificial intelligence subsystem is divided into two parts. The high-level AI takes the current game state (robot and ball positions, velocities, and game history) as input, and determines the role of each robots, such as shooter, defender, mid-fielder, goalie, etcetera. Please see [2], and the references therein, for a thorough description of the application of role based systems in robotic soccer. Once the role is determined, desired final state such as time-to-target, robot orientation and robot velocity at the final position are computed from the current state. More than 20 roles are preprogrammed. The low-level AI subsystem resides on the each roles, and generates the trajectory to the target point and computes the wheel velocities to transmit to a robot.

6.2 Trajectory Control

The task of low-level AI is to generate trajectories and to control the robot to follow the trajectories. It takes as inputs the current state of the robot and the desired ending state. The current state of a robot consists of the robot x and y coordinates, orientation, and left and right wheel velocities. A desired target state consists of the final x and y coordinates, final orientation, final velocity as well as the desired amount of time for the robot to reach the destination.

Compared to reactive control strategies, such as those in [1] for example, we perform a global trajectory optimization for each robot and take advantage of the mechanical characteristics of the robots. Two position feedback loops are employed for the robot's trajectory control. The first is a local feedback loop and the other a global feedback loop. The local feedback loop resides on the microcontroller of each robot and is in charge of controlling the motor position[4]. The global feedback control also has a position feedback loop via the global vision system and makes the robot follow the desired trajectory. These two position feedback controls improve the robot's staying performance. The performance enhancement shows up especially when the goalie is facing an opponent robot. The desired velocity of each of the robot wheels are generated and then transmitted

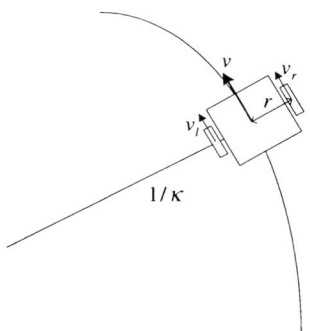

Fig. 2. Trajectory Generation

to the robots through the RF communication link at every sixtieth of a second. Fig. 1 shows the schematic diagram for the entire trajectory control loop.

The low-level AI needs to be efficient and robust to imperfections in robot movement. Currently, our algorithm can run more than 60 times per one computation cycle, which is sufficient considering it only needs to be run at about four times per cycle (for four robots excluding the goalkeeper).

This complex problem is solved by breaking the problem of trajectory generation into two parts. The first part generates a geometric path. The second part calculates wheel velocities such that the robot stays on the path and completes it in the allocated time.

Generating a Geometric Path Our geometric path is represented by two polynomials in the x and y coordinates of the robots. The x coordinate polynomial is a fourth-degree polynomial and the y coordinate polynomial is third degree.

$$x(p) = \sum_{k=0}^{4} \alpha_k p^k \qquad (1)$$

$$y(p) = \sum_{k=0}^{3} \beta_k p^k \qquad (2)$$

The task is to solve for the nine polynomial coefficients for a particular path requirement[7]. The 9 constraints on the polynomial path are: initial x coordinate, initial y coordinate, initial orientation, initial curvature (determined by the initial left and right wheel velocities), initial speed, final x coordinate, final y coordinate, final orientation, and final speed.

Generating Wheel Velocities Every point on the geometric curve has a curvature value, which defines a relationship between the left wheel velocity v_l

and the right wheel velocity v_r at that point in the curve. This relationship is:

$$v = (v_l + v_r)/2 \qquad (3)$$
$$v_l(1 + \kappa \cdot r) = v_r(1 - \kappa \cdot r) \qquad (4)$$

where κ is the curvature of the path, and r is the half distance between the two wheels and v is the forward moving velocity of the robot(See Figure 2). Thus, we simply need to choose a forward moving velocity of the robot to solve for v_l and v_r at every point on the curve, which can then be sampled at the cycle rate of our AI system. Obviously, the forward moving velocity is constrained by the time-to-target as well as mechanical limits of the robot.

Even though each run of this algorithm generates a preplanned path from beginning to end, it can be used to generate a new path after every few cycles to compensate for robot drift. The continuity of the paths generated is verified through testing. However, this algorithm breaks down when the robot is very near the target because the polynomial path generated might have severe curvature changes. In this case, the polynomials are artificially created (and not subject to the above constraints) on a case-by-case basis, and these are guaranteed to be smooth.

7 Other Team Features

7.1 Vision Calibration

The vision calibration consists of 4 main parts. They are:

- barrel distortion correction
- scaling
- rotation
- parallax correction

The barrel distortion correction is performed using a look-up table to map a point in image-coordinates into a new coordinate equidistant coordinate system. Due to the necessity for a wide angle lens barrel distortion becomes a significant problem in the image processing. Barrel distortion is a function of the lens of the camera and is radially symmetric from the center of the image. To invert the distortion, points are measured from the center of the image to the corner of the image. Since the points are measured equidistantly, a scaling can be done to convert corresponding points in image coordinates, which tend not to be equidistant. A look-up table is generated which contains the scaling factor and indexed by the distance from the center point in the image. Points between the indices are linearly interpolated from the two surrounding points. This transformation ensures that straight lines in reality are mapped back into straight lines.

The scaling is then computed such that the sides of the field are computed as accurately as possible. These values are computed so that the transformation from image coordinates to field coordinates is accurate for the center of each of

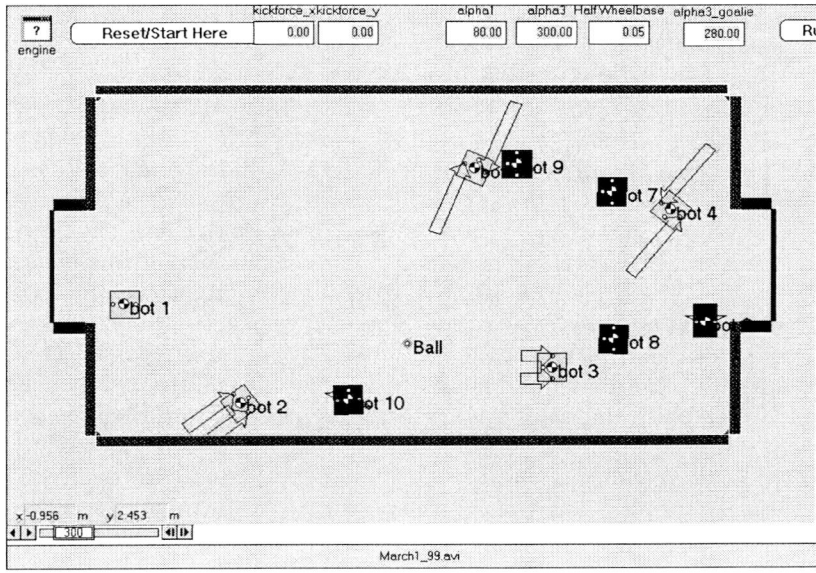

Fig. 3. Testing platform for the Simulation

the sides of the field, i.e. the x-coordinate of the center of the goals, and the y-coordinate for the centers of the lengthwise walls. This anchors the transformation to the sides of the field, ensuring that the walls of the field correspond to exactly where the artificial intelligence system expects them to be.

Since the camera cannot be mounted perfectly, the rotations about the center axis of the camera also need to be taken out. The points are rotated so that the sides of the field have constant x-coordinates along the widthwise walls and constant y-coordinates along the lengthwise walls.

Finally the parallax error that results from differences in object height is removed by scaling the x- and y-coordinates proportionally to the distance that the object is from the center of the camera projected onto the field plane.

7.2 High Fidelity Simulation

To provide a realistic testing platform for our artificial intelligence system, we have constructed a simulation of the playing field that models the dynamics of our environment.

The dynamic modeling of our system is performed by a Working Model 2D[8] rendering of the complete playing field. The model includes two teams of five individual players, the game ball, and the playing field. Real world forces and constraints are modeled, including the modeling of the motion of the tires and the inertia of the robots and ball. Additionally, the physical interactions between the players and each other, the ball, and the playing environment are all modeled in Working Model's two dimensional environment.

The simulator accepts external input and output 60 times per simulated second, the rate at which the artificial intelligence operates, and the rate at which new robot commands are issued. To simulate the time lag and noise we encounter in our real world simulation, the Working Model parameters are passed into Matlab[9], where random noise, error, and delay are introduced to model the limitations of our vision and communication systems. In addition, the kicking mechanisms and simple referee functions are managed, such as the detection of a team goal, and the reset of the playing field. This information is then passed to the artificial intelligence module. Information normally transmitted across the communications link is then passed back to Matlab from the artificial intelligence module, and is interpreted in Matlab before it is applied to the model of our system, to simulate the delay associated with our real world communications link.

The transition of our intelligence code between the simulated environment to the real playing environment is fairly smooth. However, there are some underlying differences that need to be considered when making the transition. In the simulator, the intelligence must run synchronously with the rest of the system, unlike in our final playing environment. In addition, simpler feedback control mechanisms are implemented into the physical model to increase the speed of the simulator, which are slightly different than the control methods that we use on our actual robots. Despite these limitations, we have found the differences between the environments to be small enough that we can make the transition between them without the need to significantly alter the functionality of the system intelligence and control code.

In the time before we had a fully operational real world system, the simulator provided us with a means of testing the artificial intelligence play-by-play, allowing code and strategy development to begin a full four months before the first robots were built. In addition, it allowed us to run competing strategies and full games without the need for a full complement of ten robots and two workstations running different intelligence algorithms simultaneously.

The simulator is a simple and manageable platform that allows us to recreate the real-world problems that exist in our system. At the same time, it removes several annoying physical constraints of the real-world system, such as limited battery life and operating time, providing a more convenient environment for new algorithms to be tested. Because the simulator performs more reliably than the real-world system, problems can be traced more quickly and more reliably to problems within the intelligence code. The simulator is a convenient and fairly accurate rendering of our real-world system, and an invaluable tool in the design and implementation of our artificial intelligence system.

7.3 System Delay Test

In any feedback control system, a change in the system delay can cause unwanted oscillations and loss of system control. To prevent unwanted oscillations, we have created a simple testing procedure to accurately measure our system delay. We measure the delay by sending a command to a robot to change the state of the playing field, and then we measure the time needed to detect the arrival of the change back at the place in the system where we issued the command.

There are two major reasons that we chose to implement the delay measurement in this manner. First, this method of measurement provided us with an accurate representation of the system delay that could be used when modifying the trajectory controller, which is responsible for modeling the system and properly handling the system feedback. Second, a quick measurement of the system delay allows us to easily check if the system is functioning normally.

The results of our delay measurements have reflected a total system delay of approximately 83 to 117 milliseconds. The total delay time can be attributed to specific components within the system. Our intelligence contributes a single frame delay of about 16 milliseconds (ms) to the total system delay. The communications link approximately adds an average of 13.3 ms of delay due to the time needed to buffer our 13 byte packets into the device, send the packets across the data channel, and to decode the packets at the robot.

The time that is needed for the vision system to capture the field state and relay it to the intelligence system is approximately 45 ms. This can be attributed to a delay of 16 ms to capture a new frame from the camera and the time needed to interpret a single frame, which is approximately 29 ms.

The vision information is incorporated into the next artificial intelligence cycle, which begins a new cycle every 17 milliseconds. The entire system delay breakdown gives us a minimum system delay rate of 75 ms, with the possibility of additional delay due to the asynchronous nature of the links between the camera, vision, and artificial intelligence subsystems.

8 Conclusion

Even though our team performed well at the competition last year, there are many subsystems and components that need to be improved. The main ones are outlined below:

- A more robust vision system. The current vision system performs well when operational, but does fail on occasion. In addition, it takes a very long time to calibrate the system. One of our objectives for next year is to construct a reliable vision system that can be set up in less than 30 minutes.
- Role coordination. This will allow us to implement set plays.
- More refined trajectory generation, obstacle avoidance, and trajectory control.
- Reduce the system latency.
- Innovative electro-mechanical designs.

Acknowledgment

The authors would like to thank all of the members in Cornell RoboCup team for their help in the design and construction of the system, Professors Bart Selman and Norman Tien for helpful advice and comments, and Professor Manuela Veloso for her insights and information on the CMU AI during a RoboCup lecture at Cornell University.

References

1. RoboCup-98:Robot Soccer WorldCup II, Minoru Asada and Hiroaki Kitano(Ed.), Springer Verlag. 1998
2. Manuela Veloso, Michael Bowling, Sorin Achim, Kwun Han and Peter Stone. The CMUnited-98 Champion Small Robot Team. In RoboCup-98:Robot Soccer WorldCup II, Minoru Asada and Hiroaki Kitano(Ed.), Springer Verlag. 1999
3. Eli Brookner. Tracking and Kalman Filtering Made Easy. A Wiley-Interscience Publication. 1998
4. Joseph L. Jones and Anita M. Flynn. Mobile robots: Inspiration to Implementation. Wesley. 1993
5. Rafael C. Gonzalez and Richard E. Woods, Digital image processing, Addison-Wesley, 1992
6. S. Blanchard and W.J. Fabrycky. System Engineering and Analysis. Prentice Hall, 3rd Edition, (1997)
7. G. Thomas and Ross Finney. Calculus and Analytic Geometric. Addison Wesley. 1996
8. Working Model. Knowledge Revolution, San Mateo, CA.
9. Matlab, The Mathworks Inc., Natick, MA.
10. Pro/ENGINEER. Parametric Technology Corporation, Wahlthan, MA
11. OrCAD. OrCAD Inc., Beaverton, OR.

Middle Sized Soccer Robots: ARVAND

M. Jamzad[1], A. Foroughnassiraei[2], E. Chiniforooshan[1], R. Ghorbani[2],
M. Kazemi[1], H. Chitsaz[1], F. Mobasser[1], and S.B. Sadjad[1]

[1] Computer Engineering Department
Sharif University of Technology
Tehran, Iran.
jamzad@sina.sharif.ac.ir, {chinif, chitsaz,
sajjad}@linux.ce.sharif.ac.ir, {mobasser, kazemi}@ce.sharif.ac.ir
http://www.sharif.ac.ir/~ceinfo

[2] Mechanical Engineering Department
Sharif University of Technology
Tehran, Iran.
{ghorbani, forough}@linux.ce.sharif.ac.ir
http://www.sharif.ac.ir/~mechinfo

Abstract. Arvand is the name of robots specially designed and constructed by sharif CE team for playing soccer according to RoboCup rules and regulations for the middle size robots. Two different types of robots are made, players and the goal keeper. A player robot consists of three main parts: mechanics (motion mechanism and kicker), hardware (image acquisition, processing unit and control unit) and software (image processing, wireless communication, motion control and decision making). The motion mechanism is based on two drive unit, two steer units and a castor wheel. We designed a special control board which uses two microcontrollers to carry out the software system decisions and transfers them to the robot mechanical parts. The software system written in C++ performs real time image processing and object recognition. Playing algorithms are based on deterministic methods. The goal keeper has a different moving mechanism, a kicker like that of player robots and a fast moving arm. Its other parts are basically the same as player robots. We have constructed 3 player robots and one goal keeper. These robots showed a high performance in Robocup-99: became champion.

1 Introduction

In order to prepare a suitable level for research in many different aspects involved in autonomous robots, we designed and constructed all parts of the robots by our group members in different laboratories of our university. These robots which are the 2nd generation which we made in the last two years, have a controllable speed of maximum 0.53 m/sec. In addition to the basic movements of a robot, special mechanical design of the player robot, enables it to rotate around any point in the field. In practice, the distance between ball center and robot geometrical center is calculated and the robot can be commanded to rotate around the ball center until seeing the opponent team goal. This unique mechanics, to a good extent, simplified and accelerated our playing algorithms.

The machine vision system of player robots uses a widely available home use video camera and a frame grabber. But for goal keeper we used one CCD camera in front and two small video conferencing digital cameras in sides rear. Our fast image processing algorithm can process up to 16 frames per second and recognize objects in this speed. For any recognized object, its color, size, distance and angle from robot is determined. The wireless communications between robots made it possible to test the cooperative behavior in a multi-agent system in a real-time changing environment. TCP/IP protocol was used for communication.

The software is based on deterministic algorithms, designed in object-oriented method and implemented in C++ using DJGPP compiler in MS/DOS. The reason for using MS/DOS was mainly due to the fact that we had to use a floppy disk drive for booting the system because of its reliability in a moving robot in RoboCup environment and also its low price compared to hard disk.

In the following we describe the mechanics, hardware and software systems used for goal keeper and player robots.

2 Mechanical Architecture

According to the motion complexity of a soccer player robot, proper design of its mechanics can play a unique role in simplifying its motion and as a result the playing algorithms. In this regard, different specific mechanisms were designed and implemented for player and goal keeper, that together with the motors current feedback measurement, to a good extent, guided us to the current mechanism which showed a better performance in Robocup-99.

2.1 Player Robot Motion Mechanism

Arvand consists of two motion units in front of the robot and one castor wheel in the rear. Each motion unit has a drive unit and a steer unit. A drive unit is responsible for rotating its wheel in forward and backward directions and also, a steer unit is responsible for rotating its respective drive unit around the vertical axis of the drive unit wheel. The combination of drive unit movement and proper settings of steer units angles with respect to robot front, provides the robot with a continuous rotational move around any point (this point can be selected to be inside or outside robot body) in the field in clockwise or counter-clockwise direction.

Drive unit consists of a wheel which is moved by a DC motor and a gearbox of 1:15 ratio [1]. The steer unit uses a DC motor and a gearbox of 1:80 ratio. For controlling a steer unit, an optical encoder is mounted on the respective motor shaft and its resolution is such that one pulse represents 0.14 degrees of drive units rotation. Figure 1 is from the robot top view and shows the position of drive units for rotating around point A in the field. The coordinates are as shown in the figure 2.

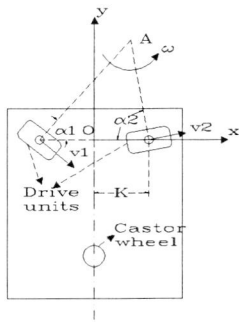

Fig. 1. The position of drive units to make the robot move around point A in the field.

The velocity vectors and angles made by steer units are calculated according to following formulas [2]:

$$v_1 = w \times \sqrt{y_1^2 + (k + x_1)^2} \qquad (1)$$

$$v_2 = w \times \sqrt{y_1^2 + (k - x_1)^2} \qquad (2)$$

$$\alpha_1 = Arctg(\frac{y_1}{k + x_1}) \qquad (3)$$

$$\alpha_2 = Arctg(\frac{y_1}{k - x_1}) \qquad (4)$$

In the above equations, x_1 and y_1 are the coordinates of point A (i.e. the rotation center); k is the distance between y axis and the drive unit rotation center; w is the angular speed of robot around point A; α_1 and α_2 are the rotation angles of left and right drive units with respect to x axis; v_1 and v_2 are the speeds of left and right drive motors, respectively. In special case, if the rotation center A is located on the y axis, equations 1 to 4 summarize to the following equation:

$$v_1 = v_2 = w \times \sqrt{y_1^2 + k^2} \qquad (5)$$

$$\alpha_1 = \alpha_2 = Arctg(\frac{y_1}{k}) \qquad (6)$$

This means that, to rotate the robot around a point $(0,y_1)$, both drive units should be set by the same angle α_1 and then , they should move by the same velocity v_1. As a result the robot will rotate with angular speed of w around the point $(0,y_1)$.

In summary, this mechanism has the following capabilities:

1. Rotating around any point in the field. Appropriate rotation of steer units can bring the drive units in desired angular positions α_1 and α_2. After these angles setting, if the ratio between the angular speed of two drive units is set according to equation 7 (extracted from equations 1 and 2), as a result the robot will rotate around point A.

$$\frac{w_1}{w_2} = \frac{\sqrt{y_1^2 + (k + x_1)^2}}{\sqrt{y_1^2 + (k - x_1)^2}} \qquad (7)$$

In the above formula, w_1 and w_2 are the angular speed of left and right drive units. By setting one of w_1 or w_2, the other is calculated according to above equation.

2. In our software system we can set the drive units to be parallel to each other while having a specific angle related to robot front. This mechanism is useful for taking out the ball when stuck in a wall corner and also dribbling other robots.

3. A kicker arm is installed in front of robot. A solenoid is used to supply it with kicking power. A simple crowbar connects the solenoid to the kicking arm. The power of kicking is controlled by duration of 24 DC voltage applied to the soleniod.

2.2 Goal Keeper Motion Mechanism

We think the goal keeper should have a complete different mechanism from player robot. Because it keeps the goal, it seems that more horizontal speed in front of goal area and deviation-less movement is a great advantage for the goal keeper. Thus, in order to guarantee a nearly perfect horizontal movement for the goal keeper, 4 drive units are installed in the robot (the castor wheel has been eliminated because it causes deviation in the robot movements). However, in practice the robot will be displaced after some movements, therefore it should have the ability to adjust itself when displaced. Horizontal movements and self adjustment can be done by a combination of the following three basic movements:

1. Move forward and backward (Fig. 2-a).
2. Rotate around its geometrical center (Fig. 2-b).
3. Move straight towards left and right (Fig. 2-c).

In order for the robot to perform these movements, 4 drive units and two steer units are installed in the robot. One steer unit rotates two front drive units round their vertical axes simultaneously in opposite directions and the other steer unit does the same on two rear drive units. The drive units wheel has a diameter of 8 Cm and a gearbox of 1/15 ratio. Measurement of the rotation angle for drive units is done by encoders installed on steer unit motor shafts.

To minimize the adjustment movements and also increase goal keeper performance, we installed a fast moving sliding arm on it, such that this arm can slide in left or right direction before the robot body itself moves in these directions. This arm can slide to its leftmost or rightmost position within less than 0.1 of

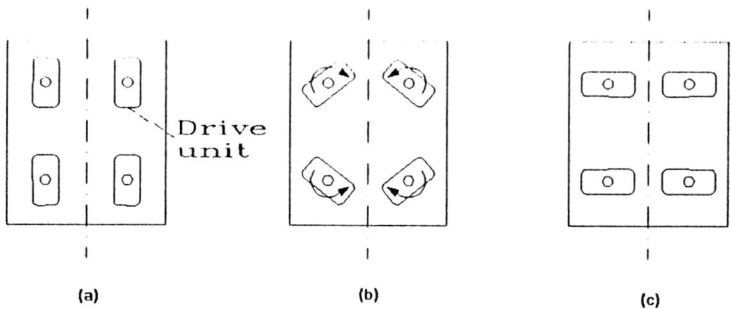

Fig. 2. Goal keeper drive units

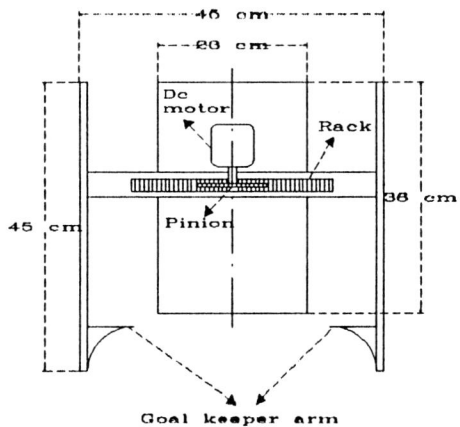

Fig. 3. Sliding arm movement

a second. Considering the front body size of goal keeper which is 23 Cm, and the arm size which is 45 Cm, the robot can cover 68 Cm which is approximately 1/3 of the goal area, within less than 0.1 of a second. Compared to goal keeper maximum speed which is 75 Cm/sec, this arm gives better protection of goal area from very fast moving balls.

Sliding arm movement is carried out by a rack and pinion mechanism, as seen in Fig. 3. To control the amount of arm sliding, an encoder is mounted on the shaft of pinion motor. It is necessary to fix the arm when goal keeper is in a stuck situation with other robots. This is done by using a solenoid which can lock the arm in its present position.

3 Hardware Architecture

The goal of our hardware architecture is to provide a control unit independent of software system as much as possible and also reduce the robots mechanical errors.

Arvand hardware system consists of three main parts: Image acquisition unit, processing unit and control unit.

The image acquisition system of goal keeper consists of a Topica PAL color CCD camera with 4.5 mm lens in front and two digital Connectix Color Quick-Cam2 for the sides rear view. For other robots we used a widely available home use video camera in front which could record the scene viewed by robot too. All robots including the goal keeper used a PixelView CL-GD544XP+ capture card which has an image resolution of 704x510 with the frame rate of 25 frames per second.

The processing unit consists of an Intel Pentium 233 MMX together with a main board and 32MB RAM. Two onboard serial ports are used as communication means with the control unit. A floppy disk drive is installed on the robot from which the system boots and runs the programs.

The control unit senses the robot and informs the processing unit of its status. It also fulfills the processing unit commands. Communication between the control unit and the processing unit is done via two serial ports with RS-232 standard[3]. Two microcontrollers 89C52 and 89C51 [4] are used in control unit. They control the drive units, steer units and kicker. Two limit switches are mounted on each steer unit. Microcontroller counts the number of pulses generated by the encoders mounted on the motor shafts to control the drive unit rotation. Each pulse represents 0.14 degrees of the drive unit rotation. The motor speeds are controlled by PWM pulses with frequency of about 70kHz. Fig. 4 shows the block diagram of the control unit. It allows distributed processing among the main board processor and two processors on this board.

This board performs the following main tasks:

– PWM generation on two drive units, two steer units and also the kicker. PWM control of drive units is done by MOSFET. A relay is used for changing motor rotation direction. In order to control the steer units not to rotate

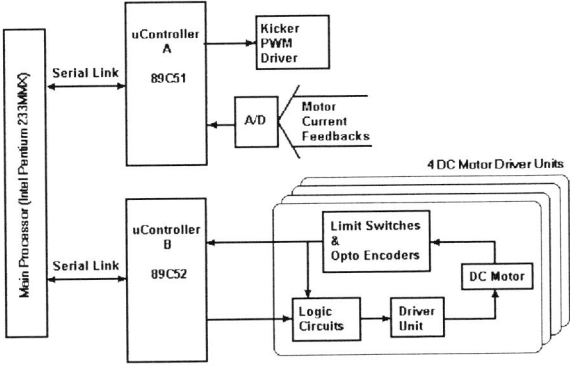

Fig. 4. Player robot hardware components interconnection diagram

beyond their angular limits, two limit switches are installed. If a steer unit touches a limit switch, no more movement in that direction will be possible.
- Measurement the current feed back of drive units. Motor current is measured by an A/D and processed by software in the main processor.
- Measurement of the batteries voltages in order to find their charging level.
- Control of goal keeper sliding arm and its relating lock.
- A pause key on control unit board allows us to stop all movements of robot and also send a signal to the main processor to bring it to suspend mode, which will save batteries.

The control unit board is designed to be robust, easy to test, fast failure finding, easy maintenance and modification, and reliable performance. In addition, it has the capability to handle extra signals which could come from new sensors on the robot, in the future.

4 Software Architecture

Software architecture of **Arvand** consists of four main parts: Real time object recognition, Motion control, Communication and Decision making module. Software which is written in C++ has an object oriented design with 5 classes as follows: Camera class (all related functions for working with frame grabber), Image class (machine vision functions), Motion class (motion functions which is the interface between software and hardware), Communication class (all TCP/IP related networking) and Decision class (all robot playing methods and algorithms).

4.1 Real Time Object Recognition

Objects are detected according to their color. We used HSI[5] color model for recognizing colors, because of its advantage in representing approximately each color in a cube in the HSI space. The color output of our frame grabber board is in RGB. To reach a near real-time speed in color processing, HSI color space is constructed from RGB in off-line. For each color to be recognized, its HSI range is determined in off-line as well (i.e. this range can be set for all colors according to the lighting condition). A one dimensional array of size 65536 that shows all possible RGB input values in our system is filled with a color name according to its HSI range, in off-line. Therefore, in real time, the color of a pixel is determined by two single array access (once for finding the RGB value of a pixel from frame grabber memory and the second time for finding the color name from the above mentioned array). Due to RoboCup regulation, each main object such as ball, ground, wall and a goal is assumed to have a single predefined color. This routine generates a segmented image matrix such that all pixels belonging to an object are assigned the same color name.

To find all objects in a scene, the image matrix is processed from top to bottom only once. To speed up this routine, instead of examining each single pixel in the image matrix, only one pixel from each subwindow of size $w_i \times h_i$ is selected and tested (i.e. w_i and h_i are the minimum width and minimum height of an object which can exist in a scene). If this pixel has the desired color, then we move upward in one pixel step until hitting a border point. At this point a contour tracing algorithm is performed and the contour points of the object are marked.

To find the next object, the search is continued from a subwindow located to the right of the subwindow in which the start point of the previous object was found. In searching for the next object, the marked points are not checked again. At the end of this routine, all objects are determined.

To overcome the possible color error of image acquisition system, during moving on the object contour, if it reaches a pixel with a color different from that of the object, but 3 of its 4 neighbors have the object color, then that pixel color is changed to the color of the object and it is considered to be a contour point. In addition, during contour tracing algorithm, the minimum and maximum x, y coordinates of contour points are calculated. The extracted object fits in a rectangle which upper left and lower right corner have the (min_x, min_y) and (max_x, max_y) coordinates. The size of this rectangle is estimated to be proportional to the real object size. If the object size is smaller than a predefined size, it is taken as a noise and eliminated. However, if because of lighting condition, one or more objects are found inside a larger object with the same color, the smaller objects are considered as noise and deleted. For any object found, its size, color, angle and distance from robot camera are passed to the decision making routine.

4.2 Object Distance Calculation

The object distance from camera can be determined in two methods. In the first method, since in RoboCup, the real size of objects are known (except that of opponent robot which can be estimated before the game), the object distance is calculated as a ratio of its real size and the size calculated in object detection routine. However, this method works only if the robot sees an object completely (there are many situations where only part of an object is visible).

In the second method, the distance is calculated from the object position in the image matrix. This method is independent from detected object size and therefore has less error. We calculated the object distance D according to the following formulas. In these formulas, X_0 is the number of pixels between the image matrix bottom position to the point that has lowest y value in the object selected from image matrix. $YSIZE$ is the image height in number of pixels. The constant parameters H, A and B are calculated off-line. Where, H is distance from camera focal lens center to ground. A is the distance such that if the object is located there, then the object bottom is seen in the lowest part of the image. B is the distance such that if the object is located at that position, then its bottom is seen in the image center. Fig. 5 shows the relation between these parameters.

$$b = Arctg(\frac{B}{H})$$

$$L = (B - A) \times Sin(\frac{\pi}{2} - b)$$

$$K = \sqrt{H^2 + A^2 - L^2}$$

$$X = L(1 - \frac{2X_0}{YSIZE})$$

$$a = Arctg(\frac{X}{k})$$

$$D = H \times Tan(a + b)$$

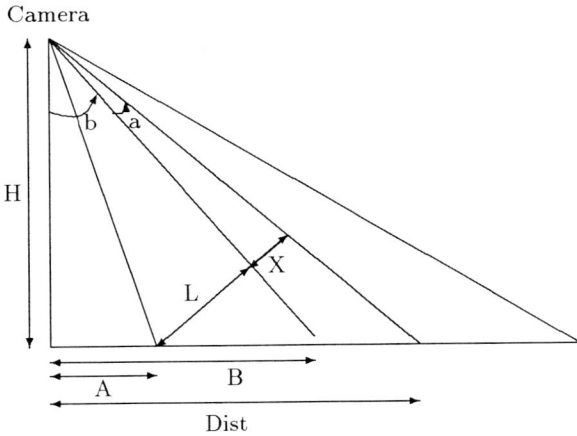

Fig. 5. Geometrical relations for finding object distance

4.3 Motion Control

This module is responsible for receiving the motion commands from the "Decision Making Module" and make the robot move. As it is mentioned in the hardware architecture section, the communication between the processing unit and the control unit is via two on-board PC serial ports using RS-232. So, just some basic computations are done in this module and commands are sent via serial ports to microcontrollers where they are executed.

For example, some commands for player robot movement are go(forward), go(backward), rotate(left), rotate(right), rotate_round(left, 10) (i.e. this stands for rotation around a point 10 centimeters straight from the robot geometrical center), kick (i.e. kicks the ball) and etc.

4.4 Communication

Communication between robots is done by wireless LAN under TCP/IP protocol. We used WATTCP whose main kernel can be downloaded from [6]. Each robot has a wireless network adapter, and there is a computer, we named it message server, outside the field which processes messages of robots and coordinates them. The server provides a useful user interface to command robots manually. Server's main responsibility is to receive the robots messages and inform them of each robot status. For example, in our multi-agent system, if one robot takes control of the ball, it will inform all others via server, and then other robots will not go for the ball.

4.5 Decision making

Principally, the decision making module is that part of **Arvand** software that processes the results of real time object recognition, decides accordingly and finally commands the motion control software. We have taken deterministic approach in these routines. This module is a finite state machine (FSM) whose inputs for changing state are machine vision results, motion control hardware feedbacks and server messages. Each robot playing algorithm kernel finds the ball, catches it, finds the opponent goal and finally carries the ball towards the goal and kicks. But there are a large number of parameters that affect this main kernel and cause interrupts in its sequence. For example, the main method for finding the ball is rotating. When our robot is moving in the field it tries not to collide with other robots. Collision avoidance is done by calculating the distance and angle of other robots and changing the speed of its motors. This capability showed its good performance in dribbling other robots.

In addition, robot ability to measure the motors current, enables it to determine stuck situations and thus making appropriate move to come out of that state.

5 Discussion

Considering the dynamic situation in a soccer game, the elementary generation of robots which are able to show the clarity of a future when we or our children will see humanoid robots playing soccer with human soccer players in a real field, should have a special mechanics, control and vision system which can simplify robots move in such environment. In this regard, it is obvious that soccer robots should be divided into two categories, players and goal keeper, each designed with different concept. The responsibility of a goal keeper and the type of its movements is basically different from other players.

Considering these facts, at the present time, we can not purchase robots which could fulfill these works in the way that we think is appropriate. Therefore, we decided to go through all difficulties of designing and constructing all parts of our robots, including mechanics, control hardware and software by our group.

In RoboCup-99 this idea showed its superiority compared to robots purchased from certain manufactures which had a general purpose design. The possibility to make changes in their mechanics and movement capability were limited, that is why the users were bounded to a certain extent to the parameters put through by the manufacturer.

We believe that one of the keys to the success in this field is designing the mechanics and control hardware which can best fit our idea of soccer player robot. Also we should concentrate on a fast and reliable vision system. It is unbelievable for a human soccer player to mistake a ball in real soccer field. Therefore our robots should be able to be improved to that ability. High resolution of CCD camera and fast and reliable frame grabbers are among the essential tools needed for the vision system.

The sliding arm of our goal keeper moves much faster than robot body to the left and right. Like a human goal keeper, when he tries to catch a ball from sides, his hands move faster and before his body. This design not only enables the robot to catch fast moving balls going from sides, but also reduces the risk of horizontal displacement of robot in fast left and right moves. Because we use two PC main boards with a frame grabber installed on one of them and other hardware boards, the height of our goal keeper is too much (i.e. 60 cm). There is not a proper balance between its width, height and weight, that is why in high accelerations the robot itself become unstable. To overcome this problem, we suggest the replacement of large size mother boards and frame grabber to some small size, so we could fit all hardware equipment in a smaller space. If this problem is solved, it is suggested to use two CCD cameras for sides view instead of digital video conferencing cameras.

At present our player robots can communicate with each other using wireless network by TCP/IP protocol. In practice we sometimes encountered communication stall. We think this is due to TCP/IP protocol when waiting for acknowledgment anytime a packet is sent, and this wait lasts because of electromagnetic noise in the environment. We think it will be more appropriate to use UDP protocol, because it is not a connection oriented protocol and does not wait for acknowledgments. A reliable real-time communication in a multi-agent system is the base for a successful team play.

6 Conclusion

Arvand is the 2nd generation of robots constructed by our team. One advantage of **Arvand** is its unique mechanics design which enables it to rotate around any point in the plane. Therefore, the robot can rotate around the ball center while simultaneously finding the goal position. Object distance measurement and motors speed control, enabled the robots to implement special individual playing techniques in dribbling, releasing themselves when stuck and taking out the ball from a wall corner.

Another advantage of our robots is the use of MS/DOS operating system, because it can be executed on a floppy disk which is a more reliable device compared to hard disk, on mobile robot. Our robots showed a good performance in real games and we are going to improve our software algorithms based on individual techniques and also team play. The wireless LAN system used in our robots provided the communication between robots resulting a cooperative behavior, specially when a robot has the control of ball. A well defined cooperative behavior in a multi-agent system, is the key to success of team play algorithms and also individual techniques. Sliding arm of goal keeper enables it to take fast moving balls from sides. The four drive units moving mechanism reduces the horizontal displacement of goal keeper during movements.

7 Acknowledgments

We like to give our best gratitude to the president of Sharif University of Technology, Dr. S. Sohrabpoor for his special support and caring, also to our university adminisitrative deputy Mr. A. Bayati and research deputy Dr. M. Kermanshah for their full support of this research work. Many thanks to Dr. E. Shadravan from department of mechanical engineering and his technicians in mechanics laboratories, and also all university staff who played a vital role at the right time to help this project.

We like to thank a lot our financial supporters in Iran: Mega Motor, SAIPA, Ministry of Culture and Higher Education, Planning and Budget Organization, Iran Industrial Expansion and Renewal Organization, Fajar Petrochemical Industries, Sanier Co., Iran Telecommunication Co. (Data group) and Advanced Industrial Systems Research Center.

Many thanks to Mr. A. Rajaiyan, Mr. A. Rajabzade and Mr. M. Seyed Razavi the students in computer engineering department, whose help and cooperation was very valuable for us in difficult and desperate situations.

References

1. Shigley, J.E., *Mechanical Engineering Design*, McGraw-Hill, 1986.
2. Meriam, J.L., *Dynamics*, John Wiley, 1993.
3. Mazidi, M.A., and Mazidi, J.G., *The 80x86 IBM PC and Compatible Computers, Volume II*, Prentice Hall, 1993.
4. MacKenzie, I.S., *The 8051 Microcontroller*, Prentice Hall, 1995.
5. Gonzalez, R.C., and Woods, R.E., *Digital Image Processing*, Addison-Wesley, 1993.
6. WATTCP http://www.geocities.com/SiliconValley/Vista/6552/l6.html

Vison Based Behavior Strategy to Play Soccer with Legged Robots

Vincent HUGEL, Patrick BONNIN, Ludovic RAULET, Pierre BLAZEVIC,
Dominique DUHAUT

Laboratoire de Robotique de Paris
10, 12 avenue de l'Europe, 78140 VELIZY France
E-mail : {hugel, bonnin, blazevic, ddu}@robot.uvsq.fr

Abstract. This work deals with designing simple behaviors for legged robots to play soccer in a predefined environment : the soccer field. Robots are fully autonomous, they cannot exchange messages between each other. Behaviors are based on information coming from the vision sensor, the CCD camera here which allows robots to detect objects in the scene. In addition to classical vision problems such as lighting conditions and color confusion, legged robots must cope with "bouncing images" due to successive legs hitting ground. This paper presents the work achieved for our second participation. We point out the improvements of algorithms between the two participations, and those we plan to make for the next one.

1 Introduction

It is not so easy to make robots play soccer in team. It is more difficult to design locomotion for legged robots than for wheeled machines But in both cases, « Vision » is essential to analyze the scene correctly, and therefore the current game action, in order to adopt the right strategy : go to the ball, go to the opposite goal with the ball, push the ball away of its goal, etc. Real soccer players would appreciate ! For the moment, the strategies seem to be selfish, but the level and the quality of the games increase.
Robots are now able to adopt some reactive behaviors and to analyze the current situation so that it can adapt itself when it is changing. Reactivity is absolutely necessary when the robot is facing unpredictable events, for example if it loses the object it is tracking or if it falls down. Reactive behaviors are like reflexes, that make the machine react always in the same way when confronted to the same situation. Adaptability is different from reactivity since it results from a kind of reasoning capability. It is also necessary for robots to be able to analyze a real situation so that they can make up their minds as for what to do next. Here the quality of the analysis determines the suitability of the behavior selected. For instance, adaptability comes in the case where reflexes lead to deadlocks, the machine should switch autonomously to

a rescue behavior. Examples of adaptive behaviors can be slowdowns to avoid obstacles or turning round a ball to bring it back to the desired location, ...
This work, describes the whole strategy used to design behaviors of legged robots which are operating in a predefined structured world. The example of a soccer field is considered here. Section 2 outlines the improvements between the exhibition held in 98 in Paris and the first competition held in 99 in Stockholm. The improvement of the visual recognition system is detailed in section 3. Section 4 is devoted to the strategy used to play soccer games on a colored field. Finally, the last section presents future improvements in the design of behaviors for soccer robot players.

2   From Paris 98 to Stockholm 99

Previous work, achieved for the First Legged Robot RoboCup Exhibition, held in Paris in July 1998, stresses upon basic tasks of locomotion and vision [HBB98(a)]. The 98 prototype shown in figure 1 was used in soccer games. For locomotion purpose, we developed our own walking patterns, and transitions between the different patterns: forward motion, left and right turns and turn-in-place modes[HBB98(b)]. For vision purpose, we developed an-easy-to use interface to tune vision parameters of the color detection hardware available on the robot, and the embedded vision algorithms [BHS99]. By lack of time, a very simple strategy which consisted in pushing the ball was implemented. This strategy was not effective: our robots were pushing the ball in our goal during the play !
Considering the new design of the 99 prototype (figure 1) of AIBO, the first work was to adapt locomotion algorithms to new leg configurations. The most important problem came from the nails at the leg tips. They are necessary for a « like a dog » design, but they limit the possible positions of the feet during the walk. Embedded vision algorithms do not need adaptation. They take advantages of the new CCD camera, and the new processor : the processing rate increases from 15 to 30 images per second.

Fig. 1. 98 and 99 Prototypes of AIBO

The improvements of the algorithms from the Paris 98 version to the Stockholm 99 one concern the vision module : embedded algorithms as well as the interface to tune parameters (see § 3). Moreover, two real game strategies have been implemented : behaviors for the goalkeeper and for the attacker / defender (see § 4).

3 Vision Improvement

3.1 No Coherent Behavior Possible without Reliable Information

It is impossible to make the appropriate decisions of action, that is to say to make the robot switch to the correct behavior, without reliable information coming from the vision module. Reliable information means identifying the scene element correctly: ball, beacons (landmarks), goals, players. Positions of scene elements are always correct.

All the behaviors we describe here to illustrate the consequences of the color confusions result from observations during the games in Stockholm. We do not think that they were implemented by teams voluntarily ! The confusion between the orange ball and red players made robots follow red players instead of going to the ball : our robots had this behavior during the training play, we corrected this problem for the competition, but we saw other robots with this behavior. The confusion between the orange ball and the shadows on both sides of the yellow goal made robots stay part of a game inside the yellow goal. During part of a game, two attackers of the opposite team were staying in our goal on each side of our goal keeper, meanwhile our attackers went with the ball.... A robot stayed many minutes in front of a beacon, blocked by the inclined wall of the soccer field.

From a practical point of view, it is possible to avoid color confusion, by choosing correctly the threshold values of the Color Detection Hardware. But it is not enough, results of the Color Detection hardware must be carefully processed in order to reject false pixel classifications.

3.2 More Autonomous Interface

Since our first participation, we developed a practical interface, aimed at tuning the threshold values of the Color Detection Hardware in the best way possible. This interface is designed for the following tasks:

1. It can display a large number of images (between 10 and 20 in practical cases), taken from different points of view of the soccer field, and their segmentation (*) results (cf. figure 2)
2. It allows to select or to remove areas of the images representing different scene objects (like the ball, the goals, the beacons, the players) viewed from several locations, under various lighting conditions. These areas can be chosen « automatically » by pointing regions on the segmentation results (improvement of this last participation), or « by hand » (as in the first version) by determining a box.
3. It computes the threshold values for each color template in the YUV color space from a set of image areas representing the same color. We add, in this new version, the possibility to process the threshold values, to make the results of the color

detection more independent of the positions of the scene object and of the robot on the soccer field.
4. It checks the coherence between color templates : a color pixel may belong to one color template at most, so in other terms, there is no intersection between color templates. If this condition is not satisfied, it is possible to apply again processing from step 2. An « error message » informs the user about confusion between colors.
5. It processes the YUV color images and displays the results of processing : the identified connected components (representing ball, goal, beacon, player). If results are not satisfying, it is possible to apply again procedures of step 2.

Finally, it writes the file containing the threshold values, which will be loaded into the robot.

This interface includes other tasks : it is in fact a more general tool for color image analysis.

(*) To be able to determine automatically a region in the image on which color parameters will be computed after, we develop a segmentation method, adapted version of the multispectral cooperative segmentation, specially developed for military applications (NATO project) of sensor fusion (Radar, Visual and Thermal Infra Red radiations) [BHP95] et [BHP96]. YUV homogeneous regions are extracted from YUV edges. In fact, the parameters of homogeneity were too strictly tuned: the ball and the goal are divided into several (really homogeneous) regions, and it is possible to select individually each of them, to avoid color confusions. Using this automatism, the tuning of the threshold values is easier to perform by the operator, and more precise because the regions on which thresholds are computed are selected by a segmentation of good quality, which gives better results than the human operator.

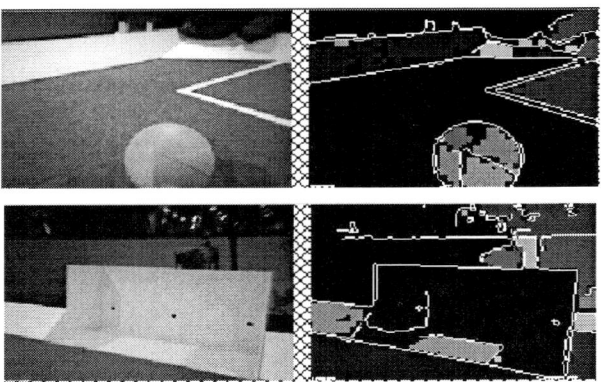

Fig. 2. Image and its Segmentation Results

After lots of experiments, we find a strategy for the choice of the regions on the segmented images, which allows for a correct tuning of threshold values.

3.3 Low Level Filtering

The previous described method was tested systematically using all coherent values of camera parameters : the mode, (4 modes : indoor, outdoor, fluorescent and user, according to the temperature of light, are available), the minimal and maximal gains. Then we choose the best combination by looking at object color appearances in several images for each set of parameters values, and by trying to find some regions in the image to tune parameters without detecting incoherence.

Some « control » images, results of the Color Detection hardware, with the best combination of camera parameters and with the tuning previously described (see § 3.2) are taken. In fact, the color confusion was noticed in only a few points which are generally isolated. So, a low level filtering, « similar to an Opening Procedure » using an isotropic 3 by 3 centered neighborhood was implemented to remove with success this wrong information (cf. figure 3).

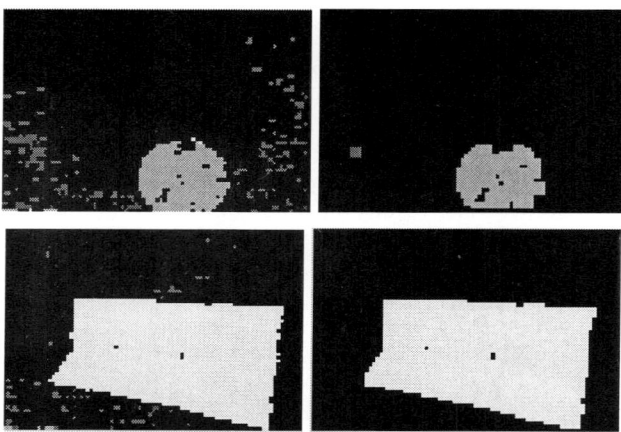

Fig. 3. The Results of the Color Detection Hardware, and the « Low Level » Filtered Image

But this filtering has a drawback : the maximal distance of detection of scene elements like the ball and the beacons decreases, which is a problem for the absolute localization of robots on the soccer field. The choice we made results from a trade-off : good identification (necessary including this filtering) / large distance detection of scene element, and seems to be good. It is better to be sure of the extracted information, in order to design a good strategy.

4 Behavior strategy to play soccer

4.1 Influence of locomotion on behavior

Locomotion is not omnidirectional. To change direction, the robot must compute the best suited turning circle and switch to a turning pattern as soon as possible. If the turning circle is below the threshold, the machine should stop and switch to the turn-in-place mode in the desired direction. Therefore, if the supervision module needs to plan the trajectory of the body within a world reference frame, it must take these characteristics into account. Moreover, it must incorporate the possible delays for a new walking pattern to trigger. A supervision module was tested on simulation and gave satisfactory results.
The problem appears when trajectory planning is carried out on the real machine. In spite of the careful design of walking patterns, some leg slippage, drifts and falls occur sometimes. Slippage is due to the nature of the interaction between the ground and the leg tip, which is changing. Since the robot does not master these interactions, drifts cannot be avoided over a long distance (4 or 5 times the length of the body). And it may happen that the robot falls down when it bumps into walls or other objects in the scene, one of the reason is that the locomotion is open-loop controlled and that the machine does not know how to deal with these situations. By adding some closed loops using exteroceptive data from force sensors, gyrometers and accelerometers, the machine should be able to detect collision, loss of balance, etc., and then react in real time to avoid falling down. In the worse case where falls cannot be avoided, the machine should be aware of it and capable of recovering by itself. Basic recovery behaviors have been implemented for real, thanks to its three-axis accelerometer, the robot can detect on which side it has fallen and chose the right standing up procedure. When confronted to real situations, absolute trajectory planning does not provide good results. This is because it relies on localizing procedures, therefore on the vision recognition system. Absolute localization is not possible while moving. The strategy adopted by our team is to use relative positioning, which can be used in real time. As precise positioning is not required, it does not matter if local motion is not accurate, the most important thing is to reach the goal. Therefore, in function of the landmarks captured while moving, the trajectory of the body is being corrected in real time by varying the turning circle.

4.2 Influence of vision on behavior

The vision recognition system has to deal with changing lighting conditions and "bouncing images". The first difficulty can be overcome using color adaptive algorithms. Color thresholds could be changed online after analyzing the results on the field. However, the vision system cannot cope with the second difficulty. It often occurs that an object is detected in the first image acquisition and lost the time after.

Since it is not always possible to track an object in the successive images captured, it is essential to memorize the object during a certain amount of time. If the object is not refreshed until the time-out expires, the robot should switch to some different behavior. The first behavior consists in searching for the object lost. If it fails, and if the robot is completely lost, the next step consists in beginning an absolute localization procedure. Since it takes a lot of time, this behavior should only be called in critical situations. In other cases, relative positioning is preferable. For instance, on the soccer field, each time the robot can spot the ball, one goal or the other, or one of the landmarks surrounding the field, it can deduce some information and adopting a more thoughtful behavior.

4.3 Two kinds of behavior

This section illustrates the strategies used to make quadruped robots play soccer successfully. Behaviors are separated into two groups. The first group includes basic behaviors that do not need analysis of the situation. The second group contains more high level behaviors that are based on interactions with the environment.

4.3.1 Basic primitives
Two kinds of basic primitives are described here.
- The first kind regroups behaviors that can be seen as reflexes. Reflexes are always triggered when the robot is confronted to the same situation. The duration of such a behavior is limited. It is a kind of quick reaction to a similar event. Examples of reflexes are all the recovery procedures in case of fall. Another reflex is the one thanks to which the robot turns its head towards the last position of the object lost. It can help save searching time.
- The second type of basic behaviors are not limited in time. In fact, these behaviors need time to reach the objective set by the supervision module. For instance, ball

Fig. 4. Online trajectory correction using landmarks captured

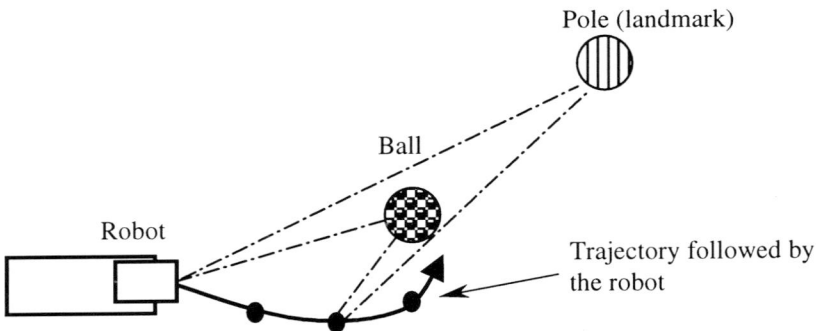

searching and ball tracking illustrate this kind of behaviors. In ball searching, the goal consists in finding the ball again. In ball tracking, the objective is to get closer to the ball. These behaviors are basic ones since no further analysis of the situation is required.

4.3.2 Higher level behaviors involving situation analysis
Four kinds of high level behaviors have been designed to play soccer.
- The first behavior consists in curving the trajectory in function of the pole (landmark) captured. The objective in the case of soccer game consists in bringing back the ball to the opponent field. Every time a landmark is spotted, it is memorized, and the turning circle is corrected in such a way that the robot goes to the ball from the right or the left side, depending on which side of the field the pole is situated. The robot must have the ball in its field of sight or have it memorized, and have captured a landmark in the same time. The turning circle of the current trajectory depends on which pole is detected. The machine must therefore analyze the situation before triggering the behavior. In soccer game, this behavior is used to defend its own goal or to attack the opponent one. In defense mode, the robot can head for the ball located in the own field and bring it back by moving behind it. In attacking mode, the robot can bring the ball back from the corners towards the opponent goal. Figure 4 illustrates this behavior.
- The second behavior is used by players to attack the opponent goal. In this case, the robot tries to align the ball and the center of the adversary goal along its longitudinal axis. To trigger this behavior, the robot must spot the ball and the opponent goal, the ball should be close enough. To this purpose, the robot uses the

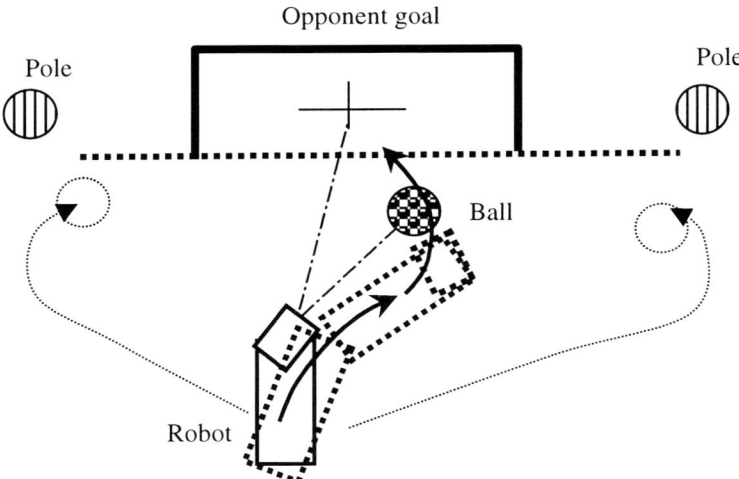

Fig. 5. Attacker behavior in front of the opponent goal

angle of the ball and the goal computed by the vision module. However, memorization should be carefully tuned. Refreshment rate must be sufficient otherwise the robot could miss the ball. Figure 5 shows the role as an attacker.

- The third behavior consists in defending the own goal when the situation is very dangerous, that is when the robot sees the ball and its own goal in the same time. When all conditions are met, the robot starts to get around the ball. The movement

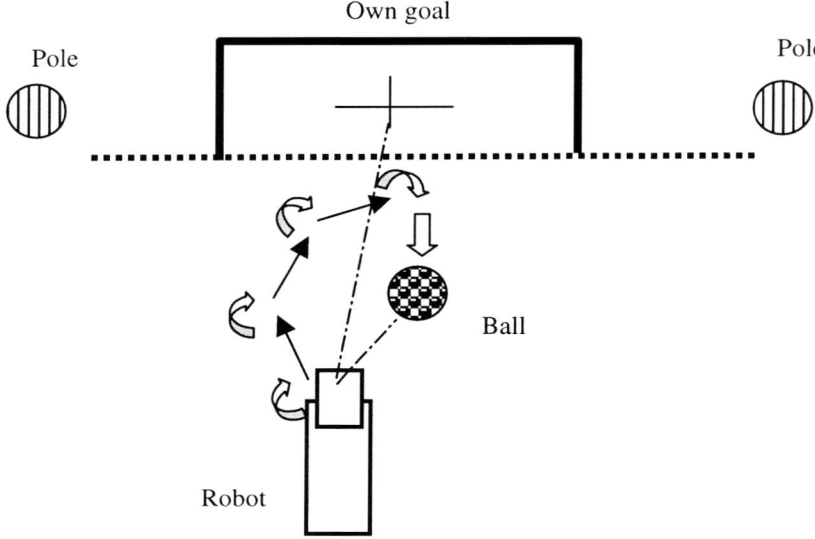

Fig. 6. Turn-round-the-ball defense. Successive turn-in-place and forward motion sequences around the ball.

is a sequence of straight line and turn-in-place motions, as shown in figure 6. During the procedure, the robot does not turn the head. If it loses the ball, the machine switches to turn-in-place motion until it sees the ball again, then it moves straight forward until losing the ball, and so on. Normally the robot makes a turn of approximately 180°, it can therefore push the ball to the opponent field.
- The last behavior is specially designed for the goalkeeper. First, the robot directs its head to the ball. If the ball gets close, the goalie aligns its body to the ball. Once the ball gets closer, the robot moves towards the ball to strike it and push it away. After that, it searches for its goal, goes to it. It detects the walls of its goal thanks to its infrared proximity sensor, and begins then to search for the opponent goal by turning in place. Once it has found it, it tries to spot the ball again. The procedure can restart. In this behavior, the robot has to estimate the distances to the ball and to the goal. Experiments must be carried out to tune them since they depend on the lighting conditions.

4.4 Behavior interaction

In the case of soccer, the interaction between the different behaviors is easily managed. In fact, some behaviors are naturally preemptive over others. The direct attack of the goal has a higher priority than others. If it were the behavior that curves the trajectory, the robot could miss the ball and the goal. In case of defense it is the turn-round-the-ball behavior which preempts all the others. This is obvious since it refers to the most critical situation. It can be seen as a rescue or last chance behavior to prevent the other team from scoring. Figure 7 summarizes the behavior interactions.

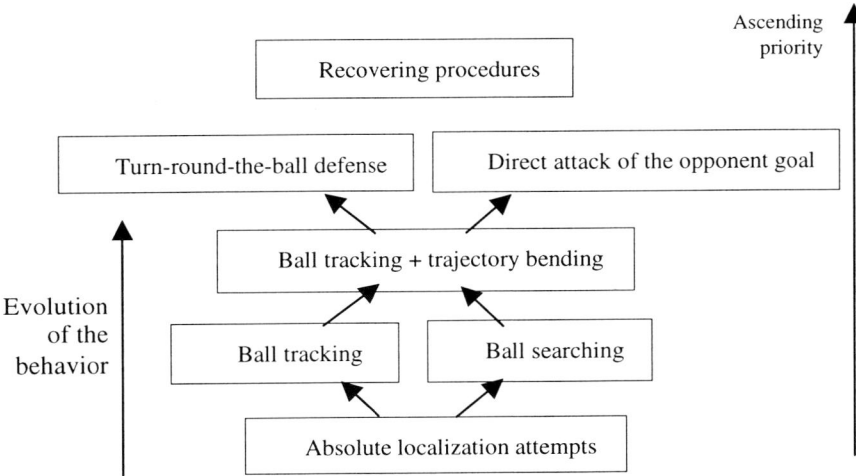

Fig. 7. Scale of behaviors priorities. Example of soccer game.

The experience brought by soccer experiments taught us that behaviors should not be limited to classic basic ones such as turn to the right, turn to the left, go to a specific location, stop to localize, achieve a fixed number of steps, and so on. It is better designing some adaptive behaviors that interact with the environment. Decisions should be made while moving, and relative positioning should be used most of the time.

5 Conclusion

The experiments carried out on quadruped robots playing soccer allow us to test behaviors in real situation. It helps us become aware of real problems of implementation. Vision and locomotion skills should be taken into account in the design of behaviors. A strategy that combines reactive and adaptive behaviors should give satisfactory results, provided that the interaction between both is well managed. We are satisfied by the improvements we added between Paris 98 and Stockholm 99.

Regarding Vision, we will try to increase the maximal distance of scene element detection without decreasing the quality of identification, which is essential. This point is necessary in order to guarantee a reliable absolute localization, on which we are currently working.

Another research subject is to embed an automatic color segmentation, like the one we use for the interface, to modify the threshold value of the color detection hardware. This will confer the vision system a better adaptation to inhomogeneous illumination.

In the first series of experiments (Paris 98 and Stockholm 99), communication between robots was not possible. However the next version of quadruped prototypes might be provided with wireless communication links. Behaviors could then be improved as robots could know the relative or absolute positions of others. Information could then be crossed to increase the accuracy of positioning.

Another way to improve the overall behavior is to use gyrodometry. This technique consists in using gyrometer data locally to correct drifts from odometry. The robot can then estimate its displacement with enough precision over a few more steps.

6 Acknowledgement

The French LRP team would like to thank SONY, which provided four robots, a development environment, and very cooperative technical support.

7 References

[BHP95] P.Bonnin, B.Hoeltzener, E.Pissaloux, A new way of image data fusion : the multi-spectral image cooperative segmentation, IEEE Int Conf on Image Processing ICIP95 Washington, October 1995

[BHP96] P.Bonnin, B.Hoeltzener, E.Pissaloux, The multi-spectral cooperative segmentation : How to Perform simultaneously image segmentation and image data fusion, Defence Optics's 96 (NATO Conference) Royal Military Academy Bruxelles Belgique 3-5 juin 1996

[BSH99] P Bonnin, D.Solheid, V.Hugel, P.Blazevic, Software Implementation of of a Real Time System inboard a mobile and autonomous robot for the SONY RoboCup Challenge Vision Interface 99, Mai 99, Trois Rivières Canada, pp 172-179

[HBB98(a)] V.Hugel, P.Blazevic, JC bouramoué, D.Solheid, P.Blazevic, D.Duhaut, Quadruped Robot Guidede by Enhanced Vision System and Supervision Modules in RoboCup-98 Robot Soccer World Cup II, M.Asada, H.Kitano (eds)

[HBB98(b)] Vincent HUGEL, Patrick Bonnin, Jean-Christophe Bouramoué, Pierre BLAZEVIC, Integration of vision and walking skills together with high level strategies for quadruped robots to play soccer, Advanced Robotics, The International Journal of the Robotics Society of Japan, Special Issue on RoboCup.

Automated Assistants to Aid Humans in Understanding Team Behaviors

Taylor Raines, Milind Tambe, Stacy Marsella

Information Sciences Institute and Computer Science Department
University of Southern California
4676 Admiralty Way, Suite 1001
Marina del Rey, CA 90292, USA
{Raines, Tambe, Marsella}@isi.edu

Abstract. Multi-agent teamwork is critical in a large number of agent applications, including training, education, virtual enterprises and collective robotics. Tools that can help humans analyze, evaluate, and understand team behaviors are becoming increasingly important as well. We have taken a step towards building such a tool by creating an automated analyst agent called ISAAC for post-hoc, off-line agent-team analysis. ISAAC's novelty stems from a key design constraint that arises in team analysis: multiple types of models of team behavior are necessary to analyze different granularities of team events, including agent actions, interactions, and global performance. These heterogeneous team models are automatically acquired via machine learning over teams' external behavior traces, where the specific learning techniques are tailored to the particular model learned. Additionally, ISAAC employs multiple presentation techniques that can aid human understanding of the analyses. This paper presents ISAAC's general conceptual framework, motivating its design, as well as its concrete application in the domain of RoboCup soccer. In the RoboCup domain, ISAAC was used prior to and during the RoboCup'99 tournament, and was awarded the RoboCup scientific challenge award.

1 Introduction

Teamwork has been a growing area of agent research and development in recent years, seen in a large number of multi-agent applications, including autonomous multi-robotic space missions [5], virtual environments for training [16] and education [8], and software agents on the Internet [15]. With the growing importance of teamwork, there is now a critical need for tools to help humans analyze, evaluate, and understand team behaviors. Indeed, in multi-agent domains with tens or even hundreds of agents in teams, agent interactions are often highly complex and dynamic, making it difficult for human developers to analyze agent-team behaviors.

The problem is further exacerbated in environments where agents are developed by different developers, where even the intended interactions are unpredictable.

Unfortunately, the problem of analyzing team behavior to aid human developers in understanding and improving team performance has been largely unaddressed. Previous work in agent teamwork has largely focused on guiding autonomous agents in their teamwork [6, 17], but not on its analysis for humans. Agent explanation systems, such as Debrief [7], allow individual agents to explain their actions based on internal state, but do not have the means for a team analysis. Recent work on multi-agent visualization systems, such as [9], has been motivated by multi-agent understandability concerns (similar to ours), but it still leaves analysis of agent actions and interactions to humans.

This paper focuses on agents that assist humans to analyze, understand and improve multi-agent team behaviors by (i) locating key aspects of team behaviors that are critical in team success or failures; (ii) diagnosing such team behaviors, particularly, problematic behaviors; (iii) suggesting alternative courses of action; and (iv) presenting the relevant information to the user comprehensibly. To accomplish these goals, we have developed an agent called ISAAC. A fundamental design constraint here is that unlike systems that focus on explaining individual agent behaviors [7, 12], team analysts such as ISAAC cannot focus on any single agent or any single perspective or any single granularity (in terms of time-scales). Instead, when analyzing teams, multiple perspectives at multiple levels of granularity are important. Thus, while it is sometimes beneficial to analyze the critical actions of single individuals, at other times it is the collaborative agent interaction that is key in team success or failure and requires analysis, and yet at other times an analysis of the global behavior trends of the entire team is important.

To enable analysis from such multiple perspectives, ISAAC relies on multiple models of team behavior, each covering a different level of granularity of team behavior. More specifically, ISAAC relies on three heterogeneous models that analyze events at three separate levels of granularity: an individual agent action, agent interactions, and overall team behavior. These models are automatically acquired using different methods (inductive learning and pattern matching) -- indeed, with multiple models, the method of acquisition can be tailored to the model being acquired.

Yet, team analysts such as ISAAC must not only be experts in team analysis, they must also be experts in conveying this information to humans. The constraint of multiple models has strong implications for the type of presentation as well. Analysis of an agent action can show the action and highlight features of that action that played a prominent role in its success or failure, but a similar presentation would be incongruous for a global analysis, since no single action would suffice. Global analysis requires a more comprehensive explanation that ties together seemingly unconnected aspects and trends of team behavior. ISAAC uses a natural language summary to explain the team's overall performance, using its multimedia viewer to show examples where appropriate. The content for the summary is chosen based on ISAAC's analysis of key factors determining the outcome of the engagement.

Additionally, ISAAC presents alternative courses of action to improve a team using a technique called 'perturbation analysis'. A key feature of perturbation analysis is that it finds actions within the agents' skill set, such that recommendations are plausible. In particular, this analysis mines data from actions that the team has already performed.

Overall, ISAAC performs post-hoc, off-line analysis of teams using agent-behavior traces in the domain. This analysis is performed using data mining and inductive learning techniques. Analyzing the teams off-line alleviates time constraints for these analysis techniques, allowing a more thorough analysis. Also, using data from the agents' external behavior traces, ISAAC is able to analyze a team without necessarily understanding its internals, allowing analysis of teams developed by different developers in a given domain.

ISAAC is currently applied in the domain of RoboCup soccer simulation [8]. RoboCup is a dynamic, multi-agent environment developed to explore multi-agent research issues, with agent teams participating in annual competitions. Agent-team analysis is posed as a fundamental challenge in RoboCup since team developers wish to understand the strengths and weaknesses of teams and understand how to improve such teams. (There are at least 50 such development groups around the world.) Indeed, ISAAC has been applied to all of the teams from several RoboCup tournaments in a fully automated fashion. This analysis has revealed many interesting results including surprising weaknesses of the leading teams in both the RoboCup '97 and RoboCup '98 tournaments and provided natural language summaries at RoboCup '99. ISAAC was also awarded the 'Scientific Challenge Award' at the RoboCup '99 international tournament. ISAAC is available on the web at http://coach.isi.edu and has been used remotely by teams preparing for these competitions.

While ISAAC is currently applied in RoboCup, ISAAC's techniques are intended to apply in other team domains such as agent-teams in foraging and exploration [2] and battlefield simulations [16]. For example, exploring actions, interactions, and global trends such as target hit rate, friendly fire damage, and formation balance, ISAAC could produce a similar analysis in the battlefield simulation domain, and use similar presentation techniques as well.

2 Overview of ISAAC

We use a two-tiered approach to the team analysis problem. The first step is acquiring models that will compactly describe team behavior, providing a basis for analyzing the behavior of the team. As mentioned earlier, this involves using multiple models at different levels of granularity to capture various aspects of team performance. The second step is to make efficient use of these models in analyzing the team and presenting this analysis to the user. Later sections delve into more specifics of these models. An overview of the entire process is shown in Figure 1.

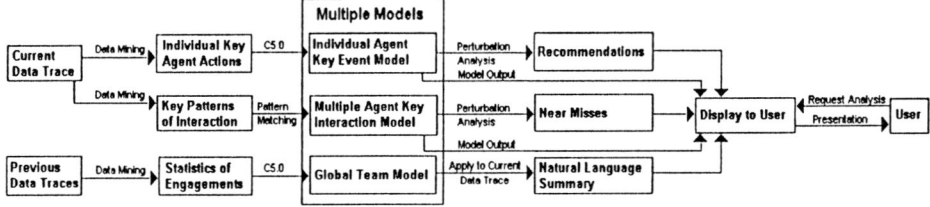

Fig. 1. Flow Chart for ISAAC Model Generation and Analysis

Input to all models comes in the form of data traces of agent behaviors. In the current implementation of ISAAC, these traces have been uploaded from users around the world through the Internet.

As shown in Figure 1, acquiring the models involves a mix of data mining and inductive learning but is specific to the granularity of analysis being modeled. Analysis of an individual agent action (individual agent key event model) uses the C5.0 decision tree inductive learning algorithm, an extension to C4.5, to create rules of success or failure [10]. For analysis of agent interactions (multiple agent key interaction model), pre-defined patterns are matched to find prevalent patterns of success. To develop rules of team successes or failures (global team model), game level statistics are mined from all available previous games and again inductive learning is used to determine reasons for success and failure.

Utilizing the models involves catering the presentation to the granularity of analysis to maximize human understandability. ISAAC uses different presentation techniques in each situation. For the individual agent key event model, the rules and the cases they govern are displayed to the user. By themselves, the features that compose a rule provide implicit advice for improving the team. To further elucidate, a multimedia viewer is used to show cases matching the rule, allowing the user to better understand the situation and to validate the rules (See figure 2). A perturbation analysis is then performed to recommend changes to the team by changing the rule condition by condition and mining cases of success and failure for this perturbed rule. The cases of this analysis are also displayed in the multimedia viewer, enabling the user to verify or refute the analysis.

For the multiple agent key interaction model, patterns of agent actions are analyzed similar to the individual agent actions. A perturbation analysis is also performed here, to find patterns that are similar to successful patterns but were unsuccessful. Both successful patterns and these 'near misses' are displayed to the user as implicit advice. This model makes no recommendations, but does allow the user to scrutinize these cases.

The global team model requires a different method of presentation. For the analysis of overall team performance, the current engagement is matched against previous rules, and if there are any matches, ISAAC concludes that the reasons given by the rule were the determining factors in the result of the engagement. A natural language summary of the engagement is generated using this rule for content selection and sentence planning. ISAAC makes use of the multimedia display here as well, linking text in the summary to corresponding selected highlights.

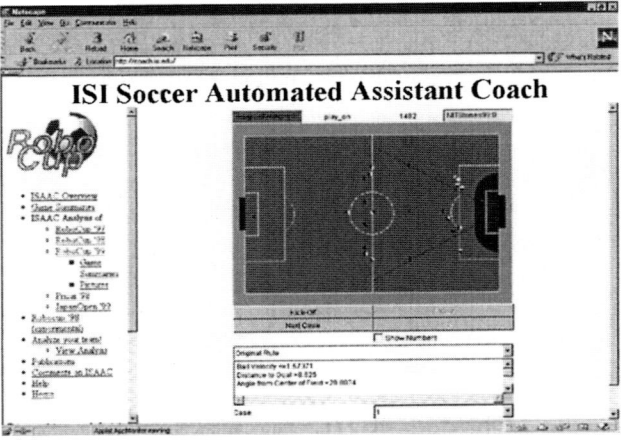

Fig. 2. Multimedia Viewer highlighting key features

ISAAC has been used in the RoboCup simulated soccer environment consisting of two opposing teams of eleven agents each. The agents do not have a centralized control, and act in a complex, dynamic, noisy environment managed by the soccer server, which acts as host and referee for the game. Figure 2 shows ISAAC's multimedia viewer, which displays the soccer field and plays from the games, and can highlight key features specific to ISAAC's analysis. For instance in figure 2, the area around the right soccer goal is highlighted.

3 Individual Agent Key Event Model

This section examines the first of ISAAC's three models, focusing on key actions taken by individual agents, and is specific to each team. In this and the following two sections, we first provide a conceptual overview of the model being analyzed and then discuss its instantiation in RoboCup.

3.1 Conceptual Overview of the Individual Agent Key Event Model

The individual agent model focuses on the analysis of critical events in a team's behavior history relevant to the team's success. There may be many critical events along the path to the team's eventual success or failure that are widely separated in time, only loosely coupled to each other, but nevertheless critical to the team's success. For instance, in a battlefield simulation, there may be many distinct attacks on enemy units, which are critical to team success, embedded in a larger history of maneuvering.

We consider critical events to be the team's intermediate successes or failures. When something occurs that directly influences the team's eventual success or failure, this is considered to be an intermediate success or failure point. At present, we assume the identification of these intermediate points is part of the domain specific knowledge available to the individual agent analysis model.

Having isolated cases of intermediate success or failure, we can now form rules of successful and unsuccessful behavior, which comprise the individual agent model. These rules are formed using inductive learning techniques over the cases of success and failure based on a set of potentially relevant features in these cases. These features, along with the decision on what are the cases of intermediate success, are the only background information or bias given to the individual agent analysis technique. The features chosen must have the breadth to cover all information necessary for the analysis, but should also be independent of each other if at all possible. In the future, a semi-automated attribute selection may be used [4].

Currently, C5.0 is used to form the rules of success and failure. Each rule describes a class of success or failure cases, based on its feature description. These rules and the cases they represent can be displayed to the user as implicit advice on how individual agents operate in critical situations.

More explicit exploration of this advice is performed using an automated *perturbation analysis*. After ISAAC has produced rules determining which circumstances govern success and failure classifications, ISAAC uses a perturbation analysis to determine which changes would produce the most benefit. Each learned rule consists of a number of conditions. We define a perturbation to be the rule that results from reversing one condition. Thus a rule with N conditions will have N perturbations. The successes and failures governed by the perturbations of a rule are mined from the data and examined to determine which conditions have the most effect in changing the outcome of the original rule, turning a failure into a success. Since these cases are mined from the original data traces, the recommended changes must already be within the agent's skill set. Perturbation analysis is explained in greater detail in Section 3.3.

3.2 Application of Individual Agent Key Event Model to RoboCup

In applying the approach to RoboCup, the domain specific information has to be identified that would be used by ISAAC as bias in its analysis. In particular, in the RoboCup domain, success means outscoring the opponent. Shots on goal are therefore key points of intermediate success or failure as these are situations that can directly affect the outcome of the game. Thus, the focus of ISAAC's individual agent analysis in RoboCup is shots on a team's goal as well as shots by the team on an opponent's goal.

Having defined shots on goal as key events, we need to determine which domain dependant features might be useful in classifying the success or failure of a shot on goal. After an initial set of experiments with a relatively large feature set, ISAAC currently relies on a set of 8 features to characterize successes and failures.

Having determined which features to use in the analysis and the key events (the cases) to examine, the task is transformed to mining the raw data and feeding it to the C5.0 decision tree learning algorithm. From the resulting decision tree, C5.0 forms rules representing distinct paths in the tree from the root of the tree to a leaf as a classification of success (goal-score) or failure (goal not scored). Each rule describes a class of similar successes or failures.

Figure 3 shows an example success rule, describing a rule where shots taken on the Windmill Wanderer team will fail to score (Successful Defense). This rule states that when the closest defender is sufficiently far away (>13.6 m) and sufficiently close to the shooter's path to the center of the goal (<8.98°), and the shooter is towards the edges of the field (>40.77°), Windmill Wanderer will successfully defend against this shot. When viewed using ISAAC, the user can see that the defender is far enough away to have sufficient time to adjust and intercept the ball in most of these cases. Thus the user is able to validate ISAAC's analysis. This rule provides implicit advice to this team to keep a defender sufficiently distant from the ball, or to try to keep the ball out of the center of the field.

> Distance of Closest Defender > 13.6 m
> Angle of Closest Defender wrt Goal <= 8.981711
> Angle from Center of Field > 40.77474
> → class Successful Defense

Fig. 3. Sample Rule from shots on Windmill Wanderer team of RoboCup'98

The application of a decision tree induction algorithm to this analysis problem must address some special concerns. The goal-shot data has many more failure cases (failed goal shots) than success cases (goals scored). However, analyzing such data using a traditional decision tree induction algorithm such as C4.5 gives equal weight to the cost of misclassifying successes and failures. This usually yields more misclassified success cases than misclassified failure cases. For example, in our analysis of shots by the Andhill team from the RoboCup'97 tournament, our original analysis misclassified 3 of 306 failure cases (less than 1%), but misclassified 18 of 68 success cases (26%). Since a much larger portion of the success cases is incorrectly classified, this produces overly specific rules that govern success cases. To compensate for this lopsided data set, the ability of C5.0 to weight the cost of misclassification is used. Specifically, the cost of misclassifying a success case is set to be greater than the cost of misclassifying a failure case [18]. ISAAC uses a 3 to 1 ratio by default, but this is adjustable.

More generally, differential weighting of misclassification cost provides a mechanism for tailoring the level of aggressiveness or defensiveness of ISAAC's analysis. Consider shots on goal against a team. If a very high cost is assigned to misclassifying a successful shot on goal, the rules produced will likely cover all successful shots, and quite a few misclassified failure cases. In this case, the rule conditions are implicitly advising to make the team very defensive. On the other hand, if a low cost is assigned, the rules may not cover all of the successful cases. Therefore, ISAAC would only give "advice" relevant to stopping the majority of shots on goal. This may not be appropriate if we consider any goal to be a serious failure. Therefore, we allow the user to adjust the weight on success case misclassifications.

3.3 Perturbation Analysis

Perturbations of a failure rule enable users to see what minimal modifications could be made to agent behaviors to convert the failures into success. Mining instances of perturbed failure rules, the developer determines steps that could be taken to move the agent from failure to successful behavior.

For example, one of ISAAC's rules states that when taking shots on goal, the Andhill97 team often fails to score when (i) ball velocity is less than 2.37 meters per time step and (ii) the shot is aimed at greater than 6.7 meters from the center of goal (which is barely inside the goal). ISAAC reveals that shots governed by this rule fail to score 66 times without a successful attempt.

Now consider the perturbations of this rule. In cases where the rule is perturbed such that ball velocity is greater than 2.37 m/t and the shot aim is still greater than 6.7m, Andhill scores twice and fails to score 7 times. In another perturbation, where ball velocity is again less than 2.37 m/t but now shot aim is equal to or less than 6.7m (i.e. shots more towards the center of the goal), Andhill is now scoring 51 times and failing to score 96 times (See figure 4). These perturbations suggest that improving Andhill97's shot aiming capabilities can significantly improve performance, while trying to improve agents' shot velocity may not result in a drastic performance increase.

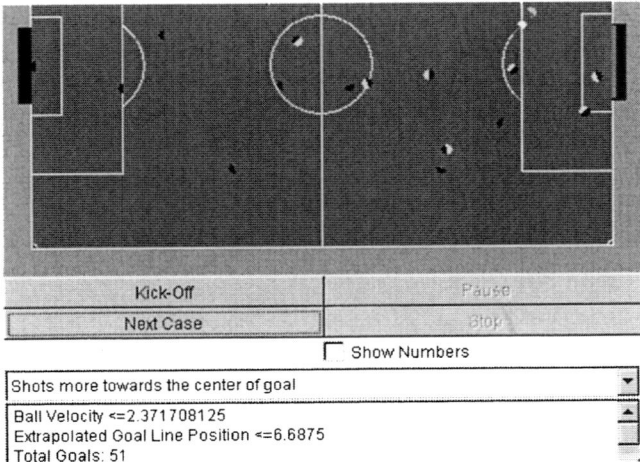

Fig. 4. Perturbation analysis showing Andhill '97 scoring using a perturbed shooting behavior

Perturbations of success rules are also useful. There are two reasons for this. First, it allows ranking of conditions contributing to success. In particular, some changes to the rule will take a team further from success than another. For example, a team may succeed in scoring 95% of the time when all conditions are met. The percentage of success may drop to 50% if the first condition is changed and down to 5% if the second condition is changed. In this case, the developer may decide that even if the first condition is not met, shooting is still the correct course of action while doing so if the second condition is not met is a bad decision. Secondly, allowing the user to see how the team succeeds or fails, more insight can be drawn as to why these conditions are important. Human oversight is important at this juncture to determine if the reasons ISAAC comes up with are truly the reasons the team is succeeding or failing.

4 Multiple Agent Key Interaction Model

4.1 Conceptual Overview of the Agent Interaction Model

To analyze agent interactions, ISAAC relies on matching predefined (possibly user-defined) patterns. These patterns consist of sequences of abstracted actions of different agents that result in intermediate successes (or failures). The patterns are matched against the data traces to find actual instantiated interactions. ISAAC then classifies the patterns into those that lead to intermediate success or failures. This approach shares similarities with meta-pattern based data-analysis [13], where a user provides 'interesting templates' to discover patterns in the data.

The perturbation analysis in this model finds near misses in the case history, i.e., patterns similar to the successful patterns (within a given threshold) that end in failure. ISAAC identifies these near misses by perturbing the successful patterns, e.g., it may find interactions resulting in failure that result from a slight change in one of the actions specified in the pattern. Near misses help the user to scrutinize the difference between successful and unsuccessful patterns.

For example, in an air combat simulation, suppose that ISAAC finds a common pattern where friendly aircraft respond to the enemy's pincer maneuver with a flanking maneuver and a missile shot, causing an enemy aircraft to be shot down. The perturbation analysis will review traces to find instances where an identical pattern, with a slight variation in the missile shot (or one of the maneuvers), does not result in an enemy aircraft getting shot down. The user is then able to view and compare the successful and unsuccessful (yet similar) patterns to determine possible differences and causes for failure.

4.2 Application of Agent Interaction Model

The first step necessary to apply the agent interaction model to the RoboCup domain is the determination of the patterns to examine and a notion of success or failure of these patterns. We again use a soccer goal as a notion of success, so any pattern leading to a goal is successful. The interaction patterns are made up of the players' actions (kicks) causing the ball to be shot in to the goal.

To further illustrate this type of analysis, we present examples from the Windmill Wanderer and ISIS teams from RoboCup '98. The ISIS team scored 20 of their 35 goals when the player kicking the ball before the shot was on the opponent team (described by a *shooter→opponent→shooter* pattern). This suggests a very opportunistic scoring behavior for ISIS, and viewing the cases shows that ISIS tends to push the ball deep into the opponent territory, and sometimes they are able to get the ball back for a quick shot. In contrast, the Windmill Wanderer team scored 17 goals from the shooter dribbling in before the shot (*shooter→shooter→shooter*) and another 9 goals from a teammate controlling the ball before passing to the shooter (*teammate→teammate→shooter*), out of a total 37 goals. Thus, this team scores more often when they control the ball all the way in to the goal, a stark contrast from the ISIS team.

For the Windmill Wanderer team from above, 27 near misses were found similar to the 17 goals from the dribbling pattern, suggesting this pattern was well defended or the team was making some mistakes. Windmill Wanderer placed third in the tournament, and the 27 near misses may have been the culprit in its third place finish. The developer can review and compare these cases in the multimedia viewer, and make the determination as to what changes would be most beneficial.

5. Automated Game Summary: Team Model

5.1 Conceptual Overview of Team Model

The purpose of the global team model is to analyze why teams succeed or fail over the course of the entire engagement (as opposed to teams' intermediate success/failure at a single time point). The assumption of this model is that there can be many different factors that impact a team's overall success or failure. In a complex environment, a team may have to perform many actions well in order to be successful. These actions may involve entirely different sub-teams, and very different kinds of events, perhaps widely separated in time, may be implicated in success or failure. Nevertheless, there may be patterns of these factors that, although not strongly related in the behavioral trace, do in fact correlate with whether a team succeeds in a domain. The global team analysis attempts to find these patterns to explain success/failure.

In designing this model, we had two options possible. One was to tailor the analysis to specific teams. In particular, by analyzing data traces of past behaviors of a specific team, it would be possible to explain why this specific team tends to succeed or fail. This approach would be similar to the one followed in Section 3, which explained why agents' critical actions tend to succeed or fail in a team-specific manner. A second option was to analyze teams in terms of why teams succeed or fail in the domain in general, in a non-team-specific manner (which does not require data traces from the particular team being analyzed, but from other teams in this domain). Despite the advantages of option 1 (team-specific explanations), this option was rejected due to the lack of large amounts of team-specific engagement data, and option 2 was used. In particular, unlike the individual agent model as in Section 3, which can obtain lots of key event data points even from a single engagement, a single engagement is just one data-point for the global team model. For instance, even a single RoboCup game provides large numbers of shots on goal to begin learning the individual agent model; yet, this single game is not enough to begin learning a global team model.

Exercising option 2 above implies acquiring the team model by examining the behavior traces of many different teams in a domain. Here again we rely on the domain expert to provide the set of overall features that lead to success or failure over an entire engagement. Again the C5.0 induction algorithm is used on these features, classifying the engagement as a success or failure for each team, and learning rules that capture the dominant features that lead to success (or failure).

A different approach is taken for using the rules learned via C5.0. When analyzing a specific engagement, we mine the features from the engagement and determine which learned rule the current game most closely matches. This rule then becomes the reasoning for why each team succeeded or failed. ISAAC uses the rule as the basis for its natural language summary generation to ease human understanding of the engagement as a whole.

The learned rules are critical in ISAAC's natural language summaries. Indeed, we initially attempted to provide a natural language summary without using such learned rules. In this case, we were only able to present all of the game statistics glossed with natural language phrases, with no ordering or reasoning behind them. This initial approach failed because all of the summaries were long, uniform, and they failed to emphasize the relevant aspects of the game.

Instead, with the current method, ISAAC generates a natural language summary of each encounter employing specific rules as the basis for content selection and sentence planning in accordance with Reiter's architecture of natural language generation [11]. Reiter's proposal of an emerging consensus architecture is widely accepted in the NL community. Reiter proposed that natural language generation systems use modules for content determination, sentence planning, and surface generation. ISAAC's NL generation can be easily explained in terms of these modules.

Starting with the raw data of the game, ISAAC mines the features it needs, and matches it to a pre-existing rule. This rule is thus used in content determination for the natural language generation, since the rule contains that which ISAAC believes pertinent to the result of the game. Furthermore, the conditions of each rule also have some ordering constraints, since the rules come from a decision tree learning algorithm, and we use this to form our sentence planning. We consider branches closer to the root of the tree to have more weight than lower branches, and as such should be stated first. Each fact is associated with a single sentence, and ordered accordingly. From these, ISAAC creates a text template of the summary for performing the surface generation. This template is augmented with specific data from the game, and links to examples of the features found earlier to be shown in the multimedia viewer.

5.2 Application of Team Model to RoboCup

To learn rules of why teams succeeded or failed in previous engagements, ISAAC reviews statistics of previous games. The domain expert must provide the domain knowledge of what statistics to collect, such as possession time and number of times called offside. ISAAC uses this information (10 features in all) to create a base of rules for use in analysis of future games.

ISAAC learns and uses seven classes of rules covering the concepts of big win (a victory by 5 goals or more), moderate win (a victory of 2-4 goals difference), close win (1 goal victory), tie, close loss (by 1 goal), moderate loss (2-4 goals), and big loss (5 or more goal loss). The motivation for such subdivision is that factors leading to a big win (e.g., causing a team to outscore the opponent by 10 goals) would appear to be different from ones leading to a close win (e.g., causing a one goal victory) and should be learned about separately. While this fine subdivision thus has some advantages, it also has a disadvantage, particularly when the outcome of the game is at the border of two of the concepts above. For instance a 2-0 game (moderate win) could very well have been a 1-0 game (close win). Thus, we anticipate that the learned rules may not be very precise, and indeed as discussed below, we allow for a "close match" in rule usage.

To use these rules, ISAAC first matches the statistics of a new (yet to be analyzed) game with the learned rules. If there is a successful match, ISAAC checks the score of the game against that predicted by the matching rule before writing the summary. If the match is exact or close (e.g. the actual game statistic matched a close win rule, although the game had an outcome of 2-0), the template is used as is. If there are multiple matches, the closest matching rule is used. However, if no match is close to the actual score, ISAAC still uses the rule, but changes the template to reflect surprise that the score did not more closely match the rule.

The matched rule discussed above provides the content selection, so ISAAC now has a template to shape the game summary. The template orders components of the rule according to their depth in the original decision tree, in accordance with our sentence planning technique. ISAAC then fills in the template, mining the features of this particular game to create a summary based on the rule. An example rule is shown in Figure 5.

> Ball in Opponent Half > 69%
> Average Distance of Opponent Defender <= 15 m
> Bypass Opponent Last Defender > 0
> Possession time > 52%
> Distance from Sideline of Opponent Kicks > 19 m
> → class Win Big

Fig. 5. Example team rule for big wins.

To see how this rule is used in creating a natural language summary, we examine one summary generated using this rule as a template. In this case, ISAAC is arguing the reasons for which 11Monkeys was able to defeat the HAARLEM team:

HAARLEM Offense Collapses in Stunning Defeat at the hands of 11Monkeys! [1]

11monkeys displayed their offensive and defensive prowess, shutting out their opponents 7-0. 11monkeys pressed the attack very hard against the HAARLEM defense, <u>keeping the ball in their half of the field for 84% of the game</u> and allowing ample scoring opportunities. HAARLEM <u>pulled their defenders back to stop the onslaught</u>, but to no avail. To that effect, 11monkeys <u>was able to get past HAARLEM's last defender</u>, creating 2 situations where only the goalie was left to defend the net. 11monkeys also handled the ball better, <u>keeping control of the ball for 86% of the game</u>. HAARLEM <u>had a tendency to keep the ball towards the center of the field</u> as well, which may have helped lead them to ruin given the ferocity of the 11monkeys attack.

The underlined sentences above correspond directly to the rule, with some augmentation by actual statistics from the game. By using the rule for content selection

---

[11] The title and first sentence of our summary does not come from the model, but is based solely on the score of the game. We used examples from (human soccer) World Cup headlines and let ISAAC randomly choose among these in categories of tie, close win, moderate win, and big win.

and sentence planning, ISAAC is able to present the user the reasons for the outcome of the engagement, and avoid presenting irrelevant data consisting of irrelevant features.

6. Evaluation and Results

To evaluate ISAAC, we evaluate each of its models in isolation and then the effectiveness of the integrated ISAAC system. We begin by evaluating the individual agent model.

A key measure of ISAAC's individual agent model is the effectiveness of the analysis, specifically the capability to discover novel patterns. Section 3.3 highlighted a rule learned about the Andhill97 team concerning their aiming behavior. This rule was one instance of ISAAC's surprising revelation to the human observers; in this case, the surprise was that Andhill97, *the 2nd place winner of '97*, had so many goal-shot failures, and that poor aim was at least a factor. Not only was this surprising to other observers, this was also surprising to the developer of the team, Tomohito Andou. After hearing of this result, and witnessing it through ISAAC's multimedia interface, he told us that he "was surprised that Andhill's goal shooting behavior was so poor..." and "... this result would help improve Andhill team in the future." [Andou, personal communication]

Another interesting result from the individual agent analysis model comes from the number of rules governing shooting behavior and defensive prowess. Figure 6 shows that in each year, the number of rules for defense decreased for the top 4 teams, perhaps indicating more refined defensive structures as the teams progress. Also, the number of rules necessary to capture the behavior of a team's offense is consistently more than that necessary for defense, possibly due to the fact that no single offensive rule could be effective against all opponent defenses. The key here is that such global analysis of team behaviors is now within reach with team analyst tools such as ISAAC.

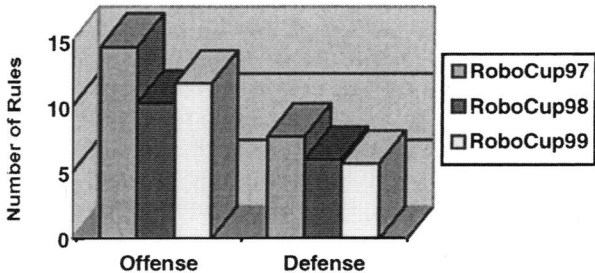

Fig. 6. Number of rules by year.

Another point of evaluation is understanding how well the model captures the shooting behaviors. To this end, ISAAC models were applied to predict game scores at RoboCup '99, a rather difficult problem even for humans. ISAAC used rules describing a team's defense and matched them with the raw averaged data of the shots taken by the other team to produce an estimate of how many goals would be scored against that team in the upcoming game. Performing this analysis for both teams produced a predictive score for the outcome of the game[2]. This prediction obviously ignores many critical factors, including the fact that some early games were unrepresentative and that teams were changed by hand during the competition. Yet in practice, ISAAC's predictive accuracy was 70% with respect to wins and losses, indicating it had managed to capture the teams' defenses quite well in its model.

To evaluate the game summary model, a small survey was distributed to twenty of the participants at the RoboCup '99 tournament, who were witnessing game summaries just after watching the games. Figure 7 shows the breakdown of the survey, showing that 75% of the participants thought the summaries were very good.

Another measure of game summaries is a comparison of number of features used in the current summaries versus those generated earlier that did not use ISAAC's approach. On average, ISAAC uses only about 4 features from its set of 10 statistics in the summaries, resulting in a 60% reduction from a natural language generator not based on ISAAC's machine learning based analysis. Thus, ISAAC's approach was highly selective in terms of content. Indeed as mentioned earlier, summaries generated without ISAAC were much longer, lacked variety, and failed to emphasize the key aspects of the game.

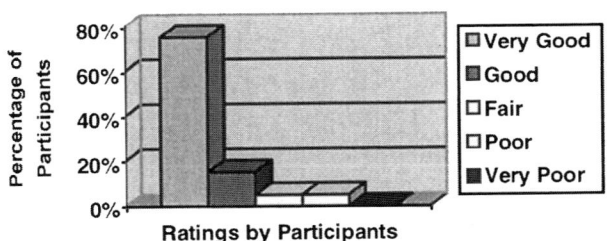

Fig. 7. Automated game summary survey.

Yet another measure of ISAAC's use of the team model for natural language generation is available by viewing the error rates from the machine learning algorithm used. These error rates tell us how accurately ISAAC's learned rules reflected the game. On the original set of games for which ISAAC's rules were learned, 87% of

---

[2] No prediction was done for the preliminary rounds as ISAAC gathered data on the teams. Prediction was performed only on the double-elimination tournament.

the games were classified correctly (70% exact match, 17% close match), resulting in an error rate of 13%. Our test set of (unseen) RoboCup '99 games produced 72% classified correctly (39% exact match, 33% close match), for an error rate of 28%. If an error does occur, ISAAC still produces a summary, but it reflects its surprise at the outcome, thus explaining the error. The high error rate on our training data could indicate that a better feature set is possible or that the data may be noisy.

Evaluating ISAAC as an integrated system is more difficult. However, some observations can still be made. ISAAC was awarded the 'Scientific Challenge Award' by the RoboCup committee. ISAAC was used extensively at the RoboCup '99 tournament in Stockholm, and received a great deal of praise and other feedback. Developers used ISAAC to analyze opponent teams after the early round matches to get a feel for the skill of upcoming opponents. Spectators and developers alike were able to view ISAAC's game summaries just minutes after a game, and there was also a great deal of speculation concerning ISAAC's predictions on future games.

7. Related Work

The research presented in this paper concerns areas of multi-agent team analysis and comprehensible presentation techniques. We compare research in each of these areas.

André et al have developed an automatic commentator system for RoboCup games, called ROCCO, to generate TV-style live reports for matches of the simulator league [1]. ROCCO attempts to recognize events occurring in the domain in real time, and generates corresponding speech output. While both ROCCO and ISAAC use multimedia presentations, ROCCO attempts to analyze events quickly to produce live reports. However, the ROCCO analysis does not use multiple models of behavior for multi-perspective analysis as in ISAAC, and its analysis is not designed to help users and developers understand teams' abilities. ROCCO also has no capability to perform perturbation analysis.

Bhandari et al's Advanced Scout uses data mining techniques on NBA basketball games to help coaches find interesting patterns in their players and opponents' behaviors [3]. Advanced scout also enables coaches to review the relevant footage of the games. Advanced Scout is able to capture statistical anomalies of which coaches can take advantage. However, Advanced Scout does not have some of ISAAC's extensions including the use of multiple models to analyze different aspects of teams, perturbations to make recommendations, and game summaries for an analysis of overall team performance.

Ndumu et al's system for visualization and debugging multi-agent systems comprises a suite of tools, with each tool providing a different perspective of the application being visualized [9]. However, the tools do not perform any in-depth analysis on the multi-agent system, and the system has no capability for perturbing this analysis. ISAAC also uses a visualization component, but only as an aid to understanding its analysis.

Johnson's Debrief system enables agents to explain and justify their actions [7]. This work focuses on agents' understanding the rationales for the decisions they make and being able to recall the situation. Debrief also has a capability for agent experimentation to determine what alternatives might have been chosen had the situation been slightly different. ISAAC performs something similar in its perturbation analysis; however, ISAAC focuses on an entire team, not just an individual, necessarily.

Stone and Veloso have also used a decision tree to control some aspects of agents throughout an entire game, also using RoboCup as their domain [14]. However, this work pertains to execution of agents rather than analysis of agent teams, and since it is internal to the agent, their work has no means of presentation.

8. Conclusion

Multi-agent teamwork is a critical capability in a large number of applications including training, education, entertainment, design, and robotics. The complex interactions of agents in a team with their teammates as well as with other agents make it extremely difficult for human developers to understand and analyze agent-team behavior. It is thus increasingly critical to build automated assistants to aid human developers in analyzing agent team behaviors. However, the problem of automated team analysts is largely unaddressed in previous work.

We have taken a step towards these automated analysts by building an agent called ISAAC for post-hoc, off-line agent-team analysis. ISAAC uses two key novel ideas in its analysis. First, ISAAC uses multiple models of team behavior to analyze different granularities of agent actions, using inductive learning techniques, enabling the analysis of differing aspects of team behaviors. Second, ISAAC supports perturbations of models, enabling users to engage in "what-if" reasoning about the agents and providing suggestions to agents that are already be within the agent skill set. Additionally, ISAAC focuses on presentation to the user, combining multiple presentation techniques to aid humans in understanding the analysis, where presentation techniques are tailored to the model at hand.

While ISAAC is intended for application in a variety of agent team domains, ISAAC has currently been applied in the context of the RoboCup soccer simulation. It is available on the web for remote use. ISAAC has found surprising results from top teams of previous tournaments and was used extensively at the RoboCup '99 tournament. ISAAC was awarded the 'Scientific Challenge Award' at RoboCup '99 where its analysis and natural language game summaries drew a crowd throughout the tournament.

Acknowledgement

We thank Intel Corp for their generous support of the work reported in this article.

References

1. André, E., Herzog, G., Rist, T. Generating Multimedia Presentations for RoboCup SoccerGames. In *RoboCup-97: Robot Soccer World Cup I*, 1997.
2. Balch, T. The Impact of Diversity on Performance in Multi-robot Foraging. In *Proceedings of the Third Annual Conference on Autonomous Agents*, 1999.
3. Bhandari, I., Colet, E., Parker, J., Pines, Z., Pratap, R., Ramanujam, K. Advanced Scout: Data Mining and Knowledge Discovery in NBA Data. *In Data Mining and Knowledge Discovery*, 1997.
4. Caruana, R., Freitag, D. Greedy Attribute Selection. In 11th *Proceedings of the 11th International conference on Machine Learning (ICML)*, 1994.
5. Dorais, G., Bonasso, R., Kortenkamp, D., Pell, B., Schreckenghost, D. Adjustable Autonomy for Human-Centered Autonomous Systems. *Working notes of the Sixteenth International Joint Conference on Artificial Intelligence Workshop on Adjustable Autonomy Systems*, 1999
6. Jennings, N. Controlling Cooperative Problem Solving in Industrial Multi-agent System Using Joint Intentions. *In Artificial Intelligence, Vol. 75*, 1995.
7. Johnson, W. L. Agents that Learn to Explain Themselves. In *Proceedings of AAAI-94*, 1994.
8. Kitano, H., Tambe, M., Stone, P., Veloso, M., Noda, I., Osawa, E. & Asada, M. The RoboCup synthetic agent's challenge. In *Proceedings of the International Joint Conference on Artificial Intelligence (IJCAI)*, 1997.
9. Ndumu, D., Nwana, H., Lee, L., Haynes, H. Visualization and debugging of distributed multi-agent systems. In *Applied Artificial Intelligence Journal, Vol 13 (1)*, 1999.
10. Quinlan, J. *C4.5: Programs for Machine Learning*. Morgan Kaufmann, 1994.
11. Reiter, E. Has a Consensus NL Generation Architecture Appeared, and is it Psycholinguistically Plausible? In *Proceedings of the Seventh International Workshop on Natural Language Generation*, 1994.
12. Sengers, P. Designing Comprehensible Agents. In *Proceedings of the International Joint Conference on Artificial Intelligence (IJCAI)*, 1999.
13. Shen, W.M., and Leng, B. A Metapattern-Based Automated Discovery Loop for Integrated Data Mining. In *IEEE Transactions on Data and Knowledge Engineering*, 1996.
14. Stone, P., Veloso, M. Using Decision Tree Confidence Factors for Multiagent Control. In *Proceedings of the International Conference on Autonomous Agents*, 1998.
15. Sycara, K., Decker, K., Pannu, A., Williamson, M., Zeng, D., Distributed Intelligent Agents. In *IEEE Expert*, 1996.
16. Tambe, M. Johnson, W. L., Jones, R., Koss, F., Laird, J. E., Rosenbloom, P.S., Schwamb, K. Intelligent Agents for Interactive Simulation Environments. In *AI Magazine, 16(1) (Spring)*, 1995.
17. Tambe, M. Towards Flexible Teamwork. In *Journal of Artificial Intelligence Research, Vol. 7*, 1997.
18. Ting, K. Inducing Cost-Sensitive Trees via Instance Weighting. In *Principles of Data Mining and Knowledge Discovery (PKDD 98)*, 1998.

LogMonitor: From Player's Action Analysis to Collaboration Analysis and Advice on Formation

Tomoichi Takahashi

Chubu University
1200 Matsumoto, Kasugai-shi, Aichi 487 - 8501, JAPAN

Abstract. This paper describes analysis results of collaboration among players of RoboCup '98 simulator teams and on-line adversarial model analysis using LogMonitor. LogMonitor is a tool for analyzing games from logfiles and displaying statistical data such as counts of soccer plays. Evaluation of collaboration in a multi-agent system is closely related with applied domains, which make it difficult to distinguish agent's universal ability from task oriented programs. In viewing simulation soccer games, play agents' skills are evaluated from the human soccer standards. This situation is assumed to be similar to collaboration among teammates, that is evaluated from human standards.

Adding to the basic actions of the player such as shooting, kicking, etc., a 1-2 pass among teammate agents is used to evaluate teams in collaboration. LogMonitor data shows that 1-2 pass may be useful to evaluate collaboration. Experiments show that adding adversarial information is very useful to make a team more robust.

1 Introduction

Sporting games are examples of a multi-agent system. In sports, team play and team tactics as well as an individual player's abilities are important. Reviewing scorebooks gives us information about which player scored a goal, which made a shot .etc. They are very useful for coaches to rank the players and to plan strategies for upcoming games, even though similar plays are evaluated differently in different situations. Various kinds of computer-aided scoring or analyzing methods have been developed to use in human games[SoftRB].

The RoboCup simulator game is a multi-agent system which is played between player agents through network communication [Kitano98]. Teams programmed based on various paradigms participate in RoboCup, and the games are recorded as logfiles. RoboCup provides test beds for evaluating agent systems. Using logfiles, Takahashi et al. reviewed RoboCup97 teams [Taka98], Letia et al. used logfiles to extract player's action model [Ial98], and Tanaka et al. made clear changes of games from RoboCup97 to RoboCup98 [Tanaka].

It is important to evaluate the multi-agent system not only from the game results but also from collaboration among agents. In RoboCup '98, an attempt was made to evaluate soccer simulation games from teamwork, not from scores. This paper describes analysis of collaborative plays using logfiles of the evaluation league at RoboCup 98 [Cup98]. The collaboration among agents is discussed from scores and statistical values such as numbers of kicks, passes, or 1-2 pass, etc. Next, the player agent's autonomy is discussed by comparing CMUnited 98, the champion team of RoboCup 98 with our team, Kasugabito-II. Experimental games were presented, and the result showed a team composed of less autonomous agents played weal by adding information on opponent teams.

2 LogMonitor

LogMonitor [1] is a tool for analayzing RoboCup simulation games from logfiles where the positions of the ball and all players of both teams at every simulation step are written [Log]. The data in logfiles are equivalent to images displayed on the monitor.

2.1 Actions Analysis by LogMonitor

We enjoy the game by seeing CRT images and may also record the game by taking note of which agent passed the ball, the quality of the pass, etc. For recording the games by a computer in the similar ways to a human scorer, it is necessary to recognize the player's actions such as passing, kicking .etc and the ball movement from time sequence data in logfiles.

The followings are methods used to recognize actions:

kick: The ball is kicked when the following conditions are satisfied.
1. the ball direction is changed or the change of the ball's speed is greater than the decreasing rate adapted in the soccer server at consecutive $\{t_i, t_{i+1}\}$ and $\{t_{i+1}, t_{i+2}\}$.
2. at least one player is within a kickable area at t_{i+1}.

When there are more players within a kickable area, the nearest player is assumed to kick the ball.

pass: Two consecutive kicks are assumed to be a pass, when two players of the same team kicked the ball.

interception: Two consecutive kicks are assumed to be an interception, when an opposing team player kicked the second time.

1-2 pass: A player kicks the ball to a teammate, runs behind an opponent player, and receives the ball that the teammate returned.

2.2 Position Analysis by LogMonitor

Human player's abilities are measured by their running speed, run length, etc. This corresponds to the allocation of stamina, and the range of moves during a game. The positioning of players is important in teamwork. The following data on positioning are displayed:

trajectory: the plot of a player's position in time sequence,

---
[1] LogMonitor is gained from our Home Page.

distance: the sum of distance which players moved when play_mode is play_on,
range: the area that a player moves during a game, by averages of the positions and their horizontal/vertical variance.

3 Statistical Analysis of Evaluation Leagues at RoboCup '98

The details of evaluation and logfiles of games are available at Dr. Kaminka's homepage [Gal98]. The following are short explanations of evaluation.

- All participating teams played four half games against AT_Humbolt97, the champion team of RoboCup 97.
- Four half games are referred to as phase A, B, C and D.

   phase A : the game is played under normal conditions.

   phase B : A manager of the evaluation assigns one player other than the goalkeeper randomly. The team disabled the assigned player and compete the game with ten players.

   phase C : A member of AT_Humbolt97 assigned another player who he thought was the most valuable player other than the goalie. The team omitted two assigned players and competed the game with nine players.

   phase D : The team also omitted the goalie and competed the game with eight players.

3.1 Discussion from Scoring Points

Table 1 shows the scores of teams that participated in the evaluation league and the statistics of AT_Humbolt97 plays. The first columns are the names of the teams that participated in the evaluation league. The number under the team name is the rank in the RoboCup '98 tournament league. The second column is the phase and scores. The left score is the points the team gained and the right score is AT_Humbolt97's points.

Most teams won the game at phase A, so it can be said that the level of RoboCup '98 is higher than that of RoboCup '97 [2]. The game conditions become harder for teams as the phase changes to B, C and D. The teams are said to be robust, when their scores do not vary as the phases change. From the table,

- The higher ranked teams, such as CMUnited, won the game in disadvantageous phases, while lower ranked teams, such as Kasugabito-II which were eliminated from the tournament, gained less points in disadvantageous phases than the normal phase and lost the game.

---

[2] At RoboCup '99, evaluation league were held with adding new half games. Games with AT_Humbolt97 shows that the level of RoboCup '99 is higher than that of RoboCup '97 and '98.

– Some teams in the middle rank are said to be robust from the difference in points scored. For example, Isis98 won the game at phase C with a better score than phase A. CAT_Finland lost the game at phase A, but won at phase B.

We don't think there is any relation between tournament ranks and robustness in play.

Table 1. Statistic data in evaluation games

team	phase	Score	AT-Humbolt97				1-2 pass	AT
			K	P	I	D		
CMUnited-98	A	7 - 0	53	7	23	2654	2(2)	0(0)
(1)	B	6 - 0	51	6	30	2395	5(2)	0(0)
	C	3 - 0	55	11	30	2327	4(0)	0(0)
	D	3 - 0	37	8	20	2006	0(0)	0(0)
AT_Humbolt-98	A	6 - 0	76	11	36	2622	2(0)	0(0)
(2)	B	9 - 0	87	11	46	3545	2(0)	0(0)
	C	3 - 0	76	17	33	2462	1(0)	1(0)
	D	5 - 0	80	15	43	2789	0(0)	0(0)
WindmillWanders	A	5 - 0	66	10	42	2481	1(0)	0(0)
(3)	B	7 - 1	58	9	33	2884	0(0)	0(0)
	C	3 - 0	64	13	29	2629	0(0)	0(0)
	D	2 - 1	55	10	29	2199	1(0)	0(0)
Isis 98	A	1 - 0	79	30	36	1976	1(0)	0(0)
(4)	B	1 - 0	60	13	34	2771	2(0)	0(0)
	C	2 - 0	62	17	32	1857	1(0)	1(0)
	D	1 - 2	72	24	35	2378	0(0)	3(0)
Rolling Brains	A	5 - 0	82	13	53	2737	0(0)	0(0)
(5-6)	B	2 - 0	81	21	42	2624	1(0)	0(0)
	C	1 - 0	60	14	33	2163	3(0)	0(0)
	D	0 - 0	70	16	44	1994	0(0)	0(0)
Andhill	A	6 - 0	101	33	53	2931	1(1)	2(0)
(5-6)	B	5 - 1	83	21	44	2568	1(1)	0(0)
	C	3 - 1	88	25	48	2469	1(0)	2(0)
	D	5 - 0	91	26	48	2551	0(0)	0(0)
CAT_Finland	A	0 - 1	62	18	27	1659	0(0)	1(0)
(7-8)	B	1 - 0	69	17	39	1669	0(0)	0(0)
	C	1 - 1	78	24	36	2151	1(0)	0(0)

continued on next page

team		Score	K	P	I	D		AT
	D	1 - 3	84	30	35	2288	1(0)	3(0)
Gemini	A	8 - 0	121	41	58	3241	0(0)	1(0)
(7-8)	B	5 - 1	104	37	50	2899	0(0)	2(0)
	C	1 - 0	119	44	52	2177	1(0)	3(0)
	D	8 - 2	107	41	45	3431	0(0)	0(0)
Aiacs	A	7 - 0	99	29	49	3008	0(0)	2(0)
(9-12)	B	0 - 0	61	17	29	2091	1(0)	0(0)
	C	6 - 1	82	24	42	2246	0(0)	1(0)
	D	4 - 4	72	24	37	2453	0(0)	0(0)
PasoTeam	A	0 - 1	68	11	44	1987	1(0)	1(0)
(9-12)	B	0 - 3	58	12	35	2175	0(0)	0(0)
	C	0 - 4	51	5	33	1996	2(0)	0(0)
	D	0 - 5	51	10	31	2164	0(0)	0(0)
AT_Humbolt-97	A	1 - 1	77	21	45	2290	1(0)	0(0)
(9-12)	B	1 - 1	80	21	43	2244	0(0)	1(0)
	C	1 - 4	81	21	49	2349	0(0)	0(0)
	D	0 - 2	80	36	34	2360	0(0)	2(1)
DarwinUnited	A	0 - 3	64	20	32	2289	2(0)	0(0)
(-)	B	0 - 6	57	16	30	2395	1(0)	1(0)
	C	0 - 3	71	23	34	2400	2(0)	2(0)
	D	0 - 1	54	16	31	1978	1(0)	0(0)
Kasugabito II	A	5 - 0	109	39	53	2807	1(0)	1(0)
(-)	B	2 - 0	78	29	37	2226	0(0)	0(0)
	C	0 - 2	81	34	35	2186	0(0)	2(0)
	D	0 - 2	81	29	38	2358	1(0)	1(0)

K=kick　　P=pass　　I=interception　　D=distance
AT=AT_Humbolt-97
rank(-)= eliminated from the tournament.

3.2　Discussion from AT_Humbolt97's Side

It is difficult to evaluate teams by their game scores. AT_Humbolt97 was used to normalize various team's ability. The second column is the numbers of kicks(K), passes(P), interception(I) and distance(D) of AT_Humbolt97's players. The values in Table 1 are calculated according to the methods in section 2.1.

Fig. 1 shows the changes of actions from A to B, from A to C, and from A to D. The vertical axis shows the ratio of data change from phase A and the horizontal axis is the team.

The players of the teams with less collaboration are thought to be weak in covering a disabled player, so we expect that players of AT_Humbolt97 can pass, kick and move more easily as phases changed from A to D. Against our expectation, the most vertical values of point in Fig. 1 are less than 1.0. This

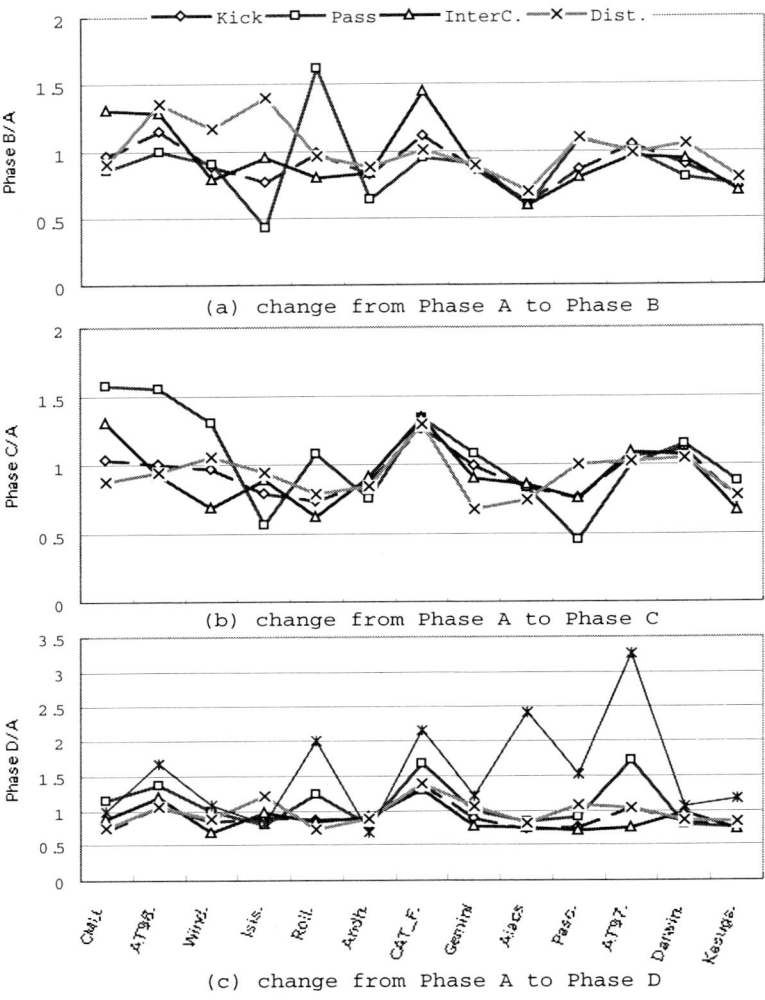

Fig. 1. Changes in actions at Phases.

means AT_Humbolt97 didn't play and move more than in phase A, and doesn't support our expectations. And the changes from phase A to B, A to C, and A to D don't show the same tendencies. For example, in the games vs. CMUnited, the number of passes of AT_Humbolt97 remained equal at phase A and B, increased 50% at phase C, and increased a little more at phase D. On the other hand, at the games vs. Kasugabito-II, they decreased in all phases following A.

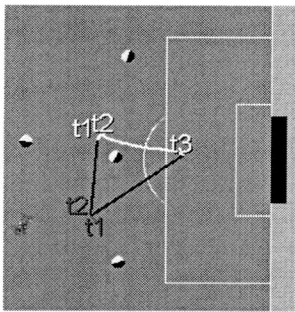

Fig. 2. 1-2 pass between CMUnited players.

3.3 Discussion from 1-2 pass as collaboration

A 1-2 pass is a collaborative actions. Fig. 2 shows a snapshot of a 1-2 pass shown at phase B of CMUnited. The white line indicates trajectories of a player who kicked the ball at t_1 and received the return pass at t_3. The black line shows the passes. The displayed players are AT_Humbolt97 players at t_1.

The last column of Table 1 is the number of the 1-2 pass. The left number is the number of 1-2 passes of teams evaluated and right number is that of AT_Humbolt97's 1-2 passes. The numbers in parentheses is the number of 1-2 passes which are connected to goals.

CMUnited players perform 1-2 passes the most at evaluations, and most of their passes scored points. AT_Humbolt97 didn't perform any 1-2 pass in games with high ranked teams, but they did in games with lower ranked teams. These findings seem to be similar to human teams that can perform well against weak teams, but perform poorly against strong teams.

3.4 Discussion of Autonomous Movement

Without communication among agents, an agent which moves by itself according to changing situations can be said to be autonomous. Fig. 3 shows the trajectories of a forward player of CMUnited 98 (left) and Kasugabito-II (right).

The figures are trajectories of the same player at phase A, B, C and D from the top. The numbers under the figure are the number of kick, distance, the variance of horizontal movement, and the variance of vertical movement.

The CMUnited player moved twice as much as the Kasugabito-II player and the range of his movement was wider. CMUnited 98 and Kasugabito-II played at preliminary games, and the score was 5-0. From the score, our Kasugabito-II played a good game. However, the CMUnited player's attack in front of the goal was superior according to the figures.

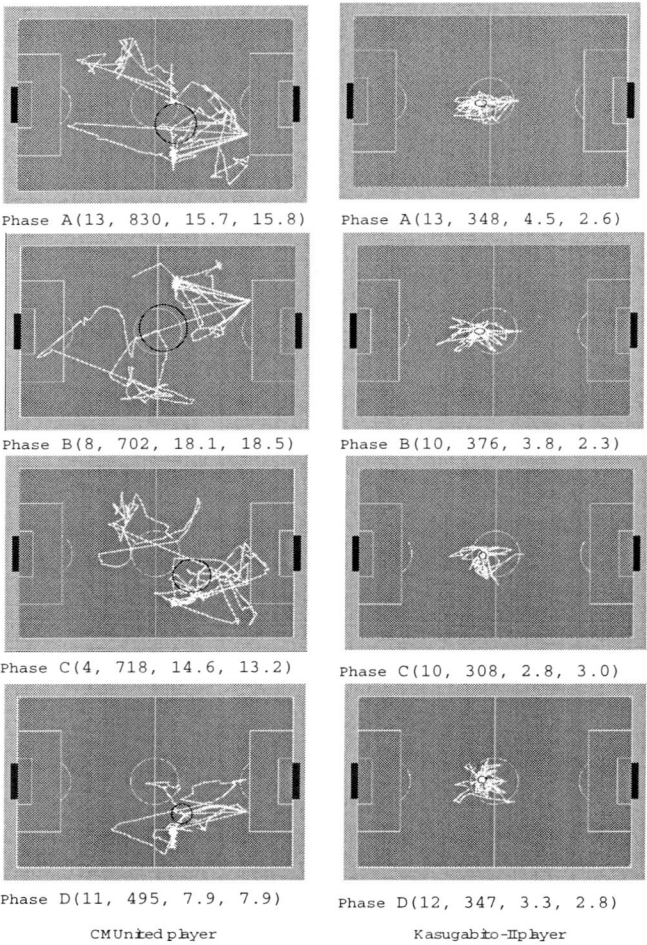

Fig. 3. Trajectories of Players.

4 Robustness and Opponent's Information

Agent programs are said to be robust when it can adapt environmental changes without hearing of the change from others. When knowing of the changes, the agent programs may modify their parameters to adapt to the changes.

4.1 Experiment for Adding Opponent's Information

At evaluation leagues, participating teams could not change their programs or parameters before the games. At regular games, the participants can modify their

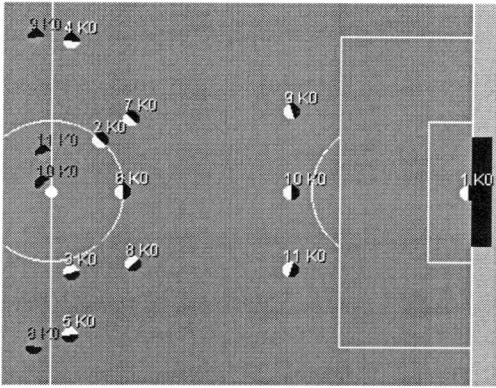

Fig. 4. Kasugabito players' initial position.

programs or tune parameters for the next game. This adjustment is equivalent to adding opponent's information gained from the previous games. We test the effectiveness of the adjustment before a game by comparing the difference of the score's between two games. One game is done with parameters modified by human, and the other games with no modifications.

experiment 1 The evaluation games between AT-Humbolt97 and Kasugabito-II were played again in our computer environment.

experiment 2 At the beginning of phase B, C and D, the initial positions of Kasugabito-II players were modified by one of our students.

Fig. 4 shows the initial positions of Kasugabito-II players in Phase A for explanation.

phase B disabled player = No.8(Defensive MF).
The No.7 player in the counter position of No.8 was moved to the center, for the purpose of defending the right and left side.

phase C disabled player = No.2 (CF).
Another forward No.3 was moved to the center, and defense positions of three attractive MDs (No.4, 5, 6) were changed 5m forward.

phase D disabled player = No.1 (goalie)
Defense positions of the center DF (No.10) was moved closer to the penalty area.

4.2 Discussion

Table 2 shows the result of the experiment games. The running environment is different from that of RoboCup98, so the values of the score are different from the values in Table 1. However, the scores in experiment 1 show a similar tendency as that shown in the evaluation games. Kasugabito-II won the game at phase A. AT-Humbolt97 became superior to Kasugabito-II as phase changed from A to D, and AT-Humbolt97 won from Phase C on. In experiment 2, Kasugabito-II continued to win the game till phase D. Judging from the scores, Kasugabito seems to have become more robust than in experiment 1.

The second column of Table 1 shows AT-Humbolt97 players' data and the last column Kasugabito-IIs' data. At phase B and C of experiment 2, the number of kicks and passes are more, and the distance is longer than in experiment 1 for both teams. At present, we cannot account for the changes in data. However, the experiments support the claim that adding the opponent's information made Kasugabito-II more robust even though the player agents were the same. This indicates that making use of opponent information as well as the agent's ability itself is important to make teams more robust.

Table 2. Results of experiments (AT-Humboldt side)

	Score	AT_Humbolt				Kasugabito-II			
		K	P	I	D	K	P	I	D
		experiment 1							
A	3 - 0	106	41	53	2553	118	27	55	2706
B	2 - 1	75	25	37	2042	72	16	34	1869
C	1 - 2	89	29	52	2265	81	17	49	2287
D	0 - 3	99	42	40	2303	81	18	34	1974
		experiment 2							
B	3 - 1	103	41	44	2333	86	20	41	2311
C	3 - 0	111	42	54	2509	103	26	51	2264
D	2 - 1	88	33	43	2249	91	20	42	1977

K=kick   P=pass   I=interception   D=distance

5 Summary

This paper presents analysis of collaborative actions in soccer simulation games. While there are many papers on how to implement multi-agent systems, there are few on evaluation of multi-agent systems. One of the reasons is that evaluation standards depend on the applied field. In soccer simulation games, we can use many of the standards used in human soccer.

In analyzing the logfiles of evaluation games at RoboCup'98, it is clear that:
- There is no direct relation with scoring to collaboration by analyzing the data of basic soccer actions, such as kicking, passing, etc.
- The 1-2 pass is assumed to be collaborative actions among agents. The number of 1-2 passes seems to be related to the ranking of teams in the tournament.

- The difference in their autonomy between CMUnited players and Kasugabito-II players seems to be bigger than the difference of the scores of their game.
- Adjusting players' initial position made Kasugabito-II as robust as CMUnited from the standpoint of scoring.

Through these analysis results, we think it is necessary to give player agents advice on games as human coaches do. We have been developing an on-line coach agent based LogMonitor that analyzes the game on-line [kIII].

We appreciate the RoboCup98 committee who planned the evaluation league, AT-Humbolt team who prepared AT-Humbolt 97, and Gal Kaminka who edited the logfiles.

References

[SoftRB] http://www.softsport.com/
[Kitano98] H. Kitano ed.: RoboCup97 Robot Soccer World Cup I, Lecture Notes in Artificial Intelligence, Vol. 1395, Spring-Verlag, 1998
[Log] http://kiyosu.isc.chubu.ac.jp/robocup/
[Gal98] http://www.isi.edu/soar/galk/RoboCup/Eval/
[Cup98] http://www.robocup.org/
[Taka98] T. Takahashi, T. Naruse: From Play Recognition to Good Plays Detection, Proceedings of the second RoboCup Workshop, pp. 129-134, 1998
[Ial98] I. A. Letia, etl.: State/Action Behavioral Classifiers for Simulated Soccer Players, Proceedings of the second RoboCup Workshop, pp. 151-164, 1998
[Tanaka] K. Tanaka, I. Frank, I. Noda, H. Matsubara: Statistical Analysis on RoboCup Simulation League (Japanese), Journal of IJSAI, Vol. 14, No. 2, pp.200-207, 1999
[kIII] Team discription: Kasugabito-III, this book

A Statistical Perspective on the RoboCup Simulator League: Progress and Prospects

Kumiko TANAKA-Ishii, Ian FRANK,
Itsuki NODA, and Hitoshi MATSUBARA

Electrotechnical Laboratory
Umezono 1-1-4, Tsukuba
Ibaraki, JAPAN 305
{kumiko,ianf,noda,matsubar}@etl.go.jp

Abstract. This paper uses statistical analysis to demonstrate the progress to date in answering the RoboCup Synthetic Agent Challenge. We analyze the complete set of log data produced by the simulator tournaments of 1997 and 1998, applying techniques such as principal component analysis to identify precisely what has improved, and what requires further work. Since the code that implements our analysis produces its results in real-time, we propose releasing a proxy server that makes statistical analysis available to RoboCup developers. We believe such a server has a crucial role to play in facilitating and evaluating research on the three specific challenge problems of opponent modeling, teamwork and learning. We also suggest that — if RoboCup is to make the most of the efforts of participating researchers — the time is ripe for the institution of a modular team based on a common model.

1 Introduction

In the year between the first and second Robotic Soccer World Cups (RoboCup-97 and RoboCup-98), the level of play of the teams has unquestionably improved. This can be seen by the way that the best teams from 1998 beat the best teams from 1997. However, this kind of high-level observation does not answer the question of *why* the teams perform better. Our principal contribution in this paper is to qualitatively demonstrate the improvement in RoboCup soccer abilities by showing how a team's play can be analyzed with statistics. We analyze the entire set of log data produced by the 1997 and 1998 simulator tournaments in terms of 32 skill-based features.

The RoboCup Synthetic Agent Challenge [Kitano et al.1997] defines three separate challenges of *opponent modeling*, *teamwork* and *learning*. Our research in particular demonstrates that significant progress has been made towards meeting the teamwork challenge. For instance, our analysis of player correlations shows that forwards (and defenders) exhibit far greater teamwork in 1998. Further, our analysis techniques are also important for directing *future* research on the Synthetic Agent challenges. Most obviously, by highlighting both the aspects of play that have improved and also those that have not, our statistics enable researchers to identify the most promising directions for further investigation. More importantly, however, the code that implements our analysis techniques produces its results in real-time, as games are running. We therefore propose offering this code to the RoboCup community in the form of a proxy server. We describe how such a proxy server can directly facilitate research on all three challenge problems, as well as on systems such as automated commentators and post-game analyzers. Further, we note that this kind of server has an obvious role in conducting the evaluation procedures set out in the challenge paper itself. This discussion leads us to suggest a new approach to RoboCup that would enable more effective evaluation and combination of the efforts of RoboCup researchers: the creation of a modular team based on a common model.

As far as we are aware, there have been few previous attempts to address the statistical analysis of soccer in an academic context. [Tanaka-Ishii *et al.*1998] have defined notions such as *pass bigrams* and suggested the use of Voronoi diagrams to measure the areas of influence of players. Also, [Takahashi *et al.*1998] and [Matsubara *et al.*1998] have carried out limited log analyses of the 1997 RoboCup and the 1998 Japan Open, respectively. To date, though, the only significant progress in the statistical analysis of soccer has been the coaching tools and analysis software produced by commercial companies (*e.g.*, see [SoftSport1998]). National teams are assumed to keep databases of statistics, but the details of such resources are a closely guarded secret.

2 The Data

Our analysis is based on all the log files of the first two RoboCups (55 games from 1997 and 81 games from 1998), and focuses on the differences between the winning and losing teams. Since team designers are allowed to manually change their team settings and code at half-time, we collect separately the statistics for the winners and losers of each *half-game* (excluding the 36 half-games that were drawn). The size of the data files for the 236 half-games that produced a winner is around 245Mbyte.

In interpreting this data, it is important to bear in mind that the simulator league also changed between the RoboCups of 1997 and 1998. Most notably:
- A goalkeeper (with a `catch` ability) was introduced.
- The stamina model was changed to make long-term resource management more important.
- An off-side rule was introduced.
- A limit was imposed on the maximum ball speed.

It is natural that client team programs change when the rules change. This is reflected in the statistics we present in the following section.

3 The Statistics

We analyzed the log files in terms of a set of thirty-two features that describe different aspects of a soccer game. These features are shown in Figure 1, along

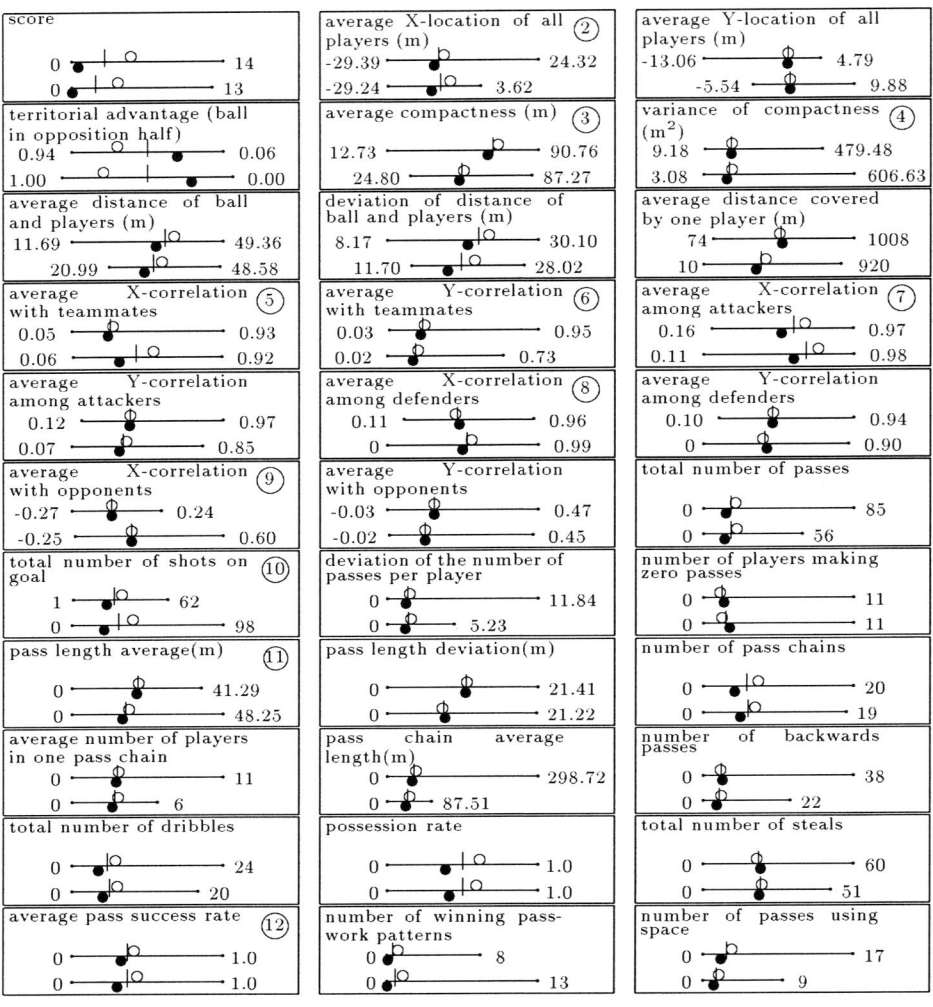

Fig. 1. The 32 evaluation features (plus 'score') that we used to analyze the RoboCup tournaments. The scales show the range of the values taken by these features for one team in one half game in 1997 (upper scale) and in 1998 (lower scale). The vertical bars denote the overall averages, and the o and • marks denote the averages for the winning and losing teams. Circled numbers refer to the figure numbers of the graphs in the text.

with some simple scales giving information on the values they took in 1997 (upper line) and in 1998 (lower line). (Note that X values are measured along the longer side of the field and Y values along the shorter side, with zero in both cases being the center spot.) Broadly, the features fall into three categories that we will discuss in detail below: 8 that represent general aspects of a team's formation (§3.1), 8 that represent formation in terms of the specific notion of player correlations (§3.2), and 16 that deal with passing and shooting skills (§3.3).

We illustrate our discussion with 11 graphs that detail the frequency distribution of selected features. In each of these graphs, the x-axis is discretized into 30 equal and contiguous intervals that span the observed maximum and minimum values of the feature in question. Considering the losers and winners in 1997 and 1998 separately, the percentage of games for which the value of the feature falls within each interval is then counted, and a smoothed curve fitted through the 30 points.

We concentrate throughout this section on highlighting the overall changes in playing skills from 1997 to 1998. We should point out, though, that the real-time nature of our code (discussed in §4) also supports the investigation of other aspects, such as the way that individual statistics change dynamically as a game develops.

3.1 Formation

Figure 2 shows the average of all the players' X-positions. Players in the opposition's half have a positive X-value, so the larger the values in the figure, the more attacking the formation. Both in 1997 and 1998, the winners were more attacking than the losers (the curves are further to the right).

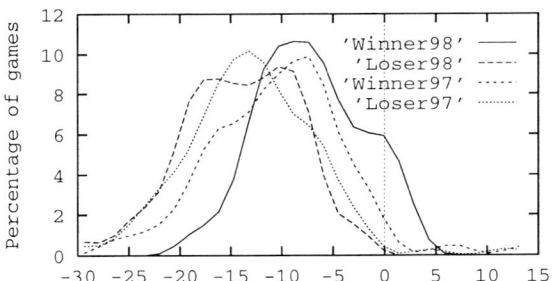

Fig. 2. The average X-location of all players (m)

In real soccer, the entire formation of a team follows the movement of the ball across the field. We define the *compactness* of a team as the X-distance between its front-most player and its rear-most player (excluding the goalkeeper). Figure 3 shows the distribution of average compactness in RoboCup, demonstrating clearly that teams in 1998 were more compact than in 1997. This improvement is probably a result of the introduction of the off-side rule. The presence of off-sides allows a team to protect itself against counter-attack if they move up-field as a unit with the ball. This becomes an aspect of the game that teams cannot afford to neglect.

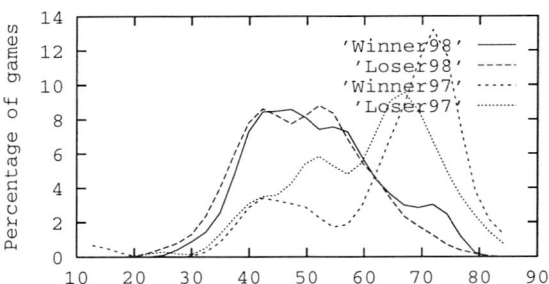

Fig. 3. The average compactness (m)

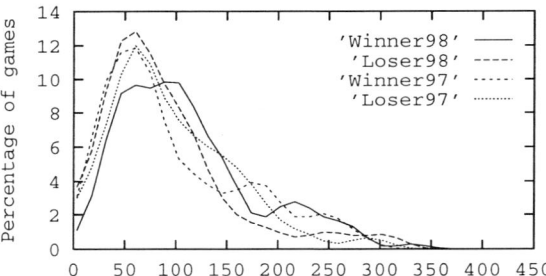

Fig. 4. The variance of compactness (m^2)

Figure 4 shows the variance of compactness during games. This does not show a remarkable difference between winners and losers or over the two years. So, although team formation has improved to dynamically follow the ball, the degree of compactness does not yet change significantly during the game. That the players follow the ball more closely in '98 can also be seen from Figure 1: the average distance between the ball and players decreases in '98. At the same time, though, the average distance covered by a single player has also decreased, indicating that teams have improved their formations whilst also handling their resources better.

3.2 Player Correlations

Another way to analyze players' movements is in terms of correlations between player locations. We calculate these correlations in the form of two 22×22 symmetric matrices, one for the correlation of players' X-locations and the other for the players' Y-locations. These matrices are generated for each time step of the game.

Figure 5 and Figure 6 show the average of the values in the X and Y correlation matrices over entire half-games. These averages are calculated for the entries in the 11×11 sub-matrices giving correlations between teammates (there were no negative elements in these sub-matrices, making the average a realistic measure). Although the Y-correlations show no significant change, the X-correlations show a substantial improvement in the movements of the 1998 winners. In fact, since this curve has two peaks it shows that there are some winners that could correlate and others that could not.

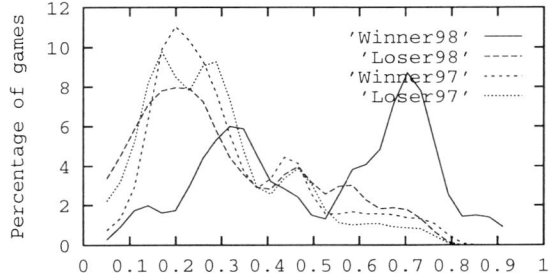

Fig. 5. The average X-correlation with teammates

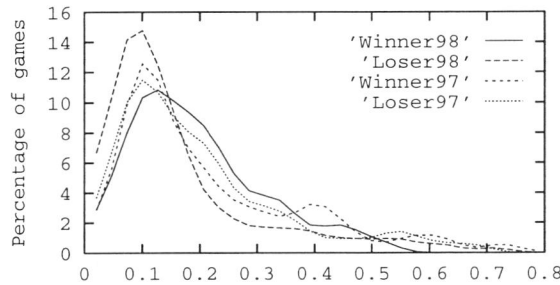

Fig. 6. The average Y-correlation with teammates

To examine teamwork in smaller groups, we also calculated the average X-correlation values for just a team's front 3 attackers and for its back 3 defenders (excluding the goalkeeper). Again, these results (Figure 7 and Figure 8) show a marked improvement in the winning teams of 1998. In contrast to this, the Y-correlations again do not
show an improvement. In fact, *all* the average values for Y-correlations in Figure 1 show a decrease in 1998. This apparent *worsening* of play is partially due to the skewing of the 1997 figures by a small number of games that exhibited

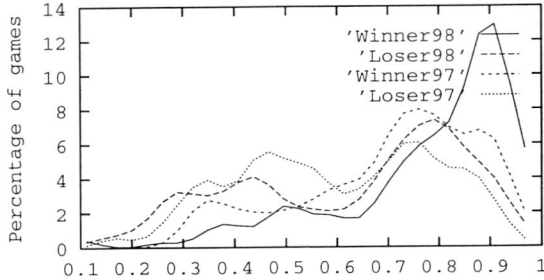

Fig. 7. The average X-correlation among attackers

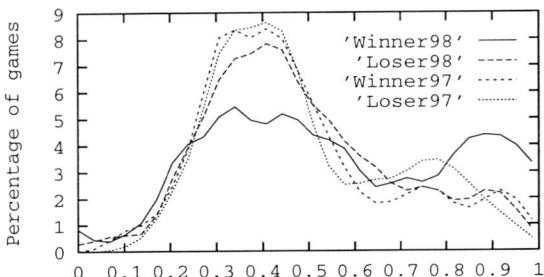

Fig. 8. The average X-correlation among defenders

the 'rugby effect': teams simply surrounding the ball with their players in a scrum that slowly moved across the field. Motivated by the introduction of the off-side rule, teams have improved their X-correlations to the extent that the 1998 average is higher than the biased 1997 value. For the Y-correlations, such improvements remain as further research.

One final use of correlations is to demonstrate the marking abilities of teams. The average of a team's X-correlations with opponents is shown in Figure 9 (the correlation matrix is symmetric, so there is only one graph for each year). This demonstrates that marking has increased from '97 to '98. In general, however, the symmetric nature of player correlations makes it difficult to identify *which* team's players are actually doing the marking. We are investigating how to collect more statistics on marking by examining the dependencies between the directions and the timings of players' movements.

3.3 Passing, Shooting & Dribbling Skills

Here, we look at the remaining sixteen features from Figure 1, which all deal with ball-handling skills. In RoboCup, client programs issue large numbers of kick commands. To distinguish kicks that have a genuine effect, we define a

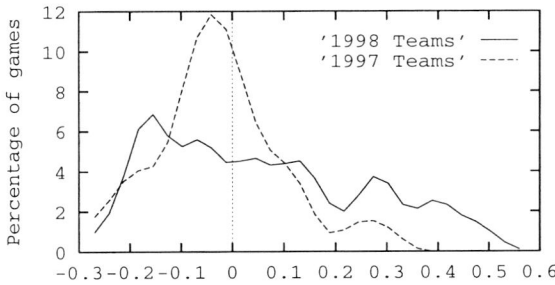

Fig. 9. The average X-correlation with opponents

genuine kick sequence as two successive successful kicks of the ball that occur with a time lag of more than 0.5s and at a physical separation of more than 0.5m. A successful pass is then a genuine kick sequence in which the two kicks are made by players of the same team; a dribble is a genuine kick sequence in which the two kicks are made by the same player.

A team's possession time is defined as starting with the first kick of a team's kick sequence, and continuing until either an opponent kicks the ball, or the ball leaves open play (a free kick, goal kick, throw in, corner, or center kick is awarded). We measure the possession of one team as the rate of the total possession time in the game. A *steal* occurs when the possession transfers from one team to another without the ball leaving open play. The pass success rate of a team is then defined as the ratio of its passes to the total of its number of passes plus the number of the opponent's steals. A *winning passwork pattern* is defined as any chain of three players A, B, C from the same team such that passes from A to B and from B to C occur at least once, and C scores at least one goal. Finally, a *pass using space* is a pass where the ball never comes within 5m of an opponent.

Let us look at some of the graphs for the ball-handling features. One of the most interesting is Figure 10, which shows the total number of shots on goal. The curves for the 1997 and 1998 winners both have multiple peaks, suggesting that there are two types of strategies for shooting. Another interesting conclusion can be drawn from Figure 11, which shows the average pass length. This suggests that shorter, more stable passwork is one of the keys for strong play. In Figure 12 we also show the average pass success rate. This is slightly better for winners than for losers, and has a small tendency to increase from 1997 to 1998.

In fact, most of the graphs for the ball-handling features show small improvements such as that of Figure 12 rather than the larger differences of Figure 10 and Figure 11. To provide a better picture of the true importance of ball-handling skills, we therefore carried out the *principal component analysis* described in the following section.

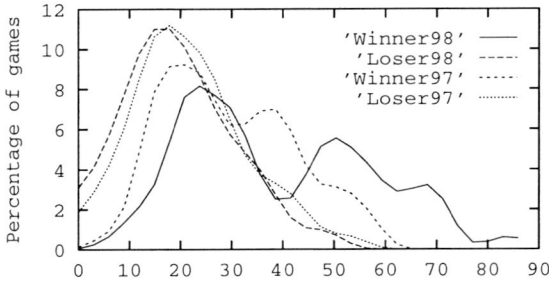

Fig. 10. The total number of shots on goal

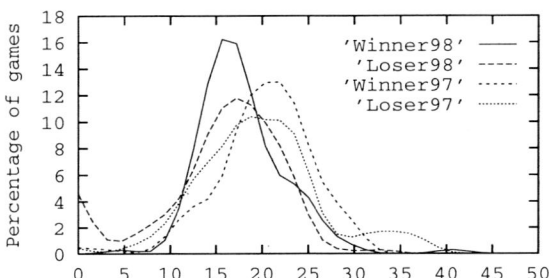

Fig. 11. The pass length average(m)

3.4 Principal Component Analysis

To use principal component analysis (PCA) in our analysis, we calculated the eigenvectors of the 32×32 correlation matrix of the features in Table 1. These eigenvectors give the directions in which the data cloud is stretched most, and so

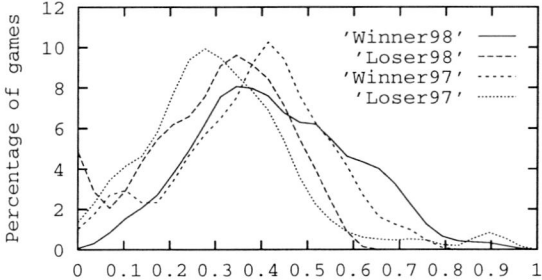

Fig. 12. The average pass success rate

allow us to identify the features that play the most important roles in RoboCup games. For the data from the 1998 RoboCup (including the drawn half-games), Table 1 shows the ten features that had the largest elements in the eigenvector corresponding to the largest eigenvalue. Note that the features related to ball handling are prominent. So, whereas the previous sections revealed that substantial progress has been made between '97 and '98 in team formation and co-ordinated movement, here we see that an underlying ability to handle the ball well is also crucial for success.

Table 1. The RoboCup98 features highlighted by PCA

Feature	Value
total number of passes	0.297
deviation of the number of passes per player	0.287
average pass success rate	0.271
number of pass chains	0.269
number of backwards passes	0.234
average distance covered by one player	0.230
number of players making zero passes	-0.222
total number of shots on goal	0.219
average X-correlation with teammates	0.214
average X-correlation among attackers	0.194

We also used principal component analysis to identify the 1998 RoboCup games that produced the best play. For each half-game, we found the inner product of the vector of the feature values (for each team) with the vector of largest eigenvalues. Since this produces larger values for teams that played well, we can identify the good teams independently of the actual tournament results. Table 2 shows the ten teams that produced the highest scoring games. Note that these teams all fared well in the actual tournament itself. Our analysis demonstrates in detail the reasons for this success.

Table 2. PCA top ten half-games played in RoboCup98

	Team (final rank)	Game
1	AT Humboldt (2)	1st half *vs* Gemini
2	CMUnited (1)	1st half *vs* Andhill98
3	CMUnited (1)	2nd half *vs* Rolling Brains
4	CMUnited (1)	2nd half *vs* Kasugabito II
5	AT Humboldt (2)	2nd half *vs* Andhill98
6	CMUnited (1)	2nd half *vs* Andhill98
7	Tumsa (Best 12)	1st half *vs* Gemini
8	Windmills (3)	2nd half *vs* AT Humboldt
9	Andhill98 (Best 6)	1st half *vs* Gemini
10	Windmills (3)	1st half *vs* AT Humboldt

3.5 Statistics Summary

The above sections have demonstrated the progress in the RoboCup simulator league from 1997 to 1998. Our overall conclusions can be summarized as follows:
- Teams work together to follow the play better in '98, despite having to manage stamina more carefully.
- Teamwork has improved significantly along the long side of the field, but not along the shorter.
- Improved teamwork can be seen in '98 in the way that forwards (and defenders) share their roles.
- Teams carry out more marking in '98.
- Ball-handling skills have not improved significantly on average, but principal component analysis shows that they are crucial for strong play.

These results demonstrate that progress has been made towards meeting the teamwork Synthetic Agent challenge. To a lesser extent, an improvement in opponent modeling is also demonstrated by our analysis of marking.

4 Prospects: A statistics proxy server

The statistical analyses we have described above answer the question of the extent to which soccer skills in RoboCup have progressed *to date*. In addition to this, however, the code that we have implemented to carry out this analysis also has a direct bearing on the *future prospects* of research on the Synthetic Agent Challenge.

In particular, our code produces its results in real-time, as a game is running. Thus, we suggest offering this code to the RoboCup community in the form of an independent proxy server. In our prototype server, we have incorporated two further techniques developed by [Tanaka-Ishii *et al.*1998]: the representation of ball-play chains as first order Markov processes, and the calculation of players' defensive areas with Voronoi diagrams. Figure 13 shows the likely uses of such a proxy server.

In addition to facilitating research on automated commentators and postgame analyzers, then, we envisage that a proxy server will directly enable research on the three challenge problems. Obvious applications are:
- facilitate off-line learning from game logs and on-line learning via the touchline coach planned for 1999,
- identify the opponent's key players, passwork patterns and styles of play (opponent modeling),
- identify players not fulfilling their assigned roles, enabling suitable adaptations to be made (teamwork).

Another application we foresee is related to the specific evaluation procedures laid out in the challenge paper itself. To our knowledge, only one of these evaluations — a test of team robustness — has so far been carried out. This test took place at RoboCup 1998, and involved 13 of the 37 participating teams. With the 1997 champions as the fixed opposition, each of these 13 teams played a single

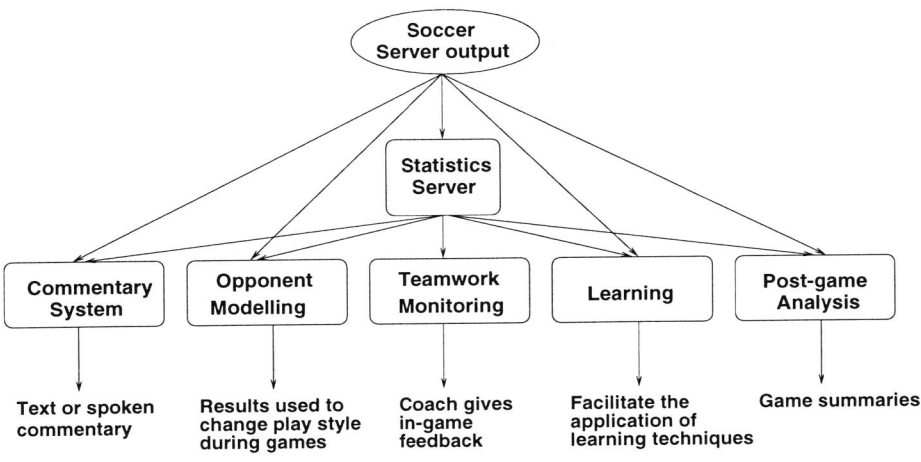

Fig. 13. Likely uses of a statistics proxy server

half-game with 11, 10, 9 and then 8 players. The logs of this test can be found at http://www.isi.edu/soar/galk/RoboCup/Eval/. We are not aware of any detailed analysis, but the logs clearly demonstrate that evaluation will not be straightforward. As might be expected, conducting just one half-game with each number of players results in noisy data. For example, the CAT-Finland team lost 0-1 when playing with 11 players, but with just 10 players it won 1-0, and with 9 players it drew 1-1.

Thus, to generate meaningful results from the robustness test, statistical analyses of the kind presented in this paper will be important. Indeed, since almost all of the evaluation procedures laid out in the challenge paper involve the playing of games, we expect to see our proxy server used in this way very often.

5 Prospects: A modular common model

We feel that what is ultimately most important in RoboCup evaluation is identifying the *AI techniques* that produce good performance; only by identifying the best techniques can they be effectively re-used in the future to advance RoboCup research. Within the current framework of RoboCup, however, this kind of evaluation is hard to carry out. Almost all teams use a combination of a number of techniques. So, even with a detailed statistical analysis of full games, it is in general difficult to identify which AI techniques worked best for particular aspects of the game. Similarly, the evaluation procedures described in the challenge paper focus on varying the conditions under which a team plays a game rather than varying the AI techniques used for particular sub-problems. The prevailing tendency in RoboCup, then, is simply to assume that teams that win are the ones that employ good techniques. This must be true to some extent, but the reverse is certainly not true: teams that lose do *not* necessarily employ bad techniques. They may actually excel at some aspects of soccer whilst being let down by others.

We suggest that the current situation in the RoboCup simulator league is similar to that in another research field often regarded as a challenge problem: Computer Go. Superficially, this field appears very healthy, with over 100 active programs being developed (including about ten that are commercial), and a number of computer tournaments (two with the status of World Championships). However, it was recently pointed out by [Müller1998] that progress is much slower than it could be. A typical program has between 50 and 100 modules, and it is generally the *weakest* of these that determines the strength of the program. Thus, the field progresses by all of the program developers simultaneously trying to improve the level of their worst modules. Müller proposed that the overall rate of progress would be significantly improved if researchers worked together on the implementation of a *common model*.

Although RoboCup is young compared to Computer Go, we believe that a similar argument applies. A common model would improve the rate of progress, provided it has the essential feature of being *modular*. Modularity is important because it is this that facilitates the isolation and direct evaluation of competing techniques for any individual aspect of the game. This in turn increases the chances of being able to identify and retain the best aspects of every RoboCup team, including those with poor overall match results.

Thus, we propose the institution of at least one modular RoboCup team based on a common model. To function successfully and to benefit most from the available manpower, we suggest that such a team should be:
- open source; anyone is free to use the code, and to contribute improved versions of any module.
- well-tested; contributors demonstrate the advantages of their modules with match results or statistical analysis such as that described in this paper.
- well-documented; the learning curve is as painless as possible.
- updated frequently; there is an efficient organization for selecting and integrating currently best modules that keeps all contributors informed of progress.

Already, the first two RoboCups have identified important sub-tasks of soccer. A small sample of these tasks includes the automatic learning of passwork skills, the pre-compilation of communication options, static evaluation functions for player positioning, and the specification of a small number of distinct 'player modes'. We believe that formalizing a common model incorporating such modules will expedite RoboCup research. Ultimately, it will also lead to a deeper understanding of soccer itself, and of the multiple challenges it represents.

6 Conclusions

We have used statistics to qualitatively demonstrate the progress to date in the simulator league of RoboCup. Our results highlighted the changes in play from 1997 to 1998 in terms of 32 features. We also demonstrated the utility of examining the correlation of players' movements and of applying principal component analysis.

Further, we emphasized that our analysis is important to the *future* of the RoboCup research. We proposed the free distribution of a proxy server that can supply in real-time the full range of statistical analyses described in this paper, and pointed out that such analyses will have a crucial role in facilitating and evaluating research on the challenge problems of opponent modeling, teamwork, and learning. Finally, we suggested that to make the best use of the efforts of RoboCup researchers, at least one modular team based on a common model should be instituted.

References

[Kitano et al.1997] H. Kitano, M. Tambe, P. Stone, M. Veloso, S. Coradeschi, E. Osawa, H. Matsubara, I. Noda, and M. Asada. The RoboCup synthetic agent challenge 97. In *Proceedings of IJCAI-97*, pages 24–29, Nagoya, Japan, 1997.

[Matsubara et al.1998] H. Matsubara, I. Frank, K. Tanaka-Ishii, I. Noda, and H. Nakashima Automatic soccer commentary and RoboCup. In *Proc. of the 2nd Intl. Wkshop on RoboCup*, pages 7–22, 1998.

[Müller1998] M. Müller. Computer go: a research agenda. In H.J. vd Herik and H. Iida, eds., *Proc. of The 1st Intl. Conf. on Computers and Games*, Japan, 1998. Springer-Verlag. LNCS series, vol. 1558.

[SoftSport1998] SECOND LOOK soccer analysis software. The makers, SoftSport Inc., can be found on the Web at http://www.softsport.com/, 1998.

[Takahashi et al.1998] T. Takahashi, T. Naruse. From play recognition to good play detection. *Proceedings of Robocup98 Workshop*, pages 129–134, 1998.

[Tanaka-Ishii et al.1998] K. Tanaka-Ishii, I. Noda, I. Frank, H. Nakashima, K. Hasida, and H. Matsubara. Mike: An automatic commentary system for soccer. In *Proceedings of ICMAS98*, Paris, France, 1998.

Real-Time Color Detection System Using Custom LSI for High-Speed Machine Vision

Junichi Akita[1]

Dept. of Elec. and Comp. Eng., Kanazawa University
2-40-20 Kodatsuno, Kanazawa, Ishikawa, 920-8667 Japan

Abstract. The quantity of image information is very large, and this is why it is difficult to process them in real time, such as video frame rate. In this paper we demonstrate the developed real-time color detection system for the vision system of RoboCup Small-Size League. The pixel's color information is converted to HSV color system at first, and judged whether it has the color information of target. The detected information is converted to gray scale NTSC signal to be captured by PC, which is faster than the full color frame grabber.

1 Introduction

The quantity of image information is very large, and this is why it is difficult to process them in real time, such as video frame rate. Pixel information from video camera is transfereed from the upper-left corner of image to the lower-right corner serially, and the color detection of pixel can be processed only according to its color information, not those of neighbor pixels.

In this paper, we describe the real-time color detection system for the vision system of RoboCup Small-Size League using specially designed LSI. We also describe the real-time color conversion system from RGB to HSV employing pipelined operation of look-up table using ROM. The original color image is fed into the developed system, containing color conversion system and color detection system, and then the gray scale image generated by color detection results comes out in 600ns, and this system can be regarded as a 'filter' of image.

The post-color-detection processes, such as labeling, recognition, are processed for these detected 'gray' image, which is much easier to process in real time than the case of both color detection and other processes are executed by software.

2 Elements of Color Detection System

Figure 1 shows the structure of the developed real-time color detection system for vision system of RoboCup Small-Size League. The video signal from color CCD camera in NTSC format is converted to RGB signal at first, and then

Fig. 1. Structure of real-time color detection system

converted to digital signal of 6×3 bits. The digitized RGB signal is converted to the pair of HSV (Hue, Saturation, and Value), and then judged whether it is a target color or not by color detection circuit. The color detection results are converted to gray scale NTSC video signal again, which PC can easily capture by frame grabber. The detection result per pixel comes out 600ns after original pixel data is fed in, which can be regarded as 'simultaneously,' and the processed image is generated by passing a kind of 'filter.'

2.1 Color conversion methodology

The video signal in NTSC format from color CCD camera is at first converted to RGB signal, with the horizontal synchronous frequency of 15.7kHz. It may be usual to express the color information of each pixel by the pair of RGB, that is the intensities of red, green, and blue, that are intrinsically acquired by CCD color imager. It will be a better way to express the color information of pixel by the pair of HSV, since the pair of HSV is more invariant than the pair of RGB for the change of lighting condition.

The conversion rule of the pair of RGB to the pair of HSV is a simple, but a non linear function. One idea to execute this conversion is to calculate this simple, but non linear conversion rule for each pixel. This procedure can be easily implemented by software program, but it needs much processing time. For example, the pixel clock cycle time is about 100ns in NTSC video signal, and the conversion from RGB to HSV should be completed in 100ns for real-time process, that is almost impossible by employing software process.

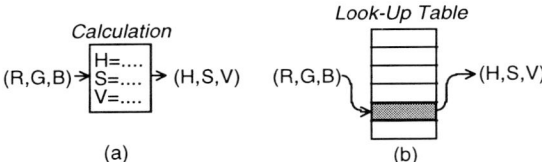

Fig. 2. Two methodologies to convert from RGB to HSV, (a)calculation per pixel, (b)employing look-up table

The other idea to execute this RGB to HSV conversion is to employ look-up table of RGB to HSV. The HSV values should be calculated for all case of possible RGB values once, and they are stored in look-up table. In conversion process, the input RGB values is used to indicate the according point of this look-up table, and the according result in this look-up table will be obtained much faster than to calculate it, as shown in Fig.2.

The look-up table can be easily implemented by ROM (Read Only Memory), where input RGB values are used as its address, and output HSV values are read out as the data in its address. Assuming the number of bits of RGB values as 6×3 bits, the total address space of ROM is equal to $2^{18} = 256K$, which is reasonable to implement by using popular ROM IC. If we employ 8×3 bits for RGB values, the needed address space is $2^{24} = 16M$, which will need more than 8 chips of ROM for one conversion look-up table bank.

2.2 Color detection criteria

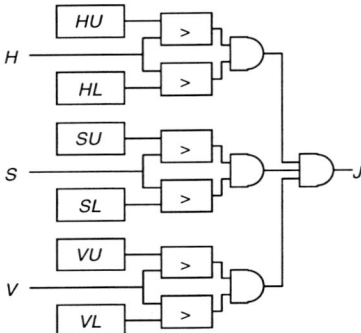

Fig. 3. Structure of color detector

The simple color detection criteria we have employed here is to detect all of hue, saturation, and value stay in the desired range, which is expressed as follows,

$$J = (H_L < H) \text{ and } (H < H_U)$$
$$\text{and } (S_L < S) \text{and } (S < S_U)$$
$$\text{and } (V_L < V) \text{and } (V < V_U),$$

where J is the judgement flag of detection, H, S and, V are the value of hue, saturation, and value, respectively, H_L and H_U are the preset lower and upper thresholds of hue, respectively, and so on.

In our developed system, the number of bits of each component is 6 for hue, 5 for both saturation and value, and the comparison operation should be executed in the accuracy of these bits. They are also needed the register where the upper and lower thresholds of each component are stored.

The whole circuit structure of the color detector is shown in Fig.3.

2.3 Design of color detection chip

We have designed the color detection chip using CMOS 1.2μm, double metal, double poly technology through LSI fabrication service of VDEC[VDEC, 1998][1].

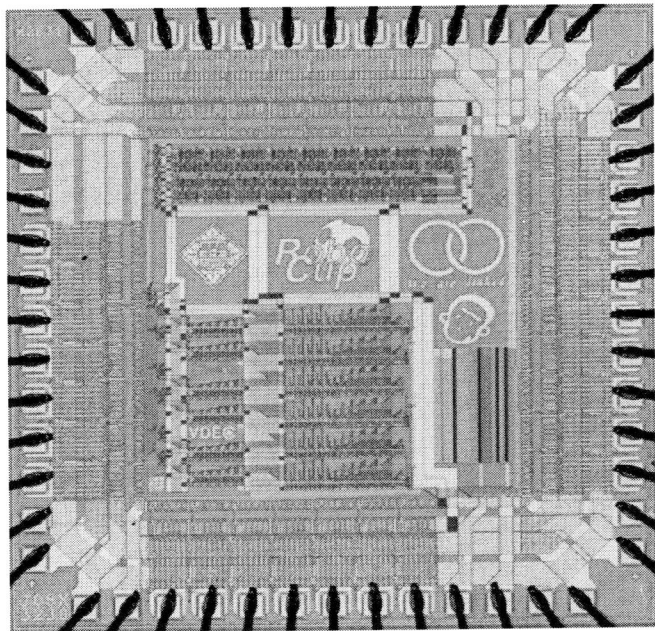

Fig. 4. Designed real-time color detection chip

The designed color detection chip consists of three blocks; 6bits magnitude comparators and 6bits registers and logic AND gates. We have designed them manually using layout editor, and gathered them in one chip with their interconnections drawn by layout editor Fig.4 shows the photograph of fabricated chip. The chip size is 2.3mm×2.3mm, and the number of transistors is 2,174. The measured operation speed of this chip is 11.2ns, which is fast enough for color detection in one pixel clock cycle, about 100ns.

The operation of color detection can be processed only according to the information of one pixel, and the detection circuit does not need to have frame memory which stores the whole image, and this is why the designed circuit consists of small size transistors and operates fast enough.

---

[1] The VLSI chip in this study has been fabricated in the chip fabrication program of VLSI Design and Education Center(VDEC), the University of Tokyo with the collaboration by Nippon Motorola LTD., Dai Nippon Printing Corporation, and KYOCERA Corporation.

3 Developed Color Detection System

The developed real-time color detection system consists of two blocks; color conversion block and color detection block.

3.1 Developed color conversion block

Fig. 5. Developed real-time RGB to HSV conversion board

Figure 5 shows the developed circuit board to convert pixel color information from RGB to HSV. The acquired RGB signal of pixel is 18bits; 6bits for red, green, and blue respectively. We have employed the total 18bits for RGB values in order to reduce the number of ROMs for look-up table in reasonable level.

The converted HSV information is assumed to be composed of 6bits of hue, H, 5bits of saturation, S, and 5bits of value, V. The conversion look-up table is implemented by 4Mbits EPROM of 27HC4096, which consists of 16bits×2^{18}. The data for each memory cell should contain according values of HSV, and this contains 6+5+5=16bits.

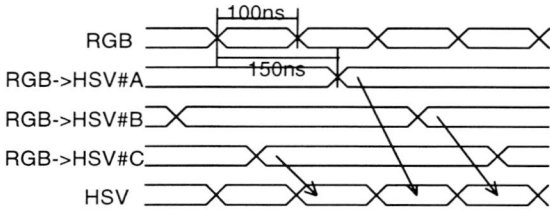

Fig. 6. Operation timing of RGB to HSV conversion

The access time of the ROM we employed is 150ns, which is longer than the cycle time of pixel clock, about 100ns. In order to maintain the average processing time per pixel within 100ns, we have employed three banks of ROM, and they are used in order, which is a kind of pipeline architecture, shown in Fig.6. The RGB is passed to ROM of bank#A, and the result will come out after 150ns. The RGB data of the following pixel is passed to ROM of bank#B, and the next to bank#C, and at this time the conversion result of bank#A is obtained, and passed to output latch of final HSV conversion result, and the result of bank#C will be obtained in the next cycle, and so on.

The control of this pipeline operation of three ROM banks is implemented by programmable logic (Complex Programmable Logic Device; CPLD) of XC95108 by Xilinx Corp[Xilinx, 1998], with using ABEL (Advanced Boolean Expression Language) as function description language.

3.2 Developed color detection block

Fig. 7. Developed real-time color detection board

Figure 7 shows the developed color detection block including the designed chip above. The HSV signal of each pixel is passed to four color detection chip at the same time, in order to detect four colors simultaneously. The four outputs of all color detection chip are gathered again, in order to generate the gray scale image according to the detected color using 4bits priority encoder. For example, it generates 'white' pixel for input of 'orange' pixel (that is the color of ball), 'bright gray' for 'blue' (that is the color of robots' marker), 'medium gray' for 'pink' (that is the color of robots' second marker to detect their directions), and 'dark gray' for 'yellow' (that is the color of enemy robots' marker). The registers of each color detectors which hold the upper and lower thresholds are set by external controller, it is set by the keyboard in this system. These thresholds are decided manually, in order to the target color can detect independently.

The generated gray scale image based on color detection is converted to NTSC video signal, which frame grabber of PC can capture. It will be a best way to send the detection results to PC by digital data, and we are currently under development of this block.

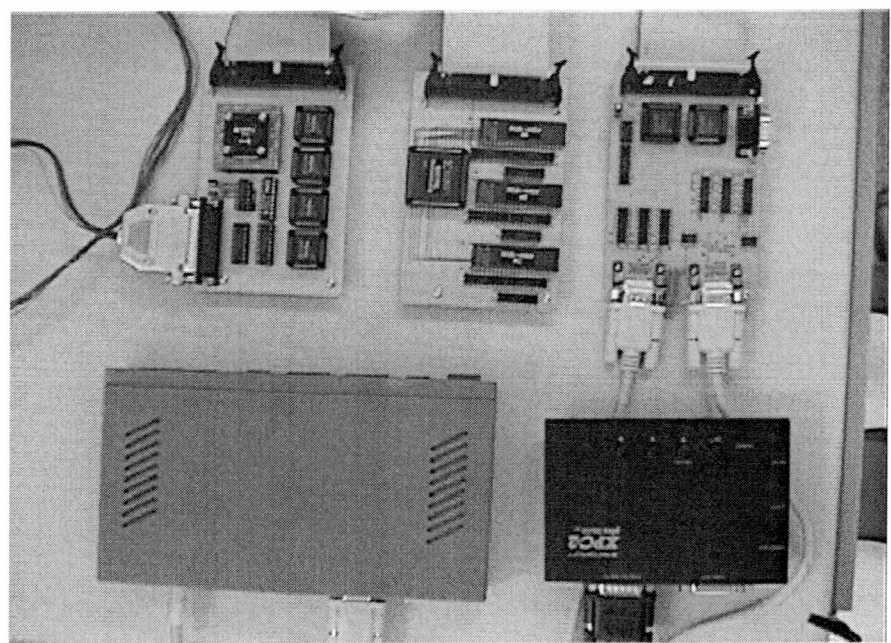

Fig. 8. Whole of the developed real-time color detection system

Figure 8 shows the whole of the developed real-time color detection system, including NTSC to RGB converter, clock generator and A/D, D/A conversion blocks, the color conversion block, the color detection block, and RGB to NTSC converter.

4 Experimental Results

We have carried out the experiments for color detection. Sample color images are captured by CCD color video camera, and this signal is fed into the developed system. The color detection results in gray-scale image is again captured by PC. Figure 9 shows one of the experimental results. The input image of Fig.9(a), where the text of 'Red' is written in red ink, 'Blue' in blue ink, and so on, is

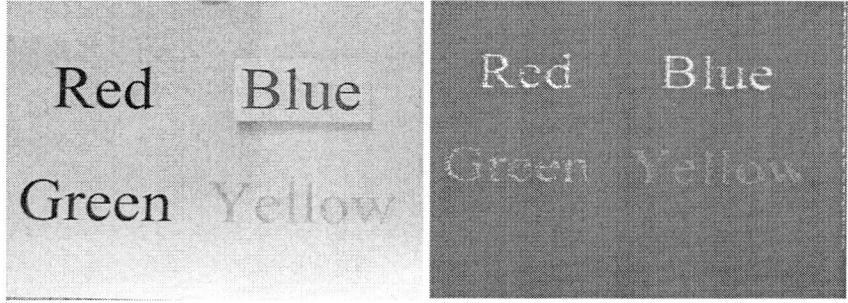

Fig. 9. The experimental result of color detection using the developed system, (a)original color image, (b)generated gray image based on color detection

processed by the developed system with the sequence of Fig.1. The generated gray image based on color detection is shown in Fig.9(b). We can recognize some lack of component pixels containing in each text, but it will be improved by tuning the threshold values of each color, and reducing the noise in A/D and D/A conversion block. It will be an another way to increase the number of bits of each color values, but it will result in the increase of the size of ROM. It will be reasonable to employ 7×3 bits for each color values, which will result in $2^{21} \times (7 + 7 + 7) \simeq 16$Mbytes in one bank of look-up table.

5 Conclusion

In this paper, we have described the developed the real-time color detection system for the vision system of RoboCup Small-Size League. The color image is fed into this system, and the result of color detection is automatically generated in NTSC video format, which can be easily captured by gray-scale frame grabber, that is faster than full-color frame grabber. This system is expected to be effective to reduce the load of image processing system; it can process just after color detection.

Acknowledgements

This study is aided by "Yugen Club Found" of Japanese Association for Mathematical Science[JAMS, 1998].

References

[Xilinx, 1998] http://www.xilinx.com
[VDEC, 1998] http://www.vdec.u-tokyo.ac.jp
[JAMS, 1998] http://www.yugen.org

A Segmentation System for Soccer Robot Based on Neural Networks

Carmelo Amoroso[1], Antonio Chella[1,2], Vito Morreale[1], and Pietro Storniolo[3]

[1] Dip. di Ingegneria Automatica e Informatica, Univ. of Palermo, Italy
[2] CERE-CNR, Palermo, Italy
[3] Dip. di Ingegneria Elettrica, Univ. of Palermo, Italy
{amoroso,chella,vito,piero}@csai.unipa.it

Abstract. An innovative technique for segmentation of color images is proposed. The technique implements an approach based on thresholding of the hue histogram and a feed-forward neural network that learns to recognize the hue ranges of meaningful objects. A new function for detecting valleys of the histogram has been devised and tested. A novel blurring algorithm for noise reduction that works effectively when used over hue image has been employed. The reported experimental results show that the technique is reliable and robust even in presence of changing environmental conditions. Extended experimentation has been carried on the framework of the Robot Soccer World Cup Initiative (RoboCup).

1 Introduction

The task of identification of objects placed at different locations in gray level images is usually difficult unless uniform lighting conditions have been achieved. Uneven illumination leads to variations in light intensity reflected from objects. This is true in the RoboCup framework [10], where light conditions vary during the day and due to different game fields. Considering this problem, in [4] a method for calibrating lighting conditions in the scene is proposed, in which a brightness map of the area is employed to select areas for which different color calibrations are needed. But in general, it is necessary to resort to additional criteria, such as size or shape in order to identify objects of interest. With this respect, color becomes a powerful descriptor that greatly simplifies the object identification and extraction tasks.

Several authors analyzed the advantages of color image analysis with respect to gray level image analysis [15]. The most widely adopted approaches are the statistical methods. They are based on the assumption that an analytic expression for the distribution functions of the pattern classes is known, such as the Gaussian probability density function. This assumption may be unrealistic, especially in the case of images acquired by a vision system placed on a mobile robot, due to the presence of illumination changes, shadowing and obstacles in the robot working environment.

The starting point of our segmentation technique is the HSI color space and particularly the hue component. In the last RoboCup edition many teams employed intensity (or value), saturation and hue components of pixels in order to label them. In [1] a real-time color detection system based on HSV information of pixels using designed LSI is described. Also the italian robot Rullit [5] adopting an omnidirectional vision sensor, performs the color classification by using the HSV color space. Alternative color models have been used by other teams during RoboCup games. For example, in [6], authors adopt a particular decision tree using the rgG color space, where r and g are red end green normalized components and G is roughly related to brightness.

To accomplish the task of a robust segmentation of color images oriented to the vision system of a mobile robot, as in the RoboCup framework, we employ a novel approach based on a feed-forward neural network that learns to recognize the hue range of the objects of interest. Unlike statistical methods, a neural network does not require the exact knowledge of the statistical distribution of input items, as it generates an internal representation of input data distribution by a suitable learning process. Furthermore, a neural network deals effectively with noisy and partial data, as it is able to generalize to new and unexpected situations. Connectionist approach has often been used to perform the segmentation step in an image analysis process providing high performances [3].

Ranges to be recognized have been extracted by using an effective technique for thresholding based on analysis of the hue histogram. Thresholding has been used by the Agilo RoboCuppers Team also in its vision system in which the YUV color space is used to distinguish different color classes: particularly color clusters are found in the UV-plane and a color label is assigned to them according to their position in that plane.

Images acquired by optical or electronic means are likely to be degraded by imperfection of sensing mechanism, that occurs in the form of a noise. Several filters have been reported in the literature to remove noise from images. As our analysis is based on the analysis of the hue component of HSI color model, we propose a novel technique to remove noise from hue images that performs well when works on angular values.

The described approach has been tested in the framework of the Robot Soccer World Championship Initiative. In this framework, a mobile robot acquires the color images by a CCD camera fixed on its top (see Fig. 1).

The images are then processed by a PC on board of the robot. Although the color of the objects in the scene are clear and well defined, several disturbing conditions, such as shadowing, nonuniform lighting and light reflections arise during the RoboCup competition, that impose tight constraints to the segmentation step. Therefore different game fields may be characterized by similar but not equal color, e.g., the green color of the training field may not be exactly equal to the green color of the game field.

The presented technique is part of the vision system adopted by the soccer player robots of the Italian robot team ART (Azzurra Robot Team) [11], a robot team in "F2000" category (middle size league). The technique is also employed as a first step of a more complex vision system for an autonomous robot aimed at the description of static and dynamic scenes [7, 8].

Fig. 1. Images of the adopted experimental framework. The soccer field is green, the walls and the field lines are near white, the goal area is blue and the ball is red.

2 Color imaging

Color images are widely used due to the superior characteristics over gray-level images. They contain object spectral information and provide valuable details for quick objects identification. Color information processing provides more reliable and accurate results for machine perception and scene analysis.

Andreadis and Tsalides [2] showed that the HSI color space compensates for illumination variations, hence contributing to simplified segmentation. According to the Phong shading model, hue has the property of multiplicative/scale invariance and additive/shift invariance [13]. According to these features, the hue parameter is our starting point for developing image processing algorithms based on color information. This choice allows us to work on a 1D space, rather than in a 2D (such as IQ, $a*b*$, $u*v*$) or a 3D (such as RGB, $CieXYZ$) color space. This simplify the search for significant clusters that identify meaningful regions of pixels.

3 Noise removing by circular blurring

The purpose of pre-processing is to improve the image in ways that increase the chances for success of the subsequent processes. The alteration of the image may be generated in a number of ways, such as interference, noise, light reflection, etc. To maximize the noise attenuation is only one aspect of image pre-processing. On the other hand, it might happen that noise is attenuated so hardly that the useful image details are lost.

Many methods for scalar image processing have been developed for gray-level image processing. A direct application of such methods to the hue image can be performed, but this generalization exhibits some drawbacks, the most important of which lies in not considering exact relationship between hue values. Hence, the development of new methods becomes necessary to realize efficient low-pass filtering of hue image. In this section, a new algorithm for hue image blurring and noise reduction is presented. This technique is based on mask processing.

Throughout this section let us denote the hue levels of pixels under the mask at any location by $h_1, h_2, \ldots, h_N$ (let N be the number of pixels within the mask) and suppose that hue values lie in the range $[0, L-1]$, due to sampling performed over this range. The only requirement on L is that it should be integer, positive, and finite.

Defining n as the largest integer value lower than k/L, that is

$$n = \left\lfloor \frac{k}{L} \right\rfloor , \tag{1}$$

we introduce the following *circular indexing function*

$$I_L(k) = k - nL \quad \forall k \in \mathbb{Z} \tag{2}$$

This operator allows us to perform any operation over the domain of the hue values considering automatically its circular nature, without changing the form of equations.

Given $I_L(k)$, the counterclockwise distance between h_i and h_j is defined as:

$$D(h_i, h_j) = I_L(h_j - h_i) \tag{3}$$

The most important properties of this function are:

$$D(h_i, h_i) = 0 ; \tag{4}$$
$$D(h_j, h_i) = L - D(h_i, h_j) . \tag{5}$$

Defining

$$d_i = \min_{\substack{1 \leq j \leq N \\ i \neq j}} \{D(h_i, h_j)\} \quad i = 1, 2, \ldots, N , \tag{6}$$

the reference value of the set $\{h_1, h_2, \ldots, h_N\}$ is the value B such that

$$B = h_b \tag{7}$$

where

$$b = \arg \max_{1 \leq i \leq N} \{d_i\} . \tag{8}$$

We can define a new sequence of relative hue values $\boldsymbol{H} = \{H_1, H_2, \ldots, H_N\}$, where

$$H_i = D(B, h_i) \quad i = 1, 2, \ldots, N . \tag{9}$$

Now, both median and linear filtering may be used to obtain the new relative hue level of the pixel located at the center of the considered mask. Median filter are best suited to double-exponential and impulsive noise but may not be effective for Gaussian noise, while linear filters are optimal in removing additive white noise (Gaussian or non-Gaussian) but will produce rather poor results in an impulsive noise environment.

Using linear filtering, the relative hue value of the pixel located at the center of the current mask is replaced by

$$m_{lin} = \mathbf{w} \cdot \mathbf{H} = \frac{1}{N} \sum_{i=0}^{N} w_i H_i \qquad (10)$$

while the median of set $\mathbf{H}$ is defined as the value m_{med} such that

$$m_{med} = \arg \min_{H_j \in \mathbf{H}} \left\{ \sum_{i=1}^{N} |H_i - H_j| \right\} . \qquad (11)$$

The absolute hue level of the center pixel in the defined window is replaced by

$$M = I_L(m + B) , \qquad (12)$$

where m is either m_{lin} or m_{med} according to the choice of the used filter.

4 Thresholding

We employed a thresholding technique to perform low-level image processing. According to the definition of the hue in the HSI model, the hue histogram is a periodic function in the same range of the hue.

Fig. 2 shows three "typical" hue histograms of color images of the chosen framework. It should be noted that the color modes (the peaks in the histograms) extend themselves within similar hue ranges. Each color mode corresponds to an object in the scene: the first one is related to the ball, the second one indicates the wall and the lines, the third one is related to the game field and the last one indicates the game goal. In the figure, the hue ranges corresponding to the color modes are highlighted.

In the literature several techniques for searching valleys and peaks of gray level histogram are presented [14, 9, 12, 16]. These methods fail when directly used for the analysis of the hue histogram, because the hue is periodic. For this reason we developed a special technique for histogram analysis, in order to find meaningful modes of the hue histogram.

Hue histogram is defined as follows:

$$q(k) = \frac{r(k)}{R \cdot S} \qquad k = 0, 1, \ldots, L - 1 \qquad (13)$$

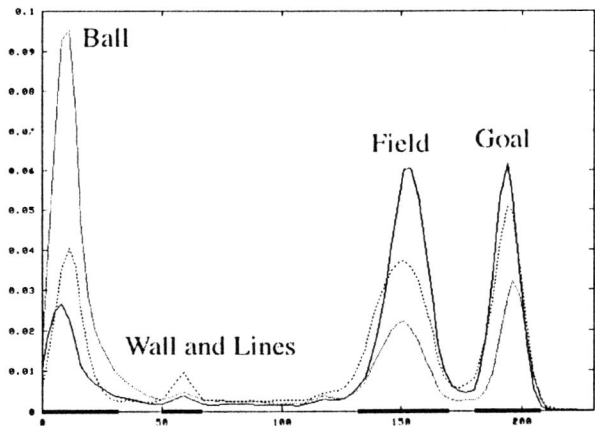

Fig. 2. Typical hue histograms related to images of the game field.

where $r(k)$ is the number of pixels with hue value equal to h_k, i.e., the k-th sampled hue value, and L is the number of hue samples. R and S are, respectively, width and height of the image.

To be able to deal correctly with hue histogram, for searching valleys and peaks, it is necessary to redefine the histogram function, according to the circular domain. Hence, starting from the histogram function $q(\cdot)$, it is possible to define a new function $p(\cdot)$, called *extended histogram function*, by means of functional composition with the function $I_L(\cdot)$ defined in (2)

$$p(k) = q(k) \circ I_L(k) = q[I_L(k)] \qquad \forall k \in \mathbb{Z} \ . \tag{14}$$

To improve the performances of the following histogram analysis, the hue histogram is first smoothed to remove noise by means of:

$$\hat{p}(k) = p(k) * w_\sigma(k) = \sum_{i=-\sigma}^{\sigma} p(k-i) w_\sigma(i) \qquad \forall k \in \mathbb{Z} \tag{15}$$

where $w_\sigma(\cdot)$ is a Gaussian smoother, defined as

$$w_\sigma(k) = \begin{cases} \frac{1}{\sqrt{2\pi}\sigma} \exp\left(\frac{-k^2}{2\sigma^2}\right) & |k| \leq \sigma \\ 0 & \text{otherwise} \end{cases} \qquad \forall k \in \mathbb{Z} \tag{16}$$

and σ is the standard deviation related to the width of the smoother.

It is therefore possible search the valley points in the histogram, according to its circular property using a *valley detection signal* $r(\cdot)$ generated from the extended histogram function by convoluting it with a *valley detection kernel* $v_P(\cdot)$:

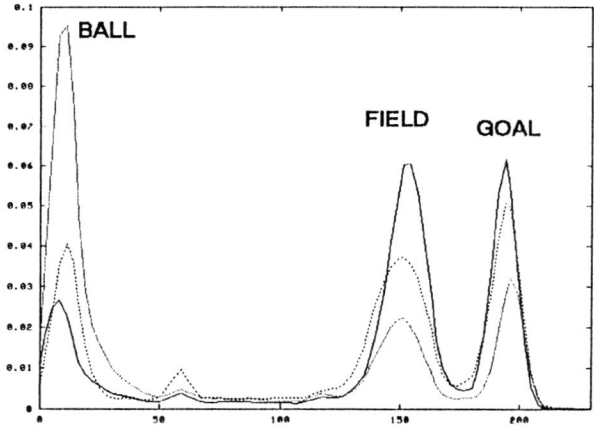

Fig. 3. An example of hue histogram and corresponding valley detection signal.

$$r(k) = \hat{p}(k) * v_P(k) = \sum_{i=-\lfloor P/2 \rfloor}^{\lfloor P/2 \rfloor} \hat{p}(k-i) v_P(i)$$
$$k = 0, 1, \ldots, L-1 \qquad (17)$$

where the valley detection kernel $v_P(\cdot)$ is defined as

$$v_P(k) = \begin{cases} \frac{k}{|k|} \frac{1}{\sqrt{\left(1-\frac{|k|}{k_0}\right)^2 + 4\frac{k^2}{k_0^2}\zeta^2}} & |k| \leq \lfloor \frac{P}{2} \rfloor, k \neq 0 \\ 0 & \text{otherwise} \end{cases} \qquad (18)$$

P is the dimension of the kernel mask, ζ and k_0 are adjustable parameters (typically, $\zeta \in [0.1 - 2]$ and $k_0 \in [1-3]$). Valleys are placed on zero crossing to negative points of the $r(\cdot)$ signal.

In Fig. 3 an example of a hue histogram and the corresponding valley detection kernel is shown: the above mentioned zero crossing points are drawn as painted black dots. Only meaningful valleys are found, due to the robustness of the kernel with respect to small and irrelevant jitters.

5 The neural networks approach

As previously explained, the hue range and the shape of a meaningful color mode of the histogram are not clearly defined because they greatly depend on illumination conditions, shadows effect and so on.

We compared several techniques for classifying the modes extracted from the hue histogram. The statistical minimum distance classifier and the optimum

Bayes classifier have not been provide satisfactory results, due to the variability of the histogram. Because of this, we have been chosen a neural network for the classification step.

To code the hue ranges of the histogram to feed them as input to the neural network, we compared three different approaches: the *bit range*, the *Gaussian* and the *value* coding. Let L be the number of the considered hue sampled levels. Each level corresponds to one of the input units of the neural network. In the *bit range* coding, we set to 1 all the input units corresponding to the hue levels belonging to the considered range, and we set the other units to 0. In the *Gaussian* coding, we consider a Gaussian distribution of probability over the considered hue range. Then, we set to 1 the input units corresponding to the hue levels belonging to the range $[m - \sigma, m + \sigma]$, where m is the mean value and σ is the standard deviation of the hue of the considered color mode. The units outside the range are set to 0. In the *value* coding, we set the input units to the effective values of the color mode of the considered hue range; the units that do not belong to the considered hue range are set to 0.

The choice of the number L of sampled hue levels corresponding to the input nodes of the network is an important issue. We have chosen $L = 32$ according to the width and the resolution of the hue segments, corresponding to the classes of objects of interest. Experiments showed that for $L > 32$ the network does not improve its performances, while for $L < 32$ the hue segments overlap and the classification ratio degrades.

In the considered application, we recognize the following four classes of object: the *field*, the *wall and lines*, the *goal* and the *ball*. Hence, we chosen the number $Oout$ of the units in the output layer so that each node represents one of the above mentioned classes; i.e., $Out = 4$.

We chose the number Hid of hidden nodes by considering the variation of performances of the network by varying Hid. The number of hidden nodes was assumed to be the lowest one that achieves the minimum error on the test set. Experiments showed that the best performances were reached with $Hid = 16$.

In conclusions, we experimented the best performances with a feed-forward neural network with 32 input nodes, 16 hidden nodes and 4 output nodes. We chose the sigmoid activation function for every node. The learning rate was 0.5.

After the classification of the hue ranges performed by the neural network, the obtained classified ranges are used to label each pixel in the image.

6 Experimental results

In this section, the performances of our segmentation technique based on neural network are analyzed. To produce the training set that has trained the network, we used a sequence of images drawn by some films realized on a soccer field during testing. In these images, there were all of the objects of interest.

Image pre-processing has been made by using the technique described in the Sect. 3: experimental results showed that a linear mean filter performs better than a median filter.

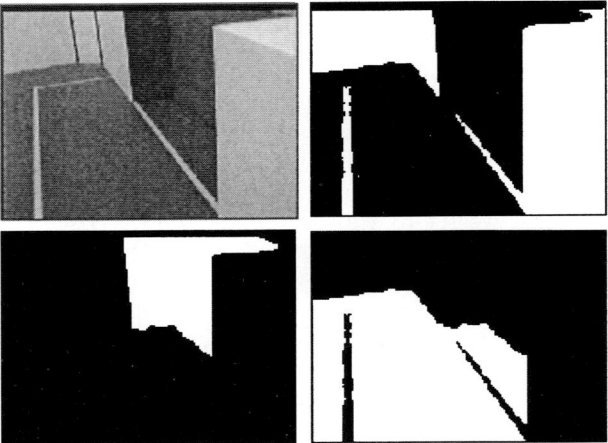

Fig. 4. An example of the operation of the thresholding (see text).

Thresholding of training and testing images has been made by using the efficient technique based on the convolution of the hue histogram with the suitable defined valley detection kernel, as described in Sect. 4.

A visual example of the obtained results are reported in Fig. 4 that shows the acquired image and *wall*, *field*, and *goal* images in which white patches represent selected regions.

To analyze in details the performances of the network, we accomplished three different tests: *Test A*, *Test B* and *Test C*, described below. In all of the tests, the training and the test sets are non-overlapping sets of patterns. We repeated the tests for all the three input coding described in the previous section.

In *Test A*, we analyzed the performances of the network by using training and testing images drawn by the same soccer field. To generate the training an the testing sets, we used 300 images acquired by the robot moving in the soccer field built in our laboratory. Different lighting conditions have been considered during acquisition in order to provide the network with a high degree of robustness.

We divided the training set into 5 mutually exclusive groups of 60 patterns. In each group, there were 15 patterns of each of the four classes. The network has been trained using 4 of these groups and tested using the fifth group. This train-and-test process has been repeated 5 times, using all groups of 60 test patterns. At the end, a mean misclassification error has been evaluated: the results are shown in the first row of Tab. 1. Results show the high reliability of the network when it works with pattern generated from images acquired in the same environment of the training phase.

To verify the capabilities of the network to work in different environmental conditions, during *Test B* we used the same training set of *Test A* but the test set was made up by images acquired by the robot during the previous RoboCup competition held in Paris in 1998. Therefore, environmental conditions, tonality

Table 1. Misclassifications of the neural network (see text).

	Bit range	Gaussian	Value
Test A (%)	0.33	0.67	0.67
Test B (%)	10.0	19.0	17.0
Test C (%)	6.17	7.83	9.33

of colors and illuminations in test set were very different with respect to the training set. The corresponding misclassification errors are shown in second row of Tab. 1. It should be noted that the error is maintained within acceptable limits. These results confirm that our method is satisfactory with respect to severe environmental changes.

To evaluate the performances of the network working on noisy data, during *Test C* we used the same training set of the previous tests, and a test set made up by noisy images. We produced the test set as follows: each pattern of the training set has been modified by operating a right shift and a left shift of the input values. Hence, for any training pattern, we produced two new test patterns. Altogether, 600 new patterns have been generated, on which the network has produced the results shown in third row of Tab. 1. Once again, experimental results highlight that the network is robust with respect to the noise.

By analyzing Tab. 1 with respect to the input coding, it should be noted that the *bit range* always performed better in terms of misclassification error with respect to the other coding. This is due to the fact that the *bit range* coding do not make any assumption on the form of the hue histogram. It therefore provides a low sensibility with respect to the form of the color modes. The result is an improvements in terms of robustness and adaptation to the environmental situations.

7 Conclusions

The problem of automatic segmentation of color images acquired by a camera placed on a mobile robot platform has been discussed. A new algorithm for noise removing in hue image has been presented. A special technique for dealing with multiple threshold selection from hue histograms has been developed. Here, a new valley detection signal is proposed. A feed-forward neural network has been employed to classify hue ranges corresponding to hue modes extracted from histograms.

Experimental results showed that the proposed approach is effective and robust with respect to changes of environmental conditions, such as shadows and illumination variations. Although this method has been applied in the RoboCup experimental framework, it is an effective intermediate level step in generic image interpretation tasks where the color is a meaningful descriptor.

We suppose that, training the network with a new training set according to the new working environment, this method is suitable to be used in different environments. This fact provides our method a high degree of adaptability redefining the correspondences between the chromatic components and objects by the neural network. Furthermore, we think that the approach based on to feed the representation of the most meaningful modes of the histogram as input to the neural network may be easily extended to N-dimensional analysis.

Acknowledgments

Authors would like to thank E. Ardizzone and S. Gaglio for stimulating discussions, and D. Nardi and all the persons involved in the ART (Azzurra Robot Team) for their encouragement. This work has been partially supported by CNR and MURST "Progetto Cofinanziato CERTAMEN" and ex 60%.

References

1. J. Akita. Real-time color detection system using custom lsi for high-speed machine vision. In *Proc. of the Third International Workshop on RoboCup*, pages 1–4. IJCAI Press, 1999.
2. I. Andreadis and P. Tsalides. Coloured object recognition using invariant spectral features. *Journal of Intelligent and Robotic Systems*, (13):93–106, 1995.
3. E. Ardizzone, A. Chella, and R. Pirrone. A neural based approach to image segmentation. In *Proc. of the International Conference on Artificial Neural Networks*, pages 1152–1156. Springer-Verlag, London, 1994.
4. J. Baltes. Practical camera and color calibration for large rooms. In *Proc. of the Third International Workshop on RoboCup*, pages 11–16. IJCAI Press, 1999.
5. A. Bonarini. The body, the mind or the eye, first? In *Proc. of the Third International Workshop on RoboCup*, pages 40–45. IJCAI Press, 1999.
6. J. Brusey and L. Padgham. Techniques for obtaining robust, real-time, colour-based vision for robotics. In *Proc. of the Third International Workshop on RoboCup*, pages 63–67. IJCAI Press, 1999.
7. A. Chella, M. Frixione, and S. Gaglio. A cognitive architecture for artificial vision. *Artificial Intelligence*, (89):73–111, 1997.
8. A. Chella, M. Frixione, and S. Gaglio. An architecture for autonomous agents exploiting conceptual representations. *Robotics and Autonomous Systems*, 25(3-4):231–240, 1998.
9. L. Hayat, M. Fleury, and A. F. Clark. Candidate functions for a parallel multi-level thresholding technique. *Graphical Models and Image Processing*, 58(4):360–381, July 1996.
10. H. Kitano, M. Asada, Y. Kuniyoshi, I. Noda, E. Osawa, and H. Matsubara. Robocup: A challenge problem for AI. *AI Magazine*, 18(1):73–85, 1997.
11. D. Nardi, G. Clemente, and E. Pagello. Art: Azzurra robot team. Technical report, DIS Universita' di Roma *La Sapienza*, 1998.
12. N. Otsu. A threshold selection method from gray-level histogram. *IEEE Transactions on Systems, Man, and Cybernetics*, 9(1):62–66, 1979.
13. F. Perez and C. Koch. Toward color image segmentation in analog VLSI: Algorithm and hardware. *International Journal of Computer Vision*, 12(1):17–42, 1994.

14. E. Saber, A. M. Tekalp, R. Eschbach, and K. Knox. Automatic image annotation using adaptive color classification. *Computer Models and Image Processing*, 58(2):115–126, 1996.
15. G. Sharma and H. J. Trussel. Digital color imaging. *IEEE Transactions on Image Processing*, 6(7):901–932, July 1997.
16. J. Yen, F. Chang, and S. Chang. A new criterion for automatic multilevel thresholding. *IEEE Transactions on Image Processing*, 4(3):370–378, March 1995.

Practical Camera and Colour Calibration for Large Rooms

Jacky Baltes

Centre for Image Technology and Robotics
University of Auckland, Auckland
New Zealand
j.baltes@auckland.ac.nz
http://www.citr.auckland.ac.nz/~jacky

Abstract. This paper describes a practical method for calibrating the geometry and colour information for cameras surveying large rooms. To calibrate the geometry, we use a semi-automatic system to assign real world to pixel coordinates. This information is the input to the Tsai camera calibration method. Our system uses a two stage process in which easily recognizable objects (squares) are used to sort the individual data points and to find missing objects. Fine object features (corners) are used in a second step to determine the object's real world coordinates. An empirical evaluation of the system shows that the average and maximum errors are sufficiently small for our domain. Objects are recognized through coloured spots. The colour calibration uses six thresholds (Three colour ranges (Red, Green, and Blue) and three colour differences (Red - Green, Red - Blue, Green - Blue)). This paper describes a fast threshold comparison routine.

1 Introduction

Our research work focuses on the design of intelligent agents in highly dynamic environments. As a test-bed, we use the RoboCup domain, which is introduced in section 2. In this domain, small toy cars play a game of soccer.

This paper describes an accurate, cheap, portable, and fast camera calibration system (Section 3). After an initial preprocessing step (which is guided by the user), it automatically computes real world coordinates for features in the image (Section 4). Section 5 discusses our algorithm in more detail. The Tsai camera calibration algorithm is briefly described in section 6.

Section 7 shows the accuracy that can be obtained by our method in a sample and a real world problem. Both the average and maximum error are sufficiently small for our application.

Section 8 discusses the blob detection used in our video server. Objects are identified using coloured spots. The colour detection uses the R-G-B colour model. Each colour is identified by twelve parameters. Six parameters identify the minimum and maximum threshold for the red, green, and blue colour channels.

Another six parameters identify minimum and maximum values for the difference channels (red - green, red - blue, and green - blue).

To be able to maintain a frame rate of 50 fields per second without special purpose hardware, the video server uses a number of optimizations described in section 9.

In section 10, we discuss ideas for further research to improve the accuracy of the calibration and to find colour thresholds automatically.

2 The Laboratory Setup

RoboCup [4] is a domain initially proposed by Alan Mackworth ([5]) to provide a challenge problem for AI researches that requires the integration and coordination of a large number of techniques. The problem is to create autonomous softbots and robots that can play a game of soccer.

RoboCup is a difficult problem for a team of multiple fast-moving robots under a dynamic environment that requires the designer to incorporate techniques such as: autonomous agents, multi-agent collaboration, strategy acquisition, real-time reasoning, robotics, and sensor-fusion. RoboCup also offers a simulation environment for research on the software aspects of RoboCup.

RoboCup is a standard problem which allows the evaluation of proposed methods to solve these problems in a friendly competition. Apart from Machine Learning, which has used databases of problems extensively in research [6], such an agreed upon evaluation method is sadly missing from lots of AI research areas. However, the importance of such test-beds has been realized in other AI fields as well. The planning community agreed on a common domain description language and held the first planning competition in 1998.

The RoboCup environment at the University of Auckland consists of a commercially available cheap video camera mounted on a tripod. The video camera is connected to a video server (a Pentium PC). The video server interprets the video data and sends position, orientation, and velocity information to other clients on the network (three PCs).

Lighting is provided by fluorescent lamps on the ceiling. All the equipment is readily available and most of the room has been unchanged. Although playing soccer is our main objective, there are other tasks that we are working on such as parallel parking and time trials on a race track. Time trials along a race track (called Aucklandianpolis [1]) proved to be very popular with students. Figure 1 shows our environment.

In contrast to all other teams in the RoboCup competition, our camera is mounted on the side of the playing field, which introduces large perspective distortions. Therefore, the geometry calibration is very important.

Since we are often asked to give demos of our system, we needed an accurate, cheap, portable, and fast method for camera calibration.

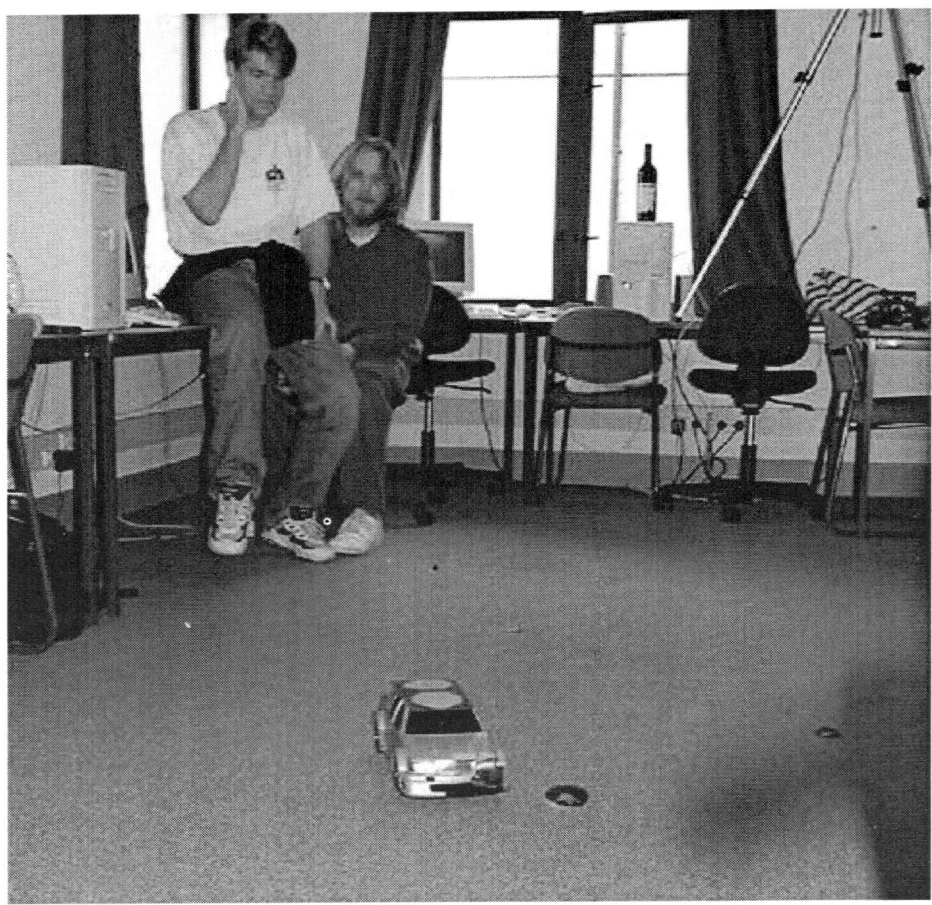

Fig. 1. Aucklandianapolis at the University of Auckland. The tripod of the vision system can be seen on the top right corner of the image. The video camera is just out of the picture. The video server determines position and orientation of the cars by bright dots on the car. As can be seen, the speed trials took their toll on our cars.

3 Camera Calibration

The problem of camera calibration is a very fundamental problem in Computer Vision. The input to a calibration method is a set of known world coordinates and their matching pixel coordinates in the image and the output is a set of external and internal parameters for a camera model. Given this calibrated camera model, it is easy to determine the real world coordinates of image points (if at least one dimension is known) or compute the image coordinates for known real world coordinates.

Traditional camera calibration relies on the availability of known image coordinates for some known world points. For example, a simple pin hole camera model requires that at the real world coordinates of at least 12 image points are known [3]. Once a sufficient number of matching points have been found, well known camera calibration algorithms can be used. For example, the Tsai calibration method uses a complex eleven parameter model with six external and five internal parameters [7]. In our work, we are using a public domain implementation of the Tsai calibration method, which is available from the WWW [9].

This paper focuses on the problem of finding a suitable set of matching points for camera calibration. The need for portability and speed of the calibration method ruled out traditional methods of using feature points inherent in the scene (since these feature points will not be available when moving to different rooms) or of painting feature points into the scene (a labour intensive and error prone task for a large set of points). The creation of a special calibration pattern of sufficient size and with a sufficient number of points was also too expensive. For example, a large wooden board with calibration points (a) would be difficult to move, (b) may not fit into rooms that do not have similar geometry (e.g., a part of the rectangle is cut out by a wall), and (c) expensive and labour intensive to manufacture.

However, we clearly needed a portable calibration pattern[1], so we decided to use readily available and light material. We looked at a number of possibilities including carpets (have a dense texture and are expensive) and linoleum carpets (accurate pattern, but expensive and has an undesirable warping property).

In the end, we decided to use a duvet cover (250x200cm) with a square pattern on it. The back half of the duvet cover was removed to reduce artifacts due to the transparency of the cloth material. The duvet cover is well suited for our environment, since it is easily portable and can be adapted to room outlays[2]. Drawbacks are that the cloth material stretches and warps. Both drawbacks can be minimized through the handy use of an iron. However, they can not be eliminated and thus introduce errors, which limit the accuracy of the camera calibration that can be obtained.

---

[1] Otherwise our overweight charges when flying to the RoboCup competitions would be even higher

[2] It is also a handy blanket for my graduate students when they get caught up in their work and end up sleeping in the lab

Figure 2 shows a picture of the calibration duvet cover as seen by the video camera.

Fig. 2. Calibration Pattern as seen by the Camera

4 Find Matching Points Algorithm

Given the picture shown in Fig. 2, our system uses a semi-automatic method for calculating the matching points. In the preprocessing step, the user removes unwanted parts of the picture, such as the table top on the left side of the calibration picture. Secondly, the colour image is converted into a gray scale image and thresholded, so that only the white squares are left in the image. Currently, we use a global threshold value on the red channel, which was sufficient for our environment.

After this initial preprocessing step, the system automatically computes the matching points. The idea is to find features in the image that can be assigned world coordinates by the known geometry of the calibration pattern (i.e., by knowing that the dimensions of the squares are $8.0 \times 8.1 cm$). A false colour image of the result of the preprocessing step can be seen in Fig. 3. The figure shows some of the practical problems in assigning real world coordinates to image features: (a) some of the squares are missing from the right side of the image, and (b) some parts of the squares are missing (e.g., in the bottom right corner).

First, the system uses a simple pattern (5 by 5 pixel squares) to find the white squares in the picture. This step ignores small artifacts and handles missing

Fig. 3. Calibration Pattern after Preprocessing

squares. The centre point of each square is computed by calculating the moments along the x and y direction. Then, the squares are sorted. This sorting step is of critical importance, since if it is done in the wrong order, the assigned real world coordinates will be wrong, which will result in unusable calibration parameters.

The following algorithm `find_real_world` is used to sort the squares and to assign real world coordinates to their centres. The algorithm takes an unsorted sequence of squares as input and assigns a real world coordinate to the centre of each square. First, the squares are sorted in increasing order of their y coordinate (line 3). This is used to repeatedly extract the next row from the sequence. A row is defined by an initial sequence of squares from `y_sort_squares`, whose y coordinates are within the tolerance limit `eps`. The system also initializes the variable `guess_y`, which is used as a guess of the distance in pixels between the previous and the current row. Lines 8-12 calculate the ratio of the actual distance between the previous and the current row to the current estimate. This ratio is used to check for missing rows in the input image. The current y coordinate `Wy`, and `guess_y` are updated in lines 13-14. Similarly to the rows, the squares within a row are then sorted based on their x coordinate (line 16) and an x coordinate is assigned (line 24) based on a guess of the distance in pixels to the next square `guess_x` (lines 20-23 and line 27).

Note that the estimates to the next row and column are adaptive, so this method will work in pictures with obvious perspective distortion (as can be seen in Fig. 3) as long as the change from one row to the next is not more than 50%.

```
1   Procedure find_real_world_coords(unsorted_squares) {
2
3     y_sort_squares=sort(unsorted_squares,y-direction);
4     guess_y=0; prev_avg_y=0;
5     Wy=0;
6
7     while (row=extract_row(y_sort_square,eps)!=empty) {
8        avg_y = average_y_coor(row);
9        if (guess_y != 0)
10           factor = (avg_y-prev_avg_y)/guess_y;
11       else
12           factor = 0;
13       Wy=Wy+factor*SQUARE_Y_DIMENSION;
14       guess_y=avg_y-prev_avg_y;
15
16       x_sort_squares=sort(row,x-direction);
17       guess_x=0; prev_square=null;
18       Wx = 0;
19       foreach square in x_sort_square {
20          if (guess_x != 0)
21             factor=(square.x-prev_square.x)/guess_x;
22          else
23             factor=0;
24          Wx=Wx+factor*SQUARE_X_DIMENSION;
25          square.realworld_x = Wx;
26          square.realworld_y = Wy;
27          guess_x = square.x - prev_square.x;
28          prev_square = square;
29       }
30       prev_avg_y = avg_y;
31  }
```

Table 1. Algorithm for finding real world coordinates

After approximate real world coordinates have been assigned to the centres of all squares, the system uses four edge detection steps to find the coordinates of all four corners. If a corner has been identified, it is assigned a real world coordinate by the geometry of the calibration pattern (for example in the first column, the first top left corner has coordinates 0.0, 8.1, the bottom left corner of the next square is 0.0, 16.2 and the top left corner of the second square is 0.0, 24.3.

This means that the assignment of the real world coordinates to the corners is independent of the assigned real world coordinates of the centres of the squares themselves. This is an important feature in our algorithm, since the centres of objects are distorted by the perspective projection and are moved to the lower end of the picture (see Fig. 4), which means that they are unsuitable for applications that require high accuracy. Of course, given an accurate camera model, this perspective distortion can be compensated for, but this leads to a chicken and egg problem, since we are using this information to calibrate the camera in the first place.

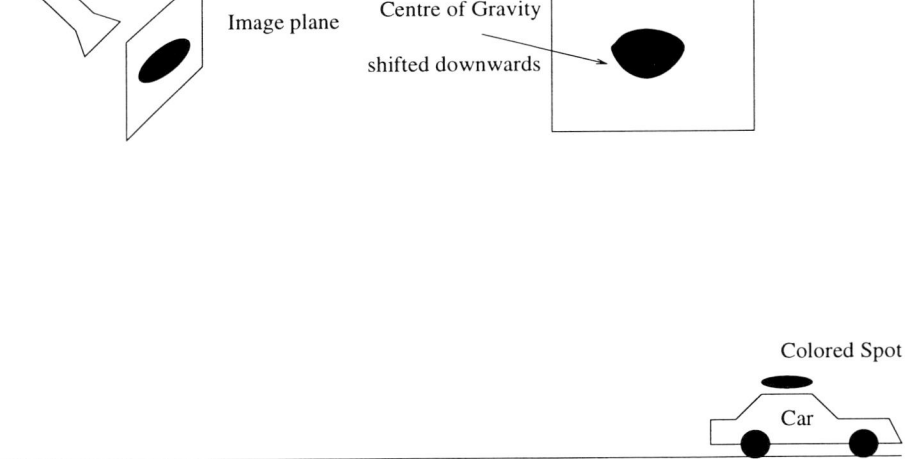

Fig. 4. Movement of the center of a circle under perspective distortion

The real world coordinates of the centres are only used for sorting the squares, which means that only their relative values are important since they are used to determine, which square is the next square in a row or column or whether a square is missing.

Also we found in our tests that this two-stage approach (sort centre of squares, find corners for each square) works better than assigning world coordinates to all corner points. Missing squares or missing data points makes this one step assignment very difficult and error prone.

5 Discussion

The `find_real_world_coors` algorithm assumes that the perspective distortion increases along the y axis. This assumption does not always hold. Sometimes, our camera has to be mounted such that the perspective distortion increases along the x axis. In this case, the user can simply rotate the image by 90 degrees. Note that rotation by 90 degrees simple involves swapping pixel coordinates and does not incur a loss of information.

Given the corners of the calibration carpet, it is also possible to compute the necessary rotation angle. We experimented with arbitrary rotation angles and found that the errors introduced through the rotations were too big and that the calibration data computed was therefore useless.

The algorithm also does not deal with missing squares in the first column. Without knowing the values for the perspective distortion it is impossible to compute where the next row starts and therefore to find out whether the first square in a column is missing.

In practice, both of these aspects are under the user's control. The carpet is manually aligned with the camera coordinate system through visual feed back. The second problem is solved by the user removing any leading columns with missing squares.

6 Tsai Camera Calibration

After the computation of the matching points, we use a PD implementation of Tsai's camera calibration to compute the extrinsic and intrinsic parameters of the camera model.

The Tsai calibration method uses a four step process to compute the parameters of a pin hole camera with radial lens distortion.

Firstly, the position (X_T, Y_T, Z_T) and the orientation (R_X, R_Y, R_Z) of the camera with respect to the world coordinate system is computed. This involves solving a simple system of linear equations. This step translates the 3D World coordinates into 3D camera coordinates and computes the six extrinsic parameters of the camera model.

In Step 2, the perspective distortion of a pin hole camera is compensated for. This step is a non-linear approximation and computes the focal length f of the camera. The output of this step are the ideal undistorted image coordinates.

Thirdly, the radial lens distortion parameters (κ_1, κ_2) are computed. These parameters compensate for the pin cushion effect of video cameras, that is straight lines along the edges of the camera are rounded. An example is seen in the top and bottom row of the calibration image in Fig. 2. The output of step 3 are the distorted image coordinates.

Lastly, the image coordinates are discretized into the real image coordinates by taking the number of pixels in each row and column of an image into consideration.

The last three steps compute five intrinsic parameters of the camera model (focal length, lens distortion, scale factor for the rows, and the origin in the image plane).

The Tsai method is a very efficient, accurate, and versatile camera calibration method and is therefore very popular in computer vision.

7 Evaluation

We evaluated the system in practice (by calibrating different rooms on a number of occasions) and quantitatively through the use of syntheticly generated and real camera pictures.

The synthetic picture was generated by computing a perfect image of all feature points (corners of squares) given our current camera setup (camera mounted on a tripod, $2.58m$ above ground). Since in this case the matching points are 100% accurate, it gives an indication of the maximum accuracy that can be obtained with an eleven parameter camera model.

Given the input image shown in Fig. 2, the corner detection finds 815 corner points. Table 2 summarizes the average error and the standard deviation of the error with increasing number of calibration points n. The data in the table was generated by averaging the results of three cross validation runs for each picture. In each test, n points were selected at random. The camera was calibrated with the data from the calibration points and then the average error, standard deviation, and the maximum error (all in millimeters) were computed.

n	Synthetic Picture			Real picture		
	avg. err	stddev	Max. err	avg err	stddev	Max. err
50	0.9936	0.0653	0.7291	15.2802	7.6748	85.0945
100	0.0964	0.0553	0.3307	17.2455	7.9873	50.0908
150	0.0931	0.0511	0.3068	13.0654	3.8769	37.0576
200	0.0939	0.0557	0.5121	13.8500	5.0923	55.2477
300	0.0904	0.0498	0.3186	13.6753	4.3130	43.3685
400	0.0901	0.0504	0.3207	13.6320	4.2632	56.5799
500	0.0899	0.0497	0.3152	13.5105	3.6942	34.5634

Table 2. Results of the Evaluation. All measurements are in millimeters.

As expected, increasing the number of calibration points improves the calibration of the camera in the synthetic picture. A similar trend can be observed in the real picture.

The differences in errors between the synthetic and the real world image are due to warping of the material and inaccuracies in determining the feature coordinates precisely.

Also, even when using only 150 points, the predictive power of the algorithm is sufficient for our purposes. The error of the calibration is less than $1.3cm$ on average and the maximum error is $3.4cm$. This data is confirmed by testing the accuracy of the coordinates in uncovered areas of the picture (on the very top and bottom of the image). Although, there were no calibration points that covered these areas, the measured error for this region is around $1.5cm$.

The system also proved its worth during competition. The camera calibration was tested at the PRICAI-98 RoboCup and RoboCup-99 competition and proved very stable and fast [2]. For example, it took us less than 15 minutes to calibrate the geometry of our camera system.

8 Colour Calibration

In the RoboCup domain, the ball is a bright orange golf ball and to simplify recognizing the cars they are marked with coloured dots (blue and yellow) or table tennis balls. These dots provide position and orientation information. Since we do not have access to special purpose video hardware, all processing must be done by the video server (Pentium 200MHz).

A simple and fast method for the colour detection is to use minimum and maximum thresholds for the three colour channels R, G, B. In this model, a colour is defined as a cube in the R–G–B cube. This method is not robust enough, since in any practical situation, the colour values will vary greatly with lighting across the field. This means that the thresholds for the colours must be made very large and only a small number of different colours can be detected. In our experience, even when spending a lot of time fine tuning the calibration, it is impossible to distinguish more than four colours with this method.

Although a change in lighting will affect the absolute colour values (e.g., the R, G, B channels are lower in a shadow), the relative distribution of colours is more stable. Therefore, our video server also computes the three difference channels R - G, R - B, and B - G and uses minimum and maximum thresholds for the three difference channels.

The addition of the difference thresholds allows us to detect eleven separate colours reliably, which is sufficient in our domain.

9 Object Tracking

To be able to maintain recognition at 50 fields/sec, the video server uses a number of optimization techniques to reduce computation time: integer threshold comparison, object prediction, and a sampling grid.

9.1 Integer Threshold Comparison

The most frequently used subroutine in our video server is the colour matching routine. Therefore, it was a natural choice for optimization.

Firstly, we tried standard improvements such as hand coding the routine in 80X86 assembly language. Secondly, we even used the special purpose MMX instructions. Neither of these approaches led to the hoped for improvement. The Assembly language implementation only led to 5% speedup. The MMX routine run somewhat faster, but stalled the FPU which slowed down subsequent computations of the real world coordinates. Both approaches, of course, have the additional disadvantage that they are specific to the Pentium CPU and are thus not portable to others architectures.

Therefore, we looked to a general solution that would make use of the following facts:

- Most modern processors support word (32 bit) operations on integer operands and word memory accesses.
- Pixels are stored as words (32 bit) in either ARGB (big endian) or BGRA (little endian) format.

The motivation for our approach is to test all three channels R,G and B against the minimum in one operation by interpreting the pixel as a 32 bit word. Similarly, our method only uses one operation in the comparison against the maximum.

Our implementation is based on the realization that subtracting two bit fields will result in a borrow if the first operand is smaller than the second operand. If bits at position i are both 0, then there can never be a borrow. Therefore, if the resulting bit is a 1, it must have resulted from a borrow at position $i-1$.

Our colour threshold routine uses the least significant bit of the alpha, red, and green channel as a stop bit to detect borrows from the red, green, and blue channel respectively. This means that the least significant bit of the colours is ignored. This does not cause a problem, since there is very little difference between for example, a red value of 110 or 111.

The algorithm for our colour thresholding routine is shown in table 3. Variable `pixel` is an integer representation of the pixel value. Variable `lower` is the concatenation of R_{min}, G_{min}, and B_{min} anded with 0x7efefeff (so that the least significant bits are cleared). Variable `upper` is the concatenation and masking of the least significant bits of the upper thresholds respectively.

The least significant bits of the red and green channel in the pixel are cleared and the lower threshold is subtracted from the pixel. Should a colour channel (R, G, B) be less than the corresponding threshold, a borrow will have resulted in bits 8, 16, or 24. If such a borrow occurred, the routine returns 0, otherwise 1.

Comparisons against the upper thresholds are done similarly by subtracting the pixel value from the maximum threshold.

We use a similar method to calculate and test the difference thresholds R-G, R-B, and G-B. This routine resulted in a 20% speedup in our code.

Table 3. The Colour Threshold Routine

```
int matchColourThreshold(int pixel, int lower, int upper) {
    int ret;

    pixel = pixel & 0x7efefeff;
    if (((pixel - lower) & 0x81010100)||
        ((upper - pixel) & 0x81010100)) {
        ret = 0;
    } else {
        ret = 1;
    }
    return ret;
}
```

9.2 Object Prediction

Another method that we use to speed up the object detection routine is to use the previous position of an object as a starting point for a new search. The video server maintains the X and Y velocities of all object. When looking for an object in the next frame, a new position for the object is predicted using these velocities and a small 32*16 pixel subarea is searched for the object.

If the object is not found within this area, the object is put on a scan queue.

Object prediction works very well in our domain. In over 90% of the times, an object can be found in the predicted region. The reason for prediction failure is most often a fast moving ball, which is deflected or occluded by a robot.

9.3 Sampling Grid

Given the current hardware, we do not have sufficient processing power to scan the whole image even once. Therefore, we use a sampling grid whose size is determined by the smallest object that we are trying to find.

In our domain, these are the yellow and blue ping pong balls, which on the far end of the field are about 6*3 pixels. Therefore, we are using a 6*3 scanning grid.

9.4 Field Mask

As can be seen in the sample picture 2, only about 2/3 of the image contains the actual playing field. The tables on the left side and the top of the picture are not used. The video server uses a mask to distinguish the playing field from the surrounding area. This has two advantages: (a) finding objects is faster since only a sub area of the image must be scanned, and (b) the video server is more robust, since if someone with blue shoes walks through the image it will not be incorrectly classified as an opponent.

10 Conclusion

This paper describes a practical implementation of camera calibration in large rooms. It combines the use of a well known calibration algorithm with a semi-automatic method for computing the matching points.

The method uses a two stage approach. Initial approximations of the centres of objects (in our example squares) are used to sort the objects, but specific object features are used to assign real world coordinates. We intend to use feature detection mechanisms with sub-pixel accuracy, such as the ones described in [8] in the future to improve the accuracy of the calibration.

Object detection is based on blob detection of coloured spots on the car and the ball. The videoserver uses three colour ranges and three difference ranges to identify different colours. Under general lighting conditions, such as the ones that exist during RoboCup, this method allows us to distinguish between up to eleven different colours. A fast integer threshold comparison is used which lead to a 20% speed-up of the video server.

Currently, only geometry and brightness information in the calibration image is used to calibrate the camera. We are currently working on extending the system to compute the colour changes for blue and white squares. This would allow us to estimate the spectrum of the light source. The goal is to compute the colour thresholds for orange, blue, and yellow balls automatically given a single calibration picture as input.

References

1. Jacky Baltes. Aucklandianapolis homepage. WWW, February 1998. http://www.tcs.auckland.ac.nz/~jacky/teaching/courses/415.703/aucklandianapolis/index.html.
2. Jacky Baltes, Nich Hildreth, Robin Otte, and Yuming Lin. The all botz team description. In *Proceedings of the PRICAI Workshop on RoboCup*, 1998.
3. K.S. Fu, R.C. Gonzales, and C. S. G. Lee. *Robotics: Control Sensing, Vision, and Intelligence*, chapter 7.4, pages 306–324. McGraw Hill, 1987.
4. Hiroaki Kitano, editor. *RoboCup-97: Robot Soccer World Cup I*. Springer Verlag, 1998.
5. Alan Mackworth. *Computer Vision: System, Theory, and Applications*, chapter 1, pages 1–13. World Scientific Press, Singapore, 1993.
6. C.J. Merz and P.M. Murphy. UCI repository of machine learning databases, 1998.
7. Roger Y. Tsai. A versatile camera calibration technique for high-accuracy 3d machine vision metrology using off-the-shelf tv cameras and lenses. *IEEE Journal of Robotics and Automation*, RA-3(4):323–344, August 1987.
8. Robert J. Valkenburg, Alan M. McIvor, and P. Wayne Power. An evaluation of subpixel feature localisation methods for precision measurement. In *Videometrics III*, volume SPIE 2350, pages 229–238, 1994.
9. Reg Willson. Tsai camera calibration software. WWW, 1995.

Path Tracking Control of Non-holonomic Car-Like Robot with Reinforcement Learning

Jacky Baltes and Yuming Lin

Centre for Image Technology and Robotics
University of Auckland, Auckland
New Zealand
j.baltes@auckland.ac.nz
http://www.citr.auckland.ac.nz/~jacky

Abstract. This paper investigates the use of reinforcement learning in solving the path-tracking problem for car-like robots. The reinforcement learner uses a case-based function approximator, to extend the standard reinforcement learning paradigm to handle continuous states. The learned controller performs comparable to the best traditional control functions in both simulation and also in practical driving.

1 Introduction

The CITR at the University of Auckland has a mobile robotics lab, which hosts the Aucklandianapolis competition ([2]). The goal of the competition is to drive car-like (non-holonomic) robots five laps around a race track as quickly as possible. The cars are simple remote controlled toy cars with proportional steering and speed controls. A parallel port micro-controller based interface ([7]) allows us to control the cars (65 speed settings, 65 direction settings). Position and orientation information for the cars is provided by a video camera mounted on top of the playing field.

A non-holonomic path planner ([3]) creates a path for the car around the race track. The path contains only three different path segments: (a) straight lines, (b) maximum turns to the right, or (c) maximum turns to the left. The toy cars do not have shaft encoders so there is a feed forward control error when driving a given path.

Therefore, we need a controller which keeps the car on the track. Note that the control function described in this paper only depends on the curvature of the path and is thus mostly independent of the path itself. This means that our results are also applicable to more dynamic environments, such as RoboCup.

Some popular methods to control a non-holonomic mobile robot in such a path tracking problem include:

1. Feedback control as described by Alessandro and Giuseppe [5].
2. A Sliding-mode controller suggested in [1], which was used during initial trials for the Aucklandianapolis. This state of the art controller performed extremely well in simulation, but performed poorly in the practice. The

motivation of this project was to improve on its performance in the real world, see section (5).
3. A Fuzzy logic controller [8] which currently holds the unofficial track record for the Aucklandianapolis. The fuzzy logic controller is able to drive a car twice as fast as the sliding mode controller mentioned above.

This paper describes another method based on dynamic programming, a reinforcement learning controller. At the core of the reinforcement learner is a value function, called Q-value, which is why it is also called Q-Learning ([9]).

The following section describes the kinematic model of the car-like vehicle, or just car for simplicity. The model is used throughout the paper. Section 3 gives a brief introduction to reinforcement learning. Section 4 describes a case-based function approximator, which is used to approximate the value function in our implementation. Section 5 describes the results of our experiments using both simulation and practical driving. Section 6 concludes the paper.

2 Kinematic Model

In this research, we use a kinematic model, which is relative to the path. The controller knows the current position and orientation errors and the curvature of the path. However, the future path is not known. This model is appropriate in highly dynamic domains such as RoboCup.

The kinematic model is shown in Fig. 2. The car is at position (x,y) and is following a path with curvature R. The point $(\hat{x},\hat{y})$ is the closest point on the path to point (x,y). The position error $\tilde{y}$ is the distance between points (x,y) and $(\hat{x},\hat{y})$. $\hat{\theta}$ is the tangent of the path at the point $(\hat{x},\hat{y})$, θ is the orientation of the car, $\tilde{\theta}$ is the orientation error of the car (that is, $\tilde{\theta} = \theta - \hat{\theta}$).

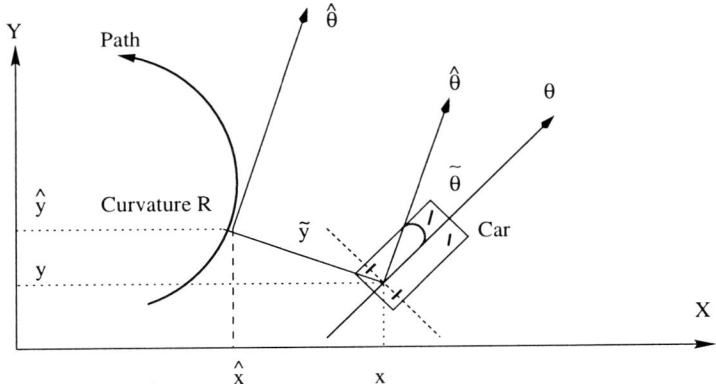

Fig. 1. The Kinematic Model

In the representation used in this paper, the current state of the system is defined by the positional error $\tilde{y}$, the orientation error $\tilde{\theta}$ and the curvature of the path R (a 3-tuple). The input to the controller is the three tuple for the current state and the outputs are desired settings for speed and direction.

3 Reinforcement Learning

This section gives a brief introduction to reinforcement learning. The most important concepts in reinforcement learning are the agent and the environment. The agent (a remote-controlled car in our case) has a number of possible actions. The agent performs some actions in the environment (which is modeled through a set of states). In some states, the agent receives feedback from the environment about how good or bad a certain state is. This feedback is called *reward*. The task of reinforcement learning is to find the action with the highest expected reward (*Q-value*) in the current state. In the path tracking problem, the reward is based on how well the agent tracked the given path.

At any time, the agent is in one state (X), it finds the optimal action (U) and executes it. Usually, the selected action is the one with the highest expected reward (*Q-value*). To prevent premature convergence on suboptimal action sequences, a reinforcement learner will sometimes not select the best action so that it can further explore the environment. This is called the exploration vs. exploitation trade-off.

After executing the action, the agent enters another state(X'). The agent may get a reward (positive or negative reinforcement) when entering certain states. The function $Q(X,U)$ is the value function for a given state(X) and action(U). It is the immediate reward r after the execution of the action(U), plus the best Q-value (discounted by a factor γ) in the following state. The reinforcement learning algorithm is shown below (Algorithm 1).

Algorithm 1 Reinforcement Learning Algorithm
---
  for each pair of $< X, U >$ **do**
    $Q(X,U) \leftarrow 0$
  end for
  Observe the current state X
  loop
    1. Select an action U and execute it
    2. Receive immediate reward r
    3. Observe the new state X'
    4. Update the table entry for Q(X,U), as follows:
        $Q(X,U) = r + \gamma * \max_{U'} Q(X',U')$
  end loop

Initially, since all function values are zero, the agent just selects an action randomly. With more and more experience, the function values may converge to the actual values [6] and the agent may use the learned function values for optimal control.

It is important to note that in general, the size of the state space determines how quickly the algorithm will converge on the correct function. The larger the state space, the longer it will take to learn the correct function.

3.1 Reward function in the car domain

The reward function is of critical importance in the design of a reinforcement learner. The reward function must accurately reflect the progress that an agent is making towards achieving a goal, since otherwise the agent will learn the wrong behavior.

In the car driving domain, we want to keep the car on the path, so it is reasonable to base the reward function on the position ($\tilde{y}$) and orientation error ($\tilde{\theta}$). Preliminary experiments, however, with Balluchi's controller ([1]) suggested that it is also important to have a smooth control function. Therefore, the reward function in this research is based on the weighted sum of normalized position error ($\tilde{y}$) and orientation error ($\tilde{\theta}$) as well as the necessary control work (Difference in control setting U at time t and time $t-1$) as shown in Equation 1.

$$r = -w_1 * (\frac{\tilde{y}}{2})^2 + w_2 * (\frac{\tilde{\theta}}{2\pi})^2 + w_3 * (\frac{U_t - U_{t-1}}{9})^2 \quad (1)$$

In a control problem, in principle a reward can be associated with every state. However, to get a better estimate of the real reward of a state, we return as reward the sum of the rewards for the last five states.

3.2 Reinforcement Learning with Continuous States

One may notice that the algorithm listed above assumes discrete states and actions. This is a problem in our path-tracking domain. Although the actions of the car (i.e. left-turn, right-turn etc) are discrete, the state, a 3-tuple vector $<\tilde{y},\tilde{\theta},R>$, is continuous. We must provide some mechanism to quantize the state space before reinforcement learning can be applied in this problem. There are at least two approaches.

The first one is to quantize the state directly and apply reinforcement learning. An example is shown in figure 2. This method is simple but inflexible and inefficient. It will unnecessarily increase the size of the state space. For example, assume that the car is facing in the right direction when following a straight line. In this case, if the car is only slightly to the right of the line, we want to turn gently left to approach the line and to not overshoot it. If the car is far away from the line, we want to turn sharply to get back onto the path. Therefore, we would require a fine quantization. But if we are following a circle, then independently of how far away we are from the outside of the circle, we want to make a sharp turn, since all circles are maximum turns. This means that the fine quantization will generate unnecessary states, which will greatly reduce the convergence speed of the algorithm.

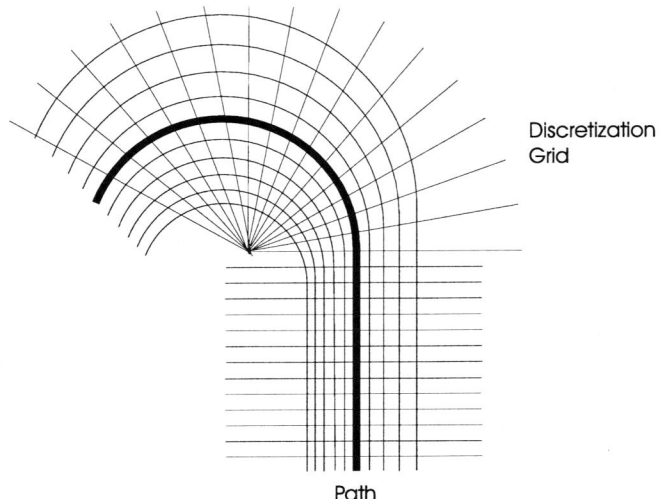

Fig. 2. Static Quantization. Each intersection of lines represents of the discretization grid a world state.

The second method is the use of a function approximator for the value function Q. In this case, the quantization is implicit and based on previous cases, which are stored in a database. Figure 3 shows an example of a case-based quantization. A state is assigned a value based on a prototypical case (e.g., close to the straight line or further away). In this simple example, all cases have the same area of influence. The key is to calculate the distance from the current state to those existing states in the database. In the example, only the nearest case determines the type of state, but in our implementation, a nearest neighbor set is computed.

In general, this distance is used to measure the contribution of all those selected states in the database in the Q evaluation. The research described in this paper uses a case-based function approximator, which is described in more detail in the following section.

4 Reinforcement learning using a Case-based Function Approximator

Function approximators are used to represent the value function(Q) for a continuous state problem. In discrete space, a finite resource can be used to store the value function, whereas in continuous space this is not the case. There are many functions approximators. For more details please see [10]. The Case-based function approximator is one of them and it is suitable for our task because of its structure. Operations are defined for the evaluation and update of the value function. Details can be found in [10].

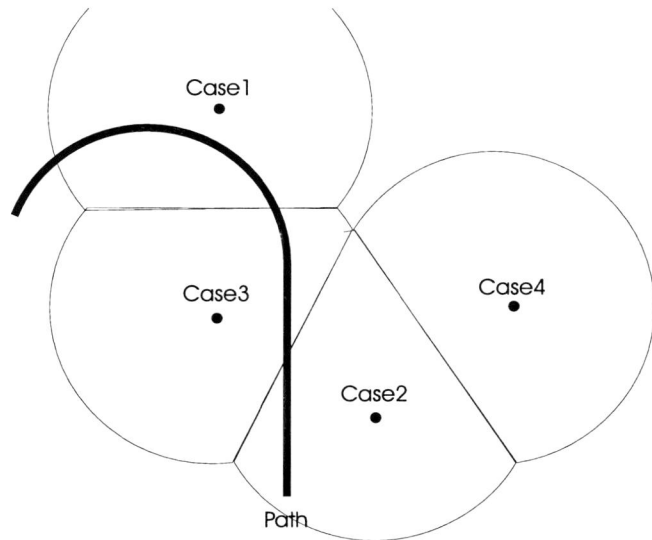

Fig. 3. Case-based Quantization

4.1 Case Structure

Every case in the database corresponds to one input point X_i that the agent has visited ($X_i = <\tilde{y}_i, \tilde{\theta}_i, R_i>$ in our case). One case C_i is:

$$C_i = (X_i, Q_i, e_i, \{U_{ij}, Q_{ij}, e_{ij}\})$$
where $i = 1 \ldots N$ is the number of cases (2)
and $j = 1 \ldots M$, M is the number of actions

From 2, it can be seen that C_i consists of two separate portions, the first portion (x_i, Q_i, e_i) is associated only with the state, the other (U_{ij}, Q_{ij}, e_{ij}) is associated with actions within the state. e_i is the eligibility trace of the state [11], while e_{ij} is the eligibility trace for action j within the state i.

4.2 Function Evaluation

To evaluate the value function $Q(X_Q, U_Q)$ for the query state X_Q, the database is searched for those states that are similar to the state in question. The distance (d_i) from an existing state (X_i) to the query state X_q can be used for the estimation of similarity $(d_i = f(\tilde{y}_i - \tilde{y}_q) + g(\tilde{\theta}_i - \tilde{\theta}_q) + h(R_i - R_q))$. After search through the entire database, a nearest neighbor set NN_q for the query state X_q is generated. NN_q consists of those states with distance to X_q less than a predefined threshold τ_k. That is

$$NN_q = \{Q_i | d_i <= \tau_k\} \quad (3)$$

The distance measure d_i is defined as:

$$d_i = \sqrt{\left(\frac{\tilde{y}_i - \tilde{y}_q}{2}\right)^2 + \left(\frac{\tilde{\theta}_i - \tilde{\theta}_q}{2*\pi}\right)^2 + \left(\frac{R_i - R_q}{2}\right)^2} \quad (4)$$

The distance is based on three parts: current distance error, current orientation error and the curvature of the path. The distance error is normalized to $-1..1$ meter, the orientation error to 360 degrees, and the curvature to $0..2$.

From NN_q in Equation (3), all existing cases in the database that are similar to the current input X_q can be found, thus the Q value for the query point $<X_q, U_q>$ can be calculated by the following formulas:

$$Q_i(U_q) = (1-\rho)Q_i + \quad (5)$$

$$\rho \left(\sum_{\forall u_{ij} \in C_i} \frac{K^u(d_{ij}^u)}{\sum_j K^u(d_{ij}^u)} Q_{ij} \right) \forall C_i \in NN_q$$

$$Q(X_q, U_q) = \sum_{\forall C_i \in NN_q} \frac{K^x(d_i^x)}{\sum_j K^x(d_j^x)} Q_i(U_q) \quad (6)$$

The value $Q_i(U_q)$ is the overall Q value for the current query action U_q in state X_i. It consists of two parts: (a) the Q value for state X_i and (b) the sum over all actions in this state.

The action having the highest $Q(X_q, U_q)$ is selected as the current action for the input X_q.

4.3 Learning Update

All Q-values in the database must be updated after a new reward is returned from the environment for the given action. The eligibility traces (e_i, e_{ij} in Equation(3)) are also updated according to their contribution to $Q(X_q, U_q)$. Based upon the distance function, a new case is created if no case near enough to the query input X_q exists [10].

4.4 An Example of Function Evaluation

This section gives an example of how to evaluate the Q-value for an input $X_q = <0.5, 0, 1>$ and to find the best action for state X_Q. For simplicity, after searching the database, only two cases are in NN_q in Equation (3), as shown in Figure 4. Table 1 shows details of the cases in NN_q. There are only three actions(0 for left-turn, 1 for go-straight, 2 for right-turn) here. The actual implementation uses nine different steering angles.

In Table 1, d_{iq} is calculated by Equation (4), The selection of K^u in Equation (5) and K^x in Equation (6) is based on the strategy of exploitation and exploration[6]. Set $\rho = 0.6$ in Equation(5), and let K^u be such that in Equation(5)

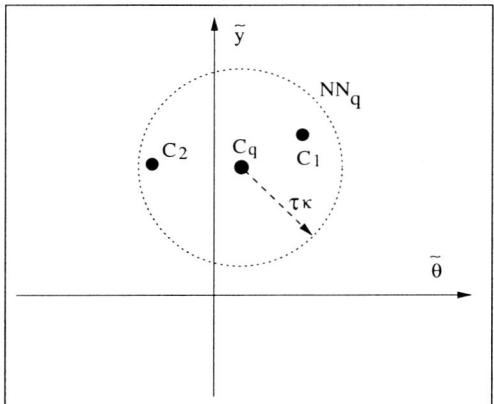

Fig. 4. Two cases in the NN_q

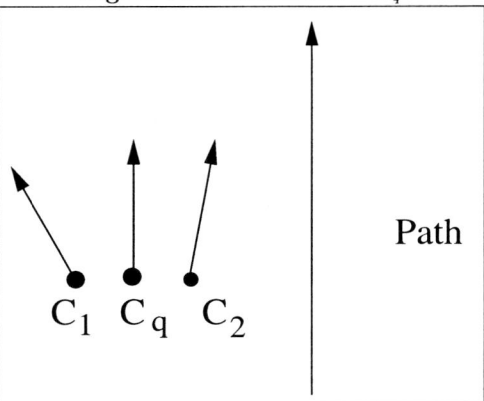

Fig. 5. One of the situations as shown in Fig. 4

Case	$\tilde{y}$	$\tilde{\theta}$	R_i	Q	Q_0	Q_1	Q_2	d_{iq}
1	0.6	0.3	1	-0.2	-0.8	-0.6	-0.3	0.069
2	0.4	-0.1	1	-0.1	-0.7	-0.1	-0.2	0.052

Table 1. The two cases in the nearest neighbor set NN_q

$Q_i(U_q) = (1-\rho)Q_i + \rho Q_{iq}$ (that is, only the value of the action that is the same as the query action is considered), $K^x(d_i^x) = d_i^x$.

The distance of case 1 and 2 to the query state are given as:

$$d_1 = \sqrt{\left(\frac{0.6-0.5}{2}\right)^2 + \left(\frac{0.3-0}{2\pi}\right)^2 + \left(\frac{1-1}{2}\right)^2} = 0.069$$

$$d_2 = \sqrt{\left(\frac{0.4-0.5}{2}\right)^2 + \left(\frac{-0.1-0}{2\pi}\right)^2 + \left(\frac{1-1}{2}\right)^2} = 0.052$$

The overall Q values for the actions in state X_1 and X_2 are computed using Equation 5.
$Q_1(0) = (1-\rho)Q_1 + \rho Q_0 = (1-0.6)*(-0.2) + 0.6*(-0.8) = -0.56$
$Q_1(1) = (1-\rho)Q_1 + \rho Q_1 = (1-0.6)*(-0.2) + 0.6*(-0.6) = -0.44$
$Q_1(2) = (1-\rho)Q_1 + \rho Q_2 = (1-0.6)*(-0.2) + 0.6*(-0.3) = -0.26$

So action 2 (right turn) has the best Q value in case 1.
$Q_2(0) = (1-\rho)Q_2 + \rho Q_0 = (1-0.6)*(-0.1) + 0.6*(-0.7) = -0.46$
$Q_2(1) = (1-\rho)Q_2 + \rho Q_1 = (1-0.6)*(-0.1) + 0.6*(-0.1) = -0.1$
$Q_2(2) = (1-\rho)Q_2 + \rho Q_2 = (1-0.6)*(-0.1) + 0.6*(-0.2) = -0.16$

Similarly, action 1 (straight) has the best Q value for case 2. As shown below, we evaluate the best action for the current state X_q by using Equation 6.

$Q(X_q, 0) = \frac{d_1}{d_1+d_2}Q_1(0) + \frac{d_2}{d_1+d_2}Q_2(0) = \frac{0.069}{0.121}(-0.56) + \frac{0.052}{0.121}(-0.46) = -0.52$

$Q(X_q, 1) = \frac{d_1}{d_1+d_2}Q_1(1) + \frac{d_2}{d_1+d_2}Q_2(1) = \frac{0.069}{0.121}(-0.44) + \frac{0.052}{0.121}(-0.1) = -0.29$

$Q(X_q, 2) = \frac{d_1}{d_1+d_2}Q_1(2) + \frac{d_2}{d_1+d_2}Q_2(2) = \frac{0.069}{0.121}(-0.26) + \frac{0.052}{0.121}(-0.16) = -0.22$

As $Q(X_q, 2)$ has the highest value, the agent will take action 2, namely turn right when the input X_q is $< 0.5, 0, 1 >$

5 Experiments

The controller described above has been implemented both in simulation and practical driving. Surprisingly, the database generated during simulation can be directly applied to practical driving. This means that the controller in a real world environment does not need to learn from scratch, which is very difficult in practice because it requires too many training episodes and because you need to put the car close to the path again if the current trial fails.)

The Aucklandianapolis race track is used as the sample path, both in simulation and practical driving.

Table 2 shows the average position and orientation errors for different numbers of learning episodes. Each trial consists of 200 steps. The data is averaged over 100 trials after the training phase.

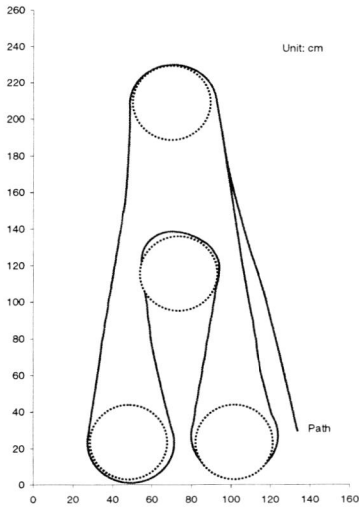

Fig. 6. Learned result in simulation after 1000 trials

Figure 7 shows the result of using the learned controller to drive the car in practice. As can be seen, the controller can use the results from simulation to *seed* the controller to drive the car in practice. Ideally, one would like the controller to improve its performance now in practice with increased experience. However, there is no noticeable improvement in practical driving. There are two reasons: (a) the reinforcement learner has settled, that is most Q values are known, on the current controller so that it is unlikely that it will explore new actions, and (b) it takes a lot longer to drive a track in practice as opposed to simulation where it is easy to drive a few thousand laps.

Experiment	Training	Avg. $\tilde{y}$(m)	Avg. $\tilde{\theta}$(radius)
1	200	0.2684	0.3202
2	400	0.2126	0.2802
3	600	0.0734	0.1381
4	800	0.0462	0.1043
5	1000	0.0509	0.1033
6	2000	0.0477	0.0943

Table 2. Average Control Errors in Simulation

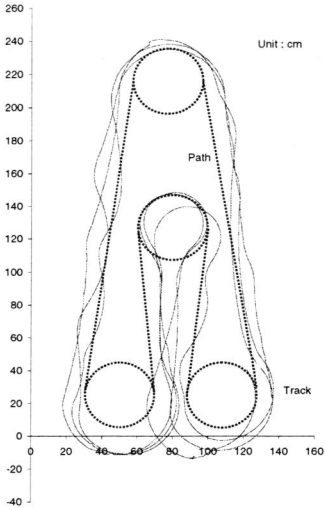

Fig. 7. Using the learned result in practical driving

6 Conclusions

In this paper we describe some aspects of reinforcement learning, such as function value representation and how it is used in the problem of path-tracking. Reinforcement learning can be adapted to control a car in path-tracking. The training can be initialized using a simulation and after the performance has stabilized, training can continue with practical driving. Since the representation is independent of the path ahead, the learned controller can be used in real time path following tasks, independently of whether the path is static (as in the Aucklandianapolis) or dynamic (as in the RoboCup competition).

The performance (average control errors) of the simulation in our experiment is satisfactory. The learned values in the simulation can also be used in the real world task of driving our toy cars.

The reinforcement learning controller has also proven itself in the Aucklandianapolis competition. It won the 1999 competition in 2 minutes 30 seconds, which is twice as fast as Balluchi's controller and comparable to the Fuzzy Logic controller ([8]). The reinforcement controller is the only controller that has been used successfully to drive cars with and without linear steering behavior. The Fuzzy Logic controller works by interpolating steering angles and thus works only for cars with at least an approximate steering behavior.

However, further work is needed to achieve significant improvement during learning (as in the simulation) in the real world with its many sources of errors e.g., noise, actuator error and slipping.

Another improvement is the use of dynamic weights in the reward function. A simple example shows that given our representation, static weights are insufficient to learn the correct control function. Assume that the car is following a straight line. When the car is far away from the path, then it will need to steer towards the path, which means that it will get a reduced reward due to the orientation error. In this case, the weight on the orientation error should be small. However, when close to the line, the orientation error is more important than the position error and should receive more weight. Full details of the practical evaluation of the reinforcement learner are given in [4].

References

1. A. Balluchi, A. Bicchi, A. Balestrino, and G. Casalino. Path tracking control for dubin's cars. In *Proceedings of IEEE International Conference on Robotics and Automation*, Centro "E. Piaggio" & Dipartimento Sistemi Elettrici e Atuomazione, Università di Pisa., 1996.
2. Jacky Baltes. Aucklandianapolis homepage. WWW, February 1998. http://-www.tcs.auckland.ac.nz/~jacky/teaching/courses/415.703/aucklandianapolis/-index.html.
3. Antonio Bicchi, Giuseppe Casalino, and Corrado Santilli. Planning shortest bounded-curvature paths for a class of nonholomic vehicles among obstacles. In *Proceedings of the IEEE International Conference on Robotics and Automation*, pages 1349–1354, 1995.
4. Yuming Lin. *Path Tracking Control of Non-Holonomic Car-Like Robots with Reinforcement Learning*. Master's thesis, University of Auckland, Auckland, New Zealand, July 1999.
5. Alessandro De Luca, Giuseppe Oriolo, and Claude Samson. Feedback control of a nonholonomic car-like robot. Technical report, Universita di Roma La Sapienza, 1996.
6. Tom Mitchell. *Machine Learning*. mcGraw-Hill, 1997.
7. Ben Noonan, J. Baltes, and B. MacDonald. Pc interface for a remote controlled car. *In Proc. Of the IPENZ sustainable city conference*, pages 22–27, 1998.
8. Robin Otte. Path following control of a nonholonomic vehicle at high speeds using a fuzzy controller. Technical report, Computer Science, The University of Auckland, rott001@cs.auckland.ac.nz, 1998.
9. Stuart Russel and Peter Norvig. *Artificial Intelligence: A Modern Approach*, chapter 20, pages 598–624. Prentice-Hall Inc., Englewood Cliffs, New Jersey 07632, 1995.
10. JC Santamaria, RS Sutton, and A Ram. Experiments with reinforcement learning in problems with continuous state and action spaces. *Adaptive Behavior*, 6:163–217, 1997.
11. Singh SP and Sutton RS. Reinforcement learning with replacing eligibility traces. *Machine Learning*, 1996.

Fast Image Segmentation, Object Recognition and Localization in a RoboCup Scenario

Thorsten Bandlow, Michael Klupsch, Robert Hanek, Thorsten Schmitt

Forschungsgruppe Bildverstehen (FG BV)
Technische Universität München, Germany
{bandlow,klupsch,hanek,schmittt}@in.tum.de
http://www9.in.tum.de/research/mobile_robots/robocup/

Abstract. This paper presents the vision system of the robot soccer team *Agilo RoboCuppers* [1] – the RoboCup team of the image understanding group (FG BV) at the Technische Universität München.
We present a fast and robust color classification method yielding significant regions in the image. The boundaries between adjacent regions are used to localize objects like the ball or other robots on the field. Furthermore for each player the free motion space is determined and its position and orientation on the field is estimated. All this is done completely vision based, without any additional sensors.

1 Introduction

The vision module is a key part of our robot soccer system described elaborately in [1, 5]. Given a video stream, the vision module has to recognize relevant objects in the surrounding world and provide their positions on the field to other modules. Each robot is only equipped with a standard PC based on a single Pentium 200 MHz processor. Consequently, we have to focus on efficient and computationally inexpensive algorithms to serve the real time constraints. This is done with the help of the image processing tool HALCON (formerly known as HORUS [3]). This tool provides efficient functions for accessing, processing and analyzing image data, including framegrabber access and data management.

In general, the task of scene interpretation is a very difficult one. However, its complexity strongly depends on the context of a scene which has to be interpreted. In RoboCup, as it is currently defined, the appearance of relevant objects is well known. For their recognition, the strictly defined constraints of color and shape are saved in the model database and can be used. These constraints are matched with the extracted image features such as color regions and line segments. Figure 1 shows a data flow diagram of our vision module.

Besides recognizing relevant objects, further tasks of the image interpretation module are to localize the recognized objects and to perform the self-localization

---
[1] The name is derived from the Agilolfinger, which were the first Bavarian ruling dynasty in the 8th century, with Tassilo as its most famous representative.

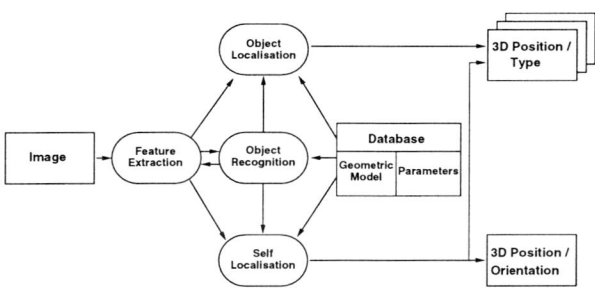

Fig. 1. Data flow diagram of the vision module.

of the robot on the field. For this the intrinsic camera parameters as well as the external ones relative to the robot were determined by calibration.

This paper is organized as follows: Section 2 describes the algorithms applied for the color based image segmentation and object extraction. In Section 3 the estimation of the position of relevant objects in the 3-D coordinate system of the observing robot is discussed. A video based self-localization algorithm using the field boundary lines is presented in Section 4. Section 5 shows the achieved results and, finally, a conclusion is given.

2 Color based Image Segmentation and Object Detection

This section describes how an YUV-image, as captured from the camera, is segmented into regions representing one color class each. The overall aim of this segmentation process is to extract one segment for each visible object of the current scene. Since all important objects of the RoboCup scenario have distinct and unique colors, color is the key feature used for object recognition.

First, color regions are determined by the image processing module using a fast classification algorithm which assigns a color class label to each pixel (see Equation 1) according to its YUV-values.

$$(y, u, v) \longrightarrow \{no\_color, black, white, \\ green, blue, cyan, \\ magenta, red, yellow\} \quad (1)$$

Then various image processing operators are applied in order to determine the regions which contain the important objects of the RoboCup scenario, e.g. ball, lines, robots and goals.

2.1 Building a robust color Classifier

Previous RoboCup events have shown that the illumination conditions differ from field to field. Therefore, an adaption to the actual lighting conditions of the field is needed.

Fig. 2. The color classifier training tool.

For building an adapted color classifier we have developed an interactive classification tool which supports supervised learning of the appropriate color classes (see Figure 2). The YUV-color space is used in order to distinguish different color classes. While the Y-channel heavily depends on the light intensity, the U- and V-values are relatively invariant regarding the brightness of the light. Since we use a camera with deactivated auto white balance the U- and V-values of a class have a quite small variance. The classification tool determines color clusters in the UV-plane according to test-samples and assigns color class labels to them. For achieving a robust classifier the training can be performed over several different images.

A color class label is assigned to a cluster in the YUV-color space as follows. An YUV-image is grabbed and the Y-channel is displayed. The user draws a region which contains only pixels of the same color class and assigns a label to it. First, the minimum and maximum brightness values according to the Y-channel are determined. Then a 2-D histogram for the U- and V-channels of the selected region is computed. This 2-D histogram is interpreted as a 256×256 image, and several threshold, closing and opening operators are applied to it. This eliminates faulty responses arising form color noise in the camera image, and provides a compact cluster representing one color class in the UV-plane.

This procedure is repeated for all color classes of interest incorporating different images. As a final result the color calibration tool saves a 256×256 color

Fig. 3. The color segmented view of a robot.

class label image for the U- and V-channels and the minimum and maximum Y-values for each color class label.

2.2 Fast color based Segmentation

During the match the interface integrating the framegrabber device driver into the image processing library performs the color based segmentation. For each pixel the U- and V-values serve as indices into the previously determined color class label image. Each pixel is assigned to one region which corresponds to a specific color class, if its brightness is within the previously determined brightness interval for that color class. It is noteworthy that this procedure determines one region for each color class and that these regions need not to be interconnected and may be distributed all over the image.

After the whole image has been processed the framegrabber interface returns the determined color regions using a runlength encoding. This encoding is the default data structure used by the image processing system HALCON for storing regions. Consequently all HALCON operators can now easily and efficiently be applied to the pre-segmented image.

2.3 From Color Regions to Object Segments

Once the color regions are known, several different operators are applied in order to determine the independent segments which contain objects such as the ball, robots or goals.

First, different morphological operators to the color regions are applied, in order to remove small disturbances and make the regions more compact. Then we determine the single object segments performing a connected component analysis using the 8-neighborhood. Each of these segments is now regarded as an object hypothesis. Finally, objects can simply be recognized through the computation of certain shape features and verified by plausibility tests. For example, the ball segment can be detected using the shape feature anisometry, goals are discovered by their size, and lines by their length and their angle of inclination. Figure 3 shows a result of our segmentation and object recognition process. The yellow goal, the goal keeper, the ball and one opponent are clearly visible.

3 Object Localization

In this section we explain how the 2-D regions, introduced in the previous section, are used to estimate the position of relevant objects such as the ball, the goals or other robots in the 3-D coordinate system of the observing robot.

The cameras of the robots are calibrated using the approach presented in [6]. With the help of the intrinsic camera parameters pixel coordinates can be converted into image coordinates. Since we use a camera with a wide viewing angle, about 90 degree, for this conversion it is important to take the radial distortions of the lens into account even if not high precision is needed and high speed is desired.

3.1 Restriction on a 2-D Localization Problem

We assume that all relevant objects are located on the ground of the field, i.e. the distance between an object and the plane defined by the ground of the field is zero. Of course for a jumping ball this is not correct. However, such cases are quite rare in RoboCup games. The restriction onto the ground provides an one-to-one correspondence between a point on the ground plane $\mathbf{E}$ and its observation in the image plane.

In order to estimate the location of an opponent, for example, we determine in the corresponding segmentation result the lowest point $\mathbf{p}$. This point and the optical center of the camera define a viewing ray $\mathbf{r}$. Its intersection with the plane $\mathbf{E}$ of the field yields an estimate of the opponents maximum distance and its direction in camera coordinates, see Figure 4. The 3-D point $\mathbf{P}$ corresponding to the observed image point $\mathbf{p} = [p_x, p_y]^T$ is given by

$$\mathbf{P} = \frac{h}{-p_y}[p_x, p_y, f]^T \qquad (2)$$

Here the focal length of the camera is denoted by f and its height, the distance to the ground, by h. The point $\mathbf{P}$ given in camera coordinates can easily be expressed in the coordinate system of the robot, since the pose of the robot's camera is given in robot coordinates.

In general the 3-D point $\mathbf{P}$ corresponding to the observed image point $\mathbf{p}$ lies not exactly on $\mathbf{E}$. However for the robots taking part in the RoboCup competition the distance $d(\mathbf{P}, \mathbf{E})$ is in general small enough.

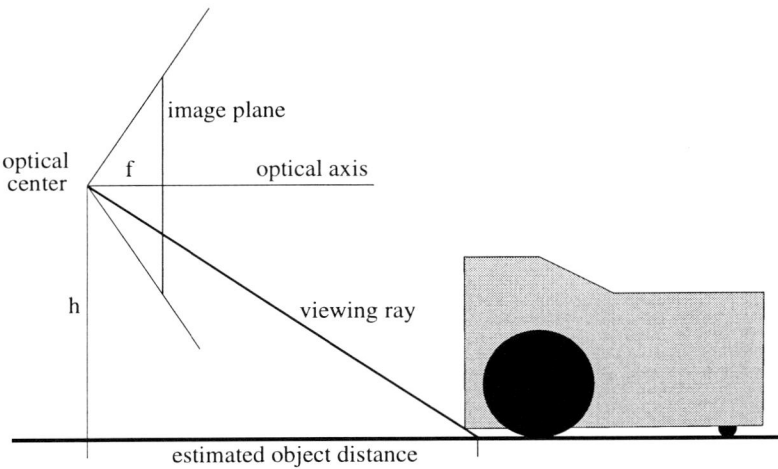

Fig. 4. An estimate of the object distance is given by the intersection of the viewing ray with the ground plane of the field.

The segmentation algorithm described in the previous section only classifies a pixel if its classification is quite sure. This means some pixels do not belong to any segment, especially at the border of two segments. Therefore, the distance computed as discussed above is an upper limit for the object distance. In order to obtain a lower limit we use the highest point **p′** under **p** classified as field. This point defines a second viewing ray **r′**. Its intersection with the plane **E** yields a lower limit for the object's distance.

3.2 Using Shape Restrictions for the Localization of the Ball

Due to noise, reflections or other disturbances the segmentation process may provide more than one red segment as candidates for the ball. In order to decide if a segment actually corresponds to the ball we use several plausibility checks. For all hypothesis given by red segments we compute the distances. Using the ball's distance we determine the radius r of the ball's projection onto the image plane. A segment of size s_x in x-direction and size s_y in y-direction is rejected as ball hypothesis if the condition

$$(s_x < b_x r) \text{ or } (s_y < b_y r) \tag{3}$$

holds. The parameter b_y is chosen to be smaller than b_x since due to reflections on the top of the ball this part quite often can not be classified as red. Furthermore we reject segments which cover a number of pixels which is too small in comparison with the calculated projection of the ball.

3.3 Free Motion Space

In order to navigate autonomously it is essential to know where the robot can move without collision. The localization approach described above yields the positions of objects localized with a relatively high probability. However, objects which could not get localized could still represent an obstacle. Therefore, we compute the free motion space independently of the object localization.

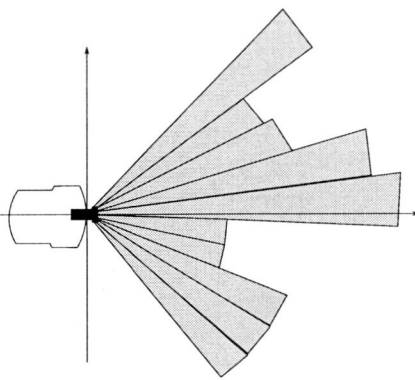

Fig. 5. Free motion space: The possible moving area of the robot is equally divided into sectors. To each of the sectors a maximal moving distance is associated, representing the space in which the robot can move without collision.

We divide the space of possible motion directions into sectors of equal angle as depicted in Figure 5. Each of these sectors on the field corresponds to a region in the image plane. Since the white lines of the field are no obstacles in contrary to the also white wall, we distinguish the white lines from the white wall by investigating the neighborhood of white regions. For determining the free motion space the green field regions and the white line regions are merged. Then morphological operations are applied in order to eliminate small artifacts. Since the ball can be moved by the robot the ball is also no real obstacle. Therefore, we compute for each sector the furthest point such that all closer points are either green or red, which means they are no obstacle. This point defines the length of a sector.

4 Self Localization

Obtaining the pose of the robots in the field coordinate system is a crucial term for the strategic planning of the robots, especially for actions, where several robots have to collaborate. Our self-localization algorithm is based on the boundaries of the field. They are easy to detect and allow a robust pose estimation. If a robot detects only one boundary line, the distance to this line and the robot's orientation can be adjusted. With two lines the robot can determine its absolute position and orientation. In the following sections we will discuss the feature extraction and pose estimation process.

4.1 Feature Extraction

The aim of the feature extraction process is to detect one or two lines of the field boundary, for which the relative 3-D coordinates are computed.

To detect the field boundaries only the border area between the field regions – green field and white lines – and the white wall regions is investigated. This area or region of interest can easily be achieved by a dilatation operation applied on both the wall and the field region followed by intersecting the two resulting regions.

Two different methods were implemented for ascertening the field boundaries. The first approach uses directly the field-wall-border region mentioned above. The skeleton is calculated and transformed into a set of contours $\mathcal{C}$. In the second method, a subpixel accurate edge filter is applied onto the Y-channel only within the previously determined region of interest, also calculating a set of contours. This method results in much more accurate contours but needs more computation time (approx. 30 ms vs. 5 ms). Both methods can be used alternatively.

The next steps remove camera distortions and approximate the contour segments with straight line segments. For this the following process is performed:

1. $\forall c_i \in \mathcal{C}$: Compute a regression line.
2. Filter lines by angle and length. Vertical lines and too short lines are discarded.
3. Join contours c_i and c_j which are collinear and closer than a maximum distance.

To achieve 3-D line segments we project the endpoints of the 2-D line segments onto the ground plane using the method described in Section 3.1. Collinear 3-D line segments are joint.

4.2 Obtaining Correspondences

For the self-localization we use a model of the field consisting of the four boundary lines. In order to estimate the pose we have to find correspondences between the 3-D model lines and the 3-D backprojection of their 2-D observations resulting from the method described in Section 4.1.

The two goals with their distinct colors are used to obtain the needed correspondences. If a 2-D line segment is adjacent to a goal segment then this line segment corresponds to the 3-D line next to the observed goal. This test is performed using a dilatation method described in [2]. If a second orthogonal 3-D line segment is given then this segment corresponds to a side line. The position of the goal segment and the 2-D line segment in the image defines whether the

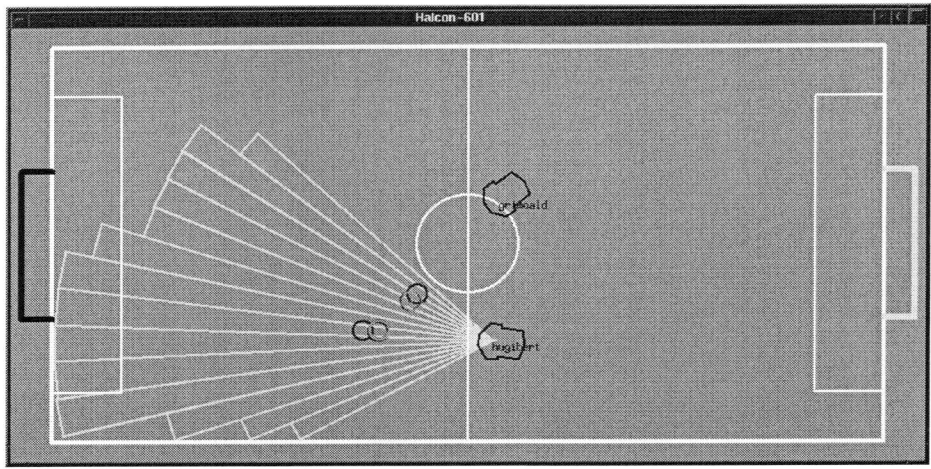

Fig. 6. The RoboCup monitor visualizes the actual positions of robots, ball hypotheses and free motion spaces.

line segment corresponds to the left or right side line. Due to noise the two back-projections are usually not exactly orthogonal. In this case we substitute the back-projections by two orthogonal lines having minimal distance to the original back-projected lines.

If no goal is extracted we can not establish correspondences as described above. In this case an absolute relocalization is not possible. Therefore, the pose of the robot at time t_i is predicted from the data at time t_{i-1} and the odometric data from the robot. We match the back-projected lines given in the robot coordinate system with the model line given in the world coordinate system such that the difference of the robot's new and predicted orientation is minimal. With this renewing method a correct match is performed, if the error of the orientation prediction is lower than 45 degree which holds most of the time.

4.3 Pose Estimation

Once the correspondences are given, we have to determine the robot's pose, such that the back-projected lines, given in robot coordinates, fit to the model lines, given in world coordinates.

The robot's pose has three degrees of freedom and is represented by a two-dimensional translation vector $\mathbf{T} := (T_x, T_y)^T$ and a rotation angle Φ. Since the feature extraction yields either zero, one or two line segments we have to distinguish three cases. In the quite rare case where no line segment is given no vision based update of the robot's pose is possible.

For a single line segment the problem is underdetermined. However, the rotation and one component of the translation can still be determined. The new orientation Φ_t and the previous orientation Φ_{t-1} are related by

$$\Phi_t = \Phi_{t-1} + \Delta\Phi_t \tag{4}$$

where $\Delta\Phi_t$ denotes the angle between the back-projected line and the corresponding model line. Furthermore the robot is able to update its distance to the observed line. Similar to the orientation, the relation between the new translation $\mathbf{T}_t$ and the old translation $\mathbf{T}_{t-1}$ is

$$\mathbf{T}_t = \mathbf{T}_{t-1} + \Delta\mathbf{T}_t. \tag{5}$$

Here the vector $\Delta\mathbf{T}_t$ is the distance vector between the model line and the back-projected line. Note, after the orientation is updated using Equation (4), these two lines are parallel and $\Delta\mathbf{T}_t$ is usually not zero. The translation along the line can not be estimated from a single line. Therefore the component of $\mathbf{T}_t$ parallel to the line is obtained from the previous translation vector $\mathbf{T}_t$.

If two line correspondences are given, the orientation can be updated again using Equation (4). Since the two back-projected lines are forced to be orthogonal (see Section 4.2), the quantity $\Delta\Phi_t$ is the same for both lines. With two lines both components of the translation can be updated. Once again the new translation is given by $\mathbf{T}_t = \mathbf{T}_{t-1} + \Delta\mathbf{T}_t$. However here $\Delta\mathbf{T}_t$ denotes the distance vector between the intersection of the model lines and the intersection of the back-projected lines.

5 Results

We have implemented the above robot vision system in an object-oriented framework that allows the implicit modeling of time within image processing systems [4]. The whole system is based on a modular design and we are able to exchange all image processing algorithms at run-time. This offers great flexibility and enables us to build rapid prototypes of sophisticated vision algorithms. The algorithms presented in this paper are currently applied and represent up-to-now the best solution for our robot soccer team.

In order to verify and measure the accuracy of the (self) localization algorithms we have developed a monitoring program that visualizes the positions, orientations, free motion spaces and ball hypotheses of the robots (see Figure 6). In a second window the states of the robots are displayed, such as the current action and role as well as the planning state (see Figure 7).

The vision system has been tested on a Linux operating system using an low-cost Pentium 200 MHz processor. An inexpensive PAL color CCD camera (Siemens SICOLOR 810) is mounted on top of the robot console and linked to the S-VHS input of the video capture card (BT 848 based with PCI interface). Gain, shutter time, and white balance of the camera are adjusted manually.

With this configuration we are currently able to process 7 to 12 frames with a size of 384×172 pixels per second. The frame rate mostly depends on the method used for determining the field boundaries in order to perform self localization.

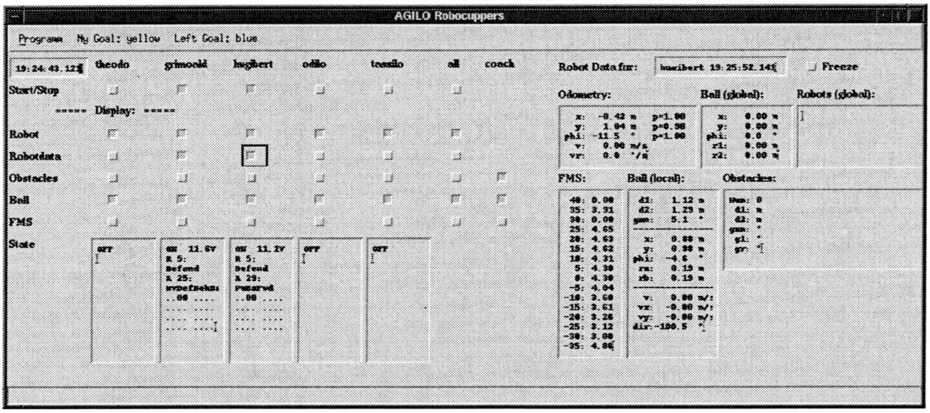

Fig. 7. A second window displays the states of the robots.

Overall the self-localization and object detection algorithms work quite robust. The robots are capable to detect the ball and score goals as well as to detect other robots and avoid collisions. However, there are still cases where a robot estimates its pose incorrectly. In general this occurs when most of the goal and border lines are hidden behind other robots. We are currently investigating further methods, incorporating lines on the field as well as global sensor fusion, to overcome this problem.

6 Conclusions

We have presented a system that segments objects by their color and generates a 3-D interpretation of the scene in real time. Our robot team has proven it's playing capabilities at several occasions (i.e. RoboCup'98 in Paris and the German Vision RoboCup'98 in Stuttgart). However a few problems remain.

The presented color classification algorithm is fast and robust, but we still have to adjust the classifier manually before each game. A more precise assignment of the color class labels to regions in the YUV-space might be one solution, but we are also considering time, position and orientation dependent solutions.

We hope to overcome the problem of incorrect pose estimations with a global sensor fusion system, that constructs a global view of the playing field from the local robot views. So far we had difficulties in exploiting this possibility, as we were relaying on a very unstable wireless radio ethernet. A further approach will also exploit the positions of field lines and the non-linear center circle in the 3-D CAD model.

References

1. BANDLOW, T., HANEK, R., KLUPSCH, M., AND SCHMITT, T. Agilo RoboCuppers: RoboCup Team Description. In *RoboCup'99 (http://www9.in.tum.de/research/mobile_robots/rob ocup)* (1999), Lecture Notes in Computer Science, Springer-Verlag.
2. ECKSTEIN, W. Unified Gray Morphology: The Dual Rank. *Pattern Recognition and Image Analysis 7*, 1 (1997).
3. ECKSTEIN, W., AND STEGER, C. Architecture for Computer Vision Application Development within the HORUS System. *Journal of Electronic Imaging 6*, 2 (Apr. 1997), 244–261.
4. KLUPSCH, M. Object-Oriented Representation of Time-Varying Data Sequences in Multiagent Systems. In *International Conference on Information Systems Analysis and Synthesis (ISAS'98)* (Orlando, FL, USA, 1998), N. Callaos, Ed., International Institute of Informatics and Systemics (IIIS), pp. 833–839.
5. KLUPSCH, M., BANDLOW, T., GRIMME, M., KELLERER, I., LÜCKENHAUS, M., SCHWARZER, F., AND ZIERL, C. Agilo RoboCuppers: RoboCup Team Dscription. In *RoboCup'98* (1998), Lecture Notes in Computer Science, Springer-Verlag.
6. LENZ, R., AND TSAI, R. Y. Techniques for Calibration of the Scale Factor and Image Center for High Accuracy 3-D Machine Vision Metrology. *IEEE Trans. on Pattern Analysis and Machine Intelligence 10*, 5 (Sept. 1988), 713–720.

Using Hierarchical Dynamical Systems to Control Reactive Behavior

Sven Behnke, Bernhard Frötschl, Raúl Rojas, Peter Ackers, Wolf Lindstrot, Manuel de Melo, Andreas Schebesch, Mark Simon, Martin Sprengel, and Oliver Tenchio

Free University of Berlin, Institute of Computer Science
Takustr. 9, 14195 Berlin, Germany
{behnke|froetsch|rojas|ackers|lind|melo |schebesc|simon|sprengel|tenchio}
@inf.fu-berlin.de, http://www.inf.fu-berlin.de/~robocup

Abstract. This paper describes the mechanical and electrical design, as well as the control strategy, of the *FU-Fighters* robots, a F180 league team that won the second place at RoboCup'99. It explains how we solved the computer vision and radio communication problems that arose in the course of the project.

The paper mainly discusses the hierarchical control architecture used to generate the behavior of individual agents and the team. Our reactive approach is based on the Dual Dynamics framework developed by H. Jäger, in which activation dynamics determines when a behavior is allowed to influence the actuators, and a target dynamics establishes how this is done. We extended the original framework by adding a third module, the perceptual dynamics. Here, the readings of fast changing sensors are aggregated temporarily to form complex, slow changing percepts.

We describe the bottom-up design of behaviors and illustrate our approach using examples from the RoboCup domain.

1 Introduction

The "behavior based" approach has proved useful for real time control of mobile robots. Here, the actions of an agent are derived directly from sensory input without requiring an explicit symbolic model of the world [1, 2, 5]. In 1992, the programming language PDL was developed by Steels and Vertommen as a tool to implement stimulus driven control of autonomous agents [8, 9]. PDL has been used by several groups working in behavior oriented robotics [7]. It allows the description of parallel processes that react to sensor readings by influencing the actuators. Many basic behaviors, like taxis, are easily formulated in such a framework. On the other hand, it is difficult and expensive to implement more complex behaviors in PDL, mostly those that need persistent percepts about the state of the environment. Consider for example a situation in which we want to position our defensive players preferentially on the side of the field where the offensive players of the other team mostly concentrate. It is not useful to take this decision based on a snapshot of sensor readings. The positioning of the defense has to be determined only from time to time, e.g. every minute, on the basis of the average positions of the attacking robots during the immediate past.

The Dual Dynamics control architecture, developed by Herbert Jäger [3,4], arranges reactive behaviors in a hierarchy of control processes. Each layer of the system is partitioned into two modules: the activation dynamics that determines at every time step whether or not a behavior tries to influence actuators, and the target dynamics, that describes strength and direction of that influence. The different levels of the hierarchy correspond to different time scales. The high-level behaviors configure the low-level control loops via activation factors that set the current mode of the primitive behaviors. This can produce qualitatively different reactions if the agent receives the same stimulus again, but has changed of mode due to stimuli received in the meantime.

The remainder of the paper is organized as follows: The next section describes the mechanical and electrical design of our RoboCup F180 league robots. Then the vision and communication systems are presented. In Section 5 we explain the hierarchical control architecture that we use to generate behaviors for the game of soccer and illustrate it using examples from the RoboCup domain.

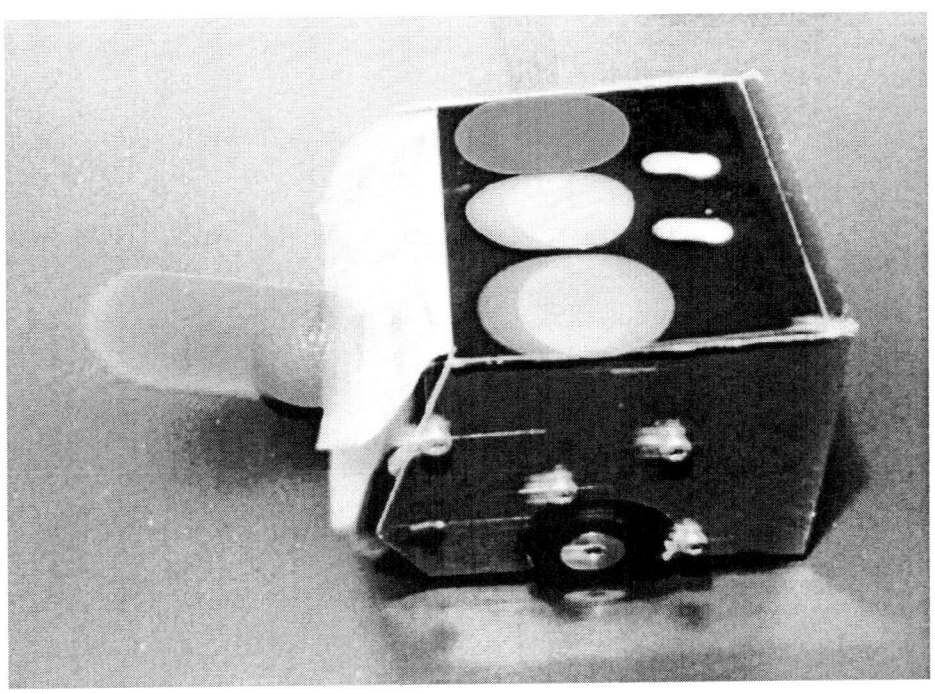

Fig. 1. A FU-Fighters robot kicking the ball. (Photo: Stefan Beetz)

2 Mechanical and Electrical Design

Our robots were designed in compliance with the new F180 size RoboCup regulations. We built four identical field players and a goal keeper. All robots have stable aluminum frames that protect the sensitive inner parts, as shown in Fig. 1.

They have a differential drive with two active wheels in the middle and are supported by one or two passive spheres that can rotate in any direction. Two Faulhaber DC-motors allow for a maximum speed of about 1 m/s. The motors have an integrated 19:1 gear and an impulse generator with 16 ticks per revolution.

One distinctive feature of our robots is a kicking device (Fig. 2) which consists of a rotating plate that can accumulate the kinetic energy produced by a small motor and release it to the ball on contact.

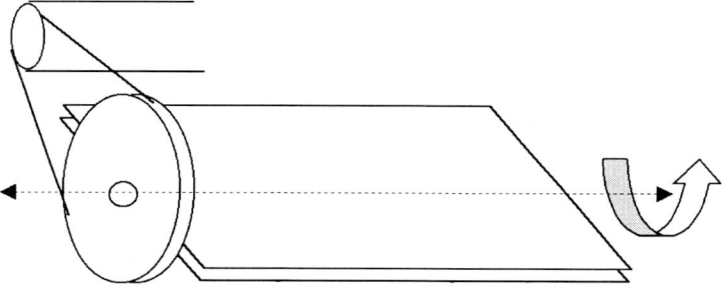

Fig. 2. Sketch of the kicking device.

We use C-Control units from Conrad electronics for local processing. They include a Motorola microcontroller HC05 running at 4 MHz with 8 KB EEPROM for program storage, two pulse-length modulated outputs for motor control, a RS-232 serial interface, a free running counter with timer functions, analog inputs, and digital I/O. The units are attached to a custom board containing a stabilized power supply, a dual-H-bridge motor driver L298, a beeper, and a radio transceiver SE200. The robots are powered by 8 + 4 Ni-MH rechargeable mignon batteries.

3 Video Input

The only physical sensor for our control software is an S-VHS camera that captures the field from above. The camera produces an analog video stream in NTSC format. Using a PCI-framegrabber, we feed images to a PC running MS-Windows. We capture RGB-images of size 640 × 480 at a rate of 30 fps and interpret them to extract the relevant information about the playing field. Since the ball, as well as the robots, are color-coded, we designed our vision software to find and track several colored objects. These objects are the orange ball and all the robots that have been marked with colored dots, in addition to the yellow or blue team ball.

To track the objects we predict their positions in the next frame and then inspect the video image first at a small window centered around the predicted position. We use an adaptive saturation threshold and intensity thresholds to separate the objects from the background. The window size is increased and larger portions of the image are investigated only if an object is not found.

The decision whether or not the object is present is made on the basis of a quality measure that takes into account the hue and size distances to the model and geometrical plausibility. When we find the desired objects, we adapt our model of the world using the measured parameters, such as position, color, and size.

4 Communication

The actions selected by the control module are transmitted to the robots via a wireless serial communication link with a speed of 9600 baud. We use radio transmitters operating on a single frequency that can be chosen between 433.0 MHz and 434.5 MHz in 100 KHz steps. The host sends commands in 8-byte packets that include address, control bits, motor speeds, and a checksum. A priority value can be used to transmit more packets to the most active players.

The microcontroller on the robots decodes the packets, checks their integrity, and sets the target values for the control of the motor speeds. No attempt is made to correct transmission errors, since the packets are sent redundantly. To be independent from the state of the battery charge, we implemented locally a closed loop control of the motor speeds. The microcontroller counts the impulses from the motors 122 times per second, computes the differences to the target values and adjusts the pulse length ratio for the motor drivers accordingly. We use a simple P-control to adapt the motor power.

5 Behavior

5.1 Architecture

Our control architecture is shown in Figure 3. It is based on the Dual Dynamics scheme developed by H. Jäger [3,4]. The robots are controlled in closed loops that use different time scales and that correspond to behaviors on different levels of the hierarchy.

We extend the Dual Dynamics concept by introducing a third element, namely the perceptual dynamics, as shown on the left side of the drawing. Here, either slow changing physical sensors, such as the charging state indicators of the batteries, are plugged-in at the higher levels, or the readings of fast changing sensors, like the ball position, are aggregated by dynamic processes into slower and longer lasting percepts. The boxes shown in the figure are divided into cells.

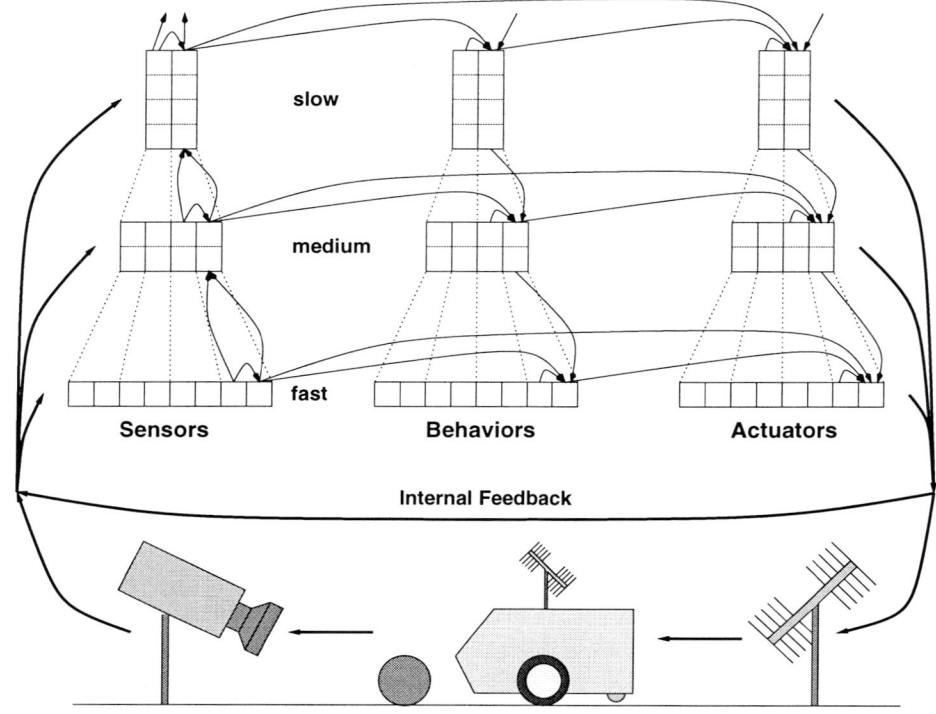

Fig. 3. Sketch of the control architecture.

Each cell represents a sensor value that is constant for a time step. The rows correspond to different sensors and the columns show the time advancing from left to right.

A set of behaviors is shown in the middle of each level. Each row contains an activation factor from the interval $[0, 1]$ that determines when the corresponding behavior is allowed to influence actuators.

The actuator values are shown on the right hand side. Some of these values are connected to physical actuators that modify the environment. The other actuators influence lower levels of the hierarchy or generate sensory percepts in the next time step via the internal feedback loop.

Since we use temporal subsampling, we can afford to implement an increasing number of sensors, behaviors, and actuators in the higher layers without an explosion of computational cost. This leads to rich interactions with the environment.

Each physical sensor or actuator can only be connected to one level of the hierarchy. One can use the typical speed of the change of sensor readings to decide where to connect a sensor. Similarly, the placement of actuators is determined by the time constant they need to produce a change in the environment. Behaviors are placed on the level that is low enough to ensure a timely response to stimuli, but that is high enough to provide the necessary aggregated perceptual information, and that contains actuators which are abstract enough to produce the desired reactions.

5.2 Computation of the Dynamics

The dynamic systems of the sensors, behaviors, and actuators can be specified and analyzed as a set of differential equations. Of course, the actual computations are done using difference equations. Here, the time runs in discrete steps of $\Delta t^0 = t_i^0 - t_{i-1}^0$ at the lowest level 0. At the higher levels the updates are done less frequently: $\Delta t^z = t_i^z - t_{i-1}^z = f \Delta t^{z-1}$, where useful choices of the subsampling factor c could be 2, 4, 8, ... In the figure, $c = 2$ was used.

A layer z is updated in time step t_i^z as follows:

$\mathbf{s}_i^z$ – *Sensor values:*
  The n_s^z sensor values $\mathbf{s}_i^z = (s_{i,0}^z, s_{i,1}^z, \ldots, s_{i,n_s^z-1}^z)$ depend on the readings of the n_r^z physical sensors $\mathbf{r}_i^z = (r_{i,0}^z, r_{i,1}^z, \ldots, r_{i,n_r^z-1}^z)$ that are connected to layer z, the previous sensor values $\mathbf{s}_{i-1}^z$, and the previous sensor values from the layer below $\mathbf{s}_{ci}^{z-1}, \mathbf{s}_{ci-1}^{z-1}, \mathbf{s}_{ci-2}^{z-1}, \ldots$.
  In order to avoid the storage of old values in the lower level, the sensor values can be updated from the layer below, e.g. as moving average.

α_i^z – *Activation factors:*
  The n_α^z activations $\alpha_i^z = (\alpha_{i,0}^z, \alpha_{i,1}^z, \ldots, \alpha_{i,n_\alpha^z-1}^z)$ of the behaviors depend on the sensor values $\mathbf{s}_i^z$, the previous activations α_{i-1}^z, and on the activations of behaviors in the level above $\alpha_{i/c}^{z+1}$. A higher behavior can use multiple layer-z-behaviors and each of them can be activated by many behaviors. For every behavior k on level $(z+1)$ that uses a behavior j from level z there is a term $\alpha_{i/c,k}^{z+1} T_{j,k}^z(\alpha_{i-1}^z, \mathbf{s}_i^z)$ that describes the desired change of the activation $\alpha_{i,j}^z$. Note that this term vanishes, if the upper level behavior is not active. To determine the new activations the changes from all T-terms are accumulated. A product term is used to deactivate a behavior, if no corresponding higher behavior is active.

$\mathbf{G}_i^z$ – *Target values:*
  Each behavior j can specify for each actuator k a target value $g_{i,j,k}^z = G_{j,k}^z(\mathbf{s}_i^z, \mathbf{a}_{i/c}^{z+1})$.

$\mathbf{a}_i^z$ – *Actuator values:*
  The more active a behavior j is, the more it can influence the actuator values $\mathbf{a}_i^z = (a_{i,0}^z, a_{i,1}^z, \ldots, a_{i,n_a^z-1}^z)$. The desired change for the actuator value $a_{i,k}^z$ is: $u_{i,j,k}^z = \tau_{i,j,k}^z \alpha_{i,j}^z (g_{i,j,k}^z - a_{i-1,k}^z)$. If several behaviors want to change the same actuator k, the desired updates are added:
  $a_{i,k}^z = a_{i-1,k}^z + u_{i,j_0,k}^z + u_{i,j_1,k}^z + u_{i,j_2,k}^z + \ldots$

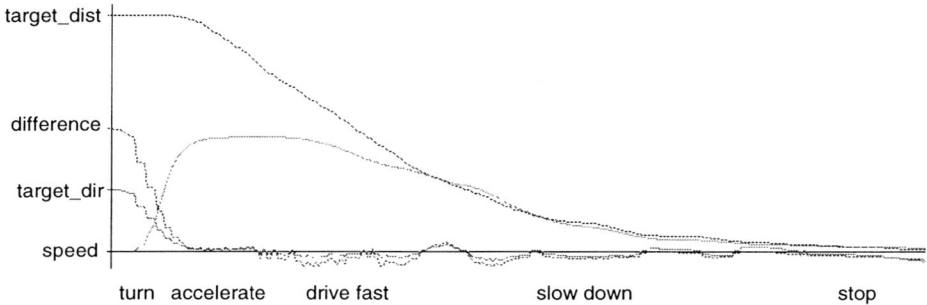

Fig. 4. Recording of two sensors (distance and direction of the target) and two actuators (average motor speed and difference between the two motors) during a simple taxis behavior. The robot first turns towards the target, then accelerates, drives fast, slows down, and finally it stops at the target position.

5.3 Bottom-Up Design

Behaviors are constructed in a bottom-up fashion: First, the processes that should react quickly to fast changing stimuli are designed. Their critical parameters, e.g. a mode parameter or a target position, are determined. When the fast primitive behaviors work reliably with constant parameters, the next level can be added to the system. For this higher level more complex behaviors can now be designed that influence the environment, either directly, by moving slow actuators, or indirectly, by changing the critical parameters of the control loops in the lower level.

After the addition of several layers, fairly complex behaviors can be designed that make decisions using abstract sensors based on a long history and that use powerful actuators to influence the environment.

In a soccer playing robot, basic skills, like movement to a position and ball handling, reside on lower levels, tactic behaviors are situated on intermediate layers, while the game strategy is determined at the topmost level of the hierarchy.

5.4 Examples

To realize a Braitenberg vehicle that moves towards a target, we need the direction and the distance to the target as input. The control loop for the two differential drive motors runs on the lowest level of the hierarchy. The two actuator values used determine the average speed of the motors and the speed differences between them. We choose the sign of the speed by looking at the target direction. If the target is in front of the robot, the speed is positive and the robot drives forward, if it is behind then the robot drives backwards. Steering depends on the difference of the target direction and the robot's main axis. If this difference is zero, the robot can drive straight. If it is large, it turns on the spot. Similarly, the speed of driving depends on the distance to the target. If the target is far away, the robot can drive fast. When it comes close to the target it slows down and stops at the target position. Figure 4 shows an example where the robot first turns around until the desired angle has been reached, accelerates, moves with constant speed to a target and finally decelerates. Smooth transitions between the extreme behaviors are produced using sigmoidal functions.

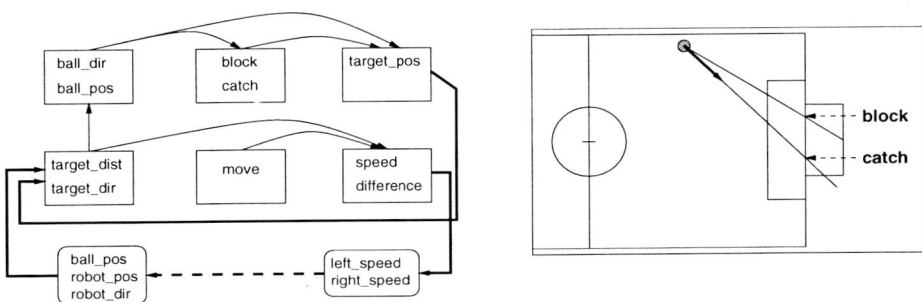

Fig. 5. Sketch of goal keeper behavior. Based on the position, speed, and the direction of the ball it decides to either block the ball or to catch it.

This primitive taxis behavior can be used as a building block for the goal keeper. A simple goal keeper could be designed with two modes: block and catch, as shown in Figure 5. In the block mode it sets the target position to the intersection of the goal line and a line that starts behind the goal and goes through the ball. In the catch mode, it sets the target position to the intersection of the predicted ball trajectory and the goal line. The goal keeper is always in the block mode, except when the ball moves rapidly towards the goal.

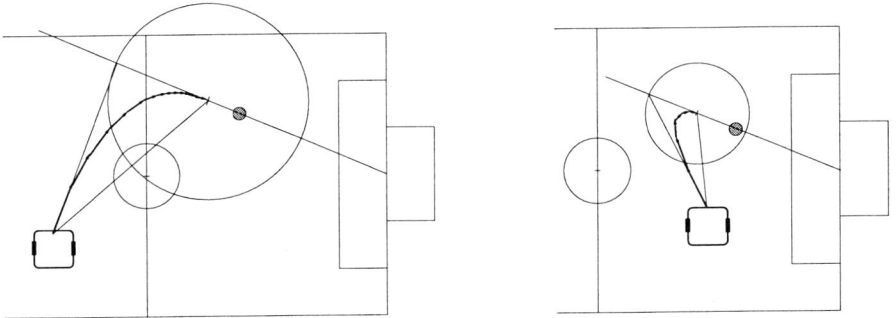

Fig. 6. Trajectories generated in the run mode of the field player. It smoothly approaches a point behind the ball that lies on the line from the ball target through the ball.

The control hierarchy of the field player that wants to move the ball to a target, e.g. a teammate or the goal, could contain the alternating modes run and push. In the run mode the robot moves to a target point behind the ball with respect to the ball target. When it reaches this location, the push mode becomes active. Then the robot tries to drive through the ball towards the target and pushes it into the desired direction. When it looses the ball, the activation condition for pushing is no longer valid and the run mode becomes active again. Figure 6 illustrates the trajectory of the field player generated in the run mode. A line is drawn through the ball target and the ball. The target point is found on this line at a fixed distance behind the ball. The distance from the robot to this target point is divided by two. The robot is heading always towards the intersection of the dividing circle and the line. This produces a trajectory that smoothly approaches the line. When the robot arrives at the target point, it is heading towards the ball target.

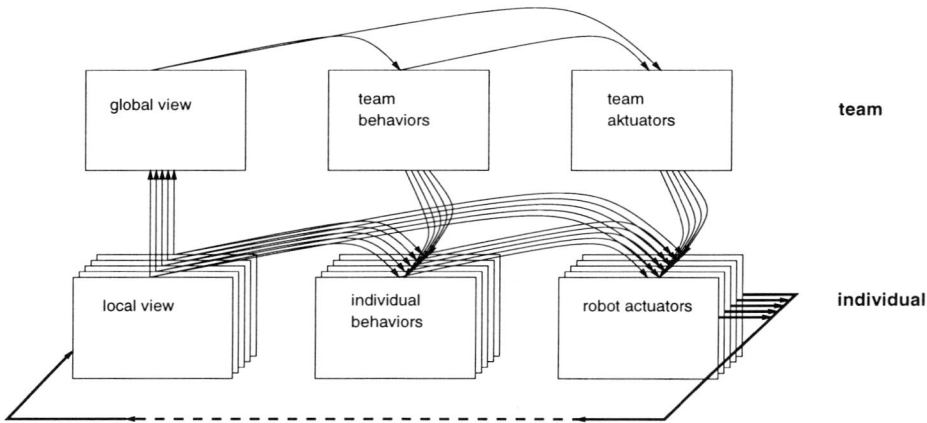

Fig. 7. Sketch of the relation between the team and the individual robots.

Each of our robots is controlled autonomously by the lower levels of the hierarchy using a local view of the world, as indicated in Figure 7. We present, for instance, the angle and the distance to the ball and the nearest obstacle to each agent. In the upper layers of the control system the focus changes. Now we regard the team as the individual. It has a slow changing global view to the playground and coordinates the robots as its extremities to reach strategic goals. For example, it could position its defense on the side of the field where the offensive players of the opponent team mostly attack and place its offensive players where the defense of the other team is weak.

We implemented some of these complex behaviors for the RoboCup'99 competition. They include for instance dynamic homing, where the home positions of our defensive players are adjusted such that they block the offensive robots from the other team, and the home positions of our offensive players are adjusted, such that they have a free way to the goal. Another example is ball interception, where we predict the ball trajectory and the time it takes for the robot to reach successive points on this trajectory. We direct the robot towards the point where it can first reach such a point earlier than the ball. This results in an anticipative behavior. We also detect when a robot wants to move, but does not move for a longer time, e.g. because it is blocked by other robots or got stuck in a corner. Then we reverse for a short time the motor speeds, in order to unstuck the robot.

6 Summary

We designed robust and fast robots with a kicking device, reliable radio communication, and high speed vision. To generate actions, we implemented a reactive control architecture with interacting behaviors on different time scales. These control loops are designed in a bottom-up fashion. Lower level behaviors are configured by an increasing number of higher level behaviors that can use a longer history to determine their actions.

This framework could be used in the future to implement mechanisms, like adaptation and learning using Neural Networks [6]. We successfully participated in the RoboCup'99 F180 league competition, finishing second, next to Big Red from Cornell University.

We thank the companies Conrad ELECTRONICS GmbH, Dr. Fritz Faulhaber GmbH & Co KG, Siemens ElectroCom Postautomation GmbH, and Lufthansa Systems Berlin GmbH for their support that made this research possible.

References

1. R.A. Brooks. Intelligence without reason. A.I. Memo 1293, MIT Artificial Intelligence Lab, 1991.
2. T. Christaller. Cognitive robotics: A new approach to artificial intelligence. *Artificial Life and Robotics*, (3), 1999.
3. H. Jäger. The dual dynamics design scheme for behavior-based robots: A tutorial. Arbeitspapier 966, GMD, 1996.
4. H. Jäger and T. Christaller. Dual dynamics: Designing behavior systems for autonomous robots. In S. Fujimura and M. Sugisaka, editors, *Proceedings International Symposium on Artificial Life and Robotics (AROB '97) – Beppu, Japan*, pages 76–79, 1997.
5. R. Pfeifer and C. Scheier. *Understanding Intelligence*. MIT press, Cambridge, 1998.
6. R. Rojas. *Neural Networks*. Springer, New York, 1996.
7. E. Schlottmann, D. Spenneberg, M. Pauer, T. Christaller, and K. Dautenhahn. A modular design approach towards behaviour oriented robotics. Arbeitspapier 1088, GMD, 1997.
8. L. Steels. The pdl reference manual. AI Lab Memo 92-5, VUB Brussels, 1992.
9. L. Steels. Building agents with autonomous behavior systems. In L. Steels and R.A. Brooks, editors, *The 'Artificial Life' route to 'Artificial Intelligence': Building situated embodied agents*. Lawrence Erlbaum Associates, New Haven, 1994.

Heterogeneity and On-Board Control in the Small Robots League

Andreas Birk and Holger Kenn

Vrije Universiteit Brussel, Artificial Intelligence Laboratory, Belgium
c/o birk@ieee.org, http://arti.vub.ac.be/~cyrano

Abstract. Versatile physical and behavioral features as well as their exploitation through computation-power onboard the robot-players are feasible and necessary goals for the RoboCup small robots league. We substantiate this claim in this paper by classifying different approaches and by discussing their potentials and limitations for research on AI and robotics. Furthermore, we present the most recent results of our approach to these goals in form of the so-called CubeSystem, a kind of construction-kit for robots and other autonomous systems. It is based on a very compact embedded computer, the so-called RoboCube, a set of sensor- and motor-modules, and software support in form of a special operating system and highlevel languages.

1 Introduction

The Small Robots League of RoboCup [KAK+97, KTS+97] allows global sensing, especially bird's view vision from an overhead camera, and restricts the size of the physical players to a rather extreme minimum. These two, most significant features of the small robots league bear an immense potential, but as well some major pitfalls for future research within the RoboCup framework.

First of all, it is tempting to exploit the set-up with an overhead camera for the mere sake of trying to win, reducing the robot-players to RF-controlled toy-cars within a minimal, but very fast vision-based closed-loop. The severe size limitations of the players in addition encourage the use of such "string-puppets" with off-board sensing and control instead of real robots. The Mirosot competition gives an example for this type of approach [Mir]. This framework would lead to dedicated solutions, which are very efficient and competitive, but only of very limited scientific interest from both a basic research as well as from an application-oriented viewpoint. If the teams in the small robots league would follow that road, this league could degenerate to a completely competition-oriented race of scientifically meaningless, specialized engineering efforts.

Though the two major properties of the small robots league, global sensing and severe size restrictions, discourage the important investigation of on-board control, they also have positive effects. First of all, the global sensing eases quite some perception problems, allowing to focus on other important scientific issues, especially team behavior. An indication for this hypothesis is the apparent difference in team-skills between the small robots league and the midsize league, where global sensing is banned.

The size restrictions as a second point also have a beneficial aspect for the investigation of team-behavior. The play-field of a ping-pong-table can easily be allocated in a standard academic environment, facilitating games throughout the year. It is in contrast difficult to embed a regular field of the midsize league into an academic environment, thus the possibilities for continuous research on the complete team are here limited. The severe size restriction of the small robots league has another advantage. These robots can be much cheaper as costs of electro-mechanical parts significantly increase with size. Therefore, it is more feasible to build even two teams and to play real games throughout the year, plus to include the team(s) in educational activities.

The rest of this article is structured as follows. In section two, different team-approaches are classified and possible implications are discussed. Section three presents the hardware aspects of the CubeSystem, i.e., the RoboCube V2.0 and the mechanical components used in our approach to the RoboCup small robots league. In the fourth section, the software aspects of the CubeSystem are discussed. First, its operating system CubeOS and highlevel language support are shortly presented. Then, it is shown with the example of path-planning that the RoboCube is indeed capable of quite powerful computations within realtime constraints. Last but not least the implications for team-coordination when using very heterogeneous systems are discussed. Section five concludes the paper.

2 Classification of Team-Approaches

For a more detailed discussion of the role of heterogeneity and on-board control in the small robots league, it is useful to have a classification of different types of teams and players.

Minoru Asada for example proposed in the RoboCup mailing-list to use a classification of approaches based on the type of vision (local, global or combined) and the number of CPUs (one or multi). He also mentioned that in the case of multiple CPUs a difference between systems with and without explicit communication between players can be made. Though this scheme is useful, it is still a first, quite rough classification. Therefore, we propose here to make finer distinctions, based on a set of crucial components for the players.

In general, a RoboCup team consists of a (possibly empty) set of host-computers and off-board sensors, and a non-empty set of players, each of which consist of a combination of the following components:

1. minimal components
   (a) mobile platform
   (b) energy supply
   (c) communication module
2. optional components
   (a) computation power
   (b) shooting-mechanism and other effectors
   (c) basic sensors
   (d) vision hardware

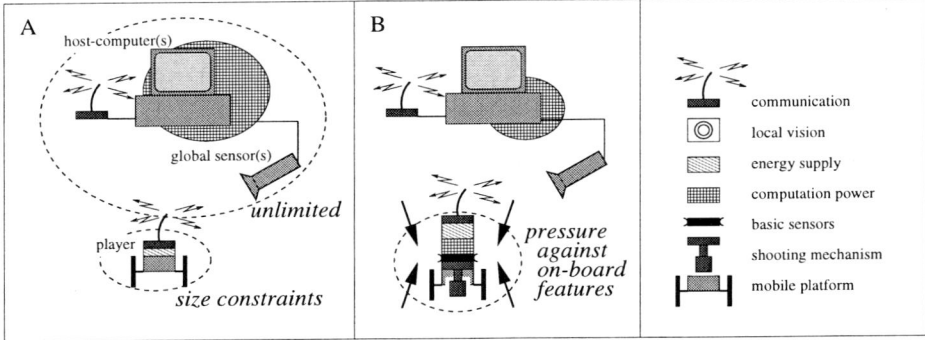

Fig. 1. There are several basic components which can be, except the minimal ones, freely combined to form a player. Situation A shows the most simple type of player, a radio-controlled toy-car, which can hardly be called a robot. Situation B shows a much more elaborated player. Unfortunately, the size-constraints of the small robots league put a strong negative pressure against the important implementation of on-board features for the players.

Note, that the most simple type of player, consisting of only minimal components, is hardly a robot. It is more like a "string-puppet" in form of a radio-controlled toy-car without even any on-board sensors or computation power (though it could well be possible that this type of device has an on-board micro-controller for handling the communication protocol and the pulse-width-modulation of the drive motors). The actual control of this type of players completely takes place on the off-board host(s).

Based on this minimal type of player, the optional components can be freely combined and added. In doing so, there is a trade-off between

– on-board sensor/motor components,
– on-board computation power, and
– communication bandwidth.

A player can for example be built without any on-board computation power at the cost of communication bandwidth by transmitting all sensor/motor-data to the host and back. So, increasing on-board computation power facilitates the use of a smaller communication bandwidth and vice versa. Increasing sensor/motor channels on the other hand increases the need of on-board computation power and/or communication bandwidth.

On-board features are important for research in robotics as well as AI and related disciplines for several reasons. Mainly, they allow research on important aspects which are otherwise impossible to investigate, especially in the field of sensor/motor capabilities. For effector-systems for example, it is quite obvious that they have to be on-board to be within the rules of soccer-playing. Here, the possibilities of systems with many degrees of freedom, as for example demonstrated in the SONY pet dog [FK97], should not only be encouraged in special leagues as e.g. in the one for legged players, but also within the small robots league. In general, a further splitting of the RoboCup activities into too many leagues seems not to be beneficial and it also seems not to be practical. Too many classifications which would justify just another new league would be possible. In addition, the direct competition and comparison of different approaches together with the scientific dialogue are one of the main features of RoboCup.

In the case of sensors and perception, the situation is similar to the one of effector-systems, i.e., certain important types of research can only be done with on-board devices. This holds especially for local vision. It might be useful to clarify here the often confused notions of local/global and on-/off-board. The terms on- and off-board are easy to distinguish, general properties. They refer to a piece of hardware or software, which is physically or logically present on the player (on-board) or not (off-board). The notions of local and global in contrast only refer to sensors, i.e., particular types of hardware, or to perception, i.e., particular types of software dealing with sensor-data. Global sensors and perception tell a player absolute information about the world, typically information about its position and maybe the positions of other objects on the playfield. Local sensors and perception in contrast tell a player information about the world, which is relative to its own position in the world. Unlike in the case of on- and off-board, the distinction between local and global is fuzzy and often debatable. Nevertheless, it is quite clear that the important issue of local vision can only be investigated if the related feature is present on-board of the player.

Hand in hand with an increased use of sensor and motor systems on a player, the amount of on-board computation power must increase. Otherwise, the scarce resource of communication bandwidth will be used up very quickly. Note, that there are many systems using RF-communication at the same time during a RoboCup tournament. Especially in the small robots league, were only few and very limited off-the-shelf products suited for communication exist, transmission of large amount of data is impossible. It is for example quite infeasible to transmit high-resolution local camera images from every player to a host for processing.

3 Towards a Robot Construction-Kit

3.1 The Motivation

Existing commercial construction-kits with some computational power like Lego Mindstorms^{TM} [Min] or Fischertechnik Computing^{TM} [Fis] are still much too limited to be used for serious robotics education or even research. Therefore, we decided to develop our own so-to-say robot construction-kit.

3.2 RoboCube V2.0

For RoboCup'98, the VUB AI-lab team focused on the development of a suited hardware architecture, which allows to implement a wide range of different robots. The basic features of this so-called RoboCube-system are described in [BKW98]. For RoboCup'99, the system was further improved and extended. A more detailed description is given in [BKW00].

The most recent version of the RoboCube boots out of a 1 MByte Flash-EPROM which holds a basic input/output operating system (BIOS) and offers space for a small file system. A huge part of the BIOS is dedicated to the efficient handling of different actuators and sensors. In the basic configuration the main memory consists of a 1 MByte low power SRAM, which can be extended by additional 12 MByte.

In its basic version, one I/O subsystem board of the RoboCube features

- 24 analog input
- 6 analog output
- 16 binary Input/Output (binI/O)
- 4 timer channels (TPC)
- 4 DC-motor controller with quadrature-encoding

The number of ports can simply be doubled by stacking a second I/O subsystem board on top of the first one. All sensor-motor-interfaces come with proper software support allowing an easy high-level usage.

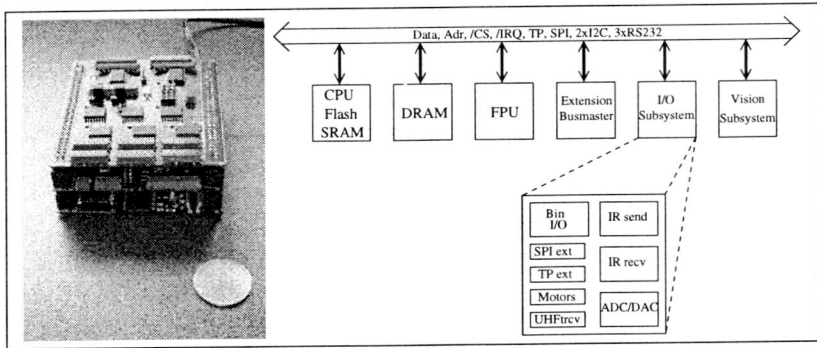

Fig. 2. A picture of the RoboCube (left) and the layout of its internal bus structure (right).

The RoboCube-system is constantly further improved, on the software as well as on the hardware side. At the moment for example, several options for inexpensive high-resolution color-vision are investigated.

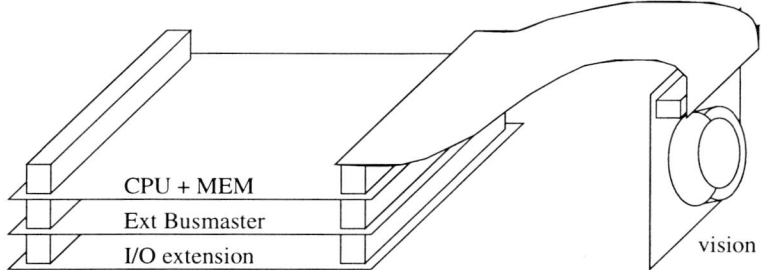

Fig. 3. Physical layout of the complete RoboCube

3.3 Mechanical Components for RoboCup

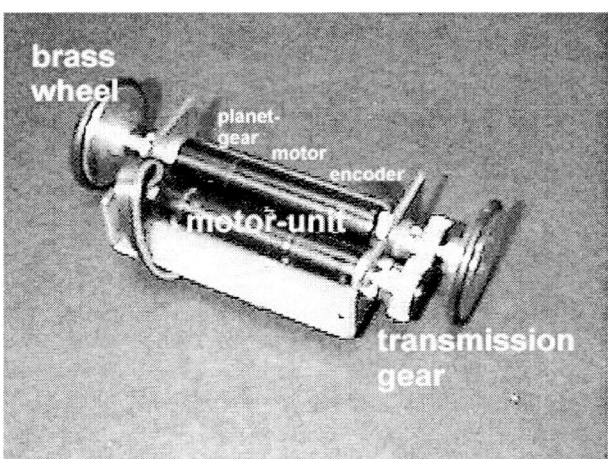

Fig. 4. The drive unit as a mechanical building-block, which can be mounted on differently shaped bottom-plates, forming the mechanical basis for diverse body-forms. Different ratios for the planetary gears in the motor-units are available, such that several trade-offs for speed versus torque are possible.

In some of our education and research activities, the RoboCube-system is combined with Lego^{TM} or Fischertechnik^{TM} components on the mechanical side. For RoboCup competitions, we developed a solid but still flexible solution based on metal components.

Keeping the basic philosophy of construction-kits, a "universal" building block is used for the drive (figure 4) of the robots. The drive can be easily

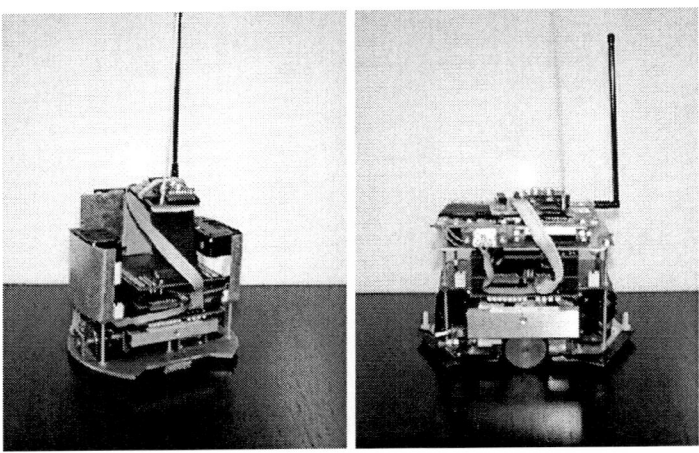

Fig. 5. A forward- (left) and a defender-type (right) robot. The mechanical set-up of the robot-players is based on a piled-stack approach such that different components, such as shooting-mechanisms and the RoboCube, can easily be added.

mounted onto differently shaped metal bottom-plates, forming the basis for different body-forms like the ones shown in figure 5. The motor-units in the drive exist with different ratios for the planetary gears, such that several trade-offs for speed versus torque are possible.

Other components, like e.g. shooting-mechanisms and the RoboCube, are added to the bottom-plate in a piled-stack-approach, i.e., four threaded rods allow to attach several layers of supporting plates.

4 Powerful On-Board Control

4.1 Operating System and Programming Languages

For the RoboCube, an embedded operating system has been developed, *CubeOS*. It provides the usual features like threads, semaphores, realtime clock, communication and I/O drivers in a small core of about 30 Kbytes. Additionally, it provides functionality that set it apart from other OS kernels and make it especially useful for robotics and autonomous systems in general:

- Drivers that handle access to the various sensor and actuator devices of the RoboCube
- Support for hardware-assisted realtime processing through the MC68332's onboard TPU
- A low-latency communication protocol engine for radio communication
- hardware-independent data encoding as defined in the *External Data Representation Standard* [SM]

CubeOS is written in C and makes use of the Free Software Foundation's Gnu C-Cross-Compiler. The host system for development is a Unix or Linux Workstation, the code is downloaded into the target via a wireless serial communication link.

On top of the CubeOS API, a simple software framework has been implemented to provide easy access within normal C-programs. In additional, there is highlevel language support suited even for novice programmers. This framework, named *NPDL* (for New Process Description Language) provides several simple constructs to create control programs for robots. Within education projects, NPDL has already been mastered by students studying economics, philosophy and architecture.

4.2 Using the RoboCube for Highlevel Control

Though the RoboCube has quite some computation power for its size, its capabilities are nevertheless far from those of desktop machines. So, it is not obvious that interesting behaviors in addition to controlling the drive-motors and shooting can actually be implemented on the RoboCube, i.e., on board of the robots. Therefore, we demonstrate in this section that for example path-planning with obstacle avoidance is feasible.

24	23	22	21	20	19	18	17	16	15	14	15	16	17	18
23	22	21	20	19	18	17	16	15	14	13	14	15	16	17
22	21	20	19	18	17	16	15	14	13	12	13	14	15	16
21	20	19	18	17	16	15	14	13	12	11	12	13	14	15
20	19	18	17	16	15	14	13	12	11	10	11	12	13	14
19	18	17	16	15	14	13	12	11	10	9	10	11	12	13
18	17	16	15	14	13	12	11	10	9	8	9	10	11	12
19	18	17	16	[X]	[X]	[X]	[X]	[X]	8	7	8	9	10	11
18	17	16	17	[X]	[X]	[X]	[X]	[X]	7	6	7	8	9	10
17	16	15	16	[X]	[X]	[X]	[X]	[X]	6	5	6	7	8	9
16	15	14	15	[X]	[X]	[X]	[X]	[X]	5	4	5	6	7	8
15	14	13	14	[X]	[X]	[X]	[X]	[X]	4	3	4	5	6	7
14	13	12	[X]	[X]	[X]	6	5	4	3	2	3	4	[X]	[X]
13	12	11	[X]	[X]	[X]	5	4	3	2	1	2	3	[X]	[X]
12	11	10	[X]	[X]	[X]	4	3	2	1	0	1	2	[X]	[X]
11	10	9	8	7	6	5	4	3	2	1	2	3	4	5
12	11	10	9	8	7	6	5	4	3	2	3	4	5	6
13	12	11	10	9	8	7	6	5	4	3	4	5	6	7
14	13	12	11	10	9	8	7	6	5	4	5	6	7	8
15	14	13	12	11	10	9	8	7	6	5	6	7	8	9

Fig. 6. A potential field for motion-control based on Manhattan distances. Each cell in the grid shows the shortest distance to a destination (marked with Zero) while avoiding obstacles, which are marked with '[X]'.

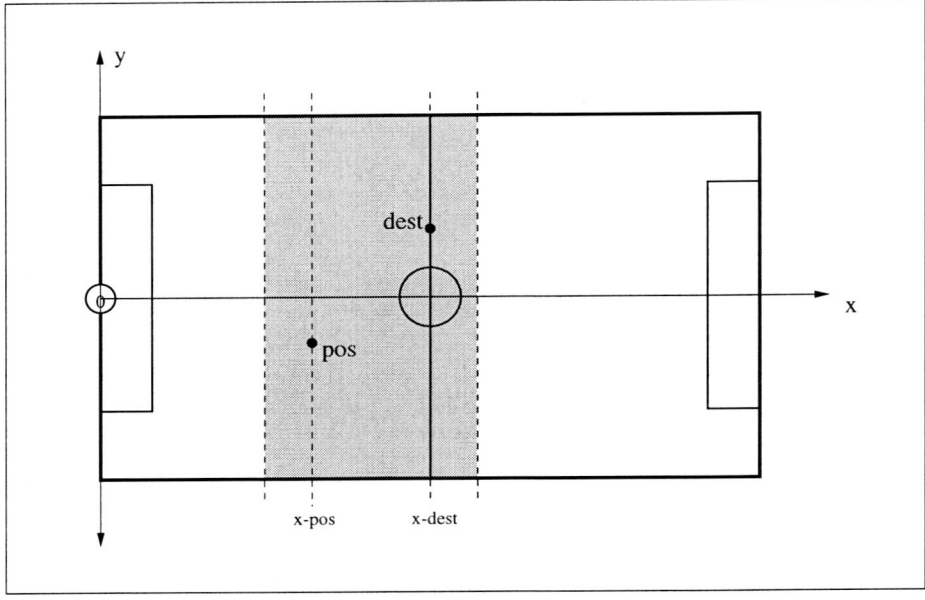

Fig. 7. The potential field (grey area) is not computed for the whole soccer-field. Instead, it is limited in the x-direction to save computation time.

Path planning is with most common approaches rather computationally expensive. Therefore, we developed a fast potential field algorithm based on Manhattan-distances. Please note that this algorithm is presented here only to demonstrate the computing capabilities of the RoboCube. A detailed description and discussion of the algorithm is given in [Bir99].

Given a destination and a set of arbitrary obstacles, the algorithm computes for each cell of a grid the shortest distance to the destination while avoiding the obstacles (figure 6). Thus, the cells can be used as gradients to guide the robot. The algorithm is very fast, namely linear in the number of cells. The algorithm is inspired by [Bir96], where shortest Manhattan distances between identical pixels in two pictures are used to estimate the similarity of images.

The basic principle of the algorithm is region-growing based on a FIFO queue. At the start, the grid-value of the destination is set to Zero and it is added to the queue. While the queue is not empty, a position is dequeued and its four neighbors are handled, i.e., if their grid-value is not known yet, it is updated to the current distance plus One, and they are added to the queue.

For the experiments done so far, the resolution of the motion-grid is set to 1cm. As illustrated in figure 7, the potential-field is not computed for the whole soccer-field to save computation time. Given a robot position *pos* and a

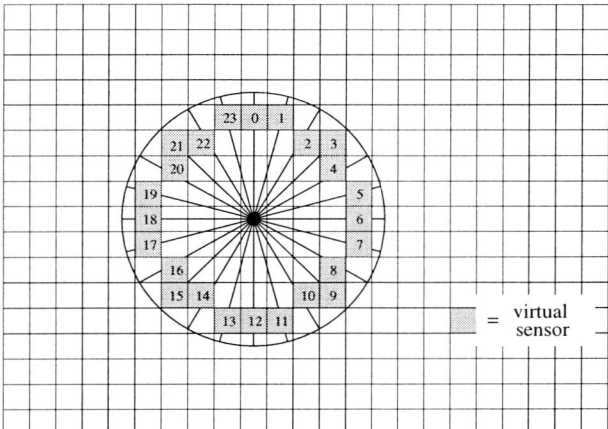

Fig. 8. Twenty-four so-called virtual sensors read the potential values around the robot position on the motion grid. The sensor values can be used to compute a gradient for the shortest path to the destination, which can be easily used in a reactive motion-control.

destination *dest*, the field is restricted in the x-direction to the difference of *pos* and *dest* plus two safety-margins which allow to move around obstacles to reach the destination.

The motion-grid is used as follows for our soccer-robots. The global vision detects all players, including opponents and the ball, and broadcasts this information to the robots. Each robot computes a destination depending on its strategies, which are also running on-board. Then, each robot computes its motion-grid. In doing so, all other robots are placed on the grid as obstacles.

Robots have so-called virtual sensors to sample a motion-grid as illustrated in figure 8. The sensor values are used to calculate a gradient for a shortest path to the destination, which is ideal for a reactive motion control of the robot. In doing so, dead-reckoning keeps track of the robot's position on the motion-grid.

Of course, the reactive control-loop can only be used for a limited amount of time for two main reasons. First, obstacles move, so the motion-grid has to be updated. Second, dead-reckoning suffers from cumulative errors. Therefore, this loop is aborted as soon as new vision information reaches the robot, which happens several times per second, and a new reactive controller based on a new motion-grid is started.

Figure 9 shows performs-results of the path-planning algorithm running on a RoboCube as part of the control-program of the robot-players. The different tasks of the control-program proceed in cycles. The execution time refers to a single execution of each task on its own (including the overhead from the operating system). The frequency refers to the frequency with which each tasks is executed as part of the player-control, i.e., together with all other tasks.

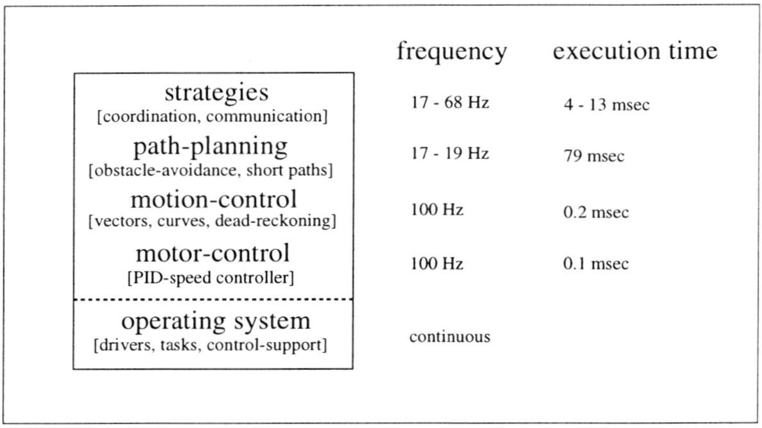

	frequency	execution time
strategies [coordination, communication]	17 - 68 Hz	4 - 13 msec
path-planning [obstacle-avoidance, short paths]	17 - 19 Hz	79 msec
motion-control [vectors, curves, dead-reckoning]	100 Hz	0.2 msec
motor-control [PID-speed controller]	100 Hz	0.1 msec
operating system [drivers, tasks, control-support]	continuous	

Fig. 9. The path-planning is part of a four-level software architecture which controls the robots players. It runs, together with the CubeOS operating system, completely on board of the RoboCube.

The control-program consists of four levels which run together with the CubeOS completely on-board of the RoboCube. The two lowest levels of motor- and motion-control run at a fixed frequency of 100 Hz. Single iterations of them are extremely fast as the TPU of the MC68332 can take over substantial parts of the processing. The strategy and path-planning level run in an "as fast as possible"-mode, i.e., they proceed in event-driven cycles with varying frequencies.

The execution of the pure strategy-code, i.e., the action-selection itself, takes up only a few milliseconds. Its frequency is mainly determined by whether the robot is surrounded by obstacles or not, i.e., whether path-planning is necessary or not. The computation of the motion-grid takes most of the 79 msec needed for path-planning. As two grids are used, one still determines the motion of the robot while the next one is computed, the cycle-frequency is at least 17 Hz. So, in a worst case scenario where the player is constantly surrounded by obstacles, the action-selection cycle can still run at 17 Hz.

4.3 Heterogeneity and Team Coordination with On-Board Control

Heterogeneity is an important feature for soccer with human players as much as with robot players. It is the main basis for adaptability of a team, either to different opponent teams within a tournament, or to the general progress of a particular game, or to very momentary situations. Heterogeneity within soccer can range from high-level roles of players in a team like forward or defender, down to different body features covering a wide-range of physical trade-offs like e.g. speed versus torque.

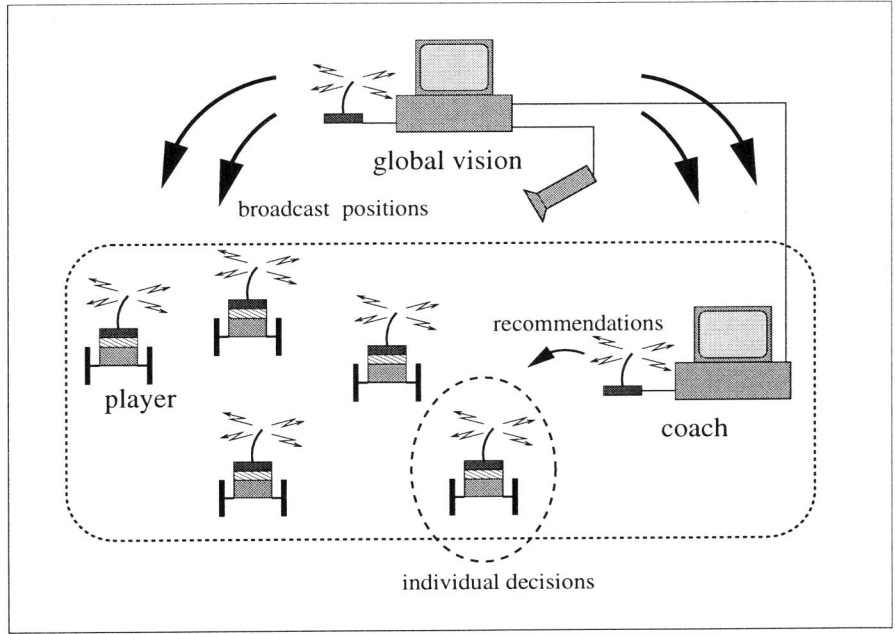

Fig. 10. A "pares inter pares" coach, residing on the same level as the players in the coordination hierarchy. This coach can be used for facilitating the coordination of heterogeneous team approaches while keeping as much as possible on-board control. The basic idea is that most of the time, the robot players decide completely on their own what to do based on their on-board control-program. Only occasionally the coach interferes as he has additional information about the capabilities of the players.

Straightforward approaches to team coordination with the expressive power of finite state automata are doomed to fail under such wide ranges of heterogeneity due to the combinatorial explosion of states. Therefore, we investigate coordination schemes based on operational semantics, which allow an extremely compact and modular way of specifying team behaviors. One step in this direction is the *Protocol Operational Semantics (POS)*, an interaction protocol based on abstract data-types and pattern matching capabilities. So far, it has only been tested in simulations, but the results are very promising. A detailed description can be found in [OBK99].

Here, we focus on the question how this approach can be integrated with a substantial exploitation of on-board control. The problem is that the expressive power of operational semantics is bought at the price of computational power. POS for example is implemented in Pizza [OW97], a super-set of Java.

A solution for this problem is a kind of additional player in form of a "pares inter pares" coach. This coach resides not above the other players in a coordination hierarchy, but resides on the same level (figure 10). The position information from the global vision is broadcasted to all players and the coach. Most of the time, all robot players individually decide what to do, based on their on-board computations. Only on rare occasions, the coach interferes as he has more background information available than the players.

To illustrate this idea, let us assume there are two heterogeneous robots of type A and A', with rather limited differences and which can be substituted against each other in the team. Both can simply run the same on-board control program, deciding most of the time the actions of the player. Only in situations when the difference plays a role, the coach interferes and provides additional information, recommending alternative actions to the player.

5 Conclusion

We claim that for serious AI and robotics research, it is necessary to work with "real" systems, i.e., heterogeneous devices with on-board control. As our contribution towards a suited infrastructure for this type of research, we develop the CubeSystem, a kind of advanced construction-kit for mobile robots and other autonomous systems. The CubeSystem consists of a special embedded hardware, the RoboCube, a set of sensors and actuators, and software support in form of a special operating system, the CubeOS, and highlevel languages.

The "string-puppet" approach of simple radio-controlled toy-cars also has its validation. It can for example serve as a rather easy and inexpensive way to enter RoboCup, it can be useful for educational purposes; shortly, it can be good for a start and to get acquainted with the basic issues of RoboCup.

But in the long run, we hope that participants in the small robots league of RoboCup cooperate to improve the options of on-board features. Only through a joint effort, it will be possible to overcome the pitfalls and to mutually benefit from the positive potential of the limited size requirements in this league.

Acknowledgments

The VUB AI-Lab team thanks Sanders Birnie BV as supplier and Maxon Motors as manufacturer for sponsoring our motor-units. Andreas Birk is a research fellow of the Flemish Institute for Applied Science (IWT); research on RoboCup is partially financed within this framework (OZM980252).

References

[Bir96]   Andreas Birk. Learning geometric concepts with an evolutionary algorithm. In *Proc. of The Fifth Annual Conference on Evolutionary Programming*. The MIT Press, Cambridge, 1996.

[Bir99] Andreas Birk. A fast pathplanning algorithm for mobile robots. Technical report, Vrije Universiteit Brussel, AI-Laboratory, 1999.
[BKW98] Andreas Birk, Holger Kenn, and Thomas Walle. Robocube: an "universal" "special-purpose" hardware for the robocup small robots league. In *4th International Symposium on Distributed Autonomous Robotic Systems*. Springer, 1998.
[BKW00] Andreas Birk, Holger Kenn, and Thomas Walle. On-board control in the robocup small robots league. *Advanced Robotics Journal*, 2000.
[Fis] The fischertechnik^{TM} website. http://www.fischertechnik.de/.
[FK97] Masahiro Fujita and Koji Kageyama. An open architecture for robot entertainment. In *Proceedings of Autonomous Agents 97*. ACM Press, 1997.
[KAK+97] Hiroaki Kitano, Minoru Asada, Yasuo Kuniyoshi, Itsuki Noda, and Eiichi Osawa. Robocup: The robot world cup initiative. In *Proc. of The First International Conference on Autonomous Agents (Agents-97)*. The ACM Press, 1997.
[KTS+97] Hiroaki Kitano, Milind Tambe, Peter Stone, Manuela Veloso, Silvia Coradeschi, Eiichi Osawa, Hitoshi Matsubara, Itsuki Noda, and Minoru Asada. The robocup synthetic agent challenge 97. In *Proceedings of IJCAI-97*, 1997.
[Min] The lego mindstorms^{TM} website. http://www.legomindstorms.com/.
[Mir] The micro-robot world cup soccer tournament (mirosot). http://www.mirosit.org.
[OBK99] Pierre-Yves Oudeyer, Andreas Birk, and Jean-Luc Koning. Interaction protocols with operational semantics and the coordination of heterogeneous soccer-robots. Technical report, Vrije Universiteit Brussel, AI-Laboratory, 1999.
[OW97] Martin Odersky and Philip Wadler. Pizza into Java: Translating theory into practice. In *Proc. 24th ACM Symposium on Principles of Programming Languages*, 1997.
[SM] Inc. Sun Microsystems. Xdr: External data representation standard. Request for Comments.

The Body, the Mind or the Eye, First?*

Andrea Bonarini

AI and Robotics Project, Dipartimento di Elettronica e Informazione,
Politecnico di Milano, Piazza Leonardo da Vinci, 32, Milano, Italy
bonarini@elet.polimi.it

Abstract. We present an approach to shape robots on their sensorial ability. We argue that the interface with the external world may strongly condition the design of a robot, from the mechanical aspects to reasoning and learning. We show the implementation of this philosophy in the RoboCup middle-size player Rullit, shaped on its omnidirectional vision sensor.

1 Introduction

We argue that the way a robot perceives the environment should strongly affect the design of each component of the robot. As it happens for animals, and human beings, the modality of interaction with the external world is strongly related to the survival behaviors, neural structures, actuators and reasoning. For instance, all the most evolved predators in natural life have eyes pointing forward, since this makes easier to follow a prey; all the preys have lateral eyes, since this is effective to be aware of the presence of predators. In some special cases (e.g., the chameleon) predators have highly movable eyes that enable cost-effective chasing strategies. We share the opinion that in a systemic perspective all the components of an agent are inter-related. However, in designing an artificial autonomous agent we should start from somewhere; here, we propose to start by considering the component that provides input, since this is the most critical to achieve the desired behavior. In the F-2000 Robocup [3] [14] environment, for instance, a black and white camera or a sonar belt, could hardly be enough to play effectively, whereas they may provide enough information to achieve many other tasks.

We have followed this approach of "shaping an agent on its sensor"(SAS) in many projects, designing both real [5] [1], and simulated [4] [8] robots by starting from the definition of the sensors the agent can exploit to operate in

---

* This research has been partially supported by the MURST Project CERTAMEN. The camera has been kindly provided by Sony Italia. I have to thank G. Borghi, who introduced me to the world of omnidirectional vision, R. Cassinis, who first implemented an Italian Robocup player around an omnidirectional sensor (the ART goalie, Saracinescu), and all the members of the Politecnico di Milano AI&R Robocup Team for their effort in the implementation of Rullit: P. Aliverti, M. Bellini, G. Contini, N. Ghironi, P. Lucioni, C. Mambretti, A. Marangon, P. Meriggi.

its environment. We have implemented Rullit, our F-2000 Robocup player (see figure 1), following this philosophy, after the first experience in Robocup98 [2]. In Rullit, the importance of a sensor adequate to the soccer playing task in the Robocup environment becomes evident, and we take it as a running example to discuss the SAS approach. In this paper, we first motivate the need for accurate

Fig. 1. Rullit, our Robocup F-2000 player.

design of the sensorial apparatus able to extract the needed information from the environment; we discuss the general issues exemplified in the Robocup environment. Then, we will describe our sensor. Finally, we discuss how a sensor may condition the design of a robot, considering the kinematics and mechanical aspects, low level control, behaviors, and learning mechanisms. This also gives us the possibility to describe the main features of our robotic agent.

2 Designing the sensor

An agent needs information to perform a task. It obtains such information by elaborating sensor data. Reasoning may somehow fulfill lacks of information by completing the available data by inference, possibly increasing uncertainty and approximation. An accurate design of the sensorial apparatus may reduce such undesired factors. We first discuss general properties that should be considered, exemplifying them in the Robocup context.

2.1 What kind of information?

The first question we have to ask ourselves is: "What kind of information will my agent need?" From the answer to this question we may decide the kind of sensor we need to acquire data, and how to extract the information from raw data. Here below, we mention some relevant aspects.

Information contents What the agent has to know? For instance, in the Robocup context, we may decide that it is interesting to know the position of the agent and the relative positions of ball, goals, and other agents, maybe discriminating between teammates and opponents, or maybe identifying each single robot. Moreover it may be interesting to know this information for any object on the field: the more the agent can perceive, the less it has to rely on inference, presuppositions, expectations, and information explicitly coming from teammates.

Information quality Should the agent be certain about the facts it is inferring from data? Can it work also with uncertain facts? Should it explicitly represent uncertainty? What kind of precision is needed? In Robocup it may be interesting to have a good precision in the neighborhood of the agent, for instance to control the ball, and to interact with close players. What happens at a great distance may be considered qualitatively, since the environment is rapidly changing, and whatever happens far from the agent do not require precise intervention before than significant changes may occur.

Information acquisition rate How frequently should the information be updated? The acquisition rate should allow to build an effective model of the events that characterize the task. The fastest event on a Robocup F-2000 field is the movement of the ball, which may run at more than 1 m/sec. The information acquisition and management should be fast enough to allow enough time to act (or react), but, at the same time, it should not be too fast since this would produce a large amount of data that has to be interpreted with the computational resources available (on board).

Information abstraction level Which kind of abstraction from raw data do we expect to need, in order to obtain the required information? Can we reason on raw data or on abstractions, and which kind of abstractions? In Robocup, we are interested in the above mentioned information, which requires a good abstraction and classification activity on raw data.

Acquisition robustness and adaptation Is the environment known, stable and static? If it is not so, data acquisition should be robust and possibly able to adapt to changing conditions. Although the Robocup rules seem to define a highly structured environment, it is not so on the real field. Illumination is never as expected, and the presence of objects in the visual field may make it changing a lot during a match.

2.2 Why omnidirectional vision?

At Robocup98, most of the robots relied on vision sensors, since the organizers seem strongly oriented to give colors a primary role in the field setting. Most of the players (included those in our team, *ART – Azzurra Robocup Team* [12]) had a fixed, color camera, that could hardly match some of the specifications mentioned above. For instance, it gives information only about a small number of objects on the field, so that, in many situations, it is hard to self-localize the agent, to know where is the ball (which runs really fast), or to understand what is happening. Our team, as others, implemented strategies to patch this lack of information, based on information exchanging, which partially failed because of transmission problems. Other teams had mobile cameras (or high mobility of the body), but most of them still seemed too slow and imprecise to keep track of the fast events on the field. The '98 winner [10] had a sensor system matching perfectly all the design requirements mentioned above, and it was one of the few teams showing really interesting behaviors. It is also to be noticed that their sensor did not follow the organizers' implicit suggestion to rely on color vision, thus avoiding most of the problems related with this type of sensor on the Paris Robocup field.

We have decided to answer to the above introduced "first question" with an omnidirectional vision sensor, which is described in details elsewhere [7]. It consists of a camera pointed upwards beneath a coaxial, revolution mirror obtained by the intersection of a truncated cone and a sphere (see figure 2). A single image contains all the objects around the agent. The data acquisition system can give with sufficient precision distance and direction from all the objects around the agent. We have designed it to exploit the camera resolution and to improve radial resolution in the peripheral areas of the circular image, containing far objects. It is thus possible to reliably detect objects such as the ball, up to 6 meters from the agent. The precision about the distance from the objects is inversely proportional to the distance itself. Uncertainty about data classification is very low, due to the image analysis system we have implemented. This is also optimized to provide all the information once every 30 ms, giving an information acquisition rate very close to the limit of the PAL European standard for video frame acquisition, which is 25 frames/sec. Since an omnidirectional image contains at the same time a large portion of the field, the average brightness is quite stable, and adaptation to light intensity is limited to the first frames, to become acquainted with a new field. In case of a standard camera pointing towards the environment, the image may contain objects with different colors (e.g., black robots, or white walls), and this requires some compensation on the average brightness, which can be obtained either by mechanical adjustment of the camera iris (slow), or electronically, requiring the analysis of at least two images unavailable for object recognition. In the next section, we give some details about the specific choices we have done in the implementation of our sensor.

2.3 Our omnidirectional vision sensor for Robocup

The sensor we are proposing is represented in figure 2. You can see that the

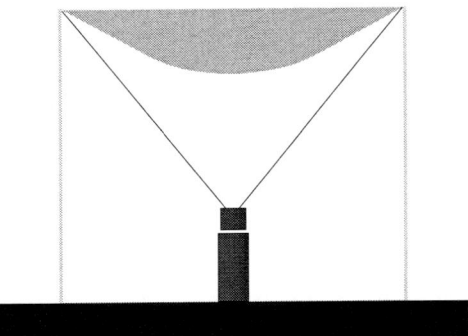

Fig. 2. The mirrror, the camera pointing upwards, the plexiglass cylinder supporting the mirror and the visual angle of the camera.

central part of the mirror consists of a truncated sphere, making it possible to see objects very close to the robot (15 cm). Tangent to this is a truncated, reversed cone, giving enhanced radial resolution from 2 m to 6 m. This design does not require adjustment of the focal length of the camera, as proposed by [17], thus avoiding the time loss due to mechanical movements. The camera with which we took the image shown in figure 3 is a low cost card camera, having 512x582 sensible elements, and a view angle of about 60^0. We are now mounting a Sony XC-999P. In figure 3, in black, on the center of the image the body of the robot, on the right the yellow goal and the goalie, on the top a ball and, at a distance, another robot and the blue goal. To implement a fast image recognition system we took hints from biology, basing it on the idea of *receptor*. The generic term "receptor" is used for any biological unit able to perceive specific stimuli from the outside world and to transform them in nervous signals, then transmitted to the central nervous system. In computer vision, image pixels are often considered as receptors. To improve computational speed, we consider specialized receptors, each consisting of a 3 by 3 pixel matrix, and characterized by the averaged HSV value. Our receptors are distributed in a pattern designed to detect the smallest object on the field (the ball) in any position of the image. Thus, we analyze only a grid of receptors on the image, thus reducing the amount of information to be considered by more than two orders of magnitude.

The vision system we have designed firstly estimates on the image the likely position of the possibly interesting objects, by classifying the receptors, and aggregating them in clusters (called *target*) by color similarity and adjacency. A target is a part of the image where it could be present an interesting objects. Once identified the targets, it is possible to operate on the part of image defined

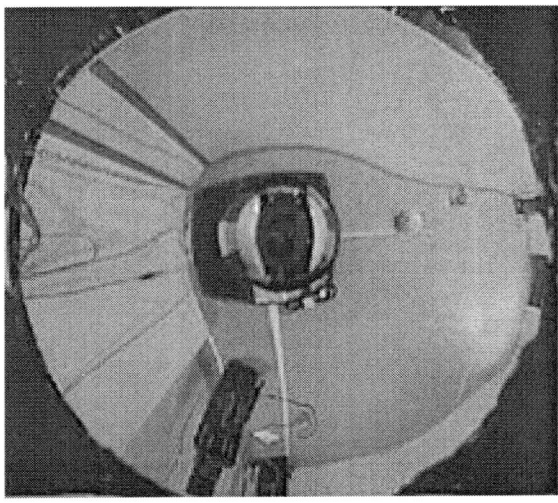

Fig. 3. A typical image taken by our sensor.

by each of them by adopting classical image processing (such as blob growing) on single pixels, object recognition and localizazion techniques. The application of these techniques significantly reduces the amount of information to handle and increases the processing speed. The image acquisition and processing requires less than 30 ms, on the on board PC, a 266 MHz AMD K6 CPU, with 32 Mb of RAM, a Matrox Meteor frame grabber, Linux RedHat 5.2 (Kernel version 2.0.36), and real-time kernel ETHNOS [13].

3 Sensors and behaviors

Now, let us discuss the impact of the type of available, sensorial information on the behaviors that can be implemented on the agent. We first consider the low-level control aspects, and then the higher level behaviors and strategies

3.1 Low-level control

In our viewpoint about robot architecture, a low-level control system may be present on an agent to provide the higher levels with reliable actions. If the higher level behavior activation cycle is long, low-level control should ensure that the desired actions are actually done as expected. For instance, if the high level control states that the agent should turn 30 degrees on the left at a speed of 0.2 m/sec, it expects that this happens; if, as usual, the actuation is imperfect, the actual action may be different. We may either have a low level control system trying to realize what the higher-level control states, or have a higher-level control designed to cope with low level problems such as imperfect actuators, and running fast enough. The information for such a kind of low-level control is different from that discussed above. Here, we need precise information about the movement of the robot wheels. We have to cover this need with other sensors, appropriate for this component of the control system, namely encoders on the wheels or on the engine axis.

Rullit, our Robocup agent, has two independent traction wheels, and we decided to attach encoders to each wheel. The precision of each measure is less than 0.1 mm, enough to implement a good, speed and jog control. We have implemented it as a fuzzy controller, so that it is also quite robust with respect to noise [11]. Notice that this same sensor (encoder) is known to be inappropriate for position control [9], and that we did not implement such a kind of controller. This is another example of the relevance to select the proper sensor to achieve a task.

3.2 Behaviors

The Robocup environment changes so rapidly that we have decided to leave to the behavioral control the decision about where to go: a plan to reach a position should be probably continuously refined, since situations change rapidly, so a controller able to bring the agent at a given position would be restarted too often.

Information provided by the omnidirectional sensor is appropriate for high level control, and influences the design of behavioral modules. We may notice that the same behavior (for instance, *Go_To_Ball*, that brings the agent on the ball) may be implemented in different ways according to the available information. If we had reliable information only about objects that are in the range of a camera pointed forward, probably the behavior can reliably trigger only when the ball is in the range of the camera, and another behavior will make the agent searching for the ball. Moreover, probably, *Go_to_Ball* will bring the agent on the ball, only while keeping it in sight; this may bring the agent in undesired situations, such as bumping the ball against the wall. Another implementation of *Go_to_Ball* with the same sensor may infer the position of the ball from past information and from information coming from other players. This may help, but may also lead to clumsy behaviors, such as that happened in the challenge during the ART-Freiburg semi-final at Robocup99, where the player didn't check often enough the ball position and originated a situation hard to manage.

By contrast, having reliable information about the ball in any position with respect to the agent from a suitable sensor, such as omnidirectional vision, a different *Go_to_Ball* may decide how to approach the ball, while keeping it in sight. In figure 4 you may see some of the ball approaching behaviors we have implemented relying on the available omnidirectional information; these include going on the ball by moving backwards (tracks 3 and 4 in figure 4), or tracking the ball by the side (track 2 in figure 4). We have implemented the behavioral control by *fuzzy behaviors* [15] [8], that is control modules that trigger on conditions consisting of fuzzy predicates. We consider two sets of such fuzzy preconditions: the *cando preconditions* enable the behavior, and the *want preconditions* give the

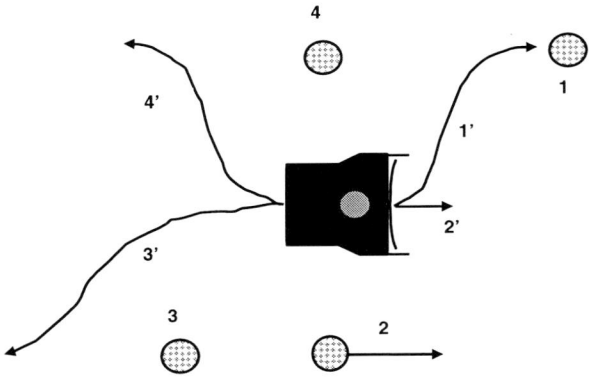

Fig. 4. Three possible trajectories (1', 3' and 4') to reach a position behind the ball (respectively in positions 1, 3, 4), and one (2') to track a moving ball(2).

amount of motivation for it. For instance, if we have the ball we can kick it, but we want to do it only if this makes sense, e.g., we are aligned with a free portion of the opponent's goal. Fuzzy predicates make it possible to classify the information coming from the sensor into higher level classes, which give it a meaning. Thus, it is possible to perform reasoning at a high level of abstraction, on a relatively small set of concepts [6], thus achieving high speed and robustness [11]. Moreover, a fuzzy interpretation gives the possibility to reason on overlapping classifications, which seems to be exactly what human beings, and some animals, do in most situations. For instance, in figure 5 we show the membership functions defining three fuzzy sets (*close*, *medium*, *far*) used to implement fuzzy predicates that classify the distance from objects. In the example, the measured distance (1 m) is classified as *close* with truth value 0.4 and *medium* with truth value 0.6. In real life, usually we adopt classifications which can be naturally represented by fuzzy predicates [11] whose definitions overlap, such as those presented in figure 5.

We associate to behaviors other two parameters: the static and the dynamic *relevance*. The first implements an a priori, partial ordering among behaviors, allowing to state, for instance, that avoiding crashes is always better than taking the ball. The dynamic relevance also implements a partial ordering, but it can be modified according to the situation faced by the agent, and it is used to implement strategies and learning mechanisms, as discussed in the next sections. At each high level control step we compute for each behavior instance whose cando preconditions are true above a given threshold, its *triggering level*, by composing: its two relevance values, the motivation coming from its want preconditions, and the possibility coming from its cando preconditions. As done by most biological beings, and in contrast with most of artificial fuzzy agents, the behavior with the highest triggering level is activated, and its actions done. We have a winner-

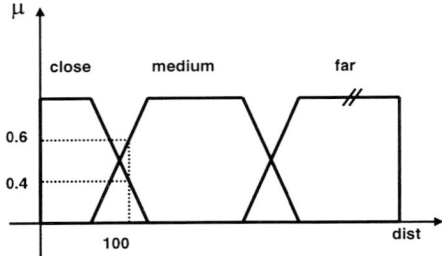

Fig. 5. An example of fuzzy classification: the distance value 100 [cm] is classified as *close* with truth value 0.4, and as *medium* with truth value 0.6.

take-all activation, instead of a composition of proposed actions typical of fuzzy systems, since this gives more coherence to behavior selection. If we decide that it is better to go on the ball instead than towards the goal expecting a passage, it does not make any sense to compose the two actions to obtain an hybrid whose success possibilities are questionable: it is better to take a decision and act coherently with it. If it was a wrong decision, it means that the activation conditions of the behavior modules have to be tuned, and this can be done also automatically, but only if it is clear which is responsible for the action taken [5]. This structure for behavioral modules is independent from the information we have decided to be needed for the Robocup task, but it can support it effectively. We adopt the same structure in other projects based on different information (and sensors). On the other side, the specific behaviors strongly depend on the available information both for their existence (a behavior needing unavailable information would have not been implemented), and for their specific implementation as discussed above for *Go_to_Ball*. The whole behavior system runs in only 7 ms on the on board PC.

3.3 Strategy

We consider that basic skills for an agent are implemented by behavioral modules. A higher level decision module may influence the behavior activation, acting on the priorities among behaviors, by modifying their dynamic relevance values. The strategic module we have implemented recognizes a situation by classifying high level data interpretations, again implemented as fuzzy predicates. According to this, it rearranges the dynamic relevance of the behaviors, to give the preference, in case of similarity of the other parameters, to a behavior or another. For instance, if a teammate has the ball, we may either protect it by hindering opponents, or follow its action expecting a passage. A deeper discussion of this topic is beyond the scope of this paper.

Also the instance of this module strongly relies on the available information. For instance, an omnidirectional vision sensor gives the possibility to detect a rich variety of situations without communicating with teammates. Probably, a fixed, front camera would not give by itself enough information to justify the existence of a strategic module, and also a mobile camera would be probably too slow to catch enough from the environment to detect the relevant facts that could make it possible to select strategies. For these reasons, most of the teams involved in the F-2000 Robocup Championship in 1998 either relied on communication to select strategies [10], or had simple, selfish strategies. In Robocup99, an increased number of teams had robots equipped by omnidirectional vision.

4 Sensors and learning

Learning and adaptation are interesting approaches to implement, or improve, control modules. We believe that, given that the present learning techniques, learning behaviors from scratch on the field is not cost-effective: a designer may develop by hand nice behaviors in less time and using less resources. The goalkeeper task [17] is simple enough to be either learnt or programmed, whereas the behaviors of the other players have to be developed on a large number of really complex situations, and have to be adapted to the behaviors of the specific opponents. Some researchers have proposed to learn more complex behaviors in the F-2000 league (such as ball passing [18]). The behaviors were learned in simulation, but hand-coded behaviors were preferred on the field. In these conditions, we consider that it is more effective to program simple behaviors and strategy modules, and then adapt them on line.

The role of sensors in learning in simulated environments is questionable. Having worked since long time on learning behaviors for simulated robots [4], we have come to the conclusion that in most cases the problems with simulation are far different from those in real world, apart from the cases where enough resources are devoted to produce sophisticated simulation environments. In particular, sensor models, and the quality of the information they provide, are usually oversimplified. Therefore, we would not like to discuss about the role of sensors in learning in simulated environments.

Adaptation is even more important than learning in applications such as a Robocup match among real robots, where the opponent's strategy is usually unknown. The quality of the incoming information is relevant for the quality of adaptation: the more and nicer information we have, the more the adaptation algorithms can exploit it to find regularities. On the other side, it would be hard to manage in real time a large amount of data.

Given the topics introduced above, we can easily imagine an adaptation mechanism it could be implemented to improve strategies to face specific opponents. For instance, an agent with the ball has to take different behaviors to contrast opponents that tend to rush on the ball, or opponents that tend to block actively any possible way to their goal. Moreover, this can only be decided on line, in real time, while playing against a specific team a specific match. We have implemented an adaptation system based on reinforcement. At present, it works only on one-to-one strategies. The strategic module classifies the situation by evaluating fuzzy predicates that interpret the incoming information, compared with information previously acquired. Then, it selects one among some predefined strategies (each corresponding to a set of relevance values for the behaviors), and provides a value of dynamic relevance for the involved behaviors. This is repeated at each high level control step, always considering the same strategy, for coherence reasons, until the situation changes. At this time, a reinforcement is computed by evaluating the new situation, and it is used to update the value $Q(s)$ of the selected strategy s according to the standard formula:

$$Q^t(s) = Q^{t-1}(s) + \alpha \left(r - Q^{t-1}(s)\right)$$

We share the motivation for using this formula, and the general background with other researchers who studied learning in the Robocup framework in other leagues [16], but we consider different models to learn, more appropriate for adaptation in the F-2000 league.

5 Conclusions

While in nature we assisted to the co-evolution of sensor, motor and neurological apparati, we claim that in robotic agent design we need to design first the appropriate sensors to get the quantity and quality of information we need to achieve a task; then, we may try to shape the other components of the robot architecture on this. We have discussed in details how we have applied this approach in the development of Rullit, our Robocup F-2000 player. We argue that most of the other components of a robot architecture could be designed in many different ways, but that the information provided by our sensor is appropriate and essential to effectively face the Robocup task. We have also shown how many problems may be solved at the sensor level, thus reducing the computational effort.

Rullit is built on our Mo^2Ro (*Modular Mobile Robot*) base, which also provides the mechanical and electronic basic modules for other robots, built around other sensors, namely: *Pop-eye*, which has a camera mounted on top of a 5 DOF ultra-light arm on board, and *RoboCOPIS*, which adopts a standard black and white COPIS sensor [19].

References

[1] G. Agazzi and A. Bonarini. Problems and solutions in acquisition and interpretation of sensorial data on a mobile robot. In *IEEE Instrumentation and Measurement Tecnology Conference – IMTC99*, Piscataway, NJ, 1999. IEEE Computer Press.

[2] M. Asada, editor. *RoboCup 98: Robot Soccer World Cup II*, Paris, F, 1998. Eurobot Project - CEC.

[3] M. Asada, P. Stone, H. Kitano, A. Drogoul, D. Duhaut, M. Veloso, H. Asama, and S. Suzuki. The robocup physical challenge: goals and protocols for phase

i. In H. Kitano, editor, *RoboCup 97: Robot Soccer World Cup I*, pages 42–61. Springer-Verlag, Berlin, D, 1997.

[4] A. Bonarini. ELF: Learning incomplete fuzzy rule sets for an autonomous robot. In Hans-Jürgen Zimmermann, editor, *First European Congress on Fuzzy and Intelligent Technologies – EUFIT'93*, volume 1, pages 69–75, Aachen, D, 1993. Verlag der Augustinus Buchhandlung.

[5] A. Bonarini. Evolutionary learning of fuzzy rules: competition and cooperation. In W. Pedrycz, editor, *Fuzzy Modelling: Paradigms and Practice*, pages 265–284. Kluwer Academic Press, Norwell, MA, 1996.

[6] A. Bonarini. Reinforcement distribution to fuzzy classifiers: a methodology to extend crisp algorithms. In *IEEE International Conference on Evolutionary Computation – WCCI-ICEC'98*, volume 1, pages 51–56, Piscataway, NJ, 1998. IEEE Computer Press.

[7] A. Bonarini, P. Aliverti, and M. Lucioni. An omnidirectional vision sensor for fast tracking for mobile robots. In *IEEE Instrumentation and Measurement TecnologyConference – IMTC99*, Piscataway, NJ, 1999. IEEE Computer Press.

[8] A. Bonarini and F. Basso. Learning to compose fuzzy behaviors for autonomous agents. *Int. J. of Approximate Reasoning*, 17(4):409–432, 1997.

[9] J. Borenstein, H.R.Everett, and L. Feng. Where am i? sensors and methods for mobile robot positioning. Technical report, The University of Michigan, Ann Arbor, MI, 1996.

[10] J-S Gutmann, W. Hatzack, I. Herrmann, B. Nebel, F. Rittinger, A. Topor, T. Weigel, and B. Welsch. The cs freiburg team. In M. Asada, editor, *RoboCup 98: Robot Soccer World Cup II*, pages 451 – 457, Paris, F, 1998. Eurobot Project - CEC.

[11] G. J. Klir, B. Yuan, and U. St. Clair. *Fuzzy set theory: foundations and applicatons*. Prentice-Hall, Englewood Cliffs, MA, 1997.

[12] D. Nardi, G. Clemente, and E. Pagello. Art azzurra robot team. In M. Asada, editor, *RoboCup 98: Robot Soccer World Cup II*, pages 467–474, Paris, F, 1998. Eurobot Project - CEC.

[13] M. Piaggio and R. Zaccaria. Distributing a robotic system on a network: the ethnos approach. *Advanced Robotics Journal*, 12(8), 1998.

[14] Robocup. The robocup initiative. http://www.RoboCup.org/, 1999.

[15] A. Saffiotti, K. Konolige, and E. H. Ruspini. A multivalued-logic approach to integrating planning and control. *Artificial Intelligence*, 76(1-2):481–526, 1995.

[16] P. Stone and M. Veloso. Tpot-rl: Team-partitioned, opaque-transition reinforcement learning. In M. Asada, editor, *RoboCup 98: Robot Soccer World Cup II*, pages 221 – 236, Paris, F, 1998. Eurobot Project - CEC.

[17] S. Suzuki, T. Kato, H. Ishizuka, Y. Takahashi, E. Uchibe, and M. Asada. An application of vision-based learning for a real robot in robocup - a goal keeping behavior for a robot with omnidirectional vision and an embedded servoing. In M. Asada, editor, *RoboCup 98: Robot Soccer World Cup II*, pages 467 – 474, Paris, F, 1998. Eurobot Project - CEC.

[18] E. Uchibe, M. Nakamura, and M. Asada. Cooperative and competitive behavior acquisition for mobile robots through co-evolution. In *Proceedings of the Genetic and Evolutionary Computation Conference – GECCO99*, pages 1406–1413, San Francisco, CA, 1999. Morgan Kaufmann.

[19] Y. Yagi, S. Kawato, and S. Tsuji. Real-time omnidirectional image sensor (copis) for vision-guided navigation. *IEEE Transactions on Robotics and Automation*, 10(1):11–22, 1994.

Motion Control in Dynamic Multi-Robot Environments

Michael Bowling and Manuela Veloso

Computer Science Department
Carnegie Mellon University
Pittsburgh, PA 15213-3890

Abstract. All mobile robots require some form of motion control in order to exhibit interesting autonomous behaviors. This is even more essential for multi-robot, highly-dynamic environments, such as robotic soccer. This paper presents the motion control system used by CMUnited-98, the small-size league champion at RoboCup-98. The team consists of five robots that aim at achieving specific goals while navigating in a limited space shared with the five other opponent robots. We introduce our motion control algorithm, which allows a general differential-driven robot to accurately reach a target point with a desired orientation in an environment with multiple moving obstacles. We describe how the features of our motion controller help to build interesting and robust behaviors. We also briefly compare our system to other motion control techniques and include descriptions and illustrations of the performance of our fully-implemented motion control algorithm.

1 Introduction

For any robotic system motion control is essential to building robust and interesting behavior. This is even more important for multi-robot systems that need to build team behaviors on top of individual behaviors. An example of such a system is robotic soccer. Here, a team of robots must coordinate their actions to push the ball into their opponents' goal. This is complicated by not only the opponent agents trying to prevent this from occurring, but also by the highly dynamic environment. This highly dynamic environment makes many traditional motion planning algorithms impractical since the environment changes before the planner can even finish its path.

This paper examines the motion control algorithm used in CMUnited-98 [6]. This team competed in RoboCup '98 in Paris in the small-size robot league. The team won four of its five games and was the league champion for the second straight year. A great deal of the success of the team can be attributed to the motion control algorithms. It not only made direct contributions by providing smooth and robust motion, but its features allowed us to build powerful individual and team behaviors. In section 2, we give a brief overview of the architecture of our team. It will describe the percepts and actuators available to the motion controller. In section 3, we describe the details of the motion control algorithm.

In section 4, we describe how the high-level attacking behaviors effectively made use of our algorithm. Finally, in section 5 we will discuss related work in this area.

2 Team Architecture

The CMUnited-98 small-size robot team is a complete, autonomous architecture composed of the physical robots, a global video camera over-looking the playing field, and several clients as the minds of the small-size robot players. Fig. 1 sketches the building blocks of the architecture. The motion controller resides in the individual client modules and bridges the gap between the output of the vision processing system and the robots' motors.

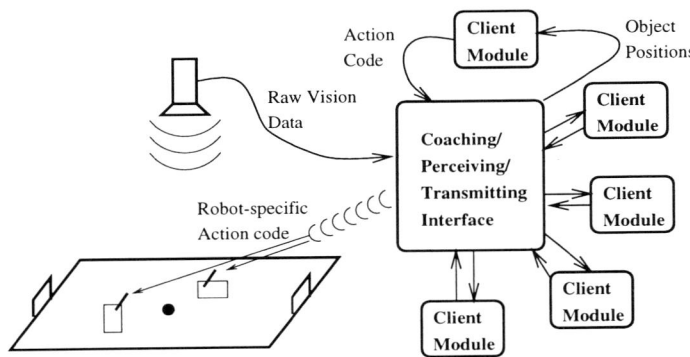

Fig. 1. The CMUnited architecture with global perception and distributed reaction.

The vision system provides the input to the motion controller. Since it overlooks the entire field, it can provide a complete view of the world. Our image processing algorithm reliably detects and tracks the position and orientation of our five robots, the position of the five opponents, and the position of the ball. Additionally it uses a Kalman-Bucy filter to provide a reasonably accurate prediction of the ball's velocity, both speed and direction. This information is computed and passed to the client modules approximately every thirtieth of a second.

The output is the motion parameters for the physical robot. The robots have two motors and use differential drive for movement. The client modules, specifically the motion controller, sends the desired wheel velocities to its physical robot using radio communication. The radio communication supports five robots each receiving over twenty commands every second. Additionally, there is no local feedback mechanism on the robots. All control is done using only visual feedback through the motion controller.

A summary of the issues the motion controller must address are given below.

– Ten moving robots ($\leq$ 12cm diameters) on a small walled field (152.5cm by 274cm). Robots are moving at speeds close to one meter per second.

- Vision system providing 30 frames per second. Radio communication able to support approximately 20 commands per second for each of the five robots.
- Must be able to push the ball towards a particular point, specifically towards a goal (50cm wide). It must also be able to intercept a moving ball and avoid fast moving obstacles.

3 Motion Control Algorithm

The goal of our motion control algorithm is to be as fast as possible while remaining accurate and reliable. This is challenging due to the lack of feedback from the motors, forcing all control to be done using only visual feedback. Our motion control algorithm is robust. It addresses stationary and moving targets with integrated obstacle avoidance. The algorithm makes effective use of the prediction of the ball's trajectory provided by the Kalman-Bucy filter.

We achieve this motion control functionality by a reactive control mechanism that directs a differential drive robot to a target configuration. The mechanism is based on CMUnited-97's motion control [7, 1], but includes a number of major improvements. The target configuration for the motion planner has been extended. The target configuration includes: (i) the *Cartesian position*; and (ii) the *direction* that the robot is required to be facing when arriving at the target position. Obstacle avoidance is integrated into this controller. Also, the target configuration can be given as a function of time to allow for the controller to reason about intercepting the trajectory of a moving target.

3.1 Differential Drive Control for Position and Direction

We begin with some basic control rules. The rules are a set of reactive equations for deriving the left and right wheel velocities, v_l and v_r, in order to reach a target position, (x^*, y^*):

$$\Delta = \theta - \phi \qquad (1)$$
$$(t, r) = (\cos^2 \Delta \cdot \text{sgn}(\cos \Delta), \sin^2 \Delta \cdot \text{sgn}(\sin \Delta))$$
$$v_l = v(t - r)$$
$$v_r = v(t + r),$$

where θ is the direction to the target point (x^*, y^*), ϕ is the robot's orientation, and v is the desired speed (see Fig. 2(a))[1]. A few aspects of these equations deserve explanation. The use of $\sin^2$ and $\cos^2$ restricts the values $(t \pm r)$ to the interval $[0, 1]$, which bounds the magnitude of the computed wheel velocities by v. These equations also do not necessarily drive the robot forward, possibly driving the robot backwards towards the target.

We extend these equations for target configurations of the form (x^*, y^*, ϕ^*), where the goal is for the robot to reach the specified target point (x^*, y^*) while facing the direction ϕ^*. This is achieved with the following adjustment:

$$\theta' = \theta + \min\left(\alpha, \tan^{-1}\left(\frac{c}{d}\right)\right),$$

---

[1] All angles are measured with respect to a fixed coordinate system.

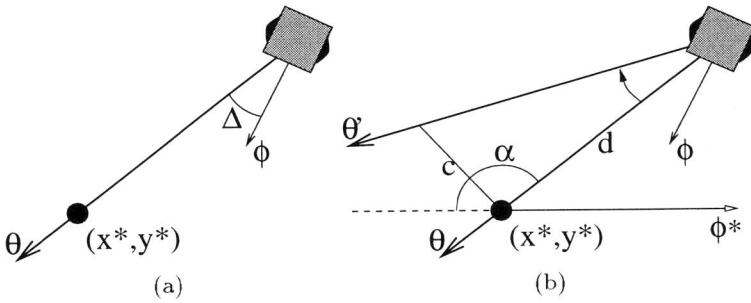

Fig. 2. (a) The parameters used to reach a target configuration (x^*, y^*), without a specified target orientation. (b) The adjustment of θ to θ' to reach a target configuration of the form (x^*, y^*, ϕ^*).

where θ' is the new target direction, α is the difference between our angle to the target point and ϕ^*, d is the distance to the target point, and c is a clearance parameter (see Fig. 2(b).) This will keep the robot a distance c from the target point while it is circling to line up with the target direction, ϕ^*. This new target direction, θ', is now substituted into equation 1 to derive wheel velocities. An example trajectory using these equations is shown in Figure 3 (a).

In addition to our motion controller computing the desired wheel velocities, it also returns an estimate of the time to reach the target configuration, $\hat{T}(x^*, y^*, \phi^*)$. This estimate is a crucial component in our robot's strategy. It is used both in high-level decision making, and for low-level ball interception, which is described later in this section. For CMUnited-98, $\hat{T}(x^*, y^*, \phi^*)$ is computed using a very simple linear function of d, α, and Δ:

$$\hat{T}(x^*, y^*, \phi^*) = w_d d + w_\alpha \alpha + w_\Delta \Delta.$$

The weights were set by simple empirical measurements. w_d is the inverse of the robot's translational speed; w_Δ is the inverse of the robot's rotational speed; and w_α is the inverse of the speed of the robot when traversing a circle of radius, c. It is interesting to note that even this crude time estimate can be incredibly useful for building more complex behaviors, which are discussed later in this paper.

3.2 Obstacle Avoidance

Obstacle avoidance was also integrated into the motion control. This is done by adjusting the target direction of the robot based on any immediate obstacles in its path. This adjustment can be seen in Fig. 4.

If a target direction passes too close to an obstacle, the direction is adjusted to run tangent to a preset allowed clearance for obstacles. Since the motion control mechanism is running continuously, the obstacle analysis is constantly replanning obstacle-free paths. This continuous replanning allows for the robot to handle the highly dynamic environment and immediately take advantage of short lived opportunities. Figure 3 (b) shows an example trajectory.

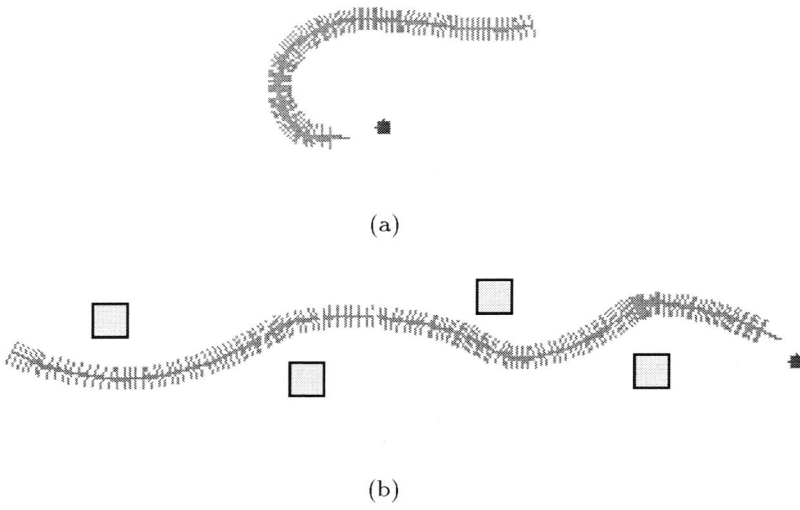

Fig. 3. Example trajectories. (a) This illustrates reaching a point with a specific orientation. The target point is the position of the ball, and the specified orientation is to the right. (b) An example trajectory illustrating obstacle avoidance.

This technique can be viewed as a path planner using only a one-step lookahead. Hence, it sacrifices completeness for the performance needed to handle the dynamic environment. Section 5 will briefly compare this technique with traditional path planning.

3.3 Moving Targets

One of the real challenges in robotic soccer is to be able to control the robots to intercept a moving ball. This capability is essential for a high-level ball passing behavior. CMUnited-98's robots successfully intercept a moving ball and several of their goals in RoboCup-98 were scored using this capability.

This interception capability is achieved as an extension of the control algorithm to aim at a stationary target. Fig. 5(a) illustrates the control path to

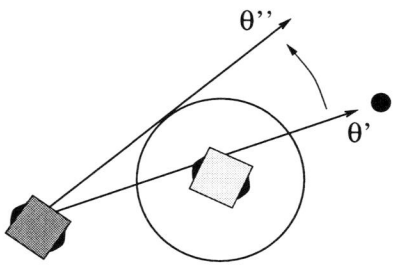

Fig. 4. The adjustment of θ' to θ'' to avoid immediate obstacles.

reach a stationary target with a specific direction, using the control mechanism described above. Our extension allows for the target configuration to be given as a function of time, where $t = 0$ corresponds to the present,

$$f(t) = (x^*, y^*, \phi^*).$$

At some point in the future, t_0, we can compute the target configuration, $f(t_0)$. We can also use our control rules for a stationary point to find the wheel velocities and estimated time to reach this hypothetical target as if it were stationary. The time estimate to reach the target then informs us whether it is possible to reach it within the allotted time. Our goal is to find the nearest point in the future where the target can be reached. Formally, we want to find,

$$t^* = \min\{t > 0 : \hat{T}(f(t)) \leq t\}.$$

After finding t^*, we can use our stationary control rules to reach $f(t^*)$. In addition we scale the robot speed so to cross the target point at exactly t^*.

Unfortunately, t^*, cannot be easily computed within a reasonable time-frame. We approximate this value, t^*, by discretizing time with a small time-step. We then find the smallest of these discretized time points that satisfies our estimate constraint. An example of this is shown in Fig. 5(b), where the goal is to hit the moving ball.

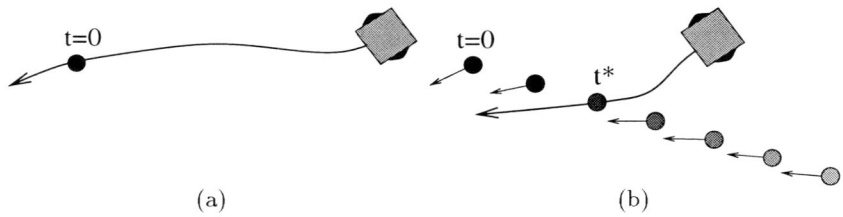

Fig. 5. (a) Control for a stationary target. (b) Control for a moving target.

The target configuration as a function of time is computed using the ball's predicted trajectory. Our control algorithm for stationary points is then used to find a path and time estimate for each discretized point along this trajectory, and the appropriate target point is selected.

4 Using Motion Control

We have described how our motion controller computes the wheel velocities to reach a target configuration, (x^*, y^*, ϕ^*), which may even be a function of time, $f(t)$. It is the responsibility of individual and team behaviors to select the appropriate target configurations for each of the robots. The features of our motion controller often simplify this problem. We will examine how these features help build two attacking behaviors, shooting and passing. Also, we will show how it contributes to our team attacking behavior, which involves a decision theoretic action selection.

4.1 Individual Behaviors: Shooting and Passing

We first developed individual behaviors for passing and shooting. For both behaviors the positional portion of the target position is the ball's position, since the goal is to push the ball. Additionally, we can use the ball's predicted trajectory to make the position a function of time, according to the trajectory.

The directional portion of the target configuration determines where the ball is pushed. The passing behavior specifies a direction that is a small amount in front of the designated receiver. For shooting, a more complex target direction is computed. Simply pushing the ball towards the center of the goal will do nothing to avoid pushing the ball into the goalie. Instead, we want to push the ball towards the largest unblocked portion of the opponent's goal. This is done by selecting the largest unblocked angular section of the goal and aiming for the angle that bisects it. Figure 6 illustrates the selected target configuration to achieve passing and shooting.

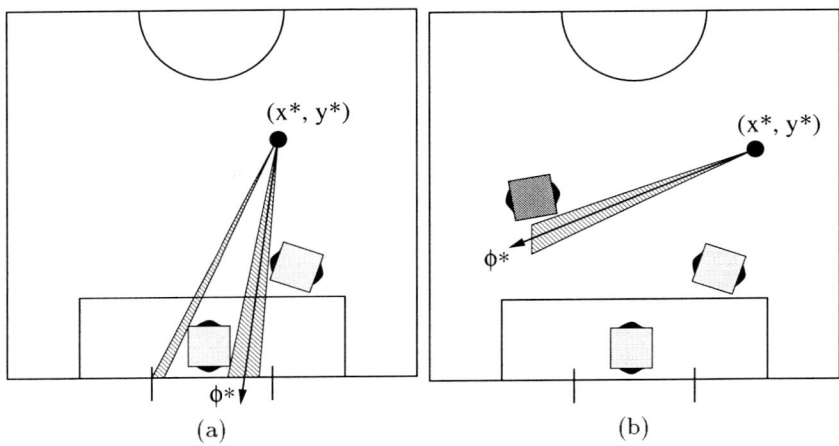

Fig. 6. An example of possible aiming positions given the position of the ball and two opponents. The largest angle is chosen and ϕ^* is the bisection of the angle.

4.2 Team Behavior: Decision Theoretic Action Selection

Given the individual behaviors, we must select an active agent, the agent that will go to the ball, and an appropriate behavior, passing or shooting. This is done by a decision theoretic analysis that uses a single step look-ahead. With n agents there are n^2 choices of actions involving shooting or a pass to another agent followed by that agent shooting. An estimated probability of success for each pass and shot is computed along with the time estimate to complete the action, which is provided by the motion controller. With these estimates, a value for each action is computed,

$$\text{Value}_{\text{pass}} = \frac{\text{Pr}_{\text{pass}} \text{Pr}_{\text{shoot}}}{\hat{T}(f_{\text{pass}})} \qquad \text{Value}_{\text{shoot}} = \frac{\text{Pr}_{\text{shoot}}}{\hat{T}(f_{\text{shoot}})}$$

The action with the largest value is selected, which determines both the active agent and its behavior. Table 1 illustrates an example of the values for the selection considering two attackers, 1 and 2. CMUnited- 98 uses a heuristic function to estimate the success probabilities of passing and shooting.

Attacker	Action	Probability of Success		Time(s)	Value
		Pass	Shoot		
1	Shoot	–	60%	2.0	0.30
1*	Pass to 2	60%	90%	1.0	0.54
2	Shoot	–	80%	1.5	0.53
2	Pass to 1	50%	40%	0.8	0.25

Table 1. Action choices and computed values are based on the probability of success and estimate of time. The largest-valued action (marked with the *) is selected.

It is important to note that this action selection is occurring on each iteration of control, approximately 30 times per second. The probabilities of success, estimates of time, and values of actions, are being continuously recomputed. This allows for quick changes of actions if shooting opportunities become available or collaboration with another agent appears more useful.

5 Related Work

An alternative to our purely reactive algorithm is to use a complex motion planning algorithm. A number of these algorithms are summarized by Latombe [3]. These techniques find complete obstacle free paths, but yet have difficulties in the robotic soccer domain. Since the environment is highly dynamic with the obstacles constantly moving, planned paths would need to be constantly reevaluated. Also, path planning often needs to be done for a large number of proposed trajectories before the high-level action can even be selected. These traditional algorithms are simply too slow for the continuous real-time execution that is demanded in robot soccer.

Another approach to motion control [4] uses a reactive mechanism with fast hardware-supported feedback, via motor encoders and on-board sensors. This makes use of a slower decision loop to provide high-level commands, and a fast control loop to perform these commands. The control loop uses the motor encoders to perform accurate movements and on-board sensing for immediate obstacle avoidance and ball manipulation. This was successfully used by CURF (Cambridge University Robot Football) in RoboCup '98. One drawback to this technique is that the fast control loop does not have access to the complete sensors (i.e. the global view of the field), and short lived opportunities, which may not be recognized by the local sensors, often cannot be exploited . ISpace [2], another team that competed in RoboCup '98, used a similar technique, but due to onboard vision could possibly overcome this drawback.

Additionally, there are also other reactive control systems for remotely controlled robots [5].

6 Conclusion

We've described the motion control algorithm used in CMUnited-98. The algorithm incorporates obstacle avoidance, and has an extended target configuration that includes orientation and can be given as a function of time. In addition to the details of the algorithm, we also described how its features simplifies the building of individual and team behaviors. The system was integral part of the team's success.

References

1. Kwun Han and Manuela Veloso. Reactive visual control of multiple non-holonomic robotic agents. In *Proceedings of the International Conference on Robotics and Automation*, Belgium, May 1998.
2. Tomomi Hashimoto, Toru Yamaguchi, and Hideki Hashimoto. Multi-vision RoboCup system using ISpace. In Minoru Asada, editor, *Proceedings of the Second RoboCup Workshop*, pages 527–537, 1998.
3. Jean-Claude Latombe. *Robot Motion Planning*. Kluwer Academic Publisher, 1991.
4. A. Rowstron, B. Bradshaw, D. Crosby, T. Edmonds, S. Hodges, A. Hopper, S. Lloyd, J. Wang, and S. Wray. CURF: Cambridge university robot football team. In Minoru Asada, editor, *Proceedings of the Second RoboCup Workshop*, pages 503–509, 1998.
5. José Santos-Victor and Carlos Carreira. Vision based remote control of cellular robots. *Journal of Robotics and Autonomous Systems*, 1998.
6. Manuela Veloso, Michael Bowling, Sorin Achim, Kwun Han, and Peter Stone. The CMUnited-98 champion small robot team. In Minoru Asada and Hiroaki Kitano, editors, *RoboCup-98: Robot Soccer World Cup II*. Springer Verlag, Berlin, 1999.
7. Manuela Veloso, Peter Stone, Kwun Han, and Sorin Achim. CMUnited: A team of robotic soccer agents collaborating in an adversarial environment. In Hiroaki Kitano, editor, *RoboCup-97: The First Robot World Cup Soccer Games and Conferences*. Springer Verlag, Berlin, 1998.

Behavior Engineering with "Dual Dynamics" Models and Design Tools

Ansgar Bredenfeld, Thomas Christaller, Wolf Göhring,
Horst Günther, Herbert Jaeger, Hans-Ulrich Kobialka,
Paul-Gerhard Plöger, Peter Schöll, Andrea Siegberg,
Arend Streit, Christian Verbeek, Jörg Wilberg

GMD – German National Research Center for Information
Technology

Abstract

Dual Dynamics (DD) is a mathematical model of a behavior control system for mobile autonomous robots. Behaviors are specified through differential equations, forming a global dynamical system made of behavior subsystems which interact in a number of ways. DD models can be directly compiled into executable code. The article (i) explains the model, (ii) sketches the Dual Dynamics Designer (DDD) environment that we use for the design, simulation, implementation and documentation, and (iii) illustrates our approach with the example of kicking a moving ball into a goal.

1 Introduction

In the RoboCup mid-size league, robots have to kick a ball into the right direction. For many reasons, this is a hard task, which calls for robotic methods from many fields:

1. The situation on the field changes rapidly and drastically. This suggests a reactive, behavior-based approach to robot control [Brooks1991].

2. Kicking a moving ball is a continuous and dynamic task. Methods from continuous-time robust control (like in [Aicardi et al.1995]) are required.

3. The meaning of "...into the right direction" also varies dynamically. A self-organising, dynamical-system realization of goals and motivations seems appropriate here [van Gelder1998].

4. Playing football involves many different kinds of actions, with complex relations and interactions between them. A hierarchical representation of actions and action selection control is a natural approach to handle this complexity [Tyrrell1993].

5. Developing complex robots is done in many iterated design-redesign cycles, often with substantial modifications both on the hardware, low-level software, and control program level. State-of-the-art co-design tools can become critically beneficial [de Micheli and Gupta1997].

This list is certainly incomplete, but it demonstrates that designing football-playing robots is a complex, interdisciplinary challenge. From a traditional engineering perspective, this cries out for a modularized, hybrid approach, where different specialized subsystems are designed by different specialists, with well-defined interfaces between them.

However, there are indications that the classical divide-and-conquer approach is not fully appropriate for football-playing robots. A fast, autonomous robot in a continuously dynamic environment must continuously construct a stream of action from a stream of sensor information. This is connected to, but transcends, the well-known action selection problem [Maes1990]: *construction is harder than selection*. The imperative of continuously "doing the right thing" can only be met by an agent that acts "holistically", or to use a more modest term, in an integrated fashion. It is difficult to conceive how a classical modular system can rise to this task, at least when it consists of subsystems that communicate with each other over relatively narrow channels according to strict protocols, hiding from each other most of what is going on inside them. Unfortunately, the notion, "to act in an integrated fashion", is as vague as the term "modular". Practical examples of robotic systems that more or less successfully construct a stream of action will help us to advance our understanding.

Building such a robotic system can only be achieved by a team of engineers that also behaves in an integrated way. At the very least, this means that there is a close, *mutually informed* collaboration – information hiding of any sort stands in opposition to the goal of building a system that can act in an integrated fashion.

Thus, a fundamental challenge for mobile robotics is to reconcile, (i) the need for *some* sort of modular design, which results from the necessity of bringing together diverse techniques and human specialists, with (ii) integratedness both in the robot and in the developing process.

At the Behavior Engineering (BE) research group in the GMD Institute of Autonomous Intelligent Systems (AiS, *http://ais.gmd.de*) we explicitly address this challenge. Our approach rests on two pillars. On the one hand, we develop a mathematical model of a behavior control system, which to a certain degree integrates the points 1 – 4 mentioned in the beginning: a behavior-based approach, robust control, a dynamical systems representation of actions and goals, and a hierarchical architecture. This is the Dual Dynamics (DD) model [Jaeger and Christaller1998]. On the other hand, we develop and utilize a design tool that fosters a close collaboration of engineers, by providing everyone with a unified access to the entire robot control system under construction. This is the Dual Dynamics Designer (DDD) tool [Bredenfeld1999].

In this article, we give a quick introduction to the DD model (Section 2), describe the DDD tool (Section 3), and demonstrate its application with the example of kicking a moving ball (Section 4).

2 The Dual Dynamics model of behavior control

The Dual Dynamics scheme is a mathematical model of a behavior control system for autonomous mobile robots. It has grown from three roots: the behavior-based approach to robotics, the dynamical systems approach to cognition, and the mathematical theory of self-organizing dynamical systems. Discussions of these foundational topics can be found in [Jaeger and Christaller 1998] [Jaeger 1998] [Jaeger 1997]. In the present article we concentrate on the mathematical and technical aspects of DD.

Behaviors are formalized as dynamical systems, using ordinary differential equations (ODEs). These dynamical systems interact through shared variables and certain control relations, yielding an complex control system, which in its entirety again is a dynamical system. The DD model specifies certain structural and dynamical constraints on admissible interactions and control relations between the various dynamical subsystems, which will be informally explained in this section. The formalism is mathematically specified in [Jaeger and Christaller 1998].

The basic assumption on which DD rests is that a situated agent can work in different *modes*. Modes are coherent, relatively stable "frames of mind", which enable the agent to tune into different situations and tasks. Specifically, agents respond to sensory signal differently in different modes. In defend mode, a football robot would react to a ball quite differently than when it is in attack mode. The DD approach rests on the assumption that transitions between modes can be formally captured by bifurcations of dynamical systems. A direct implication of casting mode changes as bifurcations is that such changes are *qualitative*, discontinuous changes, not gradual ones. Our football robots do not gradually change from defend to attack mode, they either defend or attack. However, since these transitions are regulated by dynamical systems (in contrast to finite state machines), the decision point is dynamically and continuously tuned by the full wealth of incoming sensor information.

In the remainder of this section, we explain how this basic idea becomes the ordering principle for a dynamical systems engineering approach to behavior control.

The main building blocks of a DD robot architecture are *behaviors*. They are ordered in levels (fig. 1a). At the bottom level, one finds *elementary* behaviors: sensomotoric coordinations with direct access to external sensor data and actuators. Typical examples are kick or fixateBall. At higher levels, there are increasingly comprehensive behaviors. They also have access to sensoric information but cannot directly activate actuators. Their task is to regulate modes. As a first approximation, higher-level behaviors can be seen as instantiations of modes. An example of a first-level behavior in our football robots is challenge1, which corresponds to the first video qualification task of finding a ball and scoring a goal without opponents. Second-level higher behaviors would be even more comprehensive. For instance, attack would be a second-level behavior which coincides with the attack mode.

Elementary behaviors are different from higher-level behaviors in that they are made from two subsystems (fig. 1a), which serve quite different purposes. This has given the approach its name, "dual dynamics".

The first of these subsystems is called the *target dynamics*. It calculates target trajectories for all actuators which are relevant for the particular behavior. For this calculation, the target dynamics has access to every relevant sensor information, and typically includes specific sensor preprocessing. The output of the target dynamics consists of as many variables as there are motoric degrees of freedom to be controlled.

A requirement for the target dynamics is that this system should not undergo bifurcations. This is what makes elementary behaviors elementary, and provides a very helpful criterium for deciding which behaviors are, in fact, elementary. For instance, the target trajectories of kick in our simple wheeled football robots are likely to remain qualitatively unchanged in different instances of the maneuver. Thus, kick would be a good candidate for an elementary behavior. By contrast, in an anthropomorphic football robot it is likely that there will be qualitatively different kicking maneuvers different circumstances. Each of them would thus yield a separate elementary behavior.

From an engineering perspective, the target dynamics is just a motor controller for a specific task. DD is not committed to a particular type of controller – any controller which promises success is welcome. The "no bifurcation" requirement, in this perspective, means that one has a uniform control law.

The other subsystem of an elementary behavior is its *activation dynamics*. It regulates a single variable, the behavior's *activation*. The equation ruling this variable should be written in a way that the variable displays a dynamic range between 0 and 1. Intuitively, a value of 1 means that the behavior is fully active, whereas 0 means that it is completely inhibited. High values of the activation mean that the target trajectories computed in the target dynamics are passed through to the actuators (cf. 1b).

The activation dynamics is allowed to undergo bifurcations. *The control parameters which induce these bifurcations are the activation variables of higher-level behaviors.* This is the core idea behind DD.

To illustrate this central point, consider the level-1 behaviors charge (quick advance with ball) and freeBall (liberate ball which has got stuck at wall or between robots). Consider an elementary behavior bumpRetract, a protective reflex which generally means: retract when the robot bumps into things. Standardly, the activation of bumpRetract jumps to 1 when the front bumper sensors are hit. However, this dynamical response changes qualitatively in different modes. Assume that the robot is charging and pushes the ball in front of itself. The bumper will be frequently hit by the ball. However, the activation of bumpRetract should not be triggered in this circumstance. Technically, the high activation of the level-1 behavior charge works on the activation dynamics of bumpRetract as a control parameter, pushing this dynamical system into a regime where it does not respond to bumper signals *if* the ball is seen directly

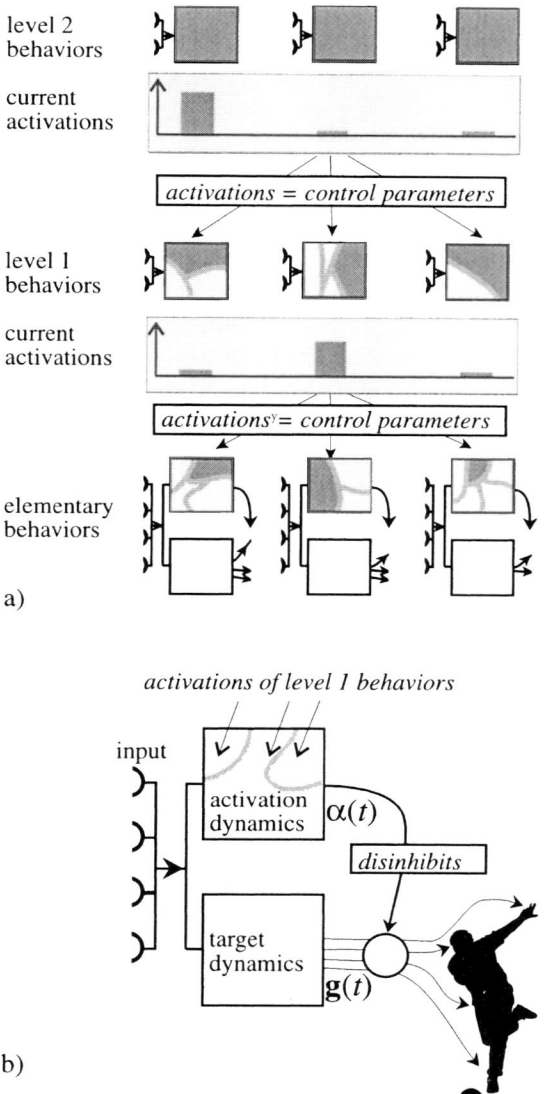

Figure 1: (a) Global structure of a DD behavior control system. At any time, every behavior has an activation. Activations of higher-level behaviors (depicted in shaded boxes) act as control parameters for the activation dynamics of lower levels. The dynamical system which maintains a behavior's activation can undergo bifurcations; this in indicated by depicting these systems as stylized "phase diagrams" (boxes with irregular partitions). A mode of the entire system is thus determined by the activations of all higher-level behaviors. (b) The target and activation subsystems of an elementary behavior.

in front. Now assume, by contrast, that the robot is trying to get the ball unstuck. Its level-1 behavior freeBall should have an activation of about 1. This value is again passed to the activation dynamics of bumpRetract as a control parameter, changing the response characteristics of this dynamical system. It should now indeed retract even when hitting the ball, since it makes little sense to try getting a ball unstuck by pushing it further into where it's been got stuck. In technical terms, the activation dynamics of bumpRetract undergoes a bifurcation when the activations of charge and freeBall change in a certain way.

These bifurcations are mathematically designed in the simplest possible way. For each relevant higher-level behavior, the activation equation is equipped with a particular additive term, which is multiplied with the concerned higher-level activation. For instance, the equation for the activation $\alpha_{\text{bumpRetract}}$ would be controlled by the activations α_{charge} and α_{freeBall} in the following way:

$$\dot{\alpha}_{\text{bumpRetract}} = \alpha_{\text{charge}} T_1 + \alpha_{\text{freeBall}} T_2 + \ldots + \text{decay}, \qquad (1)$$

where T_1, T_2 are hand-designed dynamical laws which yield an appropriate activation characteristics in the charge and freeBall modes. The decay term and other details are explained in [Jaeger and Christaller1998].

To reiterate, only the activation dynamics subsystem undergoes bifurcations in a properly designed DD scheme. The fact that bifurcations (which are inherently difficult to master from a designer's perspective) are confined to these single-variable subsystems is critical for the transparency of DD behavior control systems.

Higher-level activation variables yield control parameters for lower-level activation dynamics. Now, in the theory of dynamical systems it is assumed that control parameters change on a (much) slower timescale than the systems they control. This implies that behaviors on different levels in a DD architecture must have different timescales, with higher-level behaviors being long-term and lower-level behaviors become active/inactive on a short-term scale. This provides the designer with a formal criterium for level organization: order higher-level behaviors according to time scales.

We emphasize that an elementary behavior is not "called to execute" from higher levels. The level of elementary behaviors is fully operative on its own and would continue to work even if the higher levels were cut off. The effect of higher levels is not to "select actions", but to change the overall, integrated dynamics of the entire elementary level, by inducing bifurcations in the activation dynamics on that level.

3 The dual dynamics design tool

Programming a football playing robot is a group activity, where different researchers are occupied with designing different branches and levels of the overall robot control system. In order to achieve an "integrated" behavior, the design process must be maximally transparent for all group members. Essentially, everybody must be able to understand and use, what everybody else designs. Therefore, we have developed a unified software developing environment, the "Dual Dynamics Designer" (DDD). Specifically, DDD provides automated editing, documentation, simulation and code generation facilities.

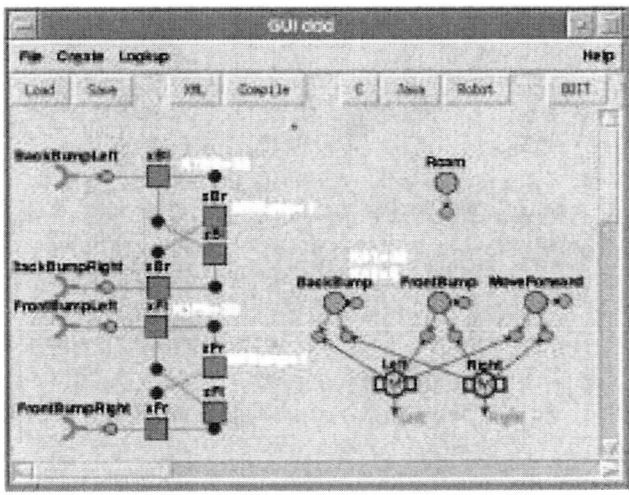

Figure 2: The primary DDD user interface. The example shows a basic roam behavior with bumper-based obstacle avoidance. Sensor filters and intermediate representations are on the left, higher-level behaviors (only roam in this case) are on the right upper part, and elementary behaviors on the right lower part of the screen.

The primary graphical user interface for designing a DD model is shown in Fig. 2. It includes icons of sensors, sensor filters and intermediate sensor representations, elementary and higher-level behaviors. Important global variables and constants (time constants, especially) appear highlighted besides the concerned icons. By clicking on the icons, context-sensitive editor windows pop up in which equations and/or ODEs can be specified in an intuitive syntax.

After designing the network of behaviors and preprocessing filters, a syntax check, global and local variable detection and checking for cyclic dependencies between equations is performed in a compilation step. Cyclic dependencies (which are unavoidable in coupled dynamical systems) are highlighted in the graphical representation on the screen. It is left to the designer to schedule a processing order for cyclically connected variables, which s/he can do by simply rearranging the icons from left to right.

By hitting the C, Java, and Robot buttons, executable standard C code, Java code, and robot C++ code is generated. The Java code can be fed into a simulation engine, which currently simulates the interaction of a single robot and a ball in an empty arena. The simulator provides a number of diagnostic traces of activations and target variables, as well as a graphical rendering of the robot's doings in the arena. We find the simulation of inestimable value in detecting "dynamo-logical" misconceptions in the designed activation and target dynamics.

The DDD tool is based on a proprietary object-oriented behavior representation, which is taken as common source for all target code generators (C, C++, Java, HTML). Therefore, the C++ code generated for our robots' onboard PCs, exactly mirrors the Java code used in the simulation. The generated documentation is common for all targets and hides language dependent implementation and syntactical details from the behavior designer. The documentation of the sensor preprocessing and DD control program allows a convenient inspection of all parts of the robot control system, ordered by various aspects.

The DDD tool itself is constructed with the Rapid Prototyping Environment APICES [Bredenfeld1998]. Readers interested in software engineering aspects can find more details on the software architecture and development process of the DDD tool in [Bredenfeld1999].

An exemplary control program (sketched in the next section), the simulator, and the documentation are available on our web server at $http://ais.gmd.de/BE/ddd/$, and can be run on Web browsers that support Java (tested on Netscape).

4 Kick a moving ball: a case study

In this section we sketch a DD behavior control system for achieving the first RoboCup-99 video qualification task. This task for a single robot consists in finding a stationary ball and scoring a goal without opponents. We made this task a bit more difficult by using a ball that rolls about while the robot tries to find and kick it.

We employ a team of custom-built 2 degree of freedom, 1-PC-3-microcontroller equipped robots that rely on the well-known Newton Lab's Cognachrome system for ball and goal detection, infrared-based distant obstacle avoidance and otherwise standard bumper ring sensors and odometry. The robots do not have concavities to guide the ball. Instead, they hit the ball with their straight front portion and rely on billiard-like ball reflection. A more detailed description of our robots is given in [Kuth et al.1998].

The difficult part of this task is kicking the moving ball into the right direction after it has been spotted. This implies hitting the ball with some appropriate (underconstrained) combination of angle, velocity and position, grounded on rather noisy estimates of ball state. This problem lies well beyond the powers of classical approaches to motor control.

We approached the task by breaking it up into various elementary behaviors, each of which comes with its own sensor-motor control strategy. The overall goal is solved by an appropriate chaining, superposition, and inhibition of the participating activation dynamics.

Fig. 3 lists the relevant behaviors (a) and depicts a typical search–intercept–kick episode (b). The latter diagram was obtained from an simulation implemented in Mathematica in the early stages of the DDD development. Initially

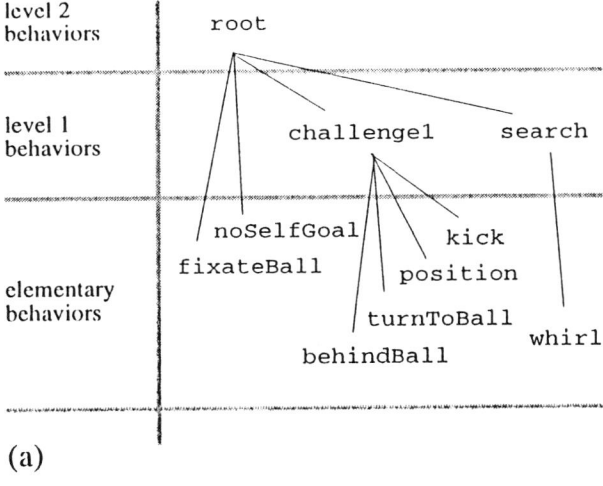

(a)

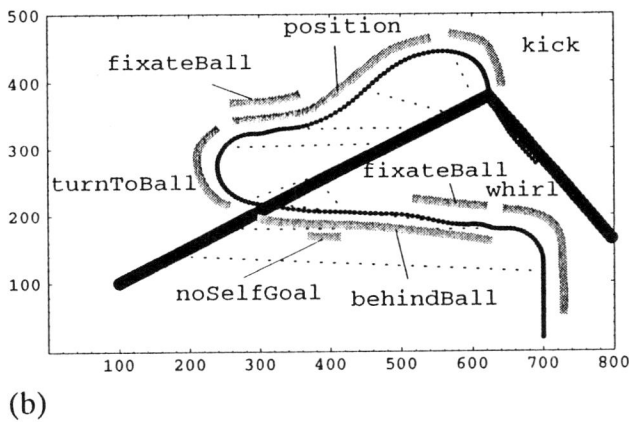

(b)

Figure 3: (a) The behaviors involved in the first qualification video task (several obstacle avoidance behaviors are omitted). (b) A successful (simulated) search – intercept – kick episode. The diagram shows head view of arena, opponent's goal on the right side. Ball (thick black line) starts at lower left corner with velocity 95 cm/sec. Robot (narrow black line) starts at lower right corner. Dotted lines connect equitemporal points on robot and ball trajectory every second. Shaded lines indicate activation periods of elementary behaviors.

the robot does not see the ball, having a vision field of only ± 33 degrees. When ignorant of the ball position, the robot falls into `search` mode. Besides some obstacle avoidance behaviors, this mode basically comprehends only the elementary behavior `whirl`. The motor commands issued by `whirl` consist of a simple alternation of straight move-forwards and circles. This pattern is active until the ball is seen after a half left turn. Seeing the ball, the robot falls into `challenge1` mode. The elementary behaviors `behindBall`, `turnToBall`, `position`, and `kick` can now potentially become activated. It depends on the situation and history which behavior is triggered. In the example in Fig. 3(b), `behindBall` is activated first. Its control law says: "move toward own goal at max velocity until robot is well behind ball". The next behavior is `turnToBall` ("turn into direction where ball is expected"), followed by `position` ("move to a position from which ball can be kicked into goal") and finally, `kick` ("bump into ball with velocity that makes it billiard-bounce toward goal"). During this sequence, there are also two activation periods of `fixateBall`. This is a mode-independent (technically: `root-mode`) elementary behavior, which tries to keep the ball inside the vision cone for about 1 sec — the time needed for sampling enough video frames for a useful estimate of the ball vector. Finally, there is a brief activation of `noSelfGoal` ("if in danger of kicking ball into own goal, avoid the ball"), which however in this case has no motor effects since the robot quickly calculates that it avoids the ball anyway.

The sensor-motor control laws of these behaviors range from trivial to tricky. For instance, the motor target trajectories generated by `whirl` are actually entirely precoded and independent of sensor input. The target dynamics of `position`, by contrast, includes mechanisms of ball prediction and a position evaluation.

The powers (and difficulties) of the DD approach, however, lie in the activation dynamics rather than in the target dynamics. Several mechanisms, all of which are locally coded into the activation dynamics laws of the behaviors, control the interaction and trigger pattern of these activations. The most important mechanisms are:

Sensor conditions. Activate or inhibit a behavior when certain sensor input conditions are satisfied. Example: `kick` gets active "opportunistically" when ball is seen roughly in line with goal.

Chaining. Activate a behavior when certain other behaviors become deactivated. Example: `kick` gets active when the activation of `position` goes down.

Inhibition. Inhibit a behavior by the activation of others. Example: most behaviors are inhibited by obstacle avoidance behaviors.

Furthermore, these activations can be gradual (e.g., `fixateBall`'s activation grows with the uncertainty of ball estimates) or almost binary (standard case); they can have a fast dynamics (typical example: protective reflexes) or a slow one (useful for behaviors whose motor trajectors blend into one another, for example the transition from `position` to `kick` is relatively slow).

Interested readers can inspect all equations of the behavior system presented here in the automatically generated DDD documentation at *http://ais.gmd.de/ BE/ddd/chall1.html*.

5 Conclusion

Identifying and coding appropriate dynamical activation schemes is decisive for the performance of a DD control system. Specifically, a simple switch-on / switch-off chaining of behaviors (like in classical action selection literature) is insufficient for a motor control task as complex and dynamic as the one in RoboCup. The phenomenology of dynamic onset, offset, and superposition of behaviors is rich and only dimly understood. We also believe that the elusive "integratedness" of situated motor control, which we mentioned in the introduction, is somehow connected to the problem of shaping appropriate activation patterns. Currently the DD framework does not spell out how the terms T in activation equations (cf. eqn. (1)) have to be written. However, certain standard terms in the activation equations begin to evolve in the BE group's everyday work. One of our major current research topics is to develop a systematic repertoire of such activation schemes, and integrate them into the DDD tool. Other robotics goups with whom we collaborate [Steinhage and Schöner1998] have started to work along the same lines.

Acknowledgments We would like to thank Karl-Ludwig Paap and his group for their generous assistance in designing and building the initial version of our robot hardware and onboard CAN communication system.

References

[Aicardi et al.1995] M. Aicardi, G. Casalino, A. Bicchi, and A. Balestrino. Closed-loop steering of unicycle-like vehicles via Lyapunov techniques. *IEEE Robotics and Automation Magazine*, March:27–35, 1995.

[Bredenfeld1998] A. Bredenfeld. APICES - rapid application development with graph pattern. In *Proceedings of the 9th IEEE International Workshop on Rapid System Prototyping (RSP 98), Leuven, Belgium*, pages 25–30, 1998.

[Bredenfeld1999] A. Bredenfeld. Co-design tool construction using APICES. In *to appear in the Proc. of the 7th Int. Workshop on Hardware/Software Co-Design (CODES'99)*, 1999.

[Brooks1991] R. A. Brooks. Intelligence without reason. A.I. Memo 1293, ftp'able at http://www.ai.mit.edu/, MIT AI Lab, 1991.

[de Micheli and Gupta1997] G. de Micheli and R.K. Gupta. Hardware/software co-design. *Proc. of the IEEE*, 85(3):349–365, 1997.

[Jaeger and Christaller1998] H. Jaeger and Th. Christaller. Dual dynamics: Designing behavior systems for autonomous robots. *Artificial Life and Robotics*, 2:108–112, 1998. http://www.gmd.de/People/ Herbert.Jaeger/Publications.html.

[Jaeger1997] H. Jaeger. From continuous dynamics to symbols. In *Proceedings of the 1rst Joint Conference on Complex Systems in Psychology, "Dynamics, Synergetics, Autonomous Agents", Gstaad, Switzerland*, 1997.

[Jaeger1998] H. Jaeger. Multifunctionality: a fundamental property of behavior mechanisms based on dynamical systems. In R. Pfeifer, B. Blumberg, J.-A. Meyer, and S.W. Wilson, editors, *From animals to animats 5: Proc. SAB-98*, pages 286–290. MIT press, 1998.

[Kuth et al.1998] A. Kuth et al. Team description of the GMD RoboCup-Team. In M. Asada, editor, *Proceedings of the 2nd RoboCup Workshop*, pages 439–450, 1998.

[Maes1990] P. Maes. Situated agents can have goals. *Robotics and Autonomous Systems*, 6:49–70, 1990.

[Steinhage and Schöner1998] G. Steinhage and G. Schöner. Dynamical systems for the behavioral organization of an anthropomorphic robot. In R. Pfeifer, B. Blumberg, J.-A. Meyer, and S.W. Wilson, editors, *From animals to animats 5: Proc. SAB-98*, pages 147–152. MIT press, 1998.

[Tyrrell1993] T. Tyrrell. The use of hierarchies for action selection. *Adaptive Behavior*, 1(4):387–420, 1993.

[van Gelder1998] T. van Gelder. The dynamical hypothesis in cognitive science. *Behavior and Brain Sciences*, 21:615–665, 1998.

Techniques for Obtaining Robust, Real-Time, Colour-Based Vision for Robotics

James Brusey and Lin Padgham[*]

Department of Computer Science, RMIT University,
P.O. Box 2476V, Melbourne 3001, Australia
{brusey,linpa}@cs.rmit.edu.au

Abstract. An early stage in image understanding using colour involves recognizing the colour of target objects by looking at individual pixels. However, even when, to the human eye, the colours in the image are distinct, it is a challenge for machine vision to reliably recognize the whole object from colour alone, due to variations in lighting and other environmental issues. In this paper, we investigate the use of decision trees as a basis for recognizing colour. We also investigate the use of colour space transforms as a way of eliminating variations due to lighting.

1 Introduction

In many domains artificial vision is the primary mechanism for a robot to sense its environment. In the domain of robot soccer it is important that vision is fast, relatively reliable in recognizing objects and is adequate for the robot to move and act appropriately in the environment. We have been exploring mechanisms to improve the reliable discrimination of objects based on colour.

In previous years we have used a vision system based on RGB (red, green and blue) pixel classification which has been manually tuned to the particular variations of colour and lighting. This year we have experimented with alternatives to RGB, such as HSL (Hue, Saturation, Luminescence) [6] and normalized RGB and evaluated the use of a decision tree learning algorithm for training the system to recognise objects from their colour in a variety of lighting conditions. This has resulted in improved object recognition.

In the following sections we describe the operating environment and the techniques explored in our efforts to improve the quality of information obtained from the vision system.

2 General and Specific Domain Information

Our work is focussed primarily on providing a vision system for use with an autonomous, mobile, soccer-playing robot, participating in the middle size league

---

[*] Part of this work was supported by CSIRO Mathematical and Information Sciences, Australia

of the Robot Soccer World Cup (RoboCup) [8]. In this scenario two teams of 4–5 robots[1] play against each other on a green carpeted field, surrounded by white walls with a blue goal at one end and a yellow goal at the other. The soccer ball is bright orange.[2] All the robots are required to be mostly black, with a light blue or purple marker to indicate team membership. Robots may be up to 80kg in weight and up to 80cm high. All sensing devices must be located on board the robots — i.e. it is not possible to have stationary cameras perceiving the environment and then transmitting information to the robots. However communication is allowed, and an off-board system may be used for computation if desired. Thus information perceived by one robot may be communicated to other robots.

Our robots currently have vision as their only sensor, and although we plan to add some additional sensors, vision will remain the primary mechanism for object recognition and an important mechanism for self-localization. While some level of self-localization is critical (e.g. to avoid kicking the ball towards the wrong goal), we are not aiming to achieve a high level of accuracy of location information. A human soccer player is able to play with only a rough knowledge of their own location, and so we feel that a rough knowledge is all that is required for a robotic soccer player.

Vision is also the only mechanism used for estimating distance from objects, though we may add tactile sensors for detecting when objects are very close.

The robots are quite fast-moving, with a top speed of over $2ms^{-1}$. Information from the vision system is the primary means used to detect an impending collision. It is essential therefore that the vision system is able to process a large number of frames per second. Consequently the speed of processing for each frame of visual data is crucial. Our initial aim for the vision processing speed was 15 frames per second. This means that the robot, when at top speed, will move up to 13cm before it has completely processed visual information.

Our robots are using Logitech QuickCam^{TM} VC cameras mounted in a fixed position on the robot at a height of 325mm from the ground and with the camera pointing at an angle of about 17 degrees downwards from the horizontal. The field of view is 45 degrees wide and we do not currently use any distorting lenses (such as a fish eye lens). The camera has a maximum resolution of 320 × 240 pixels but we currently use the 160 × 120 mode.

The main processing phases in the vision system are:

1. (grab) get frame from camera
2. (smooth) apply smoothing
3. (classify) classify each pixel according to colour as belonging to a particular object
4. (segment) divide up image into segments of the same classification
5. (filter noise) discard small segments
6. (locate) estimate distance and angle to object associated with each segment

The work we are presenting in this paper is focussed on step 3 (classification).

---

[1] There were 5 robots per team in 1997 and 1998, but only 4 per team are planned for 1999.

[2] The ball is defined in the regulations as being red, but the actual colour of the ball used is described by most people as bright orange.

3 Classification

Classification has the aim of deciding which pixel belongs to which type of object. Each pixel is 24 bits with 8 bits each for red, green and blue. This format is known as True Colour. The red, green and blue (RGB) values for an object depend mainly on the colour of the object, but also on the colour of the lighting and level of illumination, reflections from other objects, shadows and the accuracy of the camera. For a moving robot, some of these factors will change from frame to frame. Even for a single object of uniform colour, a wide variety of different RGB values will be detected.

In order to process the image as quickly as possible, and because classification has to be performed on every pixel, it is important that this step is very simple. One approach is to compare each component of the pixel (i.e. red, green and blue) to a set of minimum and maximum thresholds. Another approach is to use a look-up table (LUT) with one element for each possible pixel value.

3.1 Threshold Classification

When using thresholds [11], the threshold values can either be selected manually, or by collecting pixel samples for each object type and using the mean and variance of each pixel component to determine the thresholds. Our previous system used the latter approach, and based the threshold values on the mean, plus or minus twice the standard deviation.

The thresholds form a bounding box in the pixel feature space, aligned with the axes. Every pixel colour inside the box is considered to belong to that object. There are two main problems with this approach. The first is that it assumes a distribution that fits neatly into a box, and particularly, one aligned with the axes. Second, the boxes (and thus the thresholds) sometimes overlap.

Most objects have both specular highlights and shadowed regions which, if used in finding the thresholds, tend to expand out the threshold box so that it tends to find false positives. If they are excluded from the calibration of thresholds, these parts of the object are not detected, or in other words, they will be false negatives. We found that when using the threshold approach, the trainer had to be careful to select that part of the object that was neither too dark nor too bright as the basis for training. When this approach was used, about half the object was recognized, which was typically just enough to allow the object to be tracked, as long as the lighting remained constant.

When more extensive training was performed, the thresholds tended to overlap. The previous system allowed for this by checking the thresholds in a set order. Unfortunately, the order that was used put the ball first, which tended to result in a lot of false positives for the ball. A better approach might be to adjust the thresholds to avoid overlap.

3.2 Look-Up Table Classification

Another way to train a system to recognize a particular colour, is to use a look-up table (LUT) [5, 7], with one element for each possible colour (2^{24} bytes are required for a 24 bit pixel feature space). LUTs are often implemented in hardware and closely associated with the camera. Instead, we implemented the LUT in software, which had the benefit that the main memory was the only limitation on the size of the LUT, and in fact we initially used a 16Mb table size. (Later we dropped the least significant bit from each component, reducing the LUT size to 2Mb.) The LUT can be manually trained by collecting pixel samples for each object. Unfortunately, this typically produces a sparsely filled array and subtle variations in lighting may prevent pixels from being classified because the training didn't include that combination of red, green and blue.

Therefore, a common augmentation of the LUT scheme is to find a generalized representation of the colour associated with a particular object and then to fill the LUT based on generalized representation. Since populating the LUT only needs to occur during training, the processing time per frame is still fast.

Given the variability of the data, what is needed is a colour generalization from actual mapped data points. It is important that this generalization be at the right level for the environment. As the RoboCup environment has very distinct colours, it should be possible to have a fairly coarse grain generalization.

A common first step in generalising colour is to transform the raw RGB (red, green and blue) components into a more appropriate colour space. The problem with RGB data is that the components are highly correlated — for example, as the lighting increases, there is a corresponding increase in all components. The correlation between red and green can be seen in Figure 1. The aim of converting to a different colour coordinate space is to reduce the correlation between the coordinates.

3.3 Alternative Colour Spaces

The approach used for our past RoboCup efforts involved finding fixed maximum and minimum thresholds for each of the red, green and blue (RGB) colour components based on the tristimulus model [9]. The thresholds were determined by first establishing a training set which mapped colours (RGB triplets) to object identifiers. The variance and mean for the set of colours associated with any one object where then found, and the maximum and minimum thresholds based on this.

A number of past RoboCup teams have used variants of HSL (such as HSB or HSV [3]) for colour discrimination since it separates apparent "colour" from "brightness". The brightness values of these schemes generally do not correspond well with human perception of brightness, however this is not necessarily a problem when the aim is to discriminate colours rather than to reproduce them accurately.

Even though it is useful to separate out the measure of brightness, this coefficient cannot necessarily be discarded. In the case of RoboCup, one of the objects

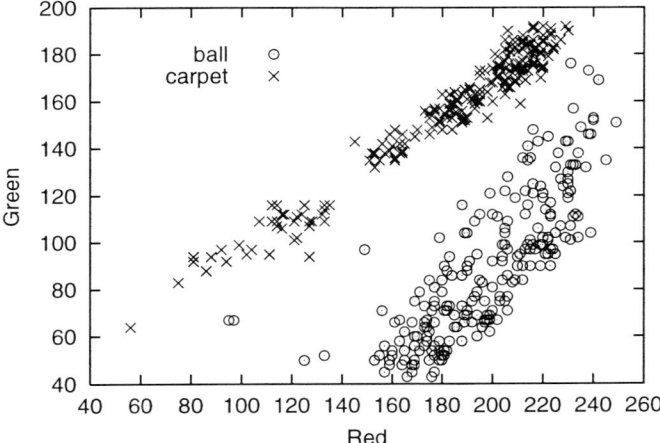

Fig. 1. Red versus Green in the RGB colour space, for sampled data for the green carpet and the orange ball.

that needs to be recognized is largely black. This object becomes impossible to differentiate from other objects if the brightness factor is ignored. To see this, we projected sampled data onto a two dimensional plane. The particular projection that we tried has previously been used with some success by Batchelor and Whelan [5] for colour discrimination of different types of apples. They refer to this projection as $\Gamma(X)$ and it is defined by:

$$U = \frac{R - B}{\sqrt{2}(R + G + B)} \tag{1}$$

$$V = \frac{2B - R - G}{\sqrt{6}(R + G + B)} \tag{2}$$

where $\Gamma(X) = (U, V)^T$. Unfortunately, in this colour space, samples from the black object are widely spread over much of the space and are not differentiable from the clusters of points for other objects. This yielded poor results when it was used as a basis for colour generalization (see Table 1).

One problem we found in using HSL coordinate space was that the hue component is measured as an angle around a colour circle and therefore wraps around to zero. This can result in different parts of the same object mapping to opposite parts of the hue axis, effectively splitting in two the cluster of data points for that object, as shown in Figure 2. Therefore, it is inappropriate to summarize hue using statistical mean and variance. Changing the mapping of the hue so that the split occurs at another point may solve the problem in some cases but is just as likely to move the problem elsewhere.

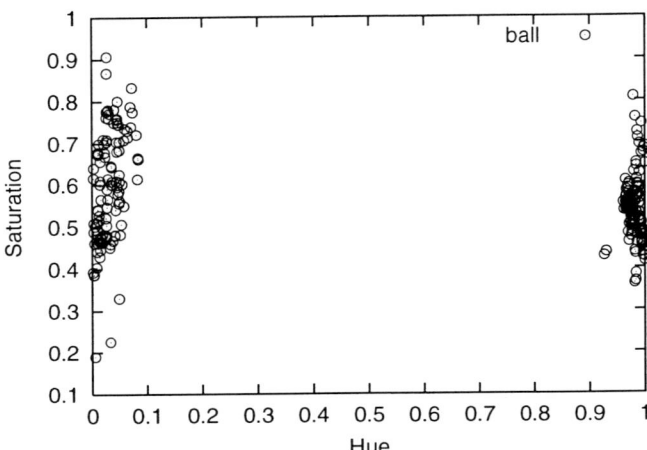

Fig. 2. Hue versus Saturation in HSL colour space, for data for the ball. Note that the data for the ball has wrapped around from 1 to 0.

This problem is similar to that found in RGB space, where the cluster of data points for the colour of an object may be of an odd shape. Changing coordinate systems doesn't necessarily remove the unusual characteristics of the shape of the cluster.

In early results, using HSL showed clear gains in robustness of object identification and the issue of colours being wrapped around the hue axis could be dealt with using a decision tree rather than a thresholding approach (described below).

An alternative method for dealing with specular highlights and shadowed areas (which cause wide variation in RGB coordinates under different lighting conditions), is to normalize the RGB coefficients [7]:

$$r = \frac{R}{R+G+B} \quad (3)$$

$$g = \frac{G}{R+G+B} \quad (4)$$

$$b = \frac{B}{R+G+B} \quad (5)$$

Since $r + g + b = 1$, one component can be dropped (usually b) as it provides no information. One of the original components, such as G, can then be added in to gain a measure of the brightness level. The transformed triplet is then (r, g, G). We refer to this colour space informally as rgG.

This mechanism is quicker to compute than HSL, and it produced colour classification results which were almost as good.

Another colour space which has been used previously to assist colour generalization [7], is HSI. HSI can be calculated from RGB as follows:

$$I = \frac{R+G+B}{3} \quad (6)$$

$$S = \frac{1 - \min(R,G,B)}{R+G+B} \quad (7)$$

$$H = \arctan\left(\frac{3(G-B)}{2R-G-B}\right) \quad (8)$$

3.4 Using Decision Trees for Learning Colour

Whichever colour coordinates are used (RGB, HSL, rgG or HSI), there is a tendency, especially where the training data is extensive, for the resulting thresholds for different objects to overlap. Thus, although thresholding is fast it can often lead to conflicts in object identification.

An alternative approach is to use a decision tree structure which allows for a more detailed partitioning of the feature space than that allowed by thresholds. The added degrees of freedom give an effect analogous to having multiple threshold cubes associated with any particular object.

If appropriate rules are known, decision trees can be designed manually. More often they are created from training data, using a learning algorithm. We used the C4.5 learning algorithm [10], which is an industrial strength, decision tree based, machine learning algorithm. It is a batch mode learning algorithm, which means that it prepares the decision tree on the basis of all the training examples, rather than adaptively changing the tree as it receives more data.

Decision trees, when based on continuous values have the advantage that they can always classify any data, unlike thresholding which may leave some data unclassified. Also no data is multiply classified as can occur with thresholding.

Figure 3 shows an example of a pruned decision tree. Pruning is performed automatically by the C4.5 algorithm, and is done by removing parts of the tree which do not significantly affect the error rate. The leaf nodes show two numbers (N/E) after the classification. The first (N) is the number of training cases which ended up at that leaf node. The second (E) is the predicted error rate for that node if N unseen cases were classified by the tree.

3.5 Classification Results

The results for the threshold and decision tree classifiers are shown in Table 1. The overall result is that the decision tree classifier is generally better than the threshold classifier and generally classifies 97% of data correctly. The HSL and HSI colour spaces wrap around and so they were inappropriate for use with the threshold classifier. The RGB colour space defeated our expectations and gave the best results. There is some margin for error in these results, but we were still surprised to see the RGB colour space rate so well.

The result for the $\Gamma(X)$ space shows that, for this type of data, three dimensional colour data is required to discriminate between the different classes.

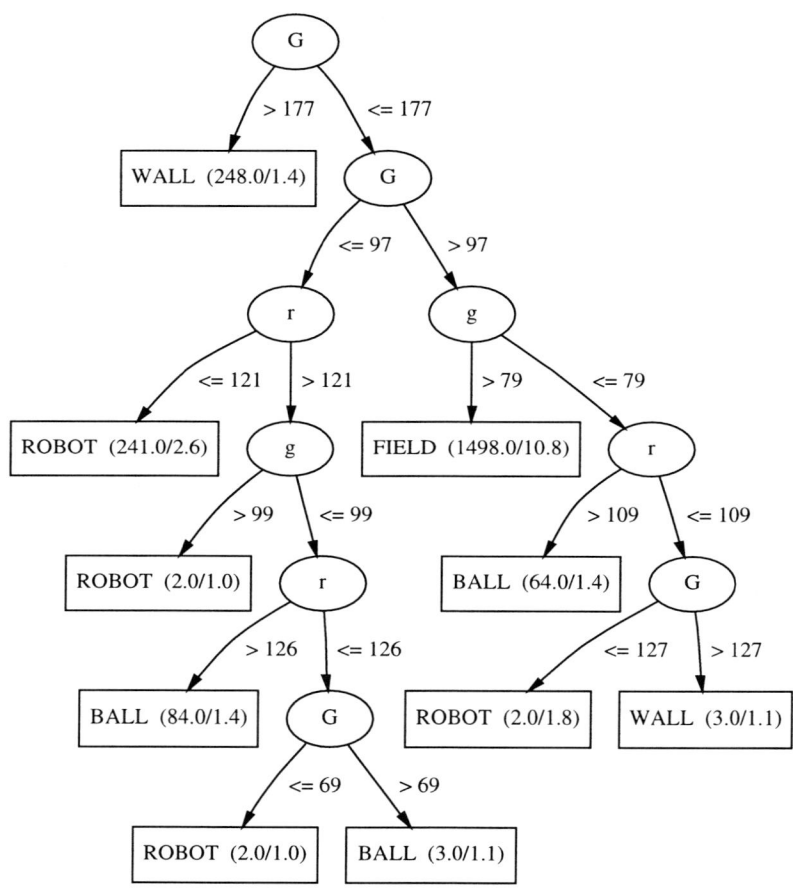

Fig. 3. An example decision tree. The colour model used in this case was rgG, where the "G" component roughly corresponds to brightness. All components are scaled in the range [0, 255].

Table 1. Summary of results for different colour models. All results were based on tests on data not seen during training (using a 50/50 split between training and test data).

Colour model	Threshold classifier			
	Correct	Incorrect	Multiply classified	Unclassified
rgG	80.0%	1.4%	8.8%	9.8%
RGB	84.9%	2.1%	6.0%	7.0%
	Decision tree classifier			
$\Gamma(X)$	87.5%	12.5%	—	—
HSI	97.5%	2.5%	—	—
HSL	97.6%	2.4%	—	—
rgG	97.3%	2.7%	—	—
RGB	97.9%	2.1%	—	—

4 Related work

Several other papers presented at the workshop dealt with the issue of colour recognition [1, 2, 4, 12]. Akita [1] focussed on the development of specialised hardware. RGB data supplied by the hardware is transformed to HSV by a look-up table before thresholds for each of the H, S, and V components are used to distinguish the colours of different objects. This classification technique appears to be the same as that described in Section 3.1.

Amoroso et al. [2] found that with the HSI transform, the different objects on the field could be discriminated using the hue component alone. Discrimination is trained by examining the hue histogram for valleys. A feed-forward neural network is used to learn the correspondence between hue "modes", extracted from the histogram, and actual objects. They also deal with the problem of having separate hue regions that correspond to a single object (as discussed in Section 3.3).

Use of hue alone may not allow for recognition of black objects, which we found hard to discriminate unless brightness was taken into account (Section 3.3). Their results show a smaller percentage of misclassifications than we were able to achieve but this may be related to our attempts at recognising black objects.

Bandlow et al. [4] start with data in the YUV colour space. A classification look-up table is trained for the UV components. For the Y component, which corresponds to the pixel brightness, minimum and maximum thresholds are determined for each object.

Terada et al. [12] mention that they use Mahalanobis distance measures to classify colours and train with samples of each colour.

One difficulty with comparing results is that the quality of the source data is subject to error that is critically dependent on the type of camera. We found that it was quite difficult to tune the QuickCam VC so that the colours looked realistic. At the boundaries between light and dark objects, lines of bright red sometimes appeared due to the way the colours are detected using a mask.

5 Evaluation and Discussion

Our goal in working with the vision system was to develop a sound perceptual footing for a cognitive system that approaches human-like behaviour within this domain. Our longer term goal is to explore issues in building, maintaining and reasoning with approximate world models in a dynamic real-time domain. Our main requirements for the vision system were therefore:

1. to achieve robust object identification (this implies virtually no false positives or negatives); when an object is present in the visual field it should be seen and recognized, the system should not "see" anything that is not actually in the visual field;
2. to process frames at a rate sufficient to judge direction of movement of target objects with some rough idea of speed;
3. to coordinate vision with movement to avoid problems such as hitting objects before it has registered that they are there, or not finding the ball due to having moved past it by the time the robot has processed that it was "seen".

The use of a decision tree colour classifier has addressed problems due to oddly shaped colour clusters, while the learning of this decision tree has ensured the appropriate amount of generalization/specialization for the environment as well as removing much of the tediousness previously associated with establishing appropriate threshold values. These combined measures have resulted in robust object identification and we appear to have eliminated problems with both false negatives and positives.

We are currently getting a frame rate of 16 frames per second, which we believe will be sufficient for judging direction of movement and for coordinating with the robot's own movement.

Our vision system does not address vision issues associated with such things as reading text or recognizing specific shapes. Such things are certain to be useful for self-localization if incorporated into the RoboCup environment as is being discussed. However, based on human behaviour it does not seem that such abilities should be necessary for this domain. We are interested rather in exploring other approaches to the issue of adequate self-localization in the absence of detailed, accurate positioning data obtained either from visual nuances or other percepts.

Two aspects of vision that we expect to address in our continued work are the use of a movable camera, and if possible, obtaining of a wider angle of vision. Both these directions aim to address the problem of seeing too little at any one time. For all players, but particularly for the goalie, it is important to have a reasonably wide visual coverage.

We are hopeful that the improvements made to date will give us adequate perceptual information to make use of our action selection and coordination mechanisms, as well as allowing us to explore dynamic teamwork.

References

[1] Junichi Akita. Real-time color detection system using custom LSI for high-speed machine vision. In Veloso [13].
[2] Carmelo Amoroso, Antonio Chella, Vito Morreale, and Pietro Storniolo. A segmentation system for soccer robot based on neural networks. In Veloso [13].
[3] Minoru Asada, editor. *RoboCup-98 : Proceedings of the second RoboCup Workshop*, July 1998.
[4] Thorsten Bandlow, Michael Klupsch, Robert Hanek, and Thorsten Schmitt. Fast image segmentation, object recognition and localization in a Robocup scenario. In Veloso [13].
[5] B. G. Batchelor and P. F. Whelan. Real-time colour recognition in symbolic programming for machine vision systems. *Machine Vision and Applications*, 8(6):385–398, 1995.
[6] A. Glassner, editor. *Graphics Gems*, pages 448–452. Academic Press, 1990.
[7] P. Jasiobedzki, B. Down, and V. Wu. Active object detection using colour. In C. Archibald and P. Kwok, editors, *Research in Computer and Robot Vision*, pages 37–53. World Scientific, 1995.
[8] I. Noda, S. Suzuki, H. Matsubara, M. Asada, and H. Kitano. Robocup-97: The first robot world cup soccer games and conferences. *AI Magazine*, 19(3):49–59, 1998.
[9] W. K. Pratt. *Digital Image Processing*. John Wiley and Sons, 1991.
[10] J. R. Quinlan. *C4.5 : programs for machine learning*. Morgan Kaufmann, 1993.
[11] M. Sonka, V. Hlavac, and R. Boyle. *Image Processing, Analysis and Machine Vision*. Chapman and Hall, 1993.
[12] K. Terada, K. Mochizuki, A. Ueno, H. Takeda, T. Nishida, T. Nakamura, and A. Ebina. A method for localization by integration of imprecise vision and a field model. In Veloso [13].
[13] Manuela Veloso, editor. *Robocup-99 : Proceedings of the Third RoboCup Workshop*, July 1999.

Design Issues for a Robocup Goalkeeper

Riccardo Cassinis and Alessandro Rizzi

University of Brescia
Dept. of Electronics for Automation
Via Branze38, I-25123 Brescia - Italy
Phone +39-030-3715.469 - Fax+39-030-380014
{riccardo.cassinis, alessandro.rizzi}@unibs.it

Abstract. This paper presents Saracinescu, the goalkeeper robot of the Italian team that was used at the Robocup '98 Paris championship. The machine features an original omni-directional vision system whose performance, enhanced by a simple but effective movement strategy, proved to be very smart and led to good results during the tournament. The paper describes the vision algorithms in detail, and discusses some issues that are still being developed and/or refined. An overview of the other components of the machine (mechanical structure and ball-kicking mechanism, computing architecture, auxiliary software routines for initial positioning, etc.) is also included.

1 Introduction

The Robocup championship offers a simple and well-structured environment, suitable for testing some innovative robot features like the visual guidance system presented in this paper. One of the simplifications that this environment introduces with respect to the real world is the small number of colors used in the playground and the rigid coding of their meaning. However, even if each game component has a unique color, problems for color matching still remain, due to illumination changes, shadows, etc.

The robot we present is the goalkeeper of the Italian team (ART), which exhibited a very good performance during the Robocup '98 Paris championship. Its main characteristics are an omni-directional vision system and a simple but effective reactive strategy.

The visual guidance system is based on color information grabbed with an omni-directional, quasi-spherical device. The intrinsic geometric complexity of the image is simplified by the use of color. Neither shape nor other geometric features are taken into account. Only the colors and the relative position of the objects surrounding the robot influence its movement. Even if the idea of using omni-directional visual devices had already been used in previous Robocup events, the presented one allows measuring not only the direction, but also the distance of relevant objects. The underlying idea has since then been adopted by several other researchers involved in the Robocup competition.

In Par. 2 it is presented the overall structure of the robot, the details of the vision subsystem are described in Par. 3, and in Par. 4 the strategy is introduced.

2 An overall robot design description

The robot is based on a modified version of the widely used, commercial RWI Pioneer 1 platform. An on-board PC, with the appropriate power supply, was added to provide the necessary computing power. The PC (an Intel Pentium II) runs LINUX operating system, and is equipped with all the necessary peripherals, that include an Intel Video Recorder frame grabber used to acquire camera images, and a Wavelan wireless networking interface.

Given the task the goalie has to accomplish, the only possibility offered by the mechanical structure was to use the robot sideways, in order to make it able to quickly reach any point of the goal area. The original castor wheel that supports the weight of the robot was replaced with a spherical device in order to eliminate lateral skids when the robot reverses its movement.

A mechanical ball-kicking device has been mounted on the left side of the robot, that always faces the playground (Fig. 1). When the ball touches the lower edge of the kicker, it activates a mechanical switch that in turn triggers a "kick and reload" mechanism. Kicking power is provided by a steel spring, while reloading is accomplished by an electric motor. After each kick, it takes about two seconds to reload the mechanism. The kicking reflex takes place locally, i.e. with no computer intervention; the main computer can, however, disable and re-enable the device.

Another significant change was the addition of a large plexiglas pipe (Fig. 1), fixed on the upper part of the robot, that supports the mirror of the vision system, as it will be described in the following paragraph.

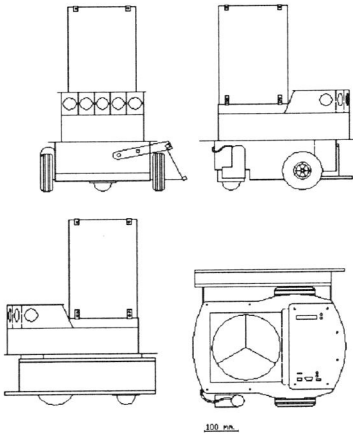

Fig. 1. Saracinescu structure design

3 The omni-directional visual device

Several kinds of omni-directional visual sensors, with different geometrical characteristics [2,3,4] and different optical pre-processing capabilities [1] have been investigated so far. They have been used for various robot navigation and self-localization tasks.

The omni-directional device developed for Saracinescu uses a mirror with a spherical sector shape and an optical grade reflecting surface, that allows a clear vision of what is happening around the robot. The spherical shape of the mirror allows the perception of a larger amount of details in the area surrounding the robot and only rough visual information of the area away from the robot. The mirror axis is vertical, and the device (actually, a 20 cm diameter stainless steel pan lid was used) is supported by a clear Plexiglas pipe, that also houses the camera. The idea is to mimic the behavior of a real goalkeeper that does not care much about what is going on in the opposite half of the field, but pays great attention when the ball comes close him.

Fig. 2, besides giving an idea of the structure of the device, shows its most interesting feature: using a quasi-spherical mirror allows measuring not only the bearing of the bjects with respect to the robot axis, but also their horizontal elevation. Since all objects in Robocup lay on the ground and have known dimensions, the system can compute objects distance as well.

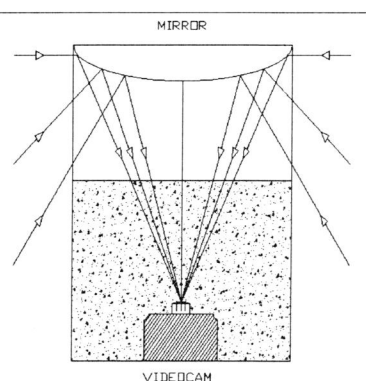

Fig. 2. The omni-directional vision system

An upward pointing CCD color camera grabs images reflected by the mirror. Its signals are processed in order to extract information about the goal posts and the ball. This information is used to keep the robot between the ball and the goal.

Fig. 3 shows an actual image grabbed from the camera. Grey scale rendering makes it hard to distinguish details, but it is quite easy to recognize the ball (red), the field walls and lines (white) and the goal (black, in the lower part of the image). The figure shows the results of processing superimposed to the original image: the recognized portion of the ball is drawn in blue, while the goal (originally yellow) is drawn in black. Careful analysis of the picture shows that all the field is shown, including the opposite goal. Far objects are poorly detailed but, as it was said, this is not important for the goalie.

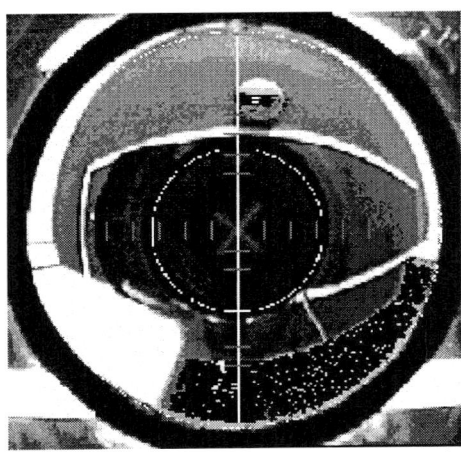

Fig. 3. An example of segmented omni-directional image

Measuring the Euclidean distance of objects from the center of the image allows estimating their distance from the robot. Since the exact shape of the mirror is not known, it would have been hard to determine the equation that gives actual distance as a function of the distance measured on the image. Instead, a look-up table was experimentally built, that lists correspondences between 50 and 230 cm (the estimated useful range), with 10 cm increments. The system uses linear interpolation for estimating intermediate values.

3.1 Color segmentation and object recognition

As it was said in the introduction, Robocup rules assign unique colors to all the objects in the playground. For this reason, no shape information is used for object recognition.

Two constraints influence the color segmentation process: the high computing speed required to effectively track the ball and the heavy varying illumination conditions. In fact, preliminary tests of the robot have been done on various training playgrounds, and have shown that apparent colors of the objects vary from field to field, due to the different light sources, materials used for building the field, etc. Additionally, our system has to deal with its very low-cost surveillance camera, whose color rendering and constancy are light-years far from perfection. Shadows produced by objects make the task even more tough.

To solve the problem of widely varying light conditions, a calibration phase has been introduced. During this calibration phase, usually performed once before each match, a supervisor has to select in the image a ball portion. The mean chromatic value among all the selected pixels is computed. This triplet is the center of the area in the RGB space that is clustered with a boundary threshold value for each chromatic channel. These boundary values are initially set equal to the relative chromatic variance of the selected pixels. This leads to a quite robust automatic clustering, but some problems still occur in presence of shadows and reflections.

Thus, a reinforcement of the color calibration phase has been introduced. In this second phase the supervisor selects a small area into which the ball is completely contained and the three RGB thresholds are incremented, thus enlarging the color clustering subspace, until a pixel that surely does not belong to the ball is selected.

After the calibration on the ball color, the same procedure is used to determine the color of the goal that can be alternatively yellow or blue. The whole procedure could be automated, but due to the strict deadline of the tournament, this has been deferred.

3.2 Ball and goal-posts recognition

During normal operation, the goal of the vision system is to recognize the ball and the goal posts, and to measure their position with respect to the robot. The speed requirements call for processing at least 10 frames per second, and some optimization has to be done to reach this goal.

The central part of the image does not contain any useful information, and is discarded. Only the external circular part of the image is considered for searching the ball and the goal-posts. With the mounted low-cost commercial camera, 5 bits per pixel for each chromatic channel have been used.

The color segmentation is performed using a region-growing algorithm which easily detects the ball. The goal posts are then detected selecting the boundary pixels between the goal color and the white field walls. Due to the geometry of the image, no confusion between the left and the right goal post can arise.

The ball is localized using the apparent center of gravity of the cluster of red pixels. To avoid localizing errors, a threshold on the minimum number of clustered pixels is used. The posts can be localized using the middle or the lower point of the connected boundary line. In the first case perspective errors due to the changing distance of the robot and the goal gate are introduced. In the second case the shadow of the robot can cover the lower post pixels introducing errors as well. The tests and the played matches proved that both methods work correctly in most practical cases. Regardless of the chosen method, what is really important for the robot movement, is not the goal-posts distance estimate, but their relative angles.

The computing speed requirements, and the restrictions of the Pioneer software system, suggested that the ball and the goal recognition should be treated as two separate activities. Moreover, since the ball moves much faster than the robot, two ball searches are performed per each goal recognition.

4 Strategy and robot movement

All the strategies described in the following paragraphs have been implemented in C using the activities functionality of the Saphira programming environment which contains all the primitives for the robot movement. It was decided to avoid using Saphira behaviors because, even if they can help solving some of the problems, they add and unacceptable computing burden, making the robot too slow for any practical use.

Using Saphira activities and micro-tasks concept, different processes can be executed sequentially and cyclically within predefined time slices. In this way, various programs can run together with a simulated parallelism.

The basic strategy obviously is to keep the robot between the ball and the goal gate at all times. In order to do that, Saracinescu lays its right side toward the goal gate and moves back and forth like in Fig. 4. In this position the kicking mechanism faces the playground.

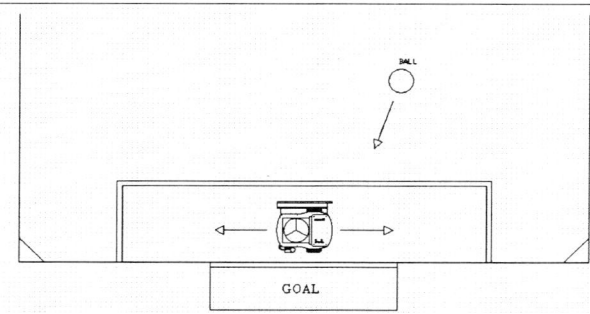

Fig. 4. Saracinescu basic strategy

Intercepting the ball only requires straight movements parallel to the goal: however, wheel skidding, encoder tolerance and crashes with other robots can result in involuntary robot rotations. To correct this problem, at regular time intervals, a process to control and correct the robot horizontal parallelism is executed.

Furthermore, since a constant angle between two reference points results in an arc, measuring the bearing of the goal posts does not allow determining the actual distance of the robot from the goal. Therefore, another task is started at regular time intervals. It controls the estimated distance of the goal posts and consequently moves the robot towards to or away from the goal.

Summarizing, there are four main logical activities running concurrently during normal robot operation:

1. grab the omni-directional image and extract distance and angle of the ball and the goal-posts,
2. keep the robot between the ball and the goal,
3. keep the robot parallel to the goal gate,
4. keep the distance between the robot and the goal constant.

All the geometrically derivable data are directly extracted from the image whenever is possible, introducing data redundancy. If, due to visual noise (occlusion, shadows, etc.) some data are missing, estimates from the data series support the robot for a limited time interval. After this period, if data are still missing, the robot enters a stall state until a sufficient amount of data is available again.

In order to obtain a not too "nervous" robot behavior and thus not to overstress the motors, a low-pass filter that averages data in time has been applied. It can be seen as an inertia increment that allows Saracinescu to keep a steady position without high frequency oscillation.

4.1 The goalkeeping strategy

The goalkeeping strategy is to maintain the line from the robot to the ball parallel to the bisecting line of the angle formed by the goal-posts and the robot (Fig. 5). In order to accomplish this strategy, this activity moves the robot only in the two directions parallel to the goal (Fig. 4).

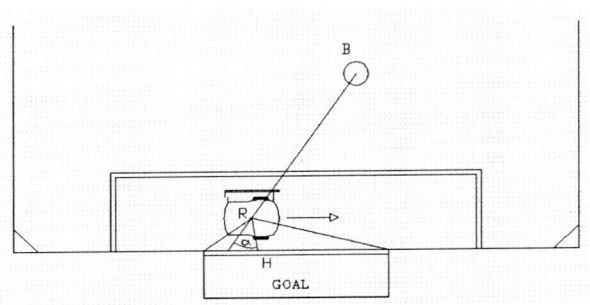

Fig. 5. The goalkeeping strategy

The bisecting line has been chosen instead of the median because it better protects the goal gate. The median would put the robot too close to the goal center. On the contrary, the bisecting line better protects the goal-post surroundings (Fig. 5), stopping also most of the bouncing kicks.

Moreover, this activity keeps the robot inside the small goal area monitoring the goal posts distance. Whenever the robot gets near a goal post, the main positioning algorithm is excluded and the robot starts to decrease velocity, so to stop at the goal-post point. Overtaking this point, in fact, is useless.

The parallelism keeping activity controls the angle between the robot heading and the goal line. The smallest angle the Pioneer can rotate is 5 degrees, thus angles below this value are not considered. Moreover, the Pioneer firmware has its own accidental rotation automatic corrector, so it is important to insert a delay before triggering this activity.

The goal distance control activity controls the robot distance from the goal. If the distance exceeds a threshold value, all the activities are stopped and a repositioning algorithm is executed. This algorithm drives the robot towards its default starting position like shown in Fig 6.

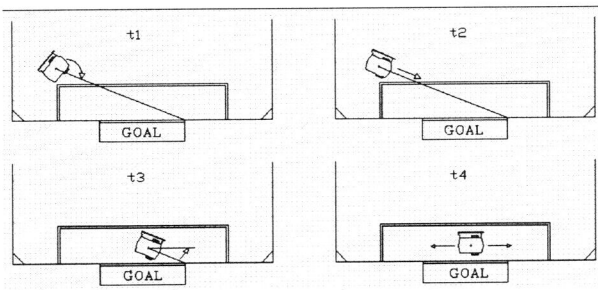

Fig. 6. The repositioning algorithm

5   Conclusions and perspectives

The performance of Saracinescu, that has been the official goalkeeper of the Italian Azzurra Robot Team at the Robocup 1998 championship in Paris, has gone beyond the most optimistic expectances. Its reactive behavior lacks any reasoning and forecast capability, but the machine was perfectly apt for the simple task it had to accomplish. The robustness of the vision system was however a determining factor for its success. It localized the ball during all the matches with good precision and regardless of the illumination, of the shadows and of reflections in the playground. This suggests that all the robots in a team, and not only the goalkeeper, should use a similar system for visual data acquisition.

It can be expected, however, that the performance of other robots will increase in future championships, and that several improvements should be made to the structure of our goalie. Besides the mechanical requirements, that suggest completely rebuilding the machine in order to have a much lighter and faster robot, it would be desirable to fully automate the color calibration procedure, using an automatic ball and goal-post detection. Object recognition should be performed using predictive algorithms, in order to speed up the process. Recognition of robots should be introduced, in order to allow reasoning about where the ball will be kicked by an opponent robot, and how it should be usefully passed to a fellow.

References

1. R. Cassinis, D. Grana, A. Rizzi, Using Colour Information in an Omnidirectional Perception System for Autonomous Robot Localization. In Proceedings of EU-ROBOT96 First Euromicro Workshop on Advanced Mobile Robots '96, Kaiserslautern (Germany), 1996
2. Y. Yagi, H. Okumura, M. Yashida. Multiple Visual Sensing System for Mobile Robot. In Proceedings of the IEEE Int. Conference on Robotics and Automation, 1994, Vol 2, pp. 1679-1684
3. Y. Yagi, Y. Nishizawa, M. Yashida. Map based Navigation for a Mobile Robot with Omnidirectional Image Sensor COPIS. In IEEE Trans. Robotics and Automation, Vol.11, No 5, 1995
4. K. Yamazawa, Y. Yagi, M. Yachida. Omnidirectional Imaging with Hyperboloidal Projection. In Proceedings of the IEEE/RSJ IROS'93, Vol. 2, pp. 1029-1034, 1993

Layered Reactive Planning in the IALP Team

Antonio Cisternino[1], Maria Simi[1]

[1] Dipartimento di Informatica, Università di Pisa
Corso Italia 40, 56125 Pisa, Italy
{cisterni, simi}@di.unipi.it

Abstract. The main ideas behind the implementation of the IALP RoboCup team are discussed: an agent architecture made of a hierarchy of behaviors, which can be combined to obtain different roles; a memory model which relies of the absolute positions of objects. The team is programmed using ECL, a Common Lisp implementation designed for being embeddable within C based applications. The research goal that we are pursuing with IALP is twofold: (1) we want to show the flexibility and effectiveness of our agent architecture in the RoboCup domain and (2) we want to test ECL in a real time application.

1 Introduction

IALP (Intelligent Agents Lisp Programmed) is a team for the simulation league of the RoboCup initiative [1, 2, 3]. The team is programmed using ECL, a public domain implementation of Common Lisp [4].

RoboCup is a real time domain task where players receive perceptions from the server and have to react within the allowed time. To make things more realistic, the environment is inaccessible (perceptions are restricted to the point of view of the player and are limited by the distance) and non deterministic (the effect of actions is not completely predictable).

For the basic architecture of IALP we have adopted a *reactive planning* approach and developed an agent architecture where the global behavior of the planner is structured in layers. The requirements we had in mind for the architecture is that it must be open and offer different levels of abstraction coping with different problems in a modular way. Moreover the architecture is meant to be general and flexible enough to allow reuse of code built for the RoboCup initiative in other domains.

The layered approach used in IALP has been inspired by agent architectures for robots, as proposed for example in [5, 6] and, in the context of multi agent applications such as RoboCup, by the idea that complex behaviors can be learned in layers of increasing complexity [7]. In our approach no learning is involved, but the complexity of behavior of the planner is obtained by defining suitable actions for each layer, by means of a language oriented to action definition built on top of Lisp.

For coping with limited perceptions, we have developed a memory model that relies on the absolute positions of objects, and offers a set of predicates allowing players to reason about the game at different levels of abstraction.

We have implemented IALP using this memory model and the planner architecture. IALP uses a model of coordination without communication [8] and a concept of role for a player that is built on top of basic abilities, common to all the agents. The layered and modular structure of the planner allows an easy reuse of the basic capabilities of the players and specialization of roles at the higher levels.

Using Common Lisp to implement IALP offers clear advantages from the AI programming point of view; in particular we have exploited the Lisp reader and the macro feature. Using the ECL implementation of Common Lisp, designed for being embeddable within C based applications, we wanted to see if such language can compete with C/C++ written teams in a real time domain such as RoboCup.

In this paper we report about the main features of the IALP team. Section 2 and 3 describe the planner architecture and the declarative language used for defining the behavior of layers; section 4 is an account of the memory model; section 5 explains how the planner and the KB have been used to program the players and the coordination model used.

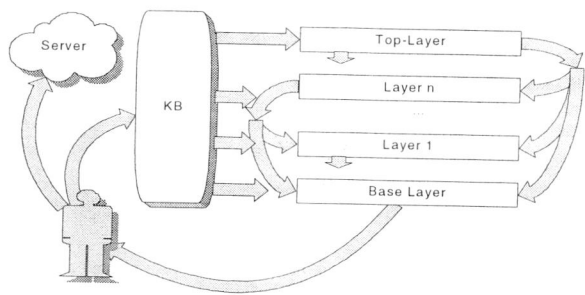

Fig. 1: The planner architecture

2 The architecture of the planner

The core of IALP is a hierarchically structured reactive planner that computes and executes plans. There is an ordered chain of layers, with a base layer and a top layer. The base layer is devoted to the communication with the RoboCup server: thus the outputs are commands like (dash *speed*) or (turn *moment*). The top layer defines the overall strategy of a player; it contains the most abstract plans and fully determines the behavior of the agent. The intermediate layers define a hierarchy of actions: each layer decides upon the implementation of an action using the actions offered by lower layers.

The overall architecture of the planner is shown in figure 1.

A plan built in a layer is a list of actions defined in one of the layers below. A *while* action can be used to repeat a sequence of actions until a specified condition is verified. Actions are expressed using the classic functional notation of LISP. A simple example of plan is the following:

```
((dash 100) (turn 90) (dash 100))
```

Another example, involving the *while* iteration construct is the following:

```
((*while* (not (can-kick? Kb))((go-ball)))
  (kick! (enemy-goal kb)))
```

The plan executor sends to the inferior layer the request to repeatedly execute the action (go-ball) until the condition (can-kick? kb) is verified; at this point the action (kick! (enemy-goal kb)) will be executed.

At each cycle, the interpreter of plans requests an action in executable form to the base layer; if this layer is executing a plan, the next action of the plan is executed. If the layer does not have a plan (has finished executing the previous one), it requests a new plan to the upper layer. This chain of requests may propagate to the top layer, which must always return an appropriate plan.

Each intermediate layer receives a plan from the upper layer and must execute the actions contained in it. The way an intermediate layer executes an action is by computing a particular function that takes into account a number of parameters and returns a plan to be executed by the lower layer. If the action is unknown to that layer the task of computing the plan is delegated to the inferior layer.

The execution of an action by a given layer may simply return a standard plan, good for any situation (in this case the system is an interpreter of plans) or involve a computation of a plan taking into account the knowledge contained in the KB. The top layer must necessarily compute a plan. Thus the planner as a whole implements a hierarchy of actions that are all available to the top layer to solve the task of writing a player for RoboCup in a suitable abstract language. Since the top level planner determines the behavior of the underlying planners, specific abilities implemented by lower levels may be reused for building different roles. In particular the layered approach is suitable for sharing low level abilities that all players should possess.

Each layer can request to reset the executing plans to upper and/or lower layers. This feature is important to implement reactive behaviors; for example, in order to react promptly to referee messages.

Another feature of the IALP planner is the possible non-determinism in the execution of actions. It is possible to define several alternative implementations for an action, all of them considered equivalent with respect to the outcome. In this case the interpreter chooses randomly the implementation to be used. With this feature, it is quite easy to introduce a richness of behavior. An advantage is that it may be difficult for an opponent team to guess the behavior of players.

The planner executes a standard perception/action cycle but we have extended the base layer to allow it to return a list of basic actions, so that all the available slots for executing actions can be exploited. Thus the basic cycle of the planner is "read a perception from the server, compute the next sequence of actions and execute them".

3 The language for defining the behavior of the planner

We have developed a simple declarative language based on LISP to define the behavior of the various layers.
Each layer contains the definition of an *update function* that, testing some conditions on the current environment, decides if some of the executing plans must be terminated. The function can return four different values: nil, UP, DOWN and ALL. When the value returned is nil the planner can continue its execution. UP means that the upper layers must abort the execution of the current plans; this capability is useful when the changes in the environment affect only the more abstract plans while lower layers can continue executing their tasks. DOWN is the dual of UP and aborts the executing plans of inferior layers; in this case it is deemed useful to continue with the overall strategy but some change in the environment make it necessary a re-planning at the lower layers. ALL is equivalent to returning both UP and DOWN and forces the planner to rebuild entirely his plans.
The way to define an update function for a layer is as follows:

```
(defupdate layer
  "Optional documentation"
  body)
```

where *body* is the body of the function. In order to define an empty update function, that is an update function that always returns nil:

```
(def-empty-update layer)
```

The possibility of aborting executing plans instantaneously is important in real time domains such as RoboCup where the environment is highly dynamic. An example is the *referee* message that changes the state of the game: each player must suddenly change his behavior to adjust to the new state. The update method that deals with *referee* messages is located in the base layer and has the following definition:

```
(defupdate basic-layer
  "Handles referee messages"
  (if (and
        (eql
          (last-percept-type kb)
          'REFEREE)
            (not (last-message-read kb)))
      (progn
            (message-read kb)
          'UP)
    nil))
```

The update function aborts all executing plans if the last perception is of type *referee* message. In this case, because the base layer is the bottommost, UP is equivalent to ALL.

In addition to the update function, a layer must define a set of actions. For each action at least one implementation must be provided. If there are multiple implementations of a given action the interpreter chooses among them with a given *policy*. For the moment the *policy* consists in choosing randomly among the implementations. In our language the list of actions defined in a layer is specified by means of the following construct:

```
(defactions layer
  (action-list action-name)
  (action-list action-name imp_1 ... imp_n)
  ...)
```

If the action-list statement is followed only by the action-name an implementation is assumed with the same name of the action. If imp_1 ... imp_n are specified, the action definitions with these names are associated to action-name.

A definition of an action is similar to the definition of a function but uses the defaction keyword:

```
(defaction name (params)
  "Documentation."
  body)
```

The name of the action must be one of those declared within the defactions construct. The parameter list allows passing some parameter to the action, for example the go-ball action must receive as a parameter the speed that must be used. The defaction must return a plan.

As an example of action definition we include a possible implementation for the run-with-ball action:

```
(defaction run-with-ball (speed dir k)
  "The power of kick is speed * k."
  (if (can-kick? kb)
      `((turn ,(ball-dir kb))
        (kick ,(* k speed) ,dir)
        (turn ,dir)
        (dash ,speed))
      '((sleep))))
```

This action checks if the player can kick (the ball is close enough) and, when this is the case, returns a plan that prescribes: "turn towards the ball, kick in the direction requested, turn and dash". If the player cannot kick, the plan ((sleep)) is returned and the player does not do anything because the action requested cannot be executed". This plan causes the immediate termination of the action and the request for a new action.

4 The memory model

A memory model is used in IALP to record basic properties of the environment used to decide which actions should be sent to the server. The memory of a

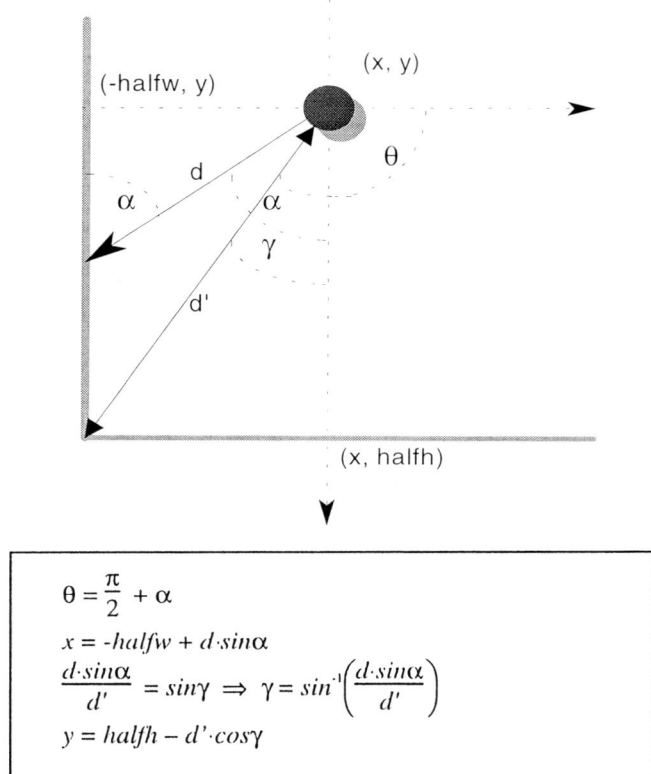

Fig. 2: Absolute position

player, or KB, keeps track of objects seen recently and is responsible for computing the absolute positions of any object and of the other players. The memory also stores the messages heard and the physical status of the player.

The IALP player executes a standard cycle: receives a perception from the server, updates the memory, computes a new set of actions and sends them to the server. In deciding the next actions the planner uses higher level predicates implemented from the information contained in the memory.

When the perception is received it is parsed using the `read-from-string` function provided by LISP. This is very convenient because the perceptions sent by the server are strings containing S-expressions and the LISP reader knows how to deal with them.

If the received perception is *see* the memory tries to update the absolute position of the player. The coordinates are the same used by the server: the (0, 0) position corresponds to the center of the field, the *x* direction is towards the enemy goal and the *y* direction is on the right of a player looking at the opponent's goal.

The absolute position of the player is computed using a borderline and a flag. When a borderline is visible the player can easily compute his distance from the line, and thus one coordinate, which is *x* or *y* depending on the line, and the direction in the coordinate system chosen. If a flag is also perceived the player can compute the second coordinate. Figure 2 shows how the absolute coordinates of the player are computed from the perception of a line (i.e. its distance d and its relative direction α) and a corner flag (i.e. its distance d' and its relative direction γ).

This method has a good precision and is fast to compute. The basic assumption is that the player movements are continuous and if the player at a given time cannot compute one or both coordinates he can assume the previous ones without making a significant error.

Once the position of the player has been computed, the absolute coordinates for each dynamic object present in the *see* perception (players and ball) are also computed using standard trigonometric calculus.

Differently from what has been proposed by other researchers [9] we have decided to maintain absolute positions for the following reasons. The absolute positions kept for all moving objects can be exploited when an object is not in the current *see* perception. For example if the player sees the ball at simulation cycle t and another player covers the ball at time $t + k$, we can assume that the ball is near to the last position recorded into memory. This assumption is reasonable only if the elapsed time k is reasonably small. Moreover, if the player changes its direction, the information stored in the memory is not affected and it is not necessary to update object positions. Also the distance among objects can be easily computed from their absolute positions.

The *hear* and *sense body* perceptions are treated similarly: they are parsed and all the information stored in appropriate structures in the memory of the agent. The *referee* messages are stored separately from other messages since they contain the status of the game and it is necessary to make sure that they are acted upon.

Given this memory model, we have defined functions and predicates and derived more abstract properties of the environment useful for defining player behaviors. Some of these predicates are used to make qualitative statements about the environment: for example the predicate can-kick? is true when the ball is near enough to the player, which in terms of lower levels means within 2 meters.

Two very important functions are distance and dir-x-y. The function distance computes the distance between the player and another point (x, y) and is fundamental for evaluating distances from objects during the game. This function is used instead of the relative distance provided in the perceptions, because, by referring to its memory, an agent is able to estimate the distance of an object even when the object is not currently perceived. This function is also exploited to evaluate the distance from a given point situated in a zone of the pitch; this is useful to implement a zone-based strategy.

The function dir-x-y allows a player to know the moment of which to turn to see a point (x, y). This function also exploits the fact that the player can estimate

the absolute coordinates of every object. Together with the distance function, dir-x-y is very useful to implement a goto-x-y action.
As an example of the flexibility of our memory model we show the implementation of the *outside* predicate:

```
(defun outside? (kb)
  (dolist p (enemies kb)
    (when (and
            (not (is-goalie? p))
            (> (pos-x kb)
               (obj-info-x p kb)))
      (return nil)))
  T))
```

5 The implementation of IALP

The Embeddable Common Lisp is an implementation of Common Lisp designed for being embeddable within C based applications [4]. ECL uses standard C calling conventions for Lisp compiled functions, which allows C programs to easily call Lisp functions and vice versa. No foreign function interface is required: data can be exchanged between C and Lisp with no need for conversion. ECL is based on a Common Runtime Support (CRS) which provides basic facilities for memory management, dynamic loading and dumping of binary images, support for multiple threads of execution. The CRS is built into a library that can be linked with the code of the application. ECL is modular: main modules are the program development tools (top level, debugger, trace, stepper), the compiler, and CLOS. A native implementation of CLOS is available in ECL: one can configure ECL with or without CLOS. A runtime version of ECL can be built with just the modules required by the application.
Using ECL has been our bet. RoboCup is a real-time domain task where system level languages like C/C++ seem to be much more effective than traditional AI languages like LISP or PROLOG. On the other hand, LISP provides a lot of advantages: no need for a parser of the messages sent by the server, automatic garbage collection, macros and closures and other high level language features traditional in AI programming were all available, so that we were able to concentrate on high level programming tasks since the beginning. Preliminary experiments have shown that LISP processes, implementing IALP players, are capable of maintaining the synchronization between server and clients.
IALP is built on top of the architecture described in previous sections: we have implemented the functions and predicates required in the RoboCup domain and defined a number of layers describing the capabilities of the different players. So far we have defined a preliminary hierarchy of layers that we intend to evolve and adjust by gaining more feedback from actual matches. Figure 3 shows the structure of the layers for the team members of the current IALP implementation. Since most of the abilities are common to all the agents, players in different roles tend to have a great number of shared layers. In fact right now they share all the

layers but the topmost. The goalie has an additional layer to implement capabilities that are specific of this role.
This homogeneity among players is justified by the definition of role that we have

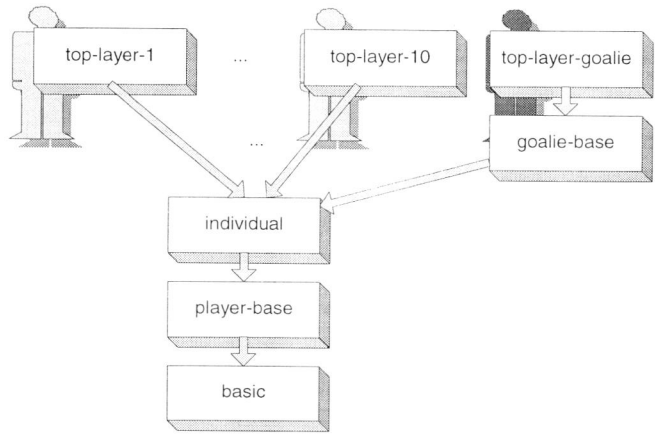

Fig. 3: Layers used in IALP

assumed: a role is *a prevalence of a behavior*. This definition of role implies that the basic capabilities of the various players must be the same and only the overall strategy of the team and the environment determine the effective behavior of a player. To understand why we have chosen this definition, consider a situation where a defender finds himself in an attack position for some reason: we want the defender to behave like an attacker for the period that he is involved in the action. An analogous situation is when all the team is forced in a situation of defence and the attackers must behave like defenders. In the real soccer it is impossible to find players able to perform top level in any role; players usually specialize in a set of tasks. Simulated soccer is different also in that it is not a problem to replicate abilities: why shouldn't we give to all the players the best capabilities in kicking and controlling the ball that we were able to develop? The only exception is the goalie that must have capabilities of his own inapplicable to the other players.

The *basic* layer is the bottom layer of the planner and its outputs are actions that are sent to the server. The actions defined at this level are basic actions such as turn, dash, kick, say, catch, move, and low level actions such as sleep, ndash and turn-ball. The sleep action is a no-action command telling the interpreter that no commands must be sent to the server until a new perception is received. The ndash action tells the interpreter that n dash commands must be sent to the server in sequence, without waiting for a new perception. The turn-ball sends a sequence of actions to the server to turn the player and the ball of a given angle.

A little preprocessing of the arguments may be done also for the basic actions: for example a turn action with a moment less than 1 is not sent to the server because it is not relevant.

The *player-base* defines a first layer of abstraction. A subset of the actions defined in this layer is the following: go-ball, see-ball, pass-ball, run-with-ball and goto. The go-ball action causes the player to find and reach the ball. The see-ball action makes the player turn until he is able to see the ball. The pass-ball action passes the ball to a given player; in this case the assumption is made that an upper layer has checked that the player can receive the ball. Finally the run-with-ball allows the player to run with the ball using the kick, dash and turn actions required to produce this complex behavior.

The *individual* layer defines individual behaviors of a player like stay-in-zone that situates the player in a given zone of the field. Another action is handle-with-ball that manages the ball and tries to move the player with the ball towards the enemy goal. The free-kick action is devoted to the execution of a free kick.

The three layers described above are shared by all the players because they correspond to abilities that all players should possess. Each member of the team has its specific top layer that distinguishes the behavior according to the role strategy. The top layer code for all players is similar and changes only in those aspects accounting for the prevalence of behavior.

The bigger differences between the goalie and other players are reflected in an additional layer, the *goalie-base*. This layer defines actions like free-kick, used to follow a far away action or catch-ball used to catch the ball.

The model of coordination used to pass the ball does not involve communication. The player possessing the ball evaluates the possible candidates for a pass and the risk of loosing the ball; if it decides to make a pass to a certain player he does so. The coordination is in the fact that a player near the action is typically interested in the ball and thus able to recognize the pass.

The overall strategy of the team emerges from role definitions. A role is substantially defined by the zone of the field assigned to a player when he is not engaged in the current action. The player is responsible for the ball and opponents in his zone. When the player has the ball he checks whether he can pass the ball or shoot into the enemy goal; if he can't, he tries to move in the direction of the opponent's goal until a pass becomes possible or he can shoot.

The flow of the ball from the defense zone to the attack zone is a consequence of the decision function used by the player to establish whether to pass the ball or proceed. For deciding whether to pass the ball or proceed, each player, depending on his role, has a number for each team mate, used for assigning a preference to the candidates for a pass. Thus defenders prefer to pass the ball to middle players and are not happy to pass the ball to the goalie.

The evaluation function also considers, for each possible target of the pass, the *gain* in case of success and the *risk* that the pass will be intercepted. The most promising target is thus chosen and its value compared with the gain and risk of advancing with the ball.

6   Conclusions and future work

In this paper we have described the basic ideas behind the implementation of the IALP team. We have adopted an agent architecture based on a layered reactive planner. Experience gained in past competitions showed that the low level (the communication layer responsible for handling communication with the server) was too slow; moreover the memory model, implemented in LISP, proved to be too heavy. This suggested rewriting in C both the communication level and the memory model for better performance.

For the future we want to experiment with different arrangements and implementations of layers and different means of coordination, besides testing the validity of the definition of role as a prevalence of behavior.

We are also interested in investigating an emotional approach to define the behavior of players [10]. Finally we want to observe the emerging behavior due to the introduction of non-determinism.

Acknowledgments

We want to thank Silvia Coradeschi for getting us interested in RoboCup and all the students of the course "Artificial Intelligence: Laboratory" in Pisa, during the a.y. 1997/98 and 1998/99, for taking up with so much enthusiasm the RoboCup challenge and for allowing us to learn and gain experience during the IAL-Cup 99.

References

1. Kitano, H. and Asada, M. and Kuniyoshi, Y. and Noda, I. and Osawa, E., "RoboCup: The Robot World Cup Initiative", IJCAI-95 Workshop on Entertainment and AI/Alife, 1995
2. Kitano, H., Asada, M., Osawa, E., Noda, I., Kuniyoshi, Y., Matsubara, H., "RoboCup: The Robot World Cup Initiative", Proc. of the First International Conference on Autonomous Agent (Agent-97), 1997.
3. Kitano, H., Asada, M., Osawa, E., Noda, I., Kuniyoshi, Y., Matsubara, H., "RoboCup: A Challenge Problem for AI", AI Magazine, Vol. 18, No. 1, 1997.
4. G. Attardi, The Embeddable Common Lisp, ACM Lisp Pointers, 8(1), 30-41, 1995.
5. Brooks, R. A., "A Robust Layered Control System for a Mobile Robot, *IEEE Journal of Robotics and Automation,* RA-2(1), March 1986.
6. Firby, R. J., "Task Networks for Controlling Continuous Processes", *Proceedings of the Second International Conference on AI Planning Systems,* Chicago IL, June 1994.
7. Stone, P., Veloso, M., "A Layered Approach to Learning Client Behaviors in the RoboCup Soccer Server", in *Applied Artificial Intelligence,* 12, 1998.
8. Franklin, S., "Coordination without Communication", http://www.msci.memphis.edu/~franklin/coord.html
9. Bowling, M., Stone, P., Veloso, M., "Predictive Memory for an Inaccessible Environment", In *Proceedings of the IROS-96 Workshop on RoboCup*", November 1996.
10. Franklin, S., McCauley, T. L., "An Architecture for Emotion", AAAI 1998 Fall Symposium "Emotional and Intelligent: The Tangled Knot of Cognition".

From a Concurrent Architecture to a Concurrent Autonomous Agents Architecture

Augusto Loureiro da Costa
loureiro@lcmi.ufsc.br

Guilherme Bittencourt
gb@lcmi.ufsc.br

Department of Automation and Systems
Federal University of Santa Catarina
CEP 88.040-900 - Brazil

Abstract. In this paper, the autonomous agent architecture used to implement the RoboCup simulator league UFSC-Team is presented. This architecture consists of three concurrent processes that encapsulate different inference engines. These take decisions in three different levels, called reactive, instinctive and cognitive. This architecture is an evolution of the concurrent architecture for cognitive multi-agents, used in the implementation of the UFSC-Team'98 that has participated in the RoboCup'98. The present implementation was designed to solve some agent synchronization and real-time response problems presented by the old architecture, due mainly to its centralized decision approach.

1 Introduction

In its first participation in the simulator league of the RoboCup'98, the UFSC-Team presented a concurrent cognitive multi-agent architecture [12]. The idea was to implement perception, action, communication, cooperation, planning and decision making exploring the concurrent programming approach [1].

The first concurrent architecture was based on three processes: interface, coordinator and expert. The interface was designed to handle perception and action. The agent/environment interaction supported by the Soccerserver consists of message exchange using a Inet Domain Socket channel. The perception information is received and the action commands are sent through this same channel. The function of the process interface was just to translate the perception and communication information into the Parla language [10] (the Agent Communication Language used by the UFSC-Team agents) and expressions from the Parla language to Soccerserver commands.

The process coordinator was responsible for the agent communication and for starting and conducting the cooperation processes. According to the original architecture proposed in the Expert-Coop environment [9], this process was responsible for the inter-process communication management, i.e., it should receive directly the messages sent by other agents and handle them. But, according to the RoboCup simulator league rules, all inter-agent communication must be done only through the Soccerserver. Because of this, the inter-process communication and the perception information are all received through the same Inet

Domain socket channel. Therefore, in this implementation, the process interface also received the inter-agent messages and forwarded them to be handled by the process coordinator, along with the perception information and referee messages.

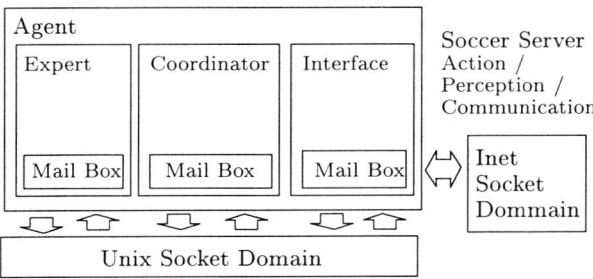

Fig. 1. The Concurrent Architecture

Finally the process expert was responsible by planing and decision making. It had a knowledge-based system encapsulated where the perception information, the messages from the referee and from other UFSC-Team agents were stored and used to infer appropriate decisions, according to the knowledge-based system rules. These three processes communicated among them by message exchange using sockets into the Unix domain (see Figure 1).

This first concurrent implementation, with a centralized decision approach, has presented some problems with agent/environment synchronization and the response time was considered too high. In fact, the best real-time responses presented by the UFSC-Team'98 agent architecture were between 70 and 80 ms, even using Case-Based Reasoning [2] to split the knowledge into different packets. Beside this, the knowledge-based system responsible for the agent decision making became very complex, because it had to include rules to treat information from high level, like what kind of collective play should be chosen in a given situation or what agents can be joined into a known play, to low level, like which power dash value or which turn value should be chosen.

To solve these problems, the agent architecture used in UFSC-Team has migrated from a concurrent approach with centralized decision making, to an autonomous agents architecture, inspired by the architecture proposed in [4], with three decision levels – reactive, instinctive and cognitive – implemented in a concurrent way (see Figure 2). The concurrent model was kept with the same three processes: interface, coordinator and expert. But now each one of these processes encapsulates a different inference engine and is responsible for one of the three decision levels. Both the first implementation and the current one were written in the C++ programming language and they integrate a partial implementation of the environment to develop cognitive multi-agent systems under real-time restrictions called Expert-Coop++.

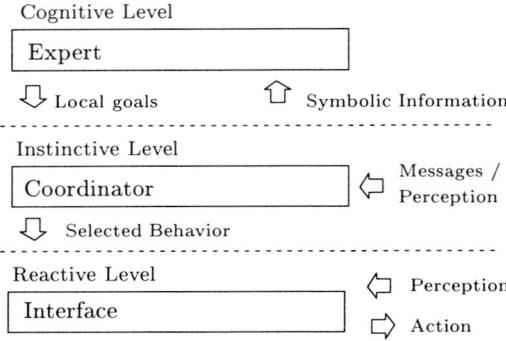

Fig. 2. The agent information flow

The reactive level inference engine is implemented in the interface process and is responsible for the real-time response of the agent, i.e., for receiving the perception information from the Soccerserver and for sending the adequate action commands to it. It consists of a set of fuzzy controllers. At any given moment, only one fuzzy controller is active and it decides which commands should be sent to the Soccerserver, along with their respective values. This choice is based on the information received from the Soccerserver and it is determined by the active fuzzy controller rules. Each one of the fuzzy controllers available in the agent represents a specific *behavior* and has some associated conditions that specify the situations in which it is effective.

The instinctive level inference engine is implemented in the coordinator process and it is responsible for updating the symbolic variables used by the cognitive level and for chosing the adequate behaviors, i.e., the adequate fuzzy controllers, that should be used in the reactive level in order to achieve a given *goal*. A goal can be achieved by a sequence of reactive behaviors that leads the agent to an intended situation. The choice of this behavior sequence is implemented through a one cycle expert system that chooses, every time the game *state* changes, the most adequate reactive behavior. Each state of the game is defined by a set of conditions that are monitored by the instinctive level. These conditions refer to perception and to the referee messages, and are used in the condition part of the rules, analogously to the reactive level. But on the instinctive level, the conclusion part of the rules are symbolic and are used either to update the symbolic information used in the cognitive level, or to select a reactive level behavior. At each moment, the chosen behavior should perform actions in the direction of the intended goal and should have its associated conditions satisfied by the game state. Once a behavior is chosen, the instinctive level keeps monitoring the conditions associated to this behavior and, if some of them are no more satisfied, it uses its rules to infer a new behavior. If this is impossible, the goal fails and a new goal should be specified. The instinctive level also handles the messages sent by the referee informing a change in the game status.

These changes are treated analogously to the game state changes, they cause the instinctive level to choose a new appropriate behavior.

Finally, the cognitive level inference engine is implemented in the expert process, and it is responsible for determining the local and global goals of the agent. The cognitive level does not have a direct effect over the reactive level, it just chooses the present goal and passes it to the instinctive level. This has the effect of changing the rules of the inference engine in the instinctive level, what indirectly will cause different behaviors to be selected. As long as a goal does not fail or succeed, the cognitive level does not interfere in the game. This idle time is used for strategic planning. This planning consists in the determination of possible future local goals, according to the result of the present one, and in the specification of cooperation requests to achieve global goals. These requests will be handled by the coordinator process and will result in other agents adopting local goals compatible with the intended global goal. The cognitive level is also implemented through an expert system, but this expert system can be much more complex than the instinctive level one, because its response time is much greater.

In the new implementation, the three processes are implemented using the multi-thread programming approach [3]. This technology allows to split a process into parts and to run these parts concurrently. In our case, each process consists of two threads. The first one is responsible for handling the Unix interruption SIGIO, used to inform that a new message has been received by the socket, and by putting this message into the mail box. The other thread, the main thread, is responsible by the process activities. The mutual exclusion between the threads is achieved by using semaphores. This implementation is a concurrent approach to the classical producer/customer problem. It avoids that the main process spend some precious time checking if there is a new message in the socket or not.

The paper is organized as follows. Section 2 describes the reactive level. The instinctive level and cognitive level are presented in Sections 3 and 4. Section 5 presents an example where this new architecture allows the agent to concurrently react to an environment stimulus in real-time and perform more sophisticated tasks like make plan, to establish new goals, open or participate into a cooperation processes, etc. Finally, in Section 6, the conclusions and future works are presented.

2 The Reactive Level

The reactive level inference engine is implemented in the process interface. This process consists of one mailbox, a set of fuzzy controllers, an input filter and an output filter (see Figure 3). The mailbox is responsible for the process message reception. All messages received by the process, including the perception information sent by the Soccerserver, will be stored in the mailbox.

The fuzzy controllers are implemented using a C++ library. This library was designed to aid implementation of fuzzy expert systems or fuzzy controllers implementation, it is called CNCL [13]. Each fuzzy controller is responsible by one reactive agent skill, called *behaviour*. At first, the following set of behaviors was chosen to be implemented into the UFSC-Team agents: *Initialize-Player, Kick-Off-Position, Move-to-Position, Move-to-Ball, Pass-Ball, Kick-to-goal, Dribble-Opponent, Drive-Ball-Fwd, Get-Ball-Control, Tackle, Follow-Opponent, Rounding-Opponent, Watch-Ball* and *Catch-Ball*. The fuzzy controller set associated with each agent depends on which agent group it belongs to: goalie, defensive players, midfielders or attack players. Of course, it does not make a lot of sense for an attack or midfielder player to have a fuzzy controller responsible by catching the ball, or for a goalie to have a fuzzy controller responsible for shooting the ball into the opponent goal.

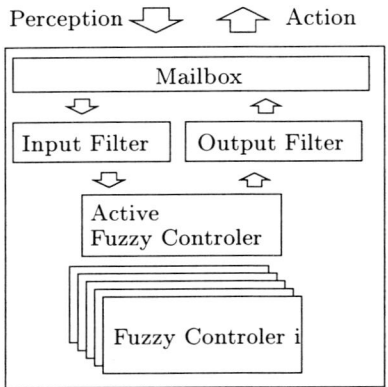

Fig. 3. The interface process

The input filter is responsible for extracting the linguistic variable values, used by the active fuzzy controller, from the perception information sent from the Soccerserver. The output filter is responsible for checking the active fuzzy controller outputs and combining them. The following criteria are observed by the output filter:

- **Null Output:** if dash power output and/or turn moment output present one null output, the respective command is not sent to Soccerserver.
- **Simultaneous turn and dash:** if the fuzzy controller presents simultaneous turn moment output and dash power output, then at first the turn command with turn moment value is sent to Soccerserver. After a 20 ms delay, the dash command with the respective dash power value is sent to the Soccerserver.
- **Kick direction and kick power:** The Kick Direction output and Kick Power output are always joined into the kick command.

Most of the fuzzy controllers have four outputs: kick-direction, kick-power, turn-moment and dash-power. The *Pass-Ball, Kick-to-goal* fuzzy controllers just have the kick-direction and kick-power output and *Move-to-Position* fuzzy controller just have turn-moment and dash-power outputs. The inputs are a set of linguistic variables, depending on which behavior is active. Each fuzzy controller has its own set of linguistic variables and the Input Filter is responsible for extracting from the perception information, the respective values that will be used to set the linguistic variables.

Using fuzzy controllers to implement the reactive level has some advantages. First of all, it is possible to synchronize the agent just adjusting the ratio between input and output, or in other words, adjusting the controller gain. This gain adjustment is made on the fuzzy set wich represent the controller input and the controller output. It is also possible to fine tune, or to get a smooth response adjusting these fuzzy sets. Figure 4 shows the fuzzy set used by the *turn-moment* output and the respective linguistic variable *ball-direction*. Note that in this case the controller gain is $0.56 = \frac{50}{90}$.

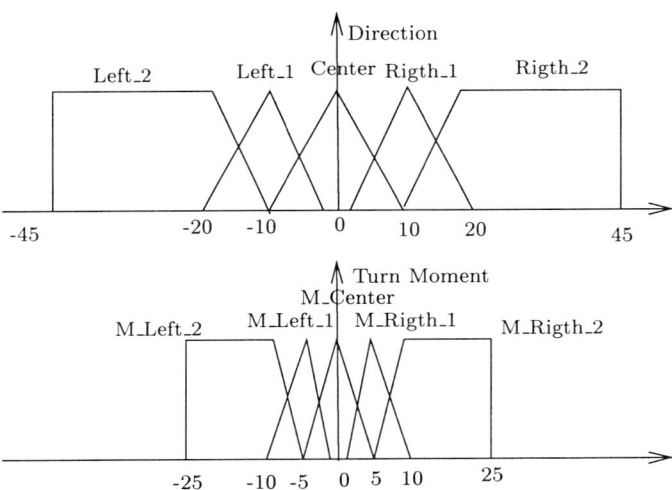

Fig. 4. The Turn Moment Fuzzy Sets

The rules used in the fuzzy controllers can be built in an intuitive way, avoiding the difficult and time consuming task of building a model of the dynamic environment (see Figure 5). It is also possible to use genetic algorithms [5] to improve the fuzzy sets used in the controllers.

Another important advantage of implementing the reactive behavior using fuzzy controllers is that it is possible to ensure that a given fuzzy controller will always be able to satisfy the real-time requirements, because the fuzzy controller is a deterministic system. Beside this, once the active fuzzy controller is the

```
rule_200.add_lhs(new CNFClause(ball_direction, left_2));
rule_200.add_rhs(new CNFClause(turn_moment, m_left_2));
rule_201.add_lhs(new CNFClause(ball_direction, left_1));
rule_201.add_rhs(new CNFClause(turn_moment, m_left_1));
rule_202.add_lhs(new CNFClause(ball_direction, center));
rule_202.add_rhs(new CNFClause(turn_moment, m_center));
rule_203.add_lhs(new CNFClause(ball_direction, right_1));
rule_203.add_rhs(new CNFClause(turn_moment, m_right_1));
rule_204.add_lhs(new CNFClause(ball_direction, right_2));
rule_204.add_rhs(new CNFClause(turn_moment, m_right_2));
```

Fig. 5. The Turn Moment Fuzzy Rules Sets

most appropriated behavior in a given situation, it releases the instinctive and cognitive levels to spend more time into more sophisticated tasks like extracting interesting symbolic features from the perception, making plans, establish goals or participating into cooperation processes.

3 The Instinctive Level

The instinctive level inference engine is implemented in the process coordinator and it is responsible for both the execution of the agent local *goals* and the generation of symbolic information to update the cognitive level knowledge base. It is implemented through a one cycle expert system that chooses, every time the game *state* changes, the most adequate reactive behavior given the current local goal. The current local goal is established by the cognitive level and it determines the set of rules to be used in the inference engine. Each state of the game is defined by a set of conditions on the perception information. These conditions usually depend on some threshold values, that must be determined experimentally.

The inputs to the instinctive level inference engine are the perception information, received from the interface process, and the messages from the referee. The perception information consists of the same synchronous perception information received by the interface from the Soccerserver, but, differently from the reactive level, the instinctive level presents a memory. This memory consists of a buffer, where perception information is stored, and whose initial size is a parameter of the implementation. It makes it possible to choose how many visual information frames can be used in one inference cycle of the inference engine. For example, assuming that the agent has been receiving visual information every 150 ms and that the buffer size is 3, in a given time t, the cycle inference process will take into account the visual information sent at times t, *t-150*, *t-300*.

The perception information is stored into the *Sync* buffer and the messages received from referee are stored in the *Async* buffer (see Figure 6). Each time

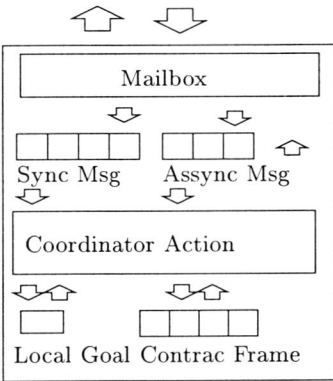

Fig. 6. The coordinator process

one of these buffers is updated or when a new local goal has been received from the cognitive level the expert system is executed. Given the input, the rules are able to recognize changes in the game state. The result of executing the rules can be either the updating of the cognitive level knowledge base or the selection of the most appropriate reactive level fuzzy controller to drive the agent from the current state to the local goal.

Suppose, for example, that the opponent team has the ball control, and our team is performing a defensive play, where the goal is *get-ball-control-back* the current behavior is *Rounding-Opponent*. Also suppose that the opponent player who has the ball control makes a mistake and kicks the ball out of the field. Then the game state is changed to *kick_in_side* and this change will be perceived by a message received from the referee informing about this new game status. In this case, a behavior can be directly selected to be performed by the reactive level, i.e., *Move-to-Ball*, and it also means that the goal *get-ball-control-back* was achieved. The cognitive level will be updated and will generate a new goal. The point here is that in a situation like that, both the new planning and the execution of the new behavior can happen concurrently.

The process coordinator is also responsible for the cooperation. A *Contract Frame Buffer* is provided to store the necessary information involved into a cooperation process that uses the *Dynamic Social Knowledge* Cooperation Strategy [11]. The coordinator process also includes a real-time policy and some management algorithms for distributed system communication, like System Fault Tolerance [6].

4 The Cognitive Level

The cognitive level inference engine is implemented in the expert process. It consists of a symbolic object-oriented knowledge-based system that handles both the symbolic information received from the instinctive level, and the asynchronous messages received from others UFSC-Team agents. It generates the local goals and the global goals. This knowledge-based system has three knowledge bases: *Dynamic KB, Static KB and Export KB*.

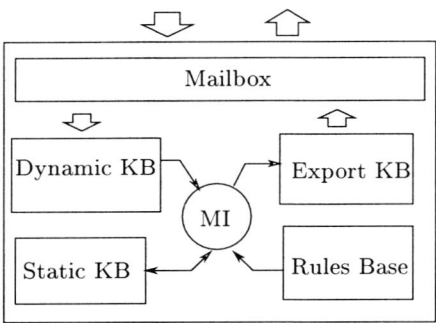

Fig. 7. The expert process

The Dynamic KB is used to store the symbolic information generated by the instinctive level and the asynchronous messages sent by other UFSC-Team agents. The Static KB stores the knowledge that has been inferred by the expert process about the game, the team, the opponent, the agent plans, goals, etc. Both, the Static KB and the Dynamic KB, are used by the inference engine to generate the agent goals. The new facts about the game, the plans and goals are stored into the Static KB. Export KB is used to store the expert process output. Basically this output consists of local goals to be sent to the instinctive level, information to be used in cooperation strategies, or messages to be sent to other UFSC-Team Agents.

Suppose that some symbolic information and/or messages have been received from the coordinator process, followed by a request. This causes the following sequence of actions:

1. The information is stored in the Dynamic KB.
2. The inference engine evaluates both the information stored in the Dynamic KB and the facts stored in the Static KB.
3. The generated new facts, plans and goals are stored into the Static KB.
4. If a new goal is chosen and/or there are some information to be sent to another agent, it is stored into the Export KB.
5. If the Export KB is not empty, its contents are sent to the coordinator.
6. The Dynamic and Export KB are cleaned.
7. A reply to coordinator is sent.

An important feature of this new architecture is that the cognitive level can spend more time planning, establishing goals, etc, once the reactive level and, in some situations, the instinctive level is responsible for real-time interaction with the environment. The cognitive level also helps the coordinator in the evaluation of the cooperation process to achieve global goals.

5 Example

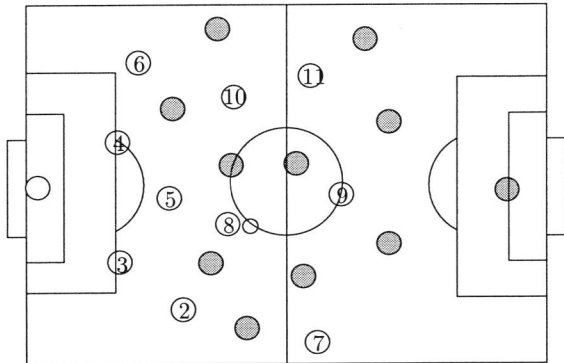

Fig. 8. An example

An example where the proposed concurrent autonomous agent architecture can be useful is presented in this section. Let's assume a situation where the opponent team has the ball control and is performing an offensive play and trying to pass the ball. In this situation the player close to the opponent player, who has the ball control, will have *get-ball-control-back* as its local goal and will be performing the behavior *Rounding-Opponent* into the reactive level. Suppose that the opponent has tried to pass the ball, makes a mistake and the midfielder, player number 8, gets the ball control back. At this situation, shown in figure 8, the agent number 8 instinctive level will recognize that the goal has been achieved and will change the current behavior to *Drive-Ball-Fwd* and inform the cognitive level of the new game state. Then the interface will perform the new behavior and the cognitive level inference engine will choose one of the play-patterns stored into the cognitive level to be performed involving the attack player. This means to select a global goal to be achieved, broadcast this goal to the involved agents and send a local goal, related with the selected global goal to be performed by the instinctive level. Another possibility in this situation is, just after the new behavior begins to be performed by the reactive level, to ask the coordinator to start a cooperation process involving the attack players and midfielders to choose which one of the known global goals is more appropriate for that situation. In the first possibility, the choice is done just taking into account one agent information, i.e., its believes about the environment and about the other agents. It is not considered whether another player has stamina enough to perform the selected play. In the second possibility, the agent will perform a cooperation process and choose which global goal is more appropriate, taking into account the perception information of all the agents involved in the cooperation process.

6 Conclusions and Future Work

A concurrent autonomous agent architecture to a simulated robot soccer team was presented in this paper. This new architecture explores the concurrent approach to implement an architecture with three levels of decision. It allows the agent to react to an environment stimulus, to make plans, to establish goals and to perform complex agent cooperation strategies concurrently and respecting real-time constraints. It also provides a memory where the perception information can be stored, allowing the agent not just to evaluate the current information, but to evaluate the current and some early past information together. This allows information about the object movement to be handled and the current and past information to be used to make inferences about the environment.

Some real-time policy is also provided to ensure that the agent is handling the newest information, like dedicated buffer to perception information. If it is assumed that the size of the buffer is two, it is sure that the two last perceptual information are stored in that buffer. Also it is possible to choose what kind of information will be handled first, perceptual or asynchronous message. Beside this, some implementation effort was done to handle the received messages, like multi-threads programming approach associated with the Unix SIGIO interrupt, and represented a significant improvement in the real-time response.

One further advantage of the proposed architecture is that the numerical parameters of the implementation are partitioned into two subsets: the limits of the fuzzy sets used in the reactive level controllers and the thresholds used by the instinctive level to calculate the logical values of the symbolic variables used in the cognitive level. In the future, we intend to use optimization methods, such as Reinforcement Learning [8] and Genetic Algorithms [5], to independently improve these two set of parameters. These parameters, because they refer only to local behaviors, can be improved using simplified situations, whith few players.

This architecture has been used to implement the UFSC-Team and it is already integrated to the partial implementation of the environment to build cognitive multi-agent system under real-time restriction, called Expert-Coop++. It will allow the UFSC-Team to employ complex cooperation strategies that use both perception information and some communication among the agents involved into the cooperation process. In a near future some cooperation strategies will be implemented and evaluated in the UFSC-Team.

Acknowledgments

The authors express their thanks to the hospitality of the staff of the IAKS (Universität Karlsruhe) where part of this work was developed. The authors are also grateful to the "Fundação Coordenação de Aperfeiçoamento de Pessoal de Nível Superior (Capes)", a Brazilian research support agency, for the partial support of this project.

References

1. Andrews, G. R.: Concurrent Programming: Principles and Practice. The Benjamim/Cumming Publishing Co. (1991)
2. Allen, B.P.: Case-Based Reasoning: Business Applications Communications of ACM (1994), March, vol. 37, Nr 3, pages 40-42
3. SunSoft: Multi Threads Programming Guide. Sun Microsystems Inc. (1994).
4. Bittencourt, G.: In the Quest of the Missing Link. Proceedings of IJCAI 15, Nagoya, Japan, August 23-29 Morgan Kaufmann (ISBN 1-55860-480-4), (1997), pages 310-315.
5. Davies, L.: Handbook of Genetic Algorithms. CVan Nostrand Reinhold, New York (1991)
6. Jalote, P.: Fault Tolerance in Distributed System. PTR Prentice Hall, Englewood Cliffs, New Jersey (1994).
7. Kitano,H.: RoboCup: The Robot World Cup Initiative. in Proc. of The First International Conference on Autonomous Agent (Agents'97). Marina del Ray, The ACM Press, (1997).
8. Kaelbling, L. P. and Littman, M. L. and Moore, A.W.: Reinforcement Learning: A Survey. Journal of Artificial Intelligence Research, (1996), vol 4, pages 237-285 .
9. Bittencourt, G. and Costa, A. C. P. L. da: Expert-Coop: An Environment for Cognitive Multi-Agent Systems in pre-printers IFAC/IFIP MCPL'97, Conference on Management and Control of Production and Logistics, vol 2, (1997), pages 492-497.
10. Costa, A. C. P. L. da and Bittencourt, G.: Parla: A Cooperation Language for Cognitive Multi-Agent Systems. EPIA'97, 8th Portuguese Conference of Artificial Inteligence, Spring-Verlag, Lecture Notes in Artificial Inteligence vol 1323, (1997), pages 207-215. Production and Logistics, vol 2, (1997), pages 492-497.
11. Costa, A. C. P. L. da and Bittencourt, G.: Dynamic Social Knowledge: A Cooperation Strategie for Cognitive Multi-Agent Systems. Third International Conference on Multi-Agent Systems, ICMAS'98,Paris, France, July 2-7 (1998), IEEE Computer Society, pages 415-416.
12. Costa, A. C. P. L. da and Bittencourt, G.: UFSC-Team: A Cognitive Multi-Agent Approach to the RoboCup'98 Simulator League. RoboCup'98 Workshop - Team description (1998), pages 371-376.
13. Junius, M. and Stepple, M.: CNCL Reference Manual. Universität Aachen, (1997), http://www.comnets.rwth-aachen.de/cnroot_engl.html

Tracking and Identifying in Real Time the Robots of a F-180 Team

Paulo Costa[(1)(2)], Paulo Marques[(2)], António Moreira[(3)],
Armando Sousa[(1)(2)], Pedro Costa[(1)]

paco@fe.up.pt, amoreira@fe.up.pt, asousa@fe.up.pt,
pamarques@riff.fe.up.pt, pedrogc@fe.up.pt

[(1)] Lecturer at the Faculdade de Engenharia da Universidade do Porto (FEUP)
[(2)] PhD. Sutdent at the FEUP, [(3)] Assistant Professor at the FEUP

Abstract - This paper describes the method employed to track and identify each robot during a Robocup match. Also, the playing ball is tracked with almost no extra processing effort. To track the robots it is necessary the use of adequate markers so that not only the position is extracted but also the heading. We discuss the difficulties associated with this problem, various possible approaches and justify our solution. The identification is performed thanks to a minimalist bar code placed in each robot. The bar code solves the problem of resolving some ambiguities that can arise in certain configurations. The procedure described can be executed in real time as it was shown in Paris in RoboCup-98.

1 Introduction

To be able to play consistently a team must, amongst other things, know its kinetic state. More, it must know the other team and the playing ball kinetic states. By kinetic state, we mean all the variables that must be known to uniquely characterize the physical position of a body.

The main source of information, for a F-180 Team, is the global camera that is typically placed above the playing field. Usually, that camera captures an image that covers all the playing field. Having that image transferred to a computer enables us to apply some kind of processing to extract the needed information. That task is very time consuming and the required robustness is sometimes difficult to obtain. Light variations in the illumination can be important disturbances the vision system and the design must be prepared to cope with that kind of problem [2].

2 The Team State

We can see a player as a rigid body constrained to have one of its faces parallel to the xy plane (the floor). With that restriction the full mechanical state of the robot can be described by the vector $(x, y, \theta, v_x, v_y, w)$, where x and y specify the robot's position, θ is the orientation, v_x and v_y are the robot's velocity in the x an y axis and finally w is the robot's rotation speed. If the robot has some kind of extra constraint in the possible trajectories then the state can be further reduced. For example, with the usual differential drive type and if we assume that the robot will not slip sideways then we can define the state only by (x, y, θ, v, w). That happens because assuming the no slippage condition the robot velocity has always the direction of its orientation.

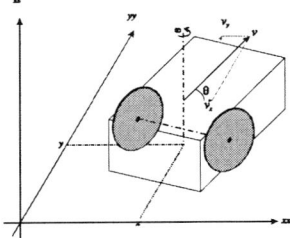

Fig. 1. Robot and state variables

From a still image it is not very easy to extract information about the velocity of an object, especially if the speed is low and we do not have any noticeable motion blur. On the other hand, that information can be obtained using information from past frames so we can concentrate on the data that can be directly extracted from the captured image.

The global kinetic state of the team it is the aggregation of all robot states [3] and [5]. The first step to extract that state from a captured image of the playing field is to get the position of all the available markers.

3 Required Markers

The Robocup Regulation enforce the presence of a Table Tennis ball placed on the top of each robot. Furthermore it is advised that the ball should be placed on the rotation center of the robot. But, using only this ball, we can only know the x, y coordinates. It gives us no information about θ. That problem can be solved using a second marker placed on top of the robots. We used another tennis table ball of a different color. That extra marker will give the robot's heading with a precision that increases as the distance between the two markers increases. That is why we placed them a few centimeters apart. The distance was not further increased because it is advantageous to have both balls surrounded by the top of the robot. The purpose is to allow the camera to get an image where the border around the ball is always black. This second ball is of the same color for all robots.

4 Image Processing

The main problem, posed in the process of extracting the required information, is the short time available and the amount of data involved: if we want to extract all the items' positions at 25 *Hz* we have only 40 *ms* to process each image. For an RGB image with PAL resolution of 384 x 288 we can have 216 *KBytes* if we are using 16 bits to hold the RGB values and 340 *KBytes* in the 24 bits case. That is a lot of data and even the simplest filter can take too long to be applied to the image. Clearly there is the need to perform some kind of data reduction in a computationally cheap way.

We choose to process the entire field in an attempt to make the processing of each frame independent from the previous ones. Another advantage is that the time consumed is more predictable as it can be more or less stable from frame to frame. We do not loose the information from previous frames. As mentioned before, the speed of the robots and the ball can only be extracted from previous frames, but the actual image processing can be done without that information.

The overall processing system and its main building blocks are shown in the next figure:

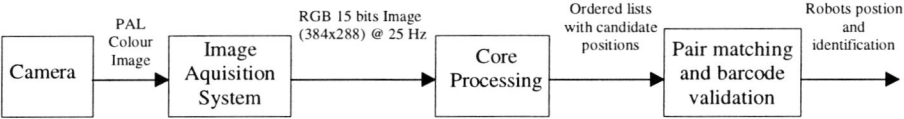

Fig. 2. Overall Image Processing System

We can use the concepts of Fuzzy Logic base our approach [7]. The core of our system relies in a function f that catalogues each pixel with its degree of membership to the set of an ideal color c. Let us can consider that each color that we want to identify is a fuzzy set Sc. Each element of this fuzzy set is a point in the RGB space. In other words it is a three dimensional vector (r, g, b). To define this set we must build a function $m(r, g, b)$ that maps each point to its set membership. To extract n colors we must define n of these sets and their respective membership functions. The function f then applies each of the p_i functions to each pixel and selects the color with the highest degree of membership. Setting a lower bound for the degree of membership we can rule out a lot of pixels that are not interesting. Almost all the pixels related to the green of the table, the white of the lines or the black of the robots cover can be dropped by this filter and we will retain only the interesting pixels. In doing that we can lower the amount of data to be processed by dropping the potentially uninteresting pixels. More, while doing that we can have now an idea of the possible color that each interesting pixel represents.

This function is locally independent and does not use any information from surrounding pixels. That is a big advantage because it can be evaluated only once for each pixel and in any order. The information inherent to the locality of each pixel will be used only in a later processing stage.

We now have a crisp set of interesting pixels and each pixel p can be represented by a vector (x, y, c, m) where x, y are the pixel coordinates, c is the color set in which the pixel showed the highest degree of membership and m is the respective degree of membership. Then, we can aggregate the pixels by their positional proximity and build a set of candidate positions for each ball in the image.

5 From Interesting Pixels to Ball Positions

We can label the n colors associated with ball markers as $c_1, c_2, ..., c_n$. After processing the image we can get n ordered lists, $cp_1, cp_2, ..., cp_n$ with the candidate positions for each ball. In each list we can have more candidates than there are balls. Each candidate is a vector (x, y, np, tf) where np and tf are figures of merit that are related with this candidate's proximity to the ideal image of a ball. More precisely, np states the number of pixels in this candidate that matched the required color and spatial proximity. The other component, tf, is the sum of each pixel fitness to the ideal color. The coordinates x and y of the candidate position are found by an weighted average of all the x, y coordinate of the points belonging to it.

The algorithm to perform the task is as follows (in pseudo code):

:Order the pixels
    Build a list lp_c of pixels that possibly represent the same color c;
      Order each list by decreasing degree of membership to the S_c
:Cluster
    for each list lp_c
      for each pixel $p(x, y, c, m)$
        if its distance to a existent candidate position of the same color is bellow a certain threshold r_b then add the pixel to that candidate position;
        else create a new candidate position and then add the pixel to that candidate position;
:Order and filter the candidate positions
    for each set of candidate positions for balls of the same color
      np_o := optimal number of pixels associated with a candidate position;
      for each candidate positions in the set
        np := number of pixels associated with the candidate position;
        if the $np < np^{min}$ then discard the candidate position
        tf := sum of each pixel degree of membership m
        if the $tf < tf^{min}$ then discard the candidate position
      order by increasing distance of np to np_o.

The core processing system and its main building blocks are shown in the figure:

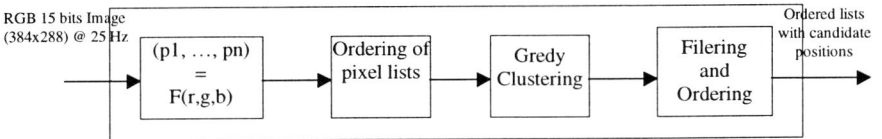

Fig. 3. Core processing system details

6 Robots' Positions and Their Identification

With best candidate positions for each ball we can pair then and extract the robot's new observed positions. A simpler procedure can be done to find the position of the playing ball. The best candidate center for the color associated with the ball is elected to give us the new playing ball position. Applying a Kalman filter or similar techniques can produce further refinements. For more on this subject see [6],[4] and [1].

This approach does not supply the identification of each robot. In earlier games we used a locality assumption to identity each robot. By assuming that each robot was the one closest to previous known location we could track each robot almost al the time. The problem with this approach was that it required the initialization by human intervention and in some cases the ambiguities could induce an error that was not easily recoverable and led the mismatched robots to become useless. Also, the pairing of the balls some times struggled with ambiguous cases that added another potential error source.

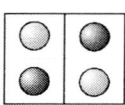

Fig. 4. Some ambiguous robot configurations

To overcome the problem we fitted each robot with a set of black and white marks that can be seen as a bar code label. The size of the markers could not be too small and in the available area on the top of the robot restricted us to use only three markers. With three markers with two states: black and white, we can only have eight different values represented. Restricting the use of all whites and all black we have six possible configurations always with, at least, one black mark and one white mark. That is enough for a five robots team.

Fig. 5. Bar code in action

Using the bar code to validate a pair of balls we can eliminate the ambiguities shown in figure 4. That can be achieved by setting a luminance threshold to accept a white mark and the same to a black mark. Only if the luminance of the white mark is above the threshold and the luminance of the black mark is bellow another threshold, we accept the code as a valid. That, coupled with the requirement that the valid bar codes have both kinds of mark, can almost ensure that a false reading is never accepted and gives us a very strong validation of the balls pair. Naturally the bar code identifies each robot without any ambiguity and we do not need anymore to rely in previous information to achieve that goal. Since we started playing with this setup, the reliability of our control increased substantially.

7 Conclusions and Future Work

In this paper we described a vision system for a F-180 team. We used this system in Paris in Robocup-98 with great success. The achieved frame rate of 25 Hz and the robustness showed by this system are its greatest strengths. It was implemented in a mix of C and C++ and the computer running it was a PC with a 266 MHz PentiumII and 64 $Mbytes$ of RAM. For the image acquisition we used a standard camcorder and a PCI frame grabber based on the Bt848 chip.

The bar code technique solved one of the worst problems, the identification of each robot and provided a tool to eliminate ambiguities and the dependence on past information.

Using vision to recognize the overall system state, consisting in the ball position and speed, our team's robots' state and the adversarial robots' state, is still a very difficult task. As the quality of the team behavior is very dependent from the accuracy of that system any improvement in this area shows dramatic results in the overall team performance.

References

1. Arthur Gelb, Joseph Kasper Jr.,Raymond Nash Jr., Charles Price, Arthur Sutherland Jr.: Applied Optimal Estimation, The M.I.T. Press, (1989)
2. J. Borenstein, H. R. Everett, L. Feng, S. W. Lee and R. H. Byrne: Where am I? Sensors and Methods for Mobile Robot Positioning, (1996)
3. J. Carvalho: Dynamic Systems and Automatic Control, Prentice-Hall, (1993)
4. Huibert Kwakernaak, Raphael Sivan: Linear Optimal Control Systems, Willey-Interscience, (1972)
5. Jean-Claude Latombe: Robot Motion Planning, Kluwer Academic Publishers, (1991)
6. Manuela Veloso, Peter Stone, Kwun Han, Sorin Achim: The CMUnited-97 Small Robot Team, RoboCup-97: The First World Cup Soccer Games and Conferences, H. Kitano (ed.), Springer Verlag, Berlin, (1998)
7. Bart Kosko: Neural Networks and Fuzzy Systems, Prentice Hall

VQQL. Applying Vector Quantization to Reinforcement Learning

Fernando Fernández and Daniel Borrajo

Universidad Carlos III de Madrid
Avda. de la Universidad, 30
28912 Leganés. Madrid (Spain)
ffernand@scalab.uc3m.es, dborrajo@ia.uc3m.es
http://scalab.uc3m.es/~ffernand, http://scalab.uc3m.es/~dborrajo

Abstract Reinforcement learning has proven to be a set of successful techniques for finding optimal policies on uncertain and/or dynamic domains, such as the RoboCup. One of the problems on using such techniques appears with large state and action spaces, as it is the case of input information coming from the Robosoccer simulator. In this paper, we describe a new mechanism for solving the states generalization problem in reinforcement learning algorithms. This clustering mechanism is based on the vector quantization technique for signal analog-to-digital conversion and compression, and on the Generalized Lloyd Algorithm for the design of vector quantizers. Furthermore, we present the VQQL model, that integrates Q-Learning as reinforcement learning technique and vector quantization as state generalization technique. We show some results on applying this model to learning the interception task skill for Robosoccer agents.

1 Introduction

Real world for autonomous robots is dynamic and unpredictable. Thus, for most robotic tasks, having a perfect domain theory (model) of how the actions of the robot affect the environment is usually an ideal. There are two ways of providing such models to robotic controllers: by careful and painstaking "adhoc" manual design of skills; or by automatically acquiring such skills. There have been already many different approaches for learning skills in robotic tasks, such as genetic algorithms [6], or neural networks and EBL [13].

Among them, reinforcement learning techniques have proven to be very useful when modeling the robot worlds as MDP or POMDP problems [1, 12, 17]. However, when using reinforcement learning techniques with large state and/or action spaces, two efficiency problems appear: the size of the state-action tables and the correct use of the experience. Current solutions to this problem rely on applying generalization techniques to the states and/or actions. Some systems have used decision trees [3], neural networks [9], or variable resolution dynamic programming [14].

In this paper, we present an approach to solve the generalization problem that uses a numerical clustering method: the generalized Lloyd algorithm for the design of vector quantizers [10]. This technique is extensively employed for signal analog-to-digital conversion and compression, which have common characteristics to MDP problems. We have used Q-Learning [18] as reinforcement learning algorithm, integrating it with vector quantization techniques in the VQQL model.

We have applied this model for compacting the set of states that a Robosoccer agent perceives, thus dramatically reducing the reinforcement table size. In particular, we have used the combination of vector quantization and reinforcement learning for acquiring the ball interception skill for agents playing in the Robosoccer simulator [16].

We introduce the reinforcement learning and the Q-learning algorithm in section 2. Then, the vector quantization technique and the generalized Lloyd algorithm are described in section 3. Section 4 describes how vector quantization is used to solve the generalization problem in the model VQQL, and in sections 5 and 6, the experiments performed to verify the utility of the model and the results are shown. Finally, the related work and conclusions are discussed.

2 Reinforcement Learning

The main objective of reinforcement learning is to automatically acquire knowledge to better decide what action an agent should perform at any moment to optimally achieve a goal. Among many different reinforcement learning techniques, Q-learning has been very widely used [18]. The Q-learning algorithm for non deterministic Markov decision processes is described in table 1 (execution of the same action from the same state by an agent arrives to different states, so different rewards could be obtained). It needs a definition of the possible states, $\mathcal{S}$, the actions that the agent can perform in the environment, $\mathcal{A}$, and the rewards that it receives at any moment for the states it arrives to after applying each action, r. It dynamically generates a reinforcement table $Q(s,a)$ (using equation 1) that allows it to follow a potentially optimal policy. Parameter γ controls the relative importance of future actions rewards with respect to immediate rewards. Parameter α refers to the probabilities involved, and it is computed using equation 2.

$$Q_n(s,a) \leftarrow (1-\alpha_n)Q_{n-1}(s,a) + \alpha_n\{r + \gamma \max_{a'} Q_{n-1}(s',a')\} \quad (1)$$

$$\alpha_n \leftarrow \frac{1}{1 + \text{visits}_n(s,a)} \quad (2)$$

where $\text{visits}_n(s,a)$ is the total number of times that the state-action entry has been visited.

Q-learning algorithm (S, A)
For each pair $(s \in \mathcal{S},\, a \in \mathcal{A})$, initialize the table entry $Q(s, a)$ to 0.
Observe the current state s
Do forever
– Select an action a and execute it
– Receive immediate reward r
– Observe the new state s'
– Update the table entry for $Q(s, a)$ using equation 1
– Set s to s'

Table1. Q-learning algorithm.

3 Vector Quantization (*VQ*)

Vector quantization appeared as an appropriate way of reducing the number of bits needed to represent and transmit information [7]. In the case of large state spaces in reinforcement learning, the problem is analogous: *how can we compactly represent a huge number of states with very few information?* In order to apply vector quantization to the reinforcement learning problem, we will provide first some definitions.

3.1 Definitions

Since our goal is to reduce the size of the reinforcement table, we have to find out a more compact representation of the states.[1] If we have K attributes describing the states, and each attribute a_i can have *values(a_i)* different values, where this number is usually big (in most cases infinite, since they are represented with real numbers), then the number of potential states can be computed as:

$$S = \prod_{i=1}^{K} values(a_i) \qquad (3)$$

Since this can be a huge number, the goal is to reduce it to $N \ll S$ new states. These N states have to be able to approximately capture the same information as the S states; that is, all similar states in the previous representation, belong to the same new state in the new representation. The first definition takes this into account.

Definition 1. *A vector quantizer Q of dimension K and size N is a mapping from a vector (or a "point") in the K-dimensional Euclidean space, R^K, into a finite set C containing N output or reproduction points, called* code vectors, codewords, *or* codebook. *Thus,*

$$Q : R^K \longrightarrow C$$

where $C = (y_1, y_2, \ldots, y_N)$, $y_i \in R^K$.

---
[1] We will refer indistinctly to vectors and states.

Given C (computed by the generalized Lloyd algorithm, explained below), and a vector $x \in R^K$, $Q(x)$ assigns x to the closest state from C. In order to define the closeness of one vector x to a state in C, we need to define a measure of the *quantization error*, which needs a *distortion measure* (analog to the similarity metric of clustering techniques).

Definition 2. *A distortion measure d is an assignment of a nonnegative cost $d(x, q)$ associated with quantizing any input vector $x \in R^K$ with a reproduction vector $q = Q(x) \in C$.*

In digital communications, the most convenient and widely used measure of distortion between an input vector x and a quantizer vector $q = Q(x)$, is the squared error or squared Euclidean distance between two vectors defined by equation 4.

$$d(x, q) = \|x, q\|^2 = \sum_{i=1}^{K}(x[i] - q[i])^2 \qquad (4)$$

However, sometimes differences in one attribute value are more important than in another. In those cases, the weighted squared error measure is more useful, because it allows a different emphasis to be given to different vector components, as in equation 5. In other cases, the values $x[i]$ and $q[i]$ are normalized by the range of values of the attribute. This is a special case of the equation 5 where weights would be computed as the inverse of the square of the range (maximum possible value minus minimum possible value).

$$d(x, q) = \sum_{i=1}^{K} w_i(x[i] - q[i])^2 \qquad (5)$$

Once a distortion measure has been defined, we can define Q as in equation 6.

$$Q(x) = \arg\min_{y \in C}\{d(x, y)\} \qquad (6)$$

In order to measure the average error produced by quantizing M training vectors x_j with Q, average distortion is defined as the expected distortion calculated among any input vector and the quantizer Q:

$$D = \frac{1}{M}\sum_{j=1}^{M} \min_{y \in C} d(x_j, y) \qquad (7)$$

Finally, we define *partition* and *centroid*, concepts needed for presenting the Lloyd algorithm for computing C from M input vectors.

Definition 3. *A partition or cell $R_i \subseteq R^K$ is the set of input vectors (old states) associated to the same (new) state in the codebook C.*

Definition 4. We define the centroid, cent(R), of any set $R \subseteq R^K$ as that vector $y \in R^K$ that minimizes the distortion between any point x in R and y:

$$cent(R) = \{y \in R^K \mid E[d(x,y)] \leq E[d(x,y')], \forall x \in R, y' \in R^K\} \quad (8)$$

where $E[z]$ is the expected value of z.

A common formula to calculate each component i of the centroid of a partition is given by equation 9.

$$cent(R)[i] = \frac{1}{\|R\|} \sum_{j=1}^{\|R\|} x_j[i] \quad (9)$$

where $x_j \in R$, $x_j[i]$ is the value of component (attribute) i of vector x_j, and $\|R\|$ is the cardinality of R.

3.2 Generalized Lloyd Algorithm (GLA)

The generalized Lloyd algorithm is a clustering technique, extension of the scalar case [11]. It consists of a number of iterations, each one recomputing the set of more appropriate partitions of the input states (vectors), and their centroids. The algorithm is shown in table 2. It takes as input a set T of M input states, and generates as output the set C of N new states (*quantization levels*).

Generalized Lloyd algorithm (T, N)
1. Begin with an initial codebook C_1.
2. Repeat
    (a) Given a codebook (set of clusters defined by their centroids) $C_m = \{y_i; i = 1, \ldots, N\}$, redistribute each vector (state) $x \in T$ into one of the clusters in C_m by selecting the one whose centroid is closer to x.
    (b) Recompute the centroids for each cluster just created, using the centroid definition in equation 9 to obtain the new codebook C_{m+1}.
    (c) If an empty cell (cluster) was generated in the previous step, an alternative code vector assignment is made (instead of the centroid computation).
    (d) Compute the average distortion for C_{m+1}, D_{m+1}
    Until the distortion has only changed by a small enough amount since last iteration.

Table2. The generalized Lloyd algorithm.

There are three design decisions to be made when using such technique:

Stopping criterion Usually, average distortion of codebook at cycle m, D_m, is computed and compared to a threshold θ ($0 \leq \theta \leq 1$) as in equation 10.

$$(D_m - D_{m+1})/D_m < \theta \tag{10}$$

Empty cells One of the most used mechanisms consists of splitting other partitions, and reassigning the new partition to the empty one. All empty cells generated by the GLA are changed in each iteration by another cell. To define the new one, another non-empty cell with big average distortion y, is splitted in two:

$$y_1 = \{y[1] - \epsilon, \ldots, y[K] - \epsilon\}, \text{and}$$
$$y_2 = \{y[1] + \epsilon, \ldots, y[K] + \epsilon\}$$

Initial codebook generation We have used a version of the GLA as explained in table 3, that requires a partition split mechanism as the one described above inserted into the GLA in table 2.

GLA with Splitting (T)
1. Begin with an initial codebook C_1 with N (number of levels of the codebook) set to 1. The only level of the codebook is the centroid of the input.
2. Repeat
   (a) Set N to $N * 2$
   (b) Generate a new codebook C_{m+1} with N levels that includes the codebook C_m. The rest N undefined levels can be initialized to 0
   (c) Execute the GLA algorithm in table 2 with the splitting mechanism with parameters (T, N) over the codebook obtained in previous step
   Until N is the desired level

Table3. A version of the generalized Lloyd algorithm that solves the initial codebook and empty cell problems.

4 Application of *VQ* to *Q-learning*. VQQL

The use of vector quantization and the generalized Lloyd algorithm to solve the generalization problem in reinforcement learning algorithms requires two consecutive phases:

Learn the quantizer. Or to design the N-levels vector quantizer from input data obtained from the environment.

Learn the Q function. Once the vector quantizer is designed (we have clustered the environment in N different states), it is needed to learn the Q function, generating the Q table, that will be composed of N rows, and a column for each action (one could also use the same algorithm for quantizing actions).

We have two ways of unifying both phases:

Off-line mode. We could obtain the information required to learn the quantizer and the Q function, and, later, learn both.

On-line mode. We could obtain data to generate only the vector quantizer, and, later, the Q function is learned by the interaction of the agent with the environment, using the previously designed quantizer.

The advantages of the first one are that it allows to use the same information several times, and the quantizer and the Q table are learned with the same data. The second one allows the agent to use greedy strategies in order to increase the learning rate (exploration versus exploitation).

In both cases, the behavior of the agent, once the quantizer and the Q function are learned, is the same; a loop that:

- Receives the current state, s, from the environment.
- Obtains the quantization level, s', or state to which the current state belongs.
- Obtains the action, a, from the Q table with bigger Q value for s'.
- Executes action a.

5 The Robosoccer domain

In order to verify the usefulness of the vector quantization technique to solve the generalization problem in reinforcement learning algorithms, we have selected a robotic soccer domain that presents us all the problems that we have defined in previous sections. The RoboCup, and its Soccer Server Simulator, gives us the needed support [16].

The Soccer Server provides an environment to confront two teams of players (agents). Each agent perceives at any moment two types of information: visual and auditorial [16]. Visual information describes a player what it sees in the field. For example, an agent sees other agents, field marks such as the center of the field or the goals, and the ball. Auditorial information describes a player what it hears in the field. A player can hear messages from the referee, from its coach, or from other players. Any agent (player) can execute actions such as run (dash), turn (turn), send messages (say), kick the ball (kick), catch the ball (catch), etc.

One of the more basic skills a soccer player must have is ball interception. The importance of this skill comes from the dependency that other basic skills, such as kick or catch the ball, have with this one. Furthermore, ball interception is presented as one of the more difficult tasks to solve in the Robosoccer simulator, and it has been studied in depth by other authors [17]. In the case of Stone's work, neural networks were used to solve the ball interception problem posed as a supervised learning task.

The essential difficulties of this skill come from the visual limitations of the agent, as well as from the noise that the simulator includes in movements of objects. In order to intercept the ball, our agents parse the visual information that they receive from the simulator, and obtain the following information:[2]

---

[2] The Robosoccer simulator protocol version 4.21 has been used for training. In other versions of the simulator, other information could be obtained.

- Relative Distance from the ball to the player.
- Relative Direction from the ball to the player.
- Distance Change, gives an idea of how Distance is changing.
- Direction Change, gives an idea of how Direction is changing.

In order to intercept the ball, after knowing the values of these parameters, each player can execute several actions:

Turn changing the direction of the player according to a moment between -180 and 180 degrees.
Dash increasing the velocity of the player in the direction it is facing with a power between -30 and 100.

To reduce the number of possible actions that an agent can perform (generalization over actions problem), we have used macro-actions defined as follows. Macro-actions are composed of two consecutive actions: turn(T), and dash(D), resulting in turn-dash(T, D). We have selected $D = 100$, and T is computed according to $A + \Delta_A$, where A is the angle between the agent and the ball, and Δ_A can be: +45,+10,0,-10,-45. Therefore, we have reduced the set of actions to five actions.

6 Results

In this section, the results of using the VQQL model for learning the ball interception skill in the Robosoccer domain are shown. In order to test the performance of the Lloyd algorithm, we generated a training set of 94.852 states with the following iterative process, similar to the one used in [17]:

- The goalie starts at a distance of four meters in front of the center of the goal, facing directly away from the goal.
- The ball and the shooter are placed randomly at a distance between 15 and 25 from the defender.
- For each training example, the shooter kicks the ball towards the center of the goal with a maximum power (100), and an angle in the range $(-20, 20)$.
- The defender goal is to catch the ball. It waits until the ball is at a distance less or equal than 14, and starts to execute actions defined in section 5, while the goal is not in the catchable area [16]. Currently, we are only giving positive rewards. Therefore, if the ball is in the catchable area, the goalie tries to catch the ball, and if it succeeds, a positive reward is given to the last decision. If the goalie does not catch the ball, it can execute new actions. Finally, if the shooter goals, or the ball goes out of the field, it receives a reward of 0.

Then, we used the algorithm described in Section 3 with different number of quantization levels (new states). Figure 1 shows the evolution of the average distortion of the training sequence. The x-axis shows the logarithm of the number of quantization levels, i.e. the number of different states what will be used afterwards by the reinforcement learning algorithm and the y-axis shows the average distortion obtained by GLA. The distortion measure used has been the quadratic error, as shown in equation 4. As it can be seen, when using 2^6 to 2^8 quantization levels, the distortion becomes practically 0.

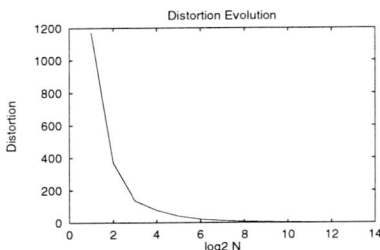

Figure1. Average distortion evolution depending on the number of states in the codebook.

To solve the state generalization problem, a single mechanism could be used, such as a typical scalar quantization on each parameter. In this case, the average quantization error, following the error quadratic distortion measure defined in equation 4, could be calculated as follows. The Distance parameter range is usually in (0.9,17.3). Thus, if we allow 0.5 as the maximum quantization error, we need around 17 levels. Direction is in the (-179,179) range, so, if we allow a quantization error of 2, 90 levels will be needed. Distance Change parameter is usually in (-1.9,1), so we need close to 15 levels, allowing an error of 0.1, and Direction Change is usually in (-170,170), so we need 85 levels, allowing an error of 2. Then, following equation 3 we need $17*90*15*85 = 1,950,750$ states!!! This is a huge size for a reinforcement learning approach. Also, the average distortion that is obtained according to equation 4 is:

$$(\frac{0.5}{2})^2 + (\frac{2}{2})^2 + (\frac{0.1}{2})^2 + (\frac{2}{2})^2 = 2.7$$

given that the quantization error on each quantization is half of the maximum possible error. Instead, using the generalized Lloyd algorithm, with many less states, 2048, the average distortion goes under 2.0. So, it reduces both the number of states to be represented, and the average quantization error.

Why is this reduction possible on the quantization error? The answer is given by the statistical advantages that the vector quantization provides over the scalar quantization. These advantages can be seen in Figure 2. In Figure 2(a), only the pairs of Distance and Direction that appeared in the training vectors have been plotted. As we can see, only some regions of the bidimensional space have values, showing that not all combinations of the possible values of the Distance and Direction parameters exist in the training set of input states. Therefore, the reinforcement tables do not have to consider all possible combinations of these two parameters. Precisely, this is what vector quantization does. Figure 2(b) shows the points considered by 1024 states quantization. As it can be seen, it only generates states that represent minimally the states in the training set. The fact that there are parts of the space that are not covered by the quantization is due to the importance of the other two factors not considered in the figure (change in distance and direction).

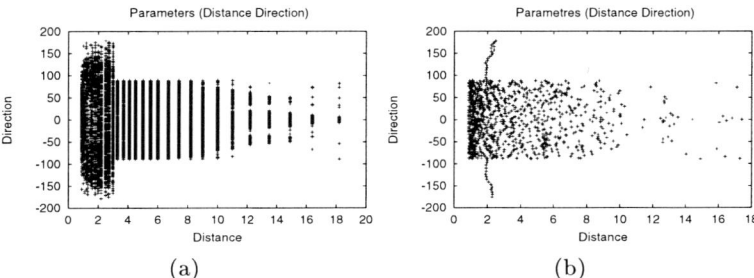

Figure2. Distance and Direction parameters from (a) original data, and (b) a codebook obtained by the GLA.

In order to test what the best number of quantization levels is, we have varied that number, and learned a Q table per obtained quantizer. We measured successful performance as the percentage of kicks of a new set of 100 testing problems that go towards the goal, and are catched by the goalie. The results of these experiments are shown in Figure 3. In that figure the performance of the goalie is shown, depending of the size of the Q table. As a reference, a random goalie would only achieve a 20% of success, and a goalie with the most used heuristic of *always go towards the ball* achieves only a 25% of successful behavior. As it can be seen, Q table sizes less than 128 obtain a quasi-random behavior. From sizes of the Q table from 128 to 1024, the performance increases until the maximum performance obtained, which is close to 60%. From 4096 states and up, the performance decreases. That might be the effect of obtaining again a very large domain (huge number of state).

7 Related Work

Other models to solve the generalization problem in reinforcement learning use decision trees as in the G-learning algorithm [3], and kd-trees (similar to a decision tree) in the VRDP algorithm [14]. Another solution is Moore's PartiGame algorithm [15] or neural networks [9]. One advantage of vector quantization is that it allows to easily define control parameters for obtaining different behaviors of the reinforcement learning technique. The main two parameters that have to be defined are number of quantization levels, and average distortion (similarity

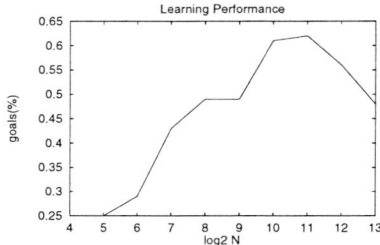

Figure3. Learning performance depending on the number of states in the codebook.

metric). Other approaches to this problem were proposed in [5] and [2] where bayesian networks are used. Similar ideas to our approach to Vector Quantization have been used by other researchers, as in [4] where Q-Learning is used as in this paper, using LVQ [8]. Again, it is easier for VQ to define its learning parameters than it is for neural networks based systems.

8 Conclusions and Future Work

In this paper, we have shown that the use of vector quantization for the generalization problem of reinforcement learning techniques provides a solution to how to partition a continuous environment into regions of states that can be considered the same for the purposes of learning and generating actions. It also solves the problem of knowing what granularity or placement of partitions is more appropriate.

However, this mechanism introduces a set of open questions that we expect to tackle next. As we explained above, the GL algorithm allows us to generate codebooks or sets of states of different sizes, each of them giving us different quantization errors. So, an important question is the relation between the number of quantization levels and the performance of the reinforcement learning algorithm. Another important issue relates to whether this technique can be applied not only to the state generalization problem, but also to actions generalization. We are also currently exploring the influence of providing negative rewards to the reinforcement learning technique. Finally, in the short term we intend to compare it against using decision trees and LVQ.

References

1. Jacky Baltes and Yuming Lin. Path-tracking control of non-holonomic car-like robot with reinforcement learning. In Manuela Veloso, editor, *Working notes of the IJCAI'99 Third International Workshop on Robocup*, pages 17–21, Stockholm, Sweden, July-August 1999. IJCAI Press.

2. Craig Boutilier, Richard Dearden, and Moises Goldszmidt. Exploiting structure in policy construction. In *Proceedings of the Fourteenth International Joint Conference on Artificial Intelligence (IJCAI-95)*, pages 1104–1111, Montreal, Quebec, Canada, August 1995. Morgan Kaufmann.
3. David Chapman and Leslie P. Kaelbling. Input generalization in delayed reinforcement learning: An algorithm and performance comparisons. *Proceedings of the International Joint Conference on Artificial Intelligence*, 1991.
4. C. Claussen, S. Gutta, and H. Wechsler. Reinforcement learning using funtional approximation for generalization and their application to cart centering and fractal compression. In Thomas Dean, editor, *Proceedings of Sixteenth International Joint Coference on Artificial Intelligence*, volume 2, pages 1362–1367, Stockholm, Sweden, August 1999.
5. Thomas Dean and Robert Givan. Model minimization in markov decision processes. In *Proceedings of the American Association of Artificial Intelligence (AAAI-97)*. AAAI Press, 1997.
6. Marco Dorigo. Message-based bucket brigade: An algorithm for the appointment of credit problem. In Yves Kodratoff, editor, *Machine Learning. European Workshop on Machine Learning*, LNAI 482, pages 235–244. Springer-Verlag, 1991.
7. Allen Gersho and Robert M. Gray. *Vector Quantization and Signal Compression*. Kluwer Academic Publishers, 1992.
8. T. Kohonen. The self-organizing map. In *Proceedings of IEEE*, volume 2, pages 1464–1480, 1990.
9. Long-Ji Lin. Scaling-up reinforcement learning for robot control. In *Proceedings of the Tenth International Conference on Machine Learning*, pages 182–189, Amherst, MA, June 1993. Morgan Kaufman.
10. Yoseph Linde, André Buzo, and Robet M. Gray. An algorithm for vector quantizer design. In *IEEE Transactions on Communications, Vol1. Com-28, N° 1*, pages 84–95, 1980.
11. S. P. Lloyd. Least squares quantization in pcm. In *IEEE Transactions on Information Theory*, number 28 in IT, pages 127–135, March 1982.
12. S. Mahavedan and J. Connell. Automatic programming of behavior-based robots using reinforcement learning. *Artificial Intelligence*, 55:311–365, 1992.
13. Tom M. Mitchell and Sebastian B. Thrun. Explanation based learning: A comparison of symbolic and neural network approaches. In *Proceedings of the Tenth International Conference on Machine Learning*, pages 197–204, University of Massachusetts, Amherts, MA, USA, 1993. Morgan Kaufmann.
14. Andrew W. Moore. Variable resolution dynamic programming: Efficiently learning action maps in multivariate real-valued spaces. *Proceedings in Eighth International Machine Learning Workshop*, 1991.
15. Andrew W. Moore. The party-game algorithm for variable resolution reinforcement learning in multidimensional state-spaces. In J.D. Cowan, G. Tesauro, and J. Alspector, editors, *Advances in Neural Information Processing Systems*, pages 711–718, San Mateo, CA, 1994. Morgan Kaufmann.
16. Itsuki Noda. *Soccer Server Manual*, version 4.02 edition, January 1999.
17. Peter Stone and Manuela Veloso. Team-partitioned, opaque-transition reinforcement learning. In M. Asada and H. Kitano, editors, *RoboCup-98: Robot Soccer World Cup II*, Berlin, 1999. Springer Verlag.
18. C. J. C. H. Watkins and P. Dayan. Technical note: Q-learning. *Machine Learning*, 8(3/4):279–292, May 1992.

Fast, Accurate, and Robust Self-Localization in the RoboCup Environment

Jens-Steffen Gutmann   Thilo Weigel   Bernhard Nebel

Albert-Ludwigs-Universität Freiburg, Institut für Informatik
Am Flughafen 17, D-79110 Freiburg, Germany
{gutmann,weigel,nebel}@informatik.uni-freiburg.de

Abstract. Self-localization is important in almost all robotic tasks. For playing an aesthetic and effective game of robotic soccer, self-localization is a necessary prerequisite. When we designed our robotic soccer team for RoboCup'98, it turned out that all existing approaches did not meet our requirements of being fast, accurate, and robust. For this reason, we developed a new method, which is presented and analyzed in this paper. We additionally present experimental evidence that our method outperforms other methods in the RoboCup environment.

1 Introduction

Robotic soccer is an interesting scientific challenge [11] and an ideal domain for testing new ideas and demonstrating existing techniques. One of our main intentions in participating in last year's *RoboCup'98* [1] was to demonstrate the usefulness of self-localization techniques that we have developed [9]. It turned out, however, that all existing self-localization techniques were not efficient enough for a dynamic environment such as robotic soccer. Furthermore, most of the techniques are not robust enough. For this reason, we developed a new technique that exploits one particular characteristic of the RoboCup environment, namely, its purely polygonal structure. Based on that we were able to come up with a very fast, accurate, and robust self-localization technique, which was most probably one of the key factors for the victory of our team *CS Freiburg* in RoboCup'98 [7,17].

Solving the self-localization problem—the problem of determining the position and orientation of the robot—is necessary for almost all tasks. In robotic soccer it seems even impossible to play an effective and aesthetic game if the soccer agents do not know where they are and how they are oriented. As a matter of fact, some of the problems displayed in the games of the middle size league at *RoboCup'97* [10] seemed to have to do with the fact that the soccer robots had the wrong idea about their positions, which led to erratic movements and a number of own goals.

The self-localization problem can be addressed using a wide range of sensors (e.g. odometry, sonars, vision, compasses, laser range finders, other sensors, or combinations thereof) and a wide range of methods. In the sequel we will only consider the combination of data from the odometry and from laser range finders (LRF), since the latter provide accurate and reliable data, which can be interpreted with much less computational effort than, say, data from a vision system.

Self-localization can be based on recognizing known *landmarks* or on *dense sensor matching*. In the first approach, features are extracted from the sensor inputs and matched with the features of the landmarks in order to determine the locations of the landmarks. However, in the RoboCup environment, there are only few natural landmarks that are always visible to the sensors and for this reason we did not consider this approach. In the second approach, all sensor inputs are matched against the expected sensor inputs for a given model. Two competing methods for dense sensor matching are grid-based *Markov localization* [3,2] and *Kalman filtering using scan matching* [5,9]. As it has been demonstrated, Markov localization is more *robust*, because it always generates some position hypotheses and because it can recover from catastrophic failures. However, self-localization using Kalman filtering based on scan matching is more *accurate* [6], since it does not rely on grids.

For robotic soccer, we need *robustness*, *accuracy*, and *efficiency*, whereby the latter property means that we want to estimate the position and orientation in a few milliseconds. Unfortunately, none of the approaches described above satisfies all three requirements. For this reason, we designed a new scan-matching approach that extracts features from the raw sensor inputs, namely, straight lines, that are matched against an *a priori* model. Using the scan match, which can be computed efficiently, the new position estimation is then derived by combining it with the odometry reading using Kalman filtering.

The rest of the paper is structured as follows. The next section sketches scan matching methods and how they can be used to estimate the position using Kalman filtering. Section 3 describes our own method and gives an analysis of the run-time complexity. Based on that, we describe in Section 4 experiments that we have made in order to compare different scan matching methods in the RoboCup environment. Finally, in Section 5 we conclude and sketch future work.

2 Scan Matching

Scan matching is the process of translating and rotating a range scan (obtained from a range device such as a laser range finder) in such a way that a maximum overlap between sensor readings and an *a priori* map emerges. Most of the scan matching methods presume an initial pose estimation that must be close to the true pose in order to limit the search space.

The robot pose and its update from scan matching are modeled as single Gaussian distributions. This has the advantage that robot poses can be calculated with high precision, and that an efficient method for computing the update step can be used, namely, Kalman filtering.

The extended Kalman filter method has the following form. The probability of a robot pose is modeled as a Gaussian distribution $l(t) \sim N(\mu_l, \Sigma_l)$, where $\mu_l = (x, y, \alpha)^T$ is the mean value and Σ_l its 3×3 covariance matrix.

On robot motion $a \sim N((\delta, \theta)^T, \Sigma_a)$ where the robot moves forward a certain distance δ and then rotates by θ, the pose is updated according to:

$$\mu_l := E(F(l,a)) = \begin{pmatrix} x + \delta \cos(\alpha) \\ y + \delta \sin(\alpha) \\ \alpha + \theta \end{pmatrix}$$

$$\Sigma_l := \nabla F_l \Sigma_l \nabla F_l^T + \nabla F_a \Sigma_a \nabla F_a^T$$

Here E denotes the expected value of the function F and ∇F_l and ∇F_a are its Jacobians with respect to l and a.

From scan matching a pose update $s \sim N(\mu_s, \Sigma_s)$ is obtained and the robot pose is updated using standard Kalman filter equations [14]:

$$\mu_l := (\Sigma_l^{-1} + \Sigma_s^{-1})^{-1} \cdot (\Sigma_l'^{-1} \mu_l + \Sigma_s^{-1} \mu_s)$$
$$\Sigma_l := (\Sigma_l^{-1} + \Sigma_s^{-1})^{-1}$$

The success of the Kalman filter depends heavily on the ability of scan matching to correct the robot pose. There are a number of methods for matching scans:

Cox [5] matches sensor readings with the line segments of a hand-crafted CAD map of the environment. He assigns scan points to line segments based on closest neighborhood and then searches for a translation and rotation that minimizes the total squared distance between scan points and their target lines.

Weiss et. al. [18] use histograms for matching a pair of scans. They first compute a so-called angle histogram for determining the rotation of the two scans and then use x and y histograms for computing the translation. Although this method seems to be well suited for the RoboCup environment it is computationally expensive and the precision of the algorithm depends on the discretization size of the histograms.

Lu and Milios [12] match pairs of scans by assigning points in one scan to points in the other scan. For finding a corresponding scan point two heuristics called *closest-point-rule* and *matching-range-rule* are applied and a combination is used for computing the rotation and translation of the two scans. This IDC algorithm (*iterative dual correspondence*) is well suited for any type of environment including non-polygonal ones.

Gutmann and Schlegel [9] use a combination of the Cox matching approach and the IDC method for combining the efficiency and robustness of the line matching method with the universal capabilities of the IDC algorithm. They call their algorithm the *combined scan matcher* (CSM).

Unfortunately all those matching algorithms possess a high computational complexity, e.g. $O(n^2)$ where n are the number of scan points, and their robustness is limited due to the small search space.

Therefore we developed a new algorithm LINEMATCH that makes use of the simple polygonal structure of the RoboCup environment and trades off generality for speed and the ability to globally localize the robot on the soccer field.

3 The LineMatch Algorithm

The LineMatch algorithm extracts line segments from a scan and matches them with an *a priori* map of line segments similar to the methods of [16,4]. We expected that this algorithm has better run-time performance and is more robust than the other scan matchers while retaining the same accuracy as the other matchers. In how far these expectations are realistic will be shown in Section 4.

In order to guarantee that extracted lines really correspond to field-border lines, only scan lines significantly longer than the extent of soccer robots are considered. The following algorithm shows how a matching between model lines and scan lines is computed by recursively trying all pairings between scan lines and model lines:

Algorithm 1. LineMatch*(M, S, P)*

Input: model lines M, scan lines S, pairs P
Output: set of positions hypotheses H
   if $|P| = |S|$ **then**
     $H := P$
   else
     $H := \emptyset$
     $s := SelectScanline(S, P)$
     for all $m \in M$ **do**
        if $VerifyMatch(M, S, P \cup \{(m, s)\})$ **then**
          $H := H \cup \{\text{LineMatch}(M, S, P \cup \{(m, s)\})\}$
   return H

SelectScanline selects the next scan line that should be matched and *VerifyMatch* verifies that the new (m, s) pairing is compatible with the set of pairings P already accepted by computing a common rotation and translation. The algorithm returns position hypotheses in the form of sets of pairs which can be easily transformed into possible locations where the scan could have been taken. For the RoboCup field the algorithm is capable of determining the global position of the robot modulo the symmetry of the field. This means that we get two position hypotheses if three field borders are visible (see Figure 1) and four hypotheses if two borders are visible.

This scan matching method is similar to the methods described by Castellanos *et al.* [4] and Shaffer *et al.* [16]. In contrast to these approaches, however, we only verify that the *global constraints* concerning translation and rotation as well as the length restrictions of scan lines are satisfied. This is sufficient for determining the position hypothesis and more efficient. Further, we do not need any initial estimation of the pose, which means that even if the robot has an extreme error in its position estimation, it may still be able to recover from that.

After matching a range scan, the most plausible position is used in the Kalman filter step for updating the robot position (see Figure 2). We use the

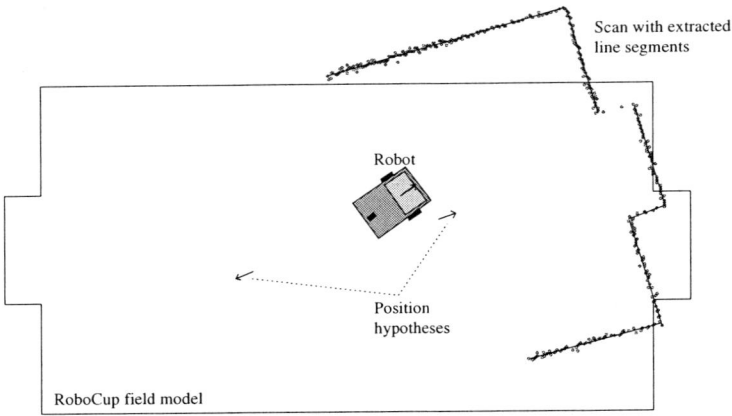

Fig. 1. The LINEMATCH algorithm returns two hypotheses for the robot position.

position information from odometry to determine the most plausible position based on a combination of closest neighborhood and similarity in heading.

For initializing the self-localization system the robot is placed at any position in the RoboCup field but roughly oriented towards the opponent goal and the mean and error covariance of the robot position are set to:

$$\mu_l := (0,0,0)^T$$

$$\Sigma_l := \begin{pmatrix} \infty & 0 & 0 \\ 0 & \infty & 0 \\ 0 & 0 & \infty \end{pmatrix}$$

This ensures global self-localization on the first scan match.

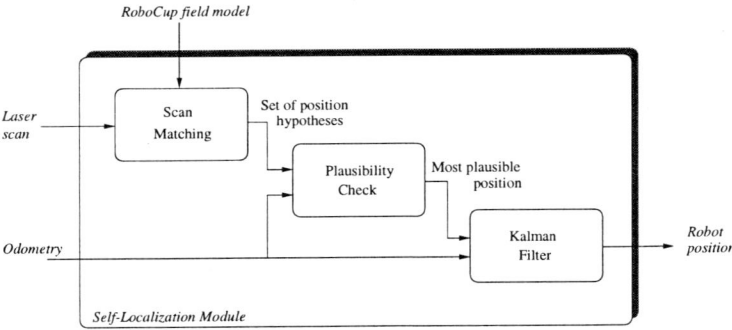

Fig. 2. Self-localization: The most plausible hypothesis is used for updating the robot position.

While it turns out that the implemented algorithm is extremely fast in the RoboCup environment (see Section 4.2), one may wonder how well it scales with the size of the set M. A first rough analysis suggests that the worst-case runtime of the algorithm is $O(|M|^{|S|})$, because the depth of the recursion is $|S|$ and in each recursive call of LINEMATCH $|M|$ different pairings are tried.

As it turns out, however, it is possible to come up with a much better runtime estimation. After the second level of recursion, when two pairings have been made, all degrees of freedom for rotation and translation have been removed (*SelectScanline* is implemented in such a way that it chooses non-parallel lines in the first two levels of recursion). This means that on deeper levels of the recursion only one pairing can be consistent, which leads to invoking another recursive call of LINEMATCH. This means that we may get $|M|^2$ possible pairings on the first two levels of recursion which are verified by further recursive calls trying $|M||S|$ different pairings. Finally, since *VerifyMatch* needs $O(|S|)$ time, we get an overall bound of $O(|M|^3|S|^2)$. In the general case, one has to live with the cubic upper bound. Nevertheless, for realistic environments where not all walls are simultaneously visible—such as is the case in office environments—preprocessing can be used to guarantee runtime almost linear in $|M|$. Such a preprocessing phase would store for each line all other lines that are simultaneously visible. Using such a data structure, the amount of lines that must be tested can be dramatically reduced and assuming a constant upper bound of simultaneously visible walls, we would get a linear complexity of the algorithm.

4 Comparison with other Scan Matchers

In order to show the advantages of the LINEMATCH algorithm we compared the Cox, CSM and LINEMATCH techniques with each other. We did not include the IDC and histogram matching methods as the properties of these algorithms are covered by the CSM algorithm [9].

Since the CSM algorithm needs a set of reference scans as its *a priori* map, we collected a small set of scans, corrected the accumulated odometry error by applying the registration method from [13], and used them as reference scans. This approach has proven to be a successful and easy way for enabling mobile robot navigation in an indoor environment without modifying the environment or creating hand-crafted maps [8].

For comparing the different methods we recorded real data with one of our mobile robotic soccer players. Each of our soccer robots is a *Pioneer I* mobile robot equipped with a *SICK* laser range finder, a *Cognachrome* vision system for ball tracking, a *Libretto 70CT* laptop with wireless ethernet connection and a custom kicking device. The laser range finder covers a 180° field of view with an angular resolution of 1° and a range resolution of $5cm$.

In order to record data of a realistic game scenario we ran the soccer robot in our RoboCup environment with several stationary and moving obstacles. From these data we computed the average run-time of the different algorithms and added different kinds of noise to the data for determining the accuracy and robustness of the methods.

Similar work has been reported by Shaffer et al. [15], who compared two scan matching methods that are similar to the Cox and LINEMATCH algorithm in this paper. However, they used only single scan matches for their experiments whereas in our experiments all data recorded during a whole robot run is taken into account. Also they only ran their algorithms in an almost static environment whereas we recorded our data in a realistic dynamic scenario with many stationary and moving obstacles that can block the robot's sensors. Therefore the results presented in this paper should give a better picture of how good the methods actually are in a dynamic environment like RoboCup.

4.1 Noise Models

There are several kinds of noise typically observed when robots operate in real-world environments. On one hand there is a typical Gaussian noise in the odometry and proximity sensors coming from the inherent inaccuracy of the sensors. On the other hand there are non-Gaussian errors arising from robot colliding with obstacles, e.g. other robot players, or from interference with the sensors.

In this paper, odometry errors coming from wheel-slippage, uneven floors, or different payloads are characterized according to the following three parameters (see left part of Figure 3).

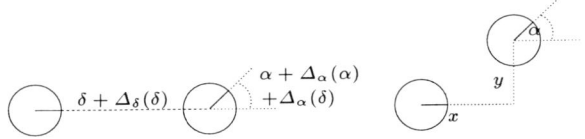

Fig. 3. Effect of adding noise $\langle \Delta_\delta(\delta), \Delta_\alpha(\alpha), \Delta_\alpha(\delta) \rangle$ (left) and bump noise $\langle x, y, \alpha \rangle$ (right) to the odometry.

Range noise: the error $\Delta_\delta(\delta)$ in range when the robot moves a certain distance δ.
Rotation noise: the error $\Delta_\alpha(\alpha) + \Delta_\alpha(\delta)$ in rotation when the robot turns a certain angle α or moves a certain distance δ.

There is another source of less frequent but much larger odometry error coming from situations in which the robot collides with obstacles. These abrupt errors can be characterized by the following parameters (see right part of Figure 3).

Error of the odometry: The error x, y, and α is added to the odometry information.
Frequency: Probability that a bump occurs if the robot travels one meter. Throughout the experiments described below, this probability was set to 0.2 per meter traveled.

4.2 Run-Time Performance

For computing the run-time performance of the scan matching techniques we measured the average time a method needed for computing the pose update before it is fused with the odometry estimate. In order to receive measurements that show the performance under real game conditions we setup a realistic game scenario in our RoboCup environment with stationary and moving objects (see Figure 4) and used our soccer robot as a right defender where it moved over the entire field a couple of times. In this run the robot moved a total distance of approximately 41 meters, turned about a total of 11000 degrees (about 30 revolutions) and collected over 3200 scans.

Fig. 4. Experimental setup: several boxes were placed in the RoboCup field to give a realistic game scenario. Noisy sensor readings are caused by moving obstacles.

Figure 5 shows run-time results performed on the robots on-board computer, a Pentium 120 MHz laptop running the Linux operating system. As expected the LINEMATCH algorithm outperforms the other competing techniques. It is 8 times faster than the Cox algorithm and about 20 times faster than the CSM method. The very low average run-time of only $2ms$ per scan match allows the processing of all incoming range finder data in real time.

Cox	CSM	LINEMATCH
$16ms$	$39ms$	$2ms$

Fig. 5. Run-time results on a Pentium 120MHz laptop.

4.3 Performance in a Game Scenario

For showing the accuracy and robustness of the LINEMATCH algorithm we used the data collected in the above run and added different kinds of noise to the odometry information. In order to measure the accuracy of the position estimates generated by the different matching methods, a set of reference positions are needed. To ease the determination of the reference positions we ran the Cox method with the recorded data and used this output as the set of reference positions.

For each set of noise values, 26 runs with different seed values for initializing a random noise generator were performed. Figure 6 shows the trajectory measured by the robots wheel encoders and a typical trajectory when adding the

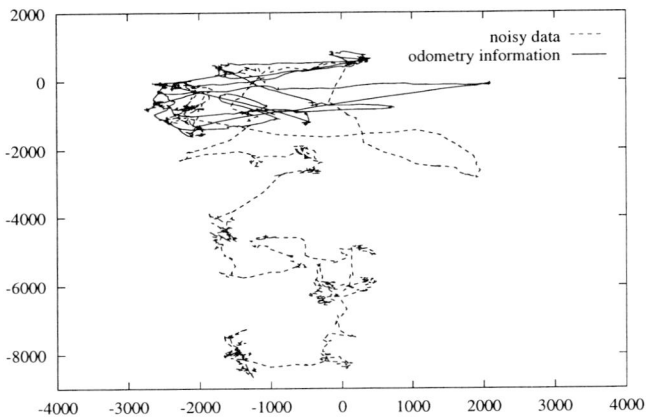

Fig. 6. Trajectory measured by the robot and typical trajectory obtained by adding large Gaussian noise with standard deviations $\langle 400, 100, 40 \rangle$ to these data.

maximum Gaussian noise $\langle 400, 100, 40 \rangle$. The values correspond to the standard deviation of the Gaussian noise $\langle \Delta_\delta(\delta), \Delta_\alpha(\alpha), \Delta_\alpha(\delta) \rangle$ with the units $\sqrt{mm^2/m}$, $\sqrt{deg^2/360°}$, and $\sqrt{deg^2/m}$.

For each scan matching method we computed the number of times the robot position was lost and the distance and heading error to the reference pose in case the position was not lost. We used a threshold of $0.5m$ for the distance and $30°$ for the heading error for determining whether or not the position of the robot was lost.

Figure 7 shows the average distance and Figure 8 the average heading error to the reference positions for five different levels of Gaussian noise. The value triples on the x-axis correspond to the standard deviation of the Gaussian noise $\langle \Delta_\delta(\delta), \Delta_\alpha(\alpha), \Delta_\alpha(\delta) \rangle$. In these and all following figures the error bars indicate the 95% confidence interval of the average mean.

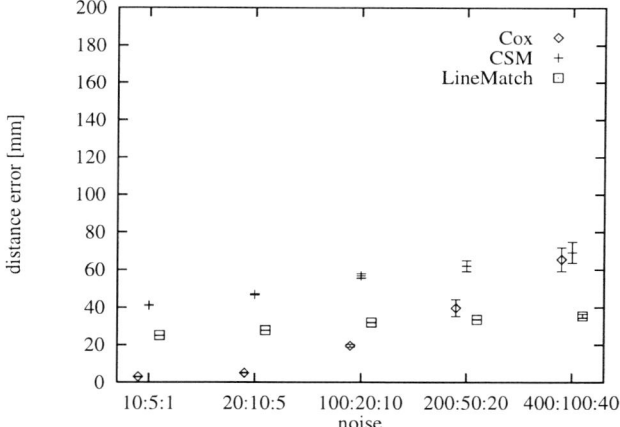

Fig. 7. Distance error to reference positions in typical game scenario for different levels of Gaussian noise.

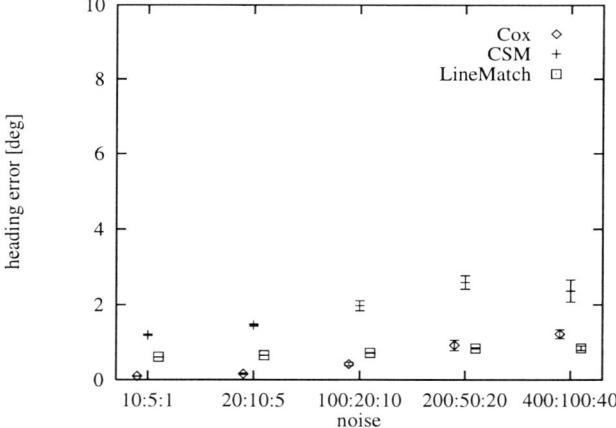

Fig. 8. Heading error to reference positions in typical game scenario for different levels of Gaussian noise.

From both figures it can be seen that all three methods have a similar accuracy usually better than $5cm$ and $2°$. Only the Cox method has a significant higher accuracy than the others when only little Gaussian noise is present but this is due to the fact that the reference positions also have been generated by the Cox method.

However, the LINEMATCH method is much more robust than the other matching algorithms. Figure 9 shows the number of times where the robot position was lost for the same levels of Gaussian noise as in the previous figures. Here the LINEMATCH algorithm shows a very good performance and keeps the

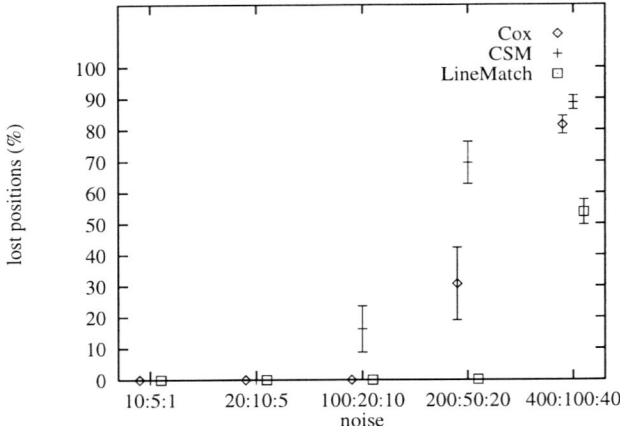

Fig. 9. Number of times where position error was above 0.5m or above 30° in typical game scenario for different levels of Gaussian noise.

robot localized even under high odometry noise. Only for the maximum level of noise, LINEMATCH also starts losing the position. We believe that the higher robustness of LINEMATCH is due to the larger search space it uses for finding matches.

In the same manner, we investigated how the methods compare given simulated bump noise. For accuracy the results were similar to the case of Gaussian noise. All three methods had a similar accuracy for the distance and heading error than in the Gaussian case. Figure 10 shows the average number of positions where the robot was lost when bump noise was added to the odometry information. The triples at the x-axis correspond to the bump noise values $\langle x, y, \alpha \rangle$ used in this experiment. The scale of these values is mm for x and y, and degrees for α. In addition to these bumps occurring with probability 0.2 per meter, we applied a small Gaussian odometry error using the parameters $\langle 100, 5, 2 \rangle$. As can be seen in Figure 10 all scan matching approaches have problems when bump noise is present. This is due to the fact that the Gaussian distribution assumption when fusing the observations with odometry in the Kalman filter does not model bump noise well. However the LINEMATCH method shows less failures than the other methods and is thus again more robust than the other ones.

In a final set of experiments, which can not be covered in this paper due to lack of space, we compared the scan matching methods in "confusing game scenarios" where a long wall was placed inside the RoboCup field. We expected that under these conditions the LINEMATCH algorithm gets irritated since the long wall is not filtered out in its preprocessing step and thus LINEMATCH produces wrong matches or relies on dead-reckoning only for the position estimation. Luckily the LINEMATCH algorithm did not suffer too much from these conditions. We suspect that this is due to the fact that there are a lot of situations where the irritating wall is not present in the range scans.

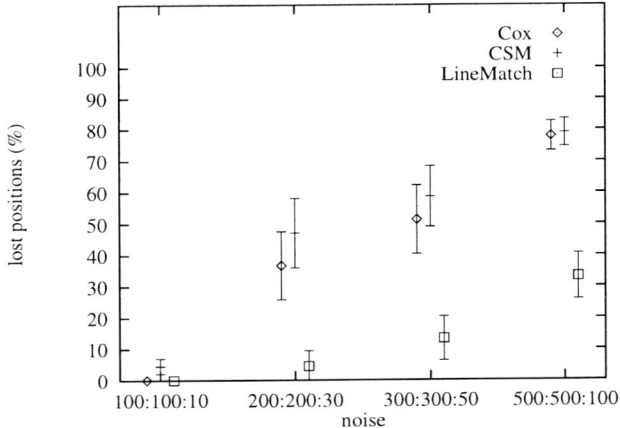

Fig. 10. Number of times where position error was above $0.5m$ or above $30°$ in typical game scenario for different levels of bump noise.

5 Conclusion and Future Work

In this paper we presented a new method for matching range scans to an *a priori* model of line segments which is well suited for localizing a mobile robot in a polygonal-shaped, dynamic environment like RoboCup. Experimental results confirm that the new method is much faster and much more robust than other existing scan matchers while retaining the accuracy of the competing methods.

The proposed method has been developed as one of the key components of the *CS Freiburg* robotic soccer team and has been proven to be fast, reliable, precise and robust. It never failed in any official or in-official game and led the team to its success at RoboCup'98 where *CS Freiburg* won the competition in the middle size league [1].

Although the method has been utilized for RoboCup so far only, it is an obvious step to use it in other polygonal-shaped environments, e.g. as a localization method in our navigation system for office environments [8]. Therefore we will extend the algorithm in various ways, e.g. to allow for partial matches where not all lines of a range scan are matched to model lines and to explore several ways to optimize the algorithm in order to deal with larger environments.

Finally we are going to explore the problem of cooperative self-localization in the RoboCup environment for allowing the reorientation of disoriented group members.

Acknowledgment

This work has been partially supported by *Deutsche Forschungsgemeinschaft* (DFG) as part of the graduate school on *Human and Machine Intelligence*, by

Medien- und Filmgesellschaft Baden-Württemberg mbH (MFG), and by *SICK AG*, who donated a set of new generation laser range finders.

References

1. M. Asada and H. Kitano, editors. *RoboCup-98: Robot Soccer World Cup II*. Springer-Verlag, Berlin, Heidelberg, New York, 1999.
2. W. Burgard, A. Derr, D. Fox, and A. Cremers. Integrating global position estimation and position tracking for mobile robots: The dynamic markov localization approach. In *Proc. of the International Conference on Intelligent Robots and Systems (IROS 98)*, 1998.
3. W. Burgard, D. Fox, D. Hennig, and T. Schmidt. Estimating the absolute position of a mobile robot using position probability grids. In *Proceedings of the 13th National Conference of the American Association for Artificial Intelligence (AAAI-96)*, pages 896–901. MIT Press, July 1996.
4. J. A. Castellanos, J. D. Tardós, and J. Neira. Constraint-based mobile robot localization. In *International Workshop on Advanced Robotics and Intelligent Machines*, Manchester, U.K., 2–3 April 1996. University of Salford.
5. I. J. Cox. Blanche—an experiment in guidance and navigation of an autonomous robot vehicle. *IEEE Transactions on Robotics and Automation*, 7(2):193–204, 1991.
6. J.-S. Gutmann, W. Burgard, D. Fox, and K. Konolige. An experimental comparison of localization methods. In *Proceedings of the International Conference on Intelligent Robots and Systems (IROS'98)*, pages 736–743. IEEE/RSJ, 1998.
7. J.-S. Gutmann, W. Hatzack, I. Herrmann, B. Nebel, F. Rittinger, A. Topor, T. Weigel, and B. Welsch. The CS Freiburg robotic soccer team: Reliable self-localization, multirobot sensor integration, and basic soccer skills. In Asada and Kitano [1], pages 93–108.
8. J.-S. Gutmann and B. Nebel. Navigation mobiler Roboter mit Laserscans. In P. Levi, T. Bräunl, and N. Oswald, editors, *Autonome Mobile System 1997*, Informatik aktuell, pages 36–47, Stuttgart, Germany, 1997. Springer-Verlag.
9. J.-S. Gutmann and C. Schlegel. Amos: Comparison of scan matching approaches for self-localization in indoor environments. In *Proceedings of the 1st Euromicro Workshop on Advanced Mobile Robots*, pages 61–67. IEEE, 1996.
10. H. Kitano, editor. *RoboCup-97: Robot Soccer World Cup I*, volume 1395 of *Lecture Notes in Artificial Intelligence*. Springer-Verlag, Berlin, Heidelberg, New York, 1998.
11. H. Kitano, M. Asada, Y. Kuniyoshi, I. Noda, E. Osawa, and H. Matsubara. RoboCup: A challenge problem for AI. *The AI Magazine*, 18(1):73–85, 1997.
12. F. Lu and E. Milios. Robot pose estimation in unknown environments by matching 2D range scans. *Journal of Intelligent and Robotic Systems*, 18:249–275, 1997.
13. F. Lu and E. E. Milios. Globally consistent range scan alignment for environment mapping. *Autonomous Robots*, 4:333–349, 1997.
14. P. S. Maybeck. The Kalman filter: An introduction to concepts. In I. J. Cox and G. T. Wilfong, editors, *Autonomous Robot Vehicles*. Springer-Verlag, Berlin, Heidelberg, New York, 1990.
15. G. Shaffer, J. Gonzalez, and A. Stentz. Comparison of two range-based estimators for a mobile robot. In *SPIE Conf. on Mobile Robots VII*, volume 1831, pages 661–667, 1992.

16. G. Shaffer et al. Position estimator for underground mine equipment. In *IEEE Transactions on Industry Applications*, volume 28, September 1992.
17. T. Weigel. Roboter-Fußball: Selbstlokalisierung, Weltmodellierung, Pfadplanung und verhaltensbasierte Kontrolle (in German). Diplomarbeit, Albert-Ludwigs-Universität Freiburg, Institut für Informatik, 1999.
18. G. Weiß and E. von Puttkamer. A map based on laserscans without geometric interpretation. In U. Rembold, R. Dillmann, L. Hertzberger, and T. Kanade, editors, *Intelligent Autonomous Systems (IAS-4)*, pages 403–407. IOS Press, 1995.

Self-Localization in the RoboCup Environment

Luca Iocchi and Daniele Nardi

Dipartimento di Informatica e Sistemistica
Università di Roma "La Sapienza"
Via Salaria 113, 00198, Roma, Italy
{iocchi,nardi}@dis.uniroma1.it

Abstract. Knowing the position and orientation of a mobile robot situated in an environment is a critical element for effectively accomplishing complex tasks requiring autonomous navigation. Techniques for robot self-localization have been extensively studied in the past, but an effective general solution does not exist, and it is often necessary to integrate different methods in order to improve the overall result.
In this paper we present a self-localization method that is based on the Hough Transform for matching a geometric reference map with a representation of range information acquired by the robot's sensors. The technique is adequate for indoor office-like environments, and specifically for those environments that can be represented by a set of segments. We have implemented and successfully tested this method in the Robo-Cup environment and we consider this a good benchmark for its use in office-like environments populated with unknown and moving obstacles (e.g. persons moving around).

1 Introduction

A general problem in mobile robot navigation is knowing the robot's pose (position and orientation) in the environment. This is a crucial feature for autonomous robots performing complex tasks over long periods of time and it is thus a main requirement for mobile robots involved in the RoboCup environment [2].

Techniques for robot self-localization (see [3] for a survey) can be distinguished according to the use of *relative* or *absolute* positioning methods. Each of these techniques provides good results as long as some assumptions are verified. For example, dead reckoning approaches are accurate only over short runs of the robot, since error in positioning constantly increases over time. Moreover, global positioning systems and artificial landmark recognition are effective as long as the environment can be appropriately structured. Since none of these techniques provides for a global solution to the self-localization problem, it is often necessary to integrate different localization methods in order to improve both the reliability and the precision of the overall result. A typical solution is to rely on dead reckoning methods (such as odometry) for a short period of time, and then to apply an absolute positioning method (e.g. map matching).

In the RoboCup context, and in particular in the F-2000 league, self-localization is one of the main problem to be addressed for robots since global positioning sensors are not allowed.

In this paper we present a self-localization method that is based on map matching by means of the Hough Transform [6] and we discuss the integration of this technique with other classical positioning approaches.

The method applies to any robot equipped with any kind of sensor that can give range information about the environment (ultrasonic sonars, laser range finders, vision and stereo vision systems, etc.), and it is quite adequate for indoor office-like environments, and specifically for those environments that can be represented by a set of segments, and has been successfully tested in the RoboCup environment. We have used this method in RoboCup-99 within the ART team [13] by making use of vision based line extraction procedures performing as a range data sensor.

The technique turned out to be sufficiently fast and accurate and the use of a vision based range sensor allows the application of the method even if current boards in the field are replaced by lines on the ground (that will be eventually adopted in the RoboCup competitions). Furthermore, we believe that the RoboCup environment is a good benchmark for the use of the method in real office-like environments, since several features are in common with the RoboCup environment: segment-based representation of the map, the possibility of using any kind of range sensor, the presence of unknown obstacles occluding part of the reference lines.

2 Self-Localization in the RoboCup Environment

The RoboCup competition consists of soccer matches between robotic teams [2]. In the F-2000 context, each soccer player is equipped with on-board acting and sensing devices.

The RoboCup environment assumes the following characteristics that must be considered for the choice of localization methods:

1. the geometry of the walls delimiting the field and of the lines drawn on the field is known,
2. the environment is highly dynamic (there are many robots and the ball moving in the field),
3. the task must be performed continuously for a "long" time (the length of each period is 10 minutes),
4. the environment cannot be modified,
5. crashes among robots are possible.

All these factors determine a difficult scenario for localization methods. Indeed, dead reckoning methods are not effective for localization, since they accumulate errors over time and they cannot deal with crashes among players. On the other hand, absolute positioning methods must consider the high noise in acquiring information from the environment due to varying conditions during data acquisition (e.g. other robots moving around).

One of the most common class of methods for absolute positioning is model matching, that is the process of determining the pose of the robot by a matching between a given model of the environment (a map) and the information acquired by the robot's sensors. Observe that these methods require an a priori knowledge of the environment (a map), but they do not require ad hoc modifications in the environment.

In the following sections we present a self-localization method that is based on matching a geometric reference map with a representation of range information acquired by the robot's sensors. We exploit the properties of the Hough Transform for recognizing lines from a sets of points, as well as for calculating the displacement between the estimated and the actual pose of the robot.

3 Hough Transform based Localization

The self-localization method we are going to describe (see also [10] for a more detailed description) is based on a matching between a known map of the environment and a local map built by the robot's sensors. The matching is performed between the Hough representation of both the reference map and the local map.

3.1 The Hough Transform

The Hough Transform is a robust and effective method for finding lines fitting a set of 2D points [6]. It is based on a transformation from the (x, y) plane (a Cartesian plane) to the (θ, ρ) plane (the Hough domain).

The transformation from (x, y) to (θ, ρ) is achieved by associating every point $P(x, y)$ in the Cartesian plane with the following curve in the Hough domain $\rho = x \cos\theta + y \sin\theta$. At the same time, a point in the Hough domain corresponds to a line in (x, y). Notice that this is a *unique and complete* representation for lines in (x, y) as long as $0 \leq \theta < \pi$.

A graphical representation of the Hough Transform can be obtained by generating a discrete grid of the (θ, ρ) plane (let $\delta\theta$ and $\delta\rho$ be the step units), and by defining $HT(\theta, \rho)$ as the number of points in (x, y) plane whose corresponding curve lies within the interval $(\theta \pm \delta\theta, \rho \pm \delta\rho)$.

Observe that it is possible to consider a Hough grid as a voting space for points in (x, y). In other words, every point in (x, y) "votes" for a set of lines (represented as points in (θ, ρ)), that are all the lines passing through that point. Notice that, in the case of a set of aligned points in (x, y), the point in the Hough domain that "receives" the highest number of votes is the one corresponding to the line passing through these points.

The Hough Transform has a number of interesting properties:

1. Given a set of input points, a local maximum of $HT(\theta, \rho)$ corresponds to the best fitting line of these points. Given a set of input points originally belonging to several lines, local maxima of $HT(\theta, \rho)$ correspond to the best fitting lines for each subset of points relative to a single line.

2. With respect to other techniques for extracting segments from a set of points, the Hough Transform is very robust to noise produced by isolated points (since their votes do not affect the local maxima) and to occlusions of the lines (since point distances are not relevant).
3. Measuring displacement of lines in the Cartesian plane corresponds to measuring distance of points in the Hough domain. Indeed, the distance between parallel lines and the angular difference between lines is given respectively by a $\Delta\rho$ and a $\Delta\theta$ between the corresponding points in the Hough domain.

3.2 Self-Localization in the Hough Domain

For applying the self-localization method in the Hough domain, we consider any sensor which returns a set of points, in the local coordinates of the robot, corresponding to a surface of an object. Observe that, in general, these sensors do not allow for simple implementation of object recognition techniques and thus they often retrieve range data from objects in the map (e.g. walls in the environment) as well as from unpredicted obstacles (such as persons moving in the world).

Given this set of points acquired by the robot's sensors and a model of the environment, we want to calculate the displacement between the estimated and the actual pose of the robot.

Under the assumption that the environment can be represented by a set of segments, and in order to exploit the properties of the Hough Transform, we address the localization problem in the Hough domain. In this way the model of the environment is represented by a set of points in the Hough domain and the range data points acquired through the sensors are transformed in the Hough plane as described in the previous section. The map matching process is thus performed over points in the Hough domain and the displacement needed for a correct re-positioning of the robot is easily calculated in the Hough plane.

Summarizing, the Hough Transform based localization method consists in the following steps:

1. extracting range information from the environment in the form of a set of points in the (x, y) plane,
2. applying the Hough Transform to the set of points generating a discrete Hough grid $HT(\theta, \rho)$,
3. determining the local maxima by a threshold,
4. finding correspondences between local maxima and reference points,
5. measuring the displacement between local maxima and the corresponding reference points in the Hough domain.

One of the most important issues that must be considered in the application of the method is that finding correspondences between local maxima and reference points (the fourth step) can lead to incorrect matching and thus to a large error in repositioning. We adopt two different strategies for dealing with this problem:

1. assuming that odometry provides for an almost correct position over a short time, (*small positioning error assumption*), the matching is performed between a local maximum and the nearest reference point;
2. in case of ambiguities, we apply a more general procedure that acquires a greater amount of data about the environment (by integrating different sensor data) and performs an overall match between the set of local maxima and the set of reference points.

It is important to notice that in some cases it is possible to detect ambiguities and thus to avoid large errors, while in other cases this is not possible. Therefore external information are required for avoiding the application of the method in those situations in which it is not accurate. We will discuss in section 5 the integration of the Hough based method with other localization method.

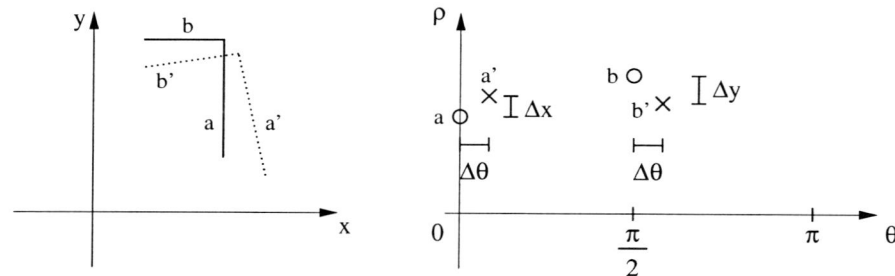

Fig. 1. Map matching in the Hough domain

Consider the example shown in Fig. 1, where the robot faces a corner. The solid segments a, b represent the map model and the set of points a', b' represent data coming from sensor devices. The four segments are also displayed in the Hough domain: a, b (indicated by a circle) are the reference points, while a', b' (indicated by a cross) represent the local maxima of the Hough Transform applied to the set of input points. In the Hough domain it is easy to calculate the displacement between the estimated and the actual pose of the robot ($\Delta x, \Delta y, \Delta \theta$).

In the example, $\Delta \theta$ is the difference $a'_\theta - a_\theta$ or $b'_\theta - b_\theta$. In ideal conditions these differences should be the same; if not, an average between these values allows for a good approximation. After the correction $\Delta \theta$ is applied to the robot's representation of the map, it is possible to calculate the other two elements $\Delta x = a'_\rho - a_\rho$ and $\Delta y = b'_\rho - b_\rho$, that are actually used for re-positioning the robot.

The accuracy of a localization method usually depends on the accuracy of the range sensor. If we consider an ideal range sensor, the noise introduced by the Hough method is due to the discretizazion of the Hough grid. Therefore the grid intervals $\delta \theta$ and $\delta \rho$ provide an upper bound on the accuracy of the Hough localization method itself.

Finally, the complexity of this method is $O(n+m)$, where n is the number of points returned by the range sensor and m is the dimension of the Hough grid, and it is thus adequate for real-time implementation.

4 Vision based range sensor

In order to apply the method described in the previous section to the RoboCup environment it is necessary to design a range sensor. The choice of a range sensor depends on the characteristics of the environment and not on the localization method.

Among other possible range sensors (ultrasonic sonars, laser range finders, and stereo-vision systems), we decided to implement a vision based range sensor that makes use of a single CCD color camera put on-board to the robot. The advantages of using a color camera with respect to other sensors are:

1. the use of colors to detect lines and boards;
2. it is possible to extract range information from the boards of the field as well as from the lines on the ground;
3. vision is a passive sensor and there are no interferences with other sensors mounted on other robots.

On the other hand the use of a vision system requires an additional effort for the calibration of the system.

We describe in the next sections the calibration techniques that have been used for tuning the system and then the range data extraction procedure.

4.1 Camera Calibration

In order to extract accurate range information from the environment it is necessary to accurately calibrate the vision system. This calibration consists in: (i) finding some critical parameters that are used for world reconstruction, that can be divided into internal and external parameters; (ii) tuning the color filter that is necessary to compensate for luminosity variability.

The method we are using (unlike many others) allows the calibration of internal and external parameters separately. This is very useful since internal calibration (that is related to the actual device) is usually performed the very first time we use a new camera, while external calibration (that depends on the position of the camera on the robot) is necessary almost before any match.

Internal Calibration. The internal parameters of the camera are mainly necessary for removing lens distortion and for computing the geometry of the camera. In other words these parameters are used to relate the ideal model of the camera (*pinhole model*) with the actual device by the following equations[1]:

$$x = (i - C_x)\, P_x\, (1 + k_1 r_d^2)$$
$$y = (j - C_y)\, P_y\, (1 + k_1 r_d^2)$$

---
[1] We make use of a simplified model of distortion that is usually quite effective with several common cameras.

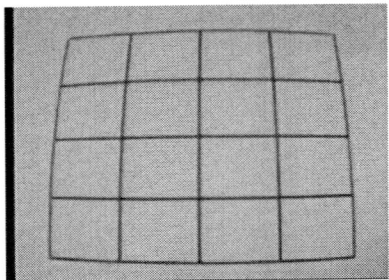 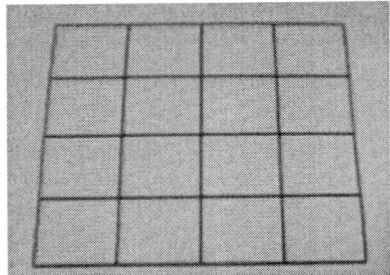

Fig. 2. Internal calibration

The above equations relate each point (i,j) in the image frame with a point (x,y) in the pinhole model of the camera. The parameters P_x and P_y are usually provided with the technical specifications of the camera, and $r_d^2 = ((i - C_x) P_x)^2 + ((j - C_y) P_y)^2$. Therefore the parameters that must be found are the first coefficient of radial distortion k_1 and the center of the distortion (C_x, C_y).

In order to calculate these parameters we make use of an easy calibration procedure that consists in putting a camera in front of a drawing containing only straight lines and starting an automatic calibration procedure that minimizes the overall distortion of the image over the three parameters k_1, C_x, and C_y (see [11] Sect. 7.3 for further details).

In Fig. 2 the original image is shown in the left side, and the undistort image computed with the internal parameters found by the calibration procedure is shown in the right side. It is important to notice that this method does not require the knowledge of the position of the lines in the world, and thus can be applied without knowing any external parameter.

External Calibration. External calibration is the task of calculating the parameters that are necessary for world reconstruction. Performing this calibration after the computation of the internal parameters, allows the use of the equations of prospective geometry referred to the pinhole model of the camera.

$$x = f\frac{X_c}{Z_c} \qquad y = f\frac{Y_c}{Z_c} \qquad (1)$$

These equations relate a point (x,y) in the pinhole camera model to a point (X_c, Y_c, Z_c) in a 3D reference system relative to the camera, with f being the focal length of the lens that is usually provided by the camera technical specifications.

Since we want to express data in a reference system relative to the robot (X, Y, Z), and not to the camera, we need to compute the translation and the rotation of the camera reference system to a robot reference system. External parameters are thus the six parameters necessary for determining the rotation and the translation of the camera reference system with respect to the robot reference system. In this way we can compute the position of a point (X, Y, Z) in the robot reference system from the point (X_c, Y_c, Z_c) in the camera reference system.

Notice that equations (1) are not sufficient for determining a point (X_c, Y_c, Z_c) (and hence a point (X, Y, Z)) in the world given a point (x, y) in the image. As described in Sect. 4.2, we add a third equation (actually $Z = 0$) since we are interested in points that are on the ground.

Color Calibration. Color calibration is necessary for compensating the variance of lumonisity in different parts of the field. We make use of a simple color filter on a HSV representation of the image that transforms the color of every pixel into one predefined color. Specifically, given a set of predefined colors $(H_i, S_i, V_i), i = 1..n$, each pixel (h, s, v) is transformed into (H_i, S_i, V_i) iff

$$|h - H_i| < \Delta H_i \qquad |s - S_i| < \Delta S_i \qquad |v - V_i| < \Delta V_i$$

The filter parameters $H_i, S_i, V_i, \Delta H_i, \Delta S_i, \Delta V_i$ for $i = 1..n$, must be accurately tuned. Many techniques can be used for finding these values (from manual search to neural network approaches [1]), in any case we found out that the effort necessary for finding a set of parameters that are adequate in many different brightness situations is deeply related with the quality of the camera.

4.2 Range data extraction

The calibration procedure described above allows for an image pre-processing that is in charge of removing lens distortion and luminosity variances. After this process we obtain an image that is in accord with the pinhole model of the camera and in which each pixel assumes one of a limited set of predefined colors.

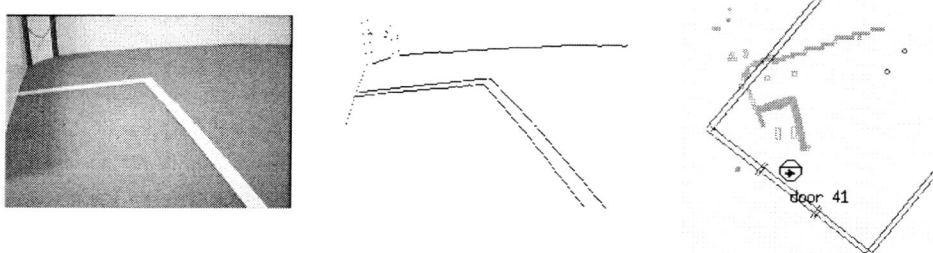

Fig. 3. Vision based range data extraction

Range data extraction is thus performed by the following steps (see Fig. 3):

1. points belonging to the lines of the field and to the base of the boards are detected by a correlation procedure identifying green-white boundaries on the image;

2. these points are transformed in the robot coordinate system by using the prospective geometry equations (1) and the equation $Z = 0$ that states that the points are on the ground.

5 Hough Transform based Localization in the RoboCup Environment

In order to provide our robot soccer players with an effective and robust localization method for the RoboCup environment, we apply the Hough Transform based localization method to the points extracted by the vision system corresponding to the boards and the lines in the field.

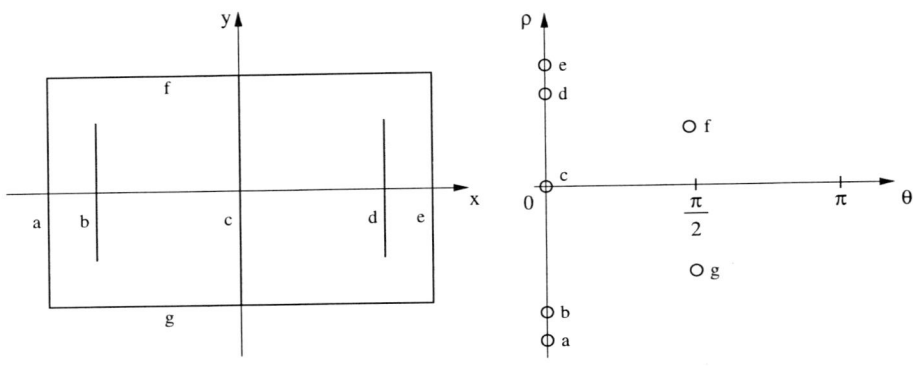

Fig. 4. The RoboCup world model

The model of the RoboCup environment is shown in Fig. 4. We consider seven segments corresponding to the four boards a,e,f,g and the three lines b,c,d. Observe that the walls are real obstacles for the robot, while lines are drawn in the field and do not correspond to obstacles. However, the vision range sensor is able to extract range information from both of them and thus we can consider both boards and lines for map matching.

A self-localization task is displayed in Fig. 5. We have verified that isolated noisy points do not affect the displacement measures and that the method is very robust to occlusion of lines, because of the properties of the Hough Transform.

The performance of the system is adequate for real-time execution with a low-cost color camera and a conventional Pentium based PC, that is on board of the robot. In fact, in our case, most of the computation time (currently around 30 ms.) is taken by the image processing procedure for line extraction, while the Hough method takes a few milliseconds of computational time. As for accuracy, when the vision system is well-calibrated, we obtain good results with a discretization of the Hough grid (and hence an average precision) of 3 degrees for θ and of 10 cm for ρ.

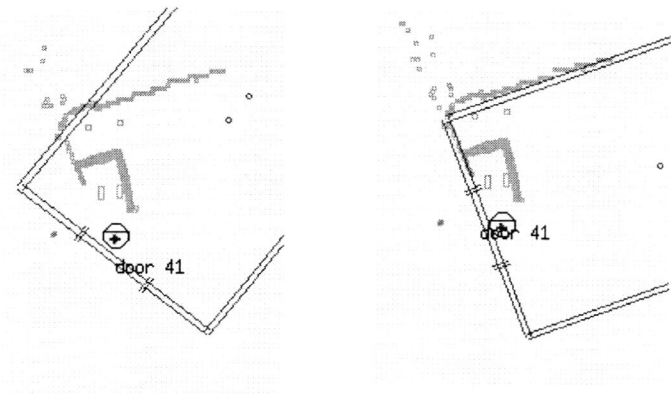

Fig. 5. A self-localization example

6 Integration of Self-Localization Techniques

Several localization techniques can be used in the RoboCup environment, such as odometry, absolute orientation with a compass, and natural landmark detection.

Odometric measures allows for a relative positioning that is reliable over a short period of time and as long as the robot is not involved in any crash. In our Pioneer robots[2] we make use of their built-in odometric system.

We also makes use of an electronic compass that provides information about the orientation of the robot. These values are often precise and reliable, unless a magnetic field produced by another robot passing by makes the values of the compass extremely incorrect.

Natural landmark detection are based on the a priori knowledge of the absolute position of some objects (landmarks) in the field and on the ability of recognizing these landmarks and place them in the environment. The effectiveness of these techniques is related to the reliability of object recognition and to the precision in object positioning. We have implemented vision based routines for detecting the poles of the goals and the corners of the field and we adopt a triangulation method for calculating their position in the field. By knowing the actual position (with respect to the robot) of a number of landmarks in the field and their predefined absolute position, it is easy to compute the absolute position of the robot in the field.

Having a number of localization techniques, we must deal with integration of their results. Observe that this is a very important issue since the process of updating the robot's pose cannot take place in an uncontrolled way. In fact we have identified two different positioning strategies. The first one, that we call *passive*, is normally used and is based on the combination of an odometric

---

[2] Pioneer robots are the commercial versions of the Erratic robot [12] and are produced by ActivMedia Inc.

method and the above Hough Transform based method. This process is continuously executed during the robot's task execution. The second one, that we call *active*, is used in special situations (e.g. a kick restart) and takes complete control of the robot to determine its position in the field without relying on any previous information about it.

The difference between the two strategies is that the passive one is performed concurrently with respect to all the other tasks of the robot, while the active one is performed by taking full control of the robot.

Moreover, during the passive positioning there exist a number of situations in which it is not convenient to apply an absolute positioning method: this is due to the possibility that absolute positioning introduces an error that is larger than the odometric error we want to correct. For example updating the robot's position when it is involved in critical tasks, such as positioning for kicking the ball towards the opposite goal, could cause an overall behavior which performs worse than the one affected by a position error.

In other words, we must provide the robot with the capability of deciding which kind of information integration is to be performed depending on its current state. To this end, we believe that numerical techniques for data fusion are not always adequate in this context. For example, consider data coming from an electronic compass: we have detected that in some situations (e.g. when another robot is moving close around the robot) these values are completely incorrect and numerical techniques for data fusion are affected by a large error provided by the compass value. It is thus necessary to make use of rules that define situations in which to apply a certain integration method, such as *"if there is a robot moving close around then do not consider compass values"*.

We are thus working on the definition of a framework for information integration based on high-level rules. Specifically, the antecedent of one of these rules specifies the conditions under which it is possible, necessary and convenient to perform a self localization, while the consequent selects those methods that are appropriate in this context and specifies how they should be integrated. This set of rules is interpreted by the robot's control system periodically and a decision about whether and how to perform a self-localization task is taken depending on the current situation in which the robot is.

An important feature of this framework is that it provides the programmer of the robot with a powerful and flexible way of dealing with sensor data fusion, and in fact it allows for defining complex criteria for information integration.

7 Conclusion

Knowing the position of a mobile robot in an environment (and specifically in the RoboCup environment) is a critical element for effectively accomplishing complex tasks requiring autonomous navigation. The localization problem has been thus addressed in the past from many different perspectives. In particular, absolute positioning methods based on map matching have been extensively studied (see [5, 14] for occupancy grid matching strategies, [9] for the angle histogram method, [4] for a probabilistic approach, [8] for scan matching techniques, and [7] for experimental comparisons).

They present different solutions that are generally robust to sensor noise, ambiguous situations, partial model description. However, in a moderately crowded and dynamic environment, map matching based localization methods must also be robust to noise given by unknown objects sensed by range sensors. The difficulty in dealing with this kind of noise, that is typical in real environments, is that it cannot be appropriately modelled.

In this paper we have presented a self-localization technique for mobile robots that is suitable with any kind of sensors able to provide range information about objects in the world. We exploit the robustness properties of the Hough Transform for defining an effective and robust self-localization method for dynamic environments. The technique is adequate for indoor office-like environments, and specifically for those environments that can be represented by a set of segments, and has been successfully tested in the RoboCup environment.

Finally, we believe that an effective integration of self localization methods should consider both the situations in which they are necessary and reliable and the current state of the robot. The definition of a high-level rule-based framework allows for describing complex criteria for information integration and for applying localization methods in a controlled way.

References

1. C. Amoroso, A. Chella, V. Morreale, and P. Storniolo. A segmentation system for soccer robot based on neural networks. In *Proceedings of 3rd RoboCup Workshop*, 1999.
2. M. Asada. The RoboCup physical agent challenge: Goals and protocols for Phase-I. In H. Kitano, editor, *Robocup-97: Robot Soccer World Cup I*, 1998.
3. J. Borenstein, H. R. Everett, and L. Feng. *Navigating Mobile Robots: Systemss and Techniques*. A. K. Peters, Ltd., 1996.
4. W. Burgard, D. Fox, D. Hennig, and T. Schnidt. Estimating the absolute position of a mobile robot using position probabilities grid. In *Proc. of 14th National Conference on Artificial Intelligence (AAAI'96)*, 1996.
5. I. J. Cox. Blanche - an experiment in guidance and navigation of an autonomous mobile robot. *IEEE Trans. on Robotics and Automation*, 7(3), 1991.
6. R. Duda and P. Hart. Use of the Hough Transformation to detect lines and curves in the pictures. *Comm. of the ACM*, 15(1), 1972.
7. J. S. Gutmann, W. Burgard, D. Fox, and K. Konolige. An experimental comparison of localization methods. In *International Conference on Intelligent Robots and Systems*, 1998.
8. J. S. Gutmann and C. Schlegel. AMOS: Comparison of scan matching approaches for self-localization in indoor environments. In *1st Euromicro Workshop on Advanced Mobile Robots (EUROBOT)*, 1996.
9. R. Hinkel and T. Knieriemen. Environment perception with a laser radar in a fast moving robot. In *Proc. of Symposium on Robot Control (SYROCO'88)*, 1988.
10. L. Iocchi and D. Nardi. Hough transform based localization for mobile robots. In *Proceedings of 3rd IMACS/IEEE International Conference on Circuits, Systems, Communications, and Computers*, 1999.

11. Luca Iocchi. *Design and Development of Cognitive Robots*. PhD thesis, Dipartimento di Informatica e Sistemistica, Università "La Sapienza" Roma Italy, `ftp.dis.uniroma1.it/pub/iocchi/publications/iocchi-thesis99.ps.gz`, 1999.
12. Kurt Konolige. Erratic competes with the big boys. *AAAI Magazine*, Summer:61–67, 1995.
13. D. Nardi, G. Clemente, E. Pagello, A. Bonarini, A. Chella, M. Piaggio, and G. Adorni. Azzurra Robot Team: ART99. In *Proceedings of 3rd RoboCup Workshop*, 1999.
14. B. Schiele and J. Crowley. A comparison of position estimation techniques using occupancy grids. *Robotics and Autonomous Systems*, 12, 1994.

Virtual RoboCup: Real-Time 3D Visualization of 2D Soccer Games

Bernhard Jung, Markus Oesker, and Heiko Hecht

Universität Bielefeld, Germany
http://www.TechFak.Uni-Bielefeld.DE/techfak/ags/wbski/3Drobocup/

Abstract. Virtual RoboCup is a real-time 3D visualization tool for 2D simulated soccer games as played in the RoboCup simulation league. Players are modeled as anthropmorphic figures and animated step-keepingly with the underlying 2D simulation. Important aspects of player animation concern the generation of natural 3D player movements and realistic player-ball interactions during kicks. A key contribution of Virtual RoboCup is its novel approach to task-level animation in which task-level directives for 3D animation of anthropomorphic characters are generated via on-line classification of fast paced 2D simulation data. As further contribution, we investigated to what extend human observers perceptually process the level of detail in Virtual RobCup animations. A psychological experiment was designed to test the effectiveness of 3D body animation. Although observers failed to notice differences in animation detail, clear effects of character animation on perceived skill were found. The experiment confirms that is is very well justified to spend valuable computational resources on naturalness and richness of detail in realtime character animation.

1 Introduction

Recent advancements in computing power and graphics rendering technology have facilitated the development of many applications (e.g. in computer games, ergonomic evaluation of virtual prototypes, multi-user virtual worlds, etc.) involving interactive animation of 3D human-like figures [7, 1]. Such animated figures are typically controled via task-level directives which are translated by the animation system into appropriate geometric transformations at the graphical level [12, 8]. Typical task-level commands for interactive animations are, for example, *walk from A to B* or *kick the ball in direction* α. Other work has explored cases where task-level animation commands originate in instructions by a human, e.g. using natural language [2, 11], or in the planning processes of autonomous agents situated in the virtual environment [10, 3, 9]. In this article, we explore a third source of input to task level control of animated 3D figures: 2D state information about a simulated soccer game.

Virtual RoboCup is a real-time 3D visualization system for simulated 2D soccer games as played in the RoboCup simulation league [5]. The 2D soccer

Fig. 1. RoboCup provides a 2D environment for simulated soccer games (top). Given 2D input from the soccer server, Virtual RoboCup generates real-time 3D visualizations where players are animated anthropomorphic figures (bottom).

simulator and a 2D visualization program are provided by the RoboCup organization (Figure 1 top). Virtual RoboCup adds anthropomorphic figures to the 2D simulation and animates the soccer game in real-time (Figure 1 bottom). Virtual RoboCup classifies 2D information about the soccer game into task-level action commands which are used to animate the 3D visualized soccer players. Special complexities of this approach arise (a) from the high frequency in which task level commands need to be generated, (b) from the relatively high number - there are 22 players to a soccer game - of animated figures, and (c) from the frequent interactions between the animated players among each other and the ball.

In developing Virtual RoboCup, high emphasis was placed on generating natural and physically plausible animations of soccer games. More concretely, we focussed on the following goals:

- Real-time animation: The 3D animation is generated on the fly; computation of 3D simulation data keeps step with 2D input from the RoboCup server.
- Natural, human-like movements of players: 3D players, as opposed to their circle representations in the 2D RoboCup server, have legs and should use them when moving on the field. Animation of players' movements should

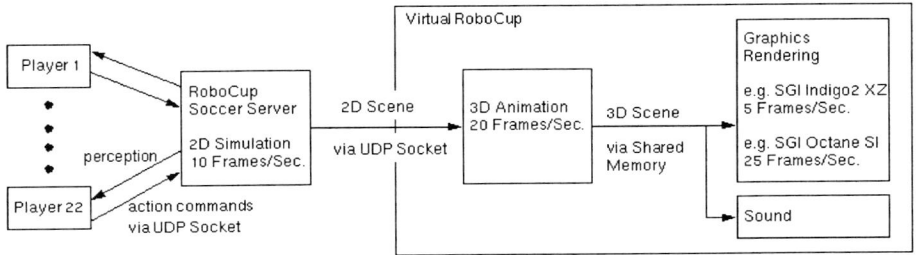

Fig. 2. The RoboCup server defines a 2D environment for simulated soccer games; player actions are controled by independent processes. Virtual RoboCup is realized through several, asynchronous processes in order to allow for a step-keeping 3D visualization of 2D RoboCup games on different hardware platforms. The 3D animation process adds intermediate states to the soccer simulation, generating 3D scenes at a fixed rate of 20 frames per second. The graphics rendering process always presents the most recent 3D scene; its update frequency depends on the underlying graphics hardware. A sound process generates acoustic feedback for successful kicking attempts.

   be fluent, without discontinuities. Furthermore, the 3D players' stepping frequency should be no faster than human stepping.
 - Physically realistic kicking actions: In order for a player to kick the ball, contact between the player's foot and the ball must be established. A kick, in contrast to its instantaneous nature in the RoboCup server, is a temporally extended action that lasts over several animation cycles.

Thus, in addition to the real-time requirement, Virtual RoboCup is mainly concerned with the addition of articulated body models and the dynamics of players' footwork when running and kicking. While the 2D information from the RoboCup server provides some constraints on these tasks, it sometimes also admits *unnatural* movement of players (see next section). Virtual RoboCup makes conservative attempts to "smoothen" unnatural player behavior into movements more consistent with human biomechanics. In general, however, the 3D animations produced by Virtual RoboCup accurately reflect the 2D game states of the RoboCop simulation.

2 System Architecture

The system architecture of Virtual RoboCup reflects the afore mentioned criteria of real-time capability and naturalness of players' running and kicking actions. Figure 2 summarizes the 3D visualization architecture of Virtual RoboCup and its relationship to the 2D soccer simulation in the RoboCup server. RoboCup soccer games are distributed simulations consisting of up to 22 players ("agents"), realized as independent control processes, and a central server that maintains the current simulation state. The simulation environment defined by the RoboCup server is two-dimensional, i.e. players and ball are represented as circles. Every 100 ms, the RoboCup server generates a snapshot of the game state describing the current positions of players and ball as well as some additional information about players' kicking actions and scored goals. These snapshots constitute the input of visualization systems such as Virtual RoboCup.

The system architecture of Virtual RoboCup consists of three asynchronous processes for 3D animation, graphics rendering, and sound generation that communicate via shared memory. The main rationale behind the asynchronous computation of 3D animation data and graphics rendering is to avoid having the (usually) faster simulation process wait for the (usually) slower rendering process. For example on an SGI Indigo 2 XZ platform, a rendering rate of 5 frames per second is achieved for Virtual RoboCup. If 3D animation and rendering were synchronized, the 3D animation step rate would also slow down to 5 frames per second, thus falling behind the input data arriving at the rate of 10 frames per second. With asynchronous animation and rendering, the rendering process visualizes the most recent 3D animation state, possibly dropping some intermediate states. If a platform with faster graphics capabilities is used, e.g. an SGI Octane SI, all frames can be rendered. The asynchronous computation thus ensures that the 3D visualization – independent from the specific graphics hardware used – keeps step with the 2D simulation.

When generating the 3D animation of the soccer game, Virtual RoboCup adds intermediate states to the original 2D RoboCup simulation, such that the internal simulation of Virtual RoboCup runs with twice the speed of the RoboCup simulation. One reason for this speed-up is that a more fluent visualization can be generated. Another, deeper reason has to do with the way that kicking actions are calculated in the RoboCup simulation: When the RoboCup soccer server establishes that, in a simulation cycle t_i, a kicking attempt of a player was successful it will also, in the same cycle, calculate a new ball position reflecting the effect of the kicking action. Thus, the soccer server might generate a new game state description for simulation cycle t_i where the ball is outside of the player's kicking range yet annotate this scene symbolically that the player has kicked the ball. Therefore, if in the kicking action contact between the player's foot and the ball is to be visualized, then the time of contact must lie before t_i, yet after the preceding simulation cycle t_{i-1}, hence in some intermediate state.

Finally, the 3D animation is slightly (including time for graphics rendering less than 0.5 seconds) delayed as compared to the original simulation. This time delay is used, for example, for continuous animation of kicks (including preparing frames for change of supporting leg, leg swinging), that are treated as instantaneous events in the RoboCup server. Also, the 2D simulation in the RoboCup server is an abstract approximation of real soccer games that sometimes allows for unnatural, or even biomechanically impossible player movements. For example, with repeated "dash-turn" commands, 2D players can zig-zag across the field with up to 5 direction changes per second. Similarly, with repeated "turn" commands, 2D players can perform up to 5 pirouettes per second. To make the

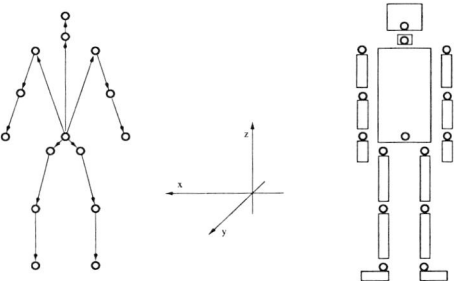

Fig. 3. The hierarchical body model of Virtual RoboCup players consists of 15 segments representing torso, legs, arms, neck and head. Body segments are shaped as boxes so as to allow for efficient graphics rendering (as welcome side effect players also appear more "robotic").

3D players' movements appear more human-like, Virtual RoboCup examines the temporal context of player movements and smoothens sharp direction changes by averaging over several simulation cycles. As a last example, with quickly repeated "kick" commands, players can 'hyper-kick' the ball in consecutive 100 ms simulation cycles. By inspecting the temporal context of kicks, such repeated, instantaneous kicks in the 2D simulation are merged into one, yet temporally extended kicking action in Virtual RoboCup.

3 Movement Animation

Virtual RoboCup tries to optimize animation of the 3D soccer players' running in two dimensions: First, players' movements should appear as natural as possible. For example, each visualized step of 3D players will extend over several cycles of the underlying 2D simulation. And second, generation of the 3D animation must occur in real time, keeping up with the 100 ms update frequency of the 2D simulation. While physical based simulations of human running result in highly realistic animations, e.g. [4], they are not yet computable in real-time. To meet the real-time requirement, Virtual RoboCup uses a keyframe based approach for animating the players' forward movements.

As foundation for the real-time movement animation, 30 locomotion keyframes are defined. The cyclic keyframe sequence shows a 3D player performing a step with the right leg followed by s step with the left leg (see Figure 4). The whole sequence will move the player 4.10 meters forward. The keyframe table also annotates each keyframe with the player's position gain as compared to the first frame of the sequence.

As the soccer players move on the field with changing speed but the animation proceeds with a fixed frequency, animation of the players' movements cannot simply consist of replaying the predefined keyframe sequence in standard

Fig. 4. Some locomotion keyframes for Virtual RoboCup players.

order. Instead, computation of the players' body postures during running animations also accounts for the player's position gain in an animation cycle: for a fast moving player, some intermediate keyframes might be dropped, whereas for a very slowly moving player, even the same keyframe might be used in consecutive animation cycles. Also, the 30 predefined keyframes are in certain cases, especially for very slowly moving players, insufficient to capture subtle changes in a player's running posture. Therefore, the players' animation postures are not limited to the predefined keyframes but calculated by interpolating between the predefined keyframes (*"inbetweening"*).

The algorithm for computing a player's body posture during running takes as input the 2D state information from the RoboCup server, or more concretely, the player's 2D position and orientation in the 100ms simulation cycle that is to be visualized as well as 2D state information from two preceding and two following simulation cycles. The animation process also maintains some additional internal state information about the players including their current posture (conceptually, a posture is represented by a float within the range [0, 4.10] that 'indexes' into the keyframe table). The output of the algorithm is the 3D player's position, orientation and body posture for the next 50 ms animation cycle. The movement animation is computed in the following way:

1. The target position and orientation of the player for the next animation cycle are calculated from the 2D input data. If the animation cycle corresponds to a simulation cycle, the target position equals the player's position in the RoboCup simulation. If the animation cycle lies inbetween simulation cycles, the target position is calculated through linear interpolation between the player's positions in neighbouring cycles. The player's orientation is calculated by averaging over several simulation cycles, thus smoothening sharp direction changes.
2. To generate the player's 3D pose, first an 'posture index' into the keyframe table is computed by adding the position gain as compared to the last animation cycle to the player's posture index in the previous animation cycle. Then two frames f_- and f_+ are selected from the keyframe table that are closest to the new posture index. These frames are weighted according to their respective proximity to the new posture index and the player's actual body posture is generated by weighted interpolation between the two frames f_- and f_+ (inbetweening).

Fig. 5. Animation of kicking actions: Preparation – contact – follow-through.

The running animations produced by this method appear especially natural in cases where 2D RoboCup players perform steady forward movements without sharp direction changes. One limitation of the current implementation is that the 3D players cannot yet come to a stillstand with both feet on the ground. In cases where the 2D simulation allows player movements that are per se impossible to perform given the constraints of human biomechanics (e.g. the zig-zagging and pirouetting behaviours described in Section 2), we had to make a choice, whether to stick with the original 2D running paths or to replace them with more natural paths. As the purpose of Virtual RoboCup is the visualization of 2D RoboCup games (and because improving on the 2D input data is in general a nontrivial, time consuming task) we decided to correct unnatural player movements only very cautiously. In particular, sharp direction changes of the players are smoothened by averaging a player's body orientation over several simulation cycles. The player's head is however always oriented according to the player's actual direction in the RoboCup server. Further improvements of the movement animation might involve the definition of several keyframe sequences for player movements in different speeds.

4 Kicking Animation

For human soccer players (and human-like 3D players), kicking a ball is a fairly complex skill, involving e.g. selection of a foot to kick the ball with, approach of the ball such that the non-kicking foot gives enough support to keep the player balanced, leg swinging and orienting of the foot such that the ball is kicked in the intended direction, and so on. In contrast, the kick-model in the 2D RoboCup simulation is rather abstract: if close enough to the ball, a player can kick the ball in any direction. Furthermore, the RoboCup simulation treats kicks as instantaneous events and it is possible for single players to kick the ball in several consecutive 100 ms simulation cycles. Virtual RoboCup aims at visualizing the players' kicking actions as natural as possible. While not all details of human movements during ball kicking are reenacted, kicking actions are visualized as extended sequences: In the preparation phase, the player's leg swings towards the ball. In the culmination phase, contact between the player's foot and the ball is established. In the final follow-through phase, the leg keeps swinging in the direction of the kick (see Figure 5).

The RoboCup simulation allows a player to kick the ball anywhere within a distance of 2 meters and in any direction. A keyframe-based approach for animation of kicking actions is thus not feasible as it would require the definition of a too large amount of keyframe sequences to cover all free parameters. Instead, animation of kicking actions uses a inverse kinematics approach to guide the foot towards the ball.

The kicking animation is triggered whenever the RoboCup simulation reports a successful kicking attempt of a player. Further constraints on the kicking task, such as timing, location, and direction of the kick are extracted from 2D game state information of several consecutive simulation cycles. The following steps are performed to generate the kicking animation:

1. The direction in which the ball is kicked is calculated from the 2D input data. This is also the direction in which the kicking foot will approach the ball.
2. Based on the kicking direction and position of the ball w.r.t. the player, the foot to kick the ball with is selected. Foot selection also ensures that the player's legs don't overlap during the kicking sequence.
3. If necessary, the player is slightly repositioned such that a kick in the correct direction is possible. This might also involve a change of the supporting leg.
4. Using inverse kinematic techniques, the player's posture at ball-foot contact as well as at preparing and follow-through postures are calculated. These postures differ in the angles of hip, knee, and ankle joints. Computation of inverse kinematics uses a simple geometric approach.

An animated kick usually lasts for four animation cycles: Two preparation cycles, during which the foot is moved towards the ball, one contact cycle, and one follow-through cycle. For the reasons detailed in section 2, the contact cycle always falls in an animation phase between the RoboCup simulation cycles. If the RoboCup simulation reports kicks of the same player in consecutive cycles ('hyper-kick', see section 2), only the first kick is animated but with an prolonged follow-through phase; thus, visualization of 'hyper-kicks' is another example, where the 3D animation slightly deviates from 2D input data in order to make the players' movements appear more natural.

In Virtual RoboCup, players can kick the ball in any direction (see Figure 5 for a sideways kick); furthermore, players can kick the ball equally well with both feet. Although player animation based on inverse kinematics is computationally more expensive than keyframe based methods, the 3D visualization still meets the real-time requirement. This is due to the geometric (i.e. closed-form, non-iterative) approach for inverse kinematics calculation but also to the fact that usually at most one player performs a kick per simulation cycle. In parallel to the visual presentation of a kick, Virtual RoboCup also generates a characteristic sound to give the observer a fuller impression of the kicking action. For the time of a kicking animation, animation of the players' running movements is suppressed. Future work might involve improving the animation of transition phases between running and kicking.

Team	Goals
Sopra1	34:00
Sopra3	14:10
Sopra4	03:10
Krislet	00:31

Table 1. Goals scored by soccer teams during a competition.

5 A psychological experiment on human perception of animation detail

Detailed animation of 3D articulated body models is in principle desirable but it is also a highly resource-intensive task. It becomes particularly critical in 3D visualizations of multiple characters in real-time game sequences, such as Virtual RoboCup. Only if human observers perceptually process (though not necessarily consciously) visually presented animation detail can it be justified to spend valuable resources on its computation. To test the influence of animation style on observers' judgments of the capabilities of RoboCup simulation league soccer teams, we designed an experiment that allowed us to contrast the level of perceived playing skill with richness of detail in character animation.

5.1 Design, Stimuli, Apparatus, and Procedure of the Experiemnt

First, a factor of objective skill level was created. Four teams were selected that span a large range of accomplishment. The teams' playing skills ranged from a tournament winner in 1998 to a team that was a few years back in evolution. Four clients (player agents) with known and heterogeneous abilities were used. A team consisted of five instances of one of the clients. Thus, each team consisted of five identical players controlled by the same algorithm but starting at different positions on the field. As main criteria for selection of clients we required that they were objectively discriminable by means of scored goals. Table 1 gives an insight into the selected teams' performance as exhibited during a round robin competition. Three of the teams were selected from a student competition that took place 1998 at the University of Bielefeld, Germany, where they made first, third and fourth place. The simple client named Krislet contributed by Kryzsztof Langner was taken from the Internet [6]. All four clients are implemented in Java and work well with Soccerserver version 3.28 that was used for the experiment.

Three competitions were recorded in which each of the four teams played against each of the others.

From the recorded soccer games we cut sequences of 20 seconds duration, which corresponds to 400 animation cycles. Each sequence showed a promising attack, which was defined as driving the ball in the direction of the opponent team's goal or at least the attempt thereof. Whenever possible we chose sequences that ended with the scoring of a goal. The selected set of sequences contained scenes of all 12 possible combinations of teams. For each team an attack against each of the other three teams was included.

Second, we created four different animation levels by suppressing some features of the animation. Conditions were as follows: 1. No animation of running or kicking, 2. animation of running only, 3. animation of kick actions only, 4. both running and kicking animated.

Each of the 12 sequences was presented 4 times using the different levels of animation detail. The resulting 48 sequences were presented in random order. The defending team was always named A, the other one B. The observer's point of view corresponded to a position near the corner to the right of the defender's goal. The direction of gaze was directed at the ball. This presentation ensured that observers could not identify teams except by the players' actions.

The experiment was carried out as a fullscreen application on an SGI Indigo II XZ machine with a 21" monitor. For the experimental session the digitally recorded sequences were presented at a frame rate of 7 Hz[1]. The refresh rate of the monitor was 72 Hz. Eight student observers (4 men, 4 women) were paid for their participation. After viewing the 20 second sequence each observer had the opportunity to review the entire sequence if so desired. Then she was asked to first decide which of the two teams was more apt and skillful in its overall play. Once this decision had been made she had to assign a grade to each team. A grade of 0 corresponded to pitifully poor skill, a grade of 12 to exceedingly adept. About 10 practice trials were randomly selected from the pool of trials and presented to familiarize observers with the task.

5.2 Results and Discussion

After all data had been collected, participants were asked what they thought distinguished the teams. They were also asked whether they had noticed any changes in animation style between trials. Amazingly, none of the observers reported changes in character animation. Even when asked directly whether in some trials players had moved or shot differently observers failed to report differences in animation style. Observers, however, recognized fairly reliably the objective skill of the teams. In 75.1 % of all cases they gave the higher ranking team (see Table 1) also a higher skill grade. The skill grades assigned to the different teams were correlated positively with their objective skill ($r = .51$,

---

[1] Rendering frequency would have been only 5 frames/second for a standard game with 22 players.

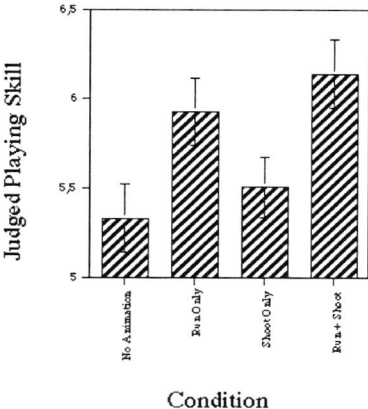

Fig. 6. Skill ratings averaged over all performance variations, plotted by degree of character animation. Error bars indicate standard errors of the mean.

$p < .0001$). Thus, as expected, the main strategic differences produced by the clients were reflected in the judgments.

For the purposes of analyzing the effects of the unnoticed changes in character animation, the grade scores were entered into a repeated measures analysis of variance (ANOVA) with the four levels of animation as independent factor. For the dependent variable, the average judged grades for both teams in a given sequence were analyzed. Note that the animation style was always the same for both teams in a given sequence. For a perfect observer there should obviously be no difference in judgement of playing skill for sequences that are identical except for animation style. Thus, the ANOVA only reflects the influence of character animation. A significant main effect was found for this factor ($F(3,21) = 3.28$, $p = .041$). As shown in Figure 6, team performance under the full animation condition was judged to be significantly more "skillful" than in absence of all character animation and also received better ratings than the animation of the shooting action only. The difference between no animation and shooting action only was not significant. Neither did the slightly better rating of full animation compared to running only reach significance.

Everything else controlled for, the fact that human observers failed to notice the manipulations of animation style did not prevent the animation style to influence their judgments. Teams whose characters are animated in their running and shooting actions are judged to be more skillful. The running action tended to be most important in this context. These findings reveal that it is very well justified to spend valuable computational resources on richness of detail in real-time 3D character animation. They also reveal that explicit judgments, such as obtained by questionnaires or by mere inspection of the displays, are insufficient to asses the importance of level of detail. Detail is processed unconsciously.

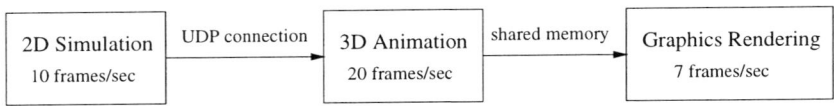

Fig. 7. The animation process generates 3D scenes at a fixed rate of 20 frames per second. In the experimental setup, a rendering speed of 7 frames per second was achieved. This information loss might contribute to the insignificant effect of kick animations on perceived playing skill.

The fact that adding shooting detail to the running animation had very little effect on the results indicates that the limit of meaningful level of detail may have been reached. The poor effect of kick animations may however not originate in the limits of observers' perceptual capacities, but may instead be caused by features of the animation itself. Soccer players are running nearly continously for the time of a match while shooting actions are performed only at times and by single players. Moreover, our animation technique may be suboptimal for visualization of short and accentuated events like kicking actions. Animation is synchronized with incoming simulation data but unsynchronized with the rendering process (see Figure 7). As a consequence the 3D animation keeps step with the 2D soccer simulation. However, some simulation steps might not be visualized if rendering is slow. This loss may result in a significant effect on observer's perception since animation of kicking actions lasts only between four and six simulation cycles and contact to the ball is established only for the time of a single step. However, as disadvantagous as multiprocessing seems to be in this context it is indispensable. Synchronization of the processes would result in a worse and flickering visualization.

6 Conclusions

We have described Virtual RoboCup, a real-time 3D visualization tool for simulated 2D soccer games. The soccer players are visualized as anthropomorphic characters whose running and kicking actions are animated in a natural fashion. Virtual RoboCup represents a novel kind of task-level animation system in which task-level commands are generated by classification of fast paced 2D simulation data. By visualizing soccer games with 22 players, Virtual RoboCup also demonstrates that real-time animation of a high number of human-like like characters is feasible with today's computing technology.

Virtual RoboCup is intended for 'live'-visualization of on-going 2D RoboCup simulation league games. Many of the design choices reflect this need for real-time capability, such as the simple body models of the 3D players, extensive use of keyframing and the asynchronous computation of 3D animation data and graphics rendering. Measurements of processing time show that computation of the 3D player animation can easily keep up with input from the RoboCup Server.

Fig. 8. A scene from the RoboCup'97 simulation league final, seen from two different perspectives.

Rendering times, on the other hand, highly depend on the underlying graphics hardware (see section 2).

Besides the real-time requirement, important design goals of Virtual RoboCup included a high degree of naturalness in the players running and kicking animations. As the computation of realistic animations is a resource intensive task, we performed an experiment to test whether the effort for detailed figure animation is worth its while. Results show clear effects of character animation on perceived skill although observers were unaware of alterations in animation detail. Especially, naturalistic animation of the virtual players' running actions proved to be an important factor.

By virtue of being a 3D visualization, Virtual RoboCup offers, as compared to 2D visualizations, a more life-like presentation of RoboCup games. Virtual RoboCup visualizes certain aspects of RoboCup games, e.g. the players' kicking attempts, that cannot be experienced in 2D visualizations. Also, soccer games can be watched from any angles, including the perspectives of the virtual players (Figure 8). As alternative to standard desk-top displays, we have also ported Virtual RoboCup to the Responsive Workbench, hardware-software unit that projects stereoscopic 3D graphics on a translucent tabletop. Another aspect of the soccer game not presented in the the 2D visualization, namely visualization of the players' changing stamina states is a desirable candidate for further development.

References

1. N.I. Badler, C. Phillips, and B. Webber. *Simulating Humans: Computer Graphics Animation and Control.* Oxford University Press, New York, NY, 1993.
2. N.I. Badler, B.L. Webber, J. Kalita, and J. Esakov. Animation from instructions. In N.I. Badler, B.A. Barsky, and D. Zeltzer, editors, *Making them Move: Mechanics, Control, and Animation of Articulated Figures*, pages 51–93. Morgan Kaufmann, San Mateo, CA, 1991.
3. B. Blumberg and T. Galyean. Multi-level direction of autonomous creatures for real-time virtual environments. In *Computer Graphics Proceedings, SIGGRAPH-95*, 1995.
4. J.K. Hodgins. Three-dimensional human running. In *Proc. of the IEEE Conference on Robotics and Automation*, 1996.
5. H. Kitano, Tambe, P. M., Stone, M. Veloso, S. Coradeschi, E. Osawa, H. Matsubara, I. Noda, and M. Asada. The RoboCup synthetic agent challenge '97. In *Proc. of the 15th International Joint Conference on Artificial Intelligence, IJACI'97*, 1997.
6. Kryzsztof Langner. Krislet, a sample client for robocup simulation league, 1997. http://ci.etl.go.jp/~noda/soccer/client/index.html, 23 March 1999.
7. N. Magnenat-Thalmann and D. Thalmann. Complex models for visualizing synthetic actors. *IEEE Computer Graphics and Applications*, 11:53–59, 1991.
8. N. Magnenat-Thalmann and D. Thalmann. Computer animation. In *Handbook of Computer Science.* CRC Press, 1996.
9. K. Perlin and A. Goldberg. IMPROV: A system for scripting interactive actors in virtual worlds. In *Proc. SIGGRAPH'96*, pages 205–216, 1996.
10. D. Terzopoulos, X. Tu, and R. Grzeszczuk. Artificial fishes with autonomous locomotion, perception, behavior, and learning in a simulated physical world. In *Artificial Life IV: Proc. Fourth International Workshop on the Synthesis and Simulation of Living Systems*, pages 17–27, Cambridge, MA, 1994.
11. I. Wachsmuth, B. Lenzmann, T. Jörding, B. Jung, M. Latoschik, and M. Fröhlich. A virtual interface agent and its agency. In W.L. Johnson, editor, *Proceedings of the First International Conference on Autonomous Agents*, pages 516–517. ACM Press, 1997.
12. D. Zeltzer. Motor control techniques for figure animation. *IEEE Computer Graphics and Applications*, 2(11):53–59, 1982.

The RoboCup-98 Teamwork Evaluation Session: A Preliminary Report

Gal A. Kaminka[1]

Information Sciences Institute and Computer Science Department
University of Southern California
Marina del Rey, CA 90292
galk@isi.edu

Abstract. Increasingly, agent teams are used in realistic and complex multi-agent environments. In such environments, dynamic and complex changes in the environment require appropriate adaptation of the teamwork (collaboration) among team-members. As RoboCup proposes to provide multi-agent researchers with a standard test-bed for evaluation of methodologies, it is only natural to use it for investigating this essential capability. During the RoboCup-98 workshop and competition a unique event took place: a comparative evaluation of the teamwork adaptation capabilities of 13 of the top competing teams. An evaluation attempt of this scale is a novel undertaking, and presents many novel challenges to researchers in the multi-agent community. This preliminary report describes the data-collection session, the experimental protocol, and some of the preliminary results from analysis of the data. Rather than proposing solutions and well understood results, it seeks to highlight key challenges in evaluation of multi-agent research in general, and of teamwork in particular..

Introduction

Agent teamwork (collaboration) is an important and challenging research area, as teams of agents are increasingly becoming a common, often required, theme in many dynamic, complex, multi-agent environments. Such application environments range from virtual environments for training (Johnson and Rickel 1997), through distributed large-scale simulations (Tambe, Johnson et al. 1995) and robotic soccer (Kitano, Tambe et al. 1997), to future space missions. These application domains present many challenges to managing teamwork (collaboration) among the agent team-members: Agents may have conflicting or incomplete local views, the environment may impose restrictions on their ability to observe each other or otherwise communicate, and individual, localized, failures may lead to global, team-

---

[1] We thank Kay Schroter and Prof. Hans-Dieter Burkhard for their help during data-collection phases.

wide difficulties. Agent teams deployed in such realistic settings must therefore be able to recognize when such situations occur and adapt individually and team-wise to the changing conditions.

Generally, evaluation of adaptive teamwork capabilities is limited to investigators using the application domain. Each application is therefore generally evaluated on absolute terms, not with respect to other adaptation techniques. Partially to address this problem, RoboCup has been proposed as a standard research domain and test-bed for multi-agent and robotics research (Kitano, Tambe et al. 1997). In particular, the RoboCup simulation environment shares many of the characteristics that make realistic domains so challenging. It involves multiple interacting agents, both in collaborative and adversarial modes, uncertainty in perception and action, random environmental effects (such as weather conditions), etc. Annual competitions, attended by researchers and programmers from across the world present an environment where the adaptive capabilities of teams may be investigated and evaluated comparatively.

Indeed, the IJCAI-97 RoboCup Synthetic Challenge (Kitano, Tambe et al. 1997) calls for rigorous scientific evaluation of teamwork techniques (among other research topics) using the simulation as the standard test-bed environment. Since 1997, over 60 different agent teams, developed independently by different groups, have been built for RoboCup. Unfortunately, most have been only evaluated on the basis of their overall performance in the annual competitions, rather than on the basis of their scientific contributions in particular areas of multi-agent research.

To remedy this situation a unique event took place during the RoboCup-98 workshop and competition: a large-scale comparative evaluation of teamwork adaptation under controlled failure condition of 13 different simulation teams. The evaluation of each team consisted of playing the team against a fixed opponent four times, as up to 3 of its players were disabled. It was not only the first evaluation session of RoboCup teams, but, to the best of our knowledge, the first multi-agent teamwork evaluation of this kind, and on this scale.

The evaluation methodology presents novel challenges to researchers concerned with teamwork. Issues such as statistical significance, evaluation protocols, comparative measures of teams, etc. all raise important questions about our understanding of teamwork in particular, and multi-agent research evaluation in general. As an example of such a difficult issue, despite the availability of an obvious measure of overall *team performance* (the score difference at the end of a game), it isn't clear how *teamwork performance* should be measured.

This short, preliminary, report provides an overview of the evaluation session from the perspective of the organizers. It describes the controlled conditions under which experimentation took place. It provides a description of the data collection protocol and the motivation for its different phases. It discusses a preliminary example how the data may be analyzed, and points at some of the challenges and questions that emerge. Rather than point out specific answers and argue for the correctness of particular methods or the preliminary results, this paper seeks to promote discussion of the underlying challenges involved in this undertaking, and the important questions that arise from them.

This report is organized as follows. Section 2 describes the data-collection controlled conditions and protocol. Preliminary results are provided in section 3. Section 4 provides a discussion of these results and points to emerging questions. Section 5 concludes.

The RoboCup-98 Evaluation Session

During the RoboCup'98 competition and workshop in July 1998 (Paris, France), a special two-day evaluation session was organized, beginning an annual tradition of rigorous scientific evaluation of RoboCup teams. The session consisted of 13 simulation teams each playing against the same fixed opponent four times, as incrementally, up to 3 of their players were disabled. The following describe the experiment protocol and controlled conditions.

Participation

Participation in the evaluation session was open to all teams who wanted to take place, but was strongly encouraged for all teams who have made it through the round-robin round to the double-elimination rounds (representing the top 16 teams in 37). All in all, 13 different teams participated in the evaluation:
- CMUnited-98 (Stone and Veloso 1998)
- AT Humboldt '98 (Burkhard, Wendler et al. 1998) and '97 (Burkhard, Hannebauer et al. 1998)
- Kasuga-Bito II (Maeda, Kohketsu et al. 1998)
- ISIS'98 (Marsella, Adibi et al. 1999)
- PaSo Team (Montesello and Pagello 1998)
- Gemini (Gemini 1998)
- Andhill'98 (Andou 1998)
- AIACS (Lubbers and Spaans 1998)
- CAT Finland (Riekki 1998)
- Darwin United (Andre and Teller 1998)
- Mainz Rolling Brains (Polani, Weber et al. 1998)
- Windmill Wanderers (Corten and Rondema 1998)

Fixed settings

All participating teams competed against a fixed opponent—the previous year's world champion "AT Humboldt'97" (Burkhard, Hannebauer et al. 1998), which was only slightly modified to accommodate changes in the simulation software made between 1997 and 1998. Note that the fixed opponent was also a strong competitor in 1998, and was also evaluated against itself.

Hardware settings (computers, network conditions, etc.) were identical to those of the actual competition: Teams were allowed to use up to 8 Sun machines each for the clients. Two different games ran in parallel, using two different machines to run the servers. Actual competition versions of the players were used. The protocol prohibited using any special versions of the code for the purpose of evaluation. Indeed, teams did not know about the evaluation session until three days before it took place, and did not know the details of the protocol until the beginning of the evaluation session. Except for the disabling of players (which was the controlled variable), the games strictly followed competition rules, with a referee and representatives of each team present during the matches. No tuning of program parameters was allowed between phases.

Evaluation Phases

Each team played four half-games (each lasting 3000 simulation "ticks", about 5 minutes) against the fixed opponent. Each such half-game constitutes an evaluation phase, in which a single change to the number of disabled players was made. These phases are denoted A through D:
- **Phase A.** The control phase. The team played against the fixed opponent under normal competition rules. No players were disabled.
- **Phase B.** The team played against the fixed opponent with a *single player disabled*. The player was randomly selected by the computer—but was not allowed to be the goalie. Thus, a different player was chosen for each team.
- **Phase C.** The team played against the fixed opponent with two players disabled:
    - The same player randomly selected in phase B, and
    - A player selected by the fixed opponent's representative with the intention of disabling the evaluated team's most valuable player. (But not the goalie)
- **Phase D.** The team played against the fixed opponent last final half-game, with three players disabled: The two players disabled in phase C and the evaluated team's goalie.

The motivation for this evaluation protocol in general was to check how well teams are able to adapt to loss of members. Phase A was intended to establish a base-line for the evaluated team's performance under normal conditions. Phases B through D provided the experimental worsening conditions. In all of these phases, the ideal would have been, for comparison's sake, to disable the same player in all teams, to see how their adaptive capabilities face to the same problem. The intention in using a randomly chosen player was to make sure that the teams could not have prepared in advance for particular evaluation settings. However, different teams assign different player numbers to the similar roles—thus randomly selecting a player number and then disabling the same player number is all teams would make little sense.

The next logical alternative is to look at the role of the players: Randomly select a role (e.g., top midfielder) and then disable the player who plays this role in each team. But again teams greatly vary in their team strategies. Thus not all teams have the same roles. The most common role was that of the goalie (disabled in phase D) but even for the goalie there was at least one team that had players take over the role of the goalie on a regular basis (Andou 1998).

Nevertheless, some random element was required, to make sure that teams did not know in advance which players were going to be disabled. We have therefore decided to randomly select a player for each team that would first be disabled in phase B, but then balance this randomness in phase C by allowing the representative of the fixed opponent to select a player that would potentially damage the evaluated team's soccer-playing ability the most. Our hope was that at the very latest, all teams would face similar difficulties when they reach phase C.

Data Collection: Experiment Execution

For each of the different phases, for each of the evaluated teams, the soccer-server log files were saved and tagged appropriately. These provide complete records of the game, with the exception of communicated messages exchanged between the players. These logs were made available publicly (Repository 1998) to any and all interested parties.

Players were disabled in their initial position and facing direction on the field at the beginning of the game, but were left on the field. The server ignored any commands sent by their respective clients, so disabled players could not communicate, move, nor turn. However, they were visible to other players from their own team and the opponent teams.

In phase B, a list of random numbers in the range 1-11 was generated by the C library's pseudo-random generator, and the numbers were assigned to the different teams in order of participation. The randomly selected player could not be the goalie: if the random number was that of the goalie for the given team, it was skipped and the next different number on the list was used instead. The randomly assigned number was not revealed to the evaluated team until it was actually disabled.

In phase C, when the representative of fixed opponent (Burkhard, Hannebauer et al. 1998) was to choose the next player that would be disabled, such that it would potentially harm the evaluated team the most, there were two potential cases of conflicts in interest: (a) when the fixed opponent was playing itself, and when the fixed opponent (AT Humboldt 97) was used to evaluate AT Humboldt 98, which was developed by the same programming team. In both of these cases, the selection of the representative of the fixed opponent was independently corroborated by a neutral party.

The players disabled for each team in each phase are presented in the Table 1. For each team, the table shows the players disabled in each evaluation phase (i.e., phases B through D).

Preliminary Results and Analysis

Very early on it became clear that actual performance of *teamwork*, rather than the *team*, is difficult to measure. With most teams, qualitative changes in team performance were not observed, and even in cases where qualitative differences were found, they were not sufficient for rigorous comparative evaluation. Quantitative measures are required which can allow us to compare teams and their performance in general, and their collaboration and coordination skills in particular.

Team Name	Disabled Players			Team Name	Disabled Players		
CMUnited 98	5			Rolling Brains	4		
	5	10			4	10	
	5	10	1		4	10	1
Darwin United	5			ATH 98	10		
	5	11			10	3	
	5	11	1		10	3	1
Windmill Wanderers	2			Kasuga-Bito II 98	8		
	2	8			8	2	
	2	8	11		8	2	1
Andhill 98	2			ISIS 98	2		
	2	8			2	9	
	2	8	1		2	9	11
CAT Finland	10			Gemini	4		
	10	9			4	9	
	10	9	1		4	9	1
AIACS	7			Paso Team	5		
	7	10			5	7	
	7	10	1		5	7	11
ATH 97	10						
	10	7					
	10	7	1				

Table 1. Disabled players for each team, in each phase (B-D).

Our expectation is that any quantitative measure used will show a trend of declining performance as more and more players are disabled. A more adaptive team would have a slower decline in performance, while a less capable team would have a sharper decline. Intuitively, a more adaptive team would have less reduction in performance when it loses team-members, as the remaining members would be able to compensate for those disabled. Of course, in practice we cannot expect team members to be successful in compensating for an arbitrary number of disabled team-members (teams, after all, are used most often when tasks are simply too complex and to big for a single agent to undertake). It is also useful to think of a purely theoretical ideal of a zero-slope performance trend, in which a theoretical team so successfully compensates for disabled players that there is no change in performance. This allows measurement of a team's performance not only relative to other teams, but also on an absolutely (0 decline being an ideal best).

At least three quantitative measures are immediately available in the domain of soccer: The number of goals scored by the evaluated team, the number of goals scored by the opponent, and the score-difference resulting from it. As an example, the score-difference results are shown graphically in figure 1. For each team, for each phase (A-D), the score-difference at the end of the half are shown. The scores are normalized for each team on the basis of the team's performance in phase A. In other words, the results of phases B through D for each team are shown relative to the team's performance in phase A, rather than on an absolute scale. The results are plotted as a function of the number of disabled players.

The results are difficult to interpret. Some teams seem to react with increased performance to loss of players, at times even outdoing their performance in the control phase (A). However, not all teams have responded in this way. Some have shown no effect at times, while others show the expected declining trend in measured performance. Plotting these means, we can see that on average, the evaluated teams show a reduction in performance as we hypothesized (Figure 2).

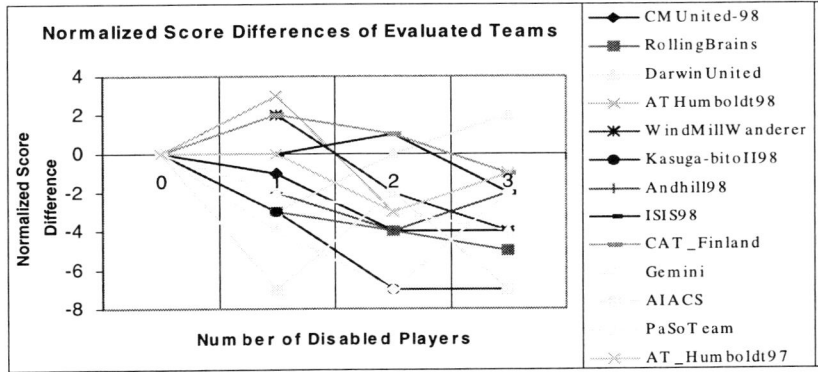

Fig. 1. Normalized Score Differences of Evaluated Teams.

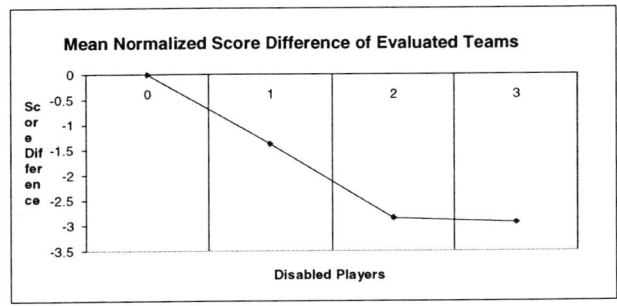

Fig. 2. Mean of Normalized Score Differences.

We use linear regression to draw a line that best represents the performance trend of the evaluated team. Figure 3 shows the computed regression slope values for each of the evaluated teams, as a function of the number of disabled players.

Maintaining performance in face of disabled players is graphically equivalent to a horizontal line, and so the more "horizontal" the performance trends of teams are, the closer they are to this theoretical ideal. Table 2 shows the rankings in our example. It should be emphasized again that these results are based on a *preliminary example analysis*. Indeed, the results are *not statistically significant*. These issues and others are discussed in the next section.

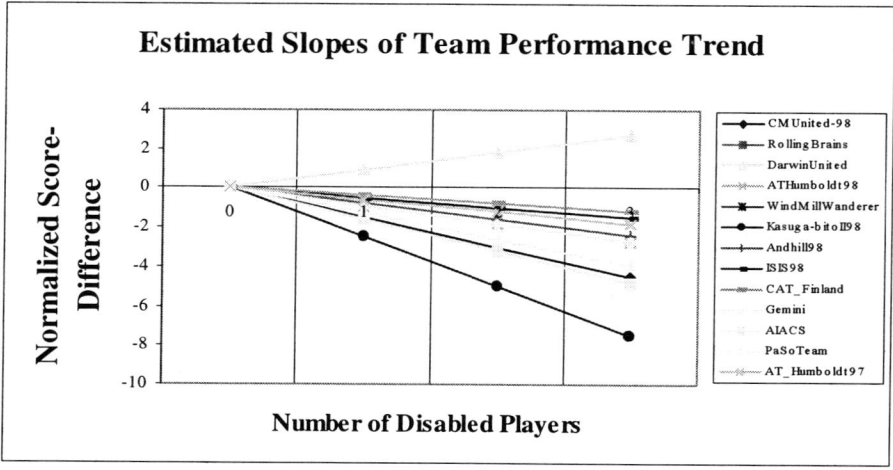

Fig. 3. A Plot of Estimated Team Performance Trends.

Team Name	Estimated Performance Trend
CAT Finland	-0.4
ISIS98	-0.5
AT_Humboldt97	-0.6
Andhill98	-0.8
ATHumboldt98	-0.9
Gemini	-0.9
Darwin United	+0.9
PaSo Team	-1.3
CMUnited-98	-1.5
Windmill Wanderer	-1.6
Rolling Brains	-1.6
AIACS	-1.6
Kasuga-bitoII98	-2.5

Table 2. Estimated ranking of teams' adaptivity (estimates n*ot* statistically significant).

Discussion

Rather than providing definitive answers to questions about actual evaluation results, or about evaluation methodology, the preliminary results and analysis seem to suggest that there is much that we have yet to understand about teamwork and its evaluation.

Measurement

Our choice of the score-difference variable as the focus of measurement raises several important issues. First, our arbitrary choice of the score-difference variable was arbitrary, and was made for demonstration purposes. The intuition behind the selection of this immediately available task performance measure is that task-performance is associated with teamwork performance. But this is an assumption on our part (see more on this issue below).

Second, even if task performance is indeed correlated with teamwork, we are still left with the problem of selecting a measurement variable. Such a variable may be more or less sensitive to the effects of teamwork. For instance, we could have chosen to measure performance by the number of passes, etc.

Third, confidence in the measurement is a serious concern. For example, the analysis results above are not, for the most part, statistically significant. Although an Analysis-of-Variance (ANOVA) of the normalized score-differences shows that the mean score-difference of each phase are significantly different ($p=0.005$), this is likely due to the normalized mean of 0 (with 0 variance) in phase A. The number of data points (4) is simply not enough to give us a clear picture of each team's response to the different evaluation phases. Although the trend suggested by the mean normalized goal scores supports our expectations of declining team performance, we need to investigate methods by which this analysis can be stated with greater confidence.

One way to do this would simply to repeat each test a relatively large number of times. For instance, if we repeat each of the phases for an evaluated team 40 times, we stand a greater chance of discovering statistically significant results. However, we should be aware that sometimes the surprising result is that even a great number of games may be insufficient to give us the level of confidence we want. Random effects in the environment and the unpredictable responses of the opponent, issues such as machine and network load, all interfere with our controlled conditions and make evaluation more difficult.

For instance, Tambe et al. (1999) report on a series of over 230 games of the ISIS'98 team against fixed opponents in which the only independent variable was the use of communications (about half the games were played with communications allowed between the evaluated team's players, and the rest were played with no communications). Despite the relatively large number of games, no statistically significant result was found. However, when a different measure was used, the average-time-to-agreement measure (ATA), very statistically significant differences were found between the communicating and non-communicating teams (the statistical significance levels exceeded 99%).

Another problem with measurement is that it may not make it clear to what extent changes in the performance (even when significant) are due to adaptation (or lack thereof) on the part of the evaluated team, rather than other factors, such as adversarial re-planning (for instance, an advanced fixed opponent might recognize that the evaluated team-member is disabled, and adapt its attack to take advantage of this weakness). It may also be that limits of performance are reached -- if an evaluated team is very strong indeed, it may be successful in showing no change in performance despite its lack of adaptations. Here, agents are able to individually compensate for the lack of adaptation on the part of the team.

Perhaps a more productive, and certainly more challenging way of approaching this problem is to change the measures used. This paper demonstrated an initial form of using the score difference to analyze *team performance*. This measure has generally been used by researchers involved in RoboCup to evaluate their teams. However, it may very well be that this measure is simply insufficient for our purposes, and that techniques that measure *teamwork performance* more directly are more useful for our purposes. The ATA measure mentioned above (Tambe et al. 1999), for example, measures the average time it takes team-members to come to agreement on chosen team-plans (tactics). But other measures can be found in the literature. Balch (1998) for instance investigated the use of Social Entropy to measure diversity in soccer, and found a positive correlation between diversity and performance. Goldberg and Mataric (1997) suggest a different measure, based on inter-agent interference.

Analysis

The preliminary results in the previous section raise many issues in our underlying intuitions and knowledge of teamwork and coordination. For instance, we have introduced a theoretical ideal of perfect adaptation as 0-slope performance trend. In other words, we have chosen to look at no change in performance as a characteristics of a perfectly adaptive team. However, this really does not take into account theoretical limits that are likely to exist on the number of agents that are required (as a lower limit) or optimal (as an upper limit) for performing a team task. For instance, perhaps soccer cannot be done with less than 9 agents, in which case disabling more than 2 players will always result in a sloping trend, even with the best possible theoretical team.

Another issue is that we assume in our earlier analysis that degradation is linear. But this may or may not have basis in reality. It could be that, as some of the results may suggest, degradation in performance is non-linear with respect to the number of disabled players. For instance, some teams show an increase in performance when some players are disabled, and a decrease as others are disabled (Figure 1). These changes may be the result of the team moving back and forth from sub-optimal organizational structures to optimal ones as a result of our disabling of clients, but they may also reflect degradation phenomenon due to other factors.

We have mentioned above that our choice (for the example analysis) of a task-performance variable assumed that task performance is correlated with teamwork. But in actuality, such correlation between teamwork and performance is an open research question. For instance, teams that have done well in the competition did not necessarily rank high in the analysis above (again, this may be due to the insignificance of the results). CMUnited'98 (Stone and Veloso 1998), which was the champion in the RoboCup-98 competition, ranks fairly low in terms of the estimated performance trend slope, while the leading team in those terms is the CAT Finland'98 (Riekki 1998) team, which proved to be a middle-level entry in terms of its placement in the competition. However, CMUnited'98 players, in particular, have been observed to send messages to the screen indicating that they correctly recognize that specific players have been disabled. The question remains why this recognition capability, which no other team was able to demonstrate, did not carry with it the implied compensation in terms of performance.

Rather than argue for the correctness of a particular technique and the implied evaluation results, we seek here to use our earlier analysis to demonstrate the possibility of evaluation on one hand, and the challenges involved in it on the other. The intent here is to stimulate discussion of evaluation methods, not of the obvious lacking of the one we have been using as a demonstration. The analysis provided earlier is a demonstration of a possible approach, and we warn again that its results are preliminary, and are not statistically significant.

Summary

The evaluation session and protocol were organized and executed with the intent of providing the multi-agent research community with a substantial data set that may be used for research in teamwork, adversarial planning, coordination, etc. The data collection and evaluation protocols were developed with this intent in mind, and we hope the research community will find them useful. However, they certainly can be improved upon. In particular, the data may not be statistically significant for certain tests, and the evaluation protocol itself can certainly be improved. We therefore welcome any and all comments and suggestions from investigators on how the evaluation methods and collected raw data may be improved.

The soccer-server log files collected during this evaluation session are to serve as the raw-data upon which the actual evaluation takes place. There are different ways of measuring the teams' performance, and these can be compared using this standard data repository. The logs present the data from the games of all 13 different participating teams, and are intended for use by the scientific community. They are publicly available for all interested parties (Repository, 1998). We hope useful, innovative measures will be developed and presented to the scientific community as a result of this data set. Perhaps most importantly, if RoboCup is truly to make an impact on AI and Robotics research, our investigations should result in evaluation techniques that generalize beyond RoboCup to other multi-agent domains.

Acknowledgements. We thank all participating teams, the RoboCup'98 Simulation-League committee (Chair: Itsuki Noda), Kay Schroter and Hans-Dieter Burkhard for their cooperation and help. Milind Tambe, Stacy Marsella, Nachum Kaminka, and Shlomit Kaminka deserve many thanks for very useful comments.

References

Andou, T. (1998). Andhill-98: A RoboCup Team which Reinforces Positioning Observability. RoboCup-98, Paris, France.

Andre, D. and A. Teller (1998). Evolving Team Darwin United. RoboCup-98, Paris, France.

Balch, T. (1998). Behavioral Diversity in Learning Robot Teams. Ph.D. thesis, Georgia Institute of Technology.

Burkhard, H.-D., M. Hannebauer, et al. (1998). AT Humboldt - Development, Practice, and Theory. RoboCup-97: Robot Soccer World Cup I. H. Kitano, Springer. **LNAI 1395**: 357-372.

Burkhard, H.-D., J. Wendler, et al. (1998). AT Humboldt in RoboCup-98. RoboCup-98, Paris, France

Corten, E. and E. Rondema (1998). Team Description of the Windmill Wanderers. RoboCup-98, Paris, France.

Gemini, 1998. A Competing entry in RoboCup-98. No team description or other reference was available for this team.

Goldberg, D. and Mataric, M. (1997). Interference as a tool for designing and evaluating multi-robot controllers. In Proceedings of the 14th National Conference on Artificial Intelligence (AAAI-97), pp. 637-642.

Johnson, W. L. and J. Rickel (1997). Steve: An Animated Pedagogical Agent for Procedural Training in Virtual Environments. SIGART Bulletin. **8**: 16-21.

Kitano, H., M. Tambe, et al. (1997). The RoboCup Synthetic Agent Challenge '97. the International Joint Conference on Artificial Intelligence (IJCAI-97), Nagoya, Japan.

Lubbers, J. and R. R. Spaans (1998). The Priority/Confidence Model as a Framework for Soccer Agents. the Second RoboCup Workshop (RoboCup-98), Paris, France.

Maeda, K., A. Kohketsu, et al. (1998). Ball-Receiving Skill Dependent on Centering in Soccer Simulation Games. RoboCup-98, Paris, France.

Marsella, S. C., J. Adibi, et al. (1999). On Being a teammate: Experiences acquired in the design of RoboCup teams. the Third International Conference on Autonomous Agents (Agents-99), Seattle, WA, ACM Press.

Montesello, F. and E. Pagello (1998). PaSo-Team'98: Learning the 'when' in RoboCup Competition. RoboCup-98, Paris, France.

Polani, D., S. Weber, et al. (1998). A Direct Approach to Robot Soccer Agents: Description for the Team MAINZ ROLLING BRAINS Simulation League of RoboCup'98. RoboCup-98, Paris, France.

Repository, (1998). The RoboCup Teamwork Evaluation Homepage. **1998**. At: http://www.isi.edu/~galk/Eval/

Riekki, J. (1998). Reactive Task Execution of a Mobile Robot. Infotech Oulu and Department of Electrical Engineering. Oulu, Finland, University of Oulu. (Ph.D. Dissertation)

Stone, P. and M. Veloso (1998). The CMUnited-97 Simulator Team. RoboCup-97: Robot Soccer World Cup 1, LNAI 1395. H. Kitano, Springer: 389-397.

Tambe, M., W. L. Johnson, et al. (1995). "Intelligent Agents for Interactive Simulation Environments." AI Magazine **16**(1).

Tambe, M., G. A. Kaminka, et al. (1999). Two Fielded Teams and Two Expert Agents: A RoboCup Challenge Response from the Trenches. International Joint Conference on Artificial Intelligence (IJCAI-99), Stockholm, Sweden.

Towards a Distributed Multi-agent System for a Robotic Soccer Team

Nadir Ould Khessal

Temasek Engineering School. Temasek Polytechnic
21 Tampines Ave 1 Singapore 529757.
nadirok@tp.edu.sg

Abstract. Many AI professionals consider RoboCup small robots league competition as an ideal platform for testing distributed artificial intelligence techniques. Among these techniques are Multi-Agent systems (MAS), which advocate collective intelligence by focusing on autonomy of agents and their intercommunication. Multi-Agent Systems have been used by computer scientists and software engineers in several disciplines such as Internet and Industry [16]. For the robotics community Multi-Agent Systems was a paradigm shift from the classical centralized approach in building intelligent machines. By the late 80's MAS were used in several multi robot systems ranging from cellular robots (Fukuda et al) [1] to a team of trash-collecting robots (Arkin et al) [2]. This paper describes a distributed approach in implementing the Multi-Agent system architecture of a robotic soccer team, Temasek POlytechnic Team(TPOT).

1. Introduction

The major characteristic of the RoboCup soccer competition is the dynamic nature of the environment in which robots operate. The only static object in the competition field is the field itself. Team and opponent robots as well as the ball can be placed anywhere in the field, be it a purposeful strategic positioning, a missed action or a forced displacement. This has led many researchers to shift from the traditional model-based top down control [3,4] to a reactive behavior based approach [5,6,7,8,9,10]. Robots need not waste a huge amount of resources building maps and generating paths that might prove useless at the time of action. Instead robots are supposed to react to the actual changes in the environment in a simple stimulus-response manner [11]. However due to the size limitations imposed by the RoboCup small robots league (15cm diameter circle) and rich visual input, on-board vision proved to be a complex and expensive task.
Due to these constraints, several RoboCup researchers [12,13] have turned to off-board global vision. Cameras are placed above the field from where relevant information about the field such as robot coordinates, identities and ball position is dumped to a stand-alone computer. This centralized approach in building the control

system has led to the adoption of a hybrid (deliberative and reactive) approach. Reactive behavior based agents are embedded in the robots for urgent time-critical actions such as obstacle avoidance and command execution. Visual data manipulation and filtering as well as high-level reasoning e.g. ball position prediction are done in the remote computer.

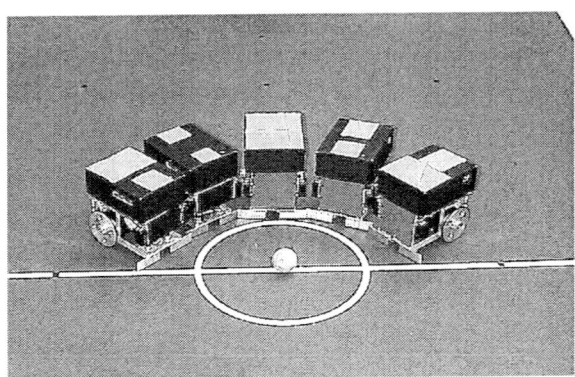

Fig. 1. Temasek Polytechnic RoboCup Team TPOT

In this paper we will present a hybrid control architecture, distributed among the robots (figure 1) and the host computer. Associated with every robot is an embedded agent, in charge of navigating the field and executing commands generated by the remote agent. Remote agents select and implement the required tasks, based on the visual data provided by the vision system and the strategy selected by the reasoning module.

2. System Architecture

The system hardware consists of a Pentium host computer, a vision system based on Newton labs Cognachrome vision card, RF transmission system and five robots (figure 2).

- *The robot* : The robot on-board controller is implemented in an 8 bit processor running at 9.216MHz with an on-board memory of 512kbyte RAM and 512Kbyte EEPROM. The board also includes a real time clock and programmable timers.

- *Sensors:* Attached to the robot are three infrared sensors mounted in the front and rear to detect obstacles whilst moving.

- *The communication module:* The host computer transmits commands to the robot via radio transceivers utilizing UHF radio waves. Each robot has its own transceiver and a unique node address. The low-powered wireless system transmits less than 1mw of power and is effective over distances of 3 to 30 meters. Two-way communication rates of up to 38.4Kbps are possible. The command set is transmitted as text code piggybacking on the transmission protocol. Commands are sent and received from the transceiver using an RS-232 interface.

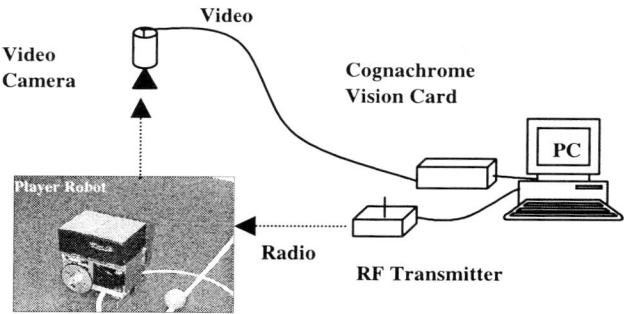

Fig. 2. System Overview

- *Vision:* A global vision system, which consists of color camcorders and a special image processor (MC68332), is used. The system is able to segment and track the robots and ball at a high frame rate. Each robot has two color pads. The image processor is trained to see the different colors and gives the locations of the center of gravity of the two color pads. Hence the orientation and robot position are known. Color pad areas are used to distinguish between different robots and minimize latency.

3. Distributed Multi-Agent System

Our approach in implementing the control architecture of the robots is based on dividing each robot controller into two parts: Embedded agent running on the on-board processor and situated in the environment (field) and Remote agent running in the off-board host computer and situated in an abstract model of the filed. The embedded agent consists of several reactive behaviors competing with each other through the use of activation levels (inhibition and suppression). The main role of the embedded agent is to execute commands issued by the remote agent and navigate safely the soccer field while avoiding other robots and obstacles.

The remote agent on the other hand implements strategies generated by the reasoning module. Based on the current score and performance of the opponent team, the reasoning module selects a strategy from a pool of pre-designed strategies and downloads it to the remote agents.

This enables the agents to select the appropriate tasks for each robot. Such tasks will enable the robot to intercept the ball, follow a target or simply move to a predetermined position. These behaviors are implemented in every robot's remote agent except the goalie. This allows the robots to swap roles e.g. from being a defender to a forward and vice versa.

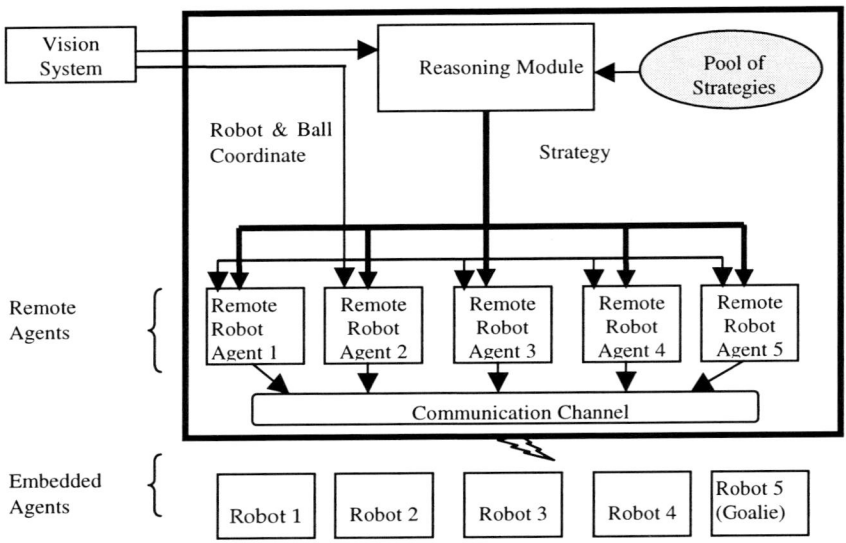

Fig. 3. Distributed Muti-Agent Architecture

3.1 Embedded Agents

An embedded agent consists of reactive behaviors designed to perform low-level navigational tasks. These behaviors are simple stimuli-response machines where the stimulus can be a sensory input or an incoming RF command. The arbitration mechanism is based on a fixed prioritization network [14]. The response of a single higher priority behavior takes over the control of the robot whenever the associated stimulus is present (figure 4).

Obstacle avoidance is the main autonomous task done by the robot. Using the three infrared sensors, the robot moves away from obstacles using the *Avoid Left* and *Avoid right* machines. The robot than moves a certain distance until a straight line path to the target position is clear. The remote agent detects that the robot is out of the previously computed path and re-computes and transmits a new path. Due to the dynamic nature of the environment obstacles are not taken in consideration while planning a path for the robot.

Remote agents transmit paths in the form of a turning angle followed by a traveling distance. The two machines on the robot i.e. *Turn* and *Move* are in charge of executing these commands. Note here that unlike the *Move* machine, the *Turn* machine is of a highest priority and therefore un-interruptible.

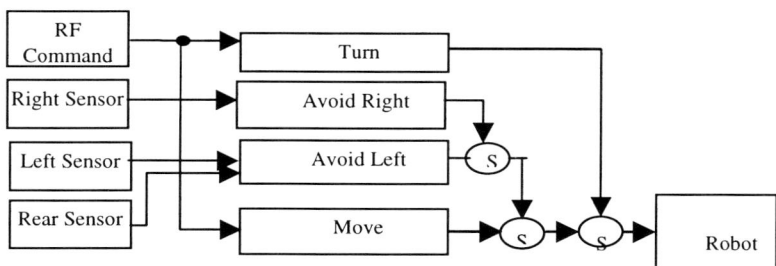

Fig. 4. A Network of Stimuli-Response Machines.

3.2 Remote Agents

For the robot to be able to play soccer it needs some basic skills such as moving towards the ball position, kicking the ball towards the goal area, intercepting the ball and passing the ball to a team member. Most of these skills could be performed by the following behaviors:
- *Intercept_ball.* This machine enables the robot to move behind a predicted ball position before kicking it towards a target area. The target area could be the opponent goal keeper area (in an attempt to score a goal figure 5.), a clear area in front of a team member (ball passing) or simply the opposite side of the field, in the case of a defending robot. This behavior is achieved by the robot first moving to an intermediate position. Once there, the robot charges towards the ball. The intermediate position is determined by computing the two lines starting at the edges of the target area and intersecting at the predicted ball position (*L1* and *L2* in figure 5). The distance between the intermediate point and the ball predicted position d is fixed (charging distance). The intermediate position corresponds to the midpoint of the base segment of the equilateral triangle b.
- *Follow:* This machine is designed to keep the robot following a target object. The target can be the ball, a team robot or an opponent robot. This is done to

keep the robot nearer to the ball and therefore in a better position to intercept the ball.
- *Homing*: Depending on the strategy being executed robots could be required to be placed at a certain position for the purpose of forming a defense wall for example. The Homing machine performs such actions.

Each of the above machines contains a path planner. This planner generates a path in the form of a straight line between the robot and its target position. After the path has been generated and transmitted the planner keeps track of the robot position. This will enable the remote agent to detect divergence of the robot from its most recent path.

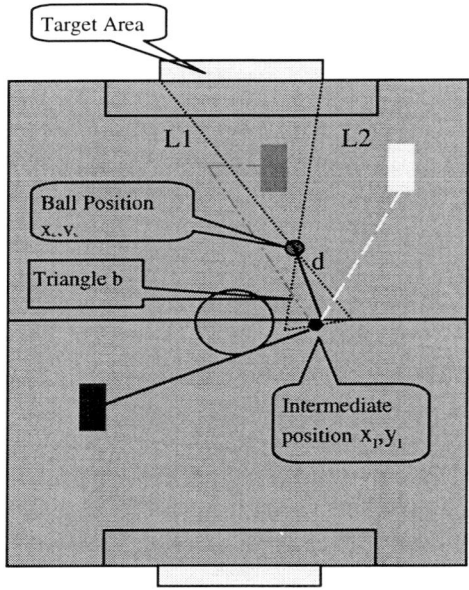

Fig. 5. To intercept the ball the robot needs to position itself in an intermediate position

Divergence of the robot could be caused by a hit from an opponent robot or a purposeful move to avoid an obstacle. In both cases the path planner computes and transmits a new path to the robot.

3.3. Behavior Arbitration

Table 1 shows a list of the behaviors implemented in each robot. The on-board agent consists of a set of reactive behaviors situated in the actual robot environment whereas the remote agents are situated in an abstract environment constructed using the visual data. Embedded behaviors use a competitive coordination mechanism; a fixed prioritization network is used to resolve behavior conflicts. The remote agent uses a supervisory coordination mechanism in the arbitration of its set of behaviors. Each robot is modeled in the following structure:

```
Robot_Player (ID)   (

            Robot_ID,    ;;each robot has an id number.

            Robot_X,     ;;the robot Cartesian coordinate.

            Robot_Y,     ;;

            Robot_Angle, ;;robot direction.

            Robot_Path,  ;;most recent path generated.

            Robot _Behavior ;;current robot behavior.

)
```

Using the Robot–Behavior parameter, the robot could be assigned any of robot remote task, i.e. *Intercept_ball, Follow and Homming*. It is therefore possible that more than one robot would be assigned the same task.

Behavior	Stimulus	Location	Environment	Coordination
Avoid_left	Infrared Sensor	Embedded	Real world –Field	Competitive
Avoid_right	Infrared Sensor	Embedded	Real world –Field	Competitive
Move	RF Channel	Embedded	Real world –Field	Competitive
Turn	RF Channel	Embedded	Real world –Field	Competitive
Intercept_ball	Visual Data	Remote	World Model	Supervisory
Follow	Visual Data	Remote	World Model	Supervisory
Homing	Visual Data	Remote	World Model	Supervisory

Table 1. Robot behaviors distributed among the robot's on-board and off-board agents.

4. Conclusion

The RoboCup small league main challenge to AI professionals is the size limitations. In contrary to the medium size league, very little computational power could be embedded in the robot itself. In this paper we summarize research done in the field of RoboCup and Multi-Agent Systems. Each robot player controller is subdivided into two agents. On-board agents composed of a collection of survival behaviors and situated in the environment. The remote agents on the other hand are situated in an abstract model of the environment. While remote agents communicate with the embedded agents in a direct manner through the RF link. On-board agents have no explicit communication link with the remote agents. The remote agent's perceived environment is therefore the only source of information on the embedded agent.

This architecture was implemented in Temasek POlytechnic Team (TPOT) during the 1998 Pacific Rim Series RoboCup Competition in Singapore [15] and the 1999 World RoboCup Competition in Sweden [17].

References

1. Fukuda, T. et al. *Structure Decision for Self Organizing Robots Based on Cell Structures- CEBOT*. In Proceeding of the IEEE International Conference on Robotics and Automation. Los Alamitos, Calif, 1998.
2. Arkin, Ronald C. and Tucker Balch. *Cooperative Multi-Agent Robotic Systems. In Artificial Intelligence and Mobile Robots*, ed. David Kortenkamp,278-296. Cambridge, Mass.: The MIT Press.1998.
3. Albus, James S. *Hierarchical Control Of Intelligent Machines Applied to Space Station Telerobots*. IEEE Transactions on Aerospace and Electronic Systems, Vol. 24. NO. 5 Sept. 1988.
4. Housheng Hu and Michael Brady, *A Parallel Processing Architecture for Sensor Based Control of Intelligent Mobile Robots*. Robotics and Autonomous Systems, vol.17,1996, pp 235-257.
5. Brooks, Rodney A.. *Achieving Artificial Intelligence Through Building Robots* A.I.Memo 899, Artificial Intelligence Laboratory, MIT, May 1986.
6. Brooks , Rodney A. *Intelligence without Reason*. Computers and thought, IJCAI-91, Also A.I. Memo 1293 Artificial Intelligence Laboratory, MIT.
7. Maja J. Mataic. *Interaction and Intelligent Behavio*. Ph.D. Thesis, Department of Electrical Engineering And Computer Science, Artificial Intelligence Laboratory, MIT, 1994.
8. Arkin, Ronald C. *The Impact of Cybernetics on the Design of a Mobile Robot System: A Case Study*. IEEE Transactions on Systems, Man , and Cybernetics Vol. 20 No 6 Nov./DEC 1990.
9. Kenneth Moorman and Ashwin Ram.. *A Case-Based Approach to Reactive Control for Autonomous Robots*. AAAI Fall Symposium " AI for real time Mobile Robots", Cambridge MA, Cot 1992.
10. Erann Gat , Rajiv, Robert Ivlev, John Loch, and David P. Miller. *Behavior Control for Robotic Exploration of Planetary Surfaces*. IEEE Transactions on Robotics and Automation, Vol. 10,NO4, August 1994.
11. Brooks, Rodney A.. *A Layered Intelligent Control System for a Mobile Robot*. IEEE Journal Robotics And Automation . RA-2, March 1986, 14-23.

12. Manuela Veloso, Peter Stone, Kwun Han, and Sorin Achim. *The CMUnited-97 Small Robot Team. In RoboCup In RoboCup-97: Robot Soccer World Cup 1,* Ed. Hiroaki Kitano, 242-256 LNAI 1395, Spriger-Verlag, 1998.
13. Gordon Wyeth, Berett Browning and Ashley tews. *UQ RoboRoos: Preliminary Design of a Robot Soccer Team.* In Proceedings of RoboCup Workshop.5th Pacific Rim International Conference On Artificial Intelligence. Ed. Hiroaki Kitano, Gerald Seet. Singapore, 1998.
14. Nadir Ould Khessal. *An Autonomous Robot Control Architecture Based on A hybrid behavior arbitration.* In The proceeding of Modeling , Identification and Control Conference. Innsbruck, Austria 1999.
15. Nadir Ould-Khessal, Dominic Kan, Sreedharan Gopalasamy, Lim Hock Beng. *Temasek Polytechnic RoboCup Pacific Rim Series Small League Team.* In Proceedings of RoboCup Workshop.5th Pacific Rim International Conference On Artificial Intelligence. Ed. Hiroaki Kitano, Gerald Seet. Singapore, 1998.
16. Jennings, N. *Cooperation in Industrial Multi-Agent Systems.* Vol 43, World Scientific Press 1994.
17. Nadir Ould Khessal, Dominic Kan, Sreedharan Gopalasamy, Lim Hock Beng. *A Distributed Multi-Agent Architecture for Temasek RoboCup* Team. Proceedings of The Third International Workshop on RoboCup. Ed. Manuela M. Veloso.1999.

A Multi-threaded Approach to Simulated Soccer Agents for the RoboCup Competition

Kostas Kostiadis and Huosheng Hu

Department of Computer Science, University of Essex
Wivenhoe Park, Colchester CO4 3SQ, United Kingdom
Email: kkosti@essex.ac.uk, hhu@essex.ac.uk

Abstract: To meet the timing requirements set by the RoboCup soccer server simulator, this paper proposes a multi-threaded approach to simulated soccer agents for the RoboCup competition. At its higher level each agent works at three distinct phases: sensing, thinking and acting. Instead of the traditional single threaded approaches, POSIX threads have been used here to break down these phases and implement them concurrently. The details of how this parallel implementation can significantly improve the agent's responsiveness and its overall performance are described. Implementation results show that a multi-threaded approach clearly outperforms a single-threaded one in terms of efficiency, responsiveness and scalability. The proposed approach will be very efficient in multi-processor systems.

1. Introduction

The creation of the robotic soccer, the robot world cup initiative (RoboCup), is an attempt to foster AI and intelligent robotics research by providing a standard problem where wide range of technologies can be integrated and examined [6][7]. Some of the fields covered include multi-agent collaboration, strategy acquisition, real-time planning and reasoning, sensor fusion, strategic decision making, intelligent robot control, and machine learning. Given the nature of the RoboCup environment, the response time of a soccer agent becomes significantly important since the soccer server operates with 100ms cycles for executing actions and 150ms cycles for providing visual sensory data [10]. In addition to that, auditory sensory data can be received at completely random intervals. It is vital that each agent has bounded response times. If an action is not generated within 100ms, then the agent will stay idle for that cycle and enemy agents that did act might gain an advantage. On the other hand, if more than one action is generated per cycle, the server will only execute one of them randomly, which might result to a non-optimal solution. An additional constraint is that Unix is not a "true" real-time system and hence real-time performance and response can only be guaranteed up to a certain resolution [12].

A real-time system is a system in which the time that output is produced is significant. In other words, a real-time system should be able to respond to stimuli from its environment within fixed time limits [4]. This is particularly true for the RoboCup simulator agents since each of them must react within an interval of 100ms, therefore achieving real-time performance. For a non real-time agent the desired behaviour is focused on the logical correctness of the result. In contrast, a real-time agent requires both logical and timing correctness. There are two broad categories of real-time agents: hard real-time agents and soft real-time agents [1]. For hard real-time agents, timing correctness is of critical importance and should never be sacrificed for other gains. In contrast, timing correctness in soft real-time agents is important but not critical. An occasional failure to meet a deadline should not result in catastrophic consequences.

Another distinction between real-time agents is whether their responses are based on events or clock times. This classification is of particular importance at the implementation level. Event-triggered agents respond to external stimuli by first detecting various conditions and then generating the appropriate reactions dynamically. Time-triggered agents operate in accordance with clock times as shown by an independent clock. In other words, clock pulses are treated as signals that generate certain actions. An appropriate action is selected based on the current state of the environment perceived by the agent.

This paper considers robotic soccer agents as hard, time-triggered real-time agents, and will only focus on how to improve the real-time performance of the soccer agents. Real-time systems are a large topic and a more detailed description of real-time systems can be found in [1][9]. In the rest of this paper, a few design issues regarding the agent architectures are discussed in the next section. Section 3 illustrates how single-threaded implementations have been developed, and then presents the proposed multi-threaded implementation. Section 4 shows experiment results and compares the two approaches. Finally, conclusions are given in section 5.

2. System Design

2.1 Agent Requirements

The robotic soccer simulator is a client/server application in which each client communicates with the server via a UDP (User Datagram Protocol) socket [10]. The server is responsible for executing requests from each client and updating the environment. At regular 150ms time intervals the server broadcasts visual information to all clients depending on their position, the quality and size of the field of their view, and their facing direction on the field. Moreover, the server sends auditory information to various clients at random time intervals.

After processing the sensory data, the clients respond by sending action requests to the server from a set of primitive actions available to them. To avoid message congestion on the server, the clients are allowed to send one request per cycle (100ms). If no message is sent within this cycle, the client will not perform any actions. If more than one message is send during the same cycle, the server executes only one randomly, which may produce undesired results. The server updates the state of the environment by serially executing each request. The results are projected on a window shown in figure 1.

It is important to mention that UDP sockets have a limited receive buffer. Messages arriving on a UDP socket will be queued until the receiving buffer is full in which case additional messages will be discarded. A client that fails to retrieve the

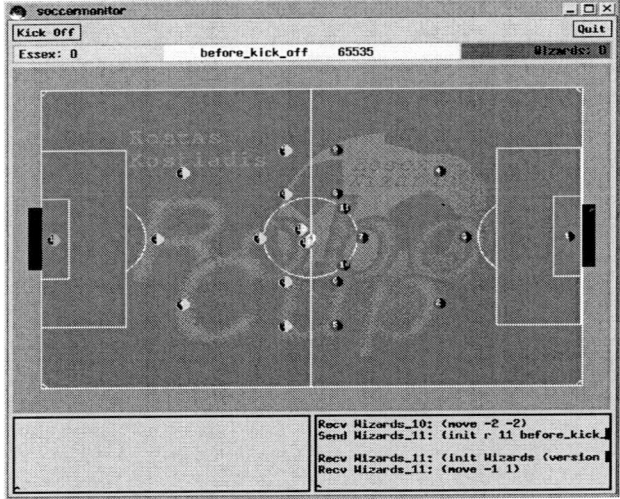

Figure 1 Robot Soccer Simulator

messages at the rate that they arrive is in danger of receiving older information from the server, since newer data will be further back in the queue. This will cause the client to create the wrong representation about the current state of the environment, leading to undesired effects when the wrong actions are executed.

The term "client", used in the client/server application context above, is the real-time agent to be built. For each cycle, the agent receives data from the server (if new data is available), processes this data, and produces an action. A very basic model of an agent's loop can be seen in figure 2. When new data is available, the agent should receive this data and update the current state of the environment. It should then "think" and send an action to the server. To be efficient an agent should satisfy the following conditions:

- To receive the newest sensory data that arrives on the socket as quickly as possible, and do not let data queue up. This enables the agent to have the most recent representation of the environment, and hence execute the most appropriate action.
- To send the action requests to the server timely, i.e. only one request/cycle. If the agent sends more than one request per cycle, the server will only execute one randomly. Otherwise, it might miss a cycle if it is too slow.
- To allow the maximum time for the thinking process and the minimum time to send or receive data.

Given the frequency of the message exchange and the timing constraints, building an agent that will satisfy the conditions described above becomes a challenging task. A brief description of the available I/O models under Unix and the agent architecture is briefly reviewed here.

2.2 I/O Models

Since the client/server communication is done via UDP sockets, there are a number of different ways to handle input and output on these sockets. Unix provides five I/O

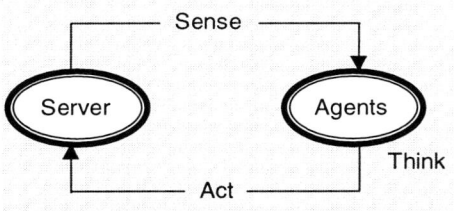

Figure 2 Agent's main loop

models: blocking I/O, non-blocking I/O, I/O multiplexing, signal-driven I/O, and asynchronous I/O. For the shake of completeness each I/O model is briefly introduced here. A more detailed description can be found in [12].

- *Blocking* I/O mode is the default mode for all sockets under Unix, which means that requesting data from a socket will not return until data is available. In other words, the whole process will be put to sleep until new data arrives or an error occurs.
- *Non-blocking* I/O mode avoids putting the process to sleep. If data is not immediately available, the kernel returns an error message, such as EWOULDBLOCK. The process will not be blocked and data can be requested at a later stage.
- *Multiplexing* I/O mode is achieved by using either the *select* or *poll* system calls. This method will block in one of these two calls rather than blocking at the actual I/O system call. When *select* returns, the socket is readable and the *rcvfrom* function is called to copy the data into the application buffer.
- *Signal-driven* I/O mode enables the kernel to notify the process with a (SIGIO) signal when a descriptor is ready. To read the data from the socket the process needs to establish a signal handler for the (SIGIO) signal. To send actions to the server, an interval timer that generates a (SIGALRM) signal when it expires can be used. By setting the timer to 100ms and using a handler that sends actions to the server the accuracy is guaranteed within certain limits.
- *Asynchronous* I/O mode enables the kernel to notify the process when the entire I/O operation is complete. The difference with the signal-driven I/O is that the kernel tells the process when an I/O operation is complete rather than an operation can be initiated. Asynchronous I/O is not very widely used primarily because of support issues.

2.2 Agent Architecture

Given this variety of I/O models supported under Unix, choosing an I/O model heavily depends upon the inner structure of the agent. In this section a novel architecture for building RoboCup agents is presented. Basically the agent contains six different modules as shown in figure 3, namely the agent's sensors, a set of behaviours, the actuators, the current play mode, a set of predefined parameters, and a memory module. The dashed arrows between the agent and the server in this diagram represent the communication links. The server sends information received by the agent's "Sensors" module and the agent sends actions back to the server through its

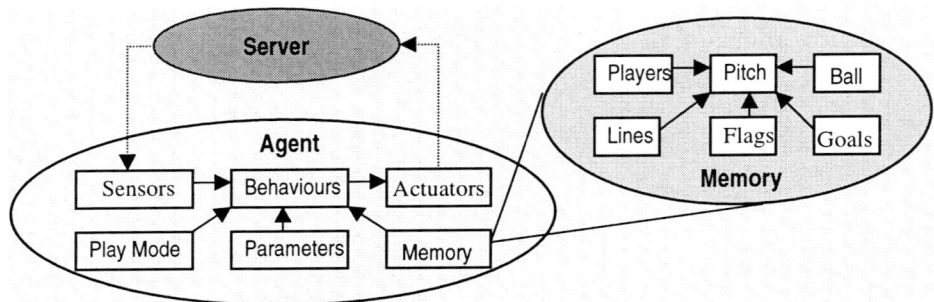

Figure 3 Block Diagram of the Soccer Agent's Structure

"Actuators" module. The solid arrows give dependence relationships between the modules. More specifically, the output produced by the "Actuators" module is directly dependent upon the input it will receive from the "Behaviours" module. The rest of the modules within the agent affect the "Behaviours" module. A brief description for each of these modules is given below.

- *Sensors* -- are responsible for receiving and analysing the visual or auditory information transmitted by the server. After receiving the data for each cycle, the agent creates a representation for the current state of the environment. Due to the limited field of view, the agent updates only part of the environment in each cycle. When new information arrives from the server, the old information is passed to the memory module, which holds a probabilistic representation for the whole environment.
- *Behaviours* -- is the most important module within the agent. It is responsible for generating actions according to the current state of the environment. The state of the environment is determined using all the modules within the agent apart from the actuators. Normally the agent needs to gather information from all other modules before an action is generated, including data regarding the agent's position and role in the team, the current formation, the position of the ball and other agents, the current play mode and so on. The "Behaviours" module should process all this information and determine the best course of action.
- *Actuators* -- are responsible for timing, and sending actions to the server. As mentioned earlier the server accepts one action each 100ms. The actuators receive an action (or a set of actions) from the "Behaviours" module, and send them to the server.
- *Play Mode* -- holds and updates the current play mode using the information received by the "Sensors" module. The current play mode directly affects the behaviour of the agent. For example the agent is expected to act differently if a free kick has been won for its team or if the opposition has won a corner kick.
- *Parameters* -- hold information regarding various settings both for the server and the agent. The parameters affect the behaviours of the agent since they include information regarding the current formation, the role of the agent in the team, and most of the server parameters including use of offside rule, player's size and maximum speed.
- *Memory* -- is a representation of the whole soccer pitch, rather than a partial representation like the one provided by the sensors. The "Pitch" contains players, lines, flags, goals and a ball. Each of these objects is associated with a confidence

value that represents the agent's confidence that an object is at the current coordinates. If an object was seen in the last cycle, its confidence value is 1. Otherwise, this value is multiplied by a confidence decay constant for every cycle that the object has not been seen. When the confidence value falls under a certain threshold the object is "forgotten".

3. System Implementation

This section is to describe the single-threaded models used so far and the multi-threaded approach being proposed for the RoboCup simulator agents. This paper focuses on POSIX threads that are the thread application-programming interface (API) specified by the standard POSIX 1003.1c-1995 [1][2]. Although POSIX threads are used in this implementation for portability reasons, the higher level design can be implemented in any system with any threads API.

3.1 Single-threaded approach

According to [2] a thread is the set of properties that suggest "continuousness and sequence" within a machine. In other words when a program is executed, a process is created. This process can be thought of as a single "thread" of execution. Of course a process has many additional properties like its own address space, file descriptors and various other data. Up to date, the majority of the implementations in the RoboCup simulation league have been single-threaded (e.g. CMUnited [13], ATHumboldt [3], Andhill). This essentially means that the initial process generated by the executable file does not create any additional threads. All computations are performed in a serial manner. If a given operation requires that the process is put to sleep (e.g. a blocking read when no data is available), then the whole execution is paused.

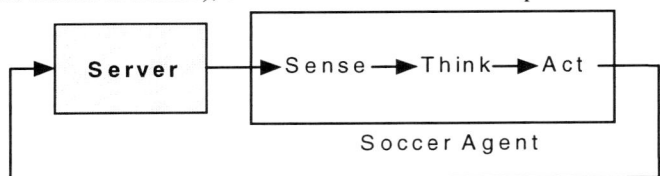

Figure 4 Single-threaded model

At the higher level, the agent is responsible for performing three individual tasks. First it needs to receive sensory information from the server via a UDP socket. Then the agent has to "think" in order to produce a desired action. Finally this action should be sent back to the server via the UDP socket. To perform all these tasks using a single thread the agent has to use a serial-processing loop shown in figure 4.

Given the nature of the simulation and the fact that two out of three operations involve I/O on a UDP socket, it is clear that a single-threaded serial implementation puts a significant limitation to the agent's capabilities. I/O operations can be extremely slow. Although an agent receives visual information every 150ms, it can send actions every 100ms. If an agent is put to sleep until new data arrives, it will miss a number of cycles, which significantly decreases the agent's overall performance. Therefore, both I/O multiplexing and signal-driven I/O are widely used in single threaded implementations.

Using I/O Multiplexing

I/O multiplexing is the capability to tell the kernel to notify the process if one or more I/O conditions are ready (i.e. input is ready to be read, or the descriptor can take more output). For the RoboCup simulator, the agent needs to check if data is available from the server, or if it is ready to send data to the server. To implement multiplexing, after the connection has been established with the server, the agent enters an infinite loop that is normally called the main-loop (or message-loop). In this main loop the *select* call is used to determine whether data are ready to be received or to be sent. If new data has arrived from the server, the agent analyses and stores the information. Alternatively if data are ready to be sent, the agent sends the required action to the server.

```
while (agent is alive)
{
    Call select to check which descriptor is ready.

    if (read descriptor is ready)
        receive and analyse data from the server.

    if (write descriptor is ready)
        send an action to the server.
}
```

Figure 5 Pseudo-code for I/O multiplexing

This approach has several disadvantages. Firstly it is difficult to accurately time each phase of the agent's cycle. The agent can only have an estimate of the time required to analyse the sensory data, perform all the thinking, and produce an action. Hence, timing correctness cannot be guaranteed for the 100ms-interval set by the server. Secondly only one I/O operation will be executed at a time. In other words, although *select* can wait on multiple descriptors, the main-loop can only perform one operation when *select* returns. Hence every time the agent enters the main-loop it will either analyse the sensory data that were received, or send an action to the server. Since only one of the two tasks can be executed, and since this task will not be interrupted before completion, the agent cannot guarantee timing correctness. Finally, the *select* function requires two system calls to receive the data. Initially *select* has to be executed to check whether data is available, and then if data exists on the socket the agent needs to call *rcvfrom* in order to receive that data. Figure 5 presents an example of a pseudo-code implementation using I/O multiplexing.

Using Signal Driven I/O

Signal driven I/O mode is used to instruct the kernel to notify the agent by generating a signal when something happens on a descriptor. To implement signal driven I/O two signals are normally used. One is generated when new data arrives from the server through the UDP socket (SIGIO), the other is generated when an interval timer expires which indicates that an action must be sent.

```
Signal handler for SIGIO          Agent's Main-Loop
{                                 {
    // SIGALRM has just been          Install SIGIO & SIGALRM handlers
       blocked.                       Set an interval timer to 100ms
                                      Start thinking process...
    Receive and analyse data
    from the server.                  /* The kernel will generate signals for
                                      the process and hence execute a signal
    // On exit, SIGALRM will be       handler if either of these two events occurs
       unblocked.                     if (100ms expire)
}                                         execute SIGALRM handler.
Signal handler for SIGALRM            If (new data arrives on the socket)
{                                         execute SIGIO handler.
    // SIGIO has just been            After the execution of the handlers has
       blocked.                       finished, the program will return
                                      precisely where it was before the
    (Since 100ms have expired...)     signal was generated. */
    Send an action to the server. }

    // On exit, SIGIO will be
       unblocked.
}
```

Figure 6 Pseudo code for signal driven I/O

Signal driven I/O can only handle one signal at a time. Although it is possible to establish signal handlers for multiple signals (e.g. SIGIO, SIGALRM, etc.) only one signal handler can be executed at any given point. Therefore, when a signal is generated while another signal handler is being executed, certain precautions have to be taken to avoid signal loss. To resolve this problem it is necessary to mask additional signals before entering the signal handler and then unmask them when the handler has finished.

However, great care needs to be taken for masking (SIGIO) signals. When a signal handler is being executed, if two more datagrams arrive on the UDP socket, the (SIGIO) signal will be generated two more times. But since the signal is blocked, when the signal handler returns, the system will only observe one (SIGIO) signal. This will force the agent to read the second datagram but the third one will remain on the queue until another (SIGIO) is generated. This clearly causes a problem since this architecture is unable to guarantee that the number of (SIGIO) generations matches the datagrams on the UDP queue. A solution would be to keep polling the queue until it is empty but this clearly consumes valuable time that could be used to perform other computations. An example of a pseudo-code implementation for signal driven I/O is given in figure 6.

3.2 A multi-threaded approach

Instead of a single-thread, a process can have multiple threads, sharing the same address space and performing different operations independently and without affecting each other. This architecture allows the agent to use a separate thread for each of the three tasks. The proposed multi-threaded model can be seen in figure 7.

Inside the agent, the three main tasks are running concurrently (or in parallel in multi-processor hardware) minimizing delays from the I/O operations. Only the "Sense" thread is responsible for waiting data from the server, and only the "Act" thread is responsible for timing and sending the actions (these relationships are

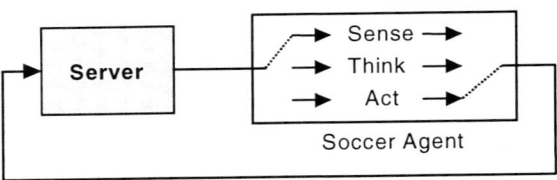

Figure 7 Multi-threaded model

indicated by the dashed arrows). In this way the agent can dedicate the maximum amount of processing power available by the processor(s) to the "Think" thread.

Firstly it is necessary to specify the I/O model that is going to be used. This now becomes much simpler since there are separate execution threads for input and output. The "Sense" thread can now use a blocking I/O connection. Since this thread is now dedicated to receiving data, it does not need to waste processing time querying the socket as to whether data is available. With a blocking connection the *rcvfrom* call will put the "Sense" thread to sleep until new data arrives on the socket. Putting the "Sense" thread to sleep does not affect the execution of the other two threads that can proceed as normal. When data arrives on the socket, the "Sense" thread will be awaken, execute a *rcvfrom* call to receive the next available datagram from the server, and analyse the new data. This approach does not allow for datagrams to be lost, or queue up. This would only happen if the server transferred data faster than the thread could analyse it, which is impossible.

Secondly, the "Act" thread needs to send any available actions to the server at 100ms intervals. By having a dedicated thread to perform this task, the accuracy of the timing is only limited by the resolution of the operating system's clock. The *gettimeofday* function and a conditional variable are used to implement the "Act" thread. In other words, the current absolute time is incremented by 100ms and then the thread waits for the conditional variable to return. Assuming spurious wake-ups will not occur, and since no other thread will signal this conditional variable, it will only return when the 100ms have passed, in which case the "Act" thread can send an action to the server. This method provides highly accurate timing comparing to the single-threaded approaches described earlier. This enables the agent to guarantee certain levels of timing correctness, which is something single-threaded approaches failed to do. In addition to that, this thread is also put to sleep while waiting for the conditional variable to return. This provides the other threads with the maximum amount of resources available.

Thirdly the "Think" thread is the only one that stays permanently awake, and consumes the majority of the available resources to perform most of the computations. Part of the "Think" thread of the current implementation can be found in [8]. If required, various scheduling policies and different priorities among the threads could be implemented. However, given the nature of the problem, no scheduling is needed. In addition to that, synchronization between separate threads can also be implemented if required. Multi-threaded programming is a large topic itself. Describing issues such as how threads compete for resources, or how often pre-emption occurs falls outside the scope of this paper. A detailed description on multi-threaded programming can be found in [2]. A pseudo-code implementation of the proposed multi-threaded approach is shown in Figure 8. An alternative multi-threaded approach based on an organic programming language can be found in [11].

Table 1 Performance testing

Cycle loss	Single-thread I/O multiplexing	Single-thread signal driven I/O	Multi-thread approach
1 agent	1651 (27.52%)	24 (0.4%)	0 (0%)
6 agents	1653 (27.55%)	1020 (17%)	0 (0%)
11 agents	1656 (27.60%)	3294 (54.9%)	4 (0.07%)

4. Results and Analysis

4.1 Experimental Results

To compare the two models a simple experiment was carried out using a PC with an Intel Pentium II 450MHz processor, running Redhat Linux 5.2. The purpose of this experiment was to evaluate only the timing correctness of the models described above, since logical correctness is based on the individual AI implementations. The test required an agent to connect to the server and send a turn command in each cycle. In addition to that the agent is required to receive and analyse all sensory information to simulate a real game situation. The sensory analysis functions are identical for all implementations. Initially, a single agent was tested on the field. The same experiment was then carried out using 6 and 11 agents to examine what would happen as computational resources decrease. The results for a 6000 cycle test (a game duration) are illustrated in table 1.

The table entries indicate the number of cycles for which an agent failed to send an action and hence did not meet the timing requirements. Adding more agents has a similar effect as extending the thinking process. In both cases system resources are

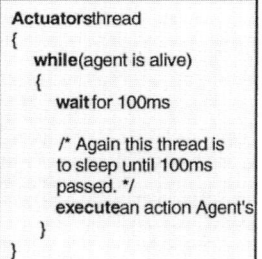
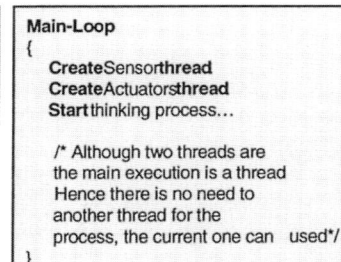

Figure 8 Pseudo code for the proposed multi-threaded approach

depleted by adding more computations. This is to test the scalability of each implementation in terms of number of operations per cycle. As can be seen, I/O multiplexing using *select* has over 27% cycle loss even when only 1 agent is used. However multiplexing is highly scalable since this percentage does not increase much more for 6 or even 11 agents. On the other hand signal driven I/O performs significantly better in the single agent trial with only 0.4% cycle loss. However this approach is not scalable since there is 17% loss for six agents and nearly 55% cycle loss for 11 agents.

From the results generated by the multi-threaded approach, it can be seen that there is nearly 0% cycle loss for all trials. This proves the superiority of multi-threaded approaches in terms of performance and scalability. It should be noticed that multi-threaded implementations are very efficient in multi-processor hardware. The use of parallel processing is significantly important in obtaining real-time performances even in highly complicated systems such as aerospace control systems [15] and advanced AGVs in industry [5].

4.2 Discussion and Analysis

Although comparing the multi-threaded approach against the single-threaded ones is not an easy task, certain points about the two approaches are briefly presented here. The main advantages of the proposed multi-threaded approach include:
- The multi-threaded model allows more efficient exploitation of the agent's natural concurrency. Various computations can be performed while waiting for slow I/O operations to complete. This provides the agent with a significant advantage considering the number of I/O operations per game.
- Using multiple-threads provides a modular programming model that clearly expresses relationships between independent events within the agent. However, designing a multi-threaded approach clearly illustrates various program dependencies and synchronization requirements between different modules within the agents. Mixing CPU-intensive processes with I/O intensive processes results in better utilization of resources of an uni-processor system. Note that multi-threaded implementations allow an agent to achieve even better performance in a multi-processor environment.
- A multi-threaded approach is highly scalable. In the current implementation, consecutive actions can be sent 100ms apart. A multi-threaded implementation could have one thread performing an expensive search algorithm, and a separate thread executing simple reactive behaviours. By combining the results from both threads, the agent can significantly improve its performance.

However, multi-threaded programming has also a few limitations. First, a multi-threaded model is normally more complicated than a single threaded one. Careful design is required to keep track of synchronization protocols and program invariants. Second, it is necessary to avoid deadlocks, races, and priority inversions. Third, the POSIX standard does not provide an interface for object oriented languages. However a solution does exist and can be found in [14].

5. Conclusions

It has been shown in this paper that the behaviour of real-time agents is not founded only on the logical correctness of their actions. Timing correctness becomes an equally important factor especially in applications where response times can significantly affect the result. In such cases the quality of the results becomes a function of both logically correct output and response time. To satisfy all the necessary timing constraints for a real-time agent, a single-threaded implementation will not suffice. This is mainly due to the low speed of network I/O operations, and the limiting serial nature of such architectures. Therefore, a multi-threaded implementation is proposed in order to overcome this problem. Based on this approach, the agents can perform various computations concurrently (or even in parallel on multi-processor hardware) and hence avoid waiting for the slow I/O operations to complete. This allows the agents to guarantee a certain degree of timing correctness that is only limited by the resolution of the given operating system. In addition, the experimental results have shown that a multi-threaded model clearly outperforms a single-threaded one in terms of responsiveness and efficiency.

Acknowledgements

We would like to thank Kaz Kylheku for his useful comments on the Unix/Linux systems and the University of Essex for the Research Promotion Fund DDP940.

References

1. A. Burns and A. Wellings, Real-time Systems and Programming Languages, Addison-Wesley, 1997
2. R.D. Butenhof, Programming with POSIX Threads, Harlow, Addison-Wesley, 1997
3. B. Hans-Dieter, et al., AT Humboldt -- Development, Practice and Theory, In Proceedings of RoboCup-97-- Robot Soccer World Cup I, H. Kitano (Editor), 1997, 357-372
4. H. Hu, M. Brady, F. Du, P. Probert, Distributed Real-time Control of a Mobile Robot, Int. Journal of Intelligent Automation & Soft Computing, Vol. 1, No. 1, 1995, 63--83
5. H. Hu and M. Brady, A Parallel Processing Architecture for Sensor-based Control of Intelligent Mobile Robots, Int. J. of Robotics and Autonomous Systems, Vol. 17, No. 4, 1996, 235-258
6. H. Kitano, RoboCup: The Robot World Cup Initiative. Proceedings of the 1st Int. Conference on Autonomous Agent (Agents-97)), Marina del Ray, The ACM Press, 1997
7. H. Kitano, M. Tambe, P. Stone, M. Veloso, S. Coradeschi, E. Osawa, H. Matsubara, I. Noda, and M. Asada, The RoboCup Synthetic Agent Challenge'97. Proceedings of International Joint Conference on Artificial Intelligence, 1997
8. K. Kostiadis and H. Hu, Reinforcement Learning and Co-operation in a Simulated Multi-agent System, Proc. of IROS'99, Korea, October 1999
9. N. Nissanke, Real-time Systems, London, Prentice Hall, 1997
10. I. Noda, Soccer Server: A Simulator for RoboCup. JSAI AI-Symposium 95: Special Session on RoboCup, 1995
11. I. Noda, Kappa: Agent Program by Gaea. Proceedings of the 2nd RoboCup Workshop, July 1998, 387-392
12. W.R. Stevens, Unix Network Programming: Networking APIs: Sockets and XTI (Volume 1), London, Prentice-Hall (1998)
13. P. Stone and M. Veloso and P. Riley, The CMUnited-98 Champion Simulator Team, In "RoboCup-98: Robot Soccer World Cup II", M. Asada and H. Kitano (eds.), Springer Verlag, Berlin, 1999.
14. B. Stroustrup and S. Lippman, Pointers to Class Members in C++. Proc. USENIX C++ Conf., Denver, 1988, 305-326
15. G. S. Virk, J. M. Tahir, P. K. Kourmoulis, Parallel Processing in Aerospace Control Systems, Proc. the 2nd Int. Conference on Applications of Transputers, UK, July 1990

A Functional Architecture for a Team of Fully Autonomous Cooperative Robots*

Pedro Lima, Rodrigo Ventura, Pedro Aparício, and Luis Custódio

Instituto de Sistemas e Robótica
Instituto Superior Técnico
Av. Rovisco Pais, 1 — 1049-001 Lisboa
PORTUGAL
{pal, yoda, aparicio, lmmc}@isr.ist.utl.pt
http://lci.isr.ist.utl.pt/projects/mrob/socrob/

Abstract. A three-level functional architecture for a team of mobile robots is described in detail, including the definition of the role assigned to each level, the main concepts involved, and the corresponding implementation for each individual robot. The architecture is oriented towards teams of fully autonomous cooperative robots, able to carry out different types of cooperative tasks. Complexity is reduced by the decomposition of team strategies into individual behaviors, which in turn are composed of primitive tasks. Relationships among robots of the team are modeled upon the joint intentions framework. An application to Robotic Soccer and some of its preliminary results are presented.

1 Introduction and Motivation

Different functional architectures have been proposed in distributed artificial intelligence and intelligent control literature to handle the complexity of controlling a fully autonomous mobile robot or a team composed of such robots. A common concept among those approaches is the existence of atomic *primitive tasks* or *behaviors* which are the kernel of the architecture. Tasks executed by the robot result from the composition of those entities.

The main difference between the existing approaches concerns the interaction among the atomic entities. While some authors allow full flexibility, so that a team behavior emerges from a negotiation between running behaviors [3], others prescribe, with different flexibility levels, the task decomposition into primitive tasks [1], to an extent which may even forbid any direct communication between primitive tasks [7].

A three-level functional architecture for a team of mobile robots is introduced in this paper. The architecture is oriented towards teams of fully autonomous cooperative robots, able to carry out different types of cooperative tasks. The

---

* This work was supported by the Science Service of the Calouste Gulbenkian Foundation and by the Portuguese Foundation for Science and Technology (ISR/IST programmatic funding).

level splitting is inspired by the work of Drogoul and his co-workers [3], but there are important differences regarding the modeling of the relational level, which describes inter-agent negotiation and role assignment. The *joint intentions framework* [8, 5, 2] provides a solid foundation for teamwork modeling, and will be used in this work to support the implementation of the relational level.

Complexity is reduced by the decomposition of team *strategies* (i.e., what should be done) into individual behaviors, which in turn are composed of primitive tasks. A set whose elements are the behaviors assigned to each robot of the team is designated as the *tactics* (i.e., how to do it) for a given strategy. An application to Robotic Soccer and some of its preliminary results developed during and after the RoboCup'98 contest are presented.

The paper is organized as follows. Section 2 describes the team and individual architectures, with details of teamwork modeling at the relational level and of the foreseen/current implementation for the introduced concepts. Section 3 maps the concepts onto a robotic soccer team. Section 4 closes the paper with some preliminary conclusions and reference to future work.

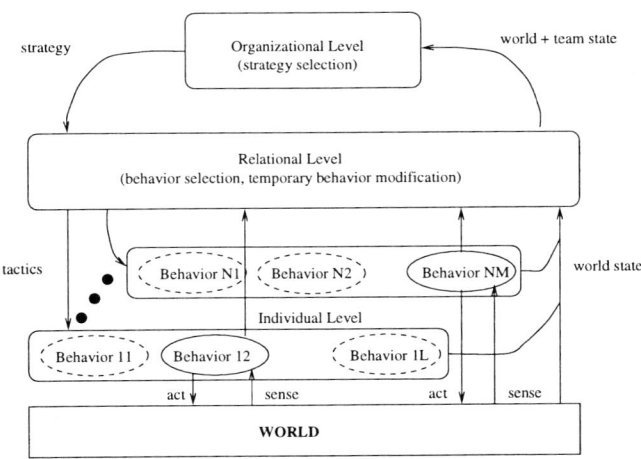

BehaviorIJ: behavior I of robot J

Fig. 1. The functional architecture of the ISocRob team.

2 The Functional Architecture

2.1 Team Architecture

The team architecture, based on the 3-level agent team architecture first proposed by Drogoul and his co-workers [4], is depicted in Fig. 1. Our interpretation of Drogoul's three-levels follows:

- **Organizational level**: establishes the strategy to be followed by the whole team, given the team and world states. The team state corresponds to the current set of behaviors under execution. The following examples, taken from the robotic soccer context, illustrate the concept of world (game) state, which is divided in two classes:
    1. game situations reached upon the application of RoboCup tournament rules (e.g., kickoff, end-of-game, penalty-for, penalty-against);
    2. team evaluation of current game status (e.g., losing & close to the end of the game, ball close to our goal).
  Strategies can be divided in, at least, two major categories:
    • pre-programmed scenarios for game situations in game state category 1 above;
    • dynamic strategies (e.g., defend, attack, counter-attack), corresponding to game state category 2 above.
- **Relational level**: at this level, relationships among robots are established. The robots negotiate and eventually come to an agreement about some team and/or individual goal. Moreover, behaviors are assigned to the individual robots, after a selection from within behavior sets representative of alternative tactics for the strategy selected by the organizational level. The selected behavior set depends on the current world plus team states. Behavior assignments may also be temporarily modified as a result of inter-robot negotiations.
- **Individual level**: encompasses all the available robot behaviors. Those include the primitive tasks (e.g., SeekBall, KickBall, RotateLeft) and their relations.

A *behavior* corresponds to a set of purposive (i.e., with a goal) primitive tasks sequentially and/or concurrently executed. A *primitive task* is a *sense-think-act loop* (STA loop), a generalization of a closed loop control system which may include motor, ball tracking or trajectory following control loops, to name a few.

STA loops are composed of the following key components:

- **goal**: the objective to be accomplished by the primitive task (e.g., moving to a given position plus orientation (pose) set-point, tracking the ball in the image);
- **sense**: sensor data required to accomplish the goal (e.g., distance to an object, object position in an image);
- **think**: the actual algorithm which, using the sensor data, does what is required to accomplish the goal (e.g., motion controller, ball visual servoing);
- **act**: the actions associated to the **think** algorithm (e.g., moving the wheel motors).

The sequence of primitive tasks is traversed as the logical conditions associated with the connections among them become true. The logical conditions are defined over a predicate set. There are two *predicate* classes:

- predicates which check the value of a given variable (e.g., the variable *goal* in lastseen(goal)=left);

- predicates which check the occurrence of a given event (e.g., the see(ball) predicate checks whether the ball became visible).

A *world model* is required to provide information to the relational and organizational levels regarding the world state. Since all computation is supposed to be distributed over the team members, with no external storage available, a distributed world model representation is required, containing all the relevant information for negotiation between agents, and in general the result of processing raw data, for primitive tasks usage. A *distributed blackboard* is proposed to implement the world model [9].

2.2 Individual Robots Architecture

Each individual robot is provided with all the three levels of the team functional architecture. However, the organizational level is only active in one of the robots, assigned as the team *captain*. The remaining robots have a dormant organization level, to ensure fault-tolerance: whenever the captain robot has a malfunction, the next robot in the list takes over as the captain. The list has no special order since, from the hardware standpoint, all robots of the team are currently homogeneous. In an non-homogeneous population, the potential captains (from a computational capacity standpoint) should be sorted according to their descending computational power.

An agent-based programming language has been specified and is currently under development [9, 10], to provide the team strategist (e.g., the coach, in robotic soccer) with the means to program the population in order to achieve the strategic objectives, embedded in the behaviors and in the primitive task STA loops.

Each of the above concepts will be implemented as follows:

- the *strategy* is determined at the organizational level by a *state-machine* whose transitions are traversed upon the matching of specific world states, and whose states define the current strategy. Therefore, strategies change when the world state (as perceived by the team) changes;
- *tactics* selection, including behavior selection, negotiation, and temporary behaviors modification, is implemented by *relational rules* at the relational level;
- a *behavior* consists of a *state-machine*, where each state corresponds to an *STA loop* and each transition has associated logical conditions defined over the predicate set described in subsection 2.1;

Team organization is necessarily a centralized operation. As such, decisions on strategies must be taken by a single agent, designated as the *captain*. Thus, the organizational state machine runs in the captain. To increase team robustness, whenever the current captain does not signal that it is alive for more than a timeout period, a new captain must take control of the team.

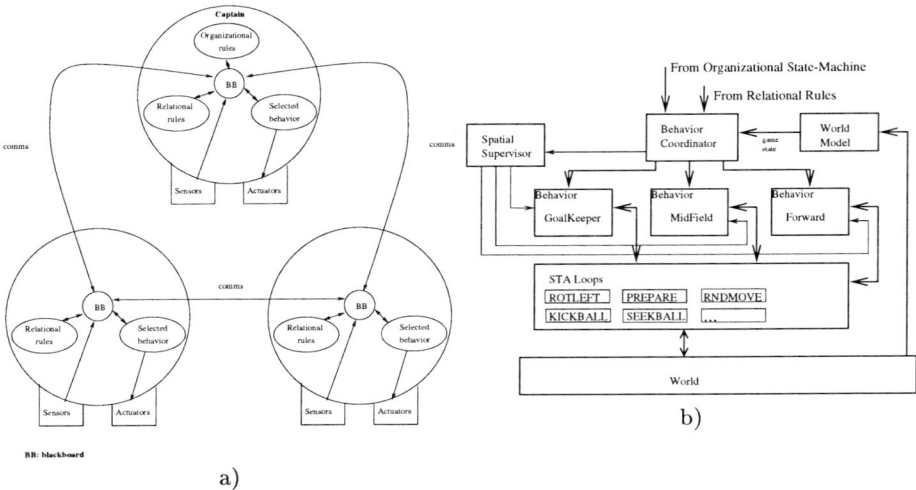

Fig. 2. Implementation of team functional architecture at each individual robot: a) Blackboard, relational rules, organizational state-machine and communications. b) Behavior selection and coordination.

A *blackboard* implements global shared memory and event-based[1] communication. The blackboard is the sole medium of communication between the agents, supporting the message exchange required for negotiation. One of the key factors will be the distribution of data per robot, which should minimize the need to communicate in order to obtain information (e.g., data obtained from processing an image should be stored in the robot where the image was acquired). The global team strategy is also stated in the blackboard, as a variable which triggers some of the relational rules.

The schematic block diagram of the implementation, at each individual robot, of the team architecture is depicted in Fig. 2 - (a). The behavior selection and spatial coordination are detailed in Fig. 2 - (b): the *behavior coordinator* selects the correct behavior for the robot, based on the world state and on the strategy and tactics provided by the organizational and relational levels of the team architecture. When a behavior is selected, the corresponding *spatial supervisor* is also activated. The *spatial supervisor* ensures that the robot always stays within the influence zone associated to its active behavior (see Section 3).

2.3 Relational Rules

Relationships among the team robots are established at the relational level of the team functional architecture. Given a strategy established by the organizational level, different tactics can be used to implement it. Tactics consist of behavior sets, whose elements are the behaviors assigned to each individual robot of the

---

[1] Event is interpreted here in the context of a computational model.

team. A tactics is chosen based on the current world state, but also on each agent's current internal state. The strategy must specify not only the goal to be attained (e.g., **attack**, **defense**) but also criteria to check how close to the goal the team is. Behaviors can be temporarily modified as a result of inter-robot negotiation, as part of the tactics to attain the goal.

An example is the situation where two teammates, both assigned a forward-like behavior, actively try to get the ball. In such a case, one of them should signal the other its intention. A negotiation process would follow, where the teammates would determine their distances to the ball, to decide which one should pursue it. After taking such a decision, the other player should temporarily modify its normal forward-like behavior.

The absence of such a relational mechanism leads to situations where team behavior is poor. Consider the case of two forward-like players with similar behaviors, that often conflict with each other while trying to reach the ball. The key to solve this problem is to endow the team members not only with individual goals, but also make them knowledgeable of the team goals. This is clearly related to concepts such as *joint persistent goal*, *joint intentions* and *joint commitment* [8, 5, 2]. Moreover, it requires *communication* between team members.

For instance, a joint persistent goal is defined in [5] as follows: *A team of agents has a joint persistent goal, relative to q, to achieve p iff: they all mutually believe that p is currently false; they all mutually believe that they all want p to be eventually true, and until they all come to mutually believe either that p is true, that p will never be true, or that q is false, they will continue to mutually believe that they each have p as a weak achievement goal relative to q.*

The example above can be interpreted under this definition. The strategy p (e.g., **attack**) is a weak achievement goal relative to the main goal q of scoring a goal. Suppose the two players both assumed the Forward behavior as part of the selected tactics. They will pursue the strong goal q (i.e., they will attempt to score a goal) by executing their Forward behaviors so as to attain p. A criterion to check whether p is attained is to determine whether the players are able to keep playing within their assigned influence zone. Both players will continue to work towards meeting this and the other criteria which define the **attack** strategy until they all come to mutually believe either that all the criteria were met, that the criteria will never be attained (e.g., after a timeout), or that 'scoring a goal' is no longer the main team endeavor (e.g., because the game is over). Working towards meeting the criteria includes temporarily modifying their behaviors to cope with the team goal (e.g., refraining from pursuing the ball). This distinguishes a group of non-cooperative agents whose individual goals just happen to be the same, from a group of cooperative agents which share a common aim. The latter exhibits cooperation and coordination, while in the former the individual agents compete when the resources are scarce [5].

The relational rules implement a *recipe* which is commonly agreed by all the agents of a team [5]. This recipe is embedded in the rules and may either be prescribed initially (i.e., before joint action is started) or evolve over time. We are currently looking at the possibility of changing the recipe over time using reinforcement learning techniques, based on a performance function which weights the reliability (i.e., the ability to meet specifications) and the cost (computational or other) of a given recipe [6]. Reinforcement learning should be able to determine the recipe (to achieve a joint intention) that best balances cost and reliability.

3 Application to a Soccer Robot Team

Some of the concepts described in the previous section will now be mapped onto a team of fully autonomous soccer robots.

Fig. 3 presents a functional division of the field in several regions. These are zones where robots try to locate themselves inside the field, according to their assigned behaviors, e.g., defenders should stay inside the D zone and Forward players should stay inside the F zone. This division helps the assignment of influence areas to players.

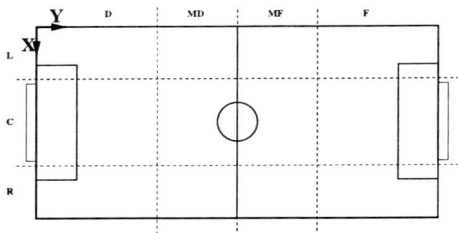

Fig. 3. The field division in actuation areas.

Besides Defense (D), MidDefense (MD), MidForward (MF) and Forward (F), further divisions are introduced to increase the field resolution. Along the field longitudinal axes, the field is divided in Left (L), Center (C) and Right (R) parts. This division is particularly useful when the team has more than one player acting in the same functional area (e.g., L and R defenders).

3.1 Player Behaviors

As explained before, behaviors are composed of primitive tasks sequentially or concurrently executed. A field influence zone is associated to each behavior. Several behaviors must be implemented in a robot soccer team. The most significant ones, whose influence zones are depicted in Fig. 4, are:

- **GoalKeeper** – Defends the goal. To do that, it continuously looks for the ball and, if necessary, leaves the goal area and kicks it away. The influence zone is defined by the goal area lines and is shown in Fig. 4 - a).

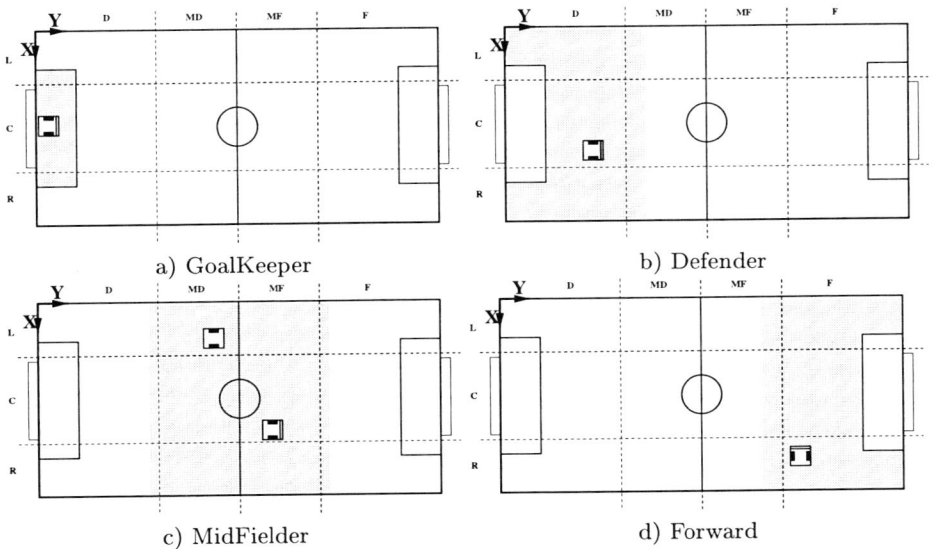

Fig. 4. Influence zones for individual behaviors.

- **Defender** – The defender mission is to move the ball from the vicinity of its team goal to the opponents field. If possible, it should try to move the ball to the vicinity of a MidFielder. It should return to its original position (D) when the ball is once again in the opponent's field (see Fig. 4 - b)).
- **MidFielder** – Such as in real soccer, the Midfielder is able to play in a variety of positions. Its influence zone lies within the MD and MF areas (see Fig. 4 - c)). This player natural ability is to receive the ball from its own team field and decide what to do, based on the other players availability. If a Forward is in the near vicinity of the opponents goal (F area), the MidFielder should try to pass it the ball.
- **Forward** – The Forward behavior induces the player to be in the F zone (see Fig. 4 - d)). If the ball goes into our field, the Forward's mission is to keep track of the ball, although it should not move out of its zone by its own initiative. When the ball moves into the F zone, it must try to take control over it and kick it into the opponents goal. Such a behavior is implemented by the state machine of Fig. 5 - a).

An alternative implementation would consist of letting the Forward players move up and down the field, using the lateral L and R corridors.

3.2 Relational Behavior Modification

Individual behaviors can be temporarily modified to allow cooperative relations between teammates, as explained in Subsection 2.3. Fig. 5 - b) depicts the state machine which implements the Forward behavior endowed with states

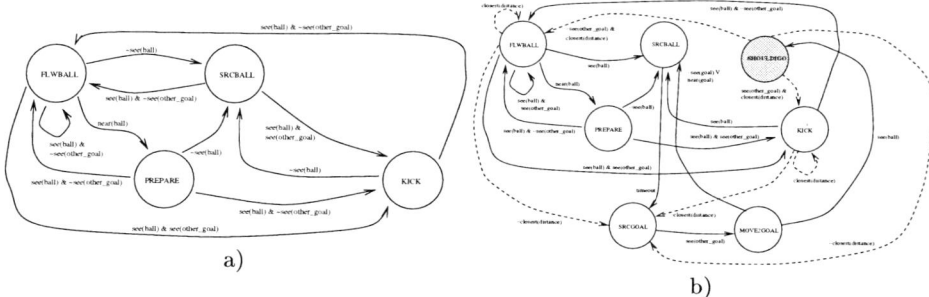

Fig. 5. State machines for the Forward behavior: a) With no cooperation. b) With cooperation. States and state transitions representing relational rules are filled and dashed in the diagram, respectively.

and state transitions representing relational rules. The negotiation implemented corresponds to the example in Subsection 2.3 concerning two Forward players who actively try to get the ball. The additional state ShouldIGo is entered by a Forward player upon its detection of the ball (predicate see(ball)). In this state, a message is broadcasted through the blackboard to all teamates stating that this Forward player saw the ball and also the estimate of its distance to the ball. The occurrence of messages from the the other teamates stating that they saw the ball and including the estimate of their distance to the ball is also checked in the ShouldIGo state. The distance to the ball of all the Forward players who saw the ball recently is continuously sorted. The sorted distances are checked by the closest(distance) predicate, associated to all the dashed arcs of the state machine in Fig. 5 - b), which are responsible by the temporary behavior modifications with respect to the non-relational Forward behavior of Fig. 5 - a). The most important modification consists of not immediately kicking (KickBall state) or following (FlwBall state) the ball upon its detection in state SrcBall, but rather moving to state ShouldIGo where distances to the ball are compared. Should the Forward be the closest the ball among all its Forward teammates (or the only one who sees the ball), the state machine execution proceeds as in the non-relational Forward behavior. Otherwise, the behavior is modified by making the player move to a location close to the other team's goal (states SrcGoal and Move2Goal).

3.3 Game State

The *game state* refers to either situations reached as a result of the application of RoboCup rules or to an evaluation of the current game status. State changes are induced by the time flow and teams actions during the game. Examples of game states are as follows:

Game situations

- game-start – This happens in the beginning of the game, after a goal or when the game restarts after a break;
- penalty-for, penalty-against;
- end-of-game – This is signaled by an external event (e.g., two whistle blows).

Evaluation of game status

- ball-our-ofield – One of our players has ball possession. The ball is in our field;
- ball-nour-otfield – None of our players has ball possession. The ball is in the other team field;
- losing & close to the end of the game;
- ball close to our goal.

3.4 Scenarios for Game Situations

Pre-defined scenarios are usually associated with the game states corresponding to game situations (see above). An example is the **game-start** situation, where the players must move to their pre-determined start positions (see Fig. 6). Self-location of the players must be accomplished at this state as they must be correctly positioned prior to the start of the game. After positioning, the players will wait for the external kickoff signal (e.g., a whistle blow) that signals the start of the game.

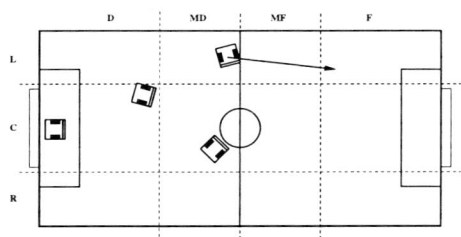

Fig. 6. Players position at game start-up.

3.5 Dynamic Strategies and Tactics

During the game the ball moves inside and outside of the team mid-field. Depending on factors such as the ball position and motion, the current game state, the current score, the number of available players and their behaviors, the opponents positions, the elapsed time, and the current strategy, the team strategy may change. This is inspired by real soccer. Possible strategies are:

- defense – The ball must be prevented from entering our field. Should that happen, it must be moved into the opponents field. Several defense tactics exist. Two examples of tactics for the Defense strategy are:
    • *Strong Defense* (SD) – This strategy points towards creating a continuous, physical barrier between the ball and our goal. It is aimed at avoiding opponent players from moving towards our goal. When re-positioning, the Defender players should try to avoid occluding the GoalKeeper visibility of the field, i.e., the DC zone should be free of players (see Fig. 7 - a)).

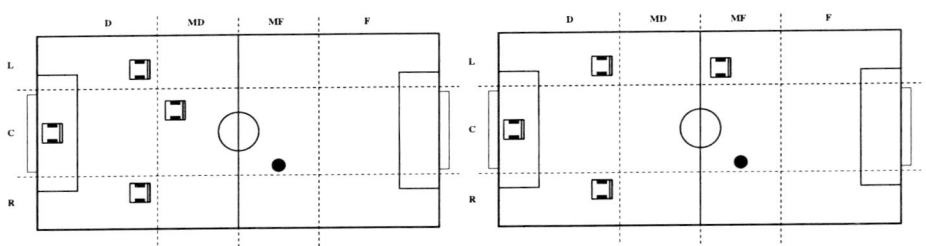

a) Strong defense initial positions.   b) Medium Defense initial positions.

Fig. 7. Strong and Medium Defense tactics for the Defense strategy.

    • *Medium Defense* (MD) – Points towards a strong defense and a good recovery mechanism, essential to counter-attack. The concept is illustrated in Fig. 7 - b). The difference between the SD and the MD is that in MD not all players are moved into our field. This makes the transition to Counter-Attack easier, as one of the players stays in the opponent field.
- counter-attack – A counter-attack happens if the team is positioned to move the ball quickly into the opponent field and score a goal. It requires a Defender, to handle a possible interception of the ball by an opponent, a MidFielder, to pass the ball into the Forward area, and a Forward player to kick the ball into the opponent goal.
- attack – Under this strategy, the whole team moves forward. Besides the GK, there is only one player in our field. This movement requires two MidFielders (one in the MD zone and another in MF zone), and one Forward. The idea is to have the ball passed from the M zone into the F zone, where a Forward player is to pick it up and kick at the goal.

4   Preliminary Conclusions and Future Work

Currently, our mid-size real robots are capable of simple but essential behaviors, composed of primitive tasks, such as following a ball, kicking a ball, scoring goals and defending the goal, using vision-based sensors (see Fig. 5). Current available behaviors include shooting at an empty goal starting from increasingly more difficult situations or defending the goal by permanently tracking the ball and kicking it out of the goal area as soon as it gets too close. One Forward *vs* one Defender and a GoalKeeper have also been successfully tested in live action.

Our current and future work is centered on four main topics:

- development of the self-localization system based on a vision camera and a mirror;
- update and tuning of the primitive tasks software;
- design and implementation of an agent-based programming language suitable for multi-agent systems;
- study and development of a teamwork model and its integration with the team functional architecture.

Among those, self-localization is perhaps the most essential. The functional architecture described in this paper relies on a (at least rough) awareness by each robot of its location in the field and, consequently, of the team current disposition in the field.

The work has been carried out in a bottom-up fashion, since we believe that many conceptual issues can be raised from and are strongly constrained by the actual implementation problems. Nevertheless, the basic framework described in the paper, concerning hardware, software and functional architectures, was designed in a top-down fashion in the beginning of the project and has been essentially kept unchanged so far.

References

1. J. S. Albus. Outline for a Theory of Intelligence. *IEEE Transactions on Systems, Man, and Cybernetics*, 21(3), May/June 1991.
2. P. Cohen and H. Levesque. Teamwork. Technical Report 504, Center for the Study of Language and Information, SRI International, March 1991.
3. A. Drogoul and A. Collinot. *Autonomous Agents and Multi-Agent Systems*, chapter Applying an Agent-Oriented Methodology to the Design of Artificial Organizations: A Case Study in Robotic Soccer. Kluwer Academic Publ., 1998.
4. Alex Drogoul and C. Dubreuil. A distributed approach to n-puzzle solving. In *Proceedings of the Distributed Artificial Intelligence Workshop*, 1993.
5. N. R. Jennings. Controlling cooperative problem solving in industrial multi-agent systems using joint intentions. *Artificial Intelligence*, 75:195–240, 1995.
6. P. U. Lima and G. N. Saridis. *Design of Intelligent Control Systems Based on Hierarchical Stochastic Automata*. World Scientific Publ., 1996.
7. G. N. Saridis. Architectures of intelligent controls. In M. M. Gupta and N. Sinha, editors, *Intelligent Control Systems*. IEEE Press, Piscataway NJ, 1995.
8. M. Tambe. Towards flexible teamwork. *Journal of Artificial Intelligence Research*, 7, 1997.
9. R. Ventura, P. Aparício, and P. Lima. Agent-based programming language for multi-agent teams. Technical Report RT-701-99, RT-401-99, Instituto de Sistemas e Robótica, IST-Torre Norte, March 1999.
10. R. Ventura, P. Aparicio, P. Lima, and L. Custódio. IsocRob — Intelligent Society of Robots. In *Collection of RoboCup99 Team Description Papers*, August 1999.

Extension of the Behaviour Oriented Commands (BOC) Model for the Design of a Team of Soccer Players Robots

C. Moreno[1] – A. Suarez[1] – E. Gonzalez[1] – Y. Amirat[2] – H. Loaiza[3]

Université d'Evry – LaMI – Bd François Mitterrand 9100 Evry France[1]
{moreno, suarez, gonzalez}@lami.univ-evry.fr
Université Paris XII – LIIA – 122, Rue P. Armangot Vitry S/Seine France[2]
amirat@univ-paris12.fr
Universidad del Valle –PayRA – Ciudad Universitaria – Cali – Colombia[3]
hloaiza@eeie.univalle.edu.co

Abstract. The Real Magicol soccer players team for Robot Cup 99 is based on a BOC architecture extension which combines reactive and deliberative reasoning by the distribution of the knowledge system into modules called behaviors. The hardware remains the same as the 1998 Robot Cup one, our research bearing on the cooperative architecture based on agent concept.

1 Introduction

Soccer players constitute a complex application of collective robotics. In fact, in such a context, the control architecture of a robot grows in complexity since this robot has to evolve in dynamic environment and cooperate with its partners in order to develop a team game. Such an application has to be considered as a set of decision–making, processing and communications, rather than a centralised and monolithic set, where little room is left to the autonomy and little importance is granted to interactions between the different robots. This agent oriented approach allows a redistribution of the decision–making among the different levels of the organisation leading consequently to a distributed, dynamical and reactive organisation.

The control architecture of a soccer player robot constitutes a complex system. This complexity is due to the necessity to make coexist both mechanisms of knowledge management (deliberative actions), and constraints of reactivity to "survive" in a dynamic environment. To process this problem, several architectures have been proposed in mobile robotics.

Deliberative architectures [MORAVEC 89] [CHATILA 85] use a centralised model of the environment to verify sensory information and to find the best response in the physical world. All actions of the robot are directed to a known final goal. These architectures allow to endow the robot with reasoning capacities that confer it the function of planning.

Reactive architectures [BROOKS 86] [CONNEL 90] are interesting to study the emergent behaviour of the robot from its primitive behaviours. On the contrary of deliberative architectures, these ones neither allow the planning of operations nor authorise elaborate reasoning. These architectures use a set of modules of behaviours that react directly to variations of the environment by using simple transfer functions. Consequently, need of a model of the world disappears.

Hybrid architectures [SIMMONS] offers a compromise between the two previously described approaches to be both reactive and capable to follow a plan.

We propose in this paper the concepts that are going to allow the modelization of an organisation of soccer playing robots from both of a local point of view (the agent robot component) and a global point of view (interactions between 'robot' agents). The objective is to propose a control architecture model for each robot, that includes a both deliberative and reactive dimension, in order to take into account, on the one hand, its interactions with the environment and on the other hand, its interactions with its partners to make emerge a strategy of group. These works concern the participation of the team "Real Magicol" (Realismo Mágico Colombiano [Realismo Màgico] to the RoboCup-99 in the category " middle size ". The hardware architecture as well as general concepts of control architecture implemented by this team have been described in a communication presented during the RoboCup-98 WorkShop [LOAIZA 98]. Given the amount of constraints of soccer playing robots, we have proposed an architecture of control based on the Behaviour-Oriented Commands (B.O.C) concept [CUERVO 96]. This architecture confers to the robot both deliberative (in order to reason on complex situations) and reactive capacities (in order to respect deadlines). More specifically, it relies on a decomposition in competitive modules which interaction allows to endow the robot of typical reactive, deliberative or hybrid behaviours.

In this paper, we present results of works started on the extension of the B.O.C model in order to endow each robot with a supplementary deliberative dimension oriented towards the group strategy. This is necessary in the case of a soccer playing robots since it allows to endow each robot with a motivation going in the senses of the satisfaction of a collective tendency. Consequently, it allows through collaboration and co-operation mechanisms between robots to make emerge a collective game at the organisation level. In the 2nd. paragraph, we introduce notions of classification of roles to define an organisational model of the team. The objective being to make emerge at the organisation level behaviours with collective tendency. We present in the paragraph 3, the extension of the BOC model and particularly deliberative behaviour units "individual strategist" and "strategist of group". The paragraph 4 describes collaboration and co-operation mechanisms between robots, used for the implementation of the distributed B.O.C. We have also introduced at the goalkeeper control architecture a BlackBoard cartographer to provide relevant information, used in the game strategy. The objective being to guarantee a temporal coherence when deliberative behaviour units oriented to the strategy of group are activated.

2   Emergent collective behaviours and organisational model

Each of robots playing on the field can be considered as an agent participating to the strategy of game. Indeed, if a game strategy can be assimilated to the convergence of a set of actions to a goal or to several ones, it is then necessary to decentralise and assign each action to each member of the team and to establish a collaboration between them. Thus, the emergent collective behaviour results from the sum of local agents behaviours.

To obtain a coherent collective behaviour, it is essential to well co-ordinate 'robot' agents. Each has to have a predetermined role to avoid a divergence of the consequent of players actions. More, the efficiency of co-operation of players results also from their good synchronisation. Without this synchronisation, the collective behaviour is inoperative, and consequently, the strategy of group is non-existent. Indeed, because of the uncertainty of the "freshness" of information from others, each robot has then to evolve individually and to act in autonomous manner. If this determinism is absent at the individual level (robot level), the realisation of a behavioural model is made impossible. Thus, behaviour units work stand by themselves and no emergence arise. To reach these objectives, a real time deterministic system managing communications is necessary.

According to the previous discussion, the fact that the global behaviour of robots is the consequent of local behaviours of each one implies a good distribution of roles to each robot.

Classification of roles, notion of formation
The chosen classification is in relation with a division of the field in zones. Each robot is assigned to a zone of activity, they are told in formation. Here is an example:
- The goalkeeper, number 1 on the figure 1, rest in the zone of goal of the team.

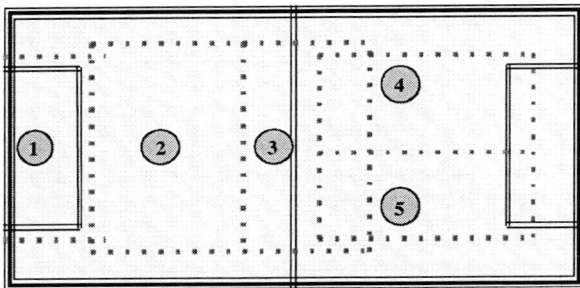

Fig. 1. The team formation

- The two defenders, numbers 2 and 3 on the figure 1, are in two distinct zones, sharing transversely half of field.

- The two attackers, numbers 4 and 5 on the figure 1, are also in two distinct zones, sharing lengthways half of field from adverse side.

Note that these zones may have a non empty intersection. Moreover, formations may be subdivided into sub formations. Each one includes then a subset of roles and interactions between these roles. Although that in a given time the robot performs only one role, it has a knowledge of all roles of the team. This allows the organisation to evolve according to two different approaches:

- The dynamic change of the role. For example, robots A and B may change their roles as well as their positions on the field if the robot A has to follow the ball in the zone of the robot B.
- The dynamic change of the formation, by taking into account the remaining time as well as the current result of the game. All the team may change its strategy leading to a more offensive or more defensive formation.

Each robot can choose between three action modes:
- It remains in its pre-defined position (the inactive state)
- It moves to the ball by tempting to shoot it to the adversary goal or, to pass it to its partner (the active state)
- It tries to intercept the ball that displaces to the goal (the auxiliary state).

These three modes of action define different behaviours according to the environment and the strategy under way of execution. Two types of behaviour can then be extracted:
- A purely opportunist behaviour, consisting in a direct answer to stimuli. This behaviour follows the reactive model.
- An intelligent behaviour computing a plan of actions in order to score at the adverse goal. This behaviour follows the deliberative model.

We present hereafter the B.O.C. model for the implementation of the control architecture of soccer playing robots.

3. BOC Architecture (Behaviour Oriented Commands)

3.1. The BOC Model

The BOC architecture is a hybrid architecture which combines reactive and deliberative reasoning by the distribution of the knowledge system into modules called behaviours. An important feature of this model is that it is easily and directly translated into a real time application.

Our architecture supplies independent entities capable of reacting directly to stimuli and also endowed with an inference system allowing deliberative planning. The knowledge of each module is encapsulated in a set of rules describing its desired behaviour. The reasoning module of each behaviour appeals to a production rules system (expert system of order 0+ or a finite state automaton) that allows to describe in a simple way complex behaviours.

Our design methodology is supported by two classical notions: abstraction and decomposition. Abstraction allows us to define a general commands without knowing its implementation while decomposition permits to map a complex set of rules into a group of less complex entities. This decomposition of the system allows the simultaneous representation of both the temporal evolution and the parallel relation of the treatments. The knowledge of the robot is represented on two levels:
- The encapsulation of a set of related rules into independent entities called behaviours.
- The establishment of association links between these entities, in order to carry out a behaviour–oriented–command (BOC) by the execution of co–operative actions.

A BOC is like a service that must be requested, executed and acquitted. It is carried out by a set of associated behaviours which can execute at the same time. Each behaviour groups a set of rules in order to achieve and maintain its own goal. It can be seen as an agent whose actions are either direct interactions with the sensors and actuators or requested services to other BOCs. The co–operative work of these Associated Behaviours (ABs) allows to solve problems beyond the scope of each one independently.

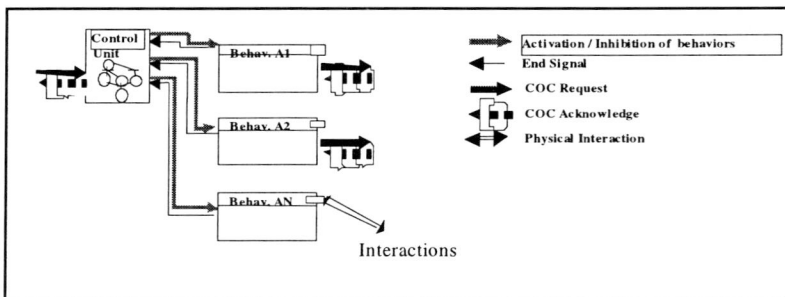

Fig. 2. Representation of a COC

A BOC is composed of a control unit and a set of associated connected behaviours which carry out a service respecting a specific set of co–ordination rules. The control unit associated to a BOC (fig. 2) includes an input port for the BOC's request, an output port for the BOC's acknowledgement and three control links between the BOC and its associated behaviours (AB): activation, inhibition and end signals.

These signals allow the co–ordination and synchronisation of the ABs. Activation signals are used to activate the necessary ABs for the correct execution of the requested BOC. End signals are sent from an AB to its BOC's control unit when it finds a BOC ending condition. Once the control unit receives the end signal, it sends inhibition signals to all the ABs which were activated. Finally, the control unit sends an acknowledgement to the behaviour that requested the BOC.

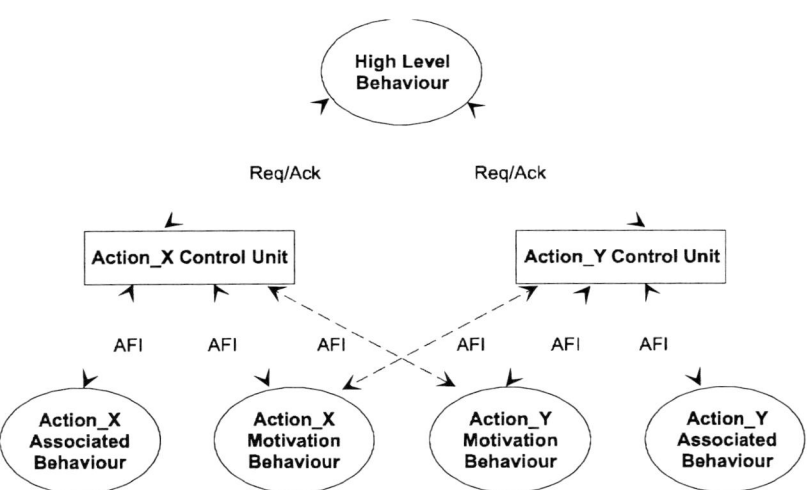

Fig. 3. Managing of concurrent actions using COC architecture

Figure 3 illustrates how the BOC architecture is used to manage two concurrent actions. Each action is modelled as a BOC which is carried out by a set of ABs. Generally, there is at least one behaviour that verifies that the conditions required to continue the execution of the BOC are valid. In a decision system, this behaviour can be seen as an entity monitoring the motivation of the action. Motivation allows to manage in a simple way the activation of concurrent actions (solid lines fig. 3).

When the motivation decreases the behaviour sends an end signal to the control unit to indicate that the action must be stopped. When the end signal arrives, the BOC is finished and acquitted. A behaviour of higher level receives this acknowledgement and uses its reasoning system to determine the next action to perform. The rules describing the behaviour and co-ordination of activities are distributed into modules, each one with a specific role.

In order to improve the performance of our decision system, the unit control of a BOC can also activate the motivation behaviour of a concurrent BOC (dotted lines in figure 3). This behaviour measures the relevance of starting the concurrent action, thus acting as an anti-motivation monitoring entity. A behaviour used as an anti-motivation sends an end signal when the motivation to perform its associated action is high enough. The trigger values are dynamically set by the high level behaviour and include an hysteresis to avoid undesirable oscillations when motivation is close to the transition value. Remark that since concurrent control units are never simultaneously active, a motivation behaviour will never be working at the same time for different control units.

Our robots use the BOC model. As a matter of fact, this model allows to easily separate into different units the robot's behaviours. It is then possible to construct the set of roles the robot can play by considering them as behaviours. For instance the attackers can be considered as an aggressive reactive behaviour and an aggressive deliberative behaviour. For instance, when the ball is found in position allowing for goal kicking, motivation of the aggressive reactive behaviour will be at its highest. On the contrary, when the ball interception is found to be impossible without obstacle avoidance the motivation of the deliberative behaviour will take over.

Currently BOCs are implemented as fully reentrant functions. Each BOC shares the source code in charge of the real time issues (communication, synchronisation, priority) allowing the programmer to concentrate in the behaviours themselves. In order to achieve even more transparency we are currently planning a future fully object oriented BOC version.

3.2. Reactive and Deliberative BOCs

The game tactics are implemented as a set of behaviours of different types.

Reflex Behaviours
This behaviours will accomplish very simple actions, as a direct response to a reduced set of pre-recorded stimuli. This allows fast "reflexive" actions. By definition, reflexive actions are simple, use only the most recently available information and have small, unambiguous rules.
Reflex behaviours have very high motivation values when their pre-conditions are met. Motivation however decreases rapidly as the expected condition is not found, allowing other more deliberative behaviours to be considered.

Deliberative Behaviours
These behaviours execute actions with a significant degree of complexity (plans). This includes a long set of movements which should finish in a better global position of each robot. No negotiation between the robots (like which one should go for an equidistant ball) has been implemented until now. Communication and synchronisation are performed through the environmental perception. Team play is mainly achieved through coherent motivation equations.

3.3. Strategy

The different roles and sub-roles are also a result of behaviour motivation evaluations.

Attack Behaviour
This is the aggressive robot's personality. Its goal is to find and keep track of kick solution.

Defence Behaviour

This is a more conservative behaviour. It looks for a backup or ball recovering position. This is achieved by trying to fill the gaps while remaining behind the ball and adverse robots.

Individual Strategy Behaviour

This is the highest level behaviour inside any single robot. It is charged of maintaining the robots global behaviour. It will therefore evaluate the motivations to play defensively or aggressively. This behaviour is activated at the beginning of the game with the parameters of initial robots position and playing attitude.

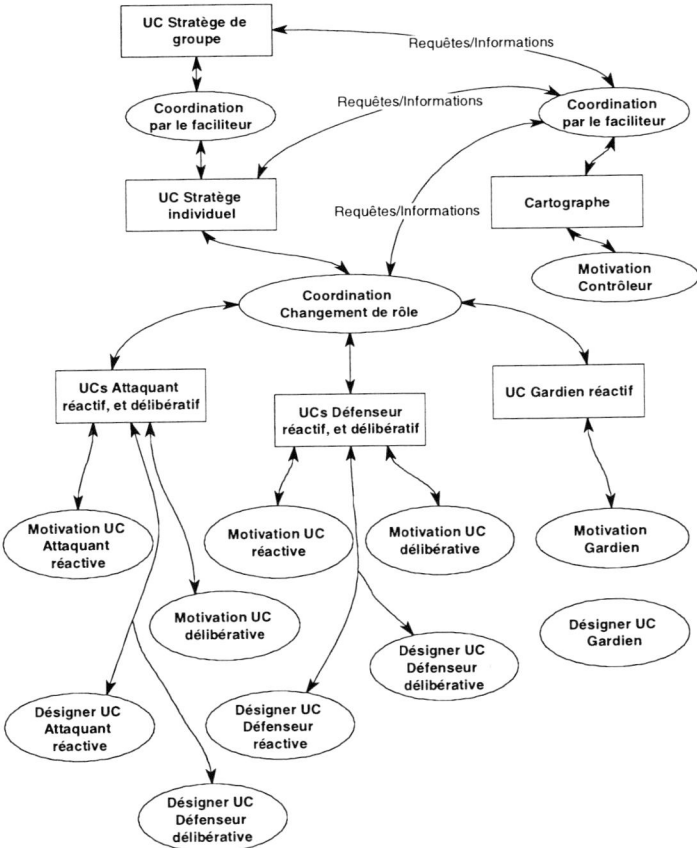

Fig. 4. C.O.C. model of each robot's role : Attacker – Defender – Keeper

Group Strategy Behaviour
This behaviour is in charge of collective play. It is the root behaviour. As such, it will start the individual strategy behaviour of all robots. It is also in charge of stopping any or all robots if necessary (dysfunction, goals, end of match...). The physical location of this behaviour can be a fixed post (used also for supervision) or any of the robots (actually the goalkeeper which has a much simpler behaviour hierarchy). Due to the network transparency of the real time kernel chosen (QNX), this can be done with no extra code. However, the physical position of this behaviour cannot be changed after starting. We intend to pass through this behaviour in order to improve the role swap of robots.

The Cartography Behaviour
Each robot saves the information gathered from different sources (camera, odometer, communications) and keeps track of a local model of the world. This allows to calculate object's future positions and to continue to evolve even when an object is not directly in sight.
A global cartography is also constructed, which takes into account information produced by several different robots. This model has to take into account local uncertainties of the different robots as well as temporal aspects. It is still in development and therefore unused for any strategic purpose. A blackboard approach is used which must fusion:
- The ball position
- The adverse goal position (Although the goal itself is a static object, we've created a dynamic goal which corresponds to the biggest goal acces and "moves away" from the opponent's goalkeeper.
- The team robots positions
- The opponents robots positions

The Blackboard control unit chooses a map building domain related with the strategic needs. Hence, if the ball's co-ordinates are necessary and a priority, they will be the first chosen domain. To illustrate the control unit's functioning let's consider the following instance :
The ball's co-ordinates belong to the designated domain, the control must choose the knowledge sources enabling it to determine them:
- The data coming from the vision modules directly from each robot;
- The data coming from the vision modules from each robot associated to make a triangulation ;
- The former ball's co-ordinates if none of the robots see the ball.

The control unit then determines the most precise source, which is to say the information coming from the closest robot to the ball or a triangulation constituted thanks to information issued from several robots. The triangulation may be chosen if the information coming from the robots is synchronous during a short time interval determined before hand according to the maximum estimated ball speed, for instance.

4. Conclusion

We have presented in this article an extension to the BOC model. This allows to endow each robot with a deliberative dimension oriented towards group strategy, while maintaining reactivity. One of our main objectives is to study and improve the emergent collective behaviours.
The notions of role and formation as well as a role classification are introduced. Defensive, offensive and individual strategy behaviours are oriented towards the role concept with a group strategy behaviour co-ordinating the whole. The entire architecture is implemented within the BOC paradigm. We currently work in an object oriented BOC implementation.

Acknowledgement

Vincent & Jean Moritz for their inconditional cooperation.

References

[BROOKS 1986] R. A. BROOKS, A Robust Layered Control System for a Mobile Robot, *IEEE Journal of Robotics and Automation*, Vol. 2, No. 1, pp 535–539, 1986.

[CHATILA 1985] R. CHATILA, J. LAUMOND, Position Referencing and consistent World Modeling for Mobile Robots, *IEEE ICRA*, 4 pp 138–145, 1985.

[CONNEL 1990] J.H. CONNEL. A Colony Architecture for an Artificial Creature, *MIT A.I. lab Technical* Report 1151, 1990.

[CUERVO 96] J. CUERVO, E. GONZALEZ, A. SUAREZ, C. MORENO, Behavior–Oriented Commands: From Distributed Kwoledge Representation to Real Time Implementation, *Proc.Euromicro Workshop on Real–Time Systems*, pp. 151–156, Jun. 1996.

[LOAIZA 98] H. LOAIZA, A. SUAREZ, E. GONZALEZ, S. LELANDAIS, C. MORENO, Real MagiCol: Complex Behaviour throught simpler Behaviour Oriented Commands, *Proceedings of the second RobotCup Workshop.* pp. 475–482, Paris 1998.

[MORAVEC 1989] H. MORAVEC. D. CHO., A Bayesian Method for Certainty Grids, *Proceedings AAAI Spring Symposium on Robot Navigation,* pp 57–60, 1989 .

[Realismo Màgico] http://artcon.rutgers.edu/artists/magicrealism/magic.html

[SIMMONS 1994] R SIMMONS, Structured Control for Autonomous Robots, *IEEE Trans. on Rob. and Automation*, Vol. 10, No. 1, 1994.

Modular Simulator: A Draft of New Simulator for RoboCup

NODA, Itsuki[1][2]
noda@etl.go.jp

[1] CSLI, Stanford University, Palo Alto CA 94306, USA
[2] Electrotechnical Laboratory, Tsukuba 305, Japan

Abstract. Soccer Server has been used as the official simulator for RoboCup Simulation League last three years. Based on this experience, I investigate the feature of Soccer Server and figure out the issue to building such kind of open simulator. Then, I propose a new design of simulator that will provide more flexible version up, easiness of maintenance, and wide application.

1 Introduction

Soccer Server has been used as the official simulator for RoboCup Simulation League last three years. The reasons why Soccer Server is chosen are open system, light weight, and widely supported platforms. These features enable many researchers to use it as a standard tool for their research. And now, we have a large community of simulation league, in which we discuss new rules, share ideas and information, and cooperate with each other to develop libraries and documents.

However, problems of Soccer Server become clear in recent years. Most of them are lied on design of Soccer Server itself. Originally, it was built just as a prototype of the simulator and modified again and again to add new features and to fix bugs. Therefore, the system became complicated and difficult to maintain.

So, it is the time we re-design a new system of simulator. In this paper, I investigate features and problems of current Soccer Server, and propose a new design of Soccer Server.

2 Issues on Soccer Server

2.1 Soccer Server

Soccer Server [5, 4] enables a soccer match to be played between two teams of player-programs (possibly implemented in different programming systems). The match is controlled using a form of client-server communication. The server (Soccer Server) provides a virtual soccer field and simulates the movements of players and a ball. A client (player program) can provide the 'brain' of a player by

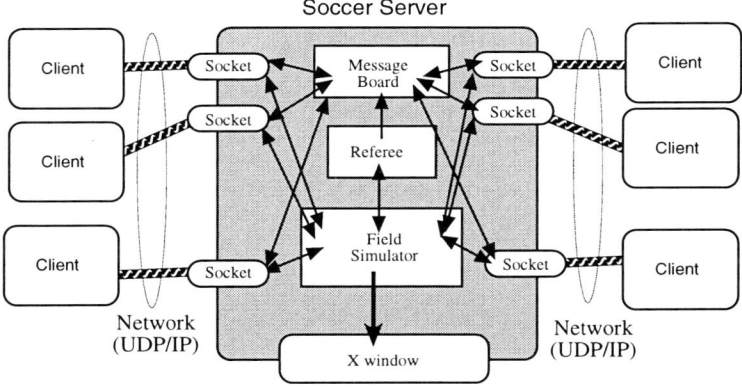

Fig. 1. Architecture of Current Soccer Server

connecting to the server via a computer network and specifying actions for that player to carry out. In return, the client receives information from the player's sensors.

A client controls only a player. It receives visual and verbal sensor information ('see' and 'hear' respectively) from the server and sends control commands ('turn', 'dash', 'kick' and 'say') to the server. Sensor information tells only partial situation of the field from the player's viewpoint, so that the player program should make decisions using these partial and incomplete information. Limited verbal communication is also available, by which the player can communicate with each other to decide team strategy.

2.2 Features of Soccer Server

Here, I like list up features of Soccer Server that are reason why it is used widely.

- Soccer Server is light. It can run entry-level PCs and requires small resources. This enables researchers to start their research from small environment. And also, in order to use it for educational purpose, it is necessary to run on PCs students can use in computer labs in schools.
- Soccer Server runs on various platforms. Finally, it supports SunOS 4, Solaris 2.x, Linux, IRIX, OSF/1, and Windows [3]. It also requires quite common tools and libraries like Gnu or ANSI C++ compiler, standard C++ libraries, and X window. They are distributed freely and used widely.

---

[3] Windows versions were contributed by Sebastien Doncker and Dominique Duhaut (compatible to version 2), and now by Mario Pac (compatible to version 4) independently. Information about Mario's versions is available from:
http://users.informatik.fh-hamburg.de/~pac_m/

- Soccer Server uses ascii string on UDP/IP for protocol between clients and the server. It enables researchers/students to use any kind of program language. Actually, participants in past RoboCup competitions used C, C++, Java, Lisp, Prolog and various research oriented AI programming systems like SOAR [8]. Version control of protocol is also an important feature. It enables us to use old clients to run in newer servers.
- The system has a separated module, soccermonitor, for displaying the field status on window systems. Simulation kernel, soccerserver, permits to connect additional monitors. While this mechanism was introduced only for displaying field window on multiple monitors, it leads unexpected activities in the different research field. Many researchers have made and have been trying to build 3D monitors to demonstrate scene of matches dynamically [6]. In addition to it, a couple of groups are building commentary systems that describe situations of matches in natural language dynamically [9, 1]. Both kinds of systems are connected with the server as secondary monitors, get information of state of matches, analyze the situations, and generate appropriate scenes and sentences.

2.3 Open Issues of Soccer Server

During past three yeas, Soccer Server was modified again and again in order to add new functions and to fix bugs. From these experience, it became to be clear that Soccer Server have the following open issues.

- huge communication:
  Soccer Server communicates various clients (player clients, monitor clients, offline-/online-coach clients) directly, so that the server often becomes a bottle-neck of network-traffic. In order to solve this problem, the server should be re-designed to enable distributed processing easily.
- maintenance problem:
  Though Soccer Server is maintained only at ETL, the source code became so complicated that it is difficult to figure out bugs and to maintain the code. The reason is that structure of classes of C++ program became not to reflect a hierarchy of required functions. Therefore, it is the time to re-design modules of the server according to required functions.
- version control:
  In order to keep upper compatibility as much as possible, Soccer Server uses version control of protocol between clients and the server. Because the current server is a single module, the server must include all version of protocols. In order to solve the problem, the server should have a mechanism that enable to connect with a kind of filter or proxy that convert internal representation and each version of the protocol.

In order to overcome these problems, I introduce a modular architecture into Soccer Server. In this architecture we divided Soccer Server into a couple of modules, which are loosely coupled via networks. These modules can run in a distributed way, so that we will be able to avoid bottle-neck problem of huge communication. Modularity also provides the way of distributed maintenance of the system. Also, it makes easy to version control by swapping modules to communicate player's clients for each version.

In the next section, I propose a new design of the simulator based on this idea.

3 A Design of New Simulator

3.1 Overview of the Design

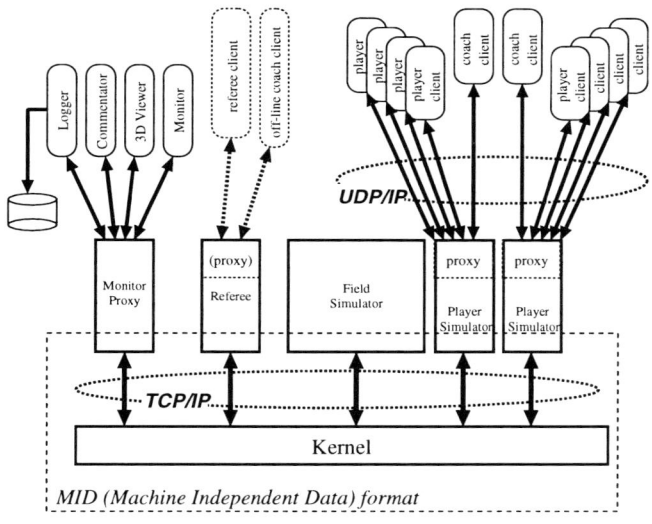

Fig. 2. Plan of Design of New Soccer Server

As mentioned in the previous section, I divided the functions of Soccer Server into the following modules:

- **Field Simulator** is a module to simulate the physical events on the field respectively.
- **Referee Module** is a privileged module to control a match according to rules. This module may override and modify the result of field simulator.
- **Player Simulators/Proxies** are modules to simulate events inside of player's body, and communicate with player and on-line coach clients.
- **Monitor Proxy** provides a facility of multiple monitor, commentator, and saving a log.

These modules are combined by a **kernel** (Fig. 2). The **kernel** manages shared data and synchronization among the modules. Each module communicates only with the kernel rather than with each other directly. In order to guarantee to run modules in various platforms, the system use **MID**, a platform-independent format, in the communication. [4]

The kernel keeps the primary data, and each module has its copy. When a module changes the data, it uploads the change to the kernel. Then other modules refer the changed data in the kernel.

The kernel treats all modules in a uniformed way. So, we can add additional modules for the system. Also, we can apply this kernel to different purposes like a rescue simulator.

3.2 Kernel

Kernel is a back born of whole system. The kernel will provide the following services to the modules:

- **Management of Shared Data**: All shared data are kept in the kernel as the primary data. Each module should have a copy of the primary data. When a module modify the data, the module must upload the data to the kernel. Other modules download the data when they use it. The kernel also provides *automatic* download mechanism. If a module is registered as a *watcher* of the data, the kernel *automatically* downloads the data when it is modified.
- **Control of Synchronization by Phase**: In order to synchronize executions of modules, the kernel provides *phase* facilities. The kernel begins a *phase* when a certain condition is satisfied. Then it notifies the beginning of the phase to all modules join the phase. Each module notifies the end of operation of the phase to the kernel. The kernel ends the phase when it receives the notification from all joined modules.

  The *phase* mechanism provides the way to give a priority to a certain operation of a certain modules. The kernel can begin a phase adjunctly before or after another phase. Therefore a user can put phases in order using the relation of adjunctness.

At the beginning of the simulation, each module connects to the kernel, and registers shared data and joining phases. Then the module start to receives services phase controls and automatic download of data. Also, the module can upload/download the data when it needs.

As mentioned above, the kernel treats all modules in a uniformed way. So there are no restriction on the number and the type of modules. This means that we can connect additional simulation units like an auditory simulator or global coaching modules. Also, we can swap the (2D-) field simulator to a 3D-simulator easily. (In this case, we may have to change other modules to adjust new data

---

[4] Currently **MID** is defined originally. However, I will move to use more common way like CORBA.

format of 3D.) The kernel is also applicable to other domains. For example, the kernel can be used in a rescue simulation. In this case, a couple of simulation modules like fire simulator, traffic simulator, and so on.

3.3 Player Simulator/Proxy

One of major problems of the current Soccer Server is management of protocol. In the Soccer Server, the protocol is implemented in various point of the whole system. Therefore, it is difficult to maintain and version-up the protocol.

On the other hand, in the new design, A player simulator/proxy receives whole information about data from the kernel, and convert it to the suitable protocol. As a result, maintainers may focus only to this module when we change the protocol.

This style brings another merit. The current system communicates with clients directly, so it the server tends to be a bottle neck of network traffic. On the other hand, this module works as a proxy that connects with multiple clients. Therefore, when we run two proxies for both teams on two machines placed in separated sub-networks, we can distribute the traffic. This also equalizes the condition for each team even if one team uses huge communication with the server.

3.4 Referee Module

The implementation of the referee module is the key of the simulator. Compared with other modules, the referee module should have a special position, because the referee module needs to affect to behaviors of other modules directly rather than data. For example, the referee module restricts movements of players and a ball, that are controlled by the field simulator module, according to the rule.

One solution is that the referee module only controls flags that specify the restrictions, and simulator modules runs according to the flags. The problem of this implementation is that it is difficult to maintain the referee module separately from other modules.

Another solution is that the referee module is invoked just before and after the simulator module and check the data. In other words, the referee module works as a 'wrapper' of other modules. The merit of this implementation is that it is easy to keep simulator modules independent from referee modules. Phase control described in Sec. 3.6 enables this style of implementation in a flexible manner.

3.5 Protocol Between Kernel and Modules

Protocol between the kernel and modules uses TCP/IP, not UDP/IP. The reason is that:

- Communication between the kernel and modules should be reliable. For example, if communication of the phase control is not reliable, the execution of

the whole system may fall into deadlock because of fault of communication. Because of the nature of UDP/IP, it is difficult to avoid such fault completely in UDP/IP during a match.
- The number of packets will be smaller than communication between Soccer Server and its clients, so that overhead of TCP/IP will be negligible.

Of course, the low-level protocol layer is designed to be independent from high-level layer. We can replace this level if more efficient protocol will be available in the future.

In order to guarantee independence of data format from machine architecture, I designed MID (Machine Independent Data) format. All data are reformed into network byte order (big-endian). In addition to it, we can use fixed decimal point format to transfer float values. It is useful because most of physical value has a limited domain of value and fixed decimal point format can reduce the size of data.

All conversion methods of data to/from MID format are defined shared header files of C++ in an object oriented manner. For example, 'position' and 'player' classes will be defined as Fig. 3.

Note that we can keep that protocol between the server and player clients is UDP/IP. The conversion of TCP/IP and UDP/IP is done by the player simulator/proxy modules.

3.6 Phase Control

The kernel controls synchronization of execution of modules by *phases*.

A *phase* is a kind of an event that have joined modules. When a phase starts, the kernel notifies the beginning of the phase by sending an **achievePhase** message to all joined modules. Then the kernel waits until all joined modules finish operations of the phase. Each module must inform the end of the operation of the phase by sending an **achievePhase** message to the kernel.

The kernel can handle two types of phases, *timer phase* and *adjunct phase*.

A *timer phase* has its own interval. The kernel try to start the phase for every interval. For example, a **field simulation phase** should occur every 100ms [5]. This phase has the field simulator as a joined module. So, the field simulator receives an **achievePhase** message for every 100ms. Then the simulator executes its operation and sends an **achievePhase** message back to the kernel.

An *adjunct phase* is invoked before or after another phase adjunctively. For example, a **referee phase** will be registered as an adjunct phase after a **field simulation phase**. Then the kernel starts the **referee phase** immediately after the **field simulation phase** is achieved. For another example, a **player phase**, in which player simulators/proxies upload players' commands, will be registered as an adjunct phase before a **field simulation phase**. In this case, the kernel starts the **player phase** first, and starts the **field simulation phase** after it is achieved.

---

[5] This interval is based on the interval of the original soccer server.

```
class FsPos {
  public:
    FsFloat x ;
    FsFloat y ;
       ...
    FsBool writeMID(FsBuffer buffer) {
      writeMID(buffer,x,10) ;
      writeMID(buffer,y,10) ;
    } ;
      // precision of x and y are 3 decimal
      // places under decimal points.
    FsBool readMID(FsBuffer buffer) {
      readMID(buffer,x,10) ;
      readMID(buffer,y,10) ;
    } ;
      ...
} ;
```
```
class FsPlayer {
  public:
    FsSide   side ;
    FsUInt   unum ;
    FsPos    pos ;
       ...
    FsBool writeMID(FsBuffer buffer) {
      writeMID(buffer,side) ;
      writeMID(buffer,unum) ;
      writeMID(buffer,pos) ;
      ...
    } ;
    FsBool readMID(FsBuffer buffer) {
      readMID(buffer,side) ;
      readMID(buffer,unum) ;
      readMID(buffer,pos) ;
      ...
    } ;
    ...
} ;
```

Fig. 3. Example of definition of data structures and their conversion to/from MID format

```
class FsPos {
  public:
    FsFloat x ;
    FsFloat y ;
        ...
    FsBool writeMID(FsBuffer buffer) {
      writeMID(buffer,x,10) ;
      writeMID(buffer,y,10) ;
    } ;
        // precision of x and y are 3 decimal
        // places under decimal points.
    FsBool readMID(FsBuffer buffer) {
      readMID(buffer,x,10) ;
      readMID(buffer,y,10) ;
    } ;
        ...
} ;
```
```
class FsPlayer {
  public:
    FsSide   side ;
    FsUInt   unum ;
    FsPos    pos ;
        ...

    FsBool writeMID(FsBuffer buffer) {
      writeMID(buffer,side) ;
      writeMID(buffer,unum) ;
      writeMID(buffer,pos) ;
        ...
    } ;
    FsBool readMID(FsBuffer buffer) {
      readMID(buffer,side) ;
      readMID(buffer,unum) ;
      readMID(buffer,pos) ;
        ...
    } ;
        ...
} ;
```

Fig. 3. Example of definition of data structures and their conversion to/from MID format

A phase may have two or more adjunct phases before/after it. To arrange them in an order explicitly, each adjunct phase has its own tightness factor. The factor is larger, the phase occurs more tightly adjoined to the mother phase. For example, a **field simulation phase** will have two adjunct phases, a **referee phase** and a **publish phase**, after it. Tightness factors of the referee and publish phases will be 100 and 50 respectively. So, the **referee phase** occurs just after the field simulation phase, and the **broadcast phase** occurs last.

Fig. 4 shows phase-control and communication between the kernel and modules in the soccer simulation. Note that the implementation of the phase control mechanism is general and flexible, so that there is no limitation on the number of phase, the duration of the interval of timer phase, or the depth of nest of adjunct phases.

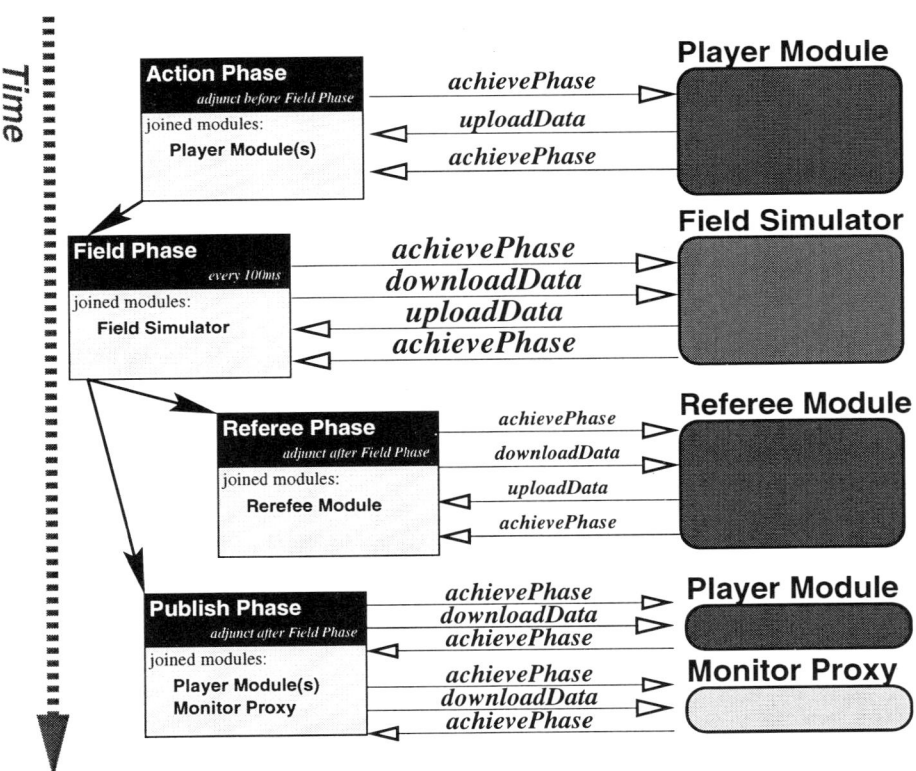

Fig. 4. Phase Control and Communication with Joined Modules

4 Related Work

4.1 Distributed Interactive Simulation

There is a series of development of distributed interactive simulation systems for military training and simulation [3, 2, 7, 11]. The most significant features of such kind of simulations is that the area of the field is very wide and objects are located relatively sparse. They use the similar architecture of the new simulator proposed here. Their main purpose is to connect simulators developed individually via network, and to enable integrated training of pilots simulate war in large scale in real-time. In addition to it, the system can connect with machine-intelligented pilots [8].

Major differences between the new simulator and these military simulators are:

- Existence of referee module is significant compared with other simulator. Referee module is tightly coupled with the field simulator, so the kernel is required to control these two modules in sequential manner. On the other hand, military simulators listed above suppose interaction between each simulation modules are localized, so that it is possible to build loosely coupled system.
- The new simulator should be light weight. One of important features is light weight and ability for researchers to run it with low cost of computer resources. In the military purpose, more realistic simulation is required even if it needs more expensive computational resources.
- The new simulator will be released under Gnu GPL. Open source policy is important in RoboCup community.

4.2 Hybrid Simulation

Hybrid simulation systems also have been investigated. [10] shows a core architecture to enable a hybrid simulation of embedded systems. Compared with the military simulations and the proposed systems, this hybrid simulation aims to simulate a system consists of more tightly coupled elements like inside of a circuit, rather than to simulate events happen in widely spread area like combat/soccer field. However, when we move to more accurate and multi-modal simulation (for example, rescue domain), we must take care the similar problem they attacked.

5 Conclusion

I investigated problems of current Soccer Server, and figure out issues that should be solved in the new simulator. Two main points are modularity and possibility of distributed simulation. Base on this investigation, I proposed a design of the new simulator. In which, simulation and communication are divided into deferent modules.

The proposed design is relatively general and is not restricted to simulation of Soccer. So, it is possible to use this design as a prototype of the kernel of other simulation of complex environment like rescue from huge disasters.

There still remain many open issues. For example, the following issues are still open:

- tradeoff between reality flexibility and computational power
- tradeoff between tight coupling and loose coupling
- generality of interaction between modules
- timing control over networks

References

1. E. Andrè, G. Herzog, and T Rist. Generating multimedia presentations for RoboCup soccer games. In H. Kitano, editor, *RoboCup-97: Robot Soccer World Cup I*, pages 200–215. Lecture Notes in Artificial Intelligence, Springer, 1998.
2. Judith S. Dahmann. High level architecture for simulation: An update. In Azzedine Bourkerche and Paul Reynolds, editors, *Distributed Interactive Simulation and Real-time Applications*, pages 32–40. IEEE Computer Society Technical Committee on Pattern Analysis and Machine Intelligence, IEEE Computer Society, July 1998.
3. Xin Li, Kien A. Hua, and J. Michael Moshell. Distributed database designs and computation strategies for network interactive simulations. *Journal of Parallel and Distributed Computing*, 25:72–90, 1995.
4. Itsuki Noda and Ian Frank. Investigating the complex with virtual soccer. In Jean-Claude Heudin, editor, *VW'98 Virtual Worlds (Proc. of First International Conference)*, pages 241–253. Springer, July 1998.
5. Itsuki Noda, Hitoshi Matsubara, Kazuo Hiraki, and Ian Frank. Soccer server: A tool for research on multiagent systems. *Applied Artificial Intelligence*, 12(2–3):233–250, 1998.
6. A. Shinjoh and S. Yoshida. The intelligent three-dimensional viewer system for robocup. In *Proceedings of the Second International Workshop on RoboCup*, pages 37–46, July 1998.
7. Stow 97 - documentation and software. WWW home page http://web1.stricom.army.mil/STRICOM/DRSTRICOM/T3FG/SOFTWARE_LIBRARY/ST% 0W97.html.
8. Milind Tambe, W. Lewis Johnson, Randolph M. Jones, Frank Koss, John E. Laird, Paul S. Rosenbloom, and Karl Schwamb. Intelligent agents for interactive simulation environments. *AI Magazine*, 16(1), Spring 1995.
9. Kumiko TANAKA-Ishii, Itsuki NODA, Ian FRANK, Hideyuki NAKASHIMA, Koiti HASIDA, and Hitoshi MATSUBARA. MIKE: An automatic commentary system for soccer. In Yves Demazeau, editor, *Proc. of Third International Conference on Multi-Agent Systems*, pages 285–292, July 1998.
10. Werner van Almsick, Thorsten Drabe, Wilfried Daehn, and Christian Müller Schloer. A central control engine for an open and hybrid simulation environment. In Azzedine Bourkerche and Paul Reynolds, editors, *Proc. of Distributed Interactive Simulation and Real-time Applications*, pages 15–22. IEEE Computer Society, July 1998.
11. Warfighters' simulation (warsim) directorate national simulation center. WWW home page http://www-leav.army.mil/nsc/warsim/index.htm.

Programming Real Time Distributed Multiple Robotic Systems

Maurizio Piaggio   Antonio Sgorbissa   Renato Zaccaria

D.I.S.T.
Department of Computer Communication and System Sciences
University of Genoa, Via Opera Pia 13, I-16145, ITALY

piaggio@dist.unige.it

Abstract. This paper presents ETHNOS-IV - a real-time programming environment for the design of a system composed of different robots, devices and external supervising or control stations. ETHNOS is being used for different service robotics applications and it is has also been used successfully used in RoboCup in the Italian ART robot team during the Stockholm 99 competition. It provides support from three main point of views which will be addressed in detail: inter-robot and intra-robot communication, real-time task scheduling, and software engineering, platform independence and code-reuse. Experimental results will also be presented.

1   Introduction

Among the positive aspects of the RoboCup [5] we believe that the emphases on multi-robot coordination and integration is particularly important. In fact most of the practical applications of mobile robotics for service tasks in civilian environments consist in systems composed of multiple robots communicating with each other, with external sensing and actuating devices, and with external supervising workstations [2]. In our opinion the research in robot architectures, currently mostly focusing on single robots and on the difficulties in the integration of different paradigms of representation (symbolic, diagrammatic and procedural) and of different types of robot-environment interactions (reactive and deliberative) [1] [3], should also address this scenario and the problems it involves. For example in the RoboCup a robot soccer player must allow a successful intra-robot integration of the different activities (visual perception, path planning, strategy planning, motion control, etc.) spanning over many different types of representation (raw sensor data, images, symbolic plans, etc.), but also guarantee a successful inter-robot integration by supporting communication and co-operation. The robot and system architecture should also allow for different levels of autonomy: the single robot, more robots in cooperation among themselves or also with external devices or supervising stations.

This paper focuses on these problems, presenting ETHNOS[1]-IV - a real-time programming environment for the design of a system composed of different robots, devices and external supervising or control stations - which has been used for different applications and which has also been successfully used in the Italian ART robot team during the Stockholm 99 competition. ETHNOS provides support from three main points of view which will be addressed in detail:

- from the communication perspective it supports and optimises transparent inter-robot information exchange across different media (cable or wireless).
- from the runtime perspective it provides support for the real-time execution of periodic and aperiodic tasks, schedulability analysis, event handling, and resource allocation and synchronisation.
- from the software engineering perspective it provides support for rapid development, platform independence and software integration and re-use.

2. ETHNOS IV

ETHNOS IV (Expert Tribe in a Hybrid Network Operating System) is the latest result (the fourth version), in a project that began more than four years ago, for the study and development of a programming environment for autonomous robotic systems. It has been designed in order to support a specific hybrid cognitive model (HEIR [8]) but, since it is sufficiently general, it has also been used in other cognitive or architectural organisations. It is composed of:

- ETHNOS IV, a dedicated distributed real-time operating system, from which the overall environment takes its name, supporting different representation, communication, and execution requirements,
- a dedicated network communication protocol designed for both the single robot and the multi-robot environment, specifically taking noisy wireless communication into account,
- an object oriented Application Programming Interface (API) based on the C++ language (and a subset based on Java),
- a set of additional development tools (a robot simulator, a Java-applet template, etc.)

The reference architecture of a single robot, and consequently of the ETHNOS operating system, is entirely based on the concept of *expert*, a concurrent agent

---

[1]In ETHNOS the expression "expert tribe" signifies an analogy between a robotic system and a tribe composed of many experts (concurrent agents) in different fields (perception, actuation, planning, etc.). The term "hybrid" indicates the support to the integration of deliberative experts with experts responsible for reactive and partially reactive behaviours; the term "network" refers to the possibility of distributing components (groups of experts) on a computer network.

responsible for a specific deliberative or reactive behaviour. Experts are members of a tribe which are distributed in separate villages (network computers) normally depending on their computational task. An example related to the HEIR cognitive model is depicted in figure 1. Three non-hierarchically organised components can be noticed, each characterised by the type of knowledge it deals with: a *symbolic* component (S), handling a declarative explicit propositional formalism, a *diagrammatic* component (D), dealing with analogical, iconic representations, and a *reactive* behaviour based component (R). The experts (depicted as circles) are also classified depending on the component they belong to: *symbolic(S)*, *diagrammatic(D)* or *reactive(R)*. Each group corresponds to a different set of computational tasks, distinguished depending on: the type of cognitive activity carried out, timing constraints, type of data managed, duration, etc.

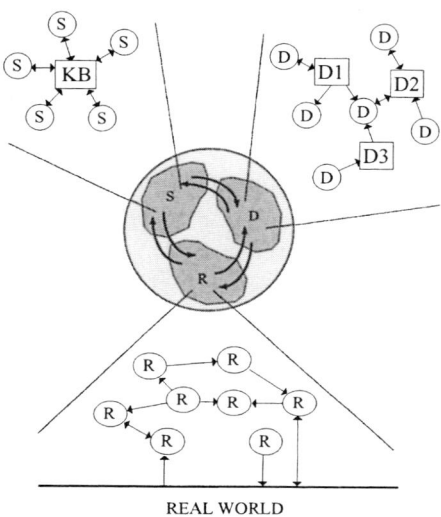

Fig. 1. HEIR Cognitive Architecture

Within this general framework, ETHNOS supports (throughout its protocols or APIs) or performs (throughout its kernel calls) different functions:

- it supports the creation of, and access to, shared representations (in the figure the KB and the different iconic instances D1, D2, etc.).
- It allows the concurrent execution and real time scheduling of different experts transparently and efficiently implemented as real-time threads, and their transparent synchronisation when accessing the above shared resources.
- It supports both asynchronous and synchronous intra-robot, inter-expert communication.
- It supports transparent asynchronous inter-robot communication.

− It allows experts to suspend, waiting for the occurrence of specific events (reception of particular message types or even combinations of them) to avoid wasting computational time. Moreover it allows each expert to specify a desired period of execution which the Posix real time scheduler will transparently be instructed to respect.

The next sections will deal with these properties in greater detail.

3. Communication

One of goals of the RoboCup competition is to encourage the comparison and exchange of methodologies, techniques and algorithms within the robotics and artificial intelligence community. A common programming environment which, without imposing significant constraints on the single components, allows to easily put together the result of different researchers within the same group (or possibly also across different groups) is certainly a contribution in this direction. In harmony with this goal, the ETHNOS programming environment allows the robots to be programmed in such a way that the different experts can be integrated, during development but also at run time, with little or no intervention in the software of the expert itself, thus facilitating both rapid prototyping and dynamic architectural reconfiguration. The first property facilitates the development of a robot application (from a set of components or behaviours) even by non highly specialised programmers (for industrial purposes but also for didactic activity in student projects, particularly relevant in RoboCup); the second property allows the system to easily scale-up to a robot capable of different complex activities and thus able to switch at run-time from a configuration in which it is performing a certain task (for example in which the active behaviours are responsible of path planning, navigation and obstacle avoidance) to a totally different configuration (for example in which the behaviours are responsible of object grasping and manipulation).

These properties are achieved by exploiting a suitable inter-expert communication protocol which comprises of three different communication methods:

− asynchronous message-based communication,
− synchronous access to an expert based on the type of service,
− access to a shared representation.

The first method is the most general of the three and it is at the base of ETHNOS applications, in particular in the communication between experts of different type (i.e. handling different types of representations). It is a message-based communication protocol (the EIEP − Expert Information Exchange Protocol [7]) fully integrated with the expert scheduler. The EIEP encourages the system developer to de-couple the different experts in execution, to reach, as close as possible, the limit situation in which the single expert is not aware of what and how many other experts are running.

In this way an expert can be added, removed or modified at run-time without altering the other components of the system.

Expert de-coupling is achieved by eliminating any static communication link. The EIEP is essentially an efficient implementation of a blackboard in a network distributed environment. In fact, the EIE protocol is mainly based on an asynchronous message-based method in which the single expert *subscribes* to the particular message types, known a priori. When another expert *publishes* any message of the subscribed types, the subscriber receives it transparently. In this way, an expert does not know, explicitly, who the sender of a message is or to which other experts, if any, its message is being distributed. Moreover, the same principles apply also if the application is distributed in a computer network: messages are distributed and received regardless of the particular machine on which they were produced or on which an expert subscribed [9]. Thus, from this point of view, the same application is programmed in the same way, whether it will run with all the experts executing on the same computer (robot soccer player) or with the experts executing on different computers connected in a network (soccer player and supervising station or remote high-level component responsible of non time-critical computationally intensive tasks). Figure 2 illustrates this concept.

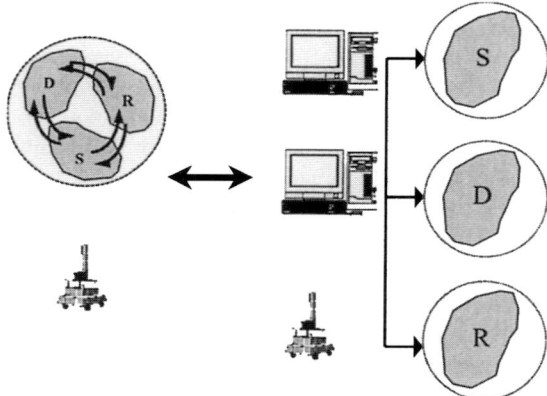

Fig. 2. Equivalence between centralised and decentralised configuration

The second method provides synchronous access in a way that is similar (from the programming point of view) to function calling. Analogously to the asynchronous communication, is based on two primitives:

– an expert declares the external services it can carry out: examples might be *get-robot-position*, *increase-robot-speed*,
– an expert requests a particular synchronous service to the operating system.

In the first step the operating system keeps the information on the types of services available, defined *a priori*, and on the experts which can carry them out. In the second step the operating system searches the previously stored information and provides the expert with the connection to the service required. It is important to notice that this connection will be maintained until the expert explicitly declares that it needs the service no further. Thus, during the normal expert execution the service access will be practically equivalent, in terms of efficiency and behaviour, to a standard function call. The benefits deriving from this approach clearly emerge only when experts are dynamically removed or modified. In this case, it is possible that a service currently in use by some other expert will no longer be available. The operating system, however, detects the critical situation and, searching again in its database, re-connects the related expert(s) to another one providing the same type of service and currently in execution.

The third and last method of communication allow the experts to share information stored in a common area of memory. It is very important when two or more experts need to work on the same, analogical or symbolic, data of relatively large dimensions. In this situation both previous methods are not applicable to real systems: the former, based on messages, is highly inefficient because of the implicit cost due to dealing with messages of inevitably large size; the latter implies the ownership of the memory by a single expert which would therefore share with it its lifetime.

ETHNOS also provides transparent inter-robot communication support. In fact, using the same EIEP principle, messages can be exchanged on the basis of their contents not only within the same (potentially network distributed) robot (soccer player) but also in-between different robots (players within the same team). However, in the single robot case, experts running on a particular machine do not have to be aware of the location of the message receivers whether on the same machine or on another one connected in the network; in inter-robot communication the situation is different and it is often important to know exactly who the sender is. Moreover it is also important to distinguish internal messages (meaning messages to be distributed within the machines implementing a single player) and external messages (meaning messages to be sent from a player to another) to avoid the explosion of unnecessary network data communication.

In ETHNOS the different experts are allowed to subscribe to *communication clubs*. For example, we may envisage a single club to which the different players belong or even different clubs, one for the reactive components, one for the diagrammatic components, etc. Message subscription and publication can thus be distinguished in internal, internal and in a specific club or internal and external in all clubs. Again, it is the responsibility of the system to dynamically transparently distribute the messages to the appropriate receivers. Figure 3 shows some example configurations that we have tested. In particular we are allowing the robots to communicate in a single club (to which all of them have subscribed) and with an external supervisor (the *coach*) which monitors the activity of all the robots.

Moreover since in the Robocup (and in general in mobile robotics) network communication is often wireless (i.e. radio link, Wavelan©, etc.), because of interference or because the robot may have moved to a blind zone, transmission packets are frequently lost. In this context, both TCP-IP and UDP-IP based communication cannot be used: the former because it is intrinsically not efficient in a noisy environment; the latter because it does not guarantee the arrival of a message, nor any information on whether it has arrived or not. For this reason we have also designed a protocol for this type of applications, called EEUDP (Ethnos Extended UDP) because, based on the UDP, it extends it with the necessary properties.

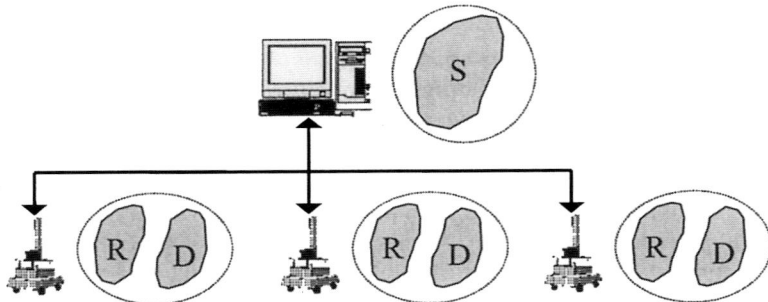

Fig. 3. Example of multi-robot ETHNOS configuration.

The EEUDP allows the transmission of messages with different priorities. The minimum priority corresponds to the basic UDP (there is no guarantee on the message arrival) and should be used for data of little importance or data that is frequently updated (for example the robot position in the environment that is periodically published). The maximum property is similar to TCP because the message is sent until its reception is acknowledged. However, it differs because it does not guarantee that the order of arrival of the different messages is identical to the order in which they have been sent (irrelevant in ETHNOS applications because every message is independent of the others), which is the cause of the major TCP overhead. Different in-between priorities allow the re-transmission of a message until its reception is acknowledged for different time periods (i.e. 5 ms, 10 ms, etc.).

4. Real-Time Expert Scheduling

An overall architecture for the development of RoboCup players should, in our view, permit the integration of reactive planning and control activities (whose execution is critical for the system and which therefore have real time requirements) with deliberative ones (whose execution is not critical for the good functioning of the system but which can increase performance in many situations).

In ETHNOS IV this is achieved in the following way. The different experts can be scheduled following different policies: real time Experts can be mapped into different Posix Threads (on one extreme) and scheduled as stated by the Rate Monotonic algorithm[6] with a Deferrable Periodic Server[11] (we chose this algorithms because of their ease of implementation with every Posix compliant OS) or executed into one single thread (on the other extreme) and scheduled following the non pre-emptive EDF algorithm [4]. In both cases the deliberative activities are performed in a lower priority thread which executes in the background.

When ETHNOS is given a particular set of experts and it is asked to schedule them, its kernel acts as follows:

- Each expert (both periodic and sporadic) is executed many times for an approximate analysis of its worst execution time.
- For each expert (both periodic and sporadic) the scheduling conditions (the Rate Monotonic conditions or the EDF algorithm ones depending on the scheduling policy we choose) are tested referring to its period (or the minimum time between two consequent runs) and its worst execution time.
- If the scheduling is possible, expert are scheduled as stated by the Rate Monotonic or the EDF algorithm.
- Should an expert miss its deadline (this is possible because of the approximate execution time analysis) a very simple recovery policy is adopted: the next deadline of the expert is obtained by adding the expert's period to the end of the last run (not to the end of the previous period). This is equivalent to changing the phase of the expert, saving the processor from computational overload and from the consequent deadline-missing by other experts.
- The background expert is scheduled when no other expert is ready for execution. Should we need a temporary intensive activity of the background expert (for example if the robot needs to perform planning activities because of a particular situation) it is a matter of the specific application to suspend or reduce real time tasks activities to leave more computational time to the background ones.

5. The Programming Interface

ETHNOS-IV has been designed to facilitate the development of complex distributed real-time applications even by non highly specialised users. This is particularly true in the RoboCup context, in which the exchange of research results between attending teams assumes a relevant interest. Moreover it should be outlined that within the ART team, in which different research groups (belonging to different Universities or Research Institutes) are involved in the same project by conducting separate research activity, these properties have been extensively exploited. In fact, different research units involved have adopted ETHNOS IV as their reference underlying architecture (for the soccer player for which they are responsible) even though they adopt different scientific solutions.

As a programming environment for the development of robotics applications, ETHNOS provides an object oriented programming interface (EPI-Ethnos Programming Interface) for the software implementation of the different components of the architecture. The EPI consists of a library of abstract C++ classes related to all the elements of the architecture, encapsulating their properties and their common behaviours. For example, the abstract class *ETExpert* is the base class for defining the behaviour of a generic expert; *ETPeriodicExpert*, *ETAperiodicExpert* and *ETBackgroundExpert* classes inherit its properties and pose further constraints on the expert timing characteristics. The user that needs to build a new expert must simply inherit the appropriate expert class.

6. Experimental Results

ETHNOS has been tested and used in service applications in hospitals and museums on different robots (TRC Labmate, GenovaRobot Snoopy robot) equipped with different sensor devices: ultrasonic proximity sensors, bumpers, laser positioning system, camera, etc. Within the RoboCup context, ETHNOS has been adopted by our laboratory in the Università di Genova, by the Università di Padova, by the Università di Parma and by the Politecnico di Milano. The first three have used as a mobile base an ActiveMedia Pioneer vision guided robot especially modified with additional devices (kicker, compass). For this purpose we have redesigned the Pioneer control libraries to directly communicate via serial interface to the robot (thus bypassing the Saphira architecture and allowing for an accurate, real time control of the actuators).

Fig. 4. Left: Mo2ro Robot developed by Politecnico di Milano, Middle: TinoZoff developed by the Università di Parma, Right: ActiveMedia® Pioneer used by Università di Genova and Università di Padova

In the Politecnico di Milano it has been implemented on a robot platform they developed (Mo2ro, additional information can be found at the internet web address: http://airlab.elet.polimi.it/projects/robocup.htm) which is equipped with kickers and an omnidirectional vision sensor. Both RoboCup robots are depicted in figure 4.
It is worth mentioning that the solutions of the different groups that adopted ETHNOS differed not only in their robots but also in the control architecture: subsumption based control, fuzzy logic control, artificial potential field based control, evolutionary learning. These diversities emphasise both the flexibility of ETHNOS and its fundamental role in the co-ordination of heterogeneous robots.

During the Stockolm99 competition the ETHNOS architecture will be fully exploited, in particular for the real time selection and execution of robot behaviours. The ETHNOS based architecture we are using in Genova is illustrated in figure 5. The lower level real time experts and middle level soft real time experts have the following functions:

– communicate with sensors and actuators

- update two diagrammatic representations: VIEW (a snapshot of what the robot currently sees suitably processed to clearly distinguish the objects in the scene) and LPS (a local perceptual space describing the situation of the field known to the robot due to its perceptions or the information coming from the other team mate robots)
- perform real-time navigation.

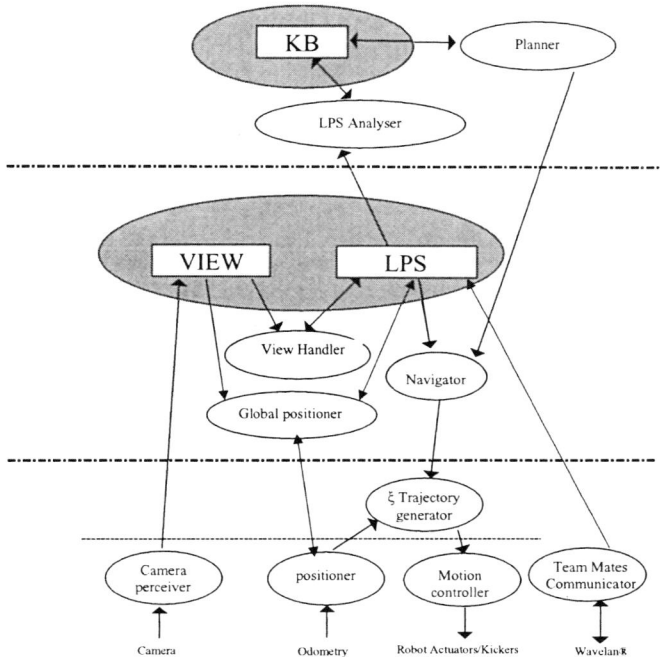

Fig. 5. ETHNOS based RoboCup Architecture

The latter in particular is responsible for ball following, ball pushing and obstacle avoidance, exploiting a non linear trajectory generator called ξ. The upper background experts analyse the information on the LPS to extract symbolic information on the match and perform high level action planning which tunes the behaviour of the real time navigator. Figure 6 depicts a photo taken during the experimenting of the architecture in which the robot has to reach the ball while avoiding the black obstacles on the floor.

The communication properties of ETHNOS have also been exploited to develop a graphical monitor for displaying and debugging the team activity during a match. In fact the monitor can receive some of the information that is being exchanged between the robots simply by subscribing to the team club and to the desired message types. In figure 7 some snapshots of the monitor during a real match have are illustrated. The small dark circle represents the ball, the larger light circles represent the opponent team robots and the other black objects are the team mate robots for which the recent trajectory is also depicted. It can be easily noticed that there are many overlapping opponent robots. This is because a circle is draw relatively to the opponent perception for each robot which slightly differs because of positioning and perceptive errors.

Fig. 6. ART Pioneer robot follows the ball while avoiding black obstacles

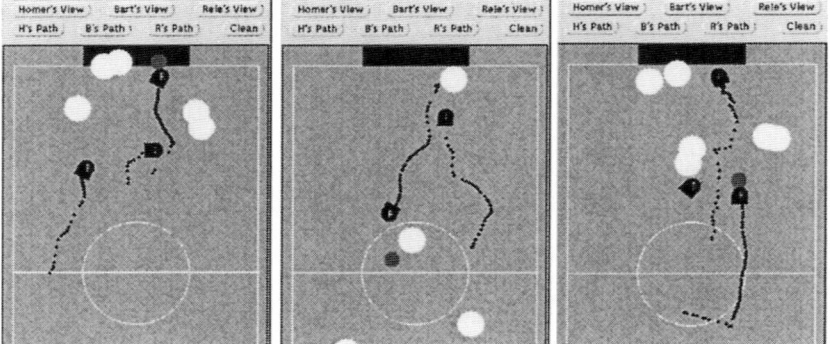

Fig. 7. Robot and ball positions perceived during a match

On the left figure the black team is scoring by pushing the ball directly into the goal. In the middle the opponent team is counterattacking and therefore the robot on the left is rapidly returning in defence. On the right figure the robot is preparing to kick the ball towards the goal while a team mate is positioning as a back up. Notice that in many cases there are sharp changes in the robot trajectory. This is either due to a change in the ball position or it is a movement to attempt to avoid an opponent (no longer depicted because out of the perceptive field of the robot).

7. Conclusions

This paper presented ETHNOS, a programming environment for real time control of distributed multiple robotic systems. Its properties have been described as well as its influences on the development of a soccer team for RoboCup. This year's competition in Stockholm has shown the importance of ETHNOS in the co-ordination of an heterogeneous team of robots independently developed. In particular it is work mentioning the ART managed to reach the finals and show effective and efficient coordination despite the fact that the robots played together less than a handful of times before the competition.

It is also worth mentioning that ETHNOS is publicly and freely available for non-commercial use to academic and research institutions. An on line HTML user guide can be found at the following address: http:// www.ethnos.laboratorium.dist.unige.it.

Acknowledgments

This work has been partially funded by the Padova Ricerche Consortium in the ART Team RoboCup project. The authors also wish to thank Prof. Bonarini of the Politecnico di Milano for the Mo2ro robot photograph and Prof. Adorni and Ing. Cagnoni of the University of Parma for the TinoZoff photograph. Many thanks also to all the ART participants that greatly contributed to the experimenting of ETHNOS in RoboCup.

References

1. R. C. Arkin, Integrating Behavioral, Perceptual and World Knowledge in reactive Navigation, Int. J. Robotics and Autonomous Syst., vol. 6, no. 1-2, pp. 105-122, 1990.
2. Special Issue on Robotics and Communication, Advanced Robotic Journal, Vol. 12, No. 8, 1998.
3. B. Hayes-Roth et al., "A domain-specific software architecture for adaptive intelligent systems", IEEE Transactions on Software Engineering, Vol. 21, No. 4., 1995.
4. K. Jeffay, D. F. Stanat, and Charles U. Martel, On Non-preemptive Scheduling of Periodic and Sporadic Tasks, *IEEE Real time Systems Symposium*, December 1991.
5. H. Kitano, M. Asada and Y. Kuniyoshi, RoboCup: A Challenge Problem for AI, AI Magazine, Vol. 18, Num. 1997.
6. C. L. Liu and J. W. Layland, Scheduling Algorithms for multiprogramming in a hard real time environment, JACM, vol. 20, no. 1, pp. 46-61, 1973.
7. M. Piaggio and R. Zaccaria, An Information Exchange Protocol in a Multi-Layer Distributed Architecture, IEEE Proc. Hawaii Int. Conf. on Complex Systems, 1997
8. M. Piaggio, HEIR - A Non-Hierarchical Hybrid Architecture for Intelligent Robots, Proc. of ATAL: workshop on agent theories architectures and languages, Paris, 1998.
9. M. Piaggio and R. Zaccaria, Distributing a Robotic System on a Network : the ETHNOS Approach, Advanced Robotic Journal, Vol. 12, No. 8, 1998.
10. A. Scalzo, A distributed multi-robot localisation algorithm, DIST Technical report, University of Genova, 1998.
11. J. K. Strosnider, J. P. Lehoczky, and L. Sha, The Deferrable Server Algorithm for Enhanced Aperiodic Responsiveness in Hard Real time Environments, *IEEE Trans. On Computers*, vol. 44, no. 1, pp. 73-91, 1995

The Attempto RoboCup Robot Team

Michael Plagge, Richard Günther, Jörn Ihlenburg, Dirk Jung, and Andreas Zell

W.-Schickard-Institute for Computer Science, Dept. of Computer Architecture
Köstlinstr. 6, D-72074 Tübingen, Germany
{plagge,guenther,ihlenburg,jung,zell}@informatik.uni-tuebingen.de
http://www-ra.informatik.uni-tuebingen.de/forschung/robocup.html

Abstract This paper describes the hardware and software architecture of the Attempto RoboCup-99 team. We first present the design of our heavily modified commercial robotic base, the robot sensors and onboard computer. Then the robot control architecture which realizes a hybrid control, consisting of a reactive behavior based component and a planner component for more complex tasks is introduced. Also the problems we currently are working on are presented, as there are a fast and reliable self localization algorithm and a robust behavior based reactive component for the hybrid control system.

1 Introduction

For building a good team of agents that can take part in the RoboCup-99 contest, ideas and results from different fields of research, e.g. artificial intelligence, robotics, image processing, engineering, multi agent systems can or even must be used and tested [1],[2].

Since we are developing a team for the mid-size contest our main objective is to build a robot system, which is able to recognize the environment in a suitable way and to build a fast and reliable control system, which is capable of solving the given task of playing football. This control system must cope with the dynamics and adverse aspects of RoboCup-99 and with complex situations, e.g. teamwork which occurs in RoboCup-99.

The remainder of the paper is structured as follows: Section 2 describes the robot, sensor and computer hardware of the Attempto team robots. Section 3 gives an overview about the three different layers of our software architecture. Section 4 focuses on our concepts to address the problems of doing a reliable and fast self localization and to develop a fast and robust reactive control component.

2 Hardware

2.1 Robot platform

As the basic robot platform for the field player we are using the Pioneer2 DX from ActivMedia Inc. (Fig. 1). This robot is equipped with a differential drive system with a free running caster wheel mounted at the back of the robot. The maximum achievable translation speed is about 1,5 m/s, the maximum rotational speed is 2 π/s. The robot can carry weights up to 20 kg and can be equipped with a maximum of three 7,2 Ah batteries, which allows an operating time with all additional hardware like PC and sensors of nearly three hours without recharging.

The two driving motors are equipped with 500 tick position encoders. With these encoders the speed and the position of the robot can be obtained. The robot is controlled by a Siemens C166 microcontroller. This device is responsible for controlling the actuators of the robot and for the calculation of the position and orientation from the motor encoder data.

Via a serial device the controller can communicate with a remote computer. This device can operate at a maximum speed of 38400 bauds. The robot sends 20 times a second a status data packet to the remote computer. It also accepts commands from the remote computer with the same rate. Therefore the minimal achievable response time for a closed loop controller is about 50ms.

As the basic platform for the goalkeeper we are using a Pioneer AT. Each of the four wheels of this robot is driven by its own motor. The wheels on each side are coupled with a belt. The battery with a capacity of 12Ah allows an operating time with our additional hardware of 1.5 hours. A custom designed board with a MC68332 CPU replaces the standard MC68HC11 board and gives faster response time, higher precision of odometry and more flexible sonar firing patterns. Despite serious problems in the preliminary rounds the goalkeeper was influential for our 1998 success at Paris reaching the final.

2.2 Sensors and actuators

As we are convinced that better sensors will result in a better situation assessment and, ultimately, in better playing capabilities, we try to employ a diversity of sensors on the robot. While the final design is not finished at the time of this writing, we are considering the use of the following sensors: Sonars, 2d laser scanner, IR sensors, colour camera, 360^0 camera, digital compass.

Sonars: The Pioneer2 DX is equipped with eight, the Pioneer AT with seven Polaroid 6500 Ultrasonic transducers, which are mounted in front and at the front side of the robot.

Laser scanner: The employed laser scanner is a LMS200 from SICK AG. It has a 180^0 field of view and a angular resolution of $0,25^0$. It can measure distances up to 15 m with an accuracy of 10 mm. With a resolution of 1^0 and a total field of view of 180^0 and 500 kbps data transfer rate over a RS422 serial device the achievable scan rate is nearly 60 Hz. This sensor, which is a successor to the device which secured Freiburg's [3] advantage last year, is currently the fastest and most precise distance measurement device. Its main drawbacks are its size (137*156*185 mm), weight (4,5 kg) and power comsumption (max. 17,5 W).

Color camera: For the task of object detection and classification we are using two vision systems. Both systems use a Siemens SICOLOR C810 CCD-DSP color camera, with a 1/3 inch CCD-chip and a resolution of 752x582 pixel. The output format is a regular CCIR-PAL signal with 625 rows and 50 half frames per second. One of the cameras is mounted at the front of the robot. This camera is equipped with a 2,8f wide angle lens. It is mainly for the detection of the ball and the objects, which lie in front of the moving robot. This camera is also responsible for distinguishing team mates from opponent robots.

360^0 camera: The second camera is mounted in an omnidirectional vision system, which is mounted at the top of the robot (Fig. 1). A 4,2f lens is mounted at this camera, to achieve a large visual field. The design of this camera has been made by Matthias Franz from the MPI for Biological Cybernetics from an earlier MPI design used for biologically inspired vision experiments. In contrast to most other omnidirectional vision systems this design has a paraboloid mirror instead of a conical mirror. This should give a better mapping of objects below the horizon.

Digital compass: This device is capable to determine the absolute orientation

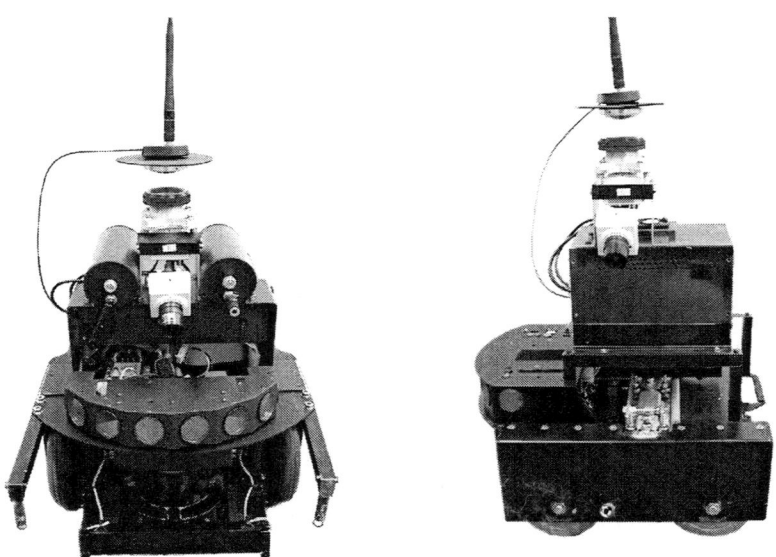

Figure1. left: The P2 robot with laser scanner in the front between the wheels, 360^0 camera on top, front camera and pneumatic kicking device.
right: The AT robot with 360^0 camera on top, front camera and electric kicking device.

of the robot, where the error in measurement does not depend on the distance traveled or on other influences the odometry suffers from. It sends heading data with 5Hz, a resolution of 1^0 and an accuracy of 2^0.

Kicker: We adapted the pneumatic kicking device used at the RoboCup-98 contest in Paris to the new robots. This kicker consists of a pneumatic cylinder, an electric valve, and a tank for compressed air. We also developed a second kicking device based on a spring mechanism wound up by a BMW car windshield wiper motor. This spring loaded kicker even shot harder than the pneumatic kicker and was successfully demonstrated at the Vision RoboCup-98 at Stuttgart, but it could not easily be fitted into the P2 chassis.

2.3 Onboard Computer

The onboard computer is the same as the one used last year in Paris, with the exception of an improved power system (the old suffered several failures in Paris and Stuttgart). It is a custom design based on standard PC parts with custom enclosure and is mounted at the rear top of the robot. Each PC has a 400 MHz AMD K6 CPU, 64 MB RAM and a 1,2 GB Hard Disk Drive. Additionally each computer is equipped with two PCI framegrabbers with a Booktree BT484 chip. These devices deliver images in YUV-format at 25 fps (PAL) and a maximum resolution of 768x576 pixels. For the connection to the laser scanner a high speed RS422 serial card was modified to achieve a data rate of 500 kbps, the highest data rate supported by the laser scanner. For the communication between the robots and to an external file server wireless PCMCIA Ethernet cards in a PCM-CIA to ISA adaptor from ARtem Datentechnik, Ulm, with a data transfer rate of 2 Mb/s are used. For this device we also developed a Linux device driver, which has now found its way back to the sponsor.

3 Software architecture

The software architecture of the Attempto team can be divided in three different layers (Fig. 2): low level data processing, intermediate level layer, high level robot control. We now describe each layer in detail.

3.1 Low level data processing

In the bottom layer different server programs organize the communication with the sensor and robot hardware, and do the first steps of data processing.
The robot server receives status data from the robot, which contains position, wheel velocity, sonar data, and battery status and sends movement commands to the robot, which are received from the Arbiter. The aim in developing the robot server was to send commands as fast as possible to the robot, under the constraint that the robot is only capable to execute 20 commands per second, to achieve a minimum duration for one control loop cycle.
The laser scanner server configures the laser scanner device at startup time with a field of view of 180^0, an angular resolution of 1^0 and a distance resolution of 10mm. The laser scanner then starts to send whole 180^0 scans with a rate of 60 Hz.

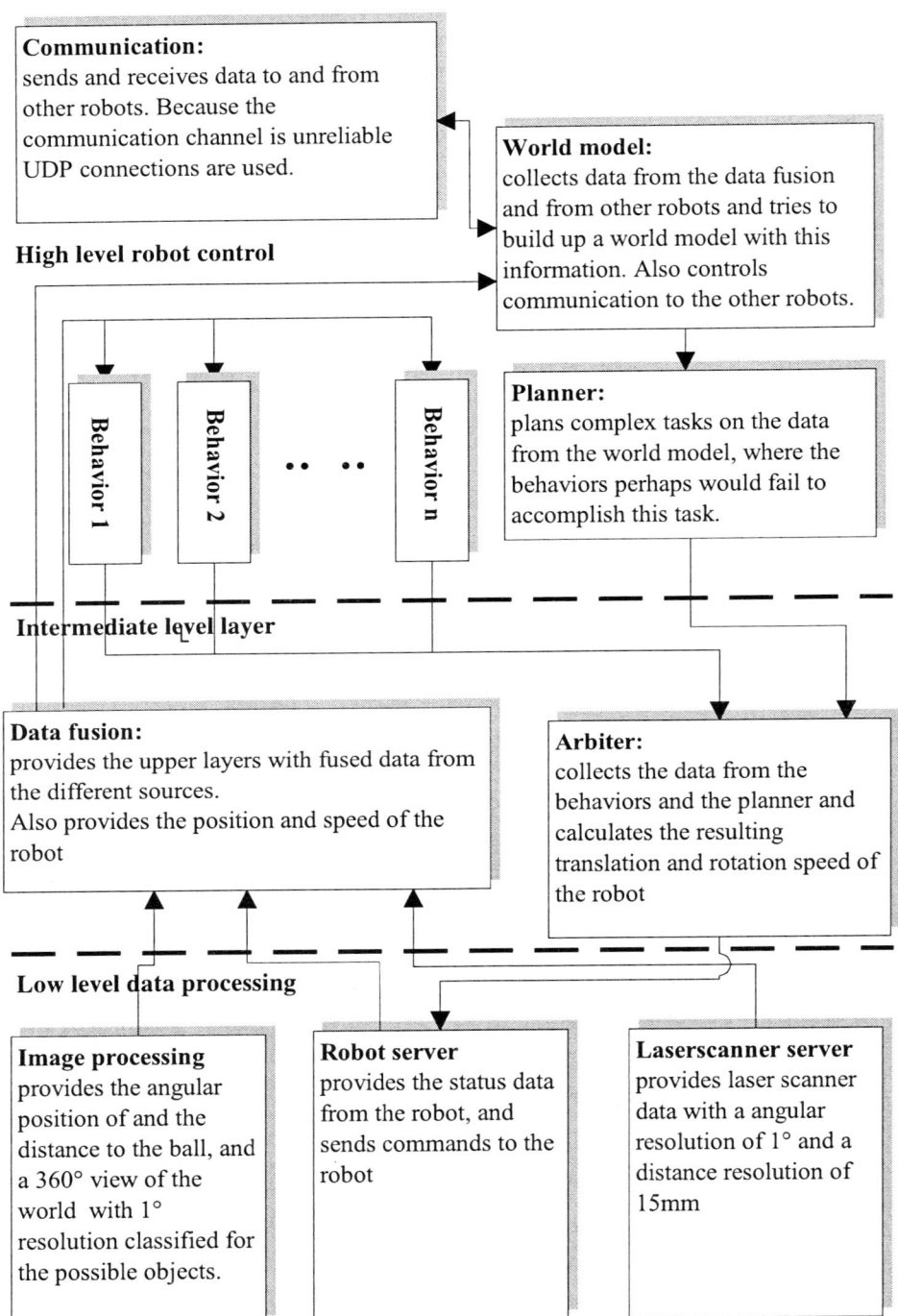

Figure 2. layered software architecture of the Attempto RoboCup-99 robots

The image processing grabs images from the front camera and the omnidirectional camera with a resolution of 384*288 pixels in YUV format. With this resolution it is possible to detect the ball with the front camera over a distance of 8 m and estimate the ball size and therefore the ball distance with an accuracy of 5 percent. The error in the angular position estimate is less than 1 degree.

To save processing time, the image processing does not search the whole image for the ball, but uses a history of ball positions in old images to predict the position of the ball in the next image. Only if the ball is not at the predicted position, the whole image is searched, starting the search at the predicted position.

Aside from the ball position the image processing provides an array data structure with 360 elements. Each of this elements represents a field of view of 1^0 and contains information about detected objects (ball, robots, goal, wall) and the determined distance and distance errors of these objects. In the field of view of the front camera the data structure additionally contains information about the type of the detected robots (own or opponent). Our high speed image processing needs only 3 ms per frame in the worst case (ball not at the predicted position). The average processing time for one frame is less than 1 ms. Therefore the image processing is capable of handling the 2 * 25 fps which the framegrabbers write to main memory in real time.

3.2 Intermediate layer

The intermediate layer consists of two different modules, the data fusion module, which fuses the data from the different low-level data processing servers and the arbiter, which receives steering and control commands from the behaviors and the planner and calculates a resulting movement command for the robot (section 4.2). The data fusion reduces the amount of information by extracting relevant object data from the raw sensor data. Objects fall in two different classes: dynamic objects like the ball and the other robots and static objects like the walls and the goals. The extracted information about an object includes opening angle in the field of view, distance and type of the object. The estimation error in the distance measurement is provided [4]. Therefore for the upper layer it is not necessary to know from which sensor source a specific distance measurement comes, because the properties of the sensor device are modeled via the measurement error. The data fusion also fuses the status data of the robot. For this reason, it receives the data from the odometry and tries to adopt the position of the robot with the information from the self localization algorithm described in section 4.1.

3.3 Top layer

The top layer realizes the hybrid robot control architecture [5]. It consists of a reactive component where a set of independent behaviors like obstacle avoidance, ball search or ball following try to fulfill their tasks. The behaviors can react quite fast on changes in the environment because they work directly on the preprocessed sensor data. This system is easy to expand because it is possible to start and stop behaviors at runtime.

The planner component is responsible for resolving more complex situations. This component is capable of suppressing or enhancing the output from specific behaviors and can also work as a special behavior with the same output to the arbiter like the other behaviors. The planner works on the data from the world model. This module fuses the data from the internal sensors and the data coming from other teammates via the wireless Ethernet connection. It tries to keep track and identify all the objects in the environment, and tries to predict the trajectories of recently undetected objects. This component is also responsible for sending data of all objects detected by the internal sensors to all the other robots.

The communication channel over the wireless Ethernet connection to the other robots is unreliable. Therefore we are using a UDP based protocol to prevent a communication action from locking while waiting for another robot to acknowledge.

4 Research Topics

In this section we give a brief overview of some of the problems we are currently working on.

4.1 Fast self localization with fused sensor data

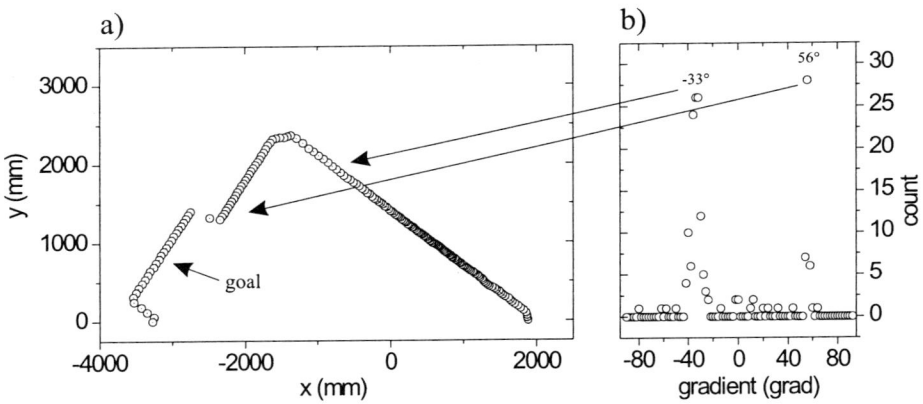

Figure3. a) laser scan of a part of our RoboCup field. b) the corresponding histogram for the directions of the difference vectors between two scan points. The two arrows point to the lines which correspond to the maximums in the histogram

Perhaps the most important information a robot needs to know to operate successfully in the RoboCup-99 is his own position within the field [3]. Therefore a fast self localization algorithm was developed, which makes use of the fused sensor data. In a first step this algorithm calculates the vectors and the directions between successive points in a laser scan. Then a histogram is calculated for these directions [6]. In a polygonal environment like the RoboCup-99 field this histogram shows usually one or more maxima which correspond to the main directions in the environment (Fig. 3). Now the laser scan is segmented into lines, by projecting each of the normalized difference vectors connecting two adjacent scan points onto the unit vectors in the main directions. If the result of this projection exceeds a certain threshold the two points cannot lie on a line with a direction equal to the main direction. After the segmentation there is a set of lines with directions according to the maxima in the histogram. The problem now is to decide, whether some of these lines correspond to a wall or a goal in the RoboCup-99 environment, because if this is the case, the distance to this wall can be used to adopt the robot position. Especially if it is possible to find two lines on different walls, the global position within the field can be calculated by trying all possibilities of matching the extracted lines against a set of lines representing the environment given as a priori information. Usually such matching algorithms possess a high computational complexity of at least second order in the number of lines. For the case that there is additional visual information from the vision systems, these matching algorithms can benefit from this knowledge. First each extracted line is classified into one of the following categories: WALL, BLUE-GOAL, YELLOW-GOAL and UNKNOWN (Fig. 4). If the classification supplies only lines in the categories WALL and UNKNOWN the number of lines which must be matched against the a priori information can be reduced by using only lines classified as WALL and so the runtime behavior of the matching algorithm improves. In the case where the classification supplies lines in the category BLUE-GOAL or YELLOW-GOAL the runtime behavior improves further, since there is only one possibility to match such a line against an environment which contains only one blue and one yellow goal. That means that by fusing the data from different sensor sources a self localization algorithm can be implemented, which works significantly faster than an algorithm working purely on range data.

At the moment we are also working on a self localization algorithm which relies only on the type classification data from the omnidirectional vision system. This algorithm works with a set of snapshots of the environment taken earlier and tries to match the current view (Fig. 4) of the environment against these snapshots to determine the actual position with respect to the positions, where the snapshots were taken [7]. The advantage of this approach is, that no prior geometric knowledge of the environment is necessary.

4.2 Hybrid robot control architecture

To play in the RoboCup-99 environment means to fulfill a quite complex task in a dynamic, adverse environment. Therefore our robots are equipped with a hybrid control architecture, existing of a reactive component and a planning component. A set of behaviors realize the fast, reactive part which is capable of

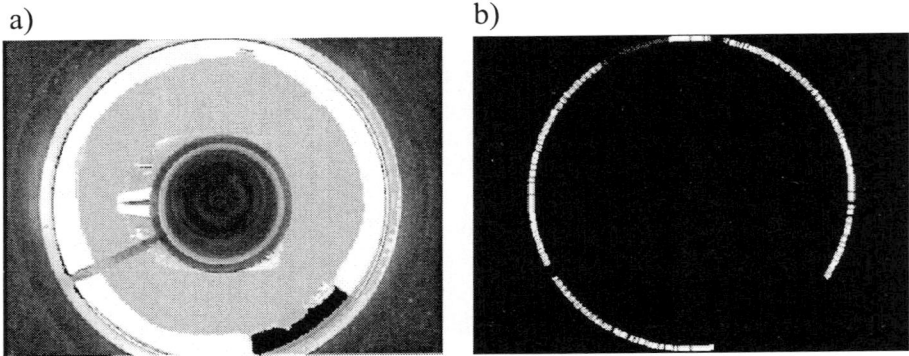

Figure4. a) Image of the omnidirectional vision system. b) Classified objects from the omnidirectional view in the left picture (white: wall, grey: the two goals)

dealing with the aspects of the dynamic and adverse environment. The planner controls more complicated tasks, where a purely reactive control could fail, e.g. team cooperation. The problem in question is to find a suitable way to merge the different outcomes of the behaviors and the planner. For this problem some solutions were proposed, e.g. the subsumption architecture [5]. We currently test and compare different ideas and proposals if they are appropriate for a scenario with the above mentioned properties of RoboCup-99.

Another problem closely connected to the mentioned one, is to find an appropriate mapping for a behavior between the sensor input and a useful response. The solutions proposed in the literature for this problem range from learned mappings to potential field methods [8].

5 Summary and Discussion

This paper described the hardware and software architecture of our RoboCup-99 robot team. Our approach so far has been hardware oriented: we tried to find the most capable robot platform within our budget and tried to maximize the number, diversity and the quality of our sensors. To this end we are using sonars, a wide angle color camera, an omnidirectional camera, a compass and a 2d scanning laser. Our underlying assumption is that at the current state of RoboCup play, improving the sensing capabilities will give a higher payoff than raising the speed of the robots or the onboard processing power or the "intelligence" of the robots. This is in contrast to the simulator or small size league, where all robots nearly have the same sensing capabilities. Our choice of sensors dictated the use of our pneumatic kicker and also the use of a larger PCI bus PC system with two frame grabbers. We use heavily specialized and optimized vision algorithms to keep the vision processing requirements low. The highlights of our software architecture are our method of sensor fusion which abstracts from individual sensors but keeps information about the reliability of sensor state with error data and the coupling of a reactive behaviour layer with additive behaviour outputs (rather than exclusive ones as in the subsumption architecture) with a planning component. We also believe we have found a good solution to update the global world model of each robot under unreliable radio ethernet communication.

References

1. H. Kitano, M. Asada, Y. Kuniyoshi, I. Noda, and E. Osawa. Robocup: The robot world cup initiative. In *Proc. of the first Int. Conf. on Autonomous Agents*, pages 340–347, 1997.
2. M. Asada, editor. *RoboCup-98: Robot Soccer World Cup II, Proceedings of the second RoboCup Workshop*, 1998.
3. J-S. Gutmann, W. Hatzack, I. Herrmann, B. Nebel, F. Rittinger, A. Topor, T. Weigel, and B. Welsch. The cs freiburg team. In *Proc. of the second RoboCup Workshop*, pages 451–459, 1998.
4. A. Mojaev and A. Zell. Sonardaten-integration für autonome mobile roboter. In Levi P., Ahlers R.-J., May F., and Schanz M., editors, *Mustererkennung 1998*, pages 556–565. Springer-Verlag, 1998.
5. R. C. Arkin. *Behavior-based robotics*. MIT Press, 1998.
6. G. Weiss and E. v. Puttkammer. A map based on laserscans without geometric interpretation. In U. Rembold et al., editors, *Intelligent Autonomous Systems*, pages 403–407. IOS Press, 1995.
7. M. O. Franz, B. Schölkopf, H. A. Mallot, and H. H. Bülthoff. Where did i take that snapshot? scene-based homing by image matching. *Biol. Cybern.*, (79):191–202, 1998.
8. J.C. Latombe. *Robot Motion planning*. Kluwer Academic Publishers, 1991.

Rogi Team Real: Dynamical Physical Agents

Josep Lluís de la Rosa, Bianca Innocenti, Israel Muñoz,
Albert Figueras, Josep Antoni Ramon, and Miquel Montaner

Institut d'Informàtica i Aplicacions
Universitat de Girona & LEA-SICA
C/Lluís Santaló s/n
E-17071 Girona, Catalonia
{peplluis, bianca, imunoz, figueras, jar, mmontane}@eia.udg.es

Abstract. Research in dynamical physical agents, consensus of proper physical decisions among physical agents, and an example of passing is shown. The interest is to introduce introspection of the dynamical behavior of each physical body so that every agent has better knowledge. This has to lead to better passes.

1 Introduction

The Rogi Team started in 1996 as the result of a doctorate course in multiagent systems in the University of Girona. The main goal of the team has always been the experimentation of *dynamical* physical agents and autonomous systems. Here, *dynamical* is understood from the automatic control background. It means dynamic temporal evolution of continuous variables of robots' physical body, which can be described by transfer functions or continuous state representation. The aim is to see the impact of dynamics of the physical agent's bodies in the co-operative world.

The micro robots have now clear dynamics for automatic control. There are good transfer functions available to describe the physical bodies of the physical agents.

Our vision hardware, result of our research, is able to process up to 50 frames/sec, locating ten robots and a ball. This sample time is enough for dynamics.

A rational physical agent's approach, result of our research, is operative for robots co-operation. It is here exemplified in terms of passing the ball joint actions.

2 Team's Hardware Description

Our micro-robot team is made up of three parts: robots, vision system, and control system. The vision and control systems are implemented in two PC. The control system is called the host. The host and vision systems are connected by TCP/IP protocol. This allows remote users to do tests on our micro-robots team and permits distant co-operative research [4]. The vision system provides data to the host that takes decisions by an agent oriented approach. The decision is converted into individual tasks for each robot and sent using a FM emitter from the host computer.

We have designed some specific hardware to perform vision, merging specific components for image processing (video converters, analog filters, etc.) with multiple purpose programmable devices (FPGAs) and using multiple color segmentation. This is a real time image-processing tool, which can be reconfigured to implement different algorithms. To locate the robots and the ball, the first step consists in their segmentation from the scene. The discriminatory properties of two color attributes, *hue* and *saturation*, are used so as to segment the objects, and different labels to pixels belonging to different color textures are assigned. Moreover, a more robust behavior under non-uniform lighting of the scenario is achieved, thanks to the stability of hue and saturation under variations on the intensity of the illuminant [2].

3 Taking Dynamics into Account for Decisions

Explicit reasoning on dynamics of the physical body of agents will improve co-operative performance of physical agents. Knowing that controllers modify (controls) dynamics of the physical body of agents, then agents have to be aware (introspection) of the set of controllers their physical body has. Control engineers need tools for developing these agents and their controllers, as stated in [3].

AGENT0 [6] is used as an agent language. In this language an agent's state consists of mental components such as *beliefs, capabilities, choices* and *commitments*. In our point of view, the *capabilities* can represent the dynamics of the agent's body. Some of the agent's capabilities have to be associated to the control of the agent's body and they are proposed as a way for the agent to be aware of what he can or cannot do. *This drives to an extension of the agent concept from physical agents* [1] *to dynamical physical agents* as follows: The physical knowledge of the physical agent contains knowledge about dynamics of its physical body, which is supported by further declarative control and supervision levels [5].

As a first example, this approach is applied to a ball passing experiment between two robots. The purpose of the example is to show the utility of inter-agent negotiation with explicit representation of dynamics and to improve the decision of *when* and *how* to do passing with respect to static knowledge. The passing experiment is here simplified as follows (see **Fig. 1**): two robots have an obstacle-free crossing trajectory and have to decide whether to pass the ball and how. The robots have several controllers to move forward in one-dimensional linear movement. The Single Input Single Output transfer functions of the robots and the ball are known. Necessary steps to do the experiment are:
- To find a model that represents dynamics of two mobile robots and a ball.
- To implement several position and speed controllers for passing. Their specification is to reach the set points with precision and stability.
- To inspect untargeted situations: not enough or too much impulse for the ball.
- A negotiation algorithm based on dynamics represented in Agent0 capabilities.

There are passes that are not physically feasible since there are not controllers to execute them. For instance, to do a short pass could be an extremely difficult task if there is no slow speed controller, and the same happens at any required speed set point where no controller exists. The physically unfeasible passes are called undesirable situations, that could be (see **Fig. 1**): (1) the robot 2 has slow dynamics, or (2) the robot 1 does not give the necessary impulse to get ball to the crossing point in some convenient time for dynamics of robot 2. This impulse has to be calculated according to dynamics of the ball.

Both robots have to agree the applicable control and moment for the pass based on the knowledge they have of their dynamics. Since passing between the robots has to be assured by proper physical decisions, then to determine whether a pass is possible or not during a football match, at least Robot 1 has to have a controller to make the ball get through the crossing point.

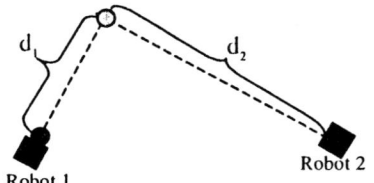

d_1 Distance between the ball and the crossing point.
d_2 Distance between the Robot 2 and the crossing point.
● Ball.
○ Crossing point of the Robot 2 with the ball.

Fig. 1: Simplified *passing experiment*

In this example the transfer functions are:

Dynamics of Each Component

Robot 1:   **Robot 2:**   **Ball:**

$$\frac{RealSpeed}{SetPoint\,Speed} = \frac{1}{20s+1}$$ $$\frac{RealSpeed}{SetPoint\,Speed} = \frac{1}{20s+1}$$ $$\frac{RealSpeed}{SetPoint\,Speed} = \frac{21}{0.035s+1}$$

Controllers of Robots

Robot 1:   **Robot 2:**   **Distance Controller Robot 2:**

$$\frac{150s+50}{s}$$ $$\frac{60s+5}{s}$$ $$Kp = 1$$

Fig. 3 shows the response of a robot to a speed set point 30 cm/s step (this step is the required speed to kick the ball in the pass). Robot 1 knows that the ball can move 10 cm away from him:
- In 0.285 s with the impulse of 32 cm/s.
- In 0.428 s with the impulse of 45 cm/s.
- And in 1.1 s with the impulse of 60 cm/s.

This knowledge is contained in its base of *capabilities*.

For doing d_1 Robot 1 can provide with three different impulse to kick to the ball as shown in the **Fig. 2**.

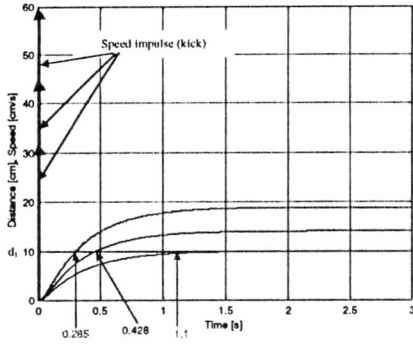

Fig. 2: *Response of the ball to three different kicks (impulses).*

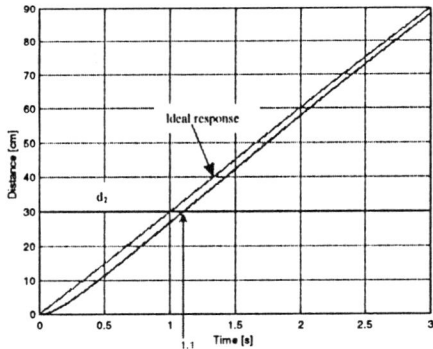

Fig. 3: *Response of Robot 2.*

Consensus Decision Algorithm

- **Step 1:** *Proposition:* With these data, Robot 1 proposes to do the pass in 0.285 seconds to Robot 2.
- **Step 2:** Robot 2 is 30 cm away the crossing point.
- **Step 3:** *Introspection:* Robot 2 looks up its base of capabilities what time is required to be 30 cm far from its actual position. The result is 1.1 s.
- **Step 4:** *Answer:* Robot 2 tells Robot 1 it can move 30 cm in 0.285s, 60% of certainty.
- **Step 5:** *Decision:* Robot 1 considers that this certainty is not enough.
- **Step 6:** *New Proposition:* Robot 1 proposes to do a pass in 0.428 s, impulse 45 cm/s.
- **Step 7:** *Answer:* Robot 2 responds 75% certainty.
- **Step 8:** *Decision:* Robot 1 considers that this certainty is not high enough.
- **Step 9:** *New Proposition:* Robot 1 proposes new time, 1.1 s with impulse 60 cm/s.
- **Step 10:** *Answer:* Robot 2 to Robot 1. It can move 30 cm in 1.1s with 90% certainty.
- **Step 11:** *Decision:* Robot 1 agrees this certainty is big enough and they do the pass.

Fig. 4 shows how the ball and Robot 2 arrive to the crossing point at the time the robots had decided. Robot 2 had to move 30 cm and the ball 10 cm. Note that Figure 3 also shows the difference between the real (dynamical) and the ideal (static) responses of the robot. If agents take decisions considering the ideal case, the decisions may be wrong. Taking into account the dynamics of their bodies in the decision, agents can assure they take the physically proper (and correct) decision.

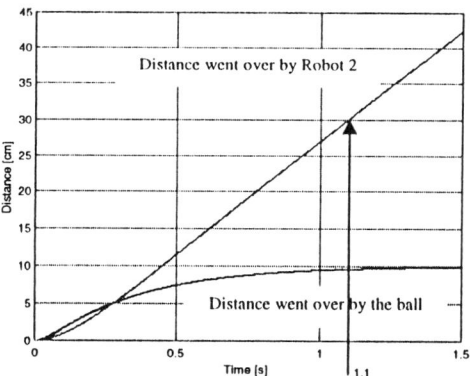

Fig. 4: *Run distances.*

Acknowledgements

This work is funded by the DAFNE CICYT TAP98-0955-C03-02 project.

References

[1] Asada M., Kuniyoshi Y., et al. "The RoboCup Physical Agent Challenge", First *RoboCup Workshop in the XV IJCAI-97*, pp.51-56, 1997.

[2] Celenk M., "A color clustering technique for image segmentation", *Computer Vision, Graphics and Image Processing*, vol. 52, pp. 145–170, 1990.

[3] de la Rosa J. Ll., et al., "Soccer Team based on Agent-Oriented Programming", *Robotics and Autonomous Systems*. Ed. Elsevier, Vol 21, pp.167-176, October 1997.

[4] Johnson J., de la Rosa J. Ll., and Kim J.H., "Benchmark Tests in the Science of Robot Football" *Proc. Mirosot98*, Paris 1998. R. J. Stonier (ed). U. C. Queensland

[5] Müller J., "The Design of Intelligent Agents: a layered approach", *Lecture Notes in Artificial Intelligence*, Vol. 1177, Ed. Springer Verlag, 1996.

[6] Shoham Y. "Agent-oriented programming", *Artificial Intelligence,* vol. 60 (1), pp.51-92, 1993 & Technical Report STAN-CS-1335-90, Computer Science Department, Stanford University, Stanford, CA, 1990.

Learning to Behave by Environment Reinforcement

Leonardo A. Scardua*, Anna H. Reali Costa**, and Jose Jaime da Cruz***

Escola Politecnica - Universidade de Sao Paulo, Av. Prof. Luciano Gualberto,
Travessa 3, 158,
05508-900 Sao Paulo, SP, Brasil
{scardua, anna}@pcs.usp.br, jaime@lac.usp.br

Abstract. This paper describes a softbot agent capable of learning to choose its actions, in order to achieve its goal when facing an opponent in a dynamic environment. The agent uses rewards gathered from the environment to assess and improve the quality of its own behavior. A multilayer perceptron neural network is assessed regarding its adequacy as a value function approximator for state-action pairs in the robotic soccer domain.

1 Introduction

The reinforcement learning approach is a very effective alternative to the standard supervised training of artificial neural networks (ANNs). Instead of having an external teacher that indicates the correct output for each given input, reinforcement learning simply provides a reinforcement signal that indicates the quality of the network output. According to [10], one can identify the following main sub-elements of a reinforcement learning system: a *policy*, a *reward function*, a *value function* and sometimes a *model of the environment*.

A *policy* is a mapping from perceived states to actions to be taken in those states.

A *reward* function defines the goal in a reinforcement learning problem, mapping each pair formed by a state s and an action a to a single number. This single number is the reward that indicates the desirability, in an immediate sense, of choosing action a when in state s.

The *value function* indicates the benefit of selecting action a when in state s, in a long-term sense. Roughly speaking, the value of a state-action pair is the total amount of reward an agent can expect to receive if it takes action a when in state s, following a policy. The only goal of a reinforcement learning agent is to maximize the total reward it receives in the long run. Methods for determining values are the most important components of almost all reinforcement learning systems.

---

* Leonardo Azevedo Scardua is supported by CNPq grant number 141802/97-9.
** Anna H. Reali Costa is partially supported by FAPESP grant number 98/06417-9.
*** Jose Jaime da Cruz is partially supported by CNPq grant number 304071/85-4(RN).

In stochastic domains, a *model* describing the state transition probabilities and the resulting immediate rewards is usually not available, at least in many relevant domains. For this reason, the reinforcement learning agent learns through direct interaction with the environment.

The main objective of our research is to develop a team of agents capable of learning cooperative behavior solely by observing the impact of their actions in the environment. To achieve this goal we have been developing an agent capable of learning to choose its actions by observing environment rewards. The first phase of the development, described in this paper, intends to assess an (ANN) as an action evaluator in the robotic soccer domain.

The RoboCup soccer server is a simulated robotic soccer domain that allows matches between two teams of up to 11 players, where each player is controlled by a single process. The soccer server works in 100 milliseconds (this is an adjustable parameter but 100 is its default value[2]) simulation cycles. It demands that agents act in real time. The environment changes are influenced by the actions of both teams of agents. This combination of features makes it a very complex and realistic domain, where decisions are made in stages, and the output of a decision can not be fully predicted. Each decision results in some immediate reward and affects the environment, influencing the reward received by the next decision to be made.

In the following section, it is first introduced the learning agent's structure and the environment state perceived by the agent. Section 3 justifies the use of an ANN as a value function approximator. It also describes the overall approach implemented, including the reward signal and the ANN. Section 4 presents the results, which are discussed in Sect. 5. Section 6 presents some related work and Sect. 7 presents the conclusions.

2 The Learning Agent

The softbot agent is in essence a reinforcement learning agent, whose main objective is to gather the maximum reward. This goal can be achieved by choosing the action that has the best value in face of the present state signal.

The agent must learn while interacting with the environment since it has no previous knowledge about the task or the domain it is about to face. Once it is not practical to use memory tables to keep the value of each state-action pair that is experienced by the agent (Sect.2.1), we decided to use an (ANN) to approximate such values (Sect.2.2).

The agent's structure is very simple, yet efficient. It works according to the following loop:

```
Sense the environment state.
If it is time to choose the best action:
  Use the neural network to evaluate each of the available actions
    in light of the state signal.
  Choose the action that has the best-estimated value.
```

```
If it is time to choose randomly:
  Choose randomly among the set of available actions.
Execute the chosen action.
Gather the reward.
Update the value estimate of the neural network.
```

2.1 The Environment State

The learning player senses the world according to the following state signal:

1. The angle between the ball and the learning agent.
2. The distance between the ball and the learning agent.
3. The angle between the opponent and the learning agent.
4. The distance between the opponent and the learning agent.

According to [2], the soccer field size is 105 x 68 where the unit is meaningless. We use the arctan function to calculate the relative angles above. We did not use any kind of generalization, in order to reduce the size of the state space. As the Soccer Server gives us float point precision, we have an enormous state space.

2.2 Set of Actions

Each of the agents is given a set of three actions, each action is implemented by a routine that has all the information it needs to perform the action.

Get the Ball.

Pre-conditions : The ball must not be within kicking distance of the agent.

Effect : This moves the agent (A) in the direction of the ball. When it reaches the ball, A stops within kicking distance from it.

Kick Towards Adversary Goal.

Pre-conditions : The ball must be within kicking distance of A.

Effect : This kicks the ball toward the adversary goal.

Turn the Ball.

Pre-conditions : The ball must be within kicking distance and behind A, with respect to the adversary goal.

Effect : This turns the ball towards the adversary goal, keeping it within kicking distance. When the ball is in front of A with respect to the adversary goal, A kicks the ball.

2.3 Decision Cycle

In this paper, we had matches between two opposing players, without goalies. One player uses the learning structure already depicted while the other chooses its actions at random. Both agents must decide which action will be taken at each simulation cycle. The action set and the short simulation cycle allow even the random player to score consistently if it does not face opposition.

3 The Learning Approach

As the main objective of this work is to assess an ANN as a function approximator in the robbotic soccer domain, we are not dealing with delayed reward.

3.1 Why a Neural Network as Value Function Approximator

It is clear that, given the chosen state representation, it is not practical to keep the actions values in memory tables, since the number of possible states is very large. This imposes the use of an approximator for the value function.

The chosen approximation architecture must attend at least, the following conditions:

1. Once it is not possible to know beforehand the features of the value function that must be approximated, the architecture must allow for good generic function approximation for at least continuos functions.
2. It must be capable of working well even in a noisy, uncertain environment.
3. It must be able to work under real-time demand.

The Kolmogorov's Mapping Neural Existence Theorem [9] gives the mathematical justification for the use of three layer perceptron networks, as universal continuos function approximators.

3.2 Coding the Agent's Goal as a Reward Signal

In the reinforcement learning framework, the reward signal is used to encode what we want the agent to do. It is the same in the present case, where the reward following the choice of each action is used to inform the agent whether it was a "good" or a "bad" choice.

In this paper, "good" means choosing an action that has all of its pre-conditions (constraints that must be valid, given the present state signal, such that the action is applicable) satisfied by the state signal the agent perceives from the environment. In this case the reward signal is positive.

If anyone of the pre-conditions of the chosen action is unsatisfied, the reward signal will be negative. In other words, the encoded goal is to learn to choose actions that have all pre-conditions satisfied by the perceived state signal. This goal and method for calculating the reward are intended just to provide an easy yet valid mean to evaluate the proposed value function approximator.

3.3 Learning the Value of a State-Action Pair

To learn the value of a state-action pair, the (ANN) is presented with the action chosen by the agent and the perceived state signal. The network output must be equal to the reward immediately received by the agent. In other words, when we talk about values in this work, we are talking about immediate rewards (given the goal of this work, that is to assess a ANN as a function approximator in the robotic soccer domain, we are not dealing with delayed rewards). If the ANN is able to learn the immediate rewards, it will be able to learn the values of state-action pairs (in the usual reinforcement learning sense) when dealing with delayed rewards.

Given the goal proposed to the agent, there is no need to estimate the true value of all state-action pairs; learning the immediate rewards will be enough.

3.4 The Neural Network

The neural network shown in Fig. 1 is a three-layer fully connected feedforward perceptron, with ten hidden neurons and one output neuron.

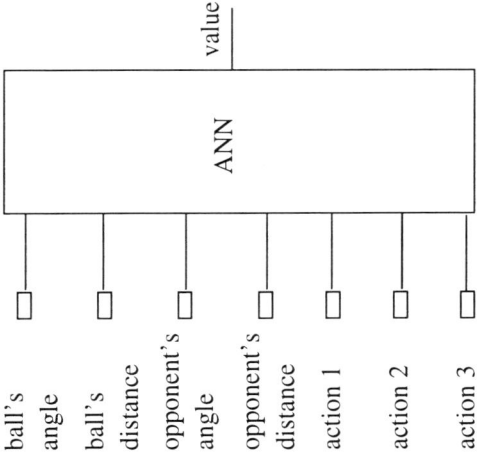

Fig. 1. ANN diagram

The hidden neurons use the hyperbolic tangent as the activation function while the output neuron just sums up all of its inputs. The training method is the backpropagation algorithm [6].

3.5 Choosing the Best Action

The choice of the best action to be taken is based solely on the state signal perceived by the agent. When it is time to choose an action, the agent feeds the (ANN) input with the current state signal and the input corresponding to the action being evaluated is made positive, while the other action inputs are kept at zero level. The ANN output is then recorded. This procedure is repeated for each existing action and the agent chooses that one that yields the biggest numerical output of the ANN. Each action is implemented by a routine and identified uniquely by a positive number.

4 Results

To assess the effectiveness of the proposed state-action pair value approximator, we conducted experiments consisting of a series of ten-minute games. The learning agent faces an opponent that selects its actions randomly.

If the learning agent effectively learns to take actions which pre-conditions are satisfied by the state of the environment, it will be able to outscore a random selecting agent, that wastes time selecting actions that may not be effective in the current environment state.

4.1 Experiment Results

Experiment 1 - The learning agent always selected a random action and did not learn from the rewards received from the environment. Therefore, the learning player actually performed just like a random player. The goal of this experiment is to assess the performance of two random players.

The results, presented in Tables 1 and 2 and Fig. 2, show that in this case, the performance is the same for both players.

Table 1. The score of each match in experiment 1

Game Number	Learning Player	Random Player
1	5	2
2	2	5
3	3	4
4	4	4
5	3	5
6	4	3
7	3	5
8	5	4
9	5	5
10	6	4

Table 2. Some statistics referring to experiment 1

Game Number	Learning Player	Random Player
Number of goals	40	41
Average goals	4	4.1
Number of wins	4	4

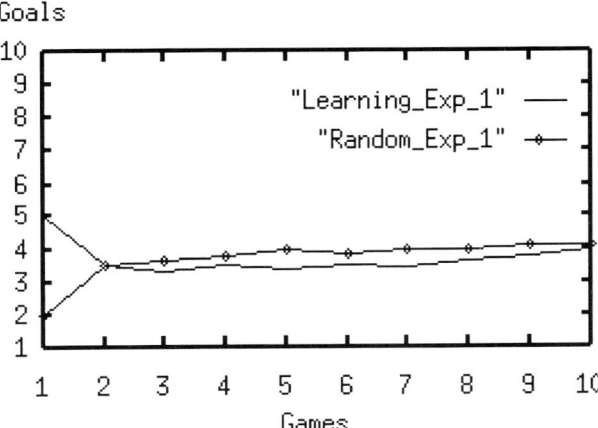

Fig. 2. Average score per game in experiment 1

Experiment 2 - At random, with probability 0.2, the learning agent selected the most valuable action, and with probability 0.8 selected randomly one action from the set of available actions.

At the beginning the ANN's weights were assigned randomly, and they changed according to the learning process throughout each game. The goal of this experiment is to assess the effect of the learning process on the player's performance.

The results, presented in Tables 3 and 4 and Fig. 3, show that in this case, the learning player is becoming superior as the number of games grow.

Table 3. Some statistics referring to experiment 2

Game Number	Learning Player	Random Player
1	5	5
2	9	3
3	2	7
4	3	5
5	6	2
6	6	2
7	2	6
8	6	3
9	7	2
10	5	2

Table 4. Some statistics referring to experiment 2

Game Number	Learning Player	Random Player
Number of goals	51	37
Average goals	5.1	3.7
Number of wins	6	3

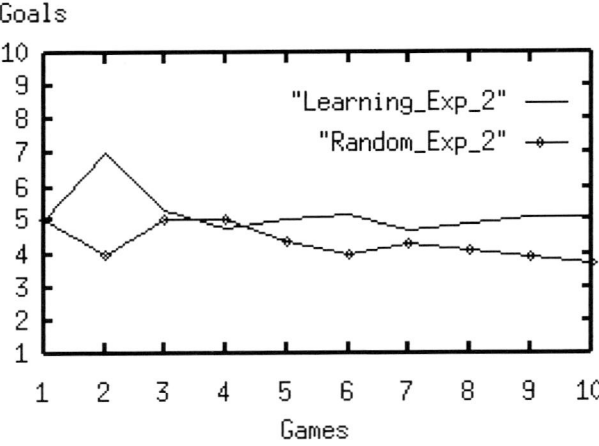

Fig. 3. Average score per game in experiment 2

Experiment 3 - At random, with probability 0.8, the learning agent selected the most valuable action, and with probability 0.2 selected randomly one action from the set of available actions.

At the beginning the ANN's weights were assigned randomly, and they changed according to the learning process throughout each game. The goal of this experiment is to assess the effect of the learning process on the player's performance.

The results, presented in Tables 5 and 6 and Fig. 4, show that in this case, the learning player becomes much superior as the number of games grow.

Table 5. Some statistics referring to experiment 3

Game Number	Learning Player	Random Player
1	6	4
2	7	2
3	5	4
4	10	2
5	7	3
6	7	3
7	7	2
8	8	4
9	9	4
10	10	2

Table 6. Some statistics referring to experiment 3

Game Number	Learning Player	Random Player
Number of goals	76	30
Average goals	7.6	3.0
Number of wins	10	0

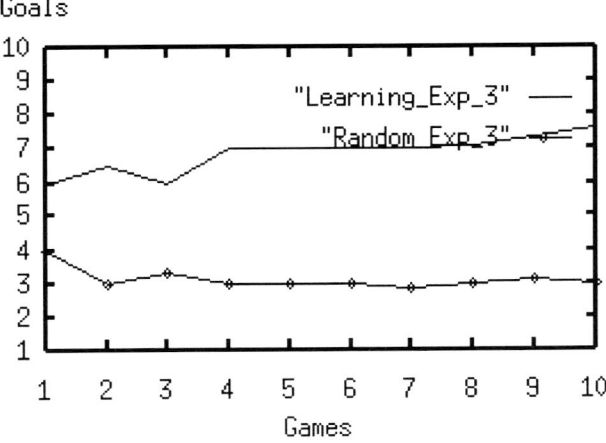

Fig. 4. Average score per game in experiment 3

5 Discussion

The simulation results show that:

1. The average number of goals per game of the random selecting player consistently decreases as the probability of selecting the most valuable action for the learning agent increases.
2. The average number of goals per game of the learning agent consistently increases as the probability of selecting the most valuable action increases.
3. The number of wins of the learning agent consistently increases, reaching 100%, as the probability of selecting the most valuable action increases.

Since the only difference between the three experiments is the learning agent's probability of choosing the best-estimated action, it is clear that this is what caused the above changes in the experiments' statistics. Because of the fact that the increase in this probability improved the performance of the learning agent and hindered the performance of the random player, it is clear that the learning agent has learned something useful. This allowed the learning agent to consistently outscore the random selecting agent.

These results make clear that a feedforward multilayer perceptron is able to approximate the value function for a reinforcement learning agent, in the robotic soccer domain, in a fast and reliable way.

6 Related Work

The work presented in this paper is inspired by Tesauro's TD-Gammon [11], an (ANN) that is able to teach itself to play backgammon solely by playing against itself and learning from the results.

The domain of backgammon, like robotic soccer, is a domain where decisions are made in stages, with a huge number of possible states, making it impossible to use a supervised learning approach for the neural network. Both domains involve playing against an unknown opponent.

Many researchers have been using learning methods in the robotic soccer domain. Among the more common methods are genetic algorithms [7], genetic programming [1], case-based reasoning [5] and reinforcement learning [3].

7 Conclusion and Future Work

Our agent learned to choose its actions using solely the rewards it obtained while interacting with the environment. It is important to stress that it did not have any previous knowledge about the robotic soccer domain in any form. It is an (ANN) that learned to play just by observing the rewards gathered while playing.

One important aspect to be emphasized is that the learning is naturally focused on the state trajectory followed by the learning agent. This greatly reduces the space state complexity, explaining the fast learning that allowed our agent to win every match in the third experiment, even having started with random weights in its neural network.

The first phase of the development of our research, described in this paper, intended to assess an (ANN) as an action evaluator in the robotic soccer domain. The results are encouraging. Our research continues with the development of a team of eleven agents that learn by being rewarded solely when they score a goal and being punished when the opponent team scores a goal. In the next step, the agents will be controlled by a version of Richard Sutton's Sarsa algorithm [10], improved with an augmented version of the ANN used in the present work, to approximate the value function.

Robotic soccer demands that the agents learn and act in real time. The output of each action is not fully predictable due to random noise imposed by the RoboCup Soccer Server. The environment changes are influenced by the actions of both teams of agents. This combination of features makes this a very complex and realistic domain. This complexity allows us to assert that the technology we have been developing can be applied to real tasks that demand learning of a control policy in dynamic huge state space environment. Good examples of real tasks are dynamic channel allocation in cellular telephone systems and elevator dispatching[10]. These two applications show the importance of a good value-function approximation architecture.

References

1. Andre, D., Teller, A: Evolving Team Darwin United. Proceedings of the Second RoboCup Workshop, Paris (1998) 317–323
2. Andre, D., Corten, E., Dorer, K., Gugenberger, P., Joldos, M., Kummeneje, J., Navratil, P., Itsuki, N., Riley, P., Stone, P., Takahashi, T., Yeap, T: Soccerserver Manual. (1999) www.dsv.su.se/ johank/RoboCup/manual
3. Andou T: A RoboCup Team wich Reinforces Position Observationally. Proceedings of the Second RoboCup Workshop, Paris (1998) 361–363
4. Bertsekas, D. and Tsitsiklis, John: Neuro Dynamic Programming Athena Scientific, Belmont, MA, (1996)
5. Burkhar, H., Wendler, J., Gugenberger, P., Schroder, K., Kuhnel, R: AT-Humboldt in RoboCup-98. Proceedings of the Second RoboCup Workshop,Paris (1998) 331–337
6. Haykin, S.: Neural Networks: a comprehensive foundation. 2nd ed., Prentice Hall, (1999)
7. Kuzuaki, E., Sadaharu, I., Yamaguchi, H., Nobui, I. and Yoshiyuki, K: Team Description for Donguri. Proceedings of the Second RoboCup Workshop, Paris,(1998), 305-308
8. Matsumura T: Description of Team Erika. Proceedings of the Second RoboCup Workshop, Paris (1998) 309–315
9. Hetch-Nielsen, R: Neurocomputing. Addison Wesley Publ. Co., New York,(1990)
10. Sutton, R. and Barto, A: Reinforcement Learning: an introduction. MIT Press, (1998)
11. Tesauro, G: Temporal Difference Learning and TD-Gammon. Communications of the ACM, (1995) Vol 38 No.3 58-68

End User Specification of RoboCup Teams

Paul Scerri and Johan Ydrén
Real-time Systems Laboratory
Department of Computer and Information Science
Linköpings Universitet, S-581 83 Linköping, Sweden
pausc,x98johyd@ida.liu.se

Institute for Computer Science, Linkpings Universitet

Abstract. Creating complex agents for simulation environments has long been the exclusive realm of AI experts. However it is far more desirable that experts in the particular application domain, rather than AI experts, are empowered to specify agent behavior. In this paper an approach is presented that allows domain experts to specify the high-level team strategies of agents for RoboCup. The domain experts' specifications are compiled into behavior based agents.
The 1999 RoboCup World Cup provided an interesting basis for evaluation of the approach. We found that for RoboCup it is not necessary to allow a user to change low level aspects of the agents' behavior in order for them to create a range of different, interesting teams. We also found that the modular nature of behavior based architectures make them an ideal target architecture for compiling enduser specifications.

1 Introduction

Simulation environments where intelligent agents play the roles of humans are used for training and testing in domains such as air combat training, fire fighting command training and large scale military simulations. The development of intelligent agents for such environments has been, and continues to be, a challenge for AI researchers. However over the years a wide range of techniques have been developed to meet many of the challenges faced. Nearly all of the developed techniques rely on the availability of agent experts to program the agents. This is highly undesirable in cases where there is either a large body of expert knowledge that needs to be incorporated into the agent or when the agent behavior needs to be changed often. Either case implies a large amount of work, better off done by a domain expert, being done by an agent expert. In this paper an approach is presented aimed at bridging the gap between domain experts and agent programming for one particular domain, namely RoboCup[3].

Our approach is to provide a graphical editor, resembling a coach's whiteboard, with which a domain expert can specify the high level strategies of a team in a manner similar to the way he might explain strategies to a real soccer team. For example, Figure 1 shows a diagram explaining one professional soccer strategy taken from the Internet. The diagram seems to be typical of the way complex, high level strategies are explained by humans for humans. The strategy editor presented here attempts to mimic this explanation style.

In our system a strategy is specified by drawing circles for player positions and arrows for the directions to pass and dribble the ball. The whiteboard style editor allows a wide range of options and flexibility providing the ability to specify complex strategies without knowledge of the underlying agent architecture. The specification, done at a team level, is subsequently compiled into eleven separate behavior-based agents, one for each player.

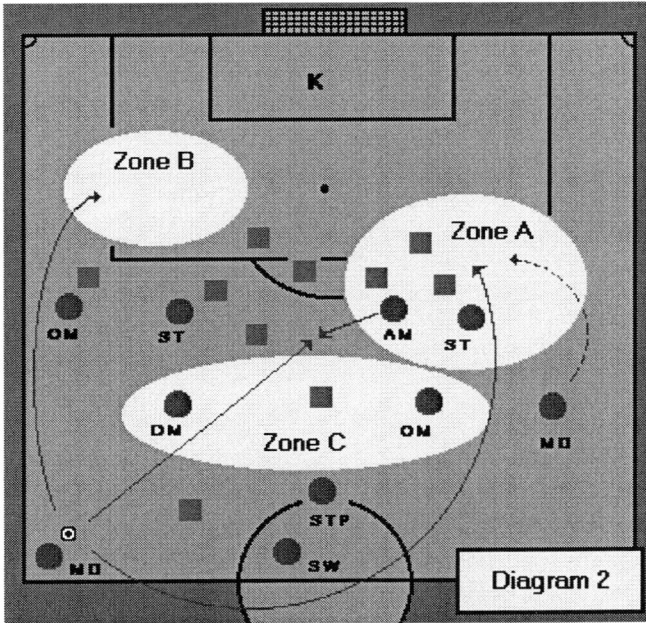

Fig. 1. A diagram from the Internet explaining a particular team soccer strategy to a human audience.

Although the system described here is intimately tied to the RoboCup domain it is hoped that lessons learned from the system can be applied generally to the problem of empowering domain experts to program agents for complex, agent-populated simulations. In particular an evaluation of the system provides insight into two questions. Firstly, are behavior based architectures good target architectures for compiling end user behavior specifications? Secondly is it sufficient in some domains to give endusers *only* the ability to specify the high level behavior of the agents?

2 The Editor

The heart of this approach to domain expert specification is a flexible, visual editing tool called the *strategy editor*. A soccer expert (henceforth referred to as the designer) indicates positions of players and intended ball movement in a style similar to drawing on a whiteboard. To create a particular team strategy the designer places all players on an image of the playing field. For each player the designer can specify an arbitrary number of positions to pass to, an arbitrary number of directions in which to dribble and directions to watch for incoming passes. The compiled agent will use its knowledge of the world (e.g. relative positions of other players) to choose at runtime which option to take when it gets the ball. Because all the players are simultaneously shown in the editor, and the designer is working on them all concurrently, it is trivial to ensure that players pass to positions where team mates are likely to be (and correspondingly that players move to positions where passes are likely to come). Soccer experts can place players and design passing and dribbling formations that lead to good RoboCup teams. Notice that once players are compiled any team behavior that occurs is "emergent", i.e. there is no explicit communication or explicit representation of passing patterns. In effect the strategy editor provides a means to visualize and specify "emergent" team behavior.

As well as control over *what* the players do, i.e. where they pass and dribble, a designer has some control over *how* the player does something. For each player the designer may choose the *style* with which the player will play (the available styles are determined from the player skeleton – see below). Some general styles are *normal play* or *shooter*, while more specialized styles are *crosser* (always tries to kick the ball to the middle) and *wait* (just watches ball, useful when referee stops play, etc.) The styles influence how the player fulfills its part of the designer's strategy. For example, the shooter style results in a player that will chase the ball in a fairly large area around its assigned position and will shoot if at all reasonable. Apart from selecting styles of play the designer has no low level control over the behavior of the players. The details of the players' behavior come from a template created in a low level individual player strategy editor.

The process of designing formations and selecting styles is repeated for each of the different game "modes" (or situations) that the player will distinguish between. Example modes are *kickoff*, *deep defense* and *transition to attack*. The modes of play that the player knows about are determined from a player "template" created in a lower level individual strategy creation tool (see below and [4]). For RoboCup99 there were about 15 different modes. At runtime the player determines the current mode mainly by the position of the ball on the ground but also considers factors such as referee calls and which team is nearer the ball. It is possible to view the strategies more than one mode at once, hence it is possible to see how a player must move when the team switches from offense to defense or vice versa.

Figure 2 shows a designer specification of how a particular defensive strategy should work. Circles represent players. Arrows with double lines indicate the direction that the player should dribble when it has the ball. Arrows with single

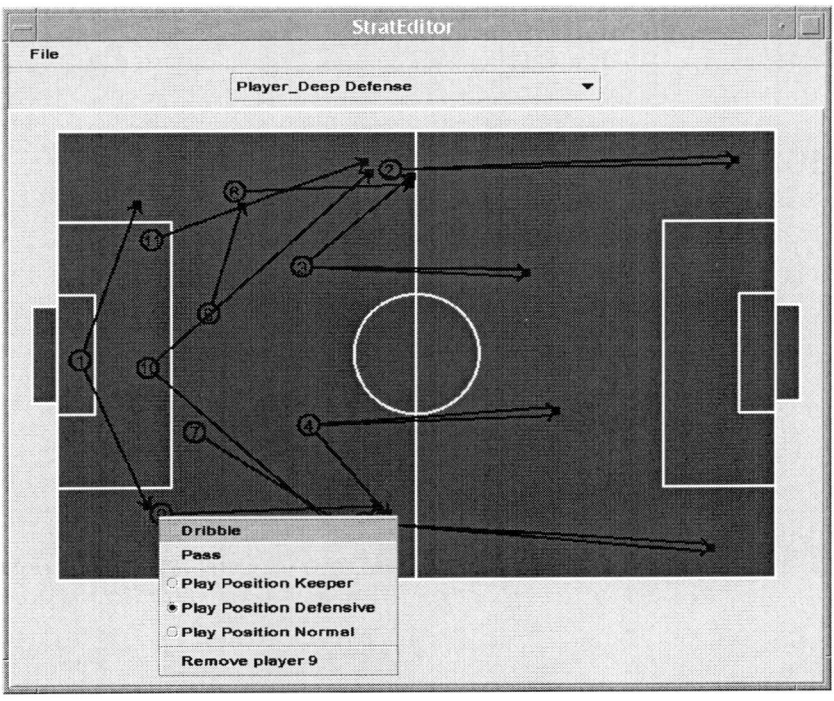

Fig. 2. A diagram of a team's defensive strategy. A player is represented by a circle. An arrow with a double line represents a direction to dribble. An arrow with a single line represents a direction to pass.

lines indicate where the player should pass when it has the ball. Priorities are not shown on this diagram. The main idea behind this particular strategy is that from defense the ball should be cleared via the sides of the ground. Notice that many of the players have multiple options when they have the ball. At runtime the agent will choose the most appropriate action that is applicable, e.g. when the player has the ball and there are opponents close the action chosen will be the pass that a teammate is most likely to receive. Notice also how the pass directions for one player are easily specified to match the positions where other teammates are likely to be situated.

3 Compiling Team Specifications to Behavior Based Trees

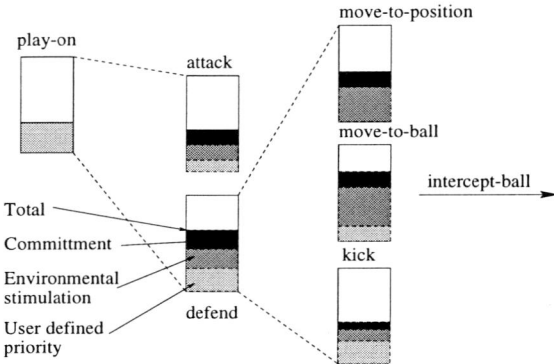

Fig. 3. A snap shot of an agent's reasoning. The grey boxes show the contributions of different components of the activation function to the total activation of the behavior.

The strategy editor is "configured" with a "skeleton" agent. The skeleton is a hierarchy of parameterized behaviors that can be subsequently instantiated into an agent. The skeleton agent can be thought of as a generic player. Three keywords are embedded in the specification of the generic player: MODE; ATTR (attribute) and ACTION. The team strategy editor uses the template (especially the keywords) to configure itself, i.e. by creating a panel for specifying each MODE, a menu item for each style (i.e. ATTR) in the mode and the ACTIONS the player can take with the style. This mechanism makes the strategy editor very flexible, quite different skeletons can be used to create teams with quite different abilities.

Once a team has been specified, partially or completely, it is compiled into a layered behavior-based agent. The process of compiling a team strategy specification consists of instantiating parameters and groups of behaviors in the skeleton hierarchy. A separate skeleton is instantiated for each player.

The architecture of the resulting agents is described in more detail in [4]. An agent is composed of a hierarchy of behaviors. At each level of the hierarchy a single behavior is chosen to act. The behavior chosen is the one with the highest activation where the activation is a function of the behavior's priority, commitment and the prevailing environmental conditions. Behaviors at higher levels of the hierarchy act by setting appropriate lower level behaviors and the lowest level behaviors act by turning on or off simple skills. Figure 3 shows a snapshot of part of a behavior hierarchy, indicating the way that activation is calculated and how it effects which behavior is selected to act.

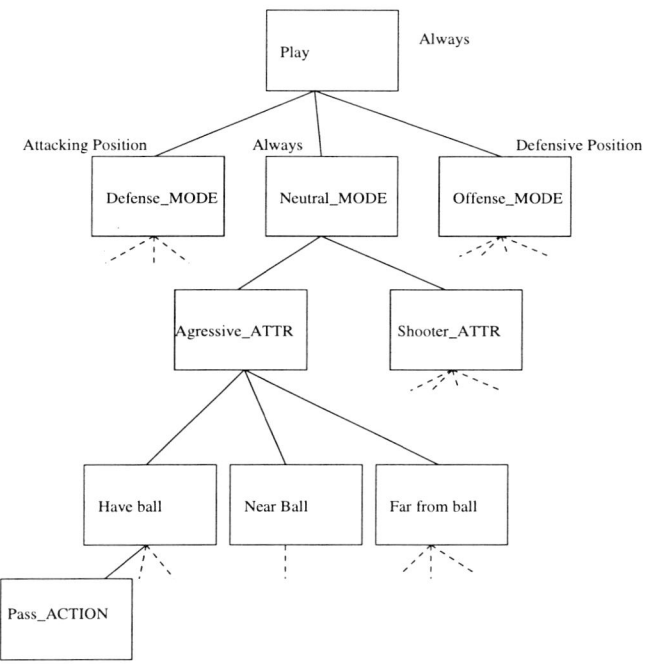

Fig. 4. Parts of the generic player's behavior hierarchy. Boxes represent behaviors. Each behavior is labeled with its name. Next to each box is the activation condition associated with the behavior.

Each team formation, i.e. one for each game mode, corresponds to one group (usually a hierarchy) of behaviors in the agent. Figure 4 shows the top part of an agents hierarchy of behaviors. Notice that there is one branch for each game mode (indicated by the keyword MODE). Appropriate conditions are associated with each hierarchy to ensure that the correct hierarchy is activated at the correct time in the game, for example the defensive strategy will be activated when the ball is in the back third of the field. A designer need not know the condition associated with the strategy.

Depending on the style chosen for the player a slightly different generic tree (indicated by keyword ATTR) is selected (by the compiler) for the agent. The hierarchies vary only slightly, mainly in terms of the functions for activating different behaviors. For example in a cautious style hierarchy behaviors for checking the location of the ball are more readily activated.

The details of the specified strategies are then used to instantiate the details of the style hierarchies. The specified position of a player on the field in a strategy is directly translated into parameters for a corresponding behavior that moves the player to a position. Figure 5 shows how one specified position has been instantiated to parameters in a behavior hierarchy. Each pass or dribble action specified for a player results in a dedicated behavior (or possibly behavior hierarchy – indicated with the keyword ACTION in the player skeleton) being added to the overall agent hierarchy. Figure 6 shows the resulting instantiated tree after behaviors for two different pass behaviors and a single dribble behavior have been added.

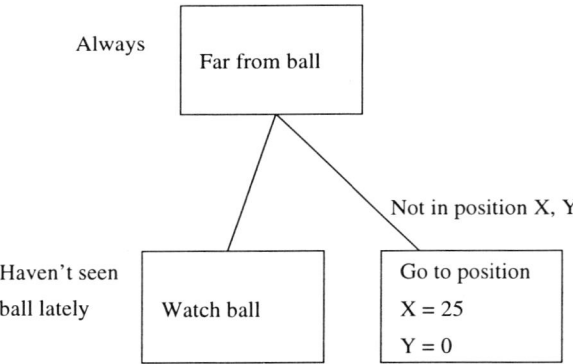

Fig. 5. This diagram shows how the position parameters of a player are incorporated into the tree. The X and Y values for the behavior and for the activation condition are taken directly from the specification.

4 Discussion

A number of different approaches have been, and continue to be, taken to meet the challenge of empowering end users to program. An evaluation of this prototype serves as an evaluation of two ideas. The first idea is that behavior based systems serve as a good target architecture for compiling end user specifications.

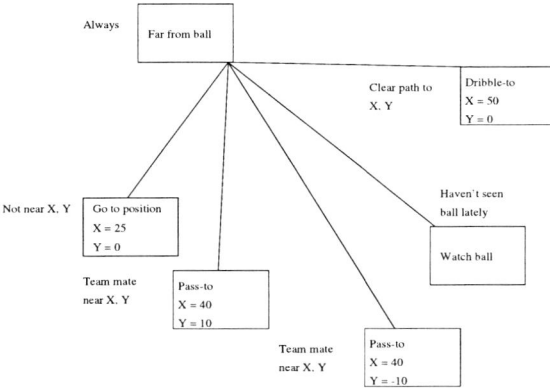

Fig. 6. This diagram shows how the behaviors have been added for two user specified pass directions and one user defined dribble direction.

The second idea is that for this domain, which we do not believe is unique, it is sufficient to give users only the ability to change the high level behavior of agents. The system described here tests both these ideas.

The first idea, that behavior based systems are an appropriate target architecture for end user programming is supported by the intuitive idea that end users describe behavior in the same way as behavior is implemented in a behavior based system. Loosely speaking, users seem to think about behavior as concurrent tasks that one switches between depending on environmental conditions. For example a soccer expert may talk about defense and attack, or dribbling and passing. In a behavior based system each of the described behaviors would be implemented as a separate logical unit (i.e. a behavior). An end user also assumes that other tasks such as maintaining situational awareness and continually evaluating whether the current behavior is most appropriate are done continuously in parallel with the current task – in a behavior based system this is standard functionality (i.e. by switching between low level behaviors).

Because the specification method of the editor is close to the usual explanation method of the designer the "translation" required by the designer is minimal. Any act of communication, even from human to human, results in some loss of information or ambiguity. The more difficult the task of communication the more the information loss and ambiguity. Hopefully by minimizing the translation required the information loss can be minimized. As a contrast consider having a user describe soccer behavior in terms of, say, hierarchical plans or beliefs and desires. Clearly an average user would need to do considerable translation from their normal way of describing what a soccer player does to express his ideas in such a formalism. Surely, it follows, that a large amount of expert knowledge would be lost in translation.

The second idea tested by our system, that a successful end user programming environment need not give programmers low level control over the behavior of agents, is perhaps less contentious. Regardless of the particular high level strategy the basic behavior of the agents does not change substantially, e.g. they still dribble, pass, run etc. in much the same way. For this reason it does not seem essential that the strategy designer be given low level control. In other domains, such as mobile robotics, the low level behavior of the agent may change dramatically with changing high level tactics rendering a high level only approach inapplicable. However simulated soccer is not the only domain where different high level strategies use the same low level behaviors, computer games and air-combat simulation being two other examples.

Unlike in domains such as manufacturing, where design, implementation and testing time may be measured in tens of man years, it is desirable that the design and implementation time of RoboCup strategies be measured in hours if not minutes. In order to support development times of this order of magnitude, the user cannot be required to make too many decisions. Even considering only high level strategies a virtually infinite design space exists. Considering that reasonable design times are unlikely to be achieved if users delve into low level behavior specifications of agents and also that at a high level there are a large range of options it seems reasonable that an end user programming system not allow the user to change the low level behavior of the agents.

During the RoboCup World Cup competition undergraduate computer science students, relatively inexperienced with agent technology, specified the behavior of HCIII and changed it (sometimes markedly) for nearly every game. Although the students became relatively experienced with the strategy editor they made very few accommodations for aspects of the underlying agent behavior and felt as though the players they specified would do as they intended. The players performed as the students intended indicating the transformation from team specification to individual agents was faithful to the designers intentions. This supported our hypothesis that behavior based architectures are a good target architecture for compiling high level specifications. The students developed a surprising range of different team strategies tailored to different oppositions. The most critical result was that the agent template rarely needed to be modified, even though the team behavior varied greatly. The students seemed more than happy to work within the confines of the team editor and managed to produce a wide range of team well tailored to particular opponents. That fact supported the hypothesis that end users did not need to specify low level aspects of the agents in order to achieve results they were looking for.

During the competition itself there was very little time for modifications of teams. However even in the shortest periods (as little as five minutes!) different team strategies, tailored to the upcoming opposition, were developed. In fact the World Cup provided a fantastic opportunity to assess the rapid development capabilities of the strategy editor and the results supported the claim that very rapid, yet effective development was possible.

5 Conclusion

In this paper we described a high level, end user programming environment for specifying team level strategies for RoboCup. The system allows a soccer expert to specify team positions and ball movement in a manner analogous to drawing on a whiteboard. The strategy editor was designed to evaluate two hypotheses, firstly that behavior based architectures were a good target architecture for compilation of high level strategies and secondly, that end users needed only high level specification capabilities when specifying complex teams. The system was put through rigorous "real-life" testing during the RoboCup World Cup. Both hypotheses were supported by the experience. Students, relatively inexperienced with agent technology, specified strategies very quickly and the resulting teams executed the strategies in accordance with the students intentions. This experience leads us to believe that the idea of high level specification by a domain expert followed by compilation into behavior based agents is worth considering in other domains.

Acknowledgments

This work is supported by Saab AB, Operational Analysis division, The Swedish National Board for Industrial and Technical Development (NUTEK), under grants IK1P-97-09677 and IK1P-98-06280, and Linköping University's Center for Industrial Information Technology (CENIIT), under grant 98.6.

References

1. Bruce Blumberg and Tinsley Galyean. Multi-level control of autonomous animated creatures for real-time virtual environments. In *Siggraph '95 Proceedings*, pages 295–304, New York, 1995. ACM Press.
2. Rodney Brooks. Intelligence without reason. In *Proceedings 12th International Joint Conference on AI*, pages 569–595, Sydney, Australia, 1991.
3. Hiraoki Kitano, Minoru Asada, Yasuo Kuniyoshi, and et. al. RoboCup: A challenge problem for AI. *AI Magazine*, 18(1):73–85, Spring 1997.
4. Paul Scerri, Silvia Coradeschi, and Anders Törne. A user oriented system for developing behavior based agents. In *Proceedings of RoboCup'98*, Paris, 1998.

Purposeful Behavior in Robot Soccer Team Play

Wei-Min Shen, Rogelio Adobbati, Jay Modi, Behnam Salemi

Computer Science Department / Information Sciences Institute
University of Southern California
4676 Admiralty Way, Marina del Rey, CA 90292-6695
U. S. A.
{shen, rogelio, modi, salemi}@isi.edu

Abstract. The annual robot soccer competition (RoboCup) provides an excellent opportunity for research in distributed robotic systems. A robotic soccer team demands *integrated robots* that are autonomous, efficient, cooperative, and intelligent. In this paper, we introduce the concept of Purposeful Behavior, to tackle the problem of achieving reactive and coordinated behavior in a team of autonomous robots. We are building a new control framework for autonomous robots to reason about goals and actions, react to unexpected situations, learn from humans and experience, and collaborate with teammates. Building such robots may require techniques that are different from those employed in separate research disciplines. We describe our experience in building these soccer robots and highlights problems and solutions that are unique to such multi-agent robotic systems in general. These problems include a framework for multi-agent programming, agent modeling and architecture, evaluation of multi-agent systems, and decentralized skill composition.

1 Introduction

The robot soccer competition (RoboCup) confronts teams of fast-moving robots that cooperatively play soccer in a dynamic environment [1]. Since individual skills and teamwork are fundamental factors in the performance of a soccer team, Robocup is an excellent testbed for integrated robots. Each soccer robot (or agent) must have the basic soccer skills—dribbling, shooting, passing, and recovering the ball from an opponent, and must use these skills to make complex plays according to the team strategy and the current situation on the field. For example, depending on the role it is playing, an agent must evaluate its position with respect to its teammates and opponents, and then decide whether to wait for a pass, run for the ball, cover an opponent's attack, or go to help a teammate.

Building these mobile autonomous robots that can function effectively in a soccer game raises a number of interesting challenges. First, autonomous robots must have an intelligent control system so that they can accomplish missions without constantly requiring humans to supply detailed instructions. This calls for the ability to not only deliberately reason about goals and plan for actions, but also reactively deal with situations and events that are not expected. Previous work on purely deliberative control systems or purely reactive control systems was limited in this respect. A purely reactive system, although fast and robust, cannot reason about actions and goals during a mission. Consider a reactive robot that is controlled by two basic behaviors, GB (go to the ball) and AO (avoid obstacles), in a goal-scoring task. When seeing the ball through an impossible gap between two idle opponents, this robot will repeatedly go back and forth on this side of the opponents without making any progress. This is because the robot is pulled back by the AO behavior to avoid the opponent robots (obstacles), but at the same time, attracted by the ball due to the GB behavior. It does not know that the correct solution in this situation is to introduce a new sub-goal to bypass the opponent robots. In a similar situation, a deliberative system, although capable of reasoning about goals, would require a detailed model of the situation in order to make a successful plan. Such a model may not be always available and there is no guarantee that any given model is complete and accurate. For example, an incomplete model may neglect the fact that the gap is impossible for the robot and thus cause the planned actions to fail in reality.

In the following sections of this paper, we will address the above issues in detail. The discussion will be organized in two parts: the description of the novel concept of purposeful behaviors, and our robot team implementation, with highlights on key components and challenges. The related work will be discussed at the end.

2 The Purposeful Behavior Approach

We are building a new control framework for autonomous robots to reason about goals and actions, react to unexpected situations, learn from humans and experience, and collaborate with teammates. Based on current research on reactive and deliberative systems, the key idea behind this approach is to make behaviors purposeful. Specifically, we propose a concept called Purposeful Behavior (**PB**) by extending a standard behavior (i.e., a pair of state $\rightarrow$ actions) with a set of purposes (i.e., state $\rightarrow$ actions $\rightarrow$ purposes). Instead of a complex mixture of heterogeneous reactive and deliberative systems, our approach is based on a uniform, and hence simpler, representation to allow a behavior-based control system to flexibly reason about goals and dynamically construct complex behaviors based on given tasks. The state in a PB is sensor-mediated so that a PB is still reactive. To ensure deliberation, the purposes of a PB are symbol-mediated so that they can serve as the desired effects of the behavior. In terms of representation, a purpose of a PB is a symbolic expression, containing a set of symbolic features that are either abstracted from the sensor data or constructed from the internal mental states. PBs differ from the operators used in traditional planning frameworks because effects directly model *all* the changes in the world that occur when the operator is executed. In contrast, a PB's purpose only describes the goal that the action achieves. A purposeful behavior provides a natural link between sensor-mediated and symbol-mediated data and integrates reactivity with deliberation without relying on separate and distinct layers of control.

The purpose of a PB can be interpreted in two ways, depending whether forward or backward reasoning is used. In the forward sense, a purpose can be interpreted as a *prediction* that specifies what is expected to be true at the end of each execution of the behavior. For example, a reasonable prediction for the GB behavior in the above example would be "reduce the distance to the ball" and it is abstracted and calculated from the sensor data on object distances. Similarly, a prediction for the AO behavior would be "increase the distance to obstacles."

The second interpretation of a purpose, when reasoning backward, is an *internal mental desire* of the robot. This is similar to the notion of desire used in the Desire-Belief-Intention (DBI) architecture. Here a desire may or may not correspond to any external state in the environment. For example, consider the desire of "staying safe" which is very difficulty to qualify in terms of descriptions of external states, but it is natural to have such an internal mental desire in order to activate a number of reactive behaviors to ensure the safety of the robot. In this light, the "staying safe" should be one of the purposes of the AO behavior. In some sense, this is similar to "goals of maintenance" (in contrast to "goals of achievement").

Fig. 1. Autonomous Soccer Robots

With the notion of Purposeful Behavior, a plan of behaviors to accomplish a given task is a *sequence of purposes or desires* (not actions) that leads and guards the accomplishment of the task. Our execution mechanism is straightforward and shares many similarities with existing execution systems. When a sequence is executed, each purpose in the sequence triggers a set of PBs that share the same purpose. These behaviors are applied reactively and in parallel until the purpose is accomplished. Upon the termination of a purpose, the next purpose in the sequence is activated, and a different set of behaviors comes into play. In this sense, behaviors are organized and played in stages and complex behaviors can be accomplished. There can be multiple sequences of purposes depending on the nature of the task. For example, the purpose of "staying safe" (avoid collisions) may be activated in parallel with other sequences of purposes so that the safety-related behaviors will be constantly active to keep the robot safe while the robot is achieving the given task.

3 Soccer Robot Implementation

The ultimate evaluation of any control software should be real autonomous robots physically functioning in the real world. In the particular case of our Purposeful Behavior approach, soccer games provide a real setting where a group of robots must cooperatively work together in a highly dynamic and unpredictable environment that contains active adversarial agents. To be able to test our ideas on the RoboCup soccer field, we have to build autonomous physical robots (see fig.1) that can function robustly in such a challenging environment. Obviously, this implies two things about our robots:

Requirement 1: They must be autonomous.
Requirement 2: They must be robust.

These requirements have significant implications on the methodology we use to build and program our robots. In particular, Requirement 1 implies that processing must be distributed and on-board. No remote computing or centralized control is allowed. Requirement 2 implies that algorithms and hardware must be simple enough to guarantee reliability. Indeed, a guiding philosophy in building these robots is to favor robustness over sophistication. Furthermore, our design philosophy for the system architecture is that we view each robot as a complete and active physical entity, which can intelligently maneuver and perform in realistic and challenging surroundings. In order to survive the rapidly changing environment in a soccer game, each robot must be physically strong, computationally fast, and behaviorally accurate [2]. Considerable importance is given to an individual robot's ability to perform on its own without any off-board resources such as global, birds-eye view cameras or remote computing processors. Each robot's behavior must be based on its own sensor data, decision-making software, and eventually communication with teammates.

3.1 Hardware Architecture

The base of each robot is a modified 4-wheel, 2x4-drive DC model car. The wheels are independently controlled, allowing in-place turning and easy maneuverability. Mounted above the base is an on-board computer. It is an all-in-one 133MHz 586 CPU board extensible to connect various I/O devices. Attached to the top of the body are twin commercial digital color QuickCam cameras made by Connectix Corp. One faces forward, the other backward. Also, we have affixed fish-eye lenses to each camera to provide a wide-angle view of the environment. The two drive motors are independently controlled by the on-board computer through two serial ports. The hardware interface between the serial ports and the motor control circuits are custom built by our team. The images from the cameras are sent into the computer through two parallel ports.

3.2 Software Architecture

Our robotic soccer team consists of four identical robots. They all share the same general architecture and basic hardware. However, they differ in their programming. We have developed three specialized roles: the forward role, the defender role and the goalie role. Each role consists of a set of behaviors organized as a state machine. For example, the forward role contains a shoot_ball behavior, dribble_ball behavior, a search_for_ball behavior, etc. The state transitions occur in response to percepts from the environment. For example, the forward will transition from the search_for_ball behavior to the shoot_ball behavior if it detects the ball and the goal from its sensory input. At game time, each robot is loaded with the program for the role it has been assigned. Note that each robot has the integrated physical abilities to play any role (i.e. detect_ball, move_forward, turn, etc.)

The software architecture of our robots is illustrated in Figure 2. The three main software components of a robot agent are the *vision module*, the *decision engine*, and the *drive controller*. The task of the vision module is to drive the camera to take pictures, and to extract information from the current picture. Such information contains an object's type, direction, and distance. This information is then processed by the decision engine, which is composed of two processing units - the internal model manager and the strategy planner.

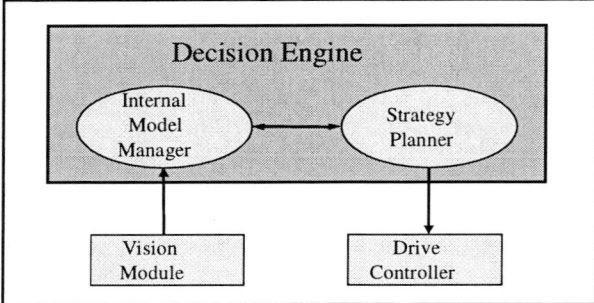

Fig. 2. The System Architecture

The model manager takes the vision module's output and maintains an internal representation of the key objects in the soccer field. The strategy planner combines the internal model with its own strategy knowledge, and decides the robot's next action. Once the action has been decided, a command is sent to the drive controller that is in charge of properly executing. Notice that in this architecture, the functionality is designed in a modular way, so that we can easily add new software or hardware to extend its working capabilities.

3.2.1 The Decision Engine

The Decision Engine receives input from the vision module and sends move commands to the drive controller. Decisions are based on a combination of the received sensor input, the agent's internal model of its environment, and knowledge about the agent's strategies and goals. The agent's internal model and strategies are influenced by the role the agent plays on the soccer field. There are three types of agent roles or playing positions: goalkeeper, defender, and forward. The team strategy is distributed into the role strategies of each individual agent. Depending on the role type, an agent can be more concerned about a particular area or object on the soccer field, e.g. a goalkeeper is more concerned about its own goal, while the forward is interested in the opponent's goal. These differences are encoded into the two modules that deal with the internal model and the agent's strategies.

In order to play a successfully soccer game, each robot must react appropriately to different situations in the field. This is accomplished by the strategy planner that resides in the decision engine on each robot. Internally, a situation is represented as a vector of visual clues such as the relative direction and distance to the ball, goals, and other players. A strategy is then a set of purposeful behaviors mapping situation to actions. For example, if a forward player is facing the opponent's goal and sees the ball, then there is a mapping to tell it to activate a Shoot@Goal PB.

3.2.2 Execution of Purposeful Behaviors

To illustrate the execution of a PB plan, consider a simple soccer task of clearing a ball out of the goal area. Assume that there exists a PB "hit ball towards opponent side" whose purposes include "push ball from own goal". To accomplish the task, a plan of desires will be set up. In this case, a single desire of "clearing ball from own goal area" will be in place. Since all PBs whose purposes match the desire will be activated, the behavior of "hit ball towards opponent side" will be triggered and applied repeatedly until the purpose is served. Compared to the traditional or augmented planning systems (e.g. [3],[4]), the advantage of this approach is that the planner does not need to specify how many times the ball needs to be hit. This relinquishes the need to have a detailed world model in order to describe how much a ball will be pushed out from the goal with one bumping action.

The purposes of PB allow the dynamic addition of new behaviors and adjustment of existing behaviors. For example, to serve the purpose of "push ball from own goal", another behavior "use kicker towards opponent goal" may be added to the system (e.g., by learning). To mediate PBs when multiple alternatives can be applied, we rely on a system of explicit preferences. These preferences are produced by the learning system, and they are constantly refined by the robot as new PBs are added. This control approach allows behaviors to be flexibly planned, grouped, and staged, as well as dynamically added or deleted. Compared to previous hybrid systems, our approach avoids the complexity inherent in programming a mixture of heterogeneous control layers. Yet it enables PBs to be dynamically activated, terminated, or prioritized during plan execution.

3.2.3 Planning

PBs can also be used to construct a plan of desires and purposes for a given task. For this objective, PBs are treated as continuous operators that have preconditions, actions, and postconditions. The state of a PB is the precondition of the corresponding operator, the action of the PB is the action of the operator, and the purpose of the PB is the postcondition of the operator. When a task is given, a plan of desires is constructed by reasoning about the purposes of individual PBs and back-propagating the goal of task through PBs' purposes (postconditions) and states (preconditions). Consider the ball-clearing example above. When the task is given, the goal of "clearing ball from own goal area" is back-propagated through the "hit ball towards opponent side" PB, whose state specifies the condition of "facing opponent goal" and "ball close ahead". If there is no ball close ahead, then the condition becomes a new purpose so a new plan is constructed with two sequential desires: "ball close ahead" and "hit ball towards opponent side". Please notice that goal-regression is only one of the existing planning methods that can be used here. In fact, most other planning methods are just equally applicable. For example, one can use a forward planner [5] and then a plan of desires is not a static sequence. In that case, the stages of desires are set up dynamically by a set of forward operators (can be implemented as a set of PBs) at the run time. We do note that the planning capabilities of our system are somewhat limited because our PB representation is simpler than traditional planning operators (not all effects may be modeled). Specifically, the system will not consider all possible behavior sequences that could accomplish a task.

The explicit representation of the purpose of a PB also enables replanning to be included in our framework. Specifically, an active PB may contribute to the decision for the next action based on the matching degree between the state description of the PB and the state of the environment. This is similar to a sub-goal mechanism used in existing planning and execution systems. It enables the robot to dynamically replan in the face of unexpected circumstances. Consider the example at the beginning where the robot fails to make any progress when seeing the ball through an impossible gap between two opponent robots. In that situation, a third PB, say ChangeTarget, may be activated because the conflict between the GB and AO behaviors (i.e., every time one behavior's purpose is satisfied, the other's purpose is violated). The ChangeTarget behavior will conclude that continuous execution of GB and AO behaviors in the current situation will not lead to any progress, thus it proposes a new desire/purpose to shift the robot's attention. This ability to dynamically modify a plan during execution is a powerful control mechanism. It is much more flexible than those fixed composition and selection methods (such as sequential, parallel, hierarchical, bidding, or arbiter) used in previous behavior-based systems. Compared to the standard behavior-based approach, where the number of complex behaviors are limited by pre-determined selection and composition methods, PB-based plans are dynamic, flexible, goal-oriented, and can be constructed and modified at the run time.

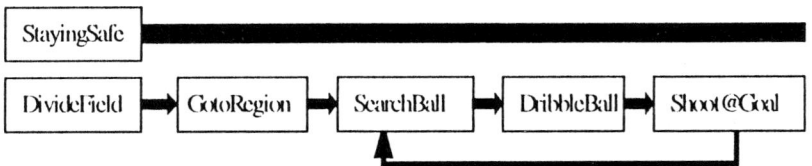

Fig. 3. The sequences of desires for a multi-robot attack mission

3.2.4 Experimental Scenarios

To illustrate the entire picture of the PB-based control architecture, let us consider a complete scenario of a simple attack mission with a group of autonomous robots. When given a specification of the designated area to be searched in the soccer field, all attack robots will construct the sequence of purposes (or stages of desires) illustrated in Figure 3.

As we can see here, the plan for the attack mission consists of two sequences of purposes, the first one has only one stage, StayingSafe, and the second one has five stages: DivideArea, GotoRegion, SearchBall, DribbleBall, and Shoot@Goal.

Each stage will have a set of PBs that are activated by their purposes. For example, at the DivideArea stage, a set of behaviors will work together to divide the area into a number of regions according to the number of robots in the team, compute the boundary of the regions, assign regions to individual robots, and so on. At the GotoRegion stage, another set of behaviors will work together enabling each robot to navigate and move into its own region, at the same time avoiding interference with other robots. At the SearchBall stage, a set of behaviors will allow robots to search for the ball in a certain pattern and get into its proximity. At this stage, behaviors such as GB mentioned above should be active. (The AO behavior is constantly active because it has the purpose of StayingSafe.) Note that some of the stages may also involve sub-stages (not shown in the picture). For example, the active behaviors at the Shoot@Goal stage may propose new (sub)purposes such as aligning ball and opponent goal, and hitting the ball as hard as possible. After shooting at goal, a robot will again enter the stage of SearchBall, and resume its search pattern.

4 Related Work

Our current approach follows an earlier, integrated system called LIVE [6] for prediction, learning, planning and action and a theory of autonomous learning in general [7]. This work also shares ideas with many cognitive architectures published in [8] that integrate planning and reacting for mobile robots, as well as recent progress in Agent research [2]. The unique feature of our approach is that a robot uses the internal model and the closed-loop control to guide its actions based on visual feedback. Our earlier work of this type includes a silver medal winner robot called YODA in the 1996 AAAI Robot competition [9].

Our approach is also closely related to behavior-based robots as described in [10]. The main difference between the PB approach and other ongoing work for autonomous robot control is that we use purposes and predictions as an explicit part of the knowledge for control. Specifically, the related ongoing work for autonomous robot control can be categorized into three basic groups in term of the structure of their control architecture: the layered, the flat, and the intermediate. The layered control architectures, see for example [11] [12] [13], use different knowledge representations to allow slower abstracted reasoning at the higher levels and faster sensor-mediated computations at the lower levels. The behaviors at the lower levels are basically black-box procedures that are not open for examination and improvement. In comparison, we use PB as a unified representation for control knowledge thus avoid the distinction between layers, and allow flexible addition, removal, and modification of control knowledge. In contrast to the flat behavior control architectures, where a fixed mechanism, such as a composer or an arbiter, is used to select or combine behaviors based on numerical values [14][15], we rely on internal desires and behavioral purposes to activate, terminate, and switch behaviors. This can avoid any built-in goal knowledge and allow reasoning and planning for goals at the run time. Compared to the intermediate approaches, where behaviors are imposed with a partial order based on the closeness to the world [16], our PBs have equal authorities so that they can decide and act on their own as long as their purposes match the current desires. This allows asynchronous and parallel execution of behaviors. Our PB approach does bear many similarities with the Soar production systems [5]. However, the major difference is the PB-based architecture allows more than one active PBs at a time, and the structures of PBs are open for examination and modification. Finally, compared to probability-based approaches [17] [18], our PB approach utilizes symbol-mediated representation for reasoning about action purposes.

5 Conclusions and Future Work

In building integrated robots that are autonomous, efficient, collaborative, and intelligent, we have demonstrated a simple but effective approach. Instead of a complex mixture of heterogeneous reactive and deliberative systems, we are introducing a new uniform, and hence simpler, representation to allow a behavior-based control system to flexibly reason about goals and dynamically construct complex behaviors based on given tasks. Moreover, it seems that the most effective approach to implement the PB concept for soccer robots is to build integrated robots using the least-sophistication to achieve the most robustness. At the present time, we are trying new sensors and we are working to improve our set of behaviors to include passing and assisting ball dribbling, and we are also adding communication to increase the robots' ability to collaborate in a wide range of situations.

References

1. Kitano H., Asada M., Kuniyoshi Y., Noda I., Osawa E. : Robocup: The Robot World Cup Initiative, in Proceedings of the first International Conference on Autonomous Agents, Marina del Rey, CA, 1997. Pp.340-347.
2. Garcia-Alegre M. C., Recio F. : Basic Agents for Visual/Motor Coordination of a Mobile Robot, in Proceedings of the first International Conference on Autonomous Agents, Marina del Rey, CA, 1997. Pp.429-434.
3. Weld, D. S. : An introduction to least commitment planning, in AI Magazine, 1994(Winter): p. 27-61.
4. Gervasio, M. : Learning general completable reactive plans, in Proceedings of the Annual National Conference on Artificial Intelligence, 1990. AAAI Press.
5. Laird, J., Rosenbloom, P. Integrating execution, planning, and learning in Soar for external environments, in Proceedings of the Annual National Conference on Artificial Intelligence, 1990. AAAI Press.
6. Shen, W. M. : LIVE: An Architecture for Autonomous Learning from the Environment., in ACM SIGART Bulletin 2(4), 1991. Pp.151-155.
7. Shen, W. M. : Autonomous Learning From Environment. W. H. Freeman, 1994. Computer Science Press. New York.
8. Laird, J.E. (ed.) : Special Issue on Integrated Cognitive Architectures. ACM SIGART Bulletin 2(4), 1991.
9. Shen, W. M., Adibi, J., Cho, B., Kaminka, G., Kim, J., Salemi, B., and Tejada, S. : YODA—The Young Observant Discovery Agent, in AI Magazine, Spring 1997. Pp.37-45.
10. Arkin, R. C. : Motor Schema-Based Mobile Robot Navigation, in International Journal of Robotics Research,1987. Pp.92-112.
11. Gat, E. : Integrating Planning and reaction in a heterogeneous asynchronous architecture for controlling mobile robots, in Proceedings of the Tenth National Conference on Artificial Intelligence. 1992. AAAI Press.
12. Miller, D., Desai, R., Gat, E., Ivlev, R., Loch., J. : Reactive navigation through rough terrain: experimental results, in Proceedings of the AAAI National Conference on Artificial Intelligence, 1992. AAAI Press.
13. Simmons, R. : Structured control for autonomous robots, in IEEE Transactions on Robotics and Automation, 1994. **10**.
14. Rosenblatt, J. K. : DAMN: A Distributed Architecture for Mobile Navigation, in Journal of Experimental and Theoretical Artificial Intelligence 9(2/3), 1997. Pp.339-360.
15. Arkin, R. C. : Integrating behavioral, perceptual, and world knowledge in reactive navigation, in Designing Autonomous Agents --- Theory and Practice from Biology and Back, P. Maes, Editor. 1990, MIT Press: Cambridge, MA.
16. Seelinger, O., and Hendler, J. : Supervenient Hierarchies of Behaviors, in Robotics, in Journal of Experimental and Theoretical Artificial Intelligence, 1997.
17. Thurn, S., Fox, D., Burgard, W. : A probabilistic approach to concurrent mapping and localization for mobile robots, in Machine Learning, 1998. **31**: p. 29-53.
18. Karen, H. Z. : Situation-Dependent Learning for Interleaved Planning and Robot Execution, in Computer Science Department. 1998, Carnegie Mellon University: Pittsburgh, PA.

Autonomous Information Indication System

Atsushi Shinjoh[1,2] and Shigeki Yoshida[1]

[1] IAMAS
3-95, Ryoke-cho, Ogaki-city,
Gifu 503-0014. Japan.
[2] Faculty of Engineering, Gifu Univ.
1-1, Yanagido, Gifu-city,
Gifu 501-1193. Japan.

{kaminari,shige}@iamas.ac.jp

Abstract. We developed an Autonomous Information Indication System for the RoboCup simulation league. This delivers and displays a three-dimensional view of the game to an audience using low-speed networks such as the Internet. Moreover, the audience has the ability to select a favorite shot from four different ones that are positioned on the field. Recenltly, our system performed succesfully at the RoboCup Japan Open 99. This paper outlines the feasibility and effectiveness of our system based on our evaluation of various experiments.

1 Introduction

The research into an infrastructure for the RoboCup simulation league has developed rapidly. Its initial motivation was to create a system that can indicate the situation of the simulation league more attractively. This research can be divided into two kinds. The first is research into a commentator system that appreciates a situation of a soccer game and makes comments similar to a on the spot broadcast in real-time[4][5]. The MIKE, developed by ETL team and demonstrated at the RoboCup-98[6] is a typical system of this type. The other is research into a three-dimensional viewer system that describes three-dimensional situations of a soccer game in real-time. Our proposal was to develop the system, actually developing two kinds of three-dimensional viewer systems, and demonstrated these systems at the RoboCup-97 and the RoboCup-98[1][2]. Several three-dimensional viewers such as RoboMon and Virtual RoboCup were also developed after our initial demonstrations[9][7].

The creation of these systems, the first stage of infrastructure domain, was demonstrated at the RoboCup-98. Each system was able to provide a vivid picture of the game situation. The infrastructure domain must be challenge to go to the next stage. The goals for this next stage are the followings:

- Using several techniques, the system can indicate the information (scene data) in several environments in real-time.
- The system can cope with the various choices of an audience in real-time.

– The system can indicate the most suitable scene data based on a prediction of the indicated situations and an evaluation of scene data.

These factors turn the infrastructure system into an information indication system for a digital broadcasting environment. An audience can watch a particular situation of a game anyplace that is connected to computer network. Furthermore, an audience does not have to control the indication system while using it. Based on these architectures, this system will be the ideal information environment for an audience.

We already started to develop an autonomous information indication system (AIIS) as a support system for personal information environments. The AIIS transmits and displays the information to each person using those methods that it has predicted as the most effective. Based on this, we developed an AIIS for the RoboCup simulation league that possessed the above features. This development has also been guided by our evaluation and analysis of our two three-dimensional viewer system. As first, we mounted two mechanisms. One was a scene data delivery mechanism that supports data delivery using low-speed networks such as the Internet. This mechanism was able to cope autonomously with rapid changes in the data delivery rate. The other was a view selection mechanism. An audience can select a favorite shot from four different cameras, including zoom. After selection, the system can indicate the scene data to several audiences. We experimented with this system at the RoboCup Japan Open 99 that held in May.

In this paper, we describe the basic concept of AIIS, and discuss its potential.

2 Autonomous Information Indication System (AIIS)

As noted above, we developed the AIIS for the RoboCup simulation league. Figure 1 shows its system architecture. The Following is a description of each part of the system.

Communication mechanism: This mechanism contains two mechanisms, a scene data deliverer and a network's traffic detector. This unit measures the immediate amount of network traffic, and continually informs the "scene selection mechanism" of its condition.

There are several methods for measuring the amount of network traffic. After evaluating the results of previous experiments, we selected one of these at the RoboCup Japan Open 99. A detailed description of each method is described in the next chapter.

Scene selection mechanism: This selects scene data for delivery based on an evaluation. First, the mechanism receives a scene request from each client system, checks the "evaluation value" added and sent by a "game condition mechanism", and receives the current scene delivery rate which has been determined by a "communication mechanism". Second, this mechanism selects the scene data based on an analysis of it.

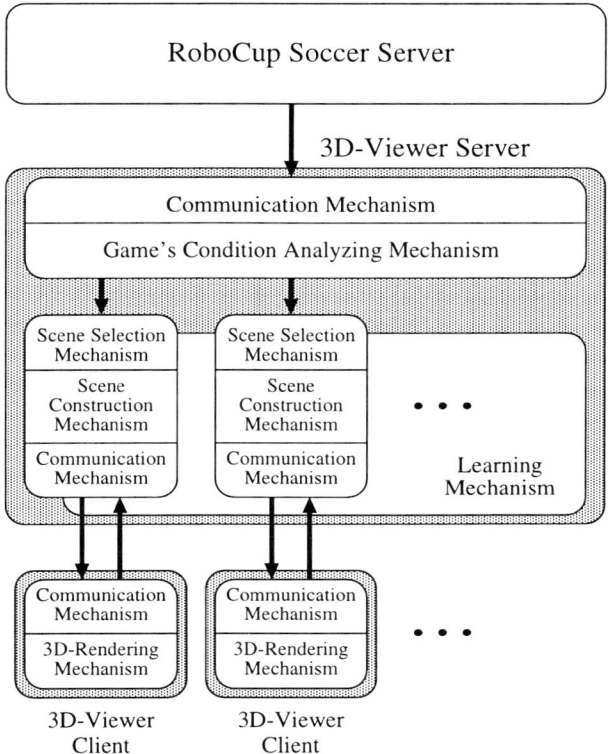

Fig. 1. The AIIS for the RoboCup.

Game condition analyzing mechanism: This mechanism evaluates the conditions of the game based on the situation of the camera and each player's positions, the formation, and the score, and adds an evaluation value to each scene. The evaluation value shows each scene's value.

3D-rendering mechanism: This renders a three-dimensional scene on the screen. This is part of a 3D-viewer client. A 3D-viewer client is composed of this and the communication mechanism. Our AIIS for the RoboCup simulation league has two 3D-viewer clients. Each client uses a different 3D-rendering mechanism. Though its detailed descriptions of a three-dimensional world are very effective in showing the situation, it needs a great deal of rendering power. As a result, this mechanism limits the number of running environments.

To cope with numerous running environments, the system adopts the two 3D-rendering mechanisms, and offers two 3D-viewer clients that use each of the mechanisms. One is a VEGA version (Figure 2). VEGA is one of the major applications of virtual reality systems for real-time processes, and supports an

Fig. 2. 3D-viewer client: VEGA version.

OpenGL library under Irix and WindowsNT environments. This version needs a VEGA runtime application. The representations of this system are quite detailed, and each player's movements are very smooth. However, this system needs a high-end graphic accelerator board with a high-performance CPU.

The other 3D-rendering mechanism is DirectX (Figure 3). This version runs under Windows95, 98, NT, and needs a DirectX library only[3]. Its description is very simple, and it can run smoothly using a mid range personal computer. This client system has six buttons for the audience to use when selecting camera positions.

3 Experimentation

3.1 Prototype 1

We developed a prototype of the AIIS for the RoboCup simulation league that has a delivery mechanism that selects the most suitable scene data that is sensitive to the network traffic. This system has the following two functions:

1. The system detects the network traffic autonomously, and controls the amount of scene data transmission corresponding to the network traffic.
2. The system selects the most suitable scene for the extraction of the "delivery scene data" in real-time.

In terms of the first function, the communication condition is judged automatically, and the scene data is selected and delivered at the appropriate rate to

---
[3] This library can be downloaded from Microsoft Web site

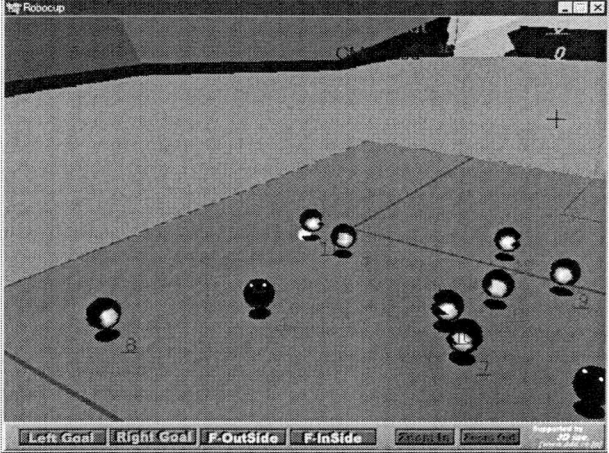

Fig. 3. 3D-viewer client: DirectX version.

the three-dimensional viewer client. This function corresponds to the communication mechanism and the scene selection mechanism in Figure 1.

As for the second function, the scene data from soccer server is analyzed. If additional scene data is required to display a game situation, this data is also selected. This function corresponds to the game condition analyzing mechanism and scene selection mechanism in Figure 1.

A basic mechanism of a traffic control At this time, we have adopted three kinds of methods that may be more effective for network control from many methods that we would have selected.

These methods are based on the following fundamental mechanism. (Figure 4)

1. When scene data is received from the soccer server, the three-dimensional viewer server adds a sequential number to each scene to distinguish each unit of scene data, and delivers this data to each client.
2. When each client receives the scene data, it makes an acknowledgment (ACK) which contains the scene data's sequential number, and returns it to the server. In one method, the client calculates the scene delivery rate, and includes it in the ACK.
3. The server or the client decides the scene delivery rate based on the return rate of the ACK or the receiving rate of the scene data.
4. The server decides whether a scene should be delivered or not, in accordance with the determination of delivery rate. Then, only after the delivery of the current scene has been decided on, does the server send scene data to the client.

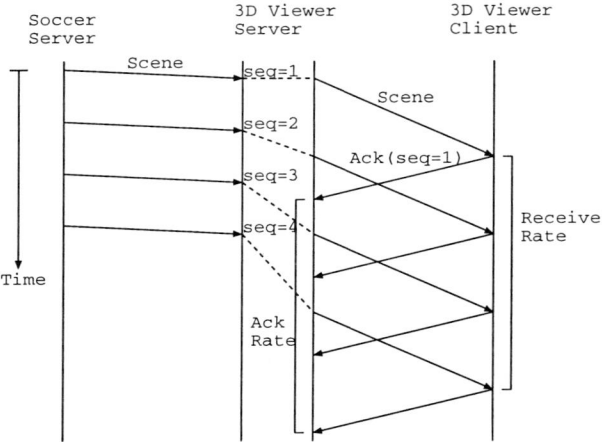

Fig. 4. The basic mechanism of traffic control.

Thus, there are two kinds of methods for deciding the scene delivery rate. The server uses one, and the client uses the other. In this prototype system, "type 1" and "type 2" adopt a method when the server decides a scene delivery rate, and "type 3" adopts a method when the client decides a scene delivery rate.

Type 1 In this method, the three-dimensional viewer server sends one scene data to the client(Please check Figure 4.). Then, the server sends a new scene data after an ACK of the scene is returned from the client.

This method is limited in that the server can't deliver the next scene before the server receives the ACK from the client. Because this situation leads to an interruption of delivery, a server sends scene data when the interval time is timed-out.

Type 2 In this method, the server calculates the delivery rate based on the rate of the ACK that is sent back from the client. The average ACK return rate is measured per "unit time". This measurement method is calculated in the following way:

- The server sends the all scene data during the first "unit time".
- When unit time has expired, the server calculates the average ACK return rate. Based on this, the server then calculates the delivery rate of the scene data for the next "unit time".
- The server delivers the scene data at a calculated delivery rate before this "unit time" has expired.
  The same procedure is then repeated.

Type 3 In this method, the client calculates the received scene data as the received scene dates rate, and then sends this data rate to the server. Based on this rate, the server determines the delivery rate.

This method adopts the two kinds of measurement methods also utilized by "type 2"

The suitable scene selection mechanism The scene data to deliver is only selected with a definite interval depends on the delivery rate, when the traffic control mechanism is used.

To select and deliver the most suitable scene data, we implemented the following mechanism into the three-dimensional viewer server.

- The server checks the scene data that was received from the soccer server. It then determines what the next scene data should be delivered when game conditions change a free kick, a goal, etc.
- The server delivers both scene data selected by the suitable scene selection mechanism and scene data selected by the traffic control mechanism.

The server judges the change in the game conditions using the "play mode flag" of the scene contained in the scene data.

3.2 The result of experimentation with the "Prototype 1"

The implementation and the method of measurement The prototype system was implemented by using a three-dimensional viewer server for UNIX, and the two-dimensional monitor that complies with the three-dimensional viewer server for UNIX and MS-Windows.

The followings are test environments.

- Substituting a logplayer for UNIX for a soccer server.
- A three-dimensional viewer server for UNIX.
- One two-dimensional monitor that complies with a three-dimensional viewer server for MS-Windows.
- A PPP connection circuit via a modem connected at 28.8 Kbps.
- A log data of the final game of the RoboCup-98 simulator league.

The measurement of the non-implemented viewer For comparison sake, we studied the case where above mentioned traffic control mechanism and a suitable scene selection mechanism were not mounted. The results of our measurements are as follows. (Table 1)

- One ACK was returned on an average of 3 - 4 scenes (0.3 - 0.4 seconds).
- After the client received 20 scenes, further scene data failed to reach to the client, and ACK wasn't returned.

Table 1. The behavior of the non-implemented viewer. (rate=scenes/sec.)

sec.	0.1	0.2	0.3	0.4	0.5	0.6	0.7	0.8	0.9	1.0	1.1	1.2	1.3	1.4	1.5	1.6	1.7	1.8	1.9	2.0
rate	10	10	10	10	10	10	10	10	10	10	10	10	10	10	10	10	10	10	10	10
sent	1	2	3	4	5	6	7	8	9	10	11	12	13	14	15	16	17	18	19	20
ACK										1				2				3		

Table 2. The behavior of "type 1". (mode: k=free kick, p=plan on)

sec.	0.1	0.2	0.3	0.4	0.5	0.6	0.7	0.8	0.9	1.0	1.1	1.2	1.3	1.4	1.5	1.6	1.7	1.8	1.9	2.0
rate	–	–	–	–	–	–	–	–	–	–	–	–	–	–	–	–	–	–	–	–
mode	k	p	p	p	p	p	p	p	p	p	p	p	p	p	p	p	p	p	p	p
sent	1	2							9				13				17			
ACK									1				2				9			

Type 1 When the method of "type 1" is implemented into a server client system, then the resulting measurements occurred. (Table 2)

- The system approaches stability, and works.
- The RTT (Round Trip Time) of one scene data is about 0.8 - 0.9 seconds (about 8 - 9 scenes).

When game condition was unchanged due to a longer delivery time of the scene data and ACK rate, compared with real possible rate (calculated by hand, so only potentially), the calculated average scene delivery rate is about 1/2.

However, when the game condition was changed, the scene data is still delivered to the client by the suitable scene selection mechanism even when ACK wasn't return to the server. Because of this behavior, the scene delivery rate increases and the game is displayed more smoothly.

Type 2 When the method of "type 2" is implemented, the following measurements result (Table 3):

1. When the unit time is 2 seconds (20 scenes), the ACK comes to a halt after about 200 seconds.
2. When the unit time is 1 second (10 scenes), the system becomes virtually stable and works.

The situation where "unit time is 2 seconds" was very similar to the situation: "non-implementation". Based on a calculation of the delivery rate of the scene data,

Table 3. The behavior of "type 2". ("unit time" = 1 sec.)

sec.	66.1	66.2	66.3	66.4	66.5	66.6	66.7	66.8	66.9	67.0	67.1	67.2	67.3	67.4	67.5	67.6	...
rate	2	2	2	2	2	2	2	2	2	3	3	3	3	3	3	3	...
mode	k	k	k	k	k	k	k	k	k	p	p	p	p	p	p	p	...
sent			663					668		670			673			676	...
ACK	600			603				606				609					...

the server changed the delivery rate after 2 seconds. However this subtle difference is accumulated at the clients step by step, until the client is unable to transmit the ACK to the server.

In this measurement, when the "unit time is 1 second", the situation becomes stable, because this unit time is quite effective for thie environment. However, when the measurement environment changed, and the speed of the connection circuit slowed down, the system had to choose a shorter unit time to deal with this environment.

Therefore, to work the system at every circuit speed, it is necessary to mount a mechanism that can find the proper delivery rate based on the measurement of the ACK information. This must be done before the system cannot handle the situation.

Type 3 In most cases, it showed the same result as "type 2".

However, when the server did not receive an ACK, the server delivered the scene data on the basis of the received scene data rate calculated by the client. Therefore, the server would not cope with changes in situation.

3.3 Prototype 2

The implementation Based on a result of the study of the Prototype 1, we developed a prototype of a system where a three-dimensional viewer client can send a request about the form of a display.

On the client side, the following items are available for selecting the position and movement of the camera. Once a selection is mode, a request is then sent to the server.

- The movements of each camera:
  - A camera positioned behind the goal of "team A".
  - A camera positioned behind the goal of "team B".
  - A camera that is moved along the side lines.
  - A camera that is moved freely around the ground.
- The motion of the camera:
  - Zooming in.
  - Zooming out.

A server has two modes to select and deliver the scene data. One mode selects a scene individually in response to each request from the clients, creating and sending the data of each scene to the clients. The other mode selects one scene as a result of a majority decision, however also creating and sending the data of the scene to the clients.

The mechanisms In this prototype system, the following methods are adopted for each mechanism in Figure 1.

A mechanism for analyzing a game condition: Any change of the game's condition is decided by the play mode flag of the scene.

A learning mechanism: Requests for a particular position or a movement of a camera concerned in response to the position of the ball or a change in the game are recorded for each client.

A scene selection mechanism: The method of prototype 1 is used for a decision about whether current scene data should be sent or not.

The decision to select a camera position and movement is made on a request from each client, and trends perceived by the learning mechanism.

The decision about which mode will be used to select and to deliver scene data is determined from out side of the system by an operator.

A mechanism for scene construction: Basically, individual scene data is created for each client. However, when a mode to select one scene for all clients is chosen, only one scene is created.

A communication mechanism: The function that adds a request from a client to the ACK data and sends it to a server is included in the mechanism implemented in the prototype 1.

3D-rendering mechanism: This mechanism rendered each scene using OpenGL or DirectX graphics library.

3.4 The result of experimentation with the "Prototype 2"

We performed an experiment of the delivery capabilities of prototype 2 using the Internet. This was a public experiment, open to everyone, was held at May 2 - 3, 1999 as a part of the RoboCup Japan Open 99. This experiment was not a replay of past simulation leagues, but an actual broadcasting of games in real-time. This experiment occurred in the following environment:

- Delivery data: The final and semi-final game of the RoboCup Japan Open 99 simulator league.
- The number of 3D-viewer clients: 10 - 20.
- The number of client machine: 10 - 20.

- Connection speed: 1.5Mbps (From conference center to the Internet)
- Operating system of clients: Not determined (MS-Windows95, 98, or NT)

Because there was a long holiday in Japan during this period, the maximum number of client accessing at same time was 20. However, in this experiment, each number of the audience could connect to a communication mechanism in low speed network environment such as 28.8Kbps. Moreover, some of there members were able to use the scene selection mechanism such as switching and zooming of cameras.

The AIIS was succeesful in managing each connection, and delivered scene data according to each client's request. Based on an analysis of this experiment, we decided to expand the "communication mechanism" and "scene selection mechanism", as well as the architecture that supports very low speed network connection such as 9.6Kbps. As you may know, this is the connection speed of mobile telephones in Japan.

4 Conclusion

In this paper, we described the basic concept of the AIIS for the RoboCup simulation league. We also showed its effectiveness and feasibility through an evaluation of the results of several experiments. Through our evaluations, we learned what areas need to be improved. We also decided to develop an AIIS for the RoboCup-rescue. The purpose of this system is to create an ideal wearable information environment for disaster rescue. Later, we will describe our developing concept of "Wearability Design", a concept of an ideal individual information environments[3].

Acknowledgement

We would like to thank Internet Initiative Japan Inc., Intergraph Corp. Japan, and 3D Inc. for their generous support. We would also like to thank many individuals who helped us. This research is one of the projects of International Academy of Media Arts and Sciences (IAMAS), which is supported by the Gifu prefecture.

References

1. A. Shinjoh: "RoboCup-3D: The construction of Intelligent Navigation System", RoboCup-97: The First Robot World Cup Soccer Games and Conferences. 179–190, Springer (1998).
2. A. Shinjoh, S. Yoshida: "The intelligent three-dimensional viewer system for RoboCup", In *proc. of the second RoboCup Workshop*, 37–46 (1998).
3. A. Shinjoh, S. Yoshida: "A development of Autonomous Information Indication System for RoboCup Simulation League", In *proc. of The 1999 IEEE Systems, Man, and Cybernetics Conference (SMC'99)*, 1999 (This will be held at October 1999).

4. H. Matsubara, I. Frank, K. Tanaka, I. Noda, H. Nakashima, K. Hasida: "Automatic Soccer Commentary and RoboCup", In *proc. of the second RoboCup Workshop*, 7–22 (1998).
5. K. Binsted: "Character Design for Soccer Commentary", In *proc. of the second RoboCup Workshop*, 23–36 (1998).
6. K. Tanaka, I. Noda, I. Frank, H. Nakashima, K. Hasida, H. Matsubara: "MIKE: An automatic commentary system for Soccer", In *proc. of Int. Conf. on Multi-agent Systems 98*, 285–292 (1998).
7. RoboMon: http://medialab.di.unipi.it/Project/Robocup/index-en.html
8. S. Yoshida, A. Shinjoh: "The prototype system of the suitable network traffic mechanism for the three-dimensional viewer for RoboCup", In *The proc. of PRICAI workshop of RoboCup*, 134–144 (1998).
9. Virtual RoboCup: http://www.TechFak.Uni-Bielefeld.DE/techfak/ags/wbski/3Drobocup/3Drobocup.html

Spatial Agents Implemented in a Logical Expressible Language

Frieder Stolzenburg, Oliver Obst, Jan Murray, Björn Bremer

Universität Koblenz-Landau, Fachbereich Informatik
Rheinau 1, D–56075 Koblenz, GERMANY
{stolzen,frvit,murray,moddy}@uni-koblenz.de

Abstract. In this paper, we present a multi-layered architecture for spatial agents. The focus is laid on the declarativity of the approach, which makes agent scripts expressive and well understandable. They can be realized as (constraint) logic programs. The logical description language is able to express actions or plans for one and more autonomous and cooperating agents for the RoboCup (Simulator League). The system architecture hosts constraint technology for qualitative spatial reasoning, but quantitative data is taken into account, too.

The basic (hardware) layer processes the agent's sensor information. An interface transfers this low-level data into a logical representation. It provides facilities to access the preprocessed data and supplies several basic skills. The second layer performs (qualitative) spatial reasoning. On top of this, the third layer enables more complex skills such as passing, offside-detection etc. At last, the fourth layer establishes acting as a team both by emergent and explicit cooperation. Logic and deduction provide a clean means to specify and also to implement teamwork behavior.

1 Introduction

Naturally, tasks to be solved by a team of autonomous agents are many-sided and complex. In order to achieve a goal, a single agent has to use a set of complementary subtasks. On the one hand, some of these actions can be performed in a purely reactive manner, meeting real-time requirements. On the other hand, tasks may require a certain amount of planning and reasoning. So, we were led to the idea of combining both the advantages of procedural and logic programming and decided on a hybrid system with a layered architecture.

1.1 Implementing Agents in Logic

In contrast to other approaches that provide an architecture for (multi-)agent systems (see e.g. [16, 24]), we use different logical and deductive formalisms not only as a specification language but also as an implementation language. Widespread in this context is the use of a Belief-Desire-Intention (BDI) architecture (see e.g. [7]), which has been originally specified by means of modal logics. A first-order axiomatization has been proposed for this kind of architecture only recently [24]. However, it seems that it is not actually used as implementation language there.

We will now describe our system architecture and show how different deductive processes—including constraint solving—can be used for the RoboCup [20]. The system combines the BDI approach with a multi-layered architecture, allowing multiple agents to perform collective actions. Nevertheless, each agent is autonomous and can be implemented in a manner similar to (Constraint) Logic Programs (CLP) [15]. This combines the advantages of being declarative and efficient to a certain extent.

The major goals of the RoboLog project, undertaken at the University of Koblenz, Germany, are the following:

- A flexible, modular system architecture should be established, meeting the various requirements for RoboCup agents. For example, on the one hand, agents have to be able to react in real-time. But on the other hand, it is also desirable that more complex behavior of agents can be programmed easily in a declarative manner.
- It should be possible to handle different representation formats of knowledge about the environment. Information may be quantitative or qualitative in nature. Therefore, we propose a deductive framework, that is expressible in plain first-order logic (possibly plus constraint technology components), that integrates axiomatic approaches in geometry, spatial constraint theories, and numerical sensor data.
- Agents should not only be able to act autonomously on their own, but also to cooperate with other agents. For this, we develop a multi-agent script language for the specification of collective actions or intended plans that are applicable in a certain situation. These scripts can be translated into logic programs in a straightforward manner.

1.2 Outline of the Approach

In the following, we discuss our layered system architecture and the functionality of the respective layers. Fig. 1 shows the complete architecture of RoboLog. The lowest layer—the RoboLog kernel, which is implemented in C++—essentially is the interface between the SoccerServer [9] and Prolog, since all other layers are implemented in this logic programming language.

The basic layer hosts reactive behavior. It is implemented in the RoboLog Prolog extension [21, 22]. This extension is an enhanced RoboCup SoccerServer interface for ECLiPSe-Prolog [14]. Time critical tasks are handled within the RoboLog module, as well as the exchange of data. The module provides the atomic SoccerServer commands and some more complex actions. Hence already at this level, logic (programming) formalisms are available. Also position determination is settled in this layer (see Sect. 2.1). It also provides more specific facilities, e.g. dribbling and ball interception. For these actions, (almost) no spatial cognition is required.

Spatial cognition is the contents of the second layer. For example, players have to recognize when passing the ball is possible or a player is offside. Many approaches (see e.g. [8, 26]) propose purely qualitative reasoning, i.e. disregarding quantitative information after it has been transferred into a qualitative representation. But this may be too inexact and too vague sometimes. Since we use logic as connecting formalism in all layers, we can access low-level data at all levels of abstraction. This implies, reasoning can be as exact as required. We will present our approach in more detail in Sect. 3.

The last two layers host complex situations, possibly requiring teamwork, i.e. single- or multi-agent plans. Nevertheless, the question remains whether teamwork should be invoked explicitly by communication or whether it is sufficient and more robust just to have implicit (emergent) teamwork. The current implementation implicitly exploits knowledge on other implementation of agents. With the exception of the goalkeeper, they are clones of each other. Cooperative behavior may be required even if the implementation details are different or not known. The problem is then, what communication language can be used in this case. See also Sect. 5.2 on this topic.

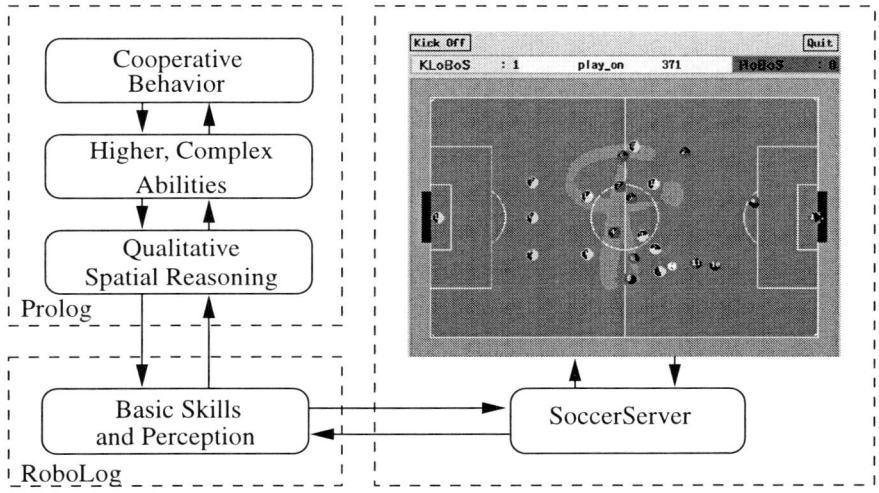

Fig. 1. System Architecture of RoboLog.

2 Basic Abilities and Actions (Layer 1)

The lowest layer in our system architecture handles basic skills and perception of the environment. The basic skills may be actions that can be performed immediately by the agent, e.g. turning around, dashing, kicking the ball etc. In addition, we will allow more complex actions in this layer, that do not need (qualitative) spatial reasoning.

Depending on the hardware used, perception of the environment, including self and object localization is a complex task, requiring more or less processing. In the simplest case, perception just means reading off the data from one of the agent's sensors. Note that we aim at having a (first-order) logic presentation for each agent. The logical description language we are going to introduce allows agent programs (scripts) to be written and interpreted in a manner similar to CLP.

Following the lines of [24], we distinguish two classes of predicates: ACTIONS a and PERCEPTIONS p. When executed successfully, a perception predicate p returns the requested data. We will assume, that this data is quantitative, i.e. some arguments of the predicate are (real) numbers. For example, a perception predicate p may return the distance to a certain landmark, measured in meters and given as a real number. The main matter of an action a is its side-effect, i.e. the performed action. Nevertheless, an action predicate (except the primitive actions of the SoccerServer) also is assigned a truth value, depending on the success or failure of the action. Note that the truth value for all predicates is dependent on the actual time t, when the action or request for data is executed.

In summary, the RoboLog interface provides the following functionality:

- For each agent, it requests the sensor data from the SoccerServer. By this, the agents' knowledge bases are updated periodically. If some requested information about a certain object is currently not available (because it is not visible at the moment), the most recent information can be used instead. Each agent stores information about objects it has seen within the last 100 simulation time steps.
- This low-level data is processed in such a way that more complex and more precise information becomes available, such as global position information (see also Sect. 2.1) or direct relations between objects with or without reference to the actual agent. The relation $is\_left(Obj_1, Obj_2)$, e.g., depends on the relative position of the agent, whereas $is\_between(Obj_1, Obj_2, Obj_3)$ is an agent independent property.
- Last but not least, Prolog predicates are provided that can be used to request the current status of sensor information on demand. The data should be synchronized with the SoccerServer, before an agent's action is initiated.

2.1 Position Determination

An important piece of information for an agent is to know its own position. Therefore, the RoboLog system provides an extensive library that makes precise object localization possible. The whole procedure implemented in the RoboLog kernel is able to work even when only little or inconsistent information is given. In particular, we employ the method for mobile robot localization using landmarks stated in [4].

2.2 Basic Skills

Agents have to be able to move in their environment without collision. This is a basic requirement for many practical robot multi-agent systems. In the RoboCup scenario agents should also be able to handle the ball. This means they must be able to run and kick to a certain position, dribble with the ball etc. Another important task is ball interception. For this, an agent has to recognize and compute the ball trajectory in advance, compute and go to the point where ball interception is possible, and stop the ball. This is a macro task, which could be executed in a certain situation without any qualitative reasoning.

A large set of low-level abilities for the RoboCup scenario is stated in [25]. There, kicking, goal-tending and—as a sub-task—getting sight of the ball among others are considered as part of the low-level architecture of an agent. Of course, such tasks may require deep computation. However, only quantitative data is used for these actions. This is the reason why it is reasonable to classify these actions as basic skills. Nevertheless, more complex actions will require (deductive) reasoning. That is the contents of the next layer (see Sect. 3).

In our system, the following basic skills (among others) are implemented (see also [18] that also describes special skills of the goalkeeper):

- The agents can search for the ball, taking into account their knowledge about the last time the ball was seen.
- Dashing and kicking to a certain position, regarding the agent's condition and avoiding obstacles is possible and (based upon these skills) also dribbling.
- Extrapolating the ball trajectory to a given time in the future enables the agents to intercept opponent passes and block shots.

3 Qualitative Spatial Reasoning (Layer 2)

During a match, a human soccer player will enter a lot of different situations, in which he has to decide what to do. In most of the cases, he will decide regarding former experience, i.e. comparing his situation to situations he already handled before. Hence, if we want to build a client, we have to provide the client with some situations and connected actions. We decided to model situations with the help of qualitative relations for two main reasons.

- The agent's situation will almost never fit exactly into a stored situation pattern (identified by its set of preconditions), so we have to parametrize and abstract the patterns. A basic set of qualities can be very easily abstracted from the visual data sent by the SoccerServer (see below). Thus the step from describing situations by quantities with tolerances to using qualitative data is easily taken.
- We think that *qualitative* spatial reasoning reflects the thoughts of a human player more clearly than the use of *quantitative* data. Consider a human soccer player who tries reaching the ball. He will think something like: the ball is *close enough*, or: a team-mate is *nearer* to the ball. Based on these *qualitative* perceptions he decides whether to run towards the ball or stay where he is. He will not calculate the trajectory of the ball and determine a set of coordinates at which he can intersect it.

What we need in order to identify situations is the abstraction of quantitative data onto a qualitative level. Therefore, we have another class of predicates—in addition to the classes mentioned in Sect. 2—, namely QUALITIES q. Qualitative predicates are defined upon the quantitative perceptions via logical rules and constraints, e.g. the *in-front-of* relation (1) (see below). But it may also be the case that there are qualitative predicates or relations based on each other. In the latter case we speak of purely qualitative predicates or reasoning, e.g. the relation *left_and_in-front-of* can be reduced to the qualitative predicates *left* and *in-front-of* (2).

$$in\text{-}front\text{-}of \leftarrow Dist > 0. \tag{1}$$
$$left \leftarrow Dir < 0.$$
$$left\_in\text{-}front\text{-}of \leftarrow left \wedge in\text{-}front\text{-}of. \tag{2}$$

For example, concerning the distance of an agent to the ball in the RoboCup scenario only a few (qualitative) aspects are interesting. Thus, in RoboLog we only distinguish few distances: *close* (the ball is in the kickable area), *near* (the agent is able to detect much detail by its sensors), *short* (maximal shooting distance), *far away* (sensor data become unreliable from this distance), *remote* (out of reach). Quantitative distance intervals can be mapped to qualities. Concerning the other direction, chosen plan schemes must be instantiated with quantitative data for the actual execution. A related work is presented in [8]. There, reasoning on the qualitative level (alone) is provided. Fig. 2 illustrates the correspondence between quantitative and qualitative distances.

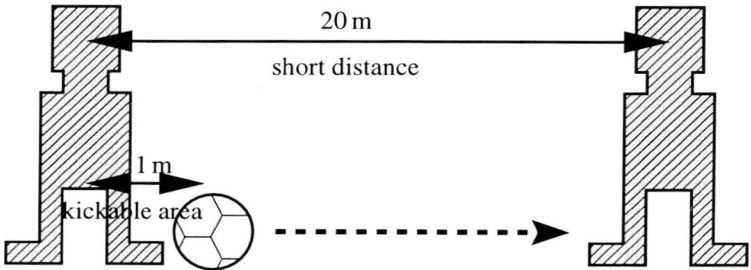

Fig. 2. Distances – quantitative and qualitative.

3.1 Constraint Reasoning

In the literature, many approaches for qualitative spatial reasoning are proposed. Most of them rely on the Region Connection Calculus (RCC), see e.g. [3, 23]. On the one hand, the advantage of qualitative information certainly is, that seemingly complex situations can be reduced to a few patterns of situations, and concentration on the relevant portion of information is possible. On the other hand, a qualitative description may be a too rough approximation of the reality, such that reasoning on a purely qualitative level may become too vague. So the question remains, how can we make use of both quantitative and qualitative information.

In most cases, if sensor data is available, it is a good idea to make use of the quantitative data by just abstracting it to a qualitative level. Only in some cases, when no more precise quantitative information is available, purely qualitative reasoning is necessary. More precise knowledge should be preferred. So, we combine real-time quantitative reasoning with qualitative spatial reasoning, that can be implemented as a constraint system (in the formal sense) and integrated in a more general deductive framework for constraint logic programming (CLP).

The process of spatial reasoning has to be seen in the context of its purpose, that is laying the basis for what action should be performed next. There are (at least) two decision problems in this context:

- If there are different sources of information (e.g. numerical sensor data, derived qualitative knowledge or conclusions thereof), there must be some control mechanism for deciding how the requested information should be obtained. In our current implementation, quantitative data is preferred: it is simply converted into a qualitative presentation. There are only very rare cases where purely qualitative reasoning is performed. This could mean applying the transitivity rule to topological relations such as *between*.
- In addition, it may be difficult to decide what should be done next in a situation where we have several options (e.g. dribbling, passing, kicking). In the current implementation, we simply make use of the backtracking facilities of Prolog for this purpose. However, it might be a good idea to employ defeasible reasoning in this decision process [11].

3.2 An Axiomatic Approach

We are also investigating the problem of modelling certain situations as patterns by means of logic programs and the full first-order theorem proving system Protein [2]. For example, passing the ball is possible in a situation where one player has the ball, another player can be reached and there is no player (of the opposite team) in between. We modelled these situations on top of the logical relations *left*, *right* and *between*. Since we use logic, the properties of the qualities have to be axiomatized. Two possibilities come into mind: we can model *between* on top of general geometric axioms [5], or use *collinear* as basic concept [12]. We believe that it is more natural to use (an ordered version of) *between* as base relation, since we can assume that the sensor data provides information about order anyway. In addition, the order information may be required for planning certain actions in detail.

However, for axiomatic approaches in general, there is one problem: how can the negative information be deduced, e.g. if we want to know that there is no opponent in between. With Prolog alone this is not possible: the built-in negation as failure sometimes causes problems if used in complex queries. So we were led to use full first-order logic with the Protein theorem prover [2]. As example for this, let us consider the problem of determining whether passing is possible. This could be checked by the following logical rule with negation in the rule body:

$$Passing \leftarrow \neg \exists Opp : Between(Me, Opp, Partner)$$

The intended meaning of this rule is as follows: passing is possible, if there is no opponent between the agent and one of its partners. The question is: how should negation ($\neg$) and existential quantification ($\exists$) be interpreted? Protein provides classical negation as usual in first-order theorem proving. Existential quantification causes problems, if treated by Skolemization, i.e. replacing existentially quantified variables by new constant or function symbols, because then we have potentially infinitely many players.

Since we need real-time behavior, we just considered the finite domain of players visible for the agent in our implementation. This is closed world or constraint domain reasoning. By this, we get a complete and terminating system. Possibly, more sophisticated kinds of non-monotonic negation can be used here in this context of decision-finding. Note that, currently, this component is not yet integrated into the actual RoboLog Koblenz implementation, but has been used for axiomatizing situations (see [6]).

4 Higher Abilities (Layer 3)

Many tasks require deeper reasoning, which can be expressed within a BDI agent architecture [24]. In our context, a BELIEF b is a qualitative predicate q, its negation $\neg q$ or a conjunction of beliefs $b_1 \wedge b_2$. A GOAL g is either an *achievement* goal $!q$ or a *test* goal $?q$, where q is a qualitative predicate. A DESIRE (or event) d is a goal or an action. Now we can build rules for a certain SITUATION in form of scripts, written $d : b - i$, where d is a desire, b is a belief (identifying the precondition of the situation), and i is the INTENTION (or, strictly speaking, the intended plan).

4.1 Intended Plans

The intended plan is a tree of desires. Edges outgoing from test goals are labeled with *yes* or *no* and possibly a time-out delay. They realize alternatives in the plan. Depending on the truth value the agent follows different paths. Edges labeled with a time-out serve to delay the predicate. The agent only follows the labeled edge, if the respective truth value holds at a time within the time-out interval. An achievement goal has to be performed actively by the actor. The actual execution of an intended plan sometimes makes it necessary to leave the abstract level of qualitative reasoning and operate on quantitative data.

If an action or achievement goal fails or an external interruption occurs (e.g. a referee message in the RoboCup scenario), the agent has to return to a *default plan*, which must be applicable without precondition.

4.2 Example 1: The Goalie Runs Home

Let us now consider an example for such an agent script. When the ball is in the opponent half of the field, the goalkeeper of RoboLog Koblenz moves to his home position and waits there in order to regain stamina. This means, if the goalie *believes* that the ball is in the opponent half, his *desire* is to be at his home position. So he executes the *intended plan* to run there. Figure 3 (a) shows the respective script. In order to execute this script, the agent has to further decompose the desire $Run\_to(home)$ as shown in Figure 3 (b).

Let us now take a deeper look at the three desires of the second intended plan in Fig. 3 (b). Each of them shows a different aspect of the language.

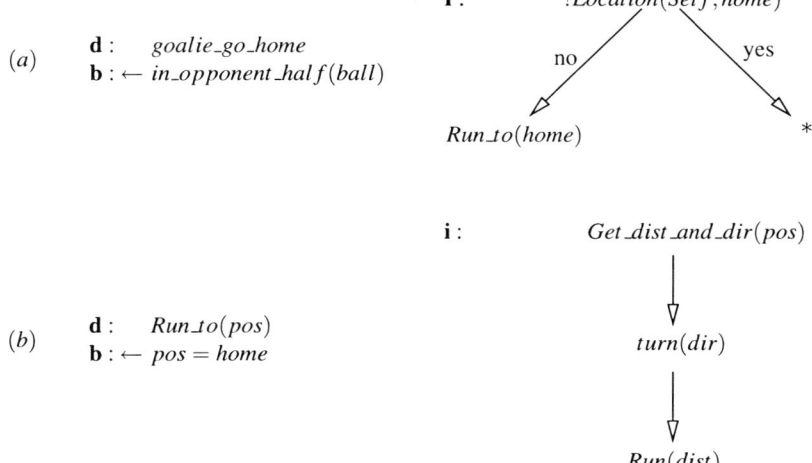

Fig. 3. Scripts for the goalkeeper.

- *Get_dist_and_dir(home)*: The satisfaction of this desire realizes the transition from the *qualitative* level to the *quantitative*. It takes a quality (*home*) as input and returns quantitative values, namely the relative distance and direction of the home position from the agent. The other desires operate on these quantities.
- *turn(dir)*: This action belongs to the lowest level of our architecture. It is atomic in the sense that it can be sent to the SoccerServer directly.
- *Run(dist)*: This, finally, is a complex action. From the point of view of our agent language, it is assumed to be *atomic*, too. But for actual execution, it has to be decomposed into a series of *dash* commands.

5 Cooperative Behavior (Layer 4)

The description language introduced in Sect. 4 is only suitable for modelling single-agent plans. But as we want to describe situations in which several agents have to cooperate, we will now extend the language to allow for the description of collective actions and multi-agent plans.

In this context a DESIRE d is a goal or an action, *indexed by a list of agents*—the actors—, which must satisfy the desire by performing some actions. Now the intended plan i becomes an acyclic graph of desires with a designated start node. Its edges are labeled with actors which must be a subset of the actors in d. Consider now all possible subgraphs wrt. edges for a certain actor. It is required that this still is a tree with the start node as root, where binary branching is only allowed after test nodes. These subgraphs represent the ROLE for the respective actor. Achievement goals are performed by the indexed actors, while non-actors wait for the achievement until a certain time limit. So for the latter such an achievement goal automatically becomes a test goal, normally labeled with a certain time-limit.

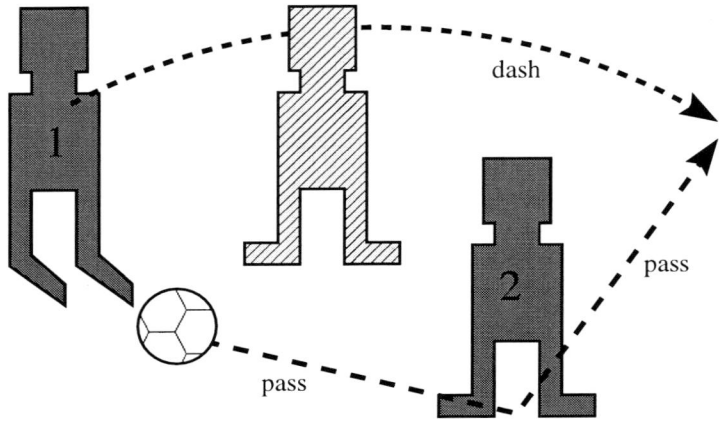

Fig. 4. Double Passing Situation.

5.1 Example 2: Double Passing

Let us now consider an example for a collective action of agents, namely double passing. There are two actors in this situation: actor 1 kicks the ball to actor 2, then actor 1 runs towards the goal, and expects a pass from actor 2. This is illustrated in Fig. 4. In order to initiate such an action, two agents simultaneously have to recognize their respective roles in the current situation in their belief state. The belief b for a double passing situation can be described as follows: 1 and 2 are nearest neighbors belonging to the same team, and player 1 has the ball. Player 2 must be clear, whereas an opponent is near to 1 such that 1 cannot dribble straight on. The intended plan i is then, that 1 passes the ball to 2 at first, then 1 runs towards the goal, and finally 2 passes the ball to 1. The respective rule can be expressed as shown in Fig. 5.

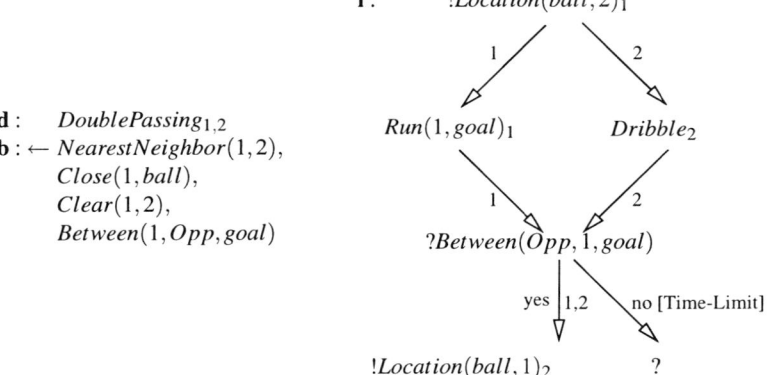

Fig. 5. Double Passing Script.

While experimenting with an implementation of double passing, we noticed that the main problem is that both actors simultaneously have to recognize their role, because one of the agents possibly does not see the other agent. In this context, communication (i.e. telling the other agent one's desire) helps a lot. A cooperating partner could tell its coordinates or even its whole own belief state. We made similar experiences with an even simpler kind of action, namely simple passing.

5.2 Communication

As we stated earlier, it makes sense to allow communication between agents. It helps them to recognize situations or their roles in them and thus reduces the complexity of the agents' reasoning and decision processes. But then another implementation decision has to be made, namely which communication language to use.

A general approach for the exchange of knowledge between agents is the Knowledge Query and Manipulation Language (KQML) [17]. However, if the domain of application is restricted, KQML may be too general. But it allows reliable communication between agents, even if their internal architecture is quite different or unknown for the other agent, by providing a common syntax. Instead, we communicate Prolog predicates directly. The advantage of this approach is, that no meta-logical interpretation of received information is necessary. A disadvantage is that for a successful communication the agents have to know each other's internal structure exactly. But this drawback can be overcome by specifying a subset of the available predicates together with their intended functionality as the communication language.

Thus, communication between the agents can be done by transmitting these predicates together with the action the recipient is expected to take on them, i.e. execute them as function. The goalkeeper, for example, could communicate his uniform-number to his teammates by saying $assert(goalie\_nr(1))$. The language is by its definition specific to the domain, thus enabling efficient communication while maintaining the flexibility of a more general language like KQML.

5.3 Translating Rules into CLP

We may distinguish several types of plans: basic plans with only one actor and complex plans where there are more than one actors. The former plans implement higher abilities (layer 3), while the latter realize teamwork (layer 4). Each BDI script can be translated into a CLP rule in a straightforward manner. For each achievement or test goal we introduce new symbols: $\underline{!P}$ and $\underline{?P}$. For each rule some default recipes are introduced:

$$P(x_1,\ldots,x_n) \leftarrow \underline{!P}(x_1,\ldots,x_n).$$
$$\underline{?P}(x_1,\ldots,x_n) \leftarrow P(x_1,\ldots,x_n).$$

The former and external events update predicates; this is the main difference to CLP. An approach that can handle external events and concurrency is *ConGolog* [10].

For each situation and for each role in it, a BDI script can be translated directly into a logic program rule, possibly with concurrent constraints (belief conditions):

$$d \leftarrow b \wedge i$$

The reader may have noticed that a situation with n roles corresponds to n CLP rules. These rules are identical wrt. their heads d. The preconditions b for the actions are also very similar; they only differ in their actor role. The last (but not least) part i is really different, because each actor plays a different role in the respective situation. For example, the instantiated plans for both actors of the double passing rule (see Fig. 5) are as follows:

<u>Role 1</u>
!$Location(ball, 2)$
$Run(Self, goal)$
?$Between(Opp, Self, goal)$
?$Location(ball, Self)$

<u>Role 2</u>
?$Location(ball, Self)$
$Dribble$
?$Between(Opp, 1, goal)$
!$Location(ball, 1)$

Recall that achievement goals are converted into test goals for non-actors. In addition, the control sequence for giving up after some time-limit is not shown here. Clearly, the translation into several CLP rules increases the time complexity for deciding which action or role therein is performed next. This problem can at least be partially overcome by communicating the next action directly to partners. In fact, we do this in our implementation by sending calls to Prolog predicates. But nevertheless, robustness of the whole system (of agents) has to be guaranteed in the case of failing actions or failing communication.

6 Conclusions

We presented a logical description language for multi-agent systems, following the lines of [24]. This language can be understood as a generalization of CLP. Both, quantitative and qualitative spatial reasoning can be built-in. With the script language proposed here, it is possible to express multi-agent plans. The RoboLog system provides a clean means for programming soccer agents declaratively. We conducted several test games with different scores on our local network—a 100 MBit Ethernet—and participated in RoboCup-99 (see also the team description *RoboLog Koblenz* in this volume).

6.1 Other Approaches with Logic Programming

Despite of the fact, that there are many logic-based approaches to agent programming in the literature, there are only few systems that are implemented with logic programming and that participated in the RoboCup. So, it seems that almost no team employs one of the well-known planning techniques in artificial intelligence (e.g. with the situation calculus [10]). *CS Freiburg*—the world champion of the middle-size league in 1997—makes use of path planning [13], but emphasizes the need of reliable basic skills. In this approach, path planning is restarted again every few milliseconds.

As mentioned in the introduction (Sect. 1), [24] proposes a framework that allows agent programs to be written and interpreted in a manner similar to that of Horn-clause logic programs. Nevertheless, only single-agent actions can be specified within this approach. The team described in [16] participated in RoboCup-98. The architecture of this system is layered (as ours) and hosts a behavior-based, a local planning, and a social planning layer. The system is implemented with *Oz*, a concurrent constraint logic programming language.

Another interesting approach is presented in [19]. There, an architecture for intelligent agents (with application to the RoboCup simulation league) is described, using the so-called organic programming language *Gaea*. It provides dynamic rearrangement of programming modules and multi-threading among other features. This, of course, is needed in a dynamic context as robotic soccer: when the system predicts or detects a change in the environment, it can swap some portion of its program accordingly.

6.2 Future Work

Further work should concentrate on the real-time requirements in exceptional situations and the concurrency of different mechanisms for information acquisition. The robustness of the decision process can be improved by means of defeasible reasoning [11] and/or organic programming [19]. Another area of research is how far logical mechanisms can be used within the lower levels of our approach. Deduction could be used to build a more complete view of the agent's world model. The application of these techniques to real robots is one of the next steps of our research activities. Finally, the specification of a flexible communication language should also be investigated.

References

[1] M. Asada and H. Kitano, editors. *RoboCup-98: Robot Soccer WorldCup II*. LNAI 1604. Springer, Berlin, Heidelberg, New York, 1999.

[2] P. Baumgartner and U. Furbach. PROTEIN: A PROver with a Theory Extension INterface. In A. Bundy, editor, *Proceedings of the 12th International Conference on Automated Deduction*, LNAI 814, pages 769–773, Nancy, 1994. Springer, Berlin, Heidelberg, New York.

[3] B. Bennet, A. G. Cohn, and A. Isli. Combining multiple representations in a spatial reasoning system. In *Proceedings of the 9th IEEE International Conference on Tools with Artificial Intelligence (ICTAI'97)*, pages 314–322, Newport Beach, CA, 1997.

[4] M. Betke and L. Gurvits. Mobile robot localization using landmarks. *IEEE Transactions on Robotics and Automation*, 13(2):251–263, Apr. 1997.

[5] K. Borsuk and W. Szmielew. *Foundations of Geometry*. North-Holland, Amsterdam, 1960.

[6] B. Bremer. Erkennung von Paß- und Abseitssituationen mit räumlichem Schließen. Studienarbeit S 570, Fachbereich Informatik, Universität Koblenz, 1999.

[7] H.-D. Burkhard, M. Hannebauer, and J. Wendler. Belief–desire–intention – deliberation in artificial soccer. *AI Magazine*, pages 87–93, 1998.

[8] E. Clementini, P. Di Felice, and D. Hernández. Qualitative representation of positional information. *Artificial Intelligence*, 95(2):317–356, 1997.

[9] E. Corten, K. Dorer, F. Heintz, K. Kostiadis, J. Kummeneje, H. Myritz, I. Noda, J. Riekki, P. Riley, P. Stone, and T. Yeap. *Soccerserver Manual*, 5th edition, May 1999. For Soccerserver Version 5.00 and later.

[10] G. De Giacomo, Y. Lespérance, and H. J. Levesque. Reasoning about concurrent execution, prioritized interrupts, and exogenous actions in the situation calculus. In M. E. Pollack, editor, *Proceedings of the 15th International Joint Conference on Artificial Intelligence*, pages 1221–1226, Nagoya, Japan, 1997. Volume 2.

[11] J. Dix, F. Stolzenburg, G. R. Simari, and P. R. Fillottrani. Automating defeasible reasoning with logic programming (DeReLoP). In S. Jähnichen, editor, *Proceedings of the 2nd German-Argentinian Workshop on Information Technology*, pages 39–46, Königswinter, 1999. To appear.

[12] C. Eschenbach and L. Kulik. An axiomatic approach to the spatial relations underlying *Left-Right* and *in Front of–Behind*. In G. Görz and S. Hölldobler, editors, *KI-97: Advances in Artificial Intelligence — Proceedings of the 21st Annual German Conference on Artificial Intelligence*, LNAI 1303, pages 207–218, Freiburg, 1997. Springer, Berlin, Heidelberg, New York.

[13] J.-S. Gutmann, W. Hatzack, I. Herrmann, B. Nebel, F. Rittinger, A. Topor, T. Weigel, and B. Welsch. The CS Freiburg robotic soccer team: Reliable self-localization, multirobot sensor integration and basic soccer skills. In Asada and Kitano [1], pages 93–108.

[14] International Computers Limited and IC-Parc. *ECLiPSe User Manual / Extensions User Manual – Release 4.0*, 1998. Two volumes.

[15] J. Jaffar and M. J. Maher. Constraint logic programming: a survey. *Journal of Logic Programming*, 19/20:503–581, 1994.

[16] C. G. Jung. Layered and resource-adapting agents in the RoboCup simulation. In Asada and Kitano [1], pages 207–220.

[17] Y. Labrou and T. Finin. A proposal for a new KQML specification. Technical Report TR CS-97-03, Computer Science and Electrical Engineering Department, University of Maryland Baltimore County, Baltimore, MD 21250, Feb. 1997.

[18] J. Murray. My goal is my castle — Die höheren Fähigkeiten eines RoboCup-Agenten am Beispiel des Torwarts. Studienarbeit S 564, Fachbereich Informatik, Universität Koblenz, 1999.

[19] H. Nakashima and I. Noda. Dynamic subsumption architecture for programming intelligent agents. In *Proceedings of the International Conference on Multi-Agent Systems*, pages 190–197. AAAI Press, 1998.

[20] I. Noda. Soccer server: a simulator for RoboCup. In *JSAI AI-Symposium*, 1995.

[21] O. Obst. *RoboLog – An ECLiPSe-Prolog SoccerServer interface: Users manual*, March 1998.

[22] O. Obst. RoboLog: Eine deduktive Schnittstelle zum RoboCup Soccer Server. Diplomarbeit D 488, Fachbereich Informatik, Universität Koblenz, 1999.

[23] D. A. Randel, Z. Cui, and A. G. Cohn. A spatial logic based on regions and connections. In *Proceedings of the third Int. Conf. on Knowledge Representation and Reasoning*, pages 165–176, San Mateo, 1992. Morgan Kaufmann.

[24] A. S. Rao. AgentsSpeak(L): BDI agents speak out in a logical computable language. In W. van de Velde and J. W. Perrame, editors, *Agents Breaking Away – 7th European Workshop on Modelling Autonomous Agents in a Multi-Agent World*, LNAI 1038, pages 42–55, Berlin, Heidelberg, New York, 1996. Springer.

[25] P. Stone, M. Veloso, and P. Riley. The CMUnited-98 champion simulator team. In Asada and Kitano [1], pages 61–76.

[26] K. Zimmermann and C. Freksa. Qualitative spatial reasoning using orientation, distance, and path knowledge. *Applied Intelligence*, 6:49–58, 1996.

Layered Learning and Flexible Teamwork in RoboCup Simulation Agents

Peter Stone[1] and Manuela Veloso[2]

[1] AT&T Labs — Research
180 Park Ave., room A273
Florham Park, NJ 07932
pstone@research.att.com
http://www.research.att.com/~pstone

[2] Computer Science Department
Carnegie Mellon University
Pittsburgh, PA 15213
veloso@cs.cmu.edu
http://www.cs.cmu.edu/~mmv

Abstract. RoboCup was introduced as a challenge area at IJCAI-97. We have been actively pursuing research in this area and have participated in the RoboCup competitions, winning the RoboCup-98 and RoboCup-99 simulator competitions. In this paper, we report on the main technical issues that we encountered and addressed in direct response to the learning and teamwork challenges stated in the IJCAI-97 challenge paper. We describe "layered learning" in which off-line and on-line, individual and collaborative, learned robotic soccer behaviors are combined hierarchically. We achieve effective teamwork through a team member agent architecture that encompasses a "flexible teamwork structure." Agents are capable of decomposing the task space into flexible roles and can switch roles while acting. We report detailed empirical results verifying the effectiveness of the learned behaviors and the components of the team member agent architecture.

1 Introduction

The RoboCup Synthetic Agents Challenge 97 [4] specifies three challenges within the simulated robotic soccer server [5]: (i) learning of individual agents and teams; (ii) multi-agent team planning and plan-execution; and (iii) opponent modeling. Many researchers are working towards these and other challenges in the robotic soccer domain [3,1], some specifically on learning, e.g. [2], and some specifically on teamwork structures, e.g. [11]. In conjunction with our own work, these pursuits have contributed to the success of the RoboCup challenge. In particular, we have addressed and successfully met the first two of the three specific challenges. Most of our specific contributions have been described in detail elsewhere. This paper serves to summarize our broad course of research within the context of the specific IJCAI challenge.

In the context of learning (challenge i), we have created three different learned behaviors and combined them hierarchically following the layered learning paradigm [8]. Given a hierarchical task decomposition, layered learning allows for learning at each level of the hierarchy, with learning at each level directly affecting learning at the next higher level.

As called for in the challenge, the three learned behaviors represent off-line skill learning by individual agents; off-line collaborative learning by teams of agents; and on-line collaborative and adversarial learning. First, individual agents learn to intercept a moving ball by training a neural network off-line. Second, they use the learned ball-interception behavior as part of the training procedure for learning to evaluate whether a pass to a particular teammate will succeed or fail. This collaborative skill is trained off-line using the C4.5 decision tree training algorithm [6]. Third, the team of agents collectively learns an effective passing and shooting policy against a particular opponent using the on-line TPOT-RL multi-agent reinforcement learning method [7].

Within the context of the teamwork challenge (challenge ii), we characterize simulated robotic soccer as an example of a periodic team synchronization (PTS) domain [9]. PTS domains are domains in which periods of limited communication and time-critical action are interleaved with periods of safe, full communication. During the limited communication periods, agents need to act autonomously, while still working towards a common team goal. Time-critical environments such as robotic soccer require real-time response and therefore eliminate the possibility of heavy communication among team agents. However, in PTS domains, agents can periodically synchronize in a safe, full-communication setting.

We implement and test a team agent architecture suitable for PTS domains. Our team member agent architecture includes a flexible teamwork structure which allows agents to decompose the task space into flexible roles and allows them to smoothly switch roles while acting. Team organization is achieved by the introduction of a *locker-room agreement* as a collection of conventions followed by all team members. It defines agent roles, team formations, and pre-compiled multi-agent plans.

The remainder of this paper is organized as follows. Section 2 presents our learning experiments within the simulated robotic soccer domain. Section 3 provides details of our solution to the teamwork challenge. Section 4 describes our complete robotic soccer team which incorporates both learning and teamwork and concludes.

2 Learning Challenge

To address the learning challenge, we have created three different learned behaviors and combined them hierarchically following the layered learning paradigm. Section 2.1 describes layered learning and Section 2.3 presents our implementation within robotic soccer.

2.1 Layered Learning

Table 1 summarizes the principles of our layered learning paradigm.

Principle 1 Motivated by robotic soccer, layered learning is designed for domains that are too complex for learning a mapping directly from an agent's sensory inputs to its actuator outputs. Instead, the layered learning approach consists of breaking a problem down into several behavioral layers. At each layer, a concept needs to be acquired. A machine learning (ML) algorithm abstracts and solves the local concept-learning task.

1. A mapping directly from inputs to outputs is not tractably learnable.
2. A bottom-up, hierarchical task decomposition is given.
3. Machine learning exploits data to train and/or adapt. Learning occurs separately at each level.
4. The output of learning in one layer feeds into the next layer.

Table 1: The key principles of layered learning.

Principle 2 Layered learning uses a bottom-up incremental approach to hierarchical task decomposition. Starting with low-level behaviors, the process of creating new ML subtasks continues until reaching high-level strategic behaviors that deal with the full domain complexity. The appropriate behavior granularity and the aspects of the behaviors to be learned are determined as a function of the specific domain. The task decomposition in layered learning is not automated. Instead, the layers are defined by the ML opportunities in the domain.

Principle 3 Machine learning is used as a central part of layered learning to exploit data in order to *train* and/or *adapt* the overall system. ML is useful for training behaviors that are difficult to fine-tune manually. It is useful for adaptation when the task details are not completely known in advance or when they may change dynamically. In the former case, learning can be done off-line and frozen during actual task execution. In the latter, on-line learning is necessary: since the agent needs to adapt to unexpected situations, it must be able to alter its behavior even while executing its task. Like the task decomposition itself, the choice of machine learning method depends on the subtask.

Principle 4 The key defining characteristic of layered learning is that each learned layer directly affects the learning at the next layer. A learned subtask can affect the subsequent layer either by:
- pruning the set of training examples;
- providing the features used for learning; and/or
- determining the actions available;

All three possibilities are illustrated in our simulated robotic soccer implementation described below.

2.2 Formalism

Consider the learning task of identifying a hypothesis h from among a class of hypotheses H which map a set of state feature variables S to a set of outputs O such that, based on a set of training examples, h is most likely (of the hypotheses in H) to represent unseen examples.

When using the layered learning paradigm, the complete learning task is decomposed into hierarchical subtask layers $\{L_1, L_2, \ldots, L_n\}$ with each layer defined as

$$L_i = (\boldsymbol{F}_i, O_i, T_i, M_i, h_i)$$

where:

F_i is the input vector of state features relevant for learning subtask L_i. $F_i = <F_i^1, F_i^2, \ldots>$. $\forall j, F_1^j \in S$.

O_i is the set of outputs among which to choose for subtask L_i. $O_n = O$.

T_i is the set of training examples used for learning subtask L_i. Each element of T_i consists of a correspondence between an input feature vector $\boldsymbol{f} \in F_i$ and $o \in O_i$.

M_i is the ML algorithm used at layer L_i to select a hypothesis mapping $F_i \mapsto O_i$ based on T_i.

h_i is the result of running M_i on T_i. h_i is a function from F_i to O_i.

As set out in Principle 2 of layered learning, the definitions of the layers L_i are given a priori. Principle 4 is addressed via the following stipulation. $\forall i < n$, h_i directly affects L_{i+1} in at least one of three ways:

- h_i is used to construct one or more features F_{i+1}^k.
- h_i is used to construct elements of T_{i+1}; and/or
- h_i is used to prune the output set O_{i+1}.

It is noted above in the definition of F_i that $\forall j, F_1^j \in S$. Since F_{i+1} can consist of new features constructed using h_i, the more general version of the above special case is that $\forall i, j, F_i^j \in S \cup_{k=1}^{i-1} O_k$.

Again, in layered learning, the task decomposition is assumed to be given a priori. Layered learning can, however, be combined with any algorithm for learning abstraction levels. In particular, let A be an algorithm for learning task decompositions within a domain. Suppose that A does not have an objective metric for comparing different decompositions. Applying layered learning on the task decomposition and quantifying the resulting performance can be used as a measure of the utility of A's output.

2.3 Layered Learning in Robotic Soccer

Table 2 illustrates our set of learned behavior levels within the simulated robotic soccer domain. We identify a useful low-level skill that must be learned before moving on to higher-level strategies. Then we build upon it to create higher-level multi-agent and team behaviors.

Layer	Behavior type	Learned behavior	Learning method	Training type
1	individual	ball interception	neural network	off-line
2	multi-agent	pass evaluation	decision tree	off-line
3	team	pass selection	TPOT-RL	on-line

Table 2: Examples of different behavior levels in robotic soccer and the learning methods used for the implemented layers in the simulated robotic soccer layered learning implementation.

L_1: **Ball Interception — an individual skill.** First, the agents learn a low-level individual skill that allows them to control the ball effectively. While executed individually, the ability to intercept a moving ball is required due to the presence of other agents: it is needed to block or intercept opponent shots or passes as well as to receive passes from teammates. As such, it is a prerequisite for most ball-manipulation behaviors. We chose to have our agents learn this behavior because it was easier to collect training data than to fine-tune the behavior by hand[1].

L_1 is defined as follows.

$F_1 = \{BallDist_t, BallAng_t, BallDist_{t-1}\}$: The agent learns what action to take based on the ball's current distance and angle from the defender, and the ball's distance a fixed time (250 msec.) in the past.

$O_1 = \{TurnAng\}$: The agent chooses an angle to turn such that it will be likely to intercept the ball.

T_1: The training procedure for ball interception involves a stationary forward repeatedly shooting the ball towards a defender in front of a goal. The defender collects training examples by acting randomly and noticing when it successfully stops the ball. Test examples are classified as saves (successful interceptions), goals (unsuccessful attempts), and misses (shots that went wide of the goal).

M_1 = **a neural network:** Ball interception is trained with a fully-connected neural network with 4 sigmoid hidden units and a learning rate of 10^{-6}. The weights connecting the input and hidden layers use a linearly decreasing weight decay starting at .1%. We use a linear output unit with no weight decay. The neural network was trained for 3000 epochs.

h_1 = **a trained interception behavior:** Table 3 shows the effect of the number of training examples on learned save percentage. With about 750 training examples, the defender is able to stop 91% of shots on goal (saves + goals: misses are omitted), a comparable save rate to that achieved when using an analytic ball interception behavior [8].

L_2: **Pass Evaluation — a multi-agent behavior.** Second, the agents use their learned ball-interception skill as part of the behavior for training a multi-agent behavior. When an agent has the ball and has the option to pass to a particular teammate, it is useful to have an idea of whether or not the pass will actually succeed if executed: will the teammate successfully receive the ball? Such an evaluation depends on not only the teammate's and opponents' positions, but also their abilities to receive or intercept the pass. Consequently, when creating training examples for the pass-evaluation function, we equip the intended pass recipient as well as all opponents with the previously learned ball-interception behavior, h_1. Again, we chose to have our agents learn the pass-evaluation capability because it is easier to collect training data than to construct it by hand.

L_2 is defined as follows.

---

[1] The learning was done in an early implementation of the soccer server (Version 2) in which agents did not receive any velocity information when seeing the ball.

| Training | | | Saves |
Examples	Saves(%)	Goals(%)	Goals+Saves(%)
100	57	33	63
200	73	18	80
300	81	13	86
400	81	13	86
500	84	10	89
750	**86**	9	**91**
1000	83	10	89
4773	84	9	90

Table 3: The defender's performance when using neural networks trained with different numbers of training examples.

F_2 = **a set of 174 continuous and ordinal features:** There are many features that could possibly affect pass evaluation. We encode a large set of attributes representing the relative positions of teammates and opponents on the field as well as statistical counts reflecting their relative positioning [8].

$O_2 = [-1, 1]$: A potential pass to a particular receiver is classified as a success with a confidence factor $\in (0, 1]$, a failure with a confidence factor $\in [-1, 0)$, or a miss ($= 0$).

T_2: The training procedure for pass evaluation involves a passer executing passes to randomly-placed teammates interspersed with randomly-placed opponents. The training scenario is illustrated within a screen shot of the soccer server in Figure 1. The dashed line indicates the region in which the teammates and opponents are randomly placed. *The intended pass recipient and the opponents all use the learned ball-interception behavior, h_1.* Trials are classified as successes (a teammate intercepts the ball), failures (an opponent intercepts the ball), and misses (no player intercepts the ball). When passing to a random teammate, 51% of passes are successful.

M_2 = **C4.5:** To learn pass evaluation, we use the C4.5 decision tree training algorithm [6] with all of the default parameters. Decision trees are chosen over neural networks because of their ability to ignore irrelevant attributes.

h_2 = **a trained pass-evaluating decision tree:** During testing, the trained decision tree returns a predicted classification as well as a confidence factor, resulting in a value between -1 and 1. Table 4 tabulates our results indicating that the trained decision tree enables the passer to choose successfully from among its potential receivers. Overall results are given as well as a breakdown by the passer's confidence prior to the pass. In this experiment, the passer is forced to pass even if it predicts failures for all 3 teammates. In that case, it passes to the teammate with the lowest likelihood of failure. 65% of all passes and 79% of passes predicted to succeed with high confidence are successful.

L_3: **Pass Selection — a collaborative and adversarial team behavior.** Third, the agents use their learned pass-evaluation capability h_2 to create the input space and output set for learning pass selection. When an agent has the

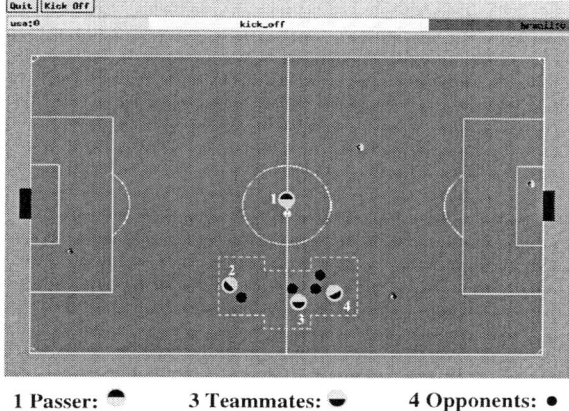

Fig. 1: The training scenario for pass evaluation. The dashed line indicates the region in which the teammates and opponents are randomly placed prior to each trial.

Result	Overall	Success Confidence: .8–.9	.7–.8	.6–.7
(Number)	(5000)	(1050)	(3485)	(185)
SUCCESS (%)	65	**79**	63	58
FAILURE (%)	26	15	29	31
MISS (%)	8	5	8	10

Table 4: The results of 5000 trials during which the passer uses the DT to choose the receiver. Results are given in percentages of the number of cases falling within each confidence interval (shown in parentheses).

ball, it must decide to which teammate it should pass the ball[2]. Such a decision depends on a huge amount of information including the agent's current location on the field, the current locations of all the teammates and opponents, the teammates' abilities to receive a pass, the opponents' abilities to intercept passes, teammates' subsequent decision-making capabilities, and the opponents' strategies. The merit of a particular decision can only be measured by the long-term performance of the team as a whole. Therefore, we drastically reduce the input space with the help of the previously learned decision tree, h_2: rather than considering the positions of all of the players on the field, only the pass evaluations for the possible passes to each teammate are considered.

L_3 is defined as follows.

$F_3 = \{PlayerPosition, O_2, O_2, O_2, \ldots\}$: The input representation consists of one coarse geographical component and one action-dependent feature [10] for each possible pass. *The action-dependent features are precisely the result of h_2 executed for each possible receiver.*

---

[2] It could also choose to shoot. For the purposes of this behavior, the agents are not given the option to dribble.

$O_3 = \{shoot\} \cup \{Teammates\}$: The result of a pass selection decision is either a shot on goal or a pass to a particular teammate.

T_3: Training examples are gathered on-line by individual team members during real games. Each individual agent learns in a separate partition of F_3 according to its position on the field. Agents learn based on the observed long-term effects of their actions [7]. *For each particular action decision, the eligible members of O_3 are pruned based on h_2: only passes predicted to succeed are considered.*

$M_3 =$ **TPOT-RL:** For training pass selection, we use TPOT-RL [10], an on-line, multi-agent, reinforcement learning method motivated by Q-learning that is applicable in team-partitioned, opaque-transition domains such as simulated robotic soccer. We use the default parameters as reported in [10].

$h_3 =$ **a distributed pass-selection policy:** We test the pass-selection learning by directly comparing two teams with identical behaviors other than their pass-selection policies. Agents on both teams begin by passing randomly, but agents on one team adjust their behavior based on experience using TPOT-RL. The other agents continue passing randomly. Figure 2 demonstrates the effectiveness of the learned passing policies.

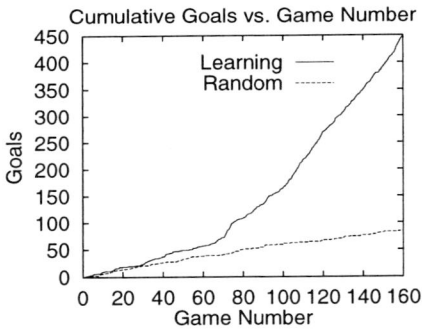

Fig. 2: Total goals scored by a learning team playing against a randomly passing team. The independent variable is the number of 10-minute games that have elapsed.

The learning methods used for each of the above behaviors are summarized in Table 2.

The three learned layers described above illustrate the principles of the layered learning paradigm as laid out in Section 2.1:

- The decomposition of the task into smaller subtasks enables the learning of a more complex behavior than would be possible if learning straight from the agents' sensory inputs.
- The hierarchical task decomposition is constructed in a bottom-up, domain-dependent fashion.
- Machine learning methods are chosen or created to suit the subtask in question. They exploit available data to train difficult behaviors (ball interception

and pass evaluation) or to adapt to changing/unforeseen circumstances (pass selection).
- Learning in one layer feeds into the next layer either by providing a portion of the behavior used for training (ball interception – pass evaluation) or by creating the input representation and pruning the action space (pass evaluation – pass selection).

3 Teamwork Challenge

To address the teamwork challenge, we characterize simulated robotic soccer as an example of a periodic team synchronization (PTS) domain and we create a general team member agent architecture suitable for PTS domains. Section 3.1 defines PTS domains and Section 3.2 lays out the agent architecture. Section 3.3 summarizes our implementation of a locker-room agreement in the robotic soccer domain. For the domain-independent formulation, see [9].

3.1 Periodic Team Synchronization (PTS) Domains

We view robotic soccer as an example of a periodic team synchronization (PTS) domain. We define PTS domains as domains with the following characteristics:

- There is a team of autonomous agents A that collaborate towards the achievement of a joint long-term goal G.
- Periodically, the team can synchronize with no restrictions on communication: the agents can in effect inform each other of their entire internal states and decision-making mechanisms with no adverse effects upon the achievement of G. These periods of full communication can be thought of as times at which the team is "off-line."
- In general (i.e., when the agents are "on-line"):
  - The domain is *dynamic* and *real-time* meaning that team performance is adversely affected if an agent ceases to act for a period of time: G is either less likely to be achieved, or likely to be achieved farther in the future. That is, consider agent a_i. Assume that all other agent behaviors are fixed and that were a_i to act optimally, G would be achieved with probability p at time t. If a_i stops acting for a random period of time and then resumes acting optimally, either:
    * G will be achieved with probability p' at time t with $p' < p$; or
    * G will be achieved with probability p at time t' with $t' > t$.
  - The domain has *unreliable communication*, either in terms of transmission reliability or bandwidth limits. In particular:
    * If an agent $a_i \in A$ sends a message m intended for agent $a_j \in A$, then m arrives with some probability $q < 1$; or
    * Agent a_i can only receive x messages every y time units.

In the extreme, if $q = 0$ or if $x = 0$, then the periods of full communication are interleaved with periods of *no* communication, requiring the agents to act completely autonomously. In all cases, there is a cost to *relying* on communication. If agent a_i cannot carry on with its action until receiving a message from a_j, then the team's performance could suffer. Because of the unreliable communication, the message might not get through on the first try. And because of the dynamic, real-time nature of the domain, the team's likelihood of or efficiency at achieving G is reduced.

The soccer server provides a PTS domain since teams can plan strategies before the game, at halftime, or at other breakpoints; but during the course of the game, communication is limited.

3.2 Team Member Agent Architecture

At the core of our agents' coordination mechanism is the locker-room agreement [9]. Based on the premise that agents can periodically meet in safe, full-communication environments, the locker-room agreement specifies how they should act when in low-communication, time-critical, adversarial environments. agreement can be hard-wired or it can be the result of deliberative automatic planning during the off-line phase of PTS domains. In our work so far, the locker-room agreement is hard-wired: we focus instead on the on-line phase.

Our team member agent architecture is suitable for PTS domains. Individual agents can capture locker-room agreements and respond to the environment, while acting autonomously. Based on a standard agent paradigm, our team member agent architecture allows agents to sense the environment, to reason about and select their actions, and to act in the real world. At team synchronization opportunities, the team also makes a locker-room agreement for use by all agents during periods of limited communication. Figure 3 shows the functional input/output model of the architecture.

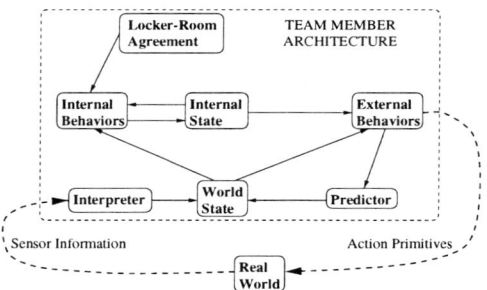

Fig. 3: A functional input/output model of the team member agent architecture for PTS domains.

The agent keeps track of three different types of state: the *world state*, the *locker-room agreement*, and the *internal state*. The agent also has two different types of behaviors: *internal behaviors* and *external behaviors*.

The world state reflects the agent's conception of the real world, both via its sensors and via the predicted effects of its actions. It is updated as a result of interpreted sensory information. It may also be updated according to the predicted effects of the external behavior module's chosen actions. The world state is directly accessible to both internal and external behaviors.

The locker-room agreement is set by the team when it is able to privately synchronize. It defines the flexible teamwork structure and the inter-agent communication protocols, if any. The locker-room agreement is accessible only to internal behaviors.

The internal state stores the agent's internal variables. It may reflect previous and current world states, possibly as specified by the locker-room agreement. For example, the agent's role within a team behavior could be stored as part of the internal state. A window or distribution of past world states could also be stored as a part of the internal state. The agent updates its internal state via its internal behaviors.

The internal behaviors update the agent's internal state based on its current internal state, the world state, and the team's locker-room agreement.

The external behaviors reference the world and internal states, and select the actions to send to the actuators. The actions affect the real world, thus altering the agent's future percepts. External behaviors consider only the world and internal states, without direct access to the locker-room agreement.

3.3 Flexible Teamwork Structure

Within the team member agent architecture, our agents are equipped with a flexible teamwork structure that allows agents to decompose the task space into flexible roles and allows them to smoothly switch roles while acting. The teamwork structure consists of flexible positions (roles), dynamically changeable formations, and pre-defined, multi-agent set-plays. The structure is "flexible" in that agents can dynamically change their collective strategy and individual behaviors within this strategy in response to changing conditions. This section summarizes the robotic soccer implementation of our general flexible teamwork structure [9].

Flexible Positions In our multi-agent approach, the player positions itself flexibly in *anticipation* of where it will be useful to the team, either offensively or defensively. The definition of a position includes *home coordinates*, a *home range*, and a *maximum range*, as illustrated in Figure 4(a). The position's home coordinates are the default location to which the agent should go. However, the agent has some flexibility, being able to set its actual home position anywhere within the home range.

Two ways in which agents can use the position flexibility is to react to the ball's position and to mark opponents. When reacting to the ball's position, the agent moves to a location within its range that minimizes its distance to the ball. When marking opponents, agents move next to a given opponent rather than staying at the default position home.

Dynamically Changeable Formations A formation consists of a set of positions and a set of units. The formation and each of the units can also specify inter-position behavior specifications for the member positions. Figure 4(b) illustrates the positions in one particular formation, its units, and their captains. Here, the units contain defenders, midfielders, forwards, left players, center players, and right players. The captain of a unit may have privileged decision-making responsibilities, for example to assign players to "mark" (stay close to) specific opponents.

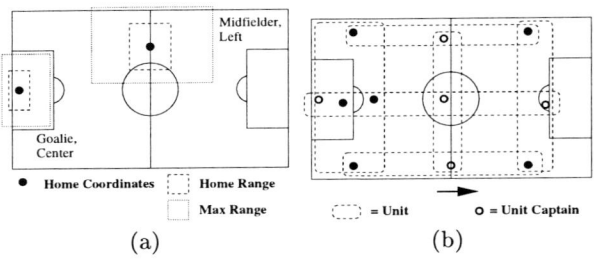

Fig. 4: (a) Different positions with home coordinates and home and max ranges. (b) Positions can belong to more than one unit.

We implemented several different formations, ranging from very defensive (8-2-0) to very offensive (2-4-4).[3] The full definitions of all of the formations are a part of the locker-room agreement. Therefore, they are all known to all teammates. However during the periods of full autonomy and low communication, it is not necessarily known what formation the rest of the teammates are using. Two approaches can be taken to address this problem:

- **static formation** - the formation is set by the locker-room agreement and never changes;
- **run-time switch of formation** - during team synchronization opportunities, the team sets globally accessible run-time evaluation metrics as formation-changing indicators.

In the RoboCup simulator competitions, our agents switched formations based on the amount of time left relative to the difference in score: the team switched to an offensive formation if it was losing near the end of the game and a defensive formation if it was winning. Since each agent was able to independently keep track of the score and time, the agents were always able to switch formations simultaneously.

---

[3] Soccer formations are typically described as goalie-defenders-midfielders-forwards.

Since the players are all autonomous, in addition to knowing its own position, each one has its own belief of the team's current formation along with the time at which that formation was adopted, and a map of teammates to positions. The players maintain consistent beliefs as to the team's state via communication and via conventions encoded in the locker-room agreement [9].

Pre-Planned Set Plays The final implemented improvement facilitated by our flexible teamwork structure is the introduction of set plays, or pre-defined special purpose plays. As a part of the locker-room agreement, the team can define multi-step multi-agent plans to be executed at appropriate times. Particularly if there are certain situations that occur repeatedly, it makes sense for the team to devise plans for those situations.

In the robotic soccer domain, certain situations occur repeatedly. For example, after every goal, there is a kickoff. When the ball goes out of bounds, there is a goal-kick, a corner-kick, or a kick-in. In each of these situations, the referee informs the team of the situations. Thus all the players know to execute the appropriate set play. Associated with each set-play-role is not only a location, but also a behavior. The player in a given role might pass to the player filling another role, shoot at the goal, or kick the ball to some other location.

3.4 Results

To test our flexible teamwork structure, we played a team using flexible positions and set-plays against one using rigid positions and no set-plays. Otherwise, the agents behaviors on the two teams were identical. Table 5 shows the results which clearly indicate the effectiveness of our teamwork structure.

(Game = 10 min.)	Flexible and Set-Plays	Default
Games won	34	1
Total goals	223	82
Avg. goals	5.87	2.16
Ball in own half	43.8%	56.2%

Table 5: Results when a flexible team plays against a rigid team. The flexible team won 34 out of 38 games with 3 ties.

To compare the different formations, we played each formation against every other several times. We used the results to help construct the formation-switching strategy used by our agents in competitions [7]. The cues for switching formations are functions of globally accessible variables (time remaining and score of the game) as defined in the locker-room agreement.

4 Conclusion

Layered learning and the team member agent architecture address the learning and teamwork components of the RoboCup Synthetic Agents Challenge 97. We leave the opponent modeling portion of the challenge as future work.

As well as being tested individually as reported above, our learning and teamwork techniques have been combined into a complete team of robotic soccer agents[4]. This team became the champion of the RoboCup-98 simulator competition, winning from among a field of 34 teams. Our team out-scored its opponents by a total of 66–0. Our subsequent team entered in RoboCup-99 also won that competition, which included 38 teams, outscoring its opponents by a total of 110–0 [7].

References

1. Minoru Asada and Hiroaki Kitano, editors. *RoboCup-98: Robot Soccer World Cup II*. Lecture Notes in Artificial Intelligence 1604. Springer Verlag, Berlin, 1999.
2. Minoru Asada, Shoichi Noda, Sukoya Tawaratumida, and Koh Hosoda. Purposive behavior acquisition for a real robot by vision-based reinforcement learning. *Machine Learning*, 23:279–303, 1996.
3. Hiroaki Kitano, editor. *RoboCup-97: Robot Soccer World Cup I*. Springer Verlag, Berlin, 1998.
4. Hiroaki Kitano, Milind Tambe, Peter Stone, Manuela Veloso, Silvia Coradeschi, Eiichi Osawa, Hitoshi Matsubara, Itsuki Noda, and Minoru Asada. The RoboCup synthetic agent challenge 97. In *Proceedings of the Fifteenth International Joint Conference on Artificial Intelligence*, pages 24–29, San Francisco, CA, 1997. Morgan Kaufmann.
5. Itsuki Noda, Hitoshi Matsubara, Kazuo Hiraki, and Ian Frank. Soccer server: A tool for research on multiagent systems. *Applied Artificial Intelligence*, 12:233–250, 1998.
6. J. Ross Quinlan. *C4.5: Programs for Machine Learning*. Morgan Kaufmann, San Mateo, CA, 1993.
7. Peter Stone. *Layered Learning in Multi-Agent Systems*. PhD thesis, Computer Science Department, Carnegie Mellon University, Pittsburgh, PA, December 1998. Available as technical report CMU-CS-98-187.
8. Peter Stone and Manuela Veloso. A layered approach to learning client behaviors in the RoboCup soccer server. *Applied Artificial Intelligence*, 12:165–188, 1998.
9. Peter Stone and Manuela Veloso. Task decomposition, dynamic role assignment, and low-bandwidth communication for real-time strategic teamwork. *Artificial Intelligence*, 110(2):241–273, June 1999.
10. Peter Stone and Manuela Veloso. Team partitioned, opaque transition reinforcement learning. In *Proceedings of the Second International Conference on Autonomous Agents*, pages 206–212. ACM Press, May 1999.
11. Milind Tambe, Jafar Adibi, Yaser Al-Onaizan, Ali Erdem andGal A. Kaminka, Stacy C. Marsela, Ion Muslea, and Marcelo Tallis. Using an explicit model of teamwork in RoboCup-97. In Hiroaki Kitano, editor, *RoboCup-97: Robot Soccer World Cup I*, pages 123–131. Springer Verlag, Berlin, 1998.
12. Christopher J. C. H. Watkins. *Learning from Delayed Rewards*. PhD thesis, King's College, Cambridge, UK, 1989.

---

[4] Robust individual agent skills also were essential for the team's performance. Thanks to Patrick Riley for his contributions to the team, developing effective individual low-level skills.

A Method for Localization by Integration of Imprecise Vision and a Field Model

Kazunori TERADA[1], Kouji MOCHIZUKI[1], Atsushi UENO[1],
Hideaki TAKEDA[1], Toyoaki NISHIDA[2], Takayuki NAKAMURA[1],
Akihiro EBINA[1], and Hiromitsu FUJIWARA[1]

[1] Graduate School of Information Science
Nara Institute of Science and Technology
8916-5 Takayama, Ikoma, Nara 630-0101, Japan
{kazuno-t, kouji-m, ueno, takeda, takayuki,
akihi-e, hiromi-f}@is.aist-nara.ac.jp
[2] Department of Information and Communication Engineering
School of Engineering
The University of Tokyo
7-3-1 Hongo, Bunkyo-ku, Tokyo 113-8656, Japan
nishida@kc.t.u-tokyo.ac.jp

Abstract. In recent years, many researchers in AI and Robotics pay attention to RoboCup, because robotic soccer games needs various techniques in AI and Robotics, such as navigation, behavior generation, localization and environment recognition. Localization is one of the important issues for RoboCup. In this paper, we propose a method of robot's localization by integrating vision and modeling of the environment. The environment model that realizes the robotic soccer filed in the computer can produce an image of robot's view at any location. In the environment model, the system can search and appropriate location of which view image is similar to the view image by the real robot. Our robot can estimate location from goal's height and aspect ratio on the camera image. We search the most suitable position with hill-climbing algorithm from the estimated location. We programmed this method, and tested validity. The error range is reduced from 1m~50cm by robot's estimation from 40cm~20cm by this method. This method is superior to the other methods using dead reckoning or range sensor with map because it does not depend on the field size on precision, and does not need walls as landmark.

1 Introduction

In the domain of the robotic soccer, there are various classes of problems; navigation, behavior generation, recognition of the environment, and localization. In these problems, localization is especially indispensable technique for robotic soccer. In order to generate cooperative behavior, an agent needs to know its position within the environment. Therefore, the problem of estimating the position of a mobile robot is one of the fundamental problems in the field of mobile robots. In this paper we propose a method of localization by integration of imprecise vision and a 3D environmental model. We will show that this method is also effective to identify opponent robots.

Dead reckoning and range finder are often utilized to the localization problem [7] [5] [3]. Dead reckoning uses odometry (i.e. counting wheel rotations) to determine the robot's position. Since, errors from slip of wheels accumulate over time, estimation of its position becomes increasingly inaccurate. The range finders such as sonar and IR sensors are not useful in the environment in which there are no walls which reflect the ultrasonic wave and infrared rays. In order to overcome this problem we need to use the method of localization there are less dependent on its environment.

In this paper we propose the system which uses both visual sensor and the soccer field model. By making comparison between images from the visual sensor and provided by the field model, the robot can estimate its position.

In the domain of vision-guided mobile robot research, there are a lot of techniques of localization using visual sensors. There have been various methods which use 3D model of environments to estimate the robot's position and orientation. In this approach, 3D model is employed to generate an expected image and it is compared with an image captured by the robot [2] [9] [6] [4] [1].

If the 3D model of environment is already known and we can measure several feature points of an object in a captured 2D image, the distance to object may be derived by geometrical model. However, it is hard to measure such feature points because of the following two reasons. One reason is that if there is an unknown object in the captured image and it obstructs the view, these feature points would not be observed. The other is that the image processing system is often confronted with change of lighting, therefore accurate measurement is impossible. In order to overcome this problem, we add a method of matching two images to distance estimation from several feature points.

2 System Architecture

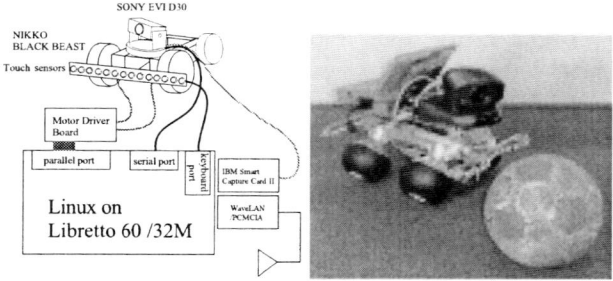

Fig. 1. Our soccer robot.

Our robot consists of four major hardware components; a portable PC (Libretto 50 or 60, 100, Toshiba), a vision system including a camera and a video capture card, a tactile sensor system, and a chassis including motors and a motor drive board. The detail of our system is described in [8].

An image captured by the vision system is processed by the vision module. The vision module provides information about objects in the field and the image is segmented into 7 colors (See Table1) regions in the vision module which runs on PC on the robot.

Table 1.

object	color
ground	green
one goal	yellow
another goal	blue
wall	white
ball	red
outside of field	black
robot	gray

Every pixel in an image is classified to 6 colors except gray by means of a discrimination rule which uses the Mahalanobis distance, and this rule is learned by sample color data. A pixel which does not classified to any colors labeled to gray.

3 localization

The procedure of our method of localization consists of the following two steps; 1) Estimate the current position by calculating distance to a landmark from an image which is sent from camera on the robot, 2) Revise the position by comparing the camera image and vision images that are generated by the field model.

We select a goal as a landmark to estimate the position. The reasons of selecting the goal as a landmark are as follows. 1) The robot can always see either goal everywhere in the soccer field. 2) The robot can always see a complete view of either goal since it can rotate its camera.

3.1 Estimation of the position

We use distance and angle from the goal estimate position of the robot in the field. Distance and angle are calculated from width and height of the goal that are identified in an image captured by the camera on the robot. The relations between distance, angle and width, height of the goal are as follows;

- The height of a goal in the image is inversely proportional to the real distance to the goal.
- The angle to the goal concerns to the ratio of the width to the height of goal.

The former relation can be easily formulated;

$$d = \frac{a}{h} \tag{1}$$

Where the h is height of the goal in the image. We select the constant a by calculating the mean value from the real data. The latter is derived from the geometrical model of the relation between a goal and a robot(See Figure 2).

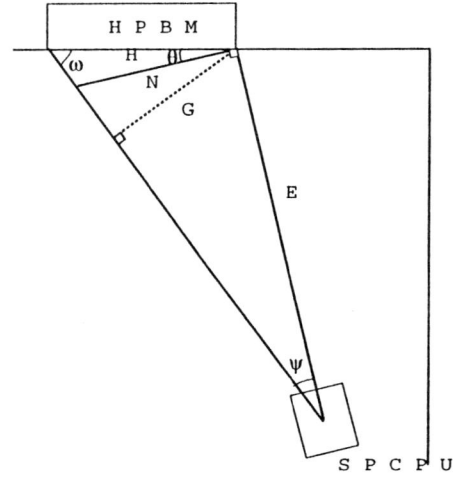

Fig. 2. The geometrical model of the relation between a goal and a robot

According to this model, angle to the goal is;

$$\theta = 90 - \arcsin(\frac{d}{g}\sin(\arctan(\frac{m}{d}))) - \arctan(\frac{m}{d}) \tag{2}$$

We define the origin of the coordinate axes as center of center circle in the soccer field. The position of robot is;

$$\begin{cases} x = 4110 - d\cos(\theta) \\ y = \phantom{4110 - {}} d\sin(\theta) \end{cases}$$

The position calculated this method has about 60cm error in average. This error is caused by limit of image precision (64 × 48 pixels) and error of color segmentation. Image precision is limited because all kinds of processing including image processing are done in a single computer in our robot and it requires real-time response.

In order to reduce this error, we revise the position with the field model.

3.2 Position Revision

The position revision system consists of a server which revises the position, and clients each of which is on a robot and send estimated positions to the server. The server communicates with multiple clients and holds the position of each robot. Figure 3 shows the client-server system applied in our work.

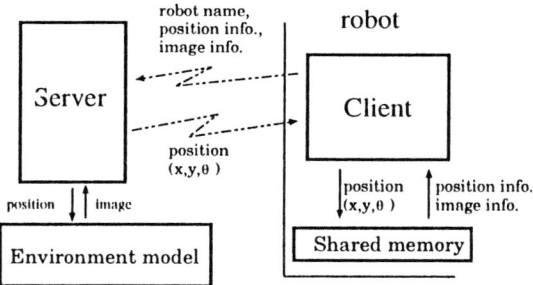

Fig. 3. Server Client System

The role of the server and the client is described as follows. Server receives both data of the position which is estimated in the computer on the robot, and images captured by the camera on the robot through wireless ethernet. The server then revises the position by comparing the two images; one is captured by the robot, another is generated in the server by means of a 3D field model.

3.3 Field Model

The server has a 3D field model. The features of this model are;

1. The size of every part of field such as wall and goal can be changed.
2. This model can generate any image of view that the robot ought to see from coordinates (x, y) and angle θ.
3. We can place multiple robots in the field.
4. The shape of every robot can be changed.

3.4 Method of Position Revision

We use two types of images for position revision. One type of images that is captured by the camera on a robot. The image is segmented to 7 colors region by the vision module on the robot and sent to the server through ethernet. The other type of images is an ideal image of the field. This image is generated by the server and is calculated from value of position; (x, y, θ) that is sent from a client to the server. The strategy of position revision is to find an optimal point in the field model. In this point the field model generates the most similar image to the captured image. The method of finding this point is to search the neighborhood point of starting point which is estimated from geometrical model. We use the hill-climbing algorithm as a search algorithm.

The evaluation function is;

$$func(a,b,c) = \frac{2c}{a+b} \qquad (3)$$

a is the number of pixels of the goal area in an image which generated by the 3D field model. b is the number of pixels of the goal area in an image which is captured by camera on the robot. c is a number of intersection pixels of both areas when two images are overlapped in order to fit center of gravity of both areas.

By means of hill-climbing algorithm search of optimal point is carried out by the following steps.

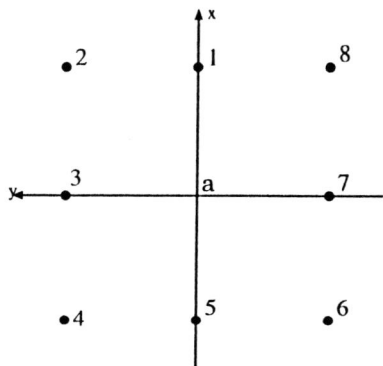

Fig. 4. Neighbor point of (i, j)

1. Step1: Evaluation of beginning of search.
   (a) Calculate the value of beginning point $p_t(i,j)$ by evaluation function.
2. Step2: Calculate the value of the next candidate points
   (a) Select the neighbor 8 points of $p_t(i,j)$ (See Figure 4).
   (b) Generate the image of each point and calculate a value by the evaluation function.
   (c) Select the point $p_{t+1}(i,j)$ in which the evaluation value is the greatest of 8 points.
3. Step3: Comparison.
   (a) If the value in $p_t(i,j)$ is greater than that in $p_{t+1}(i,j)$, finish searching.
   (b) If the value in $p_t(i,j)$ is greater than 0.99, finish searching.
   (c) If the value in $p_{t+1}(i,j)$ is the same as that in $p_{t-1}(i,j)$, finish searching.
   (d) If the value in $p_{t+1}(i,j)$ is greater than that in $p_t(i,j)$, return Step2.

4 Experimental result

We had experiments on the proposed method. We put the robot on the point of the field that is randomly selected, and executed the program. Experimental results are shown in Table 2.

Table 2. Comparison of errors

position	estimated	revised
(2000,0)	659mm	206mm
(1850,0)	639mm	308mm
(1300,0)	1113mm	320mm
(1150,0)	1111mm	381mm
(1000,0)	927mm	335mm
(1000,1000)	279mm	260mm
(0,1000)	486mm	172mm
(-1000,1000)	606mm	243mm
(1000,-1000)	341mm	141mm
(0,-1000)	258mm	215mm
(-1000,-1000)	404mm	295mm

The mean value of the error before revision is about 600mm and it is reduced to 260mm after revision. While the error before revision becomes larger with the robot goes away from the goal, the error after revision becomes less sensitive to the distance. This fact suggests that our proposed method is tolerant of the changing position in the field. But, the error after revision becomes also smaller when the robot is closer to the goal. The reason for this is that we use the number of pixel in an image as a parameter of evaluation function. Since error rate does not depend on the amount of these pixels, the larger amount of these pixels contributes to make position more accurate.

5 Identification of Opponent Robots

In this section we explain a method of identification of opponent robots. This method is realized by our proposed localization method and field model. To distinguish team mate robot from opponent robot is important ability for soccer robot. An innocent robot may pass the ball to an opponent robot. Without explicit mark, to do that is considerably hard.

Our basic idea is that there are only team mate robots in the field model constructed in the server. The server know all positions of each team mate robot by communicating with each robot, and can generate an image in which there are only team mate robots. The procedure of distinguishing a team mate from opponent robots is follows.

1. Preparation of two images; one is generated by the server, and another is sent from one of the robot in the field.
2. Counting the pixels which are colored gray. Each image is segmented to 7 colors and the robot is colored gray in each image.
3. Calculation of remainder of each number of pixels; r.
4. If r is greater than a threshold, we define that there is at least an opponent robot in the image which is captured by robot. We select the threshold as 100 pixels.

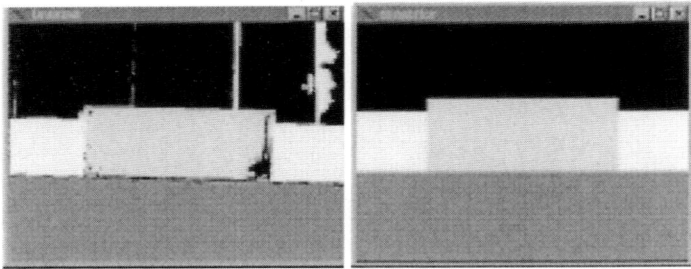

Fig. 5. There are no robots.

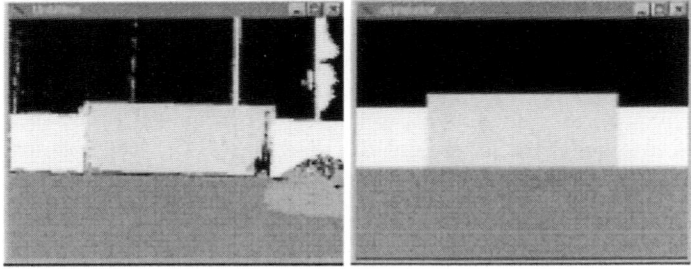

Fig. 6. There is only opponent robot.

4 experimental result are shown in Figure 5 ∼ Figure 8. In each figure, (a) is captured by a camera on a robot and (b) is generated by server. In figure (b) a team mate robot is expressed by a gray cone.

In Figure 5(a), the number of gray pixels is 66 and in Figure 5(b) 0. Therefore, the server decides that there are no robots. In Figure 6(a), the number of gray pixel is 184 and in Figure 6(b) 0. Therefore, the server decides that there is a team mate robot. In Figure 7(a), there is a robot in the left side of the goal and

Fig. 7. There is only a team mate robot.

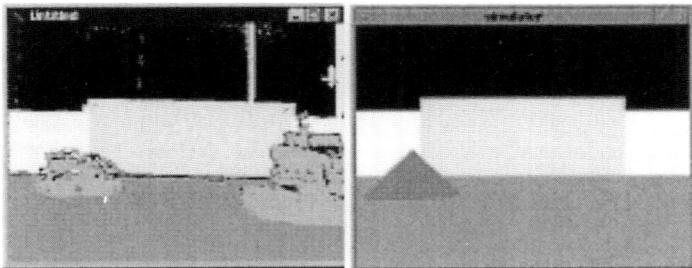

Fig. 8. There are both a team mate and an opponent robot.

the number of gray pixel is 155. In Figure 5(b) there is a corn as a team mate robot. The server know each position of each team mate robot, and can generate an image in which there is a team mate robot. The number of gray pixel is 104. Therefore the server decides that there is a team mate robot. While in Figure 8(a), there are two robots, in Figure 8(b), there is one robot, the server decides that there are both a team mate and an opponent robot.

6 Conclusions

In this paper, we propose a method of robot's localization by integrating vision and modeling of the environment. We also propose a method of identification of opponent robots by using vision and the field model. By means of our method, accuracy of the localization of the robot in the soccer field is improved.

In the method of using geometrical model and image's feature points, a noise of image prevents sampling of the feature value and directly effects the accuracy of the localization. On the other hand, in the method of comparing an image captured by robot and an image generated by the field model, starting point of search is important to reduce the computational cost. Combining these two methods, we reduced the cost of search and improved the accuracy of localization.

There are two advantages of our localization method against the other method such as a method utilized sonar or IR sensor and a method of dead reckoning. One is that our method does not depend on the size of the field. We utilize the view of goal in order to estimate and revise the position of the robot, therefore if only the goal is visible, localization is possible. If the size of the field become greater, errors from dead reckoning are proportionally increased. The other is that our method does not depend on the structure of the field. In the method utilized sonar or IR sensor, if there are no walls, the robot could not know its position.

In the method of identification of the opponent robot, we compare an image of a field model in which there are only a team mate robot to an image captured by the robot in order to distinguishing a team mate robot from opponent robots. Our experimental result shows that we can identify the opponent robot without explicit marking.

In our approach a problem still exists; If the robot can not see landmark, the estimation of position by matching two images is impossible. In order to overcome this problem we can use multiple landmarks and multiple images captured by a camera while the camera is rotating.

References

[1] H. Christensen, N. Kirkeby, and et al. Model-driven vison for in-door navigation. *Robotics and Autonomous Systems*, 12:199–207, 1994.

[2] C. Fennema, A. Hansen, E. Riseman, J.R. Beveridge, and R. Kumar. Model-directed mobile robot navigation. *IEEE Trans. on Systems, Man, and Cybernetics*, 20:1352–1369, 1990.

[3] J.-S. Gutmann, W. Hatzack, I. Herrmann, B. Nebel, F. Rittinger, A. Topor, T. Weigel, and B. Welsch. The cs freiburg robotic soccer team: Reliable self-localization, multirobot sensor integration, and basic soccer skills. In *RoboCup-98: Robot Soccer World Cup II*. Springer-Verlag, Berlin, Heidelberg, 1998.

[4] A. C. Kak, K. M. Andress, and et al. *Hierarchical evidence accumulation in the PSEIKE system and experiment in model-driven mobile robot navigation*. M. Henrion and R. Shachter and et al., Elsevier, 1990.

[5] L. Kleeman. Optimal estimation of position and heading for mobile robots using ultrasonic beacons and dead-reckoning. In *Proceedings 1992, IEEE International Conference on Robotics and Automation*, pages 2582–2587, 1992.

[6] A. Kosaka and A. C. Kak. Fast vision-guided mobile robot navigation using model-based reasoning and prediction of uncertainties. *Computer Vision, Graphics and Image Processing -IU*, 56:271–329, 1992.

[7] E. Krotkov, L. Chrisman, F. Cozman, and G. Heredia. Dead reckoning and visual landmarks for a prototype lunar rover. In *Proc. of IEEE/RSJ International Conference on Intelligent Robots and Systems 1995 (IROS95)*, 1995.

[8] T. Nakamura, K. Terada, and et al. The robocup-naist: A cheap multisensor-based moblie robot with on-line visual learning capability. In *Proceedings of the second RoboCup Workshop.*, pages 291–303, 1998.

[9] T. Tsubouchi and S. Yuta. Map-assisted vision system of mobile robots for re-choning in a building invironment. In *Proceedings of 1987 IEEE International Conference on Robotics and Automation*, pages 1978–1984, 1987.

Multiple Reward Criterion for Cooperative Behavior Acquisition in a Multiagent Environment

Eiji Uchibe and Minoru Asada

Dept. of Adaptive Machine Systems, Graduate School of Eng., Osaka University, Suita, Osaka 565-0871, Japan

Abstract. An extended value function is discussed in the context of multiple behavior coordination, especially in a dynamically changing multiagent environment. Unlike the traditional weighted sum of several reward functions, we define a vectorized value function which evaluates the current action strategy by introducing a discounted matrix to integrate several reward functions. Owing to the extension of the value function, the learning robot can estimate the future multiple rewards from the environment appropriately not suffering from the weighting problem. The proposed method is applied to a simplified soccer game. Computer simulations are shown and a discussion is given.

1 Introduction

Recently, the realization of cooperation among multiple robots has been studied by many researchers. RoboCup [3] is an increasingly successful attempt to promote the full integration of AI and robotics research, and many researchers around the world have been attacking a wide range of research issues, especially multiagent problems in a dynamically changing environment. Among them, behavior learning in a multiagent environment has been attacked based on reinforcement learning [7, 11, 12]. One of the issues in applying reinforcement learning to this domain is that the agent has to cope with multiple tasks and is needed to make a decision any time which behavior should be taken.

There are two typical coordination approaches of multiple behaviors. The first one is to coordinate multiple behaviors in sequence [6, 9]. Recently, the multi-criteria sequential decision making problems are considered based on the vectorized value function [2]. However, it seems difficult to determine the order of behaviors based on the value of the utility because it depends on the discount factor and the state space both of which should be carefully designed in real robot applications.

The second approach is a mechanism of decision making which behavior should be performed according to the current situation, but it seems difficult to decide before encountering the situations. A typical example is a task of shooting a ball into the goal avoiding collisions with other robots. The robot has to consider a tradeoff between shooting and avoiding behaviors because it depends on the situation which behavior has higher priority to other one. To cope with such a tradeoff, several modular architectures have been proposed based on the independently acquired results of reinforcement learning. For example, an arbiter coordinates behaviors based on the majority [13]. While modular reinforcement learning [11] has been proposed to coordinate behaviors in a case that the behaviors interfere with each other. However, these methods need to separately apply the learning algorithm into subtasks in advance.

As one alternative to coordinate behaviors which achieve multiple tasks simultaneously, the extension of the reward function seems promising. Simple realization is based on the weighted sum of multiple rewards [4]. In this case, fundamental behaviors have been embedded as a form of subsumption architecture, which makes learning itself simple. However, the methods of the weighted sum of reward functions are faced with the essential problem of weighting itself, that is, how to decide the weights.

In this paper, we propose a vectorized value function to cope with multiple tasks. In other words, the values of multiple behaviors are estimated by the separate value functions. We implement an architecture of an actor-critic type as a learning mechanism. The critic is a state value function. After each action selection, the critic evaluates the new state to determine whether it has become better or worse than expected.

2 Vectorized Reinforcement Learning

2.1 Temporal Difference

Before explanation of the proposed method, we show Temporal Difference (hereafter, TD) method [8] briefly for the reader's understanding. We consider the state-outcome sequences of the form $\boldsymbol{x}_t, \boldsymbol{x}_{t+1}, \cdots, \boldsymbol{x}_{t+n}, r$, where each $\boldsymbol{x}_t$ is a state vector at time t in the sequence, and r is the outcome of the sequence.

The given task is to predict the future reward to receive at each state. TD(λ) [8] maximizes (or minimizes) scalar cumulative discounted sum

$$v_t(\boldsymbol{x}_t) = \sum_{n=0}^{\infty} \gamma^n r_{t+n}, \qquad (1)$$

where γ is a discount factor between 0 and 1.

2.2 Problems of Scalar-valued Reward Function

In case of realizing cooperative behaviors in a multiagent environment, the learning robot has to consider the tradeoff between the individual and the team purposes as much as possible. Suppose that the learner has N tasks to accomplish. The multiple rewards from the environment are given to the robot as follows:

$$\boldsymbol{r}^T = \begin{bmatrix} r^1 & r^2 & \cdots & r^N \end{bmatrix},$$

where r^i denotes the reward for the i-th task.

In order to cope with multiple rewards, one of the simplest implementation of them is a weighted combination of the rewards like

$$r = \sum_{i=1}^{N} w^i r^i, \qquad (2)$$

where w^i is a weight for the reward r^i. The objective is to maximize (or minimize) the weighted sum. This method reduces the problem to the case of scalar-valued reinforcement values. However, this combination has the following deficits.

- The discount factor γ is common to the all tasks.
- The estimated value function is unstable when we give both the positive and negative rewards.

Since γ controls to what degree rewards in the distant future affect the total value of a policy, it is desirable to set γs to the appropriate values for the corresponding tasks, respectively. A typical example is "collision avoidance" which has different property (negative reward) from that of goal directed behaviors (positive reward). That is, any action can be allowed to be taken unless it causes collisions with other objects. In order to learn such a behavior, γ should be much smaller so that the utility for the distant future cannot be affected.

2.3 Extension to the Vectorized Value Function

Considering the above mentioned issue, we extend scalar value function to a vectorized value function. The discounted sum of the vectorized value function can be expressed by

$$\boldsymbol{v}_t(\boldsymbol{x}_t) = \sum_{n=0}^{\infty} \boldsymbol{\Gamma}^n \boldsymbol{r}_{t+n}, \qquad (3)$$

where $\boldsymbol{\Gamma}$ is $N \times N$ matrix. We call $\boldsymbol{\Gamma}$ discounted matrix. If the eigenvalue of the matrix $\boldsymbol{\Gamma}$ exists within the unit circle, the value expressed by Eq.(3) converges.

It is important to design $\boldsymbol{\Gamma}$. As described above, the behavior obtained by reinforcement learning depends on the value of the discount factor. Although we discussed the affect of γ in our previous work [11], it is not clear to design γ. So, we utilize the principal angles between two subspaces in order to implement $\boldsymbol{\Gamma}$. Figure 1 shows a basic idea to design $\boldsymbol{\Gamma}$. The merits of this method are regarded as follows:

- If there is a bias when the rewards are given to the robot, we can reduce the dimension of the reward space.
- Taking into account the reward space, we can modify the state space appropriately. In *LPM*, the state space is constructed based on only the sequences of observation and action.

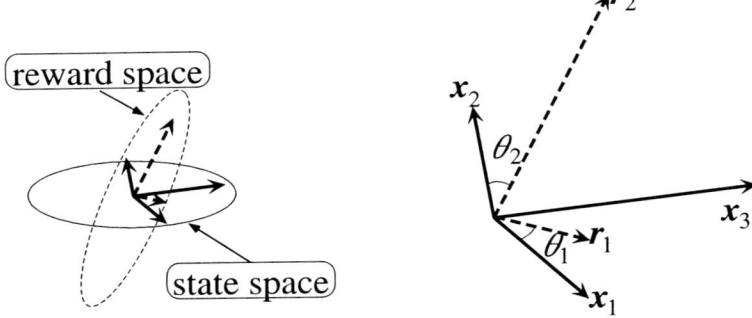

Fig. 1. Basic idea to design the discounted matrix from the relation between two subspaces

In order to calculate the principal angles, we have to compute the following singular value decomposition

$$\hat{\Sigma}_{rr}^{-1/2} \hat{\Sigma}_{rx} = U_2 S_2 V_2^T, \tag{4}$$

$$U_2 U_2^T = I_N, \quad V_2 V_2^T = I_n.$$

Then, we set $\boldsymbol{\Gamma}$ as

$$\boldsymbol{\Gamma} = S_2 = \text{diag}\left[\cos\theta_1, \cos\theta_2, \cdots, \cos\theta_N\right], \tag{5}$$

where θ ($\theta_1 \leq \theta_2 \leq \cdots \leq \theta_N \leq \pi/2$) denotes the principal angle between two subspaces ($S(\boldsymbol{R})$ and $S(\boldsymbol{X})$). Furthermore, we modify the new reward and state spaces as follows:

$$\begin{aligned} r' &= U_2^T \hat{\Sigma}_{rr}^{-1/2} r \quad \text{and,} \\ x' &= V_2^T x. \end{aligned} \tag{6}$$

The robot learns its behavior based on the new reward r' and the new state vector x'. Because the eigenvalues of $\boldsymbol{\Gamma}$ in Eq.(5) are less than one, the value expressed by Eq.(3) converges.

2.4 Behavior Learning

In order to acquire the policy (the mapping function from the state space to the action space based on the vectorized value function, we utilize the actor-critic methods [8] which are TD ones that have a separate memory structure explicitly. Let $q_t(x', u)$ be a vector at time t for the modifiable policy. The TD error can be used to evaluate the action u taken in the state x'. Eventually, the learning algorithm is shown as follows:

Initialization:
1. Collect the observation, action, and reward $\{y, u, r\}$, and construct the state space by *LPM*. Compute $\boldsymbol{\Gamma}$ by Eq.(4).
2. Initialize the value function $v(x')$ and the policy function $q(x', u)$ arbitrarily and an eligibility function $e(x') = 0$ for all states.
Repeat forever:
1. Calculate the current state x'_t by LPM and Eq.(6).
2. Execute an action based on the current policy. As a result, the environment makes a state transition to the next state x'_{t+1} and generates the reward r_t.
3. Calculate the TD error by

$$\delta_t = r'_t + \boldsymbol{\Gamma} v_t(x'_{t+1}) - v_t(x'_t). \tag{7}$$

and update the eligibility by

$$e_{t+1}(x') = \begin{cases} 1 & \text{if } x' = x'_t, \\ \lambda e_t(x) & \text{if } x' \neq x'_t, \end{cases} \tag{8}$$

where λ is the *trace-decay* parameter.
4. Update the value function and the policy function. For all $x'_t \in X'$ and $u_t \in U$,

$$\begin{aligned} v_{t+1}(x'_t) &= v_t(x'_t) + \alpha \delta_t e_t(x'_t) \\ q_{t+1}(x'_t, u_t) &= q_t(x'_t, u_t) + \beta \delta_t, \end{aligned} \tag{9}$$

where α and β $(0 < \alpha, \beta < 1)$ are learning rate.
5. Return to **2**.

In this algorithm, v and q are represented by a look-up table, where continuous state and action spaces have to be quantized appropriately. So long as the learning robot utilizes a tabular action value function which is segmented by the designer, the robots are haunted by the segmentation problem of the state and action spaces. To overcome this problem, several quantization methods such as Parti game algorithm [5], Asada's method [1] and Continuous Q learning [10] might be promising.

The learning robot has to select several actions to explore the unknown situations. Then, we use ϵ-greedy strategy [8], meaning that most of time the robot chooses an optimal action, but with probability ϵ it instead selects an action at random. We summarize the action strategy as follows.

(a) Execute the random action with probability ϵ, or goto step (b).
(b) Initialize $rank(u) = 0$ for all $u \in U$. For $i = 1, \cdots N$,
    (1) Calculate the optimal action corresponding to each q_i ($q^T = [q_1, \cdots, q_N]$),

$$u^*_i = \arg\max_{u \in U} q_i(x', u). \tag{10}$$

    (2) Increment $rank(u^*_i) = rank(u^*_i) + 1$.

(3) Execute the optimal action,

$$u^* = \arg\max_{u \in U} rank(u). \tag{11}$$

After all, since the all actions are ordered with respect to the *Pareto optimality*, the learning robot can select the action which satisfy the *optimal* action as possible.

3 Tasks and Assumptions

3.1 Environment and Robots

We apply the proposed method to a simplified soccer game including three learning mobile robots in the context of RoboCup [3]. RoboCup is an increasingly successful attempt to promote the full integration of AI and robotics research, and many researchers around the world have been attacking a wide range of research issues, especially multiagent problems in a dynamically changing environment.

The environment consists of a ball and two goals, and a wall is placed around the field except the goals (see Figure 2). The sizes of the ball, the goals and the field are the same as those of the middle-size real robot league in the RoboCup Initiative.

Each robot does not know the locations, the sizes and the weights of the ball and the other robot, any camera parameters such as focal length and tilt angle, or kinematics/dynamics of itself. Each robot has a single color TV camera and observes output vectors (image features) shown in Figure 3. The dimensions of the observed vectors for the ball, the goal, and the other robot are 4, 11, and 5, respectively.

Figure 2 (d) shows the real robot used for modeling. The robots have the same body (power wheeled steering system) and the same sensor (on-board TV camera). As motor commands, each mobile robot has a 2 DOFs. The input u is defined as a 2 dimensional vector:

$$u^T = \begin{bmatrix} v & \omega \end{bmatrix} \quad v, \omega \in \{-1, 0, 1\},$$

where v and ω are the velocities of translation and rotation of the robot, respectively. In this experiments, v and ω are quantized into three levels, which are uniformly distributed. Totally, the robot can select one from nine actions at each state.

3.2 Experimental setup

We show two experiments to verify the proposed learning algorithm using a simplified soccer game.

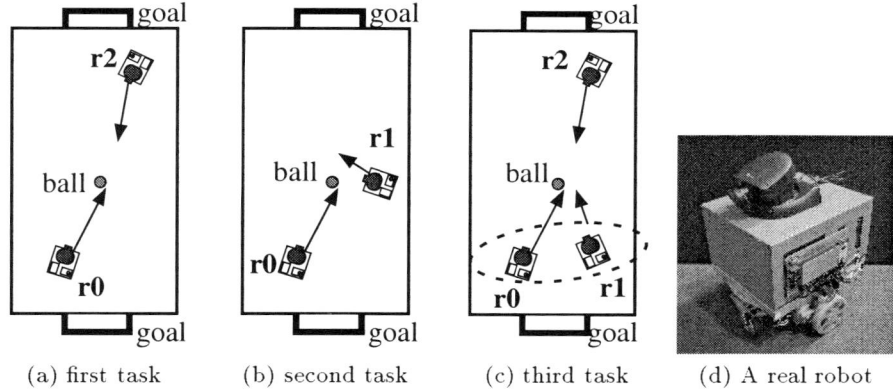

Fig. 2. An environment and three robots

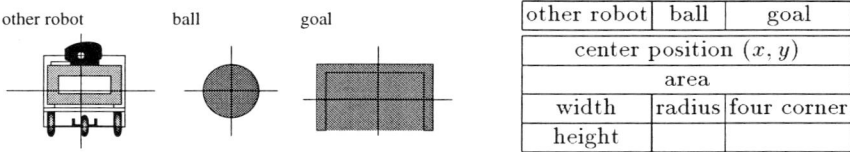

other robot	ball	goal
center position (x, y)		
area		
width	radius	four corner
height		

Fig. 3. Image features of the ball, goal, and other robot

- Shooting a ball into the goal without collisions with other robot (Figure 2(a)):
  In this experiment, there are two robots in the environment. One is a shooter **r0** which learns to shoot a ball into a goal avoiding collisions with the other robot as possible, and the other **r2** is a defender to disturb the learning robot.
- Shooting and passing a ball without collisions with each other (Figure 2(b)):
  There are two learning agents (**r0** and **r1**) in the environment. The setting of this experiment is as the same as that of one described our previous work [12] except that the role of each robot has not been pre-specified. That is to say, the reward function is common between two learning robots.
- Simplified game among three robots (Figure 2(c)):
  We perform a simplified soccer game by three learning robots. **r0** and **r1** are teammates while **r2** is a competitor against them. The difference from the experiments described above is involvement of competition.

One trial is terminated if one of the robots shoots a ball into the goal or the pre-specified time interval expires. The trial still continues even if a passing behavior is achieved, the robots make collisions, or the robot pushes the ball.

4 Experimental Results

4.1 Shooting a Ball into the Goal without Collisions

In this experiment, the tasks for the learning robot are (i) to shoot a ball into the goal, (ii) to avoid collisions with the other robot, and (iii) to kick the ball. The defender **r2** has a fixed policy of chasing the ball, and its motion speed is a 30 % of the maximum speed of the learning robot.

We assign the negative reward $r^c(=-1.0)$ when **r0** makes a collision with **r2**. In addition, we assign the positive rewards $r^s(=1.0)$ and $r^k(=1.0)$ when **r0** shoots the ball into the goal and kicks the ball, respectively. Then, the reward function is defined as a three dimensional vector. The other parameters are set to $\lambda = 0.4$, $\alpha = 0.25$ and $\beta = 0.25$, respectively. In order to compare the proposed method with the method described in Section 2.2, we prepare the following linear weighting reward function

$$r = r^c + r^s + w^k r^k, \qquad (12)$$

where w^k is a weight. We check two values: (a) $w^k = 0.0$, (b) $w^k = 1.0$. The discount factor γ is set to 0.9.

Figure 4 shows that the robot with $w^k = 1.0$ tends to push the ball more frequently than other robots. Then, the success rate of shooting behavior is worse than the case of the proposed method and no weight. The reason is that the reward function used in Task (b) is a mixture of the positive and negative value. Then, the value function based on Eq.(12) may not estimate the value appropriately. Also, the problem of the estimator based on Eq.(12) is to use the same discount factor among the given tasks. Suppose that the learning robot makes a collision with other robot. The learning robot receives r^c more frequently than r^s, which makes the robot to acquire the avoiding behavior at the beginning. As a result, the robot can not seek the feasible solutions to acquire shooting behavior.

4.2 Shooting and Passing a Ball without Collisions

In this experiment, there are two learners. The tasks for both learners are (i) to shoot a ball into the goal, (ii) to avoid collisions, (iii) to kick the ball, and (iv) to pass the ball to the teammate. The first three can be regarded as evaluation for individual robot and the fourth as cooperative one. Then, we test the coordination of shooting, passing and avoiding behaviors. We use the reward function common to them. The reward function is

$$\boldsymbol{r}^T = \begin{bmatrix} r^c & r^s & r^p & r^k \end{bmatrix}, \qquad (13)$$

where r^p denotes the rewards for passing the ball toward the teammate.

We assign the positive reward $r^p(=1.0)$ when the pass behavior is accomplished. The definition of the pass behavior in this experiments is that the agent receives r^p after the other agent touches the ball which is pushed by itself in

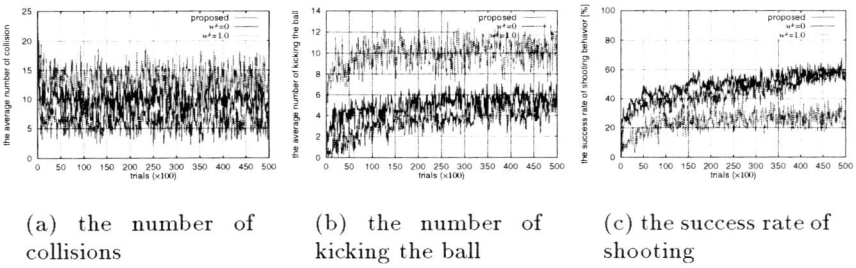

(a) the number of collisions (b) the number of kicking the ball (c) the success rate of shooting

Fig. 4. Experimental results of the acquired performance in a case of shooting and avoiding behaviors

a short time. In the same manner, other rewards $r^c(=-1.0)$, $r^k(=1.0)$ and $r^s(=1.0)$ are given when the collisions happen, the agent kicks the ball, and the agent shoots the ball into the goal, respectively.

In this experiment, since there are multiple agents that learn behaviors, simultaneous learning may cause poor performance, especially in the early stage of learning. Then, we apply the learning schedule as follows:

- **period A** (trial number $0 \sim 2.5 \times 10^4$) : **r0** is a learner while **r1** is stationary,
- **period B** (trial number $2.5 \times 10^4 \sim 5.0 \times 10^4$) : **r1** is a learner while **r0** moves around.
- **period C** (trial number $5.0 \times 10^4 \sim 7.5 \times 10^4$) : **r0** is selected as a learner again while **r1** moves around based on the result of the period **B**, and
- **period D** (trial number $7.5 \times 10^4 \sim 10.0 \times 10^4$) : **r1** construct the *LPMs*, and learn the behaviors again.

Figure 5 shows the learning curves with respect to the frequencies of the reception of the rewards r^s and r^p. As we can see from Figure 5(a), the success rate of shooting behavior gradually increased through the interactions. Until the end of period **A**, only **r0** shot the ball into the goal because **r1** did not move in this period.

From Figure 5(b), we can see that the frequencies of passing behaviors also increased. The passing behaviors not only from **r0** to **r1** but also from **r1** to **r0** are reinforced in the period **A**. Even though **r1** was stationary, the ball was sometimes run into **r1** by accident. Then, r^p was given to **r1** after **r0** pushed the ball.

4.3 A Simplified Soccer Game among Three Robots

Next, we perform three-robots' experiments. This experiment involves both cooperative and competitive tasks. **r0** and **r1** are teammates. We add new rewards

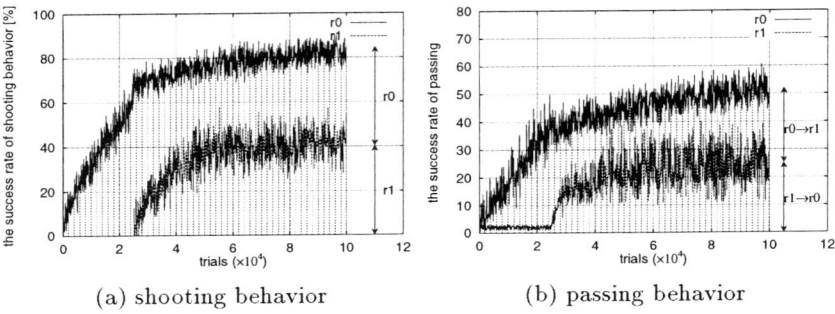

(a) shooting behavior   (b) passing behavior

Fig. 5. Experimental results of the acquired performance in a case of passing and shooting behaviors

r^l, r^m and r^p. The six dimensional reward vector is prepared for this game. We assign the negative reward $r^l(=-1.0)$ when the team to which the robot belongs lose the goal. Although the positive reward $r^p(=1.0)$ is given when the robot passes the ball toward its teammate, the reward $r^m(=-1.0)$ is given when the robot passes the ball toward its opponents. Other rewards r^c, r^s and r^k are the same as before. The other parameters are set to $\lambda = 0.4$, $\alpha = 0.25$ and $\beta = 0.25$, respectively.

We prepare the learning schedule [12] to make learning stable in the early stage. As described in Section 4.2, each interval between change of learning robots is set to 2.5×10^4 trials. In this experiment, we set up the following three learning schedules,

- case (A) : r0 → r1 → r2,
- case (B) : r1 → r2 → r0,
- case (C) : r2 → r0 → r1.

After each robot learned the behaviors (all the robot was selected at once), we recorded the total scores in each game. We perform 5 sets of the experiments, and show the histories of the game in Figure 6.

As we can see from Figure 6, the result depends on the order to learn. Although this game is two-to-one competition, **r2** won the game if we selected **r2** as the first robot to learn (case (C)). Otherwise, a team of **r0** and **r1** defeated **r2**. This scheduling is a kind of teaching, help the agents to search the feasible solutions from a viewpoint of the designer. However, the demerits of this method is also revealed when we apply it to the competitive tasks. That is, the learning schedule often leads the competitive game to the undesirable results. In other words, since one side overwhelmed its opponents, both sides reached to one of stable but low skill levels, and therefore no change happens after this settlement. Therefore, we need to extend the learning schedule to the one including competitive situations.

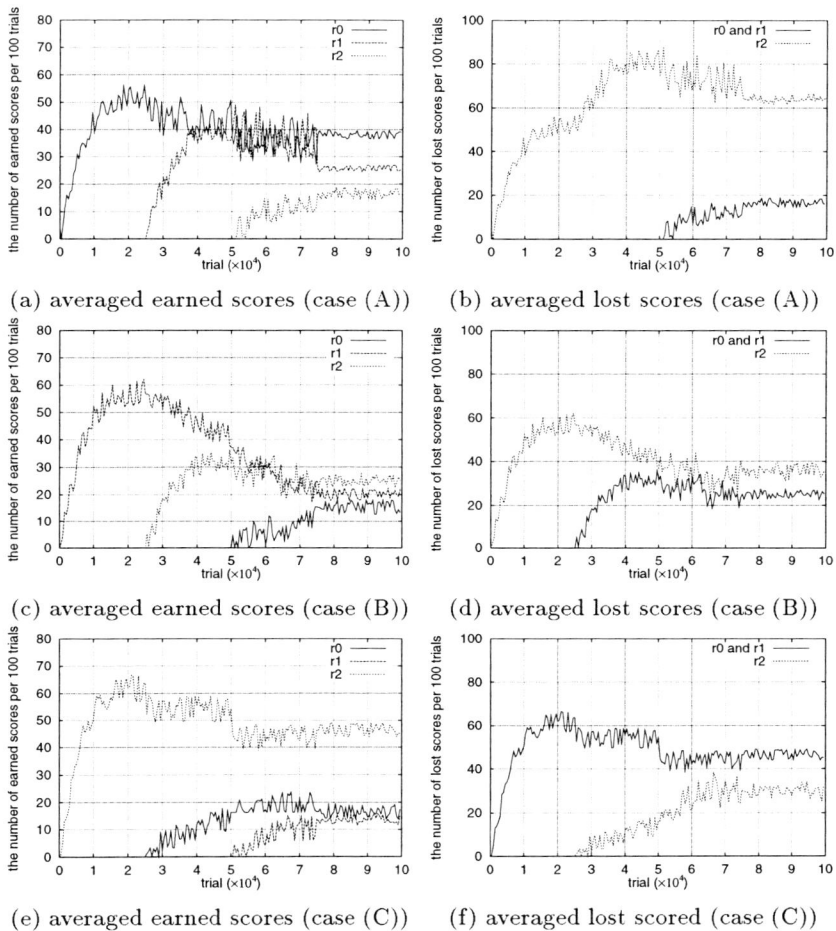

Fig. 6. Experimental results of the acquired performance in case of a simplified soccer game

5 Conclusion

This paper has shown how the learning robots cope with the multiple tasks in a multiagent environment. In order to realize cooperation, global evaluation factors are added to the reward function. In other words, the task for the robot is to consider the tradeoff between the individual evaluation and the global one. We have applied the proposed method to several simplified soccer games, and demonstrated that the learning robots can acquire the purposive behaviors. Now, we are planning to implement real experiments to check the validity of the proposed method and the obtained behaviors.

Acknowledgment

This research was supported by the Japan Society for the Promotion of Science, in Research for the Future Program titled Cooperative Distributed Vision for Dynamic Three Dimensional Scene Understanding (JSPS-RFTF96P00501).

References

1. M. Asada, S. Noda, and K. Hosoda. Action based sensor space segmentation for soccer robot learning. *Applied Artificial Intelligence*, 12(2–3):149–164, 1998.
2. Z. Gábor, Z. Kalmár, and C. Szepesvári. Multi-criteria reinforcement learning. In *Proc. of the Fifteenth International Conference on Machine Learning*, 1998.
3. H. Kitano, ed. *RoboCup-97 : Robot Soccer World Cup I*. Springer Verlag, 1997.
4. M. Mataric. Learning to Behave Socially. In *Proc. of the Third International Conference on Simulation of Adaptive Behavior: From Animals to Animats 3*, pp. 453–462, 1994.
5. A. W. Moore and C. G. Atkeson. The Parti-game Algorithm for Variable Resolution Reinforcement Learning in Multidimensional State-spaces. *Machine Learning*, 21:199–233, 1995.
6. S. P. Singh. Transfer of Learning by Composing Solution of Elemental Sequential Tasks. *Machine Learning*, 8:99–115, 1992.
7. P. Stone and M. Veloso. Team-Partitioned, Opaque-Transition Reinforcement Learning. In M. Asada ed., *Proc. of the First RoboCup-97 Workshop at ICMAS'98*, pp. 221–235, 1998.
8. R. S. Sutton and A. G. Barto. *Reinforcement Learning*. MIT Press/Bradford Books, March 1998.
9. R. S. Sutton, D. Precup, and S. Singh. Intra-option learning about temporally abstract actions. In *Proc. of the Fifteenth International Conference on Machine Learning*, 1998.
10. Y. Takahashi, M. Takeda, and M. Asada. Continuous Valued Q-learning for Vision-Guided Behavior Acquisition. In *IEEE/SICE/RSJ International Conference on Multisensor Fusion and Integration for Intelligent Systems*, pp. 255–260, 1999.
11. E. Uchibe, M. Asada, and K. Hosoda. Behavior Coordination for a Mobile Robot Using Modular Reinforcement Learning. In *Proc. of the IEEE/RSJ/GI International Conference on Intelligent Robots and Systems*, pp. 1329–1336, 1996.
12. E. Uchibe, M. Asada, and K. Hosoda. State Space Construction for Behavior Acquisition in Multi Agent Environments with Vision and Action. In *Proc. of International Conference on Computer Vision*, pp. 870–875, 1998.
13. S. D. Whitehead, J. Karlsson, and J. Tenenberg. Learning Multiple Goal Behavior Via Task Decomposition And Dynamic Policy Merging. In J. H. Connel and S. Mahadevan eds., *Robot Learning*, chapter 3. Kluwer Academic Publishers, 1993.

BDI Design Principles and Cooperative Implementation in RoboCup

Jan Wendler[1,*], Markus Hannebauer[2], Hans-Dieter Burkhard[1], Helmut Myritz[1], Gerd Sander[1], and Thomas Meinert[1]

[1] Humboldt University Berlin, Department of Computer Science,
Artificial Intelligence Laboratory, Unter den Linden 6, D-10099 Berlin, Germany
email: wendler/hdb/myritz/sander/meinert@informatik.hu-berlin.de,
WWW: http://www.ki.informatik.hu-berlin.de
[2] GMD – German National Research Center for Information Technology,
Research Institute for Computer Architecture and Software Technology (FIRST)
Planning and Optimization Laboratory, Kekulestr. 7, D-12489, Germany
email: hannebau@first.gmd.de

Abstract. This report discusses two major views on BDI deliberation for autonomous agents. The first view is a rather conceptual one, presenting general BDI design principles, namely heuristic options, decomposed reasoning and layered planning, which enable BDI deliberation in real-time domains. The second view is focused on the practical application of the design principles in RoboCup Simulation League. This application not only evaluates the usefulness in deliberation but also the usefulness in rapid cooperative implementation. We compare this new approach, which has been used in the Vice World Champion team AT Humboldt 98, to the old approach of AT Humboldt 97, and we outline our ideas for further improvements, which are still under work.

Conditions faced by deliberation in multi agent contexts differ significantly from the basic assumption of classical AI search and planning. Traditional game playing methods for example assume a static well-known setting and a fixed round-based interaction of players by a finite set of actions. Additionally, players have a rather long time for deliberation. In contrary to that, many real-world domains are characterized by a continuous action space and an environment that is permanently changed not only by the agent itself, but also by parallel events and actions of other agents. Domains with the need for real-time computing demand in addition time-bounded deliberation processes. The RoboCup Simulation League [9] is an artifical soccer testbed for the international evaluation of approaches that aim at agent deliberation in such real-time dynamic domains.

The *Belief-Desire-Intention (BDI) architecture* founded on Cognitive Science (refer e. g. to [1]) has been applied to deliberation in Artificial Intelligence by several researchers (e. g. [3, 8, 14]). It claims to be highly suitable for domains

---

[*] This research has been partially supported by the German Research Society, Berlin-Brandenburg Graduate School in Distributed Information Systems (DFG grant no. GRK 316).

that are characterized by a non-deterministic environment, competing desiderata, local information and bounded rationality. We have recently proven this to be true in RoboCup [2, 11]. We have also reported that structuring agents according to BDI is advantageous in the implementation process of RoboCup teams [7].

This report discusses two major views on BDI deliberation for autonomous agents. The first view given in section 1 is a rather conceptual one, presenting design principles, that can help to structure deliberation in real-time domains. In section 2, these design principles are directly evaluated by applying them to RoboCup. The development of RoboCup agents does not only impose constraints and difficulties on the deliberation design, but also on the implementation process. Usually, RoboCup development teams consist of several people with heterogeneous skills. Hence, section 2 also gives a glimpse on time- and resource-bounded software engineering, which is supported by the BDI design principles. In the conclusions we briefly evaluate and compare the BDI approach of AT Humboldt 97 to the presented approach, which has been used for RoboCup 98. Moreover, we outline our ideas for further improvements, which are still under work.

1 BDI Design Principles

The presented principles examine the following design tasks: **What** options of acting may an agent have in a given situation and what is their heuristic utility? **Which** of these options shall a rational agent choose as desired and intended? **How** can the intended options be pursued efficiently? These tasks directly correspond to consecutive phases in the deliberation design process.

1.1 Heuristic Options

As well as classical AI search is not sufficient for real-time dynamic domains, its set of terms, including "state" for the given situation and "operand" for an atomic action, is not sufficient to describe the agent's environment and its abilities to act. We therefore have been inspired by the notions *world* and *option* as introduced by Rao and Georgeff in [13] to substitute these terms.

A world is a timed snapshot of all environmental information which may be of use for the agent. Since we assume a situated agent's view on the environment, a world represents always local and incomplete (and even partially incorrect) knowledge about the real environment. Hence, our understanding of world is equivalent to that of *believed world*. Since the agent has usually more than one option to act, there are different following worlds. Some of these worlds fulfill certain conditions to be desirable for an agent. This subset of possible worlds is called *desired worlds* in classical BDI theory. Only a subset of these desired worlds may be achievable and consistent with respect to the given circumstances, and a rational agent may choose some of them to become *intended worlds*.

The other important notion is "option". In terms of possible future worlds, each option corresponds to a set of future worlds, where some conditions (e. g. ball control by the player) are fulfilled. Some of these worlds may be reachable by a related plan of the agent. But such a plan might also fail according to the non-determinism mentioned above, e. g. it might end in a world, where the desired condition does not hold.

The development of the world depends on the actions of others players, and of the uncertainty of the environment, too. The actions of cooperating agents are predictable to some extend. But the behavior of opponents is nearly unpredictable: We might assume that they behave in their best way, but then again we need to know about best plans (from opponents' view in this case). Thus, the outcome of a plan is non-deterministic from an agent's point of view. In [11] a theoretical model using Utility Theory (e. g. [12]) is used to describe this situation. In any case, it is impossible to compare all available plans by related calculations in a reasonable time.

This forces an approach according to the principles of "bounded rationality" [1] in the spirit of BDI. Our approach is based on *heuristic options*, which are the domain for *desires* and *intentions* (corresponding to a class option in object-oriented programming). Heuristic options are chosen from typical short-term goals, e. g. ball interception in soccer. It is important, that such goals consider a planning horizon which is restricted, but not in a uniform manner (a ball interception might e. g. last 3 or 30 steps). It is possible to enlarge this horizon by medium-term goals.

Options are realized by *skills*, which are configurable plans — or parameterizable procedures from the programming viewpoint, respectively. They implement typical basic capabilities of the agent, e. g. *running, kicking, dribbling* of RoboCup agents. As their names imply, they are in close relation to the options.

The task of deliberation is now the choice of a promising skill with appropriate parameters. This is basically the same problem as the above choice of a best plan, but our BDI-setting allows for useful heuristics. At first, we choose desires from the set of all available options: The options are ordered by approximations of their utilities, and the best scoring are considered as desires. Then it is proven, whether there really exists a plan for the achievement of such a desire. If it does, then the desire is chosen as an intention, which is successively refined by determining useful parameters for a related skill. According to BDI theory, the intention sets a screen of admissibility for the refinement of the plan, and in some cases for the consideration of conflicting future desires, too. Our utility approximation tries to determine a useful option, for which sufficiently reliable plans hopefully exist. When choosing such an option as a desire, we can dramatically restrict our search in the space of plans resp. possible future worlds. Further heuristics (including learning approaches) can be used for the determination of the concrete plan.

In principle, we can deal with multiple concurrent intentions – and we will use it in the future. Up to now additional intentions occur only in the special form of *constraints*: Several options may have certain properties in common. This holds especially in the case of resource consumption, which could be interpreted as costs for the execution of an option. Resource control is done by constraints, which can be applied to the heuristic utility calculation of an option. Constraints have also utilities. They can increase the overall expected utility of the option, which considers to fulfill them.

Concerning software technology, we state, that up to now agent oriented programming — and especially the implementation of BDI — has been mostly considered in the tradition of logic and rule based programming (e. g. [17, 6, 18]). Our approach uses agent oriented techniques as a structuring method in an object-oriented environment. Traditional programs are related to well understood control flows: The programmer knows in advance, which *procedures with which parameters* are to be called under which *concrete conditions*. This is implemented using the control structures of procedural languages. The situation changes in the case of autonomous agents in highly dynamic and complex worlds: The programmer does not know in advance all the conditions to call a procedure with appropriate parameters. Only the *criteria* of such calls can be described to some extend (as in a chess program: the programmer does not know about all concrete moves, only some decision criteria can be implemented). Instead of a fixed procedure control flow, an agent program has to implement a reasoning process for the choice of procedures. We will discuss some more details on this flexibility of control in the following subsection.

1.2 Decomposed Reasoning

BDI reasoning means to us the rational choice of promising options to become desires and intentions. This process should show the following properties, which are related to bounded rationality inside the agent as well as to software technological requirements:

Time-boundedness — In a real-time environment the reasoning process is bounded by time restrictions. Either the environment may enforce a timely decision (e. g. in applications with security demands) or late decisions may lead to suboptimal behavior (i.e. missing an opportunity to act).

Distinction of Control and Knowledge — The control of the reasoning process itself should be generic, such that it is independent from the domain-specific options and remains flexible.

Independence of Options — Alternating, deleting or adding an option should influence the reasoning process and other options only marginally or even not at all. This property could be called *scalability*.

The main idea that guarantees the fulfillment of the above mentioned demands is the decomposition of the reasoning process into modular heuristic options. In our model, every option implements a standardized interface, which defines an efficient utility estimation, an attainability predicate, a layered planner and a continuation enforcement predicate. This interface is used by the detached reasoning process as shown by figure 1. The concepts of *decision points, continuation enforcement* and *layered planning* will be described in the next subsection.

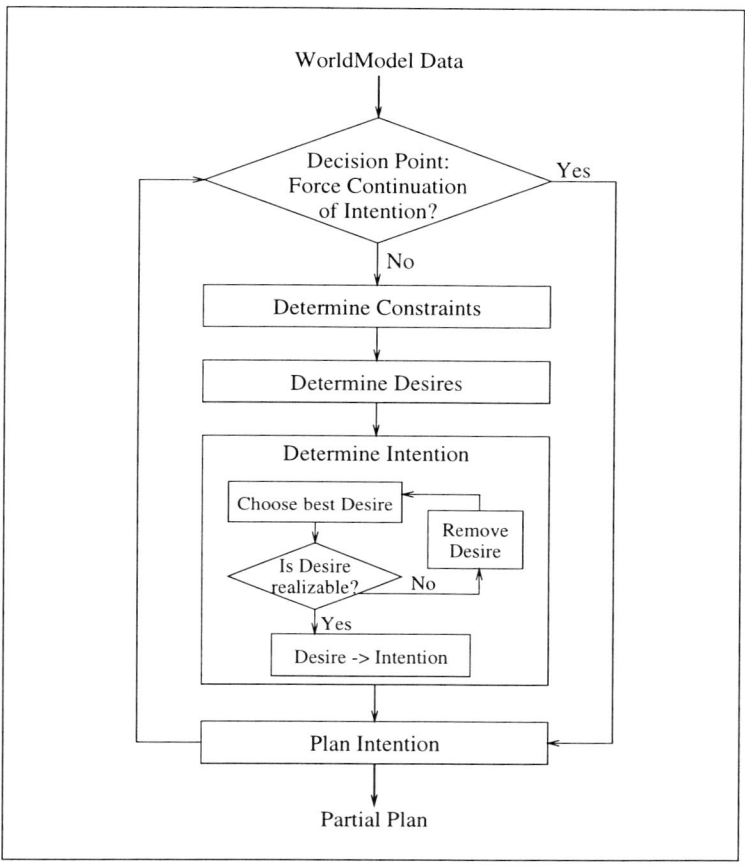

Fig. 1. BDI Reasoning Process

An initial reasoning process starts with the determination of applicable constraints. Given a domain-specific set of possible constraints PC and a utility threshold MIN_CONSTRAINT_UTILITY, the set of currently applicable constraints C can be determined by

$$C = \{c \in PC \,|\, c.\text{utility} > \text{MIN\_CONSTRAINT\_UTILITY}\}.$$

Given a domain-specific set of possible options PO and a utility threshold MIN_DESIRE_UTILITY, the set of current desires D can be determined by

$$D = \{o \in PO \,|\, o.\text{utility}\,(C) > \text{MIN\_DESIRE\_UTILITY}\}.$$

When choosing only one intention from the desires, the current intention i can be determined under consideration of the recent intention ri by

$$i = \underbrace{(\,\arg\max\nolimits_{d \in D} \{d.\text{utility}\,(C) \,|\, d.\text{isPossible}\})}_{\text{possible desire with maximal utility}}.\text{adapt}\,(ri)\,.$$

Usually there is only one desire with a maximal utility, which is then chosen as intention. If there are two or more desires with the same maximal utility no further destinction between the desires can be done, so the intention is chosen arbitrary among these desires.

Let's look at the demanded properties of a BDI reasoning process in dynamic environments. The time-boundedness of the presented reasoning process depends on the efficiency of the constraints' and options' utility estimations. The best solution for utility estimations are approximation algorithms, which deliver the better estimations, the more computing time is available. The control process itself is highly efficient, since it follows a simple greedy approach and calculates the costly `isPossible` method only in case of promising desires. Control and knowledge is fully detached, because all utility and planning heuristics are encapsulated in options. This also supports the demanded scalability of the reasoning process, since options estimate their utility independently from other options.

1.3 Layered Planning

In making a rational choice of a single intention from the given desires, the agent has decided, what to do. The planning task is then to determine, how to do it. In dynamic environments there is always a trade-off between short-term reactive control and long-term deliberative planning [4]. Reactive control has the advantage of being always well-informed about the environment and the disadvantage of a highly restricted horizon. Just the opposite holds for long-term planning. To adjust the agent's planning horizon properly, we are experimenting with *layered planning*, which tries to incorporate the advantages of both reactivity and planning (related to abstractions [10, 16] and Hierarchical Task Network (HTN) planning [5, 15]). Figure 2 shows the different layers of planning, which include coarse-grained planning on the intention layer, fine-grained planning on the skill layer and execution on the atomic actions layer. Following the principle of decomposed reasoning, all the functionality described here lies within the intention, chosen by the reasoning process.

The topmost layer shows exactly one (abstract) intention that describes the intended transition from the current world to a new world satisfying the intended conditions. The coarse-grained planning horizon directly corresponds to the estimated length of the intention. Thus, an intention corresponds to a single compound task in HTN planning. There is always a special problem in choosing the time points for monitoring the progress of intention execution and for reconsidering the intention. Too few monitoring and reconsideration might lead to a behavior, which is not appropriate to the current situation, too much of it could overload the deliberation process. Our concept of monitoring and reconsideration involves the use of so-called *decision points*. They are time points, at which the agent monitors the environment and reconsiders its choices.

Decision points usually enclose several steps for atomic actions. At this point, fine-grained planning is needed. Fine-grained planning is done by the agent's skill that is associated to the chosen intention. Compared to HTN planning, a skill is similar to a method leading to primitive tasks. Since the distance between

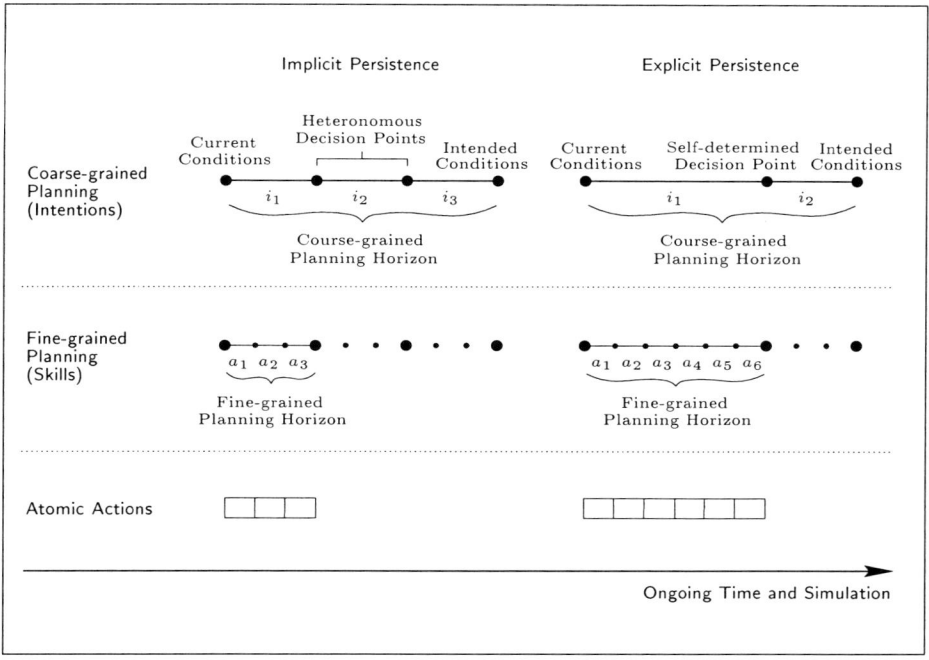

Fig. 2. Layered Planning

decision points might be dynamic and non-deterministic, fine-grained planning must have an anytime-property. That means, that these short-term plans should need no or only short initialization sequences. Otherwise, there would be much initialization overhead in case of near decision points. The approach of fine-grained planning is domain-dependent and can vary from intention to intention. For example it could use classical planning approaches, reactive planning or precompiled plan skeletons. Additionally, it should regard the constraints put onto the given intention. The action sequences planned by the fine-grained planner can directly be executed by atomic actions.

To guarantee stability of committed intentions we propose two different strategies, which influence the whole layered planning approach. The first is *implicit persistence*. Following this strategy, the agent considers all time points, at which the environment's behavior (e. g. an input of sensory data) implies reconsideration, as (heteronomous) decision points. In case of an environment, that has changed as expected, the intention chosen at one of these decision points will be most likely the same as the recent unfinished intention. In contrary to that, *explicit persistence* means, that the agent chooses on its own, on what time point to reconsider its intention. This is implemented by the intention's continuation enforcement predicate, which is illustrated in figure 1. In general, all planning layers could be subject to protections against unwanted changes: We might forbid a change of the intention while allowing a modification of the related plan, and we might even forbid changes of the plan itself (which we call "fanatism"), respectively.

2 Cooperative Implementation in RoboCup

Giving a brief example of a typical option in artificial soccer, in this section we will apply the presented generic BDI principles to the development of RoboCup agents. We will show, that these principles can not only be successfully used for agent deliberation, but also for rapid cooperative implementation. A more detailed report on this issue can be found in [7].

2.1 Constraints on Time and Staff

Artificial soccer is very useful for practical exercises in the area of Distributed Artificial Intelligence. Soccer is well-known and interesting for students. Results of work can directly be seen. In a practical exercise attached to the course 'modern methods of AI' in summer semester 1998 several students had to be included in the development process of AT Humboldt 98. In opposite to the year before, there was only few time between the start of the practical exercise (April) and the World Championship 98 (in the beginning of July).

Because of these hard time constraints, the project management was a critical task in this project. The development team of AT Humboldt 98 consisted of one core team including four persons and four feature teams with altogether eleven developers. Additionally, one of the major aims was to establish a code base, which is understandable, reusable and which can be extended and improved by additional work of students.

Due to the short time amount, much work had to be done in parallel: The students of the practical exercise had to be introduced into the new domain. Then the students had to pick up one component of the agent to specify, implement and test. After that, these components had to be integrated in the final agent and to be tested again to stabilize the behavior of the agents. The design principles of modular heuristic options and decomposed reasoning helped a lot in structuring the implementation work and breaking up the deliberation into small and to some degree independent pieces.

With AT Humboldt 99 we started to extend AT Humboldt 98 by improved skills, new options and partly by a larger planning horizon. These extensions are also subject to further improvements in the next versions of AT Humboldt.

2.2 Application of the BDI Design Principles: An Example

Our RoboCup agents know several *active options*, which may be only desired, if the agent has possession of the ball. They include such options like GoalKick, DirectPass, ForwardPass, Dribbling and so on. If the agent does not control the ball, *passive options* will be desirable. They include CollectInformation, InterceptBall, GoToHomePosition, DefendGoal and others. All these options have their own utility estimations, continuation enforcement predicates and planners. There are also two constraints, that have impact on the reasoning on option as described above. They are ConserveStamina and AvoidOffside.

One of the passive options used for our agent team is the option InterceptBall. If useful, it will represent the desire of intercepting the ball to gain ball control. The utility of intercepting the ball directly depends on the expected success in gaining ball control. The ball may be intercepted in different ways. Directly running into the way of the ball and waiting for it is secure but suboptimal, since an opponent may reach the ball first. Otherwise, if the agent tries to intercept the ball as fast as possible, it may miss the ball because of unforseen delays in its run. For a complete analysis of such a problem all possible plans have to be taken into account. As mentioned before, this seems impossible or at least not manageable. Therefore, we use a heuristic to calculate the utility of this option.

To estimate the utility of InterceptBall the agent calculates for itself and every team mate, how many time steps it will need to gain ball control if it moves optimally. For this calculation the agent uses the position of the team mate and the position and the speed of the ball. Furthermore this utility estimation can be influenced by the constraint ConserveStamina, which may be put onto InterceptBall. After having estimated the utility of this option, it may be chosen as a desire or even as an intention by the decomposed reasoning process.

In the latter case, the rough area for intercepting the ball is already known by the utility estimation. Hence, when fixing InterceptBall as intention, we determine the precise destination region at which the ball can be intercepted early but also relatively secure. The horizon of this coarse-grained plan corresponds to the time needed to gain ball control. After that, the coarse-grained plan has to be refined by a fine-grained plan reaching to the next decision point, which is done by a corresponding skill. While planning the first atomic actions towards the destination region according to the fine-grained planning horizon, the skill also considers the ConserveStamina constraint and avoids obstacles in the agent's path.

If an agent has chosen the option InterceptBall as intention and has determined a corresponding region, it may loose sight of the ball, while running to the destination region[1]. If the player reconsidered its intention every time a new sensor information arrives, it might try to look for the ball again and as a consequence loose time. To avoid this, the agent uses the continuation enforcement predicate of InterceptBall. During the calculation of the destination region, the agent also determines a "don't care" interval, in which no reconsideration is allowed. This interval is given by the minimal time, the fastest player needs to intercept the ball. Nevertheless, if the agent gets new information about the ball without additional actions, it is able to successfully reconsider its intention. In this case, the intention is not enforced to be continued. This behavior im-

---

[1] Players can use necks in Soccer Server Version 5, thus they can now turn his neck to observe the ball while running. Nevertheless, the example explains the idea of intention stability.

plements the mentioned explicit persistence and guarantees the stability of the
InterceptBall intention.

3 Conclusion

To give a brief conclusion and overall evaluation of the presented BDI principles
and their implementation, we compare them to the deliberation approach of the
former soccer team AT Humboldt 97.

The development of the deliberation process in AT Humboldt 97 was rather
intuitive. The team's knowledge of BDI design and implementation was just
evolving and did not directly guide the development. Due to this, the resulting
code did not precisely reflect the use of mental categories like belief, desires or
intentions. For example, the choice of desires and intentions was done by a fixed
hard-coded decision tree, which mixed up control and domain-specific knowledge.
Hence, its development lay within the hands of only one person and it was hard to
maintain. Writing the papers published later, helped the team to theoretically
reconsider the techniques, which had been used informally. Ideas, which were
already present in 1997, like layered planning, implicit and explicit persistence
and others, underwent a strict review by a widened group of developers. This
reconsideration lead to a full re-implementation of AT Humboldt.

The presented BDI principles have supported this re-implementation. The
decomposition of control and knowledge has shown to be highly valuable for
transparency and for rapid cooperative implementation. The notion of a modular heuristic option has played a central role in this context. By encapsulating
the domain-specific knowledge, an option could be designed and realized almost
independently from other options. Though, this modularity has certain drawbacks. The heuristic utility of a given option can not be found trivially. Another
serious problem is the global normation, such that the utilities of different options remain comparable. We consider this to be a great challenge for learning
techniques in our future work.

Since the reasoner always chooses only one intention to be pursued, the team
had to introduce constraints to allow parallel influences. This concept has not
fully paid off, since only a few of them could be identified in the real application. A better approach for future work would be the introduction of parallel
intentions. In contrary to that, the principle of layered planning and persistence
showed encouraging results. It allowed to balance the trade-off between adaption
and stability very well, especially in case of passive options.

The BDI deliberation process of AT Humboldt 98 has proven to be flexible,
scalable and maintainable. It provides a useful base for further improvements,
which are still under work. The main goal is the introduction of longer planning
intervals. A "cascade of intentions" will be used to establish raw sequences of
subsequent future steps, which can be refined according to the development of the
environment. "Emergent cooperation" in AT Humboldt 98 and 99 resulted from
the programmer's knowledge about the implementation. Now we want to extend

this behavior by explicitly plannable cooperation. On the lower level, additional and improved skills are to be developed (e. g. using learning techniques).

References

1. Bratman, M. E.: *Intentions, Plans, and Practical Reason*. Harvard University Press, 1987.
2. Burkhard, H.-D., Hannebauer, M., and Wendler, J.: *Belief-Desire-Intention Deliberation in Artificial Soccer*. AI Magazine 19(3): 87–93. 1998.
3. Cohen, P. R. and Levesque, H. J.: *Intention is choice with commitment*. Artificial Intelligence 42: 213–261. 1990.
4. Dean, T. L. and Wellman, M. P.: *Planning and Control*. Morgan Kaufmann Publishers, 1991.
5. Erol, K., Hendler, J. and Nam, D. S.: *HTN Planning: Complexity and Expressivity*. In Proc. of the 11th Nat. Conf. on AI (AAAI-94). AAAI Press, 1994.
6. Fisher, M.: *A survey of Concurrent* METATEM — *the language and its applications*. In Gabbay, D. M. and Ohlbach, H. J. (eds.): Temporal Logic — Proceedings of the First International Conference: 480–505. LNAI 827. Springer, 1994.
7. Hannebauer, M., Wendler, J., and Müller-Gugenberger, P.: *Rapid Concurrent Software Engineering in Competitive Situations*. In Chawdhry, P. K., Ghodous, P. and Vandorpe, D. (eds.): Advances in Concurrent Engineering (CE-99): 225–232. Technomic Publishing, 1999.
8. Jennings, N. R.: *Specification and Implementation of a Belief-Desire-Joint-Intention Architecture for Collaborative Problem Solving*. Int. Journal of Intelligent and Cooperative Information Systems 2(3): 289–318. 1993.
9. Kitano, H., Kuniyoshi, Y., Noda, I., Asada, M., Matsubara, H., and Osawa, H.: *RoboCup: A Challenge Problem for AI*. AI Magazine 18(1): 73–85. 1997.
10. Knoblock, C. A.: *Automatically generating Abstractions for Planning*. Artificial Intelligence 68(2). 1994.
11. Müller-Gugenberger, P. and Wendler, J.: *AT Humboldt 98 — Design, Implementierung und Evaluierung eines Multiagentensystems für den RoboCup-98 mittels einer BDI-Architektur*. Diploma Thesis. Humboldt University Berlin, 1998.
12. Neumann, J. V. and Morgenstern, O.: *Theory of Games and Economic Behavior*. Princeton University Press, 1944.
13. Rao, A. S. and Georgeff, M. P.: *An Abstract Architecture for Rational Agents*. In Nebel, B., Rich, C. and Swartout, W. (eds.): Proc. of the Third International Conference on Principles of Knowledge Representation and Reasoning: 439–449. Morgan Kaufmann Publishers, 1992.
14. Rao, A. S. and Georgeff, M. P.: *BDI agents: From theory to practice*. In Lesser, V. (eds.): Proc. of the First Int. Conf. on Multi-Agent Systems (ICMAS-95): 312–319. MIT-Press, 1995.
15. Tate, A.: *Generating Project Networks*. In Allen, J., Hendler, J. and Tate, A. (eds.) Readings in Planning: 291–296. Morgan Kaufmann, 1990.
16. Tenenberg, J.: *Abstraction in Planning*. In Allen, J., Kantz, H., Pelavin, R. and Tenenberg, J. (eds.): Reasoning about Plans. Morgan Kaufmann, 1990.
17. Thomas, S. R.: *PLACA, an Agent Oriented Programming Language*. Technical Report STAN-CS-93-1487. Standford University, 1993.
18. Wooldrige, M. J.: *This is* MYWORLD: *The Logic of an Agent-Oriented DAI Testbed*. In Wooldrige, M. J. and Jennings, N. R. (eds.): Intelligent Agents: 160–178. LNAI 890. Springer, 1995.

AT Humboldt in RoboCup-99
(Team description)*

Hans-Dieter Burkhard, Jan Wendler**, Thomas Meinert, Helmut Myritz, and Gerd Sander

Humboldt University Berlin, Department of Computer Science,
Artificial Intelligence Laboratory, D-10099 Berlin, Germany
email: hdb/wendler/meinert/myritz/sander@informatik.hu-berlin.de,
WWW: http://www.ki.informatik.hu-berlin.de

1 Introduction

Our agent team AT Humboldt 99 (AT stands for "Agent Team") was developed as extension of our former team AT Humboldt 98, which became vice champion at RoboCup-98. We started to extend it by improved skills, new options and a larger planning horizon, respectively. So the most features of our current team were already part of AT Humboldt 98 which has been briefly described in [3] and extensive described in [5]. A description of our first soccer team AT Humboldt 97, which became world champion at RoboCup-97, can be found in [1].

We are interested in virtual soccer for the development and the evaluation of our research topics in artificial intelligence which concern the fields of

- Agent oriented techniques,
- Multi-Agent Systems,
- Case Based Reasoning.

The results of our research in these areas can be found in [1, 2, 4, 7, 8]. Thus many aspects of our soccer program are heavily influenced by these fields, but it is important not to consider these fields in isolation: to create our soccer agents, we also needed a lot of contributions from other fields of computer science (e.g. programming techniques, synchronisation, concurrency) and from mathematics. Thereby we gain deeper insights for integration AI techniques in software development. This aspect is especially important for the education of our students.

2 Team Development

The virtual soccer teams "AT Humboldt" are implemented by our AI group at the Department of Computer Science at the Humboldt University Berlin. The

---

* This work was partly sponsored by TecInno GmbH Kaiserslautern, Daimler Chrysler AG Research & Technology Berlin and PSI AG Berlin
** This work has been partially supported by the German Research Society, Berlin-Brandenburg Graduate School in Distributed Information Systems (DFG grant no. GRK 316).

work is done by groups of students in practical exercises during the summer semester. A core group of up to three students maintains the coordination and the programs. Besides the experiments with AI methods, the project is also a challenge for software development by changing teams. The source code of AT Humboldt 99 has about 28.000 lines of code. It is a non trivial task to maintain the introduction of new ideas during extremly short time intervals by changing teams. To support the concurrent development we use the freely available source code management system CVS [9] and the documentation system doc++ [10].

Team Leader: Prof. Hans-Dieter Burkhard
Team Members:
   Prof. Hans-Dieter Burkhard
   – leader of the AI group
   – did lead the development and did consulting
   – did attend the competition
   Jan Wendler
   – PhD student
   – did consulting and some debugging
   – did attend the competition
   Thomas Meinert, Helmut Myritz and Gerd Sander
   – undergraduate student
   – did the design, implementation and debugging of the new ideas
   – did attend the competition
Web page: http://
   www.ki.informatik.hu-berlin.de/RoboCup/RoboCup99/index_e.html

3 World Model

Our world model uses object representations of situations, implemented by a class called Situation. For any given time an object of this class will be generated which consists of object representations for teammates, opponent players, the ball and the agent itself. Flags are only used to determine the own absolute object representation which consists of the absolute player position, speed, body direction and of the relative face direction. With this data the absolute objects representations of the other players and the ball are calculated. So the agent can get a new Situation-object from sensor information. Another way to get a new Situation-object is the simulation of the own actions of the agent, as it is done in the *SoccerServer*. So after every sensor information we have two concurrent Siuation-objects which are merged together by finding correspondig player and using the best information of both situations for the new one.

4 Communication

We have done only small efforts into communication. For the cooperation among the teammates no communication is neccessary because our agents model their teammates and can predict their behaviour very well. The agents only use communication to broadcast their world model among their teammates .

5 Deliberation and Strategy

Our agent architecture uses a mental deliberation structure which is best described by a belief-desire-intention architecture (BDI) [6]. Distinct from other (e.g. logically motivated) approaches our approach is closely related to procedural thinking, and we use object oriented programming for the implementation.

BDI reasoning means to us the rational choice of promising options to become desires and intentions. The main idea of our reasoning process is its decomposition into modular heuristic options and a generic control process.

Our BDI structure uses a set of independent options which can be differentiate into active and passive options. Each option returns a value of its utility based on the current situation of the world and obeying certain constraints as *ConserveStamina* and *AvoidOffside*.

When the agent has control over the ball one of its active options will have the highest utility and therefore he will try to pass to a teammate or to dribble in a preffered direction or to kick a goal. It depends on the utilities of the active options which one of these behaviour is chosen by the agent.

If the agent doesn't have control over the ball the passive options are candidates for the highest utility. The agent may decide to intercept the ball, or try to get a good position to get a pass from a teammate, or go back to his home region, or just collect data if already there.

Precise information about our deliberation structure can be found in our paper "BDI Design Principles and Cooperative Implementation in RoboCup" [7], which can be found in this book.

6 Skills

After a soccer agent has decided what to do, he can use its skills to fullfil its intention. All skills generate a plan which can be of any size up to the fullfilment of the intention. The most important skills are:

Goto Region: With this skill an agent is able to run to a destination region, which is given by a circle. Because we are invoking all skills with a copy of the current situation, obstacles will be avoided exactly. If a stamina limit is set, the agent will never fall below this limit of stamina. This skill can be also used to generate a plan to run backwards.
For the interception of the ball this method is used as well. The determination of the target region is done in the corresponding option.

Kick Ball: Up to now this skill kicks the ball with a demanded power into a demanded direction. Depending on the speed and the direction of the ball, it is kicked several times to reach the demanded values. This skill was implemented by using mathematical calculations.

Dribbling: We have implemented two kinds of dribbling. The simple dribble method just kicks the ball with low power to the destination direction and runs after the ball. In the second implementation, the agent puts the ball beside his body and dribbles forward, never loosing control over the ball.

Whereas the first method is stable against synchronisation problems it lacks at safety over ball control. Just the opposite holds for the second method, so we used the first one in defense and the other one in midfield and attack.

7 Conclusion

In this paper we gave a glimpse of our agent team AT Humboldt 99, which is an extension of our team AT Humboldt 98. More information about our teams can be found in [1–5, 7, 8].

Our main goals for further extensions are: the introduction of longer planning intervals (we already introduced a longer planning interval, but only for the option FreeKick), the extension of the "Emergent cooperation" of our agents by explicitly plannable cooperation, the use of learning methods for ball kicking and dribbling skills, and the modelling of opponents in the frame of the BDI architecture to predict their behaviour in advance.

Furthermore we want to support our sony legged robot team, the Humboldt Hereos, with techniques already used by our simulation team.

References

1. Burkhard, H. D., Hannebauer, M., and Wendler, J.: AT Humboldt — Development, Practice and Theory. RoboCup-97: Robot Soccer World Cup I, Springer 1998, 357–372.
2. Burkhard, H.-D., Hannebauer, M., and Wendler, J.: Belief-Desire-Intention Deliberation in Artificial Soccer. *AI Magazine* 19(3): 87–93. 1998.
3. Gugenberger, P., Wendler, J., Schröter, K., and Burkhard, H. D.: AT Humboldt in RoboCup-98 (Team description). RoboCup-98: Robot Soccer World Cup II, Springer 1999, 358–363.
4. Hannebauer, M., Wendler, J., Gugenberger, P., and Burkhard, H. D.: Emergent Cooperation in a Virtual Soccer Environment. RoboCup Papers at ICRA-98 and DARS-98, 1998, 72–81.
5. Müller-Gugenberger, P. and Wendler, J.: *AT Humboldt 98 — Design, Implementierung und Evaluierung eines Multiagentensystems für den RoboCup-98 mittels einer BDI-Architektur*. Diploma Thesis. Humboldt University Berlin, 1998.
6. Rao, A. S. and Georgeff, M. P. : BDI agents: From theory to practice. In V. Lesser, editor, *Proc. of the First Int. Conf. on Multi-Agent Systems (ICMAS-95))*, pages 312–319. MIT-Press, 1995.
7. Wendler, J., Burkhard, H. D., and Hannebauer, M.: BDI Design Principles and Cooperative Implementation in RoboCup. RoboCup-98: Robot Soccer World Cup III (this book), Springer.
8. Wendler, J. and Lenz, M.: CBR for Dynamic Situation Assessment in an Agent-Oriented Setting. In D. Aha and J. J Daniels, editors, *Proc. AAAI-98 Workshop on Case-Based Reasoning Integrations, Madison, USA, 1998*.
9. CVS: http://www.loria.fr/ molli/cvs-index.html
10. DOC++: http://www.zib.de/Visual/software/doc++/index.html

Cyberoos'99: Tactical Agents in the RoboCup Simulation League

Mikhail Prokopenko, Marc Butler, Wai Yat Wong, Thomas Howard

Applied Artificial Intelligence Project
CSIRO Mathematical and Information Sciences
Locked Bag 17, North Ryde, NSW 1670, Australia

Abstract. This paper describes a framework for formalising tactical reasoning in dynamic multi-agent systems, populated by synthetic (software) agents. The proposed framework is based on a hierarchy of synthetic agent architectures and is expressive enough to capture a subset of desirable properties from both the situated automata and subsumption-style architectures, while retaining the rigour and clarity of logic-based possible worlds semantics. This framework is successfully realised in the RoboCup Simulation League domain. Not only did it provide a solid design approach to object-orientation, but it also enabled incremental implementation and testing of software agents and their modules. In particular, the framework allowed us to correlate enhancements in the agent architecture with tangible improvements in team performance. Cyberoos98 was 3rd place winner of the Pacific Rim series at PRICAI-98. Cyberoos99 finished in the top 18 of the RoboCup-99.

1 Introduction

The principal aim of this paper is to illustrate how declarative agent specifications may facilitate design and implementation of software agents capable of tactical reasoning. Some of the architectures are well-known - for example, variants of tropistic and hysteretic agents, enabling typical perception-action feedback loop and behavioural subsumption are discussed in [2, 5, 7]. We attempted to extend these results by including new agent types (task-oriented and process-oriented agents), facilitating reasoning about tactical activities in context of multi-agent teamwork. These agent architectures allow us to capture a number of desirable properties (reactive plans, ramifications, task-oriented and, potentially, goal-directed behaviour) by embedding them in situated behaviours.

2 Situated Agent Architecture

2.1 Tropistic Agent

Following [5], we formally describe a *Tropistic* agent as a tuple A_T

$$<C, S, E, sense, tropistic\text{-}behaviour, response>,$$

where S is a set of agent sensory states, E is a set of effectors, and C is a communication channel type. A sensory function is defined as $sense: C \rightarrow S$. Activity of a *Tropistic* agent is characterised by $tropistic\text{-}behaviour: S \rightarrow E$. By allowing the set E to include composite effectors $e1; e2$, where $e1 \in E$, $e2 \in E$, we can implicitly account for the case of reactive planning - when a situated agent reacts to stimuli S with an n-length sequence of effectors. The *Tropistic* agent executes its behaviour as a sequence of commands encoded by *response*: $E \to C$ and sent to the simulator.

The *Tropistic* agent must manage two concurrent and asynchronous activities - one corresponding to receiving sensory information from the Soccer Server, and the other related to sending appropriate atomic commands to the Server. We chose to implement the required parallelism via threads, as they appeared to be intuitively appropriate for the requirements, and (assuming implementation in C or C++) have native support in the Solaris operating system. Threads allowed us to use non-blocking I/O, eliminating the necessity of timing the sensory thread. In addition, we were not limited by the number of available signals, provided by the operating system. The nature of tropistic behaviour and asynchronous character of client-server communication make precise timing imperative but problematic. Some of the observed difficulties arise due to 1) receiving outdated sensory information and 2) sending too many commands in a given simulation cycle. Non-blocking I/O network access complemented by an appropriate technique for processing of pending messages addresses the first challenge, rescuing the agents from the "living-in-the-past" syndrome. The second difficulty can be rectified if the acting thread independently schedules agent responses.

The most important examples of tropistic behaviour exhibited by a Cyberoos99 agent are obstacle avoidance, ball chasing, and (a goalkeeper) catch.

2.2 Hysteretic Agent

A *Hysteretic* agent is defined here as a reactive agent maintaining internal state I and using it as well as sensory states S in activating effectors E; i.e. its activity is characterised by *hysteretic-behaviour*: $I \times S \to E$. A memory update function maps an internal state and an observation into the next internal state, i.e. it defines *update*: $I \times S \to I$. A *Hysteretic* agent reacts to stimuli s sensed by *sense(c)* and activates effectors e according to *hysteretic-behaviour(i, s)*. This class extends its superclasses by adding *hysteretic-behaviour* and *update* functions, while retaining all previously defined functions (i.e., it is a sub-class of the *Tropistic* agent). So the *Hysteretic* agent is defined as a tuple A_H

<C, S, E, I, sense, tropistic-behaviour, **hysteretic-behaviour, update**, response>

where the **bold** style indicates newly introduced functions.

The *hysteretic-behaviour* is implemented as a (temporal) production system (TPS). Whenever the TPS fires a rule (a behaviour instantiation, expressed in terms of partial sensory and internal states, ie., in the form similar to situated automata condition-action pairs [3]), a sequence of atomic commands is placed into a queue for subsequent and timely execution. Implementation of the TPS enables monitoring of currently progressing actions, thus providing an explicit account of temporal continuity for actions with duration [7, 8], and allows us to embed actions ramifications and interactions [4, 6, 7]. For example, a dribbling action will not be invoked during shooting or passing.

It is worth pointing out that the architecture A_H can be viewed as a subsumption architecture [1] as well. It allows us to easily express desired subsumption dependencies between the *Hysteretic* and *Tropistic* levels. More precisely, resolution of a possible conflict between behaviour instantiations *e1* = *tropistic-behaviour(s)* and *e2* = *hysteretic-behaviour(i, s)* triggered by the same sensory input s, is dependent on the internal state i - leading to inhibition of the simpler level behaviour, if necessary. For instance, tropistic chase is suppressed if a team-mate has possession of the ball. A Cyberoos99 agent displays quite a few interesting instantiations of hysteretic behaviour, eg., dribble around opponents toward an enemy goal, intercepting a fast moving ball, controlling and turning with the ball, resultant-vector passing, shooting at goal along a non-blocked path, etc.

3 Tactical Agent Architecture

3.1 Task-Oriented Agent

The behaviour functions of the situated agents, described in the previous section, are uniformly defined across their respective domains and ranges. This means that the set of all behaviour instantiations $H = \{(i, s, e): e = hysteretic\text{-}behaviour(i, s)\}$ is not partitioned or structured otherwise. In other words, all agent's behaviour instantiations (action rules) are always enabled. Sometimes, however, it is desirable to disable all but a subset of behaviour instantiations - for example, when a tactical task requires concentration on a specific activity. The following agent class - *Task-Oriented* agent - is intended to capture this feature, while retaining properties of the *Hysteretic* agent (i.e., it is a sub-class of the latter). We define the architecture of a *Task-Oriented* agent as the tuple A_{TO}

<C, S, E, I, T, *sense, tropistic-behaviour, hysteretic-behaviour, update,* **decision, combination,** *response*>

A *Task-Oriented* agent incorporates a set of task states T. It uses the functions *decision*: $I \times S \times T \to T$ and *combination*: $T \to 2^H$ to decide which subset of behaviour instantiations (a task) is appropriate at a particular internal state, given sensory inputs.

Implementation of task-orientation requires some adjustments to the TPS. The TPS traces action rules whose actions may be in progress, and checks, in addition, whether a rule is valid with respect to a current task. The hysteretic behaviour instantiations mentioned in the previous section (dribbling, intercepting, etc.) are *combined* in corresponding tasks and can be selected by a Cyberoos99 agent in real-time. For example, sharp-dribbling is implemented as a task, enabling relevant (hysteretic) twists and faint moves.

3.2 Process-Oriented Agent

The *Task-Oriented* agent is capable of performing certain tactical combinations ("set pieces") in real-time by activating only a subset of its behaviour functions, and thus concentrating only on a specified task. Upon making a new decision, the agent switches to another task. In general, there is no dependency or continuity between consecutive tasks. This is quite suitable in complex and/or unexpected situations requiring a swift reaction. However, in some cases it is desirable to exhibit a more coherent agent behaviour.

A new notion, a process, is intended to capture this kind of coherent behaviour - when an agent is engaged in an activity requiring several tasks. A process constrains a set of possible tasks without specifying an exact sequential or tree-like ordering - an appropriate tactical scheme comprising a few tactical elements may simply suggest for an agent a possible subset of decisions, leaving some of them optional. For example, a penetration through centre of an opponent penalty area may require from agent(s) to employ a certain tactics - a certain set of elementary tactical tasks (dummy-runs, wall-passes, short-range dribbling) - and disregard for a while another set of tasks. It is worth noting that whereas a team's tactical formation is typically a static view of responsibilities and relationships, process-orientation is a dynamic view of how this formation delivers tactical solutions.

The architecture of a *Process-Oriented* agent can be defined as the tuple A_{po}

<*C, S, E, I, T, P, sense, tropistic-behaviour, hysteretic-behaviour, update, decision, combination, **engage**, **tactics**, response*>

A *Process-Oriented* agent maintains a process state P. It uses the functions *engage*: $I \times S \times T \times P \rightarrow P$ and *tactics*: $P \rightarrow 2^T$ to select a subset of tasks, given current internal, sensory, task and process states. Several tactical processes can be selected by Cyberoos99 agents in real-time: *Advance, Dispatch,* and *Penetration*. Importantly, tasks encapsulated in a process are not temporally ordered - instead, they make up a (currently) relevant *tactical* selection, arranged in appropriate decision-making order - for instance, an agent *engaged* in the *Dispatch* process considers the passing task before dribbling tasks, while the *Penetration* process prefers dribbling to crossing and disregards passing at all.

4 Conclusions

The notions of task and process specified for each agent open a way to formally introduce a group of agents sharing the same task or the same process, and enable formal tactical reasoning about multi-agent teamwork. It is important to realise that teamwork is formalised on the system level. In other words, the overall team behaviour/tactics emerges only as a result of agent interactions.

The described hierarchical framework provided a solid design approach to object-orientation, and enabled rigorous incremental implementation and testing of software agents and their modules. We used C++ as the implementation language; the development environment was Solaris 2.5 and GNU g++ 2.8.2 on SPARC workstations. The team took 3rd place at the PRICAI-98 Pacific Rim series. Cyberoos99 qualified for the RoboCup-99 finals, where the team played 10 games (both qualification and consolation), winning 5 of them with a total score 97:15.

References

1. Brooks, R.A. Intelligence Without Reason. In Proceedings of the Twelfth International Joint Conference on Artificial Intelligence, 569-595 Morgan Kaufmann, 1991.
2. Genesereth, M.R., and Nilsson, N.J. *Logical Foundations of Artificial Intelligence.* Morgan Kaufmann, 1987.
3. Kaelbling, L.P. and Rosenschein, S.J. Action and planning in embedded agents. In Maes, P. (ed) *Designing Autonomous Agents: Theory and Practice from Biology to Engineering and Back*, pages 35-48, 1990.
4. Prokopenko, M., Jauregui, V. Reasoning about Actions in Virtual Reality. In Proceedings of the IJCAI-97 Workshop on Nonmonotonic Reasoning, Action and Change, 159-171. Nagoya 1997.
5. Prokopenko, M., Kowalczyk R., Lee M., Wong, W.-Y. Designing and Modelling Situated Agents Systematically: Cyberoos'98. In Proceedings of the PRICAI-98 Workshop on RoboCup, 75-89. Singapore 1998.
6. Prokopenko, M. Situated Reasoning in Multi-Agent Systems. In AAAI Technical Report SS-96-05, the AAAI-99 Spring Symposium on Hybrid Systems and AI, 158 - 163. Stanford 1999.
7. Prokopenko, M., Butler M. Tactical Reasoning in Synthetic Multi-Agent Systems: a Case Study. In Proceedings of the IJCAI-99 Workshop on Nonmonotonic Reasoning, Action and Change. Stockholm, 1999.
8. Sandewall, E. Logic-based Modelling of Goal-Directed Behaviour. *Linköping electronic articles in Computer and Information science*, (2):19 1997.

11Monkeys Description

Shuhei Kinoshita, Yoshikazu Yamamoto

Yamamoto Lab. Faculty of Science and Technology Keio University, Japan

1 Introduction

The major purpose of this research is to study cooperative planning for multi-agent systems in time-critical environment. The RoboCup simulator league is the most interesting target for our research.

In Artificial Intelligence, problem solving is to transit state from initial state to goal state. There are two types of planning, namely deliberative planning and reactive planning. The former is to find all series of action before it really acts. So deliberative planning requires much computation resources. And it has also a problem of poor adaptability to dynamic environment. On the other hand, reactive planning has a good adaptability, but in many cases, it doesn't select the best choice, and it needs more optimization. So there is a potent trade-off problem in this two types of planning.

To solve this problem, we propose three layers planning, Strategy, Group, and Individual. Strategy Layer planninng determines global team strategy dependent to opponent team model. For example agents select team formation and action algorism, and it also determines a policy of management to use stamina. In Group Layer, an agent makes cooperative planning among a few teammates. In this layer agents are assigned a dynamic role, such as ball handler or support player. Dynamic role change are triggered by their own recognition of the current state, because of taking account robustness. The agent which finds a chance becomes a planner. In Individual layer, agents behave reactively according to upper layer decision, such as team formation, dynamic roles or cooperative plans.

As a result, our 11Monkeys utilizes this three layers planning and we won the championship of the simulator league Japan Open '99. And we finished the simulator league of the RoboCup'99 Stockholm in 4th place.

2 Team Development

Team Leader: Kinoshita
Team Members:
    Shuhei Kinoshita
      – Keio University
      – Japan
      – Master Student
      – attend the competition

3 Three Layers Planning Approach

In Artificial Intelligence, problem solving is to transit state from initial state to goal state. There are two types of planning, namely deliberative planning and reactive planning. The former is to find all series of action before it really acts. So deliberative planning requires much computation resources. And it has also a problem of poor adaptability to dynamic environment. On the other hand, reactive planning has a good adaptability, but in many cases, it doesn't select the best choice, and it needs more optimization. So there is a potent trade-off problem in this two types of planning.

To solve this problem, we propose three layers planning, Strategy, Group, and Individual.

3.1 Strategy Layer

Strategy Layer covers all teammates. In this layer Agents select their formation, tactics, and decide the policy of resource management. These must be decided depending upon opponent model. Static role assignment is done in this layer. Static Role is a role set, like Goalie, Defensive Half, etc. There are many types of team styles, indeed. So we need to adapt them effectively, but this is not implemented now.

3.2 Group Layer

Group Layer planning include about three or four teammates in local state near ball. In group layer agents are assigned a Dynamic Role like, ball handler, supporter.

A agent who finds the chance, can be a reactive cooperative planner. If there are no fatal condition to execute the plan, agreement will be done, and plan in group level can be executed.

3.3 Individual Layer

Individual Layer planning covers only 1 vs. 1 state. Agent selects most suitable pre-planned module. There are fatal condition, which agents withdraw his plan in every simulation step. For example agent cannot find pass course in defense area, he makes a decision of clearing ball.

4 System

The system of each agent is represented in Fig. 1. When an agent is created, he gets a common formation and is assigned one of static role in Formation/Static Role Module. In Interpreter Module he parses many sensor information from soccer server[1]. And then he updates world model. Next in Offense/Defense, Dynamic Role Module he make a decision of offense or defense, and he get a

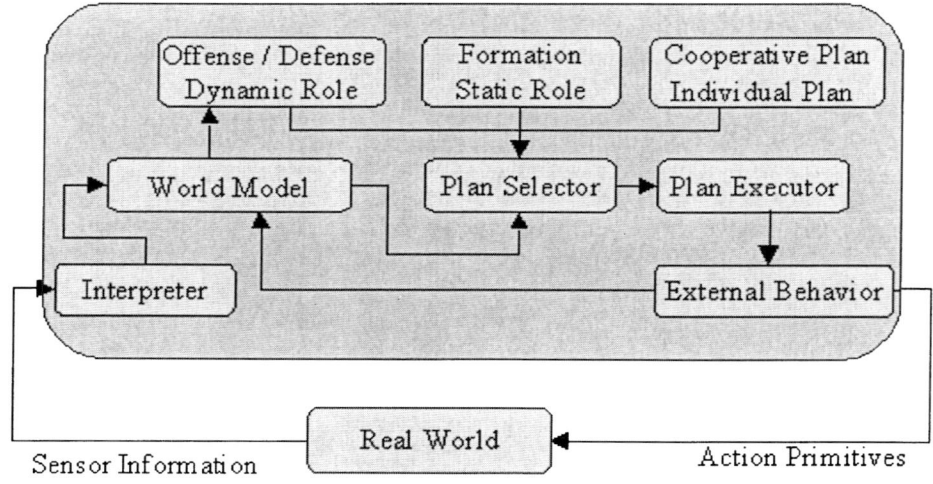

Fig. 1. Agent system

dynamic role, like ball handler. Next he selects one of abstract plan out of cooperative plans and individual plans. Then the plan is transformed into a primitive command in Plan Executor Module. At last in External Behavior Module the primitive command is sent. We uses the low level skill of CMunited'98[?, cmu]

5 Dynamic Role Assignment

Role is necessary for autonomous agents. If agents don't have any role, agents behave selfish. In many cases roles were allocated statically. It was not so flexible. We use dynamic role assignment system. We call dynamic roles as 'Player with ball" or "Player for support". In our team, dynamic role number is six. In offense mode there are three roles, and in defense mode there are also three roles. At first agents autonomously make a judgment of offense mode or defense mode. Next agent most suitable player to catch ball is allocated "Player near the ball". If in defense mode "Player near the ball" is named "First Defender". If in attack mode "Player with ball" is named "First Attacker".

```
Offense Mode
First Attacker   Player with Ball
Second Attacker  Player who supports First Attacker
Third Attacker   Other Players

Defense Mode
First Defender   Player near ball
```

Second Defender   Player who supports First Defender
Third Attacker   Other Players

That is agent selects autonomously the most suitable role of the six. He does not depend on others in making the decision. In order to take into accounts robust. The moment one player catch the ball, the others' roles automatically change.

6 Reactive Planning Mechanism

There are little time for deliberative planning, because there exists an opponent team. So an agent should select his plan reactively.

In this research, there are two steps in reactive planning.

– abstract-planning
– re-planning & plan-execution

7 conclusion

11Monkeys utilizes three layers planning and we won the championship of the simulator league Japan Open '99. And we finished the simulator league of the RoboCup'99 Stockholm in 4th place.

Table 1. The Result of 11Monkeys in RoboCup '99 Stockholm

Opponent	Affiliation	Score	Possession(%)	Side(%)
Gongeroos	University Of Wollongong, Australia	4-0	66	89
Sibiu Team	Lucian Blaga University of Sibiu, Romania	17-0	80	70
FCFoo	Link ping University, Sweden	18-0	-	73
UvaTeam	University of Amsterdam, The Netherlands	1-0	54	39
CMUnited 98	Carnegie Mellon University, USA	1-0	55	43
YowAI	University of Electro-Communications, Japan	8-0	60	39
HCIII	Link ping University, Sweden	1-0	53	37
EssexWizards	University of Essex, England	0-1	-	39
CMUnited 99	Carnegie Mellon University, USA	0-8	35	13

References

1. Itsuki Noda. *Soccer server a simulator of RoboCup* a draft distributed at IJCAI'95 Workshop: Entertainment and AI/Alife, 1995.
2. Peter Stone and Manuela Veloso *The CMUnited-98 Champion Simulator Team* In "RoboCup-98: Robot Soccer World Cup II", M. Asada and H. Kitano (eds.), Springer Verlag, Berlin, 1999.

Team Erika

Takeshi Matsumura

Department of Information and Computer Science, Waseda University

1 Introduction

Team Erika's main focus is on the facilatation of the design of agent behavior. The behavior code is generated by a graph editor which process transition diagram like graph. Since the concept is represented visually, high design efficency can be achieved. Besides, people other than computer scientist can design the behavior easily without understanding the underlying stucture.

Future work is to improve the graph editor to process graph which represents a cooperation among a few agents in single graph.

2 Team Development

Team Leader and only a member: Takeshi Matsumura
- affiliation: Waseda University
- country: Japan
- position: graduate student
- did or did not attend the competition: attended

Web page http://www.futamura.info.waseda.ac.jp/~matsu/erika

3 World Model and Communication

World model which is situated in the lower layer of the two layer structure agent was created by the sensor. The internal functions inside the agent invoked every 100ms update the world model by using the information received from the server. Agent processes all of the visual information and update the proper object inside the world model by examining the time stamp. Newly created object always supersedes the old one which is found by comparing the time stamps of each objects. (e.g. If agent had a ball object with time 14 and he got a ball information when time 16, then the time 14 ball object information is thrown away.) An agent stop its current computation immediately, reset the states and restart decision making when it receives a referee message in order to increase reactivity.

Agent used in Stockholm had no communication. Communication is being installed in current work to synchronize the computation of agents when a cooperation, like one-two pass or attack from corner kick, starts.

4 Skills

Agent has the ball prediction routine which use the world model update routine to predict the ball's future postion. The prediction is not very accurate because only the last information of the ball was being used. In future work, history of the ball position and the sequence of the ball movement will also be used in order to improve accuracy.

5 Strategy

All agents always try to monitor the ball's action even when the agent is far away from the ball because important events usually occur around the ball but basically only the agent closest to the ball goes to get it. The only exception is that when the agent is returning to his home position. In this case he will monitor the ball every 5 seconds.

6 Special Team Features

6.1 Graph editor for one player

An transition diagram like graph shown in fig.1 is used to represent each agent's play style. Each agent's computation moves from node to node, depends on the conditions described on each arcs.

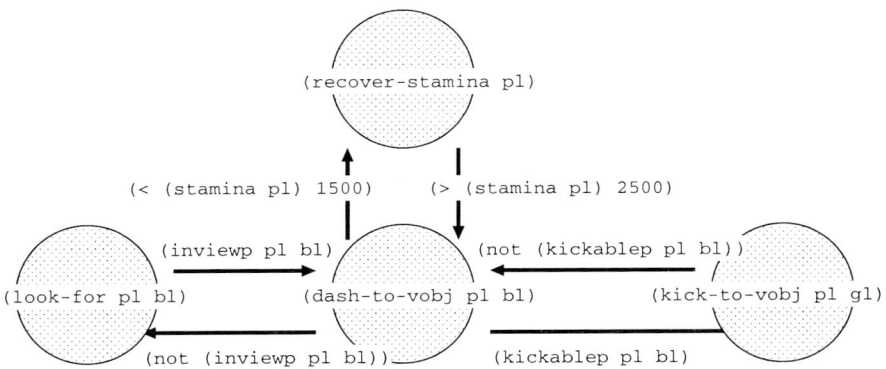

Fig. 1. graph styled flow-chat represents dango-play

Agent's computation begins at the leftmost node, for instance, node 1. At first the agent does the action "(look-for pl bl)", where the variables pl and bl are bounded to the agent himself and a ball object respectively. Then he checks the condition "(inviewp pl bl)" on the arc derived from node1. If the condition "(inviewp pl bl)" is true, which means that ball come into his view, then his computation will be moved to the center node. Otherwise, he takes the "(look-for pl bl)" action and attempt to evaluate the condition again. If the computation is at the node which has no arcs, then the action on it is taken and the computation stops. In the example, however, the computation continues permanently because there are no nodes without any arcs if there is no referee message.

After moved to another node, the action which is described on the node is absolutely taken before evaluation of conditions.

We us a graph editor to create this kind of graph and use it to generate lisp lists which are embedded inside an agent. The interpreter inside each agent's upper layer analyse the diagram and then the agent behavior is set.

6.2 Graph Editor for cooperation

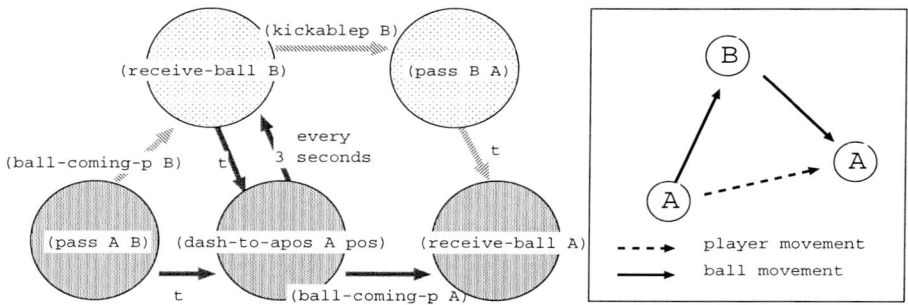

Fig. 2. one-two pass cooperation graph between two agents

Cooperation between agents can also be represented by a transition diagram like graph. In Fig.2, the left graph shows an example of the cooperation between two agents acting a one-two pass as shown in the right figure.

There are two roles. Role A who kicks a ball to role B and receives it later at another position; role B who receives the pass from role A and kicks it back towards A immediately.

The computation shown in Fig.2 will is similar to Fig.1. The most major difference is that each node and arc has a a color which is assigned to each role. We assume that an agent's computation is on a node N colored c_n. If the agent is playing the role r_n, which is assigned to the color c_n, then he does the action described on N. Otherwise, he observes another agent playing the role r_n who are assigned the color c_n instead of acting it himself.

An agent evaluates the conditions on the arcs colored the same color as the agent. If there are no such an arc, he stops his computation.

For instance, the graph of Fig.2 has two colors black and gray which are assigned to the role A and B respectively. Assumed that both A and B's computations are on the left most node.

A's computation sequence will be as follows:

1. act "(pass A B)" on the black node (node1)
2. evaluate "t" (it means true in lisp) and move to the center node (node2).
3. act "(dash-to-apos A pos)", where variable pos is a proper position.
4. if ball comes towards A, go to 6.
5. computation moves to upper left gray node (node3) every 3 seconds. Because his color is not gray, he does not act "(receive-ball B)" but observe the other agent who plays role B. A's computation returns to node2 immediately because the evaluation "t" on the black arc is derived from the node3.
6. he receives the ball and stops computation.

B's computation goes in the same manner. One-two pass cooperation will go well because of the observations of other players.

7 Conclusion and future work

With the introduction of the graph editor, behavior design of the agent is facilitated. Future work will include the improvement of graph editor to generate eager execution code for the evaluation of conditions in order to shorten the response time.

Essex Wizards'99 Team Description

H. Hu, K. Kostiadis, M. Hunter, M. Seabrook

Department of Computer Science, University of Essex
Wivenhoe Park, Colchester CO4 3SQ, United Kingdom
Email: {hhu,kkosti,mchunt,seabmn}@essex.ac.uk

Abstract: This paper describes the Essex Wizards team participated in the RoboCup'99 simulator league. It is mainly concentrated on a multi-threaded implementation of simulated soccer agents to achieve real-time performance. Simulated robot agents work at three distinct phases: sensing, thinking and acting. POSIX threads are adopted to implement them concurrently. The issues of decision-making and co-operation are also addressed.

1. Introduction

In the RoboCup simulator environment, the response time of a soccer agent becomes significantly important since the soccer server operates with 100ms cycles for executing actions and 150ms cycles for providing sensory data [12]. Moreover, auditory sensory data can be received at random intervals. It is vital that each agent has bounded response times. If an action is not generated within 100ms, then the agent will stay idle for that cycle and enemy agents that did act might gain an advantage. On the other hand, if more than one action is generated per cycle, the server will only execute one of them chosen randomly. An additional constraint is that Unix is not a "true" real-time system and hence real-time performance and response times can only be guaranteed up to a certain resolution [13]. A more detailed description of real-time systems can be found in [2,11].

In addition to the responsiveness, the ability to cope with changes in the agent's environment provides a significant advantage, especially when the environment is noisy, complex, and changes over time [5]. One of important issues for an agent therefore is to learn from its environment and past experience in order to autonomously operate without the need of human intervention. Another important issue in multi-agent systems is co-operation. It has been shown that groups of agents can derive more efficient solutions in terms of energy, costs, time and quality [6,7]. A common feature in co-operative frameworks is that of distribution of responsibilities and multiple roles. Each agent in a group has an individual role and therefore a set of responsibilities in the team [4]. In this article a form of emergent co-operation through reinforcement learning is presented. Using the back-propagating nature of Q-learning, co-operation is achieved by linking the intermediate local goals.

In section 2, a description of the agent requirements is presented. The agent architecture for Essex Wizards is illustrated in section 3. Then how multiple threads have been implemented to improve the agent's responsiveness is explained in section 4. Section 5 illustrates how machine learning is used in our team for decision making and co-operation of multiple agents. Finally conclusions and future work are briefly presented in section 6.

2. Agent Requirements

The robotic soccer simulator is an instance of a client/server application in which each client communicates with the server via a UDP socket [12]. The server is responsible for executing requests from each client and updating the environment. At regular time intervals (150ms) the server broadcasts visual information to all clients depending on their position, the quality and size of the field of their view, and their facing direction on the field. In addition to that, the server sends auditory information to various clients at random time intervals.

After processing the sensory data, the clients respond by sending action requests to the server from a set of primitive actions available to them. To avoid message congestion on the server, the clients are allowed to send one request per cycle. A cycle in the current implementation is 100ms. If no message is sent within this interval, the client will not perform any actions. If more than one message is send during the same cycle, the server executes only one, chosen at random, which might produce undesired results. The server updates the state of the environment by serially executing each request.

Since UDP sockets have a limited receive buffer, messages arriving on a UDP socket will be queued until the receiving buffer is full in which case additional messages will be discarded. A client that fails to retrieve the messages at the rate that they arrive is in danger of receiving older information from the server, since newer data will be further back in the queue. This will cause the client to create the wrong representation about the current state of the pitch, which will lead to undesired effects since the wrong actions might be executed. The term "client", used in the client/server application context above, is the real-time agent to be built. For each cycle, the agent receives data from the server, and produces an action. When new data is available, the agent should receive this data and update the current state of the environment. For efficiency the following conditions should be satisfied:

- To receive the newest sensory data that arrives on the socket as quickly as possible and no data queue up. This enables agents to have the most recent representation of the environment, and execute the most appropriate action.
- To time the execution of requests to the server accurately. If more than one request is send by an agent per cycle, the server will only execute one at random, which might be non-optimal. If the agent is too slow, it might miss a cycle and then give an advantage to the enemy agents.
- To allow the maximum time and resources for the thinking process. Since an agent has a fixed amount of time per cycle, the longer it waits to send or receive data, the less time it has to think.

Given the frequency of the message exchange and the timing constraints, building an agent that will satisfy the conditions described above becomes a challenging task.

3. Agent Architecture

Given the variety of I/O models supported under Unix, it becomes difficult to choose the most suitable one for the soccer agents. In addition to that, choosing an I/O model heavily depends upon the inner structure of the agent. As it can be seen in figure 3, the agent contains six different modules, including the agent's sensors, a set of behaviours, the actuators, the current play mode, a set of predefined parameters and a memory module. The dashed arrows in this diagram represent the communication links between the agent and the server. The server sends information received by the agent's "Sensors" module, and the agent sends information back to the server through its "Actuators" module. A brief description is given below to show the relationships among these modules:

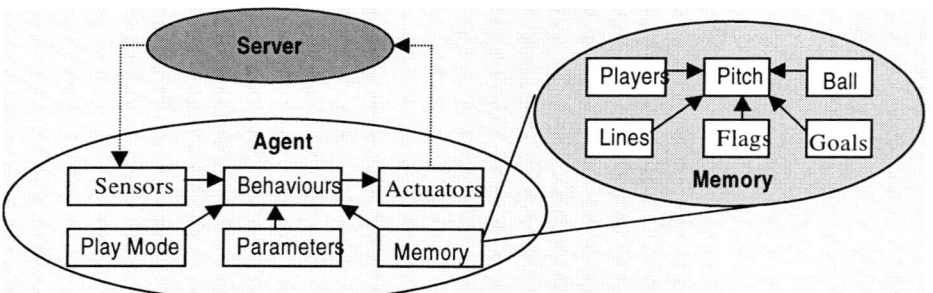

Figure 1 Block diagram of the soccer agent's architecture

- Sensors -- are responsible for receiving and analyzing the visual or auditory information transmitted by the server.
- Behaviours -- is the most important module within the agent. It is responsible for generating actions according to the current state of the environment.
- Actuators -- are used for timing and sending actions to the server every 100ms.
- Play Mode -- holds and updates the current play mode using the information received by the "Sensors" module. The current play mode directly affects the behaviour of the agent
- Parameters -- hold information regarding various settings both for the server and the agent.
- Memory -- is a representation of the whole pitch, rather than part of it.

4. Multi-Threaded Implementation

Instead of a single-thread, a process can have multiple threads, sharing the same address space and performing different operations independently. This architecture allows the agent to use a separate thread for each of the three tasks. Inside the agent, the three main tasks are running concurrently (or in parallel in multi-processor hardware) minimizing delays from the I/O operations. Only the "Sense" thread is responsible for waiting data from the server, and only the "Act" thread is responsible for timing and sending the. In this way the agent can dedicate the maximum amount of processing power available by the processor(s) to the "Think" thread.

Firstly the "Sense" thread is dedicated to receiving data using a blocking I/O mode connection in which the *rcvfrom* call will put the "Sense" thread to sleep until new data arrives on the socket. Putting the "Sense" thread to sleep does not affect the execution of the other two threads. When data arrives on the socket, the "Sense" thread will be awaken, analyze the data, and repeat a *rcvfrom* call for the next available datagram from the server.

Secondly, the "Act" thread needs to measure 100ms intervals, and sends any available actions to the server. The *gettimeofday* function and a conditional variable are used to implement the "Act" thread. In other words, the current absolute time is incremented by 100ms and then the thread is also put to sleep to wait for the conditional variable to return. Since no other thread will signal this conditional variable, it will only return when the 100ms have passed, in which case the "Act" thread can send an action to the server. This method provides highly accurate timing and enables the agent to guarantee certain levels of timing correctness.

Thirdly the "Think" thread is the only one that stays permanently awake, and consumes the majority of the available resources to perform most of the computations. The details on the "Think" thread in the current implementation can be found in [8]. More details on the multi-threaded implementation and the results are presented in [9]. The issues on multi-threaded programming can be found in [3].

5. Reinforcement Learning

Reinforcement learning (RL) addresses the question of how an agent that senses and acts in its environment can learn to choose optimal actions in order to achieve its goals. There are many systems that use RL for learning with little or no a priori knowledge and capability of reactive and adaptive behaviours [1,10,14,15]. The main advantage of reinforcement learning is that it provides a way of programming agents by reward and punishment without needing to specify how the task is to be achieved. On each step of interaction the agent receives an input i which normally provides some indication of the current state s of the environment. The agent then chooses an action to generate as output. The action changes the state of the environment and also provides the agent with a reward of how well it performed. The agent should choose actions that maximize the long-run sum of rewards.

To fully utilize the power of the learning scheme used (Q-learning), the state space is divided by assigning different roles for each individual agent. For example, the goalkeeper need not worry about how to score a goal. This task is indeed numerous stages away given the goalkeeper's responsibilities. It would take numerous iterations before the goalkeeper's training can yield acceptable levels of performance. On the other hand, a goalkeeper can easily learn to pass the ball safely to a nearby defender since this task has a goal-state that is near the goalkeeper's region. In a similar manner a defender can learn to clear or pass the ball to a midfielder and so on.

The agents in the current implementation not only have different roles and responsibilities, but also have different subgoals in the team. Hence, every individual agent in turn tries to reach its own individual goal state without worrying about the performance, or the goals of the other agents. By linking the different goals of each agent, co-operation emerges. Although each agent tries to optimize its actions and reach its own goal-state, since these goal-states are related, the agents co-operate. The ultimate goal, which is to score against the opposition, becomes a joint effort that is distributed between the members of the team. A detailed description of the Essex Wizards RL implementation can be found in [8].

6. Conclusions and Future Work

To satisfy all the necessary timing constraints for real-time agents in general, football-playing robots in particular, a single-threaded implementation will not suffice. This is mainly due to the low speed of network I/O operations, and the limiting serial nature of such architectures. Therefore, a multi-threaded implementation is presented in this article to overcome this problem. Based on this approach, the agents can perform various computations concurrently and avoid waiting for the slow I/O operations. A multi-threaded model clearly outperforms a single-threaded one in terms of responsiveness and efficiency. A decision-making mechanism based on reinforcement learning is briefly described, which can also be used to enable co-operation between multiple agents by distributing their responsibilities. By gathering useful experience from earlier stages, an agent can significantly improve its performance.

The future work for the Essex Wizards team is to focus on cooperative behaviours, team formations, sensor fusion and machine learning capability.

Acknowledgements: We would like to thank the University of Essex for the financial support to the project by providing the Research Promotion Fund DDP940.

References

1. Balch T. R., Integrating RL and Behaviour-based Control for Soccer: Proc. IJCAI Workshop on RoboCup, 1997.
2. Burns A. and Wellings: A., Real-time Systems and Programming Languages, Addison-Wesley, 1997.
3. Butenhof R.D: Programming with POSIX Threads, Harlow, Addison-Wesley, 1997.
4. Ch'ng S., Padgham L.: From Roles to Teamwork: A Framework and Architecture, Applied Artificial Intelligence Journal, 1997.
5. Hu H., Gu D., Brady M.: A Modular Computing Architecture for Autonomous Robots, Int. Journal of Microprocessors and Microsystems, Vol. 21, No. 6, pages 349-362, 1998.
6. Hu H., Kelly I., Keating D., Vinagre D.: Coordination of Multiple Mobile Robots via Communication, Proc. SPIE'98, Mobile Robots XIII, Boston, pp. 94-103, Nov. 1998.
7. Kitano H., RoboCup: The Robot World Cup Initiative, Proceedings of the 1st International Conference on Autonomous Agent (Agents-97), Marina del Ray, The ACM Press, 1997.
8. Kostiadis K. and Hu H.: Reinforcement Learning and Co-operation in a Simulated Multi-agent System, Proc. of IEEE/RJS IROS'99, Korea, Oct. 1999.
9. Kostiadis K. and Hu H.: A multi-threaded approach to simulated soccer agents for the RoboCup competition, IJCAI'99 workshop on RoboCup, 1999.
10. Mataric, J. M., Interaction and Intelligent Behaviour, PhD Thesis, MIT, 1994.
11. Nissanke N.: Realtime Systems, London, Prentice Hall, 1997.
12. Noda I.: Soccer Server: A Simulator for RoboCup, JSAI AI-Symposium 95: Special Session on RoboCup, 1995
13. Stevens W.R.: Unix Network Programming: Networking APIs: Sockets and XTI (Volume 1), London, Prentice-Hall International, 1998.
14. Stone Peter and Veloso Manuela: Team-Partitioned, Opaque-Transition Reinforcement Learning, Proc. 15th Int. Conf. on Machine Learning, 1998.
15. Uchibe E., Asada M., Noda S., Takahashi Y., Hosoda K.: Vision-Based Reinforcement Learning for RoboCup: Towards Real Robot Competition, Proc. of IROS 96 Workshop on RoboCup, 1996.

FCFoo99

Fredrik Heintz, `frehe@ida.liu.se`

Department of Computer and Information Science, Linköping university

1 Introduction

The emphasis of FCFoo was mainly on building a library for developers of RoboCup teams, designed especially for educational use. After the competition the library was more or less totally rewritten and finally published as part of the Master Thesis of Fredrik Heintz [4].

The agents are built on a layered reactive-deliberative architecture. The four layers describes the agent on different levels of abstraction and deliberation. The lowest level is mainly reactive while the others are more deliberate. The teamwork is based on finite automatas and roles. A role is a set of attributes describing some of the behaviour of a player. The decision-making uses decision-trees to classify the situation and select the appropriate skill to perform. The other two layers are used to calculate the actual command to be sent to the server.

The agent architecture and the basic design are inspired by the champions of RoboCup'98, CMUnited [6, 7]. The idea of using decision-trees and roles is inspired by Silvia Coradeschi et al [2, 3].

FCFoo99 did not do very well in the competition, but still better than expected since FCFoo was a less than six month one man project. FCFoo lost its first game against the 11 Monkeys with eighteen nil, but won its second by five nil agains Sibiu. In the third game Pardis forfeited since they were not able to run their agents properly. The fourth game against the Gongeroos was a very exciting game. FCFoo scored two quick goals but then the Gongeroos adapted its playing style and managed to score four goals and make it to the elimination round.

2 Team Development

Team Leader: Fredrik Heintz
Team Members:
 Fredrik Heintz
   − Department of Computer and Information Science, Linköping university
   − Sweden
   − Graduate student
   − Attended the competition
Team Web Page http://www.ida.liu.se/ frehe/RoboCup/FCFoo/
Library Web Page http://www.ida.liu.se/ frehe/RoboCup/RoboSoc/

3 World Model

FCFoo used an object-oriented approach to do the world modelling. Each object in the simulation, the ball, the agent, the game and so on are represented by an object. All the objects are stored in a large knowledge structure, representing the memory of the agent. Each object is responsible for all the modelling of the corresponding game object. For example the ball object is responsible for updating itself after each step of the simulation. Each object also contains the history of the last few cycles, so that previous states of the object can be used in determining the current or the next state of the object.

FCFoo also used an object-oriented approach to model the server actions available to the agent. Each primitive action is a separate object which knows how to update the state of the agent after the skill has been executed and if the primitive action is applicable, according to the world model of the agent. Before sending the actual string to the soccer server the action object makes sure the parameters used are valid, according to the parameters of the simulation and according to the current world model of the agent.

4 Communication

Inter agent communication was not used by FCFoo. Instead it used roles to coordinate the behaviour of the team.

5 Skills

The skills was one of the major weaknesses of FCFoo. The skills implemented includes score, pass, dribble, catch and intercept ball. For example the score skill only takes the position of the agent into account, not the position of the opponents, and kicks the ball just inside the closest goal post. The other skills are equally simple in their implementation.

FCFoo used a special goal keeper agent, but since the algorithm for calculating the speed of the ball did not work very well the goal keeper did not perform very well since it could not estimate where to run to catch the ball.

6 Strategy

The strategy of FCFoo is based on roles. Each player is given an initial role. The role define where it should position itself on the field, when to take freekicks and so on. The roles can be assigned and changed during the game. The roles are mainly used when making a decision on what to do next.

6.1 Roles

A role is a set of attributes that defines the behaviour of a player. A player can have only one role. But since it is possible to define new roles as combinations of roles it is actually possible for a player to have several roles. Problems with this approach are among others that one have to define a specific role for each role-type. Attributes can have different values for each individual having that role and therefore specialisation is supported.

Examples of attributes are: freekick-area, home-position, and home-area. They define where the player with this role should be and when it should do the freekicks. Examples of roles are: goalkeeper, defender, midfielder and attacker. Examples of specialisations are: left inner midfielder and right outer defender.

6.2 Decision making

The decision making of the agent is based on decision-trees (DT) and finite automatas (FA). The automatas are used in the strategy layer to decide what DT to use. The automata describes when to change states, and DT. The triggers used by the FA are referee-calls, the positions of the players and the position of the ball. Figure 1 shows a part of the FA used by the agents. The italic words on the arrows are referee-calls, the normal words are predicates and the words inside the boxes are the name of the DT to use when in that state.

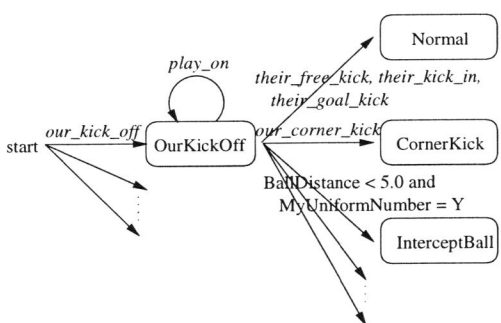

Fig. 1. Part of the finite automata used by an agent making a decision.

The DTs are used in the decision layer to classify the situation and decide what skill to execute. When the decision layer has found a skill to perform it sends that information to the skills layer. The skills layer calculates what primitive action to do based on the current position and internal state of the agent. The primitive action layer takes the action with its parameters and calculates the actual command to be sent to the server.

7 Conclusion

The purpose of FCFoo has been to test and support the development of RoboSoc, a system for developing RoboCup agents for educational use. FCFoo will not compete in RoboCup2000, instead the author and a couple of students from an AI programming course given at Linköping university [1] will create a new team, called NOAI, based on RoboSoc and on the strategy editor from the Headless Chickens III developed by Paul Scerri et al [5]. The emphasis will be on extending and evaluation RoboSoc and on improving the efficiency and expressiveness of the behaviour-based decision trees used by the Headless Chickens.

References

1. AI-Programming course web-page. http://www.ida.liu.se/~TDDA14/, March 2000.
2. Silvia Coradeschi and Lars Karlsson. A role-based decision-mechanism for teams of reactive and coordinating agents. In Hiroaki Kitano, editor, *RoboCup-97: Robot Soccer World Cup I*. Springer Verlag, Berlin, 1998.
3. Silvia Coradeschi, Paul Scerri, and Anders Törne. A User Oriented System for Developing Behavior Based Agents. In Minoru Asada and Hiroaki Kitano, editors, *RoboCup-98: Robot Soccer World Cup II*, pages 173–186. Springer Verlag, Berlin, 1999.
4. Fredrik Heintz. RoboSoc a System for Developing RoboCup Agents for Educational Use. Master's thesis, IDA 00/26, Linköping university, Sweden, 2000.
5. Paul Scerri and Johan Ydrén. End User Specification of RoboCup Teams. In Manuela Veloso, Enrico Pagello, and Hiroaki Kitano, editors, *RoboCup-99: Robot Soccer World Cup III*. Springer Verlag, Berlin, 2000.
6. Peter Stone and Manuela Veloso. The cmunited-97 simulator team. In Hiroaki Kitano, editor, *RoboCup-97: Robot Soccer World Cup I*. Springer Verlag, Berlin, 1998.
7. Peter Stone, Manuela Veloso, and Patrick Riley. The CMUnited-98 Champion Simulator Team. In Minoru Asada and Hiroaki Kitano, editors, *RoboCup-98: Robot Soccer World Cup II*. Springer Verlag, Berlin, 1999.

Footux Team Description*
A Hybrid Recursive Based Agent Architecture

Francois GIRAULT, Serge STINCKWICH

GREYC – CNRS UPRESA 6072
Université de Caen
Francois.Girault@info.unicaen.fr
Serge.Stinckwich@info.unicaen.fr

Résumé This document describes the software architecture of the Footux-99 team (simulation league). It is now well known that purely reactive (resp. cognitive) agents are out of date. An agent must be able to respond reactively when necessary, but it should have a general behaviour guideline. strategy. The most classical approach consists in using a hybrid architecture.

The architecture we are introducing in this article is a hybrid one. It combines vertical and horizontal hybrid approachs where each layer is based on a subsumption architecture.

The aim of our approach is to study the possibility to obtain a cooperative behavior within a multi-agents system without using centralized control, and thus to observe the emergence of potential relations between an agent and the society to which it belongs.

1 Introduction

The RoboCup challenge forces us to have several points of view about the agent. In a first way, the agent must be fast, clever and accept a not well known world. In a second way a soccer player must have a coherent collective behaviour. Combining both approaches implies hybrid architecture and an anticipatory approach.

In order to take both points of view about the agent into account, we need to split the basic agent model into two distinct modules. On one hand, a reactive module which acts according to reflexes. On the other hand, a set of cognitive modules making more or less long term plans. It is clear that if we do not wish to fall into the classical approaches *(i.e. reactive vs cognitive)*, both modules shall not run one after the other but shall have a complementary and parallel existence.

The hybrid architecture ([4]) approach tries to combine two antagonistic models *(reactive vs cognitive)*. A layered model such as INTERRAP([3]) or a Touring Machine divides the different levels : Basic Behaviour Level (BBL), Local Planification Layer (LPL) and Social Planification Layer (SPL). There are two different hybrid architectures. The first one is vertical : a higher layer

---

* A more recent version of this document is always available at the URL: http://www.info.unicaen.fr/~girault/footux-99

is called when the previous layer needs it. The second one is the horizontal architecture : all layers are running simultaneously, when a layer finds a solution it is the next one. Our model combines vertical and horizontal hybrid approach in the same architecture : the perception flow is vertical while the action flow is horizontal.

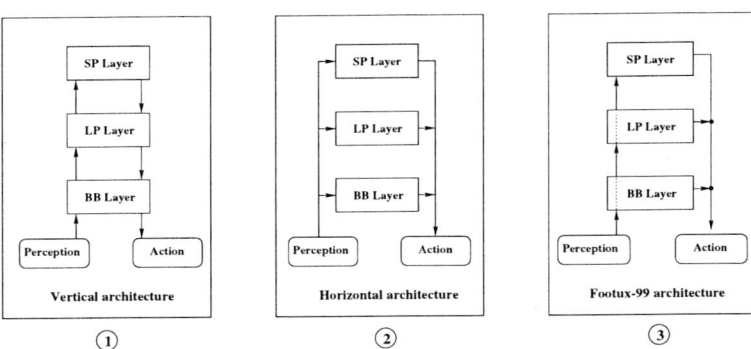

Fig.1. Hybrid architecture

2 The Footux-99 architecture

As shown in figure 1 p.2, there are two types of hybrid architectures according to the orientation of the control flow. Our model is at the intersection between these architectures. In fact, in a vertical architecture, the response delay of a high level layer such as the LPL can take long enough to give a response outdated to the call situation. In the horizontal architectures, each module computes, taking no account of the other modules results. This implies a solution management by a fairly problematic subsumption architecture as well as a potential incoherence between the different solutions.

In order to try and avoid these problems, we introduce a recursive anytime architecture : Augmented Reality by Anticipation (ARA).

2.1 Diagonal approach

We wish to have an agent which cognitive modules can always bring a relevant solution without any action conflict nor global planning reorganization. In that respect, we have to avoid the punctual aspect of reorganization, which implies a full-time communication between the modules that have to work in parallel.

This short description looks like a horizontal architecture, except that in such architectures, each module produces a solution autonomously without taking into account the other modules? behavior. Taking this into account implies stopping the cognitive mechanism except where the module works in anytime

[1]. Furthermore, this produces an incremental planning allowing us to avoid a complete new computation of the plans.

Module behaviour As just shown, a module functions anytime, and in that perspective, it is necessary for it to receive and produce a continuous data flow.

In order to simplify these data flows, we have decided to use a single data type : the REALITY. This is opposed to more classical approaches which are highly influenced by expert systems, which each level introduces a new data type, meta-data, thus adding an supplementary system to handle this meta-level.

Each module takes a reality as input and produces as output a reality modified or augmented. We call augmented reality a representation of the world in a more or less near future, reality in which some elements have been added. These augmented realities are computed, according to the module, as a function of the relative knowledge of the current situation. It seems fairly improbable that a module be able, in a real-time environment, to face all situations. It is thus important that some modules have a learning capacity.

Just as a child discovering the world, the agent placed in an very dynamical environment must be able to learn from its mistakes, to anticipate the effect of its action and to predict the behavior of objects or other agents surrounding it.

Inter-module communication The inter-module communication is the key to this model. As a matter of fact, just as the environment, the functioning of the different modules is very dynamical. Resulting from this, taking other modules intermediary results into account is as important as the modules functioning itself.

The data flow begin continuous, the interpretation of the world given by a module is used as a representation of the world by the higher level. As we can see in figure 2, the local module takes a representation P0 of the world and gives an interpretation I1. Every modification brought to this interpretation will be directly transmitted to the global module. The way the information goes down again is a crucial point in the inter-module communication. In fact, it is the choice of the right augmented reality which will place the reactive module in an emergency situation. We find again the problem of the choice of the action to do happening in horizontal architectures, to the difference that now we have a continuous information flow that has to be transmitted. This continuity allows us to chose quickly or to concatenate realities produced by two modules from neighboring levels and not all possible realities. This operation is done by the Fi control function.

A control function is an entity taking two realities in input (augmented or not) and producing an augmented reality corresponding to the ?best ? possibility. This best reality is the partial or complete concatenation of both inputs. The concatenation consists in retaining, for each element of each reality, the one being in the further future.

---

[1] the anytime process being defined such that un module can supply a solution at anytime, even if it has to be incomplete or degraded

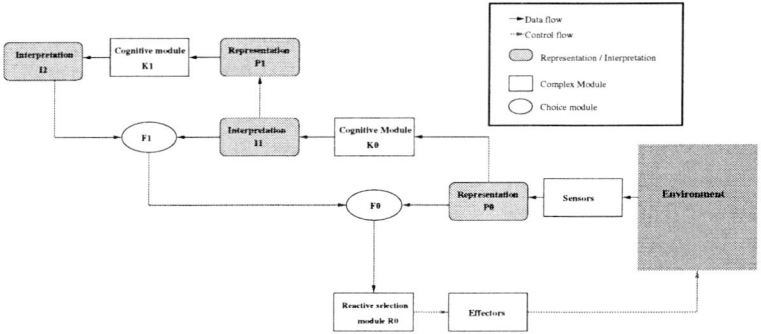

Fig.2. Cognitive model diagram

3 Discussion and perspectives

The inspiration of this model comes from the idea of anticipatory systems ([1]). Indeed, anticipation plays a significant role in the taking into account of actions unforeseen and external with the agent. It makes it possible to give an account of total and complex phenomena not apprehended by a reactive layer. To anticipate the actions of the adversary or its fellow-members, to anticipate the movements of the ball or the conflicts between team-members constitute some of the forms of anticipations which we wish to take into account in our system.

In order to ground this concept of anticipation, we are inspired here by theoretical work, in particular those of Rosen, which gives the following definition of an anticipatory system : " ... a system containing a predictive model of itself and/or its environment which allows it to change state at an instant in accord with the model's predictions to a latter instant".

An anticipatory system thus uses the knowledge of future states of the system to decide actions to take in the present moment.

This model is merely a draft. It is thus not definitively set. We can however summarize that :

In a general way :
- Every task can be split in a set of simpler sub-tasks
- Any action can be split in a set of elementary and complementary sub-actions.
- Every set of elementary actions can be split in actions running in parallel.

This model is far from being perfect : the problem with horizontal models is that they are slow at thinking. The problem with vertical models is that they do not stop thinking and thus good ideas may arrive too late. In our approach, the cognitive modules keep thinking all the time, but the lower layer can modify the representation of the world for the higher layer at any time in order to submit new information that may be important. Each module (cognitive or reactive) being built according to a parallel execution of actions, each modification of the representation of the world is taken into account as soon as it is instantiated.

The reactive module have been partially implemented in C++ programming language. Further policy managements for the layers are experiment. What needs to be improved as soon as possible is a better reactive cooperation.

References

1. Paul DAVIDSSON A Linearly Quasi-Anticipatory Autonomous Agent Architecture : Some preliminary experiments, in *Distributed Artificial Intelligence Architecture and Modelling (Lecture Notes in Artificial Intelligence 1087)*, C. Zhang and D. Lukose (ed.), pages 189-203, Springer Verlag, 1996.
2. Bertil EKDAHL, Eric ASTOR et Paul DAVIDSSON Towards Anticipatory Agents. In M. Wooldridge and N.R. Jennings, editors, *Intelligent Agents – Theories, Architectures, and languages, Lecture Notes in Artificial Intelligence 890*, page 191-202, Springer Verlag, 1995.
3. Jorg P. MULLER The design of intelligent Agents, volume 1177 of *Lecture Notes in Artificial intelligence*. Springer-Verlag, 1996.
4. Christoph G. JNG Experimenting with Layered, Ressource-Adapting Agents in the RoboCup Simulation in M. Asada (ed.), RoboCup-98 : Robot Soccer WorldCup II, *Lecture Notes in Artificial Intelligence*, Springer, 1999.
5. Hayes-Roth B. An architecture for adaptative intelligent systems. *Artificial Intelligence,* 72(1-2) :pp. 329-365, january 1995.

Gongeroos'99

Chee Fon Chang, Aditya Ghose, Justin Lipman, Peter Harvey

Department of Information Systems, University of Wollongong, NSW 2522, Australia
{c03, aditya, jl06, pah06}@uow.edu.au

1 Introduction

The Gongeroos'99 team involves agents built within the broad framework defined by the BDI agent architecture [3] with novel features involving the application of notions from team-oriented programming [2] and multi-hop ad-hoc communication networks [1] from the area of mobile computing. Gongeroos'99 achieved a 9th place ranking in RoboCup-99's software simulation league.

2 Team Development

Team Members:
  Chee Fon Chang
    – Decision Systems Laboratory
    – Australia
    – Graduate Student
    – did attend the competition
  Aditya Ghose
    – Decision Systems Laboratory
    – Australia
    – Senior Lecturer
    – did not attend the competition
  Justin Lipman
    – Decision Systems Laboratory
    – Australia
    – Graduate Student
    – did attend the competition
  Peter Harvey
    – Decision Systems Laboratory
    – Australia
    – Research Assistant
    – did attend the competition
Web page http://budhi.uow.edu.au/robocup

3 World Model

The Gongeroos'99 agent architecture is a variant of the well-known BDI architecture. As with the BDI architecture, our agents are based on the following core data structures: beliefs B, desires D, intentions I and plans PL [3]. The belief state B of an agent is denoted by a triple $< E, S, M >$ where E denotes an agent's *environment* or *world model*, S denotes an agent's *skill set* and M denotes the *domain theory* used by agent. The environment E is denoted by a collection of sensory parameters including positions of moving and stationary objects (ball, players, field), time and information accuracy. The skill set S denotes the repertoire of actions available to an agent. These skills include atomic actions such as kicking and turning as well as complex skills that make use of other skills within the same set. M is the domain theory (or *domain invariants*) that describes how actions affect states of the environment. It contains causal rules that determine the effects of actions in various states of the environment as well as (possibly partial) specifications of how actions map from one state of the environment to another.

4 Communication

The communications regime used by Gongeroos'99 agents relies on the theory of *multi-hop ad-hoc networks* [1]. Ad-hoc networks are wireless networks consisting of multiple mobile hosts. They form an unstructured, dynamic and temporary mobile network. In situations where a source host wishes to send a packet to a destination host, which is not within broadcast range, the packet may be sent via other hosts within broadcast range. Hosts forward this packet until it reaches its destination.

5 Skills

The set D of desires of an agent is partitioned into two categories: *offensive* and *defensive* desires. These are treated differently depending on how far a match has progressed. Given a certain state of the environment E, a mapping determines a unique desire to be adopted (which may be either offensive or defensive). The mapping function takes into account how far the match has progressed as well as the team *role* assigned to the agent at that point in time (team-oriented aspects are described later in this paper). The set of *intentions* or sub-goals I is partitioned into two categories: *reactive intentions* I_R and *tactical intentions* I_T. Direct mappings exist from the state of the environment E to reactive intentions in I_R, which may supersede intentions in I_T (thus, reactive behaviours take over when the ball is perceived to be within a certain radius of the player). Adoption of tactical intentions is determined, again, by an agent's role in a team at a given point in time.

The Plan Library PL is a collection of plans. Each plan is a 4-tuple denoted by $< Pr, A, F >$ where Pr denotes the plan pre-conditions, A is a sequence of actions to be peformed if Pr is satisfied and F is a measure of how successful the plan was in the past (denoted by the ratio of successful invocations to total invocations). All the actions contained in A are drawn from the skill set S. A limited form of learning can be implemented using the F measure attached to every plan. In general, multiple plans may match both the current trigger and have their pre-conditions satisfied. In such situations, plans with a higher value of F are selected. The plan library is stratified into priority levels. These levels assist in plan interruption and recovery. A plan with a higher priority can interrupt a plan with a lower priority. When a plan is selected for execution, an expiration time is assigned to the plan. When a plan is interrupted, the agent will hold the current plan and execute the higher prioritized plan. At the completion of the plan, the agent will try to recover the previous plan and try to complete it within the given expiration time. Should the time expire, the plan will be discarded.

Ball interception by an agent is a reactive behaviour and the relevant plans are invoked via direct mappings from the state of the environment (usually when the ball is within a certain radius of an agent). Dribbling is achieved via small kicks forward, followed by re-interception of the ball.

6 Strategy

Our approach to team coordination is to implement agent teams. We have adopted the concept of mutual belief, joint goal and joint intention [2] as well as dynamic role assignment [4] and team plans.

A team τ represents a set of a finite individual agents [2] and the social structure σ represents the agents' belief of belonging to a τ. The social structure σ as defined by Tidhar [2] denotes it as a pair of command and control team which we denote as $< St, Ta >$ where St denotes the command team and Ta denotes the control team for a τ, however,in our approach, there are no distinct command or control teams. Every agent would perform both the St and Ta functions. These functions are however emphasised when the agent becomes a leader. A leader is an agent that take charge of other agents' and are responsible for the their behaviours. The majority of the St component's authority is contained within the leader. Thus, a team's goal is Ta is concerned with an individual agent's contribution towards the team goal. That is, Ta takes into account the team goal. Ta could be considered the component of the agent that handles team intentions. Ta also communicates with St, which in turn communicates within the team to synchronise certain movements of players. St holds the concept of social structure and responsibilities where the agents know who is in command of it and who it is in command of. Ta adopts the role of coordinator. This approach allows the dynamic assignment of sub-teams, which we will denote as task groups. In theory, there can multiple encapsulation of sub-teams however with the limited number of agents involved we are limiting to only one level of sub-teams. These individual task groups can only receive task from the agent that initiated their formation, in other words, if the Captain initiates the formation, then only the Captain can assign tasks to the task group however transfer of task group between the Captain and the Goalie are permitted. This result in τ that can adopt two or more non-contradicting desires simultaneously. In our approach, we have adopted a social structure where the Coach is at the top of the chain, followed by the Captain and Goalie. The agent structure of the coach is a variance of the Case-Based BDI agents [5]. Before the start of the game, the Coach plays an important role where it will make decisions regarding the type of strategy to deploy as well as the composite of the initial team plan library similar to the locker room agreement [4]. Due to the time constants, we have limited the number of team plans that are available to the team. The team plan is a 3-tuple similar to the make up of the agent plans. The team plan library selection will be done based upon the success frequencies. During the game, the coach will analyse the game, determine where the plans are failing and make modifications or in extreme cases generating new team plans. These plans will then be uploaded to the player when allowed.

7 Conclusion

The Robocup environment provides an interesting range of challenges within the domain of multi-agent systems. We expect a new and improved Gongeroos2000 team based on a design philosphy that emphasizes explicit representations of both the beliefs and plans of agents as well as the trade-offs made to achieve effective behaviour in time-bounded situations.

8 References

1. Johnson, D.B. "Routing in AD Hoc Networks of Mobile Hosts", Proceedings of the IEEE Workshop on Mobile Computing Systems and Applications, December 1994.
2. Tidhar, G. "Team-Oriented Programming: Preliminary Report"
3. Rao, A.S. and Georgeff, M.P. "BDI Agents: From Theory to Practice", In Proceedings of the First International Conference on Multi-Agent Systems (ICMAS-95), San Francisco, USA, 1995.
4. Stone, P. and Veloso, M. "Task Decomposition and Dynamic Role assignment for real-time Strategic Teamwork", (ICMAS-98).
5. Chang, C.F. Ghose, A. Enguix, C.F. Olivia, O. "Case-Based BDI Agents: An Effective Approach For Intelligent Search On the World Wide Web", In Proceedings of AAAI Spring Symposium on Intelligent Agents, Stanford University, USA, 22-24 March 1999.

Headless Chickens III

Paul Scerri, Johan Ydrén, Tobias Wiren, Mikael Lönneberg and Pelle Nilsson

Institute for Computer Science, Linköpings Universitet

1 Introduction

The development of the Headless Chickens III emphasized a high level team specification environment, called the Strategy Editor, that was intended for use by endusers, rather than computer programmers[2]. Using the strategy editor consisted of placing players on a image of the ground and indicating the direction(s) the player should kick and/or dribble when they get the ball. Different player formations and passing/dribbling patterns could be specified for different game situations. The designer could also specify the style of play for each of the players, e.g. defensive or inclined to shoot or dribble.

The Strategy Editor proved to be an effective tool primarily because a "player template" could be loaded into the editor. The template specified the different modes of play the players knew about, the different styles of play the player could play and the different actions the player could take. The mechanism allowed parallel development of the low level aspects of the players behavior (developed by agent experts) and high level strategies (developed by domain experts).

Specifications made with the Strategy Editor were "compiled" into separate behavior based agents. The core of the behavior based agents runtime engine had been previously developed for earlier versions of the Headless Chickens[1].

The Headless Chickens III finished equal 5th in the 1999 World Cup competition. They were involved in some of the more exciting games of the competition including the first ever World Cup overtime game which HCIII eventually won two to one. However HCIII were clearly inferior to the best teams losing seventeen nil to CMUnited99 and eleven nil to MagmaFreiburg.

2 Team Development

Team Leader: Paul Scerri
Team Members:
   Paul Scerri
      − Linköpings Universitet
      − Sweden
      − Graduate Student
      − Attended Competition
   Johan Y'dren
      − Linköpings Universitet
      − Sweden

- Masters Student
- Attended Competition

Tobias Wiren, Mikael Lönneberg and Pelle Nilsson
- Linköpings Universitet
- Sweden
- Undergraduate Students
- Attended Competition

Web page http://www.ida.liu.se/~pausc/RC99/main.html

3 World Model

The agent architecture is split into two layers, one for skills and one for strategies. Although there is some basic information processing that is used at both layers, e.g. calculating the velocity of the ball, the way world information is presented to the different layers is quite different.

The skills layer uses world information almost directly from the sensors. Some low level calculations are done to ensure the players view of the world remains reasonably accurate between sensing cycles and after acting cycles. There is some very simple reasoning done so that objects that have been seen previously but are no longer in view are maintained in memory unless the player is looking at where the object was last seen.

The strategy layer uses world information only as abstracted fuzzy predicates. A separate Java class is associated with each fuzzy predicate. The class uses sensor information to assign a value between one and one hundred to predicates such as *the ball is close* or *defensive position*. The higher the value of the predicate the more the predicate seems to be true. Some of the predicates, for example *near ball*, have some "memory" so that the value of the predicate is still reasonable when sensor values are unable to determine its value, e.g. when the ball can be no longer seen the predicate *near ball* retains it's previous value.

4 Communication

The HCIII do not use communication between agents. It was not found to be necessary either from an individual player perspective or from a team perspective.

The reactive nature of the agents, i.e. a behavior based architecture, is well suited to having limited local information, hence there is little need for inter-agent communication about object locations.

From a coordination point of view it is the responsibility of the team designer at design time to ensure that players will be in appropriate positions at particular stages of a game to ensure that team behavior "emerges". The "emergent" team behavior does not require communication. The team designer also specifies the preferred directions for players to pass and dribble so communication is not required for that either.

5 Skills

The skills of HCIII are relatively simple. Every cycle one skill is called, perhaps with some parameters, and gets the chance to execute one action. A flag is passed to the skill indicating whether or not it was the skill that executed the previous action but it has no guarantees that it will get to execute the next action. Niether does the skill have any idea of the higher goal that the skill is part of achieving.

The algorithm for intercepting the ball looks for the closest position where the player can meet the ball. A loop calculates the expected position of the ball for subsequent cycles and for each cycle checks whether the player can reach that position within that time. There are special cases for when the ball is coming directly at the player or moving very slowly.

Because a skill can only execute one action at a time the dribble is also very simple. The agents action depends only on the position of the ball and the point that the agent should dribble to. The kick parameters are calculated by first working out where the agent wants the ball to be in two cycles then working out the power and direction to get the ball there.

The goalie is a special agent that attempts to maintain a position a certain distance along a line between the center of the goal and the ball. Any time the ball comes into the penalty box the goalie chases the ball. Once he has the ball he will wait (spinning around to watch as much of the field as possible) until finding a good player to pass to. If no good passing option is found within some time the ball is kicked hard towards the sideline.

6 Strategy

The strategy of the HCIII can vary greatly from game to game. The strategy of the team is defined in two parts, using two different graphical development systems, the individual strategy editor and the team strategy editor.

The individual strategy editor, as the name suggests, defines the strategies of a single player. In effect it defines a template of a player which will be instantiated for a particular team strategy. At a high level of abstraction the individual strategy determines the different "modes" of play that the player will react to. Example modes are *Before kick off*, *Deep defense* and *Transition to attack*. Within each of the modes the individual strategy determines the styles of play the agent has for that mode, for example *waiting* before kickoff or *crossing* from attack. Several different "styles" of behavior can be defined for one mode, for example a defensive style and an attacking style. Which particular style the player will have is determined in the team strategy editor. At a lower level of abstraction the individual strategy defines aspects of a player such as its preference for kicking with respect to dribbling, how long the player is willing to lose sight of the ball before searching for it and how keen a player is to attempt to intercept an opponent.

The team strategy editor allows an enduser to quickly instantiate the templates created in the individual strategy editor into a team configuration. In the strategy editor aspects such as where each player should be in each mode, the positions to kick to and the directions to dribble as well as the style of play for each of the players is specified.

The combination of the two strategy editors allows a great deal of flexibility. During the World Cup our development team would watch logfiles of prospective opponents and create specialized strategies for each team. A prime example was the opponent YowAI. It was realized that there were certain team tactics that would keep most of YowAI's team offside most of the time. The tactics were quickly specified in the team strategy editor without recourse to the individual behavior editor. Using the newly created tactics and having players prefer dribbling meant that HCIII had a large amount of ball possesion (bad stamina management meant that high ball possesion was not turned into a good score). A less successful strategy was against CMUnited99. It was clear that CMUnited99 were far better than HCIII and a loss was inevitable so a very aggressive strategy was employed in a (vain) attempt to be the first team to score against CMU in two years. Alas the aggressive strategy resulted only in making CMUnited's margin of victory more pronounced.

7 Conclusion

There are two teams planned for RoboCup2000 based on the HCIII. One of the teams will use the same team specification system and a similar agent architecture but port the agent runtime architecture to C++ (from Java). Porting the agent code is aimed at improving the efficency of the team. The other team planned for RoboCup2000 will also use the same team strategy editor but will use a slightly different, though still very reactive, agent architecture. The new agent architecture is aimed at providing facilities for more intelligent agent decision making, in particular the ability to simultaneously attend to multiple high level goals.

References

1. Paul Scerri, Silvia Coradeschi, and Anders Törne. A user oriented system for developing behavior based agents. In *Proceedings of RoboCup'98*, Paris, 1998.
2. Paul Scerri and Johan Ydrén. *RoboCup-99: Robot Soccer World Cup III*, chapter End User Specification of RoboCup Teams. Springer, 1999.

IALP

Antonio Cisternino[1], Maria Simi[1]

[1] Dipartimento di Informatica, Università di Pisa
Corso Italia 40, 56125 Pisa, Italy
{cisterni, simi}@di.unipi.it

1 Introduction

IALP is a team for the simulation league of the RoboCup initiative [4]. The team is programmed using ECL, a public domain implementation of Common Lisp [1].
The core of the IALP team is a reactive planner whose behaviour is structured in layers. The requirements we had in mind for the architecture is that it must be open and offer different levels of abstraction coping with different problems in a modular way. Moreover the architecture is meant to be general and flexible enough to allow reuse of code built for the RoboCup initiative in other domains.
For coping with limited perceptions, we developed a memory model based on the absolute positions of objects.
IALP uses a model of coordination without communication [3] and a concept of role for a player that is built on top of basic abilities, common to all the agents. The layered and modular structure of the planner allows an easy reuse of the basic capabilities of the players and specialisation of roles at the higher levels.
Using Common Lisp to implement IALP offers clear advantages from the AI programming point of view; in particular we have exploited the Lisp reader and the macro feature.
The team played four games at RoboCup '99: three games have been lost and one was a draw. The poor performance was due to the fact that the low level (the communication layer responsible for handling communication with the server) was too slow. Moreover the memory model, implemented in Lisp, proved to be too heavy.

2 Team Development

Team Leader: Maria Simi
Team Members:
  Maria Simi
      Dipartimento di Informatica, Università di Pisa
      Italy
      Associate Professor of Artificial Intelligence
      Attended the competition

Antonio Cisternino
    Dipartimento di Informatica, Università di Pisa
    Italy
    Undergraduate student
    Attended the competition
Web page: http://medialab.di.unipi.it/Project/Robocup/IALP/

3   World Model

A memory model is used in IALP to keep track of objects and players seen recently in terms of their absolute positions. The memory also stores the messages heard and the physical status of the player.
The IALP player executes a standard cycle: receives a perception from the server, updates the memory, computes a new set of actions and sends them to the server. In deciding the next actions the planner uses higher level predicates implemented from the information contained in the memory.
If the received perception is *see* the memory tries to update the absolute position of the player. The coordinate system is the same used by the server. The absolute position of the player is computed using a borderline and a flag. When a borderline is visible the player can easily compute his distance from the line, and thus one coordinate, which is x or y depending on the line and the direction in the coordinate system chosen. If a flag is also perceived the player can compute the second coordinate. This method has a good precision and is fast to compute. The basic assumption is that the player movements are continuous; if the player at a given time cannot compute one or both coordinates he can assume the previous ones, without making a significant error.
Once the position of the player has been computed, the absolute coordinates for each dynamic object present in the *see* perception (players and ball) are also computed using standard trigonometric calculus.
The choice of recording absolute coordinates (see [4] for a different choice) allows us to focus memory updating on the moving objects because static objects are recorded in a stable form using their position.
The *hear* and *sense body* perceptions are treated similarly.

4   Communication

The model of coordination used does not involve communication [3]. For instance if the player possessing the ball decides to pass, he simply does so. The coordination is in the fact that the target player is typically looking at the ball and close enough, thus able to see the ball coming and to intercept it.

5 Skills

The skills of the players are built using our layered architecture. We have hand coded basic skills like go to ball, run with ball, pass ball and so on. The skills of the players need to be improved in future versions of the team. For this purpose we plan to use some form of learning.

6 Strategy

All players are equivalent in IALP team: they differ only for the definition of the role they play. This homogeneity among players is justified by the definition of role that we have assumed: a role amounts to *prevalence of behaviour*. This implies that the basic capabilities of the various players are the same, and only the overall strategy of the team and the environment account for differences in behaviour. When all the team members are forced in a situation of defence, for example, we would like the attackers be able to behave like defenders.

The overall strategy of the team emerges from role definitions. The zone of the field assigned to a player, when he is not engaged in the current action, essentially defines the role. The player is responsible for the ball and opponents in his zone.

The ball flows from the defence zone to the attack zone according to a decision function used by each player. When the player has the ball, he checks whether he can pass the ball or shoot into the enemy goal; if not, he tries to move forward with the ball until a pass becomes possible or he can shoot. For deciding whether to pass the ball or proceed, each player, depending on his role, keeps a number for each teammate, assigning a preference score to the candidates for a pass. Thus defenders prefer to pass the ball to middle players and are not happy to pass the ball to the goalie. The evaluation function also considers, for each possible target of the pass, the *gain* in case of success and the *risk* that the pass will be intercepted. The most promising target is thus chosen and its value compared with the gain and risk of advancing with the ball.

7 Special Team Features

The core of IALP is a hierarchically structured reactive planner that computes and executes plans. There is an ordered chain of layers, with a base layer and a top layer. The base layer is devoted to the communication with the RoboCup server: the output are commands like (dash *speed*) or (turn *moment*). The top layer defines the overall strategy of a player; it contains the most abstract plans and fully determines the behaviour of the agent. The intermediate layers define a hierarchy of actions: each layer decides upon the implementation of an action using the actions offered by lower layers.

A plan built in a layer is a list of actions defined in one of the layers below. A *while* action can be used to repeat a sequence of actions until a specified condition is verified.

At each cycle, the interpreter of plans requests an action in executable form to the base layer; if this layer is executing a plan, the next action of the plan is executed. Otherwise (it has finished executing the previous one), it requests a new plan to the upper layer. This chain of requests may propagate to the top layer, which must always return an appropriate plan. The architecture of the planner is shown in figure 1.

Since the top-level planner determines the behaviour of the underlying planners, specific abilities implemented by lower levels may be reused for building different roles. In particular the layered approach is convenient for sharing low level abilities that all players should possess.

Each layer can request to reset the executing plans to upper and/or lower layers. This is important to implement reactive behaviours and in particular to react promptly to referee messages.

Another feature of the IALP planner is the possibility of defining several alternative implementations for an

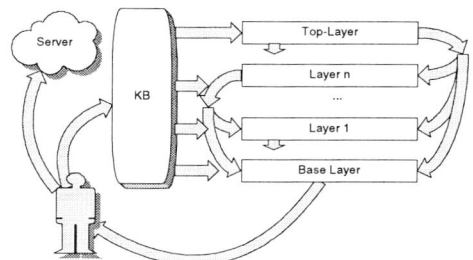

action, all of them considered equivalent with respect to the outcome. In this case the interpreter chooses randomly the implementation to be used. This feature introduces richness of behaviour and makes it difficult for an opponent team to guess the behaviour of players.

8 Conclusion

We are working on a newer version of the IALP team built on the current architecture and we hope to compete in the next European event. Experience with past competitions suggested rewriting in C both the communication level and the memory model for better performance. Moreover we plan to use a machine learning approach for improving individual and coordination skills of the players.

References

1. Attardi G., The Embeddable Common Lisp, ACM Lisp Pointers, 8(1), 30-41, 1995.
2. Bowling, M., Stone, P., Veloso, M., "Predictive Memory for an Inaccessible Environment", In *Proceedings of the IROS-96 Workshop on RoboCup*", November 1996.
3. Franklin, S., "Coordination without Communication", http://www.msci.memphis.edu/~franklin/ coord.html
4. Kitano, H., Asada, M., Osawa, E., Noda, I., Kuniyoshi, Y., Matsubara, H., "RoboCup: A Challenge Problem for AI", AI Magazine, Vol. 18, No. 1, 1997.

Kappa-II

NODA, Itsuki

CSLI, Stanford / ETL

1 Introduction

In order to realize flexible strategic planning in multi-agent systems that are working in dynamic environment, it is necessary to provide a mechanism to integrate hierarchical planning (include team planning) and reactive behavior. The main issues of this integration are:

- how to switch the context of plan
  In the dynamic environment, it is important how to terminate making and executing a plan when the environment changes so that the plan is not useful any more.
- how to organize multiple planning
  It is better that agents in a complex environment can have ability to making multiple planning, because such agents may have multiple goals in the same time. For example, in the case of soccer, while agents have an obvious goal "win the game (or score goals)", also the agent should have another instinctive goals, that is "not to miss their position", "follow the rule", and so on, in the same time. The similar requirement will be happen when agents try to make a consensus by communication during they were acting something. So, it will make the problem simple that the agent has parallel planning process, that is action planning and communication planning.

In order to attack these issues, we are proposing a programming language called Gaea and programming methodology on it.

2 Team Development

Team Leader: NODA, Itsuki
Team Members:
  NODA, Itsuki
   – Stanford Univ. / ETL
   – USA
   – visiting scholar
   – attended
  Damon Otoshi
   – Stanford Univ.
   – USA
   – under graduate student

- not attended
Stanley Peters
- Stanford Univ.
- USA
- Professor
- not attended
Web page http://ci.etl.go.jp/ noda/research/kyocho/kappaII/

3 World Model, Communication, and Skills

The World Model of kappa-II is simple. They scan visual information from the server and calculate their global positions using a revised version of `libsclient`. Then, it calculate positions of visible objects. In order to handle the ball smoothly, they also simulate movements of the ball in few steps ahead. Using this world model, a player calculates suitable parameters of turn, dash and kick commands.

Our players use the communication to require passes to a ball player. When a team is offending (keeping the ball) and a player A is keeping the ball, another player B requires if player B is locating a good position to receives the ball. Then player A can decide whether he should pass the ball to player B or not by itself.

For lower level of skill to handle the ball, we refered the code of CMUnited. Especially, circle-kick skill and intercept skill is based on the code. For higher level, we use multi-thread logic programming to behave reactively to the dynamic environment. The details are described below.

4 Strategy

As discussed in the introduction, the main issue of agent programming is:

> how to combine hierarchical planning and reactive behavior

In order to solve this issue, SOAR[5], that is used STEAM system in ISIS team[6] uses hierarchical structure of operators. In SOAR, the hierarchical structure are realized by asserting structured state data to identify where a system is extracting the plan in a decision tree. Then each rule recognizes the state of decision process by checking the state in its condition part.

Out approach takes the different realization using the similar representation of hierarchy. When a operator is selected, then the system forks a thread for execution of the operator. The original thread remains and continue to check the condition is satisfied. The new thread checks conditions of all sub-operators in the case the operator is complex operator, or execute action. In other words, All operators on a path in a decision tree becomes threads and keep alive while the path is active.

In addition to it, the system allows a parallel planning. If a operator is denoted as *multiple*, then all sub-operators are forked immediately, and keep alive until the mother operator is alive.

The merits of this architecture:

- We can realize quick and slow planning by controlling the priority of execution of each thread. Basically, high level operator does not change so quickly, while low level operator should be checked frequently. In order to realize such features, we can specify sleeping durations of each cycle of execution of thread.
- Using *multiple* operators, we can easily realize instinctive processes. Instinctive processes should always be alive and continue to check the condition. *multiple* operators enables to realize such instinctive processes in a hierarchical manner. In order to reduce the cost of such checking,

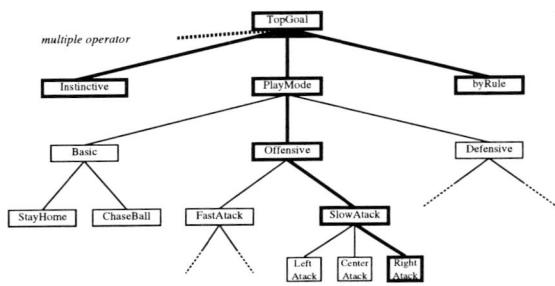

Fig. 1. Example of a tree of operators

5 Special Team Features

We use Gaea to implement our players. Gaea is a logic programming system [2, 3, 4], which has the following features:

- prolog style *logic programming* system
- *multi-thread* with flexible control mechanisms
- *dynamical program manipulation*

Because Gaea is *logic programming*, it is easy to make a translator that decomposes a hierarchical planning description to lower reactive rules. Moreover, such rules can be evaluated in parallel using *multi-thread* features. Therefore, it is easy to realize parallel processes to behave reactively or intensionally like subsumption architecture [1]. *Dynamical program manipulation* enables to combine program modules for each thread and to change the behavior of agent.

Compared with the previous RoboCup competition, Gaea has been improved in the following features:

- light weight multi-thread generation
- rich control mechanism of waiting event

6 Discussion

Compared with reactive production systems, the proposed architecture has an advantage in parallel process. In both systems, all behaviors are described in reactive rules. In the usual production systems like SOAR, these rules are checked uniformly in a single selection-application cycle. So, in every cycle, all conditions of operator should be checked. On the other hand, in the proposed architecture, the system can have multiple processes for each node of the operator tree. This enables to realize *lazy checking mechanism*. Usually, low-level operators should be tested almost every execution cycle, while conditions of high-level operators may not checked so frequently, in other words, the conditions can be checked *lazily*. In the proposed architecture, users can specify the length of interval that each process take a sleep in each cycle. So, we can realize such *lazy checking mechanism*.

Compared with the subsumption architecture[1], the proposed system has an advantage in dynamics of architectures. The problem of subsumption architecture is its static structure to subsume lower behavior modules. Because of the static structure, all combination of planning should be generated before the execution, and define the subsuming relation of each modules. In Gaea, we can realize similar subsuming mechanism by overriding definitions of predicates by manipulation of program in real-time. So, we need not prepare all combination of plan before running the agents.

References

1. Rodney A. Brooks. Intelligence without representation. Technical Report Tech. Rep., MIT, 1988.
2. Hideyuki Nakashima, Itsuki Noda, and Kenichi Handa. Organic programming for complex systems. In *Proc. of Poster Session of Fifteenth International Joint Conference on Artificial Intelligence*, page 76. IJCAI, Aug. 1997.
3. Itsuki NODA. Agent programming on gaea. In *Proc. of The First International Workshop on RoboCup*, pages 147–150. IJCAI, Aug. 1997.
4. Itsuki NODA. Kappa: Agent program by gaea. In Minoru Asada, editor, *RoboCup-98 (Proc. of second RoboCup Workshop)*, pages 387–392. The RoboCup Federation, July 1998.
5. Soar manual. http://bigfoot.eecs.umich.edu/~soar/.
6. Milind Tambe. Towards flexible teamwork. *Journal of Artificial Intelligence Research*, 7:83–124, Sep. 1997.

Karlsruhe Brainstormers - Design Principles

M. Riedmiller, S. Buck, A. Merke, R. Ehrmann, O. Thate, S. Dilger, A. Sinner,
A. Hofmann, and L. Frommberger

Institut für Logkik, Komplexität und Deduktionssyteme
University of Karlsruhe
D-76128 Karlsruhe, FRG

Abstract. The following paper describes the design principles of decision making in the Karlruhe Brainstormers team that participated in the RoboCup Simulator League in Stockholm 1999. It is based on two basic ingredigents: the *priority - probability - quality (PPQ)* concept is a hybrid rule-based/ learning approach for tactical decisons, whereas the definition of goal-orientented *moves* allows to apply neural network based reinforcement learning techniques on the lower level.

1 Introduction

The main interest behind the Karlsruhe Brainstormer's effort in the robocup soccer domain is to develop and to apply machine learning techniques in complex domains. Especially, we are interested in Reinforcement Learning methods, where the training signal is only given in terms of success or failure. So our final goal is a learning system, where we only plug in 'Win the match' - and our agents learn to generate the appropriate behaviour. Unfortunately, even from very optimistic complexity estimations it becomes obvious, that in the soccer domain, both conventional solution techniques and also advanced today's reinforcement learning techniques come to their limit - there are more than $(108 \times 50)^{23}$ different states and more than $(1000)^{300}$ different policies per agent per half time. The following describes the modular approach of the Brainstormer's team to tackle this complex decision problem.

2 The Decision Module

The task of the decision module is to compute in each time step a new basic command (i.e. *kick, turn, dash*) that is sent to the server. This command depends on the current situation s_t, which is provided by the world model module (not discussed here). As already discussed in the introduction, it is a very hard problem to do decisions at the level of basic commands.

An obvious approach - which is used in most of the known approaches in various variations (e.g. [3]) - is to introduce two levels of the decision making process. The lower level implements some useful basic skills of an individual player (for example, intercept a rolling ball). In our framework, such basic skills are called *moves* (in analogy to other strategic games as chess or backgammon). The *second level* is realized by the *tactics module*. Its task is to select one of the moves, depending on the situation. An appropriate choice of moves should finally lead to success in terms of scoring a goal. Also, aspects of team play are realized here.

2.1 The Moves

A move is a sequence of basic actions, that transforms a current situation $s(0)$ into a new situation $s(t)$ some time steps later. The resulting situation is one of a set of terminal states $\mathcal{S}^f$, which might be either positive/ desired outcomes ($\mathcal{S}^+$) or negative/ undesired situations ($\mathcal{S}^-$). The move ends, if either a terminal state is reached ($s(t) \in \mathcal{S}^f$), or the time exceeds a certain limit ($t > t_{max}$).

For example, the move *intercept-ball* terminates if either the ball is within the player's kickrange ($\mathcal{S}^+$) or if it encounters a situation, where it is no more possible for the player to reach the ball ($\mathcal{S}^-$).

Since each move has a clearly defined goal, it is now possible to find sequences of basic commands, that finally reach the defined goal. This can be done either by conventional programming, or, as it is the case in our approach, by reinforcement learning methods. In both cases, it is important that the goal of a move is reasonably chosen, that means that the solution policy is not too complex (e.g. a move 'win that game' would be desirable but its implementation will be as complex as the original problem).

Clearly, the quality and the number of different moves eventually determines the power of the individual player and therefore the whole team respectively.

2.2 Reinforcement Learning of Moves

The question now is how to implement a closed-loop policy that, after emitting a sequence of basic commands finally reaches the specified goal of the move? The above move definition directly allows to formulate the problem of 'programming' a move as a (sequential) Reinforcement Learning (RL) problem. The general idea of reinforcement learning is that the agent is only told, what the eventual goal of its acting is. The agent is only provided with a number of actions, that it can apply arbitrarily. In course of learning, it should incrementally learn a (closed-loop) policy, that reaches the final goal increasingly better in terms of a defined optimization criterion. Here we apply Real-Time Dynamic Programming methods [1], that solve the problem by incrementally approximating the optimal value function by repeated control trials. A feedforward neural network is used to approximate the value function [2].

In the current version which was used in Stockholm, the *kick*-move was learned by reinforcement learning. Several other teams have reported tricks how to implement a kick-routine by conventional programming using various heuristics. The problem with this approach is that it can be very time-consuming to find the right heuristics and to tune several parameters by hand. Instead, our reinforcement learning approach is much more convenient to handle - the work of looking for an appropriate policy is done by the agent/ computer itself. The agent is provided with a (finite) number of basic kick commands. The goal is defined in terms of a target ball direction and a target ball velocity. The agent receives costs for every kick command it uses until the ball has left its kickrange. If the ball is lost during the sequence, maximum costs occur; in case of success, the sequence is terminated with 0 costs [2]. This formulation results in a timeoptimal policy: the number of kicks until successful termination is minimized. After about 2 hours of learning, the resulting policies were quite sophisticated - similar to the proposed heuristics, the agent learned to pull the ball back and to accelerate it several times in order to produce high speeds. It was able to learn to accelerate the ball to speeds up to 2.5 m/s. Of course, it is too difficult for a single policy (i.e. a single neural net) to manage to kick in all situations to all directions with all imaginable velocities. Instead, the problem was divided into 54 subproblems; therefore the neural kick-move now is based on 54 neural networks (each of them using 4 inputs, 20 hidden and 1 output neuron).

2.3 The Tactics Module and the PPQ approach

The task of the tactics module is to select one out of the set of available moves. The difficulty with this decision is, that in general, a complex sequence of moves has to be selected, until the final goal is achieved, because normally a single move will not lead to scoring a goal. In the soccer framework, this problem becomes even worse, since the success also depends on the behaviour of the whole team - a successful sequence can only be played, if all the agents involved make the correct decisions. Although we already started some promising experiments applying reinforcement learning to this decision level also, we are still some theoretical and practical steps away from a convincing practical solution (other teams also work on this topic [3]).

For our Stockholm competition team, we therefore worked on a different solution for the tactics module, which we call the *priority - probability - quality (PPQ)* approach. The idea origins in the observation, that some parts of the problem can be elegantly solved by simple rules, whereas other aspects are not so easily judged. The PPQ approach tries to combine the worlds of programmed and learned parts.

Priority Classes and Qualities The moves are partitioned into a number of classes, e.g. the class of *goal shots*, the class of *pass plays*, the class of *dribblings*. It is now relatively easy to define a reasonable priority ordering between these classes. For example, in our Stockholm approach, we used the following priorities: *1. shoot to goal, 2. pass forward, 3. dribble, 4. pass backward, 5. hold ball.*

If there are several choices of moves within a priority class, a *quality* function decides which move to chose. Conceptually, this quality function typically follows a very simple decision rule, for example pass to the player that is closer to the goal.

Learning of success probabilities

Each move has a certain probability of success, which depends on the current situation. The idea now is to learn this probability by a simple trial and error training procedure. The agent is set into various situations, executes a certain move and notes the success or failure of the move. The learning task now is to associate each situation with its outcome, for example in terms of a '1' for success and a '0' for failure. A feedforward neural network is used here to learn the training patterns. After training, the neural network, gets a certain situation as its input and outputs the *expected* value for success/ failure.

The decision algorithm

Each priority class has a set of available moves, $\mathcal{M}(i)$. The algorithm works through all the priority classes, until it finds one, where there is a move that has a higher probability of success than a certain threshold. This set is called $\mathcal{M}^+(i) := \{m | P_{NN}(m) \geq \theta_i, m \in \mathcal{M}(i)\}$. The threshold is selected such that a reasonable chance of success is given, for example $\theta_i = 0.8$. If there is more than one such move, one of them is selected by judging its quality. Note that this final judgment can be treated very relaxed, since it already is a nearly maximal useful move (determined by the priority of its class) and its also very likely a successful move (determined by its high success probability).

To guarantee termination of the algorithm, at least one class must exist, where $\mathcal{M}^+(i)$ is not empty. This is done by the definition of a default move that is always possible.

3 Conclusion

The Stockholm version is an intermediate step within our Brainstormers' concept of a learning agent. The final goal is to have an agent, which has learned its fundamental decision behaviour by reinforcement learning. However, until then a lot of work has to be done in the field of multi-agent RL, on Semi-Markov Decision Processes, partially observable domains (POMDPs) and on large-scale RL problems. Some of very recent RL ideas have already been successfully realized. For example, the moves-concept is closely related to Sutton's et.al 'options'-framework [4]. Therefore our work can be regarded as realizing and testing some conceptual ideas in a practical environment. The Brainstormer's Stockholm agent used an ensemble of 67 feedforward neural networks, 54 for the neural kicking (RL) routine, and 13 as probability networks in the tactics module.

References

1. A. G. Barto, S. J. Bradtke, and S. P. Singh. Learning to act using real-time dynamic programming. *Artificial Intelligence*, (72):81–138, 1995.
2. M. Riedmiller. Concepts and facilities of a neural reinforcement learning control architecture for technical process control. *Neural Computing and Application*, 1999.
3. P. Stone and M. Veloso. Team-partitioned, opaque-transition reinforcement learning. In M. Asada and H. Kitano, *RoboCup-98: Robot Soccer World Cup II*.
4. R. S. Sutton, D. Precup, and S. Singh. Between mdps and semi-mdps: A framework for temporal abstraction in reinforcement learning. *Artificial Intelligence*, 1999. to appear.

Kasugabito III

Tomoichi Takahashi

Chubu University, 1200 Matsumoto, Kasugai, Aichi 487-8501, Japan

1 Introduction

Its on-line coach agent characterizes Kasuga-bito III. Kasuga-bito III is composed of LogMonitor [LogM]and Kasuga-bito II, which was runner-up in JapanOpen'98 and the champion in the RoboCup Pacific Rim Series '98. The LogMonitor advises their position to players. The positioning strategy is authenticated by analysis of the logfiles of evaluations at RoboCup'98. Our player agents are advised by the on-line coach and changes their formation.

2 Team Development

Team Leader: Takahashi Tomoichi
Team Members:
   Kazuaki Maeda, Shinji Futamase, Akinori Kohkestu
   – Chubu University
   – Japan
   – lecturer, master course student, master course student
   – did not attend the competition
Web page http://kiyosu.isc.chubu.ac.jp/robocup/

3 World Model

Player agents do not have a global field model. They have only a local world model created from *see* information from the soccer server. The model is updated as they receive *see* information. The on-line coach agent has a global field model that holds data for the most recent simulation steps (default 100 steps).

4 Communication

There is no communication between player agents. The on-line coach agent advises player agents when the ball is out of bounds.

5 Skills

Our player agents are basically the same as Kasugabito-II used at RoboCup '98. They predict the ball trajectory for the next steps (default 15 steps). If an agent can catch the ball within the steps, it tries to intercept the ball. It dribbles the ball when there is no teammate to which it can pass the ball. Dribble is combination of a kick and a dash. We do not have a special goalie. Our goalie's parameter are values gained from off-line training [Pricai98]. Defender moves to the centerline to get an offside trap.

6 Strategy

The on-line coach is introduced from soccer sever version 5.06. The coach agent can see all objects on the field and gives advice to the players at *play_off* time. The coach's advices are generated by the followings factors:

defender (offside trap advice): When opponent players are within 10 m and their direction are toward our goal, the coach advises the defenders move 10m forward.
(position back advice): After the defender moved, if there is no opponent within 10m in their back, the coach advises them to return to their default position.
forward The top forward is advised to move 5m behind the opponent offside line.
others The other players are advised that they are proportionally spread between the top forward and the defenders.

The positions of opponent players are calculated in the following steps:

- Players far from the ball are in their default positions. The players more than 30m far away from the ball are objects to be calculated, because players near the ball are assumed to catch it. The catching motion is not a usual one.
- The newer data is supposed to be the more certain. (The players may change their position as the game goes on.) The average of the most recent 100 points which satisfies the above conditions are set as the opponent players' position.

Table 1 shows the scores of the games against three team without and with the coach agent. The three teams are AT_Humbolt97, Andhill98 and Kappa. Andhill98 is known as the team that changes players' position during the game by reinforcement learning. The version with a coach won games for Andhill98 more than the version without a coach. This result supports the coach advice's effective in robustness.

7 Special Team Features

Kasugabito-III's feature is its on-line coach. Kasuga-bito III's coach is the im-

Table 1. Scores of games without /with coach.

AT_Humbolt		Andhill98		Kappa	
without	with	without	with	without	with
1 - 6	0 - 7	3 - 1	7 - 1	0 - 6	0 - 10
0 - 7	0 - 11	2 - 2	3 - 4	0 - 9	0 - 7
0 - 8	1 - 8	3 - 1	1 - 4	0 - 8	0 - 13
0 - 10	0 - 6	3 - 3	1 - 4	0 - 8	0 - 5
0 - 8	0 - 8	1 - 2	4 - 1	0 - 5	0 - 5

right score is KasugabitoIII's score.

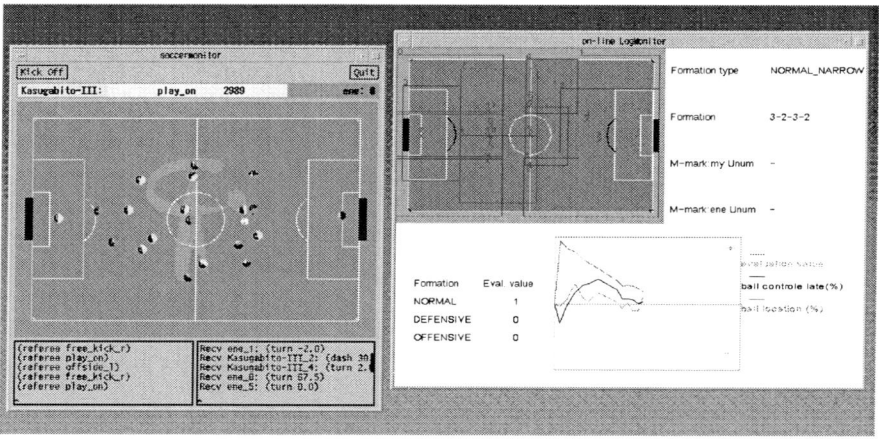

Fig. 1. On-line LogMonitor screen image

proved LogMonitor which analyzes the game on simulation cycle (Fig.1) [LogM]. The left window is the normal viewer, and the right window displays the on-line coach's status. Blocks on the field displayed upper left of right window show the range of players' movement. Line graphs show the ball location, the ball control rate, and heuristics values that indicate the coach agent thinks how the game is going.

The coach agent used at Stockholm generated advices by the follows principles:

formation The coach agent directs teammate agents to take an offensive formation, a normal formation and a defensive formation. It evaluates the game's trend based on the number of kicks and scores. When the gap of scores is more than one, the coach advises team mates for the first time. The first advice is to change to an offensive formation if the team leads the opponent, otherwise to change to a defensive one.

The coach's advice from the second time is evaluated its effectiveness by heuristics.

vertical spread The three formations - offensive, normal and defensive - have three variations - narrow layout, normal layout and wide layout. The coach directs the layout based on the average position and vertical variance of the ball during the previous period.

8 Conclusion

We were eliminated at Group H, where CMUnited 99 and Essex passed the league. The reason is the difference of agents' basic ability between out team and others are bigger than the coach's advices.

A new coach agent has been researched for improving the old one. The games at RoboCup '98 indicated that the player agent ability is required to make use of the coach advice.

References

[Pricai98] K. Maeda, A. Kohketsu, T. Takahashi: Goal-Keeping Skills in Soccer Simulation Games, Proc. RoboCup Workshop, PRICAI'98, pp.96-101, 1998

[LogM] T. Takahashi: LogMonitor: from player's action analysis to advice of adversarial model, this book.

RoboCup-99 Simulation League: Team KU-Sakura2

Harukazu Igarashi, Shougo Kosue, Takashi Sakurai

Kinki University, Higashi-Hiroshima, Hiroshima, 739-2116, Japan

Abstract. In this paper we describe our team, KU-Sakura2, which is to participate in the simulation league of RoboCup-99 Stockholm. KU-Sakura2 is characterized by soccer agents that make tactical plays and passes using communication between players.

1 Introduction

Robot soccer is one of the relevant candidates for the standard challenging problems in Artificial Intelligence. Our two teams, Team Miya and Team Niken, participated to the simulation league of RoboCup97 Nagoya (Japan, August 1997) [1]. Moreover, we sent Team Miya2 to the simulation league of RoboCup98 Paris (France, July 1998) [2]. Team KU-Sakura2 is an improved version of Miya[3] and Miya2[4]. In this short paper, we give a brief technical description of our Team KU-Sakura2.

Team Miya was characterized by individual tactical play[3]. Individual tactical play do not require communication between players, so the speed of passing was rapidly increased in RoboCup 97 games, and the team sometimes behaved as if it had been taught some tactical play. Team Miya proceeded to the quarterfinal match and was one of the best eight teams in the simulator league.

In Team Miya2, a kind of communication between players is realized by using a "say" command so that a passer can make a pass to a receiver without looking around for receivers[4]. Consequently, Team Miya2 was one of the best sixteen teams in RoboCup98.

However, more tactical play is required for the following two reasons. First, top teams of RoboCup98 showed very high-level skill in individual play. For examples, we observed a speedy dribble keeping the ball near the player's body and a safety pass without being intercepted by the opponent players. Second, the offside rule was introduced at RoboCup98. Thus forward players have to check whether they are in an offside position or not at all times. Some tactics is necessary to avoid the opponent's offside trap and succeed an offside trap against the opponent team. We use communication between players for realizing the tactics in Team KU-Sakura2.

2 Hierarchy of Actions

In Team KU-Sakura2, there is a hierarchy of actions. Actions are generally classified into four levels: strategy, tactics, individual play and basic commands(Table 1). A higher-level action includes more players and requires information in a wider range of time and space than a lower-level action. Coradeschi et al.[5] and Tambe[6] expressed the relationship between actions as a decision tree. We call such a decision tree an *action tree*. A soccer agent selects an action from the action tree at each action cycle by analyzing visual and auditory information and by considering the agent's current state. The action is then *compiled* into a series of basic commands: kick, turn, dash, catch and say.

As shown in Table1, "individual tactical play" is introduced to reduce the delay time between decisions and actions. The *individual tactical play* is defined as an action that an individual plays in a specific local situation without communication from a teammate. However, an agent expects some cooperation from a teammate in individual tactical play. For Team KU-Sakura2, we implemented three actions as individual tactical play: the safety pass, the centering pass and the post play. These three plays speed up the tactical actions of the safety pass between two players, the centering pass from a wing player, and the post play of a forward player.

Table 1. Hierarchy of actions

	Action	Definition	Examples
Level 4	Strategy	Cooperative team action	Rapid attack, Zone defense
Level 3	Tactics	Cooperative action by a few players for a specific local situation	Pass with communication
Level 2	Individual tactical play	Action of an individual player for a specific local situation without communication, but expecting cooperation from a teammate	Safety pass, Post play, Centering pass
	Individual play	Individual player skill	Pass, Shoot, Dribble, Clear
Level 1	Basic command	Basic commands directly controlling soccer agents	Kick, Turn, Dash, Catch, Say

3 Action Tree

According to the role given to the agent, each agent has its own action tree based on the hierarchy shown in Table 1. An agent's next action is specified by prioritized rules organized into its own action tree. An example of an action tree, which is used in Miya2, is shown in Fig. 1. Here, if the node offense is selected, the firing conditions of action nodes at levels 2 and 3 are checked. The knowledge of selecting actions at levels 2 and 3 are expressed as if-then rules in a C program. Examples of the firing conditions include whether there are opponent players nearby, whether the player can kick the ball, whether the ball is moving, whether the player can estimate his position correctly, and whether the player can see the ball. In addition to the if-then rules, some actions at levels 2 and 3 are prioritized.

4 Safety Pass and Safety Kick

The actions of level 2 are not unrelated to one another. The actions, shoot, centering pass, post play, dribble and clear, consist of two basic skills[4]: the safety pass and the safety kick. The *safety pass* is a skillful pass to a receiver so that it is not easily

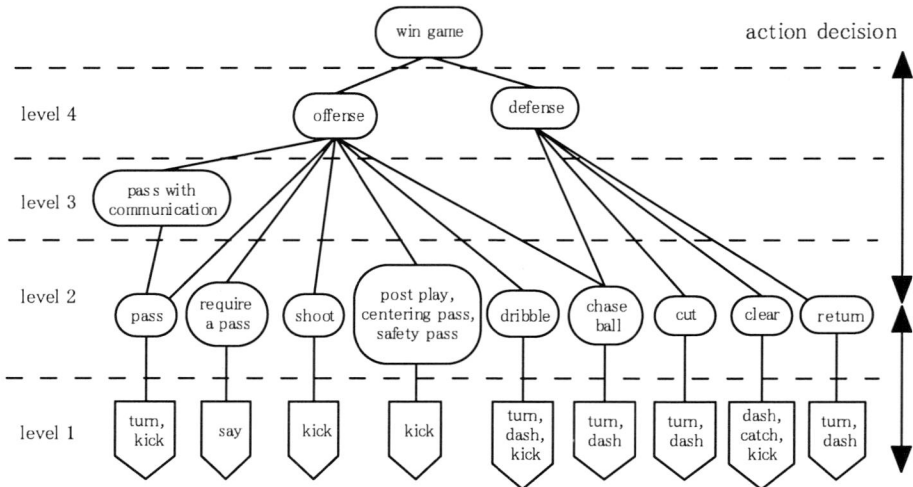

Fig. 1. Action tree used in Team Miya2

intercepted by the opponents. The *safety kick* is a skillful kick, which eludes interception by the opponents, in the direction of the objective. We call these two kinds of play *individual tactical play*.

5 Team Play Using Communication between Players

5.1 Team play in defence

The goalkeeper is a commander who orders defence players to go forward or backward. In Fig.2, the goal keeper is denoted by G and defence players are denoted by D. The optimal position of the defence line is determined by the goalkeeper taking positions of the opponent forward players into account.

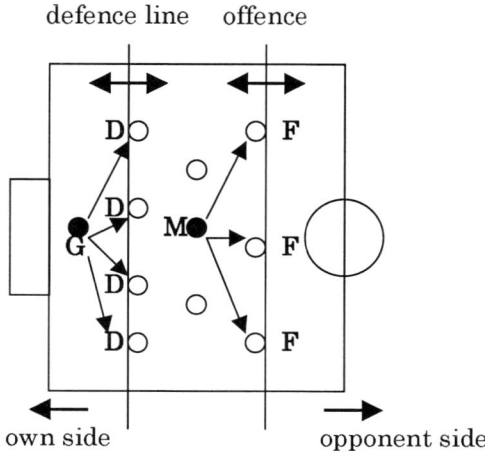

Fig.2. Movement of the defence line and the offence line.

5.2 Team play in offence

A midfielder in the center of a team, denoted by M in Fig.2, is a commander who orders forward players to go forward or backward. The optimal position of the offence line is determined by the midfield player taking positions of the opponent defence players into account.

5.3 Experiment

We ran 30 simulation games between Team KU-Sakura2 and Team Miya2. No team play mentioned in 5.1 or 5.2 is implemented on Miya2. The time length of each game is 3000 simulation-cycle steps. The results of the games are shown in Table 2.

In Table 2, one finds that the frequency of offside in Miya2, 6.77 per game, was reduced to 1.43 in KU-Sakura2. Moreover, this reduction of offside contributed to increase of scoring from 0.40 to 0.67 per game, and winning from 5 to 10 wins. The results prove that our team play is effective in actual simulation games.

Table 2. Experimental results of 30 games between KU-Sakura2 and Miya2

	Offside		Score		Win	Loss	Draw
	total	avr.	total	avr.			
KU-Sakura2	43	1.43	20	0.67	10	5	15
Miya2	203	6.77	12	0.40	5	10	15

6 Summary

Team KU-Sakura2 has a hierarchy of actions. According to the role given to the agent, each agent has its own action tree based on the hierarchy. KU-Sakura2 is characterized by individual tactical plays at level 2 and tactical plays using communication between players at level 3 of the hierarchy of actions.

References

[1] http://www.robocup.v.kinotrope.co.jp/games/97nagoya/311.html
[2] http://www.robocup.v.kinotrope.co.jp/games/98paris/312.html
[3] Igarashi, H., Kosue, S., Miyahara, M., Umaba, T.: Individual Tactical Play and Action Decision Based on a Short-Term Goal -Team descriptions of Team Miya and Team Niken-. In: Kitano, H.(ed.): RoboCup-97: Robot Soccer World Cup I, Springer-Verlag(1998)420-427
[4] Igarashi, H., Kosue, S., Miyahara, M.: Individual Tactical Play and Pass with Communication between Players -Team descriptions of Team Miya2-. Proc. of RoboCup98(submitted).
[5] Coradeschi, S., Karlsson, L.: A decision-mechanism for reactive and cooperating soccer-playing agents. Workshop Notes of RoboCup Workshop, ICMAS 1996
[6] Tambe, M.: Towards Flexible Teamwork in RoboCup. Workshop Notes of RoboCup Workshop, ICMAS 1996

The magmaFreiburg Soccer Team

Klaus Dorer

Centre for Cognitive Science
Institute for Computer Science and Social Research
Albert-Ludwigs-University Freiburg, Germany
klaus@cognition.iig.uni-freiburg.de

1 Introduction

The main interest of our research concerns **m**otivation **a**ction control and **g**oal **m**anagement of **a**gents (magma). Action Control of the magmaFreiburg team is based on extended behavior networks, which add situation-dependent motivational influences to the agent, extend original behavior networks to exploit information from continuous domains and allow concurrent execution of behaviors. Advantages of the original networks, such as reactivity, planning capabilities, consideration of multiple goals and its cheap calculations are maintained.

magmaFreiburg has been very successful in the competition finishing at second place. We scored 59:0 goals and 12:0 points in the four games of the round robin and 30:11 goals in the six games of the elimination round with all goals against us scored by the winning team CMUnited.

2 Team Development

Team Leader: Klaus Dorer
Team Members:
    Markus Plewinski, Marc Haas
    – Fachhochschule Furtwangen
    – Germany
    – students
    – did not attend the competition
Web page http://www.iig.uni-freiburg.de/cognition/members/klaus/robocup/magmaFreiburg.html

3 World Model

Each time an agent receives a perception from the server the information is entered into a *local map* containing the distances and directions of visible objects. After self-localization, the global position and direction of the agent and all visible objects are inserted into a *global map*. Information of non-visible objects gained by communication (see next section) is also entered into the map. The information in the map is updated before action selection, taking into account information on the expected effects of the agents previous actions and the inertia of the ball and the agent. Moveable objects like other players and the ball are removed from the map after three seconds or if they are not seen at the expected position. Functional objects are calculated from the information in the global map, using indexical-functional aspects of the situation [1]. This reduces the number of competence modules needed (see section 7).

4 Communication

Communication is used by the magmaFreiburg agents to share information on visible objects and the agent's internal information. This allows agents to improve information on visible players and to know about the position of players the agent can not see. Knowledge about the stamina of other agents allow players to replace tired teammates. To coordinate communication between agents, a locker-room agreement is used [6].

5 Skills

Players anticipate the future positions (20 cycles) of the ball with respect to its current velocity. They calculate the possible intersection point by taking into account the number of cycles the agent needs to approach the corresponding ball position. When receiving a ball, the agent takes into account the ball's speed to calculate the proper power vector for kicking. If the velocity of the ball is low and the agent decides to do a hard kick, the agent tries to place the ball in a position, where it can kick the ball twice. Due to poor dribbling abilities, the ball is lost in one third of all attempts. The goalie always keeps the ball in view. This is done by turning the neck (as other agents do) if moving sideways to the ball and by moving backwards if moving away from the ball. The goalie also tries to position itself in a way to minimize the attacker's angle to the goal.

6 Strategy

The player with the ball has choice of four behaviors: it can dribble with the ball, pass the ball to a teammate, kick towards the goal or just clear the ball in any direction. The mechanism used for behavior selection is described in the next section. Movement of defenders and midfielders without the ball is restricted to moving forward in order to create an offside trap if the ball is in front of the offside line, and moving backward if the ball is behind them. Midfielders and offenders without the ball try to keep themselves onside and try to keep at the level of the ball within the opponents half when attacking.

7 Special Team Features

Behavior selection of our agents is controlled by *extended behavior networks* [2, 3] that are based on work by Maes [4, 5]. Extended behavior networks consist

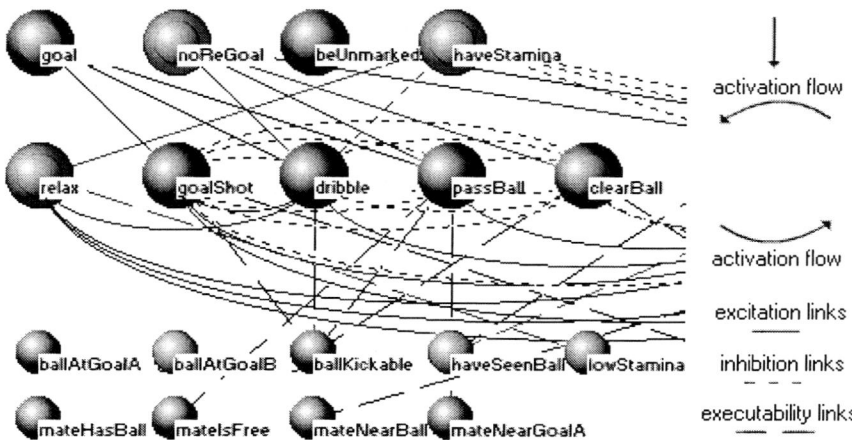

Fig. 1. Part of a behavior network used by a soccer-agent. The goals of the agent are at the top level, in the middle the competence modules and at the bottom level the situation propositions (perceptions). (The complete network contains 14 competences).

of the goals of an agent, a set of behavior rules called competence modules, the perceptions of the agent and resource nodes (see fig. 1).

Goals represent the utility of propositions that are part of the goal condition. Goals can be statically prioritized by their importance and can be dynamically, i.e. situation-dependent, prioritized by their relevance condition. The utility of a goal is calculated as the product of its importance and its relevance.

A competence module consists of the preconditions that have to be satisfied for the module to be executable, a corresponding behavior, the effects expected after behavior execution, the resources used by the behavior and an activation value. The activation of a competence module can be interpreted as the expected utility of the module's behavior with $eu = \sum_i a_i \cdot ex_i$, where a_i is the utility of effect i and ex_i is the probability of effect i to become true. The utility of effects that are part of a goal condition can be directly accessed by links from the goal to the competence module. The utility of propositions that are not part of a goal condition can be calculated by utility propagation using links between competence modules. Any unsatisfied proposition of a precondition is assigned a utility corresponding to the expected utility of the competence module. Preconditions of competence modules with a high expected utility get important subgoals of the network. For a more detailed description on utility propagation see [2]. The execution of a competence module depends on its executability, the expected utility and the availability of needed resources. Modules with high expected utility are preferred.

Perceptions represent the truth values of propositions in a domain. To improve the quality of perception within continuous domains, real-valued propositions have been introduced by extended behavior networks. This has implications on the executability of competence modules, which becomes real-valued, and on the relevance of goals, which can have continuous values. Empirical results show that real-valued propositions improve the quality of behavior selection in the RoboCup domain [2].

Resource nodes are used to coordinate the selection of multiple concurrent behaviors. Competence modules are connected with the resource nodes that correspond to the resources they use. Using these links a competence module can make sure that enough resources are available to execute the corresponding behavior and that it is the module with the highest utility requesting the resource. Concurrent actions like speaking, turning the neck and dashing have been realized by this domain independent mechanism for concurrent behavior selection.

Behavior selection in extended behavior networks is extremely cheap to calculate. All eleven agents of the magmaFreiburg team have been run on a single PC while other teams used up to five PCs. Since behavior selection can be calculated locally in each competence module, calculations can be done in parallel and could even be improved if each node of the network were run on its own processor. Besides being reactive, extended behavior networks also prefer goal-directed behavior by calculating the expected utility of a behavior with respect to the goals. In contrast to purely reactive approaches, goals of an agent can be explicitly specified.

8 Conclusion

The success in RoboCup99 has encouraged our team to take part in Melbourne 2000. Besides improvements to existing behaviors (especially ball handling and dribbling), we plan to add new behaviors to improve positioning of players not dealing with the ball and to introduce situation-dependent team strategies.

References

1. Agre, Ph., and Chapman, D. (1987). Pengi: An Implementation of a Theory of Activity. In: *Proceedings of the Sixth National Conference on Artificial Intelligence, AAAI-87*, Morgan Kaufmann, Los Altos.
2. Dorer, K. (1999). Behavior Networks for Continous Domains using Situation-Dependent Motivations. *Proceedings of the 16th International Joint Conference on Artificial Intelligence*, pages 1233-1238, Morgan Kaufmann, San Francisco.
3. Dorer, K. (1999). Extended Behavior Networks for the magmaFreiburg Soccer Team. In: S. Coradeschi, T. Balch, G. Kraetzschmar and P. Stone, *Team Descriptions Simulation League*, Linköping University Electronic Press, Stockholm.
4. Maes, P. (1989). The Dynamics of Action Selection. In *Proceedings of the International Joint Conference on Artificial Intelligence-'89*, Morgan Kaufmann, Detroit.
5. Maes, P. (1990). Situated Agents Can Have Goals. In *Journal for Robotics and Autonomous Systems*, Vol. 6, No 1, pages 49-70, North-Holland.
6. P. Stone, M. Veloso, und P. Riley. (1999). The cmunited–98 champion simulator team. In M. Asada und H. Kitano, editors, *RoboCup-98: Robot Soccer World Cup II*. Springer, Berlin.

Mainz Rolling Brains

Daniel Polani and Thomas Uthmann

Institut für Informatik, Johannes Gutenberg-Universität,
D-55099 Mainz, Germany
{polani,uthmann}@informatik.uni-mainz.de
http://www.informatik.uni-mainz.de/PERSONEN/{Polani,Uthmann}.html

1 Introduction

Our agent team is the result of a development which had to take place under tight time limitations. The total development time available was slightly less than three months where over most of the time the team developers could invest no more than a few hours per week. The code was developed from scratch to improve over the design and quality of last year's code. Thus one of the challenges was to keep a smooth development line and to avoid dead ends in the development, as well as to maintain a development environment in which a larger number of developers could work productively.

Our main challenges were twofold: to design an agent architecture which enables robust development of strategically operating agents with a larger developer team; and a lightweight implementation of agents (e.g. running the whole team on a single Ultra 1 Sun machine in Stockholm caused only a quite moderate degradation of the team's performance).

The team Mainz Rolling Brains (MRB) participated in Stockholm for the second time. Like the last year, it reached the fifth position in the total ranking, among that defeating the last year's champion team CMU '98, but succumbing to this year's winner CMU '99.

2 Team Development

Team Leaders: Daniel Polani and Thomas Uthmann
Team Members:

Christian Meyer (graduate, R)	Peter Dauscher (PhD)
Erich Kutschinski (graduate)	Tobias Jung (graduate)
Axel Arnold (graduate)	Sebastian Oehm (graduate, R)
Götz Schwandtner (graduate, R)	Frank Schulz (graduate)
Manuel Gauer (graduate, R)	Achim Liese (graduate)
Birgit Schappel (graduate, R)	Michael Hawlitzki (graduate)
Tobias Hummrich (graduate, R)	Peter Faiß (graduate)
Ralf Schmitt (graduate)	

(Each student's position is marked *graduate* or *PhD*, respectively; *R* indicates attendance at RoboCup '99)

Web page: http://www.informatik.uni-mainz.de/ANGEW/robocup.html

3 World Model

The agents are aware of their own position, of the ball position and the position of other agents, their own speed, the velocity of other agents and objects and of the age (i.e. de facto reliability) of data from objects. This is particularly important for objects not being updated by current data. The world model contains a time window of past and future world states. The past ones are filled in during the run when they are received from the server, the future ones are generated on request of the strategy unit from the most reliable known data. For every object in the game (ball or soccer player) an absolute position representation is used. The player determines his own position using the flags and reconstructs the positions of other objects combining the sensor data giving their relative positions and his own absolute position. Future positions are extrapolated by simulating the known soccer server dynamics for the ball and for the current player. Other players are simulated as not moving. The libsclient library is not used in our team.

4 Communication

Our agents only transfer minimal information about status and intention of a player that enables to improve playing quality. The fundamental doctrine of our team using communication is that it should enhance the quality of the play, but the players should in no way depend on it. Disabling communication, our players are still able to act independently, maintaining coherence of the team's play on a purely behavioural level.

Our team's messages are endowed with a checksum and slightly encoded to make sure that only our own team's messages are parsed. Though sending intentionally confusing or forged messages is deemed unfair play in RoboCup, it can still happen that without encoding the agent parser can be trapped on an innocuous message string emitted by the opponent team.

The agents send messages saying whether: they want the ball, they do not want the ball, they want to pass the ball, they are going to a given position, they are low on stamina or they estimate the offside line at a certain field value

An agent without the ball communicates that he wants the ball depending on its estimate of the situation. It can be issued by an agent if a team member is assumed to be in control of the ball. High stamina, far away opponents and being closer to the opponents goal are factors which favor an agent asking for a ball. Also the existence of a broad enough corridor between the player controlling the ball and the agent asking for the ball. The opposite factors (low stamina, close by opponents, lack of a safe corridor between sender and receiver) cause an agent to ask for not receiving the ball.

If a player intends to pass the ball, he communicates this intention, so that another player can issue a request for a ball. No handshake, however, is performed, i.e. the player does not necessarily wait for a reply.

After performing a *catch*, the goalie has the possibility to freely move to a position. By sending the *pass* message the team is informed that the goalie is indeed going to kick the ball away now. Another important situation where an agent emits a *pass* message is when a situation is considered defensive and opponents are close to the player controlling the ball.

When the available stamina is low (our agents stay always above the stamina limit which yields full recovery) *and* an opponent is close to the ball (this is a condition determined by a fuzzy rule), an agent emits the *stamina low* message. For any team player that receives the message this strongly reduces the priority of passing to the player who broadcasted it.

A player who intercepts a ball communicates his estimated interception point to his team by the *going to ball* message. This information is useful first to avoid other team members clustering around the ball and second for giving them the opportunity to optimize their positioning.

A player which has just passed should not immediately request the ball back again; this is a constellation which would be favoured by mainly reactive architectures as ours. Thus, after passing, the priority of requesting the ball back is lowered for a certain time. This, however, does not affect however the respective players behaviour: if there is danger that the ball would be lost, the player will intercede and try to secure the ball, following the doctrine mentioned above.

5 Skills

The most important higher skills available to our agents are *compound kicking*, *intercepting*, *dribbling* and *prepare kick-in*. Compound kicking includes all types of kick combinations which allow a player in reach of the ball to kick it, given a position and velocity, to some other given position and velocity (if possible). As intercepting and dribbling it may include a combination of kick and move commands to realize the desired effect. Our agent design has a strong emphasis on reactive behaviour. The mentioned compound skills, however, require processes that take place during a larger number of time steps. So, one of the most important aspects of our skill design is the realization of *pseudo-multithreads*. On the one side this yields the possibility to carry out compound actions for several time steps in a consistent way. On the other side the agents are able to interrupt a running action to either choose a different one if the situation requires a policy change or to adjust a currently running compound action.

The pseudo-multithread concept works the following way: every time step the agent considers to perform some action according to its strategy. When the agent calls a compound skill method, this calls an elementary action (like e.g. *kick*, *turn*, *dash*). The skill then returns a status value indicating whether: 1. the action was completed by the current call, 2. whether it requires further calls to be completed or 3. whether it is not possible at all to perform this action. In case 1, the skill called in the next time step can be chosen independently from the current step. In case 2, the status value indicates that the skill method has to be called again in the next time steps for the action to be complete. It also gives a reason which can be *wait* (e.g. if player intercepting a ball has not reached its target yet), *collision* which indicates that an intermediate step has been taken to avoid a collision of player and ball or *placing* which indicates that the intermediate step serves to place the ball or agent at a position useful for later kicking.

Due to this mechanism the skills always return control to the agent strategy level. This, in turn, may consider either performing an action thread until it is completed or interrupting it, as considered useful. On a *collision/placing* return status, it is usually a disadvantage to interrupt the action thread. In these cases the intermediate states may leave the ball at a position which are only adequate in a predetermined sequence of actions. For the new action sequence this position may be inconvenient.

The compound kick skill in the current version consists of a sophisticated hand-optimized sequence of ball kicks which can not be described in detail here. It attempts to perform a kick that will — from a given ball position and velocity — kick the ball to a given ball velocity in the smallest possible number of kicks. The dribbling uses a variant of the compound kick skill to kick the ball some steps ahead and follow it. It is possible to specify the average number of time steps that the player is dashing before he emits a kick. The higher this number is, the less contact the player has with the ball. It is also possible to specify a preferred direction with respect to the player where the ball is supposed to be kept.

6 Strategy

The strategy is behaviour oriented, i.e. communication is useful but not necessary for the capability of teamplay. Fuzzy predicates characterize a given situation. The predicates are then combined with fuzzy rules resulting in the respective action choice. If an agent controls the ball, it estimates the situation, i.e. the pressure by opponents, the support by team members and its own stamina state. Depending on these parameters, it will try to move the ball ahead in the field (preferably at the wings) and to kick the ball towards the centre near the opponent goal. Opponent pressure or low stamina will increase its priority to pass the ball.

If the players do not have the ball, their action is determined in part by their role (giving a preference position), in part by the current situation. A player whose *responsibility* (a fuzzy predicate) for the ball is high will dash to the ball. If his responsibility is low he will try to acquire a good position (e.g. a position giving a good opportunity for passes by a partner). If opponent players are close to the ball, players will feel responsible more easily, so that more agents will try to intercept the ball.

7 Special Team Features

In the current stage no parts developed via machine learning methods are used. Neither are opponent models used. After completion of the server communication and world model modules several independent lines of development for the strategy were taken while the agent skills module was optimized in parallel. About three weeks before the tournament a decision was taken in favour of one of the strategy development lines to be used in the official agent team. Strategical ideas from the other lines of development were then incorporated in the main line.

8 Future Work

The agent team presented here has been developed in such a way to consider the needs of a large developer group; it includes clear and well-documented module interfaces, in particular of world model and skills. This makes it amenable for further development and a participation at RoboCup 2000 is envisaged. Relevant aspects of our future work include the improvement of ball control and the positioning tactics via machine learning techniques.

NITStones-99

Kouichi Nakagawa, Noriaki Asai, Nobuhiro Ito, Xiaoyong Du, and Naohiro Ishii

Department of Intelligence and Computer Science, Nagoya Institute of Technology,
Gokiso-cho, Showa-ku, Nagoya 466, JAPAN
kouichi@egg.ics.nitech.ac.jp

1 Introduction

Since the offside rule was adopted in RoboCup-98, many teams without teamwork ability got offside penalty many times in their matches. Those teams who have dribble skill won, because most of other teams have not efficient defence strategy.

Our team focuses on a special teamwork strategy called line defence to against the offside rule. We use some basic teamwork abilities for implementing the line defence.

2 Team Development

Team Leader: Kouichi Nakagawa
Team Members:
Nobuhiro Ito
- Nagoya Institute of Technology
- Japan
- Research Associate

Kouichi Nakagawa
- Nagoya Institute of Technology
- Japan
- graduate student

3 Communication

When the ball is being close to a defender, the other defender should notify the defender that the ball is coming. It is also used by the player who is close to the ball to tell the other players the location of the ball. In addition, each player broadcast his position at regular intervals. This ability is helpful for passing the ball from a defender to a midfielder and catching the ball from the opponent offence.

4 Strategy

The offside rule makes the soccer agents have to consider new offensive strategies and new defensive strategies.

The line defence is an efficient defensive strategy which means that a group of defenders locate at a line (called offside-line) paralleled with the bottom line between the goal and the ball. we consider some ability to implement line defence.

Each defender has a home position. Usually, each defender keeps at his home position if the ball is not close to. The home position of each defender is changed dynamically depending on the position of the ball. We call the behavior of an agent to move back to its home position area as positioning. This ability is a necessity for the line defence.

Since the home positions of defenders are possibly overlaped, it is possible that a player comes into a collision with others. Moreover, a player chasing the ball may collide with the backbone of the player dribbling the ball. Hence, agents need an ability to avoid collision with others. This special behaviour is called avoiding.

Each defender should pay attention to the opponent players who are close to the offside-line,because it may break through the offside-line with the ball. That is the most dangerous situation. The defender should move to the cross point of the offside-line and the line connects one of the opponent player and the goal(see Fig.1). We call the behavior of the agent to move to that cross point as marking.

Without this ability of marking, the offside-line is possibly broken through by opponent players who have a dribble skill.

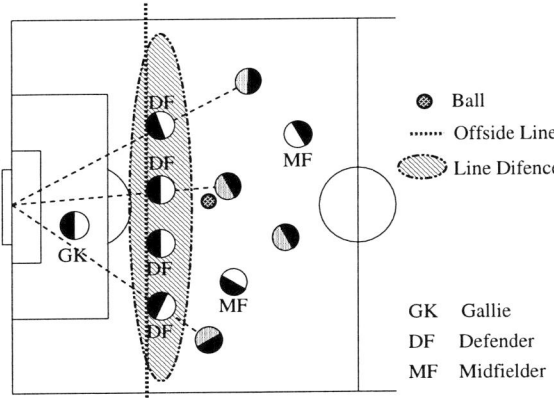

Fig. 1. Marking

The defence line varys with the position of the ball. When the ball is in the field of opponent side, the line is up. Otherwise, the line is down.

5 Conclusion

The line defence is efficient strategy against a team which using a simple pass strategy. But now, some teams have a practical dribble skill which is speedy and controllable. So our line defence strategy is not efficient against these teams. In this case, a sweeper is useful in real soccer.

We plan to vary a defence formation by modeling opponent team and using online lerning.

Oulu 99

Jarkko Kemppainen et al.

University of Oulu, Finland

1 Introduction

Oulu99 team was formed by students of University of Oulu, Finland as a part of student's Software Project course. Entire software was designed and written from scratch, even though there was source code available from previous Oulu teams. as a result, Oulu99 finished on 13th place in Robocup'99.

2 Team Development

Team Leader: Jarkko Kemppainen
Team Members:
  Mr. Jarkko Kemppainen
    – University of Oulu, Finland
    – graduate student
    – attended the competition
  Mr. Jouko Kylmäoja
    – University of Oulu, Finland
    – graduate student
    – attended the competition
  Mr. Janne Räsänen
    – University of Oulu, Finland
    – graduate student
    – attended the competition
  Mr. Ville Voutilainen
    – University of Oulu, Finland
    – graduate student
    – attended the competition
Web page http://ee.oulu.fi/~mysti/robocup

3 World Model

Clients had two coordinate systems:

– Absolute coordinates, where 0,0 is in the middle of own goal, and angle 0 is from own goal to opponent goal.
– Relative coordinates, where 0,0 is the player position, and angle 0 is directly forward from players body position.

All objects had speed vectors which are estimated to any point of time in past or in future.

Client had a memory of estimated position for each object in the field and updated this information when sensory information arrived. Otherwise the position was estimated and the probability of this information decreased as a function of time.

Ball position and referee calls played most important part in determining the current world state. Clients had different behaviourial modes depending on current play mode.

For shooting purposes, a special module calculated optimum shoot paths from own position, ball position and enemy player positions. Also, enemy goal posts played an important role in this algorithm.

To achive the most efficient learning of Soccer Server system, Team Oulu99 did not use any of the available source codes or libraries.

4 Communication

Players marked the free ball by applying movement towards it. This was a signal to other clients but one to ignore the ball. By this algorithm, ball catcher and backup player were deployed.

5 Skills

Clients had several different moving modes:

- Move with ball
- Move without ball
- Stay between two objects
- Shoot to point
- Shoot to goal
- Catch ball (goalie)

Dribble was used only when turning, it was not used for protecting the posession.

6 Strategy

High level strategy was fast point-to-point passing. Each client had a weighted position depending on ball position and play mode. All clients knew all other clients' should-be-position and could give blind passes to these coordinates.

7 Special Team Features

Team Oulu99 used different approach to determining own position than most of the other teams. Position was determined from all seen flags using distance as a weight of reliability. With extensive testing optimum values were found and the position determination was very accurate. It was also very unsensitive to noise and view blocking objects.

8 Conclusion

The final result, 13th, was very inspiring for the whole team and our sponsors, University of Oulu, Finland. We wish to thank Jukka Riekki, our inspirator.

Pardis

Shahriar Pourazin

Computer Engineering Department,
AmirKabir University of Technology, Tehran, IRAN
Tel: +(9821) 6419411, Fax: +(9821) 6413969
pourazin@ce.aku.ac.ir

1 Introduction

Pardis, was one of the entries in RoboCup-99, simulation league. It had a optimistic timing in communication with the server. And lost most of the cycles in the real league, because of relying on the enough network bandwidth. So unfortunately it had chance to be only in the first round robin. It used an experimental model, consisting of finite set of categories for each player. Each softbot in Pardis team, was a player acting as designed in a specific category. The coach had the ability to map each player in the opponent team with one of the same categories. It dynamically changed the characteristics (category) of the facing teammate to be effective against the analyzed opponent player. Although in the real league, there was no chance to see the use of the coach and it was never activated. The players read their behavioral configuration once at the start of the game and kept playing that way.

2 Team Development

Pardis was the result of 9 man-month development effort, mostly done by three people, the team leader and two undergraduate students.

Team Leader: Shahriar Pourazin
Team Members:
   Ali Ajdari Rad
   – team member, coding and representation of categories
   – IRAN
   – undergraduate student
   – attended the competition
   Houman Atashbar
   – team member, coded the low level parts, communication etc.
   – IRAN
   – undergraduate student
   – attended the competition
Web page http://www.pnu.ac.ir/~pourazin/rc99

3 World Model

The spatial model of the field consisted of separated squares making some regions. Players had patterns which described their behavior in each region. According to (say) their distance to the opponent's goal, they selected their action among the choices to pass, dribble, kick, etc. The more the player approached to the opponent's goal, the more eager it would become to kick to goal. The desire to kick, dribble, pass and other actions, had been stored as fuzzy values in an array called the *desire array*. Having no ball, near our own goalie, the player had to notice the ball and kick it away, and at the middle regions, it keeps trying not to let opponents receive any pass and if gets the ball passes it to teammates.

The real calculations on the array of fuzzy values for actions, depended on the exact region the player is in, the result of the game so far, the Boolean flag indicating that the player has the ball, and the position of other players. The player will do the action with the highest value.

All the codes were written in C++ from the scratch without any use of external prewritten libraries such as libsclient.

4 Communication

We had designed a method for message passing between players, by doing some special actions in front of the teammate, e.g., if the player does four 45 degree turns each one in the opposite direction of the previous, means that, received ball will be sent back soon. This mechanism was designed to reduce the SAY messages, but we had no chance to see its effect in the qualification of teams when no SAY is possible. They were supposed to receive from the coach some information (making the inter-player messages unnecessary), and also commands (to change their desire array).

5 Skills

Pardis has the goalie, as the player which receives the number one. It is different from the other players, such that, has no desire to take the ball toward opponent's goal, etc. It stays near the goal, uses fast moves when has the ball and the opponents are near the goal. So the goalie has no special difference in structure. All it has, is a different desire array.

6 Special Team Features

The players were designed to be as single threaded processes (i.e. no parallel processing in the players). The coach had to have a huge parallelism. It should have a plenty of models instead of each opponent, looking which model plays the same as the real opponent. This lets the coach determine the characteristics of the opponent player.

7 Conclusion

Our approach to opponent modeling had a severe assumption that the roles of the players in the opponent team are static or at least not rapidly changing. The unstructured program was not suitable for maintenance. That's why we had no chance to come up with the slow network in Stockholm. So the right thing to do is, throwing away the code and writing a new structured one.

After analyzing the timing of the whole system, we learned that working with the server could be very complicated. Even in the case that the server remains unchanged, the effect(s) of the networking problems could not be easily predicted. So how could we start our work? Is it necessary to think of all anomalies in all of our processes?

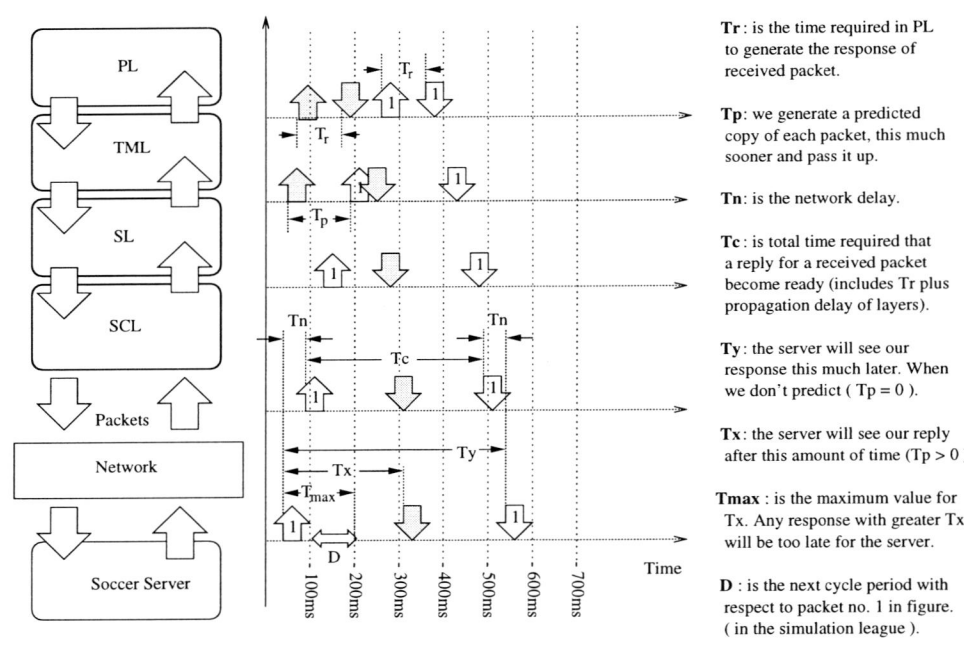

Fig. 1. The timing of an exchanged packet.

The answer could be found in the isolation of the anomalies. To do so we have to design a layered architecture for our next generation player. The structure has four layers: (1) Sense/Control Layer, (2) Synchronization Layer, (3) Task Management Layer and (4) Planning layer from bottom to top. The lowest layer (SCL) is to be modified to let us use upper layers in other leagues (F2000,

F180 or Legged). SL is to isolate the timing anomalies from the upper layers. TML is to translate abstract plans (of PL) into strings of smaller activities. This approach lets us, try to work on the (say) intelligent part of our player without being mixed up with unreliable data. Anyone who wants the corrupted data in the PL, could modify SL later. Figure 1 shows the four layers, receiving packets from server and sending back the responses. It is also shown that the response should be in the D interval. The grayed arrows stand for predicted packets. T_p should be large enough to shift the gray arrows to the left, making the player on time. Now we are working on the knowledge representation in the PL and the dynamic estimation of T_p.

PaSo-Team'99

Carlo Ferrari, Francesco Garelli, Enrico Pagello

Department of Electronics and Informatics, The University of Padua, Italy
Via Gradenigo 6a. I-35131 Padova, Italy
{carlo,pascal,epv}@dei.unipd.it

Abstract. PaSo-Team is a Multi-Agent system for playing soccer game in the Simulation League of the RoboCup competition. This paper describes the ideas and the technical structure of PaSo-Team'99, that played at RoboCup-99, in Stockholm during IJCAI'99. The main goal of the 1999 project was about the integration of a reactive model with some kind of high-level reasoning. Obstacle avoidance and motion reasoning are encapsulated at the behavior level. They use a proper world model (built from the sensed data), that focus on the relevant objects. The choice&evaluate problem is performed through an utility function over a proper coding of the prototypical game arrangements.

1. Introduction

Following the experiences done in 1997, and 1998 competitions, [2], [3], it has been developed the '99 release of *PaSo-Team* (The University of PAdua Simulated Robot SOccer Team), namely PaSo-Team'99. The PaSo-Team project has been conceived as a Multi-Agent System that must be able to play soccer game. While maintaining its historical origins of being a multi-agent reactive software architecture based on Brooks' Subsumption Architecture, PaSo-Team'99 tries to overcome the major limitations of pure reactive systems, introducing a more abstract level of learning and reasoning used to proper differentiate the current behaviour of the various component agents of the system, according to the actual phase of the game.

One of the big scientific challenge of this project, is to resolve a dilemma between two basic paradigms that mark the past and current history of Artificial Intelligence and Robotics, i.e. planning vs. reaction. The experience done at previous competitions seems to confirm the better quality of reactive-based systems, for soccer games. But it also demonstrated that a pure reactive schema suffers some major problems. In fact, it must be considered that the player is not alone in the environment. Thus, the agent cannot be recommended to execute only pure lowest-level reactive actions, due to the presence of obstacles such as other players, (team-mates or opponents) that surely interfere with the development of its strategy to score the goal. Then, obstacle avoidance routines against the opponents agents, and cooperative routines to get collaboration with the team-mates, are necessary. In the current project the emphasis is about finding a good balance between reaction and reasoning in the design of a player.

2. Team Development

Team Leader: Enrico Pagello (Associate Professor of Computer Science)
                epv@dei.unipd.it
                Dept. of Electronics and Informatics, The University of Padua,
                Via Gradenigo 6a, 35131 Padua, Italy
                and
                Institute LADSEB of CNR, C.so Stati Uniti 4, 35100 Padua, Italy,
Team Member: Carlo Ferrari (Assistant Professor of Computer Science)
                Francesco Garelli (Undergraduate Student)
                Stefano Griggio (Undergraduate Student)
                Andrea Sivieri (Undergraduate Student)

                Dept. of Electronics and Informatics, The University of Padua,
                Via Gradenigo 6a, 35131 Padua, Italy and
                {carlo, pascal, btenia, tigre}@dei.unipd.it

WEB page: http://www.dei.unipd.it/~robocup

3. The World Model

Complex data fusion procedures do not help reaction because their major role is to build a (maybe local) world model that represent a situation in a more compact form and with a focus on some aspects that are dependent from the actual task. Moreover, the continuos updating of the world model can be better performed if the validity of the acquired information can be maintained along more than one sensing cycle, trying to compute the evolution over time of the model itself. In simulated soccer game, each player is given with different information about its team-mate and opponents, as well as about ball dynamics and about the game field. Because of the nature of the problem (a team game), each single player cannot move simply reacting from raw positions and velocities information of the opponent players (like in a cat&mouse play), both because the presence of its team-mates can produce a variety of physical arrangements that it is real hard to classify, and also because its global attitude can switch very quickly from being a defender or an attacker. Moreover its focus of attention can change as well. In fact a player that has the ball maybe want to defend it against one ore more opponent or it want to drop it in the adversary net. Run-time motion planning and reasoning is the main intelligent activity at an intermediate level, for a single player. This activity is greatly based on obstacle avoidance that becomes a primary tasks in the most game situations. As obstacle avoidance is a primary task, the player world model must easily support a fast kind of obstacle avoidance reasoning. Moreover we want a model that can be produced and updated in a standard way, and that it is applicable in all the game situations

With this major requirements in mind, we developed a model that we call the *SVM* model, where SVM stands for Synthetic Visual Maps. SVM are a concise representation of the free space around the player. The SVM maps each movement direction of the player with a boolean value that says whether that direction is free or prohibited. Only the nearest elements are considered while building the SVM. In fact the influence on a SVM due to a game element (team-mate, opponent or field element) becomes as less important as its distance of the element itself is increasing. The SVM can be seen as a polar representation of the free space in a proper disk centered in the player. This representation can be easily updated at each sensing cycle, in order to consider new game elements that become important, either because they are moving towards the player or because the player itself is moving towards them. The SVM are computed from the sensed position of those game elements, that are directly reachable from the current player position. These elements form the *in-focus* set. The SVM for a single player is obtained as a composition of the SVM due to each element in the *in-focus* set. While building a SVM it is possible to explicitly take into account some correction factors that can enlarge the actual dimensions of an opponent player, in order to be able to make a choice about the movement direction that shows to be robust in spite of all the possible movement of the opponent player itself. Figure a,b and c show how to use the SVM for a dribbling action.

4. Communication

In the current PaSo-Team project we confirmed the idea of not realizing the coordination via explicit communication. In [4], it has been shown how it is possible to obtain an appropriate global team behaviour, via implicit communication., for a generic multirobot system. With implicit communication the agents can communicate looking at the current status of the environment. The interpretation procedure has been tuned to measure how the team is carrying its global task.

5. Skills

In PaSo-Team'99 we developed in C++ language, a single program that contains the features of all kind of players. This is because the simulation context is different from real soccer, where the human players have different physical abilities. In our team all the players are clones of a prototype player. The only thing that distinguishes one player from another is the role, that slightly makes changes in its reasoning strategy. The technical structure of each player is based on encapsulating the interface with the Soccerserver, and to represent the different player capabilities at different levels of abstraction. At the lowest level the PaSo-Team'99 clients act via *B-Actions* (Basic-Action), that code the information to be sent to the Soccerserver. A proper "talker" module is devoted to explicitly send information to the server, using the information prepared by the B-Actions.

B-Actions form a catalogue of primitives that can be used to model a kind of more complex action: the *Skills*. Skills are used to model atomic actions, like kicking the ball or looking around. Using Skill and B-Action let possible to separate the logical capabilities of a player from its particular implementation due to a specific server. Information from the server are handled by the "listener" module. The listener module solves the synchronization problem with the server and it receives the data items related to the game development. The data interpretation activity is devoted to a "parser" module, that can update the player local status, that is, the player "memory". The memory module is used as a data container able to answer to every request from the higher reasoning level modules. Among these modules there are the *behaviour modules* (or simply *Behaviours*). They realize the intelligent manoeuvring of a player, according to the different game situations. *Behaviours* are coded using Skills, and they also compute the SVM for motion reasoning. Some typical behaviours are devoted to control the dribbling, or to stop an opponent player, or to kick the ball away.

6. Strategy

While playing, every player can activate a particular behaviour (its *running* behaviour) at a time, and, usually, all the player of a team do not have the same running behaviour. In this way it is possible to have a global team emergent behaviour, that it is due to the compound effects of all the players behaviours. In the current approach, the PaSo-Team player do not code explicitly the coordination procedures for obtaining complementary actions, that arise, instead, because of a proper interpretation of the perceptions sent by the soccerserver. The high level reasoning is devoted to select the proper behaviour, that becomes the running behaviour. This kind of reasoning requires to solve a two-fold problem: the choice&evaluate problem.

At each game cycle each player must either confirm its running behaviour or decide to change it with another one that looks more appropriate to the game development. Moreover, it should maintain and update the evaluation of its past behaviour choices, with respect to the different game situations. Recognizing failures, and learning from them is a key point for high level reasoning. So far, the choice&evaluate problem is solved first looking for the most promising behaviour, that is, the behaviour that maximize an utility function over a proper coding of the game situations, and then updating this utility function measuring how good the behaviour has been in reaching its goal.

7. Conclusions

In the PaSo-Team'99 project we experimentally investigate how much the reaction schema for intelligent agents team must be integrated with some kind of high level reasoning. In PaSo-Team'99, sensed data are used in the decision process after they are filtered through a proper mechanism (the Synthetic Visual Maps), that builds a world model to support fast and efficient procedure for motion planning. Obstacle avoidance and motion reasoning are encapsulated at the behaviour level, while higher levels deal with decision and learning. The choice&evaluate problem is performed separately by each player, using an utility function over a proper coding of the game situations

Acknowledgements

This research work could not be done without the enthusiastic participation of the students of Electronics and Computer Engineering Undergraduate Division of Padua University Engineering School. Financial support has been provided by both CNR, under the Special Research Project on "Real-Time Computing for Real-World" and Murst, under the 60% and 40% Grants.

References

1. Burkhard, H.D, Hannebauer, M., Wendler, J.: Belief-Desire-Intention Deliberation in Artificial Soccer. AI Magazine, Fall 1998.
2. Pagello, E., Montesello, F., D'Angelo, A., Ferrari, C.: A Reactive Architecture for RoboCup Competition. In Kitano H. (ed.): Proc. First Int. Workshop on RoboCup, Lecture Notes in Artificial Intelligence, Springer -Verlag, 1997.
3. Pagello, E., Montesello, F., Candon, F., Chioetto, P., Garelli, F., Griggio, S., Grisotto, A.F.,: Learning the when in RoboCup Competition. Proc. Second Int. Workshop on RoboCup, Lecture Notes in Artificial Intelligence, Springer -Verlag, 1998.
4. Pagello, E., D'Angelo, A., Montesello, F., Garelli, F., Ferrari, C.: Cooperative behaviors in multi-robot systems through implicit communication. Robotics and Autonomous Systems, 29 (1999), 65-77
5. Stone, P.: Layered Learning in Multi-Agent Systems". Ph.D. Thesis, School of Computer Science, Carnegie Mellon University, December 1998.

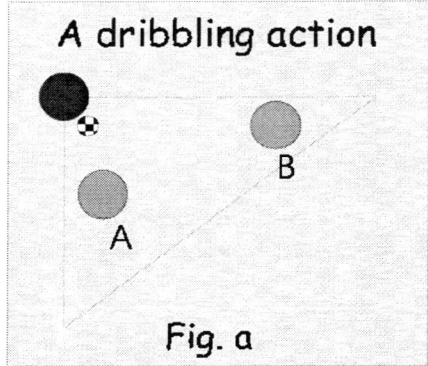

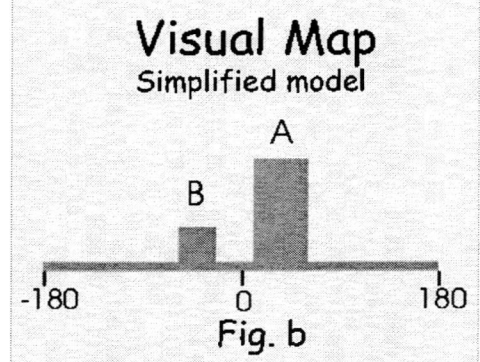

Fig. a. A typical action of dribbling: our player has the ball, A and B are opponents

Fig. b. A simplified map for situation in fig. a without filters and complex shape maps

Fig. c. The real map created by our client; note the effects of filters in promoting the forward direction.

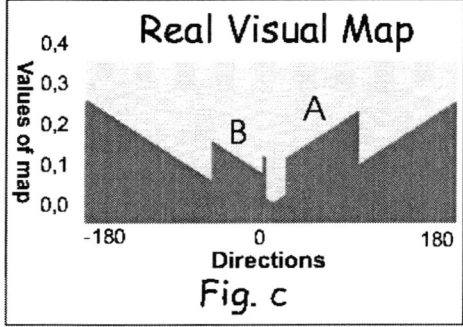

PSI Team

Alexander N. Kozhushkin

Program Systems Institute of Russian Academy of Sciences

1 Introduction

Our team – PSI was developed at Program Systems Institute of Russian Academy of Science. This paper is a short description of the dynamical refinement planning method that we use to construct our software agents.

Basic skills and roles of every agent are presented by means of the set of elementary plans. The purpose of the planning process is to compose the extended plan defining the behaviour of the agent from elementary ones.

The planning system (or just a planner), built in the agent, modifies extended plans depending on external conditions and the internal state of the agent. It adds new elementary plans to the extended one refining it and controls the execution of elementary plans in a body of the extended plan. Namely, planner can temporarily suspend (interrupt) execution of some elementary plan in favor of another one or abort execution of the inappropriate elementary plan. For its work planning system uses the family of basic relations (*interruptability, priorities* and *plan levels*) described below.

Although results of RoboCup'99 are not very well for the PSI team, we think that our method proved to be flexible and convenient for the programming the complex behaviours of the agents working in the unpredictable environment.

2 Team Development

Team Leader: Alexander N. Kozhushkin
Team Members:
   Alexander N. Kozhushkin
   - Program Systems Institute of Russian Academy of Sciences
   - Russia
   - Ph.D. student
   - did attend the competition
   Alexei P. Lisitsa
   - Program Systems Institute of Russian Academy of Sciences
   - Russia
   - senior researcher
   - did not attend the competition
Web page http://www.botik.ru/~soccer/ANK.html

3 Model of the planning system

Consider a model of the planning system built in each player. It is defined by the discrete set of time moments $T = \{0, 1, 2, 3, ...\}$, a set O of elementary actions which soccer server can execute, a set I of input (or external) states defined by all possible values of data available from soccer server and the space Ω of internal states of the player. $I \times T$ and O represent input information (*input*) and actions (*output*) of the player, respectively. Play history, or just a play, from the point of view of a player may be defined as a map $H : T \to I \times \Omega$, where $H(t)$ consists of the information from soccer server and of the internal state of the player at the moment t. Let $B(i, \omega)$, $C(i, \omega)$ be some logical conditions on the set of full states $I \times \Omega$, ϕ and ψ be some maps, $\phi : I \times \Omega \to \Omega$, $\psi : I \times \Omega \to O$. Define *elementary plan* of the player as a tuple $p = <B, C, \phi, \psi>$, where B and C are, respectively, its beginning and continuation conditions.

For our work we use some fixed finite set of *basic* elementary plans π. Define the family of subsets $D_i \subset \pi$, $i \in \{1, 2, ..., n_{lev}\}$, where $n_{lev} > 1$ the number of the hierarchic levels, such that $\cup_i D_i = \pi$ and $D_i \cap D_j \neq D_i$ for $i < j$. For every D_i define partial order relation $Prior_i$ (*priority*). By means of plan hierarchic levels introduce the binary interruptability relation Int defined on $\pi \times \pi$, with the constraint on it: $Int(p, p') \wedge p \in D_i \Rightarrow p' \in D_{i+1}$. In the rest of the paper we suppose that if the set π is defined then families $\{D_i\}$, $\{Prior_i\}$ and the Int relation are defined as well.

Explain the meaning of introduced notions. Each elementary plan from π is designed to determine a skill or a role of the player or to solve a particular task. For example, there are plans designed to solve the task of the ball interception, overtaking another player and so on. Both B and C are applicability conditions of the plan. But in practice these conditions are different. The aim of the distinction between them is a making behaviour of agents more robust and stable.

The family $\{D_i\}$ i.e. plan levels determine the hierarchy on the plans. Levels divide all plans into main and auxiliary ones. Note, that same plan can fall onto different levels. This is in agreement with the intuition that same skill can be auxiliary to some main skill and, at the same time, auxiliary to another auxiliary one. The interruptability relation presents the further refinement concerning which plan can be suspended temporarily (interrupted) in favour of another one, auxiliary to it. The priority relations serve to resolve ambiguities in the cases when several elementary plans can be applied to the current situation. Thus $\{Prior_i\}$ relations play the role of evaluation functions.

For every time t of the play H there are two elementary plan subsets $App_t^H = \{p \in \pi| \models B_p(H(t))\}$ and $Con_t^H = \{p \in \pi| \models C_p(H(t))\}$ of π whose beginning and continuation conditions are satisfied at the moment t, respectively. We would like to name the set App_t^H as a set of *applicable* and Con_t^H as a set of *executable* elementary plans.

Suppose that the set π is defined, we build its extension π^* by the following rules:
(i) $\pi \subset \pi^*$;
(ii) if $p = <B_p, C_p, \phi_p, \psi_p> \in \pi^*$ then plan $< C_p, C_p, \phi_p, \psi_p >$ is in π^*;

(iii) if $p =< B_p, C_p, \phi_p, \psi_p >\in \pi^*$ and $p' =< B_{p'}, C_{p'}, \phi_{p'}, \psi_{p'} >\in \pi^*$ then plan $< B_p \wedge C_{p'}, C_p \wedge C_{p'}, \phi_p(i, \phi_{p'}(i, \omega)), \psi_p >$ is in π^*, where $i \in I$, $\omega \in \Omega$.

Thus π^* the set of all elementary plans which an agent can use, "tuning" the basic ones to current conditions.

We define *extended* plan P as a word of a plan language L in the alphabet $A = (\pi^* \cup \{*\}) \times (\pi \cup \{*\}) \times \{0, \ldots, n_{lev}\}$. We use the symbol "$*$" as a abbreviation of the $(*, *, 0)$. L is defined by the next rules (where Q, Q_1 are letter sequences without "$*$", possibly empty):

1. $* \in L$;
2. if $Q* \in L$ and $p \in D_1$, then $Q * (p, p, 1) \in L$;
3. if $Q * lQ_1 \in L$ where $l \in \pi * \times \pi \times \{0, \ldots, n_{lev}\}$ and $p \in \pi^*$ then $Ql * Q_1 \in L$;
4. if $Q * (< B, C, \phi, \psi >, p, k)Q_1 \in L$, where
$(< B, C, \phi, \psi >, p, k) \in \pi^* \times \pi \times \{1, \ldots, n_{lev}\}$, and there exists
$< B', C', \phi', \psi' >\in \{p'|Int(p, p')\}$, then
$Q (< B, C, \phi, \psi >, p, k) * (< B' \wedge C, C' \wedge C, \phi'(i, \phi(i, \omega)), \psi' >, p', k+1) \in L$.

Every extended plan represents some play history and possible evolution of the play history. The letters lying before "$*$" represent elementary plans, which the agent has used in the past. The first elementary plan after "$*$" is a current one, i.e. behaviour of the agent is determined by the ψ map of this plan at present. All other elementary plans are those which the agent is going to execute in the future.

Introduce the notion of the *planning function* as a map $F : L \times 2^\pi \times 2^\pi \to L$. We use the special kind of the planning function defined by the following manner.

Let functions $f : A \to \pi \cup \{*\}$ and $g : A \to \{0, \ldots, n_{lev}\}$ are projections, $P \in L$ $S_B \subseteq \pi$, $S_C \subseteq \pi$ and Q is a letter sequence without "$*$", possibly empty. Define the planning function F, representing the step of our planning system work, by the following clauses:

1. If $P = Q*$ and there exist the largest element $p \in S_B \cap D_1$ wrt the $Prior_1$ relation, then $F(P, S_B, S_C) = Q * (p, p, 1)$
2. If following conditions hold:
   a) $P = Q * l_1 \ldots l_n$ where $l_1 = (< B_1, C_1, \phi_1, \psi_1 >, p_1, k_1)$, $n \geq 1$;
   b) for all i such that $1 \leq i \leq n$, $f(l_i) \in S_C$ and $f(l_i)$ is the largest element of the set $(S_B \cup \{f(l_i)\}) \cap D_{g(l_i)}$ wrt the $Prior_{g(l_i)}$ relation;
   c) there exists the largest element $p =< B', C', \phi', \psi' >\in \{p|Int(p_1, p)\} \cap S_B$ wrt the $Prior_{k_1+1}$ relation;
   then $F(P, S_B, S_C) = Ql_1 * (< B' \wedge C_1, C' \wedge C_1, \phi'(i, \phi_1(i, \omega)), \psi' >, p, k_1 + 1)(< C_1, C_1, \phi_1, \psi_1 >, p_1, k_1) \ldots l_n$.
3. If $P = Q * l_1 \ldots l_n$ where $l_1 = (< B_1, C_1, \phi_1, \psi_1 >, p_1, k_1)$, $n \geq 1$ and there exists i $(1 \leq i \leq n)$ such that:
   a) for all k such that $i \leq k \leq n$, $f(l_k) \in S_C$ and $f(l_k)$ is the largest element of $(S_B \cup \{f(l_k)\}) \cap D_{g(l_k)}$ wrt the $Prior_{g(l_k)}$ relation;
   b) if $i \neq 1$ then $f(l_{i-1}) \notin S_C$ or $f(l_{i-1})$ is not the largest element of $(S_B \cup \{f(l_{i-1})\}) \cap D_{g(l_{i-1})}$ on the $Prior_{g(l_{i-1})}$ relation;
   then $P = Q * l_1 \ldots l_i \ldots l_n$. If such i does not exist, then $P = Ql_1 \ldots l_n *$.

4. For all other cases $F(P, S_B, S_C) = P$.

Consider arbitrary history H. For every moment t' we can built the extended plan by means of the rule: $P(0) = *$, $P(t'+1) = F(P(t'), App_{t'}{}^H, Con_{t'}{}^H)$. Thus, for every time t extended plan can be built by applying consequently the planning function to the initial plan $*$.

4 How the planning system does work

Let us consider one illustrative example. We have the very simple player which has only five basic elementary plans in the basic elementary plan set: (*1*) "Go to the Ball" ($p_1 = < C_1, C_1, \phi_1, \psi_1 >$) with the application condition "I see the Ball *and* I do not possess the Ball", (*2*) "Intercept of the resting Ball" ($p_2 = < C_2, C_2, \phi_2, \psi_2 >$) – "I see the ball *and* Speed of the Ball is not equal to 0 *and* I can intercept the Ball before other players *and* I do not possess the Ball", (*3*) "Intercept of the moving Ball" ($p_3 = < C_3, C_3, \phi_3, \psi_3 >$) – "I see the Ball *and* Speed of the Ball is not equal to 0 *and* I can intercept the Ball before other players *and* I do not possess the Ball", (*4*) "Overtaking" ($p_4 = < C_4, C_4, \phi_4, \psi_4 >$) – "There is a player which prevents to the movement", (*5*) "Kick into the opponent Goal direction" ($p_5 = < C_5, C_5, \phi_5, \psi_5 >$) – "I possess the Ball". We suppose for the simplicity that $B_i \equiv C_i$ for every elementary plan. This entails $App_t{}^H = Con_t{}^H$ for every moment of time. Thus the behaviour of the agent is determined by such simple skills and the basic elementary plan set – π is $\{p_1, p_2, p_3, p_4, p_5\}$.

There are two levels in the plan hierarchy: $D_1 = \{p_1, p_2, p_3, p_5\}$ and $D_2 = \{p_4\}$. Define Int and $Prior_1$ relations as sets of pairs: $\{Int(p_1, p_4), Int(p_2, p_4), Int(p_3, p_4)\}$ and $\{Prior_1(p_2, p_1), Prior_1(p_3, p_1)\}$, respectively. The $Prior_2$ relation is obvious.

At the initial moment the agent behaviour is presented by the extended plan $P_0 = *$. Suppose that the motionless ball is far from the agent and there is another player which can intercept the ball earlier. In this case $App_0{}^H = \{p_1\}$ and planner makes extended plan $P_1 = *(p_1, p_1, 1)$ by using rule (1) of the planning function definition. At the time t_1 our agent have seen the opponent player on its way to the ball. Now $App_{t_1}{}^H = \{p_1, p_4\}$ and P_1 will be transformed to the plan (see rule (2)) $P_2 = (p_1, p_1, 1) * (p_4{}^*, p_4, 2)(p_1{}^*, p_1, 1)$.
Here, $p_1{}^*$ and $p_4{}^*$ are defined by rules (2) and (3) of the definition of π^*, respectively.

Consider two of the possible scenarios. The first – the agent overtakes the opponent player successfully, but still can not intercept the ball before some other player. In this case we have $P_3 = (p_1, p_1, 1)(p_4{}^*, p_4, 2) * (p_1{}^*, p_1, 1)$.

The second one: the agent believes that it can capture the ball before other players. In this case $App_{t_2}{}^H = \{p_1, p_2, p_4\}$. The planning system transforms P_2 into the extended plan P_4 by the next two steps (see rules (3) and (1)):
$P_2' = F(P_2, App_{t_2}{}^H, App_{t_2}{}^H) = (p_1, p_1, 1)(p_4{}^*, p_4, 2)(p_1{}^*, p_1, 1)*$,
$P_4 = F(P_2', App_{t_2}{}^H, App_{t_2}{}^H) = (p_1, p_1, 1)(p_4{}^*, p_4, 2)(p_1{}^*, p_1, 1) * (p_2, p_2, 1)$.

If the plan P_4 completes successfully at the time t_3, the agent captures the ball. Then $App_{t_3}{}^H = \{p_5\}$ and $P_4' = (p_1, p_1, 1)(p_4^*, p_4, 2)(p_1^*, p_1, 1)(p_2, p_2, 1)*$. $P_5 = (p_1, p_1, 1)(p_4^*, p_4, 2)(p_1^*, p_1, 1)(p_2, p_2, 1) * (p_5, p_5, 1)$.

5 Conclusion

Our approach is somewhat analogous to that presented in [1, 2], with one essential difference: our planning system works *on-line*, and plans refinements are being made dynamically in case of need. More detailed comparisons deserve the further investigations and are to be presented elsewhere. The further development of the method itself and the more precise formulation of its essence is a goal of further work.

References

1. M. Ginsberg. Modality and Interrupts. *Journal of Automated Reasoning* (14)1 (1995) pp. 43–91.
2. M. Ginsberg, H. Holbrook. What defaults can do that hierarchies can't. *Fundamental Informaticae* (21) (1994) pp. 149–159.

RoboLog Koblenz

Jan Murray, Oliver Obst, Frieder Stolzenburg

Universität Koblenz-Landau, Fachbereich Informatik
Rheinau 1, D–56075 Koblenz, GERMANY
{murray,frvit,stolzen}@uni-koblenz.de

1 Introduction

The RoboCup scenario yields a variety of fields of research. The main goal of the RoboLog project, undertaken at the University of Koblenz in Germany, is the specification and implementation of flexible agents in a declarative manner. The agents should be able to deal with the real-time requirements but also be capable of more complex behavior, including explicit teamwork. To this end, we develop a *declarative multi-agent script language* for the specification of collective actions or intended plans that are applicable in certain situations. The agents should be able to recognize such situations by means of *qualitative spatial reasoning*, possibly supported by communication.

The RoboLog team is based on an architecture with four layers, where layer 1 deals with the synchronization with the SoccerServer and realizes the low-level skills. Layer 2 handles qualitative spatial reasoning. More complex actions and teamwork are realized in layers 3 and 4. The focus of this paper is laid on the first layer, especially on position determination (see Sect. 3). For a more detailed description of the higher layers see the paper *Spatial Agents Implemented in a Logical Expressible Language* in this volume.

At the RoboCup-99 competition RoboLog Koblenz played in Group C in the Simulator League. The team lost only one match and managed to achieve a draw in the other three. Unfortunately, this did not suffice to enter the elimination round. The match RoboLog Koblenz vs. IALP (1–1) was the most interesting match we played, in so far as it was the *only* drawn match in the whole competition, that did not end with a score of 0–0.

Team Leader: Frieder Stolzenburg
Team Members:
    Dr. Frieder Stolzenburg
    – Universität Koblenz-Landau
    – Germany
    – Researcher
    – attended the competition
    Oliver Obst, Jan Murray
    – Universität Koblenz-Landau
    – Germany
    – Master Students
    – attended the competition
    Björn Bremer, Michael Bruhn, and Bodo van Laak
    – Universität Koblenz-Landau
    – Germany
    – Master Students
    – did not attend the competition

Web page: http://www.uni-koblenz.de/ag-ki/ROBOCUP/

Fig. 1. The RoboLog Koblenz team.

2 Team Development

The RoboLog Koblenz players were implemented by a team of 3 to 5 people. Most of them were students preparing their course work or diploma theses. We were able to conduct several test games with different scores on our local network—a 100 MBit Ethernet, sometimes using a Sun Ultra-Enterprise with 14 processors à 336 MHz and 3 GB main memory. Fig. 1 lists the team members of RoboLog Koblenz.

3 World Model

For each agent, the RoboLog interface—written in C++—requests the sensor data from the SoccerServer. By this, the agents' knowledge bases are updated periodically. If some requested information about a certain object is currently not available (because it is not visible at the moment), the most recent information can be used instead. Each agent stores information about objects it has seen within the last 100 simulation steps. So we can think of it as the agent's memory or recollection. The passing of time can be modeled in several ways with RoboLog. It provides various means for creating snapshots of the world and defining an event-driven calculus upon them, translating the agent's view of the world into a propositional, qualitative representation in *Prolog*.

The RoboLog system provides an extensive library that makes precise *position determination* possible. The whole procedure is able to work even when only few or inconsistent information is given. First of all, an agent has an egocentric view of the world. The actual sensor data provide more or less precise information about the positions of other agents, landmarks and border lines relative to the agent's position and orientation. The RoboCup scenario yields a frame of reference with absolute coordinates, given by the geometry of the playing field. Knowing the absolute position allows the agent to identify its own location wrt. other objects on the map, even if they are not visible at the moment. Secondly, an absolute frame of reference is helpful in order to communicate with other agents in situations where cooperative actions are appropriate.

The first localization method in RoboLog—introduced in [1]—requires (only) three or more directions to visible landmarks relative to the orientation of the agent to be known. Provided that at least three of them and the position of the agent neither form a circle

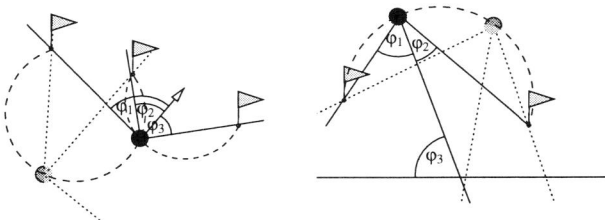

Fig. 2. Self-localization by angle information.

nor lie on a straight line, the absolute position and orientation of the agent can be computed with a time complexity that is only linear in the number of landmarks. If the corresponding equation system in complex numbers is over-determined, and the data is noisy, the procedure estimates the position applying the least squares method. Without reference to a third landmark, the actual position in question lies on a segment of a circle. This is shown in Fig. 2.

However, we also need procedures that are able to work, if only limited sensor data is available. If less than three landmarks or border lines are visible, there are two cases where the position of the agent is uniquely determined even then. Firstly, knowing the distances and angles to two landmarks is sufficient. Secondly, the position can be determined if the relative position of one landmark and of a point on a border line where the center line of the view axis of the agent crosses this line is given, provided that both points are not identical. In addition, by keeping track of the movements of the agent, it is also possible to estimate the current position. All these methods are implemented in the RoboLog module, in order to lay a solid quantitative basis for the qualitative reasoning in the higher layers.

4 Communication

The RoboLog agents use communication in order to clarify situations. Very frequently an agent is not able to recognize a situation or its own role therein, because of insufficient information. These drawbacks can be overcome by the use of communication. If an agent recognizes a situation in which a multi-agent script can be executed, it sends this fact to the others by communicating a Prolog predicate, that is executed by *all* RoboLog agents receiving it. If this command turns out to be relevant for a player, i.e. the agent has to take part in a collaborative action, it will be executed. Otherwise the agent just ignores the message. For this communication to work the agents rely on the fact, that their internal structure is the same, and more that they are all implemented in Prolog. We use communication especially for initiating teamwork such as double passing.

5 Skills

The RoboLog agents are equipped with several basic skills like *dribbling* or *kicking* to a certain position (both programmed in Prolog). These skills are not very sophisticated, which proved to be a disadvantage during the RoboCup-99 competition. All agents are clones of each other except for the goalie, which we will now describe in greater detail.

The goalie's main objects are to keep between the ball and the goal and not to lose sight of the ball. Its behavior depends mainly on the movement and (qualitative) position of the ball. The agent partitions the playing field into several regions, e.g. *opponent half* or *penalty area*. Based on the region the ball is currently in and the extrapolation of its trajectory, the goalie chooses a position to move to. It also takes into account, if the ball is owned by the opponent team or in the middle of a shot. If the ball is in a shot, the agent tries to intercept it, but otherwise it just blocks the way from the ball to the center of the goal line. After any movement it turns its neck towards the expected position of the ball, thus keeping it in sight.

6 Strategy

The underlying basic strategy of the RoboLog Koblenz team simply is moving towards the opponent goal (possibly avoiding obstacles) and trying to score. If a RoboLog agent has the ball, it tries to recognize the current situation as one in which a (multi-)agent script is applicable. In this case the script will be executed, yielding behaviors like (double) passing. Otherwise the agent sticks to the default strategy until a script can be applied.

When the other team has the ball, the players try getting it back in the following manner. Each agent checks, if at least two team mates are nearer to the ball. If that is not the case, it tries to intercept the ball. This "double attack" proved useful as it prevents situations in which no agent goes to the ball due to sensor uncertainty. Players, that are too far away to be involved, return to their special home positions.

7 Special Team Features

The RoboLog Koblenz agents make use of explicit teamwork during a game. In many other teams multi-agent cooperation emerges just as a consequence of the behaviors of the single agent. RoboLog agents, however, actively try participating in collaborative behavior specified in a multi-agent script language. The agents explicitly make use of communication to tell other players to take part in a collective action. A more detailed description of the multi-agent language can be found in the paper *Spatial Agents Implemented in a Logical Expressible Language* in this volume.

8 Conclusion

The RoboLog system provides a clean means for programming soccer agents declaratively. Cooperative Behavior can be modelled explicitly by means of multi-agent scripts. A qualitative spatial representation allows agents to classify and abstract over situations. However, the imperfection of the low-level skills turned out to be a major disadvantage of RoboLog Koblenz. The current approach only allows to react *ad hoc* to external events or interrupts. Future work therefore includes reimplementing a part of the RoboLog interface in order to enhance the basic skills as well as extending them by a number of additional abilities.

Promising areas of further research are the specification of a more sophisticated communication paradigm and the use of logical mechanisms within the lower levels of our approach. Deduction could be used to build a more complete view of the agent's world. In addition, the robustness of the decision process can be improved by means of defeasible reasoning and the use of any-time reasoning formalisms. The application of these techniques to real robots is one of the next steps of our research activities.

References

[1] M. Betke and L. Gurvits. Mobile robot localization using landmarks. *IEEE Transactions on Robotics and Automation*, 13(2):251–263, Apr. 1997.

[2] A. S. Rao. AgentsSpeak(L): BDI agents speak out in a logical computable language. In W. van de Velde and J. W. Perrame, editors, *Agents Breaking Away – 7th European Workshop on Modelling Autonomous Agents in a Multi-Agent World*, LNAI 1038, pages 42–55, Berlin, Heidelberg, New York, 1996. Springer.

Rational Agents by Reviewing Techniques

Josep Lluís de la Rosa, Bianca Innocenti, Israel Muñoz and Miquel Montaner

Institut d'Informàtica i Aplicacions
Universitat de Girona & LEA-SICA
C/Lluís Santaló s/n
E-17071 Girona, Catalonia
{peplluis, bianca, imunoz, mmontane }@eia.udg.es

Abstract. This paper describes the research in a new Rogi Team conceived by simulation in Java. It is a development of ideas for rational agents that co-operate and use revision of exchanged information and consensus techniques.

1 Introduction

This research is implemented in JAVA. Next it will be applied to real platform of robots and especially to the 11x11-soccer server. A type of rational agents is implemented by techniques inspired from consensus techniques [Chi 92] and according to new trends of agents research [Jennings 98] for emergent co-operation design.

2 Rational Agents

2.1 Reactive Decisions

In a first step of reasoning, every agent decides a private action. This first decision is considered a BELIEF of the Agent0 language [Shoham 93]. This belief depends on local environment configuration defined by two parameters: distance player-ball (DPB), and distance player-goal (DPG). The belief contains a degree of certainty.

"*Fig. 1*a" shows an example of configurations of robots and the ball in the field. Decisions are SHOOT at 'Zone 1', GET at 'Zone 2', and FORW or BACK at default 'Zone 3' depending on DPG value. Thus, reactive reasoning is the following rule:

```
BEL ( AgentX, DPB, ZONE2 )  ⇒
INFORM ( to_any_agent, AgentX, SHOOT, 0.8 )
```

Similarly, at 'Zone 3' in point 'M' (see "*Fig. 1b*"), reasoning would be the following:

```
BEL ( AgentX, DPB, ZONE3 ) ∧ BEL ( AgentX, DPG,FAR )  ⇒
INFORM ( to_any_agent, AgentX, FORW, certainty )
```

where 'certainty' is a value obtained by fuzzy inference by operating the certainties of ZONE3 ∧ FAR Agents communicate their beliefs (INFORM in terms of AGENT0 language) to the playmates. Thus reactive reasoning creates rough intentions.

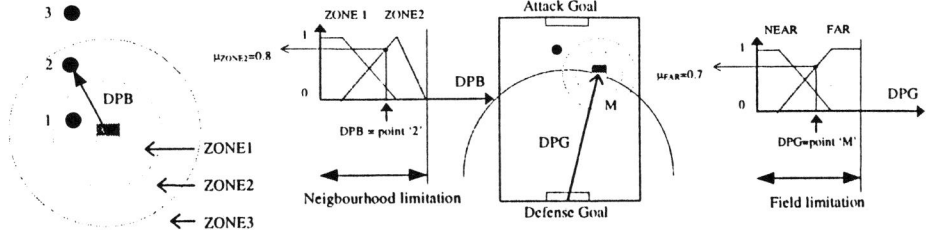

Fig. 1: Reactive beliefs of agents by fuzzy sets (a) DPB, (b) both DPB and DPG variables

2.2 Rational (Co-operative) Decisions

Rational reasoning in the sense of [Busetta 99] is implemented by communicating the former reactive beliefs. It begins with a REQUEST (a communication) action, so that every agent can know the beliefs set that contains the reactive belief, the certainty of this belief and the identification of the player *(reactive_belief, certainty, ID_player)* of all other playmates. Therefore, when two playmates realise they have conflictive beliefs then the certainty of their beliefs is taken into account and one of the playmates changes its mind by reconsidering its former reactive beliefs.

"Fig. 2" shows a situation where both Agent1 and Agent2 belief they can *GET the ball*. After REQUEST each other, Agent1 will change its belief because its DPB parameter brings lower certainty than those obtained by Agent2.

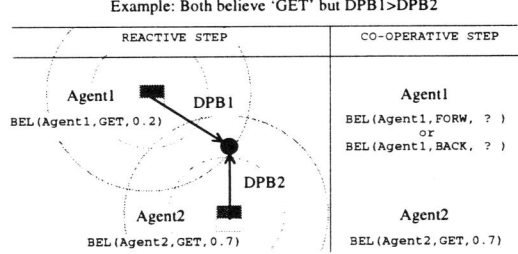

Fig. 2 Example of rational decision

Note that the exchange of beliefs and their certainties requires of revision [de la Rosa 92a]. This means that the subjective certainties associated to beliefs that are incoming from other agents have to be filtered (reviewed) at every agent. This process of revision is developed using extra knowledge about the co-operative world by means of some perception of quality and reliability of mates and of oneself [de la Rosa 92b, 93] [Acebo 98]. This reasoning procedure could be expressed as:

```
INFORM ( Agent2, Agent1, BEL ( Agent2, SHOOT, 0.7) ) ∧ BEL (
Agent1, SHOOT, 0.2 ) ⇒ BEL ( Agent1, SHOOT, f(0.7,0.2) )
```

where $f(c1,c2) = \begin{cases} 0 & , c1 > c2 \\ c2 & , \text{otherwise} \end{cases}$

In the example of figure 2 since f (c1, c2) = 0 then Agent1 will change its belief to FORW or BACK action using the here described rules.

3 Implementation of the Team in the Javasoccer

COMMUNICATION of AGENTS' BELIEFS. When players already have their reactive beliefs about possible actions to do, then this belief is communicated to the playmates. Next, every agent reviews the incoming certainty of the incoming beliefs.

3.1 Communication of beliefs

In the initialisation phase a broadcast communication channel is open. An Enumeration object is used to send beliefs. This object has two attributes: one is a emitter player's identifier, and the second attribute is the belief and certainty to send.

Every player receives the beliefs from the others. The incoming certainty of beliefs, *certesa_company* is reviewed by means of the following rules that contain the perception of every agent of the community of agents in the co-operative world. This is implemented in prestige and necessity rules. This couple of parameters, defined as follows, describes the perception of the co-operative world, that is the perception that every agent has of the rest playmate agents:

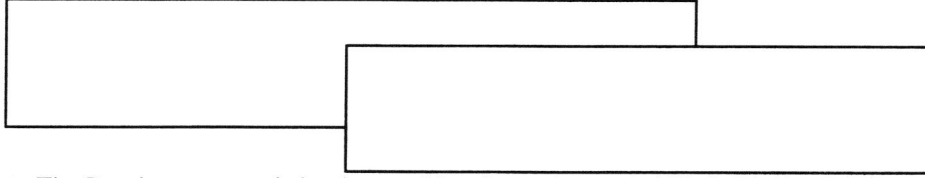

The Prestige operator is implementable when using probabilistic $P_{ij}(\varphi) = P_{ij} * \varphi$, or Sugeno's $P_{ij}(\varphi) = \min (P_{ij}, \varphi)$ implementation of the **and** operator.

3.2 Conflicts

This is the set of conflicts that agents should solve by means of the rational decision.

Player \ Mate	SHUT	ATACK	GO TO BALL	CALL BACK
SHUT	Conflict	Conflict	Conflict	
ATACK	Conflict	Conflict	Conflict	
GO TO BALL	Conflict	Conflict	Conflict	Conflict
CALL BACK			Conflict	Conflict

Alternative possible actions ATURAR (stop), DEFENSAR (defence), COBRIR (?), etc... aren't conflictive in this example. In the case of no conflict then every agent decides to convert the reactive belief into an action.

3.3 Rational decision implementation in Java

An algorithm called consensus [de la Rosa 92] [Chi 92] that recovers the perception of the co-operative world develops the reviewing process. The formulas are implemented using probabilistic implementation of the **and** logical connective.

```
if ( decisions[decidit] * consensus[RobotID] <
        certesa_company * consensus[id_company] )
  {    // lets change the belief of this agent
  }
```

4 Methods for Changing Perception of the Co-operative World

Java implementations of our rational agents show trends of better behaviour of the overall play of the teams, but there are still some lacks, as for example conflicts between defenders and goal keepers, are not properly solved. Our improvement (novelty) is to modify the perception of the co-operative world to make the consensus algorithm more adaptive to changing environments: every agent modifies its perception of the co-operative world. Two methods are proposed: (1) a positional method and (2) a reinforcement method for *winners in conflicts* to increase persistence.

4.1 Method 1: positional method.

Players are specialised. One possible effect of their specialisation is that they prefer to stay in certain position in the playground. Agents will take advantage of this feature and will modify their vision of the co-operative world by assigning the values of prestige and necessity according to positions of players.

For example, the perception of the co-operative world from a forward-player could be: *'I have big necessity of the middle-forward players and not much necessity of the goal-keeper'*. However, this perception has to be completed by more information according to the positions of the other playmates. This is the assignment of the prestige and necessity parameters:

a) The Necessity N is calculated as follows: *'We have the more necessity of information provided by a playmate when the closer of his positional area we are'*. For instance, a forward-player that helps in defence will take into account more the beliefs of goalkeeper and defenders than of others and himself.

b) The Prestige P is assigned as follows: *'The more prestige I have when the closer to my specific position I am'*. For example, a forward-player that is waiting for the ball in its position has big prestige.

Results of the method 1 Collisions in decisions are reduced compared to non-adaptive perception of the co-operative world but not eliminated. Prestige is assigned within the interval [0.5, 1] because every playmate deserves minimum credibility. Necessities vary in the interval [0, 1] but normally are low. Here follows that the behaviour of agents is as follows: when a player is far from the ball it will be passive or conservative and when the ball is closer it will be more active and aggressive.

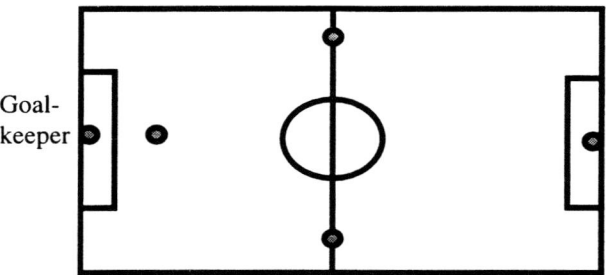

4.2 Method 2: a positional method with reinforcement of winners in conflicts.

Necessity

This is understood as the confidence any agent has on its own possibilities. This is an auto-perception. Necessity could be thought as the need of going to the ball an agent has. For example, if a defender sees the ball in the attack zone (in the opponent field) then the necessity of this player could be very low because it is not its responsibility to go to fetch the ball. This necessity will be different depending upon the perception of the world that every agent contains because of its specialised view and role.

For example, necessity of going to the ball could be maximum (1) if the ball is placed in the defence half field within the scope d distance from the origin. Progressively this necessity decreases towards the opponent half field till 0.

Prestige

Prestige is the perception of the co-operative world. It is the confidence on other playmates. Prestige that a player i is seen from a playmate j is based on using the necessity that player j has of going to the ball. This prestige, that it is initialised at a random value (0.5), will change during the game at every conflict:

- The agent that has to modify its belief because of a conflict, and happens that its reviewed certainty is lower than the reviewed certainty of the playmate. We write down the identifier of the playmate who won the conflict and its decision.
- At any moment again the agent has to modify its belief because of a conflict, then it will consider whether the conflict is with the same previous playmate. In

this case, if the conflict is solved in the same way as previously then reinforcement learning will be used, to reinforce, by means of modifying the prestige, the persistence of the rational decisions of the agents.

Results of method 2. The improvement of this method is significant and highly adaptive. Almost collisions in terms of co-operative decisions are eliminated.

5 Results

- The exchange of beliefs and application of consensus algorithm for rational reasoning improved the performance of the team by reducing the collisions.
- The change of each agent's perception of the co-operative world improved the exchange of beliefs and minimised the number of collisions almost to null.
- Prototyped is in Javasoccer. An 11x11 official RoboCup soccer simulator will be available this summer.
- The results of this research in Javasoccer are being applied to the real implementation of small-size RoboCup and FIRA robots and fields specifications.

(The results are downloadable from web page http://rogiteam.udg.es/robots/simulation.html)

6 References

[Acebo 98]   Acebo E., Oller A., de la Rosa J. Ll., and Ligeza A., "Static Criteria for Fuzzy Systems Quality Evaluation", in Tasks and Methods in Applied Artificial Intelligence, Vol. 2, pp: 877-887, *Lecture Notes in AI N°1416*, Angel Pasqual del Pobil, Jose Mira and Moonis AI Eds. Springer, 1998.

[Busetta 99]   Busetta P., Rönnquist R., et al., "JACK Intelligent Agents-Components for Intelligent Agents in Java", *Agentlink Newsletter*, No.2, pp. 2-5, January 1999.

[Chi 92]   K. Chi Ng and B. Abramson, "Consensus Diagnosis: A Simulation Study", *IEEE Transactions on Systems, Man, and Cybernetics*, Vol. 22, n° 5, September/October 1992

[de la Rosa 92a]   de la Rosa J. Ll., and Delgado A. "Fuzzy Evaluation of Information Received from other Systems", *Proceedings of the IPMU'92*, pp. 769-778, Palma de Mallorca, July 6-10, 1992, ISBN 84-7632-142-2

[de la Rosa 92b]   de la Rosa J. Ll., Serra I., and Aguilar-Martin J., "Outline of a Heuristic Protocol among Cooperative Expert Systems", *IEEE on Systems, Man, and Cybernetics*, Vol. 2 pp:905-910, ISBN 0-7803-0720-8, CHICAGO (USA) Septembre 1992

[de la Rosa 93]   de la Rosa J. Ll., Ignasi Serra and Josep Aguilar-Martin, "Applying a Heuristic Co-operation Framework to Expert Control", *IEEE International Symposium on Intelligent Control*, Chicago, August 25-27, 1993, ISBN 0-7803-1206-6

[Jennings 98]   Jennings N., Sycara K., and Wooldridge M. "A Roadmap of Agent Research and Development", *Autonomous Agents and Multi-Agent Systems*, Kluwer Academic Publishers. Vol. 1, n. 1, pp. 7-38, 1998.

[Shoham 93]   Shoham Y. "Agent-oriented programming", *Artificial Intelligence*, vol. 60 (1), pp.51-92, 1993 & *Technical Report STAN-CS-1335-90, Computer Science Department*, Stanford University, Stanford, CA, 1990.

The Ulm Sparrows 99

Stefan Sablatnög, Stefan Enderle, Mark Dettinger, Thomas Boß, Mohammed Livani, Michael Dietz, Jan Giebel, Urban Meis, Heiko Folkerts, Alexander Neubeck, Peter Schaeffer, Marcus Ritter, Hans Braxmeier, Dominik Maschke, Gerhard Kraetzschmar, Jörg Kaiser, and Günther Palm

University of Ulm, James-Franck-Ring, 89069 Ulm, Germany

1 Introduction

THE ULM SPARROWS ROBOCUP team was initiated in early 1998. Among the goals of the team effort are to investigate methods for skill learning, adaptive spatial modeling, and emergent multiagent cooperation [1]. We develop both a middle-size robot league *and* a simulation league team. Based mostly on equipment and technology available in our robot lab, we implemented a first version of both teams for ROBOCUP-98 in order to gain practical experience in a major tournament. Based on the these experiences, we made significant progress in our team effort in several areas: we designed new robot hardware, extended our vision processing capabilities and implemented a revised and more complete version of our soccer agent software architecture. In particular, we added Monte Carlo localization techniques to our robots, enhanced environment modeling, and started to apply reinforcement learning techniques to improve basic playing skills.

These improvements allowed our simulation team to consistently beat the qualification teams back home, e.g. the ROBOCUP-97 simulation champion AT Humboldt. In the ROBOCUP-99 simulation tournament, however, we had two very strong teams (the later champion CMUnited-99 and Headless Chickens) in our group and suffered some ugly defeats. Things went better for our robot team, although we still had to debug various hardware and software problems even during the tournament. In a middle size tournament with 20 teams playing the preliminaries in three groups, we finished second in our group and qualified directly for the playoffs. We then lost in our quarterfinal match against the Italian team, which advanced to the final and finished second overall. Altogether, we made substantial progress this year and laid a more solid foundation for future team development.

2 Team Development

Team Leader: Gerhard Kraetzschmar
Team Members: (Graduate students are in a M.S. program.)

- Stefan Sablatnög, PhD student, simulation team coordinator

- Stefan Enderle, PhD student, robot team coordinator
- Mark Dettinger, Thomas Boß, Mohammed Livani, all PhD students
- Michael Dietz, technician
- Jan Giebel, Urban Meis, Heiko Folkerts, Alexander Neubeck, Peter Schaeffer, Marcus Ritter, Hans Braxmeier, Dominik Maschke, all graduate student students
- Gerhard Kraetzschmar, research assistant professor
- Jörg Kaiser and Günther Palm, professors.

Web page http://smart.informatik.uni-ulm.de/SPARROWS

3 Robots

For RoboCup-99, we designed a new, modular robot which currently consists of five modules (see Figure 1): base, kicker, sonar, CPU, and camera. The latter four are common to all players. Only for the base module, which provides mobility, different designs are used for field players and the goalie: field players have a standard differential drive, while the goalie has a special four-wheel drive permitting very fast left/right movements.

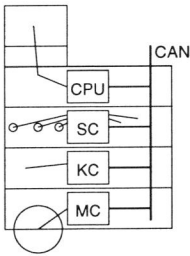

Fig. 1. A photo of the SPARROW 99 robot and a sketch of its hardware architecture. MC=motor controller, KC=kicker controller, SC=sonar controller, CPU=PC104+ with PCI framegrabber.

The design of our computer hardware for the robots follows the *smart device architecture* approach. It fosters modular, distributed designs by bringing computation closer to the data. We combine sensors and actuators with microcontrollers. These *smart sensors/actuators* perform local computation on data and thereby reduce communication and computation load on the central CPU. These *smart devices* communicate via a CAN bus with each other and the main CPU (see Figure 1). Modules can be connected using only four lines: two for power supply, and two for the CAN bus.

4 Software Architecture

We continued the implementation of a *common soccer agent software architecture* [1] (see Fig. 2) for both the simulation and the real robot teams. This year we

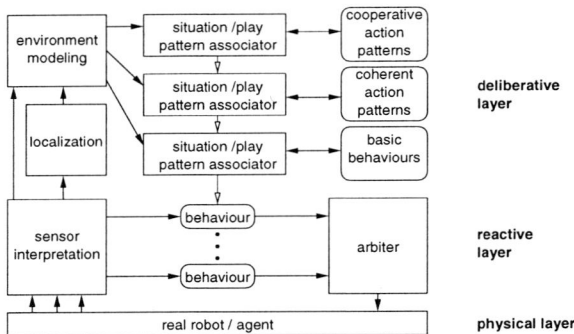

Fig. 2. THE ULM SPARROWS soccer agent architecture

provided a C++ software library that allows to quickly implement and modify the reactive layer of the architecture. It provides behavior and arbiter classes, which are easy to instantiate and ensure safe execution of behaviors and arbiters as parallel threads. A graphical policy editor can be used to specify temporal sequences of behavior sets based on events and signals; it automatically generates program code.

5 Perception

Our new robot hardware now permits us to grab frames faster and with higher resolution. We extended our behavior-based vision architecture with line-detection routines in addition to color blob trackers, and can now detect the ball, both goals, and the corners quite robustly.

6 World Model

Our soccer agents apply multi-layered spatial representations for modeling the environment. Currently, a two-layer approach that is derived and adapted from the DYNAMO spatial representation architecture (see [2]) developed in the SMART project is employed. The lower layer consists of an *egocentric feature map* representation. Relevant features include position and distance of the ball, the goals, and field landmarks. The egocentric representation is mainly used by low-level behaviors for rapid action selection.

The upper layer of the environment model is an *allocentric* spatial representation, which is constructed and maintained by integrating over time information present in the egocentric representation.

7 Skills

As demonstrated very convincingly e.g. by CMUnited-98, it is absolutely necessary to provide robust low level behaviours. Although behaviors can be hand-crafted, this is a very tedious process that must possibly be redone every time some system parameter changes. Such parameter changes have occurred almost every year in both leagues we compete in, e.g. in the soccer server the size of the physical agent and stamina model parameters have been changed, as well as the field size and lighting conditions in the middle size league. All our behaviors used in competitions so far have been hand-crafted, but we have started to apply reinforcement learning techniques to this problem. Preliminary results are quite encouraging and we will extent our efforts in this area.

8 Strategy

Our decision to **not use** communication between players during games is quite unique. We also focus on building strong individual players first before seriously pursueing cooperative team play. Our approach builds upon strong situation assessment capabilities and a broad reportoire of basic skills (behaviors). Decision making is then the association and instantiation of the right set of behaviors with particular situations. Cooperative team play will be a natural extension. If it occurs, it will be emergent behavior, which arises when two or more players independently of each other, but in temporal synchrony classify a game situation and their role in it in a consistent manner.

9 Conclusions

THE ULM SPARROWS team follows the lines of research set out in our initial team description paper [1]. We have made substantial progress and performed well in RoboCup-99. Our plans for the future focus on improving our software in almost all areas, in particular for situation assessment and skill learning.

References

1. Gerhard K. Kraetzschmar, Stefan Enderle, Stefan Sablatnög, and other. The Ulm Sparrows: Research into sensorimotor integration, agency, learning, and multiagent cooperation. In Minoru Asada, editor, *RoboCup-98*, volume TBD of *Lecture Notes in Artificial Intelligence*. Springer, 1999. to appear.
2. Gerhard K. Kraetzschmar, Stefan Sablatnög, Stefan Enderle, and Günther Palm. Application of neurosymbolic integration of environment modeling in mobile robots. In Stefan Wermter and Ron Sun, editors, *Hzbrid Neural Symbolic Integration*, volume TBD of *Lecture Notes in Artificial Intelligence*. Springer, 1999. to appear.

UBU Team

Johan Kummeneje, David Lybäck, Håkan Younes, and Magnus Boman

The DECIDE Research Group, Department of Computer and Systems Sciences,
Stockholm University and the Royal Institute of Technology, Sweden

1 Introduction

The aim of developing UBU is to subject a series of tools and procedures for agent decision support to a dynamic real-time domain. These tools and procedures have previously been tested in various other domains, e.g., intelligent buildings [2] and social simulations [6]. The harsh time constraints of RoboCup requires true bounded rationality, however, as well as the development of anytime algorithms not called for in less constrained domains (cf. [3]). Artificial decision makers are in the AI and agent communities usually associated with planning and rational (as in utility maximising) behaviour. We have instead argued for the coupling of the reactive layer directly to decision support. A main hypothesis is that in dynamic domains (such as RoboCup), time for updating plans is insufficient. Basically depending on the size requirements of agents, and on the communication facilities available to the agents, we have placed decision support either in the agents, or externally. In the former case, deliberation is made in a decision module. In the latter case, a kind of external calculator which we have named pronouncer provides rational action alternatives. The input to the pronouncer is decision trees or influence diagrams. The structure and size of these models are kept small, to guarantee fast evaluation (cf. [7]). The pronouncer can be made into an agent too, e.g., by using a wrapper. The coach function is particularly interesting in this context, since it is "free" and since it could hold the pronouncer code. An important problem here is the uncertainty and space constraints on the communication with the coach. The concept of norms as constraints on agent actions has also been investigated [1]. A team in which each boundedly rational player maximises its individual expected utility does not yield the best possible team: Group constraints on actions must be taken into account (see, e.g., [4]). Norms is our way of letting the coalitions that an agent is part of play a part in the deliberation of the agent.

The participation in RoboCup'99 was not successful as there were problems with the server-timing. UBU was among the least successful teams, ending up among the last in our group. It is not an issue whether we win or loose, it is for the scientific results we are participating.

2 Team Development

The work has been done over a period of more than two years, with two different versions of the team. The current version is the product of the latter six months.

Team Leader: Johan Kummeneje
Team Members:
  Magnus Boman (Associate Professor)
  David Lybäck (M.Sc.)
  Håkan Younes (Graduate Student)
Web page http://www.dsv.su.se/~robocup

All of the team members are connected to the DECIDE research group, and all attended RoboCup-99.

In addition, Johan Sikström, Jens Andreasen, Helena Åberg, and Åsa Åhman have made significant contributions to the different parts of the team.

3 World Model

Our players are always aware of the overall state of the game, according to the referee's messages, by internally representing the last known state.

In addition to the states given by the referee, our agents also express a degree of the certainty of their "belief" in that their team has the ball, i.e. they can determine the state, and act accordingly when they have the ball. The referee's messages in combination with the belief of having the ball, yields the situated automaton shown in Fig. 1.

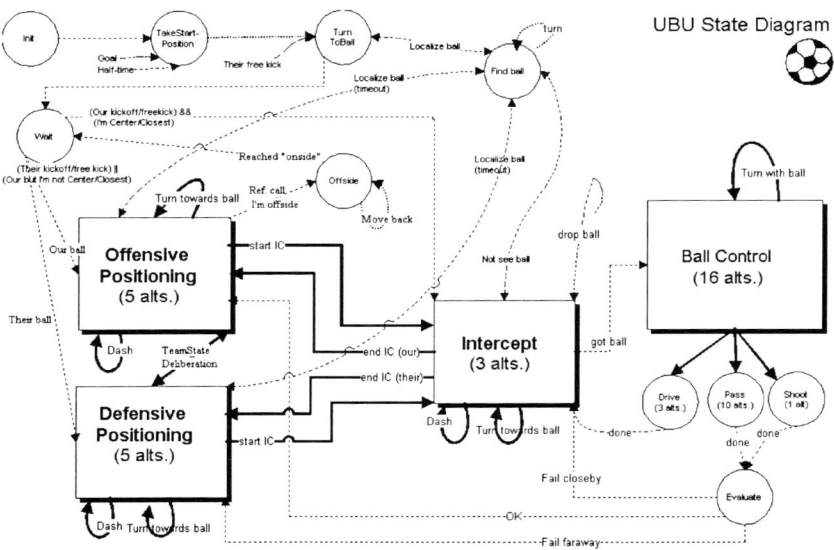

Fig. 1. The situated automaton of UBU.

The states consist of several options for what to do next, of which some are reactive (reacting to referee messages mostly) and some are deliberative. The options determine the state transitions, and through the careful design a player will never be "between" states.

We are not using the *libsclient* as we have developed the basic functionality by ourselves.

4 Communication

The inter-agent communication in the team has been very limited, and is not of any practical use yet. Each player has, however, built-in support for the communication of formation swapping, e.g., the team is able to change the formation from 2-4-4 to 4-4-2 based on messages from the libero.

Every formation swap is propagated when the libero decides to change the formation, and when a kickoff occurs.

5 Skills

The players are simple in their behaviour, as they do not make use of any special skills. When intercepting the ball a player does not calculate an intersection with the ball path. The player instead uses two rules to follow the ball. If the ball is more than 5 degrees to the left or right of the center of the player's viewcone, then turn to face the ball, else run straight forward. This works surprisingly well in most situations.

As the players have a degree of obstacle avoidance built in they can usually, by driving the ball in front of them, dribble past fairly simple opponents. We have not spent a lot of time on creating or training the dribble-behaviour of the team.

The goalie is in almost every aspect the same player as the rest of the team, with the small difference of having the capability to catch the ball. The main difference between the different roles of the team (i.e., the goalkeeper, defenders, midfielders, and forwards) is a variable that controls the defensiveness of the players.

6 Strategy

The foundation of our team is the idea that there is insufficient time for planning, and thereby the team does not in any way (yet) plan. We have instead used a lot of decision situations in which we have identified what is reasonable to do. Extending the team, we will incorporate the concept of norms, i.e. mutually agreed-upon constraints or heuristics, which each agent can decide to follow or not. When possessing the ball, each agent has three distinct choices to perform, pass, shoot at the goal, or dribble. This choice is made with respect to the situation, i.e. the position of each player, and several other factors. When not possessing the ball, the main task for each of the agent is to optimise its position on the field in order to be able to intercept the ball easily or be close to their home position.

7 Special Team Features

Our main aim of developing UBU is to subject a series of tools and procedures for agent decision support to a dynamic real-time domain. As stage, we use neither machine learning, nor opponent modelling.

We have in our team created effective thread-scheduling, i.e. by having 10-15 threads running in each player every cycle. Besides the multi-threading we have used several concepts inherent in Java, such as events and listeners. By using standard components we have been able to create a flexible and easily extensible basic foundation to build higher level functionality upon [5].

8 Conclusion

During the summer of 2000 our team will enter both the European championships and the world cup. We will be working on the improvement of the functionality of the team, and initiating several smaller projects in which we will investigate the benefits of machine learning in our team. Besides the already mentioned projects we will also incorporate concepts from the social sciences, such as incentives for coalition formation, and norms.

References

1. Magnus Boman. Norms in artificial decision making. *Artificial Intelligence and Law*, 1999. 7:17-35.
2. Magnus Boman, Paul Davidsson, Nikolaos Skarmeas, Keith Clark, and Rune Gustavsson. Energy saving and added customer value in intelligent buildings. In Hyacinth S. Nwana and Divine T. Ndumu, editors, *Proceedings of PAAM98*. The Practical Application Company, 1998. pages 505-516.
3. Magnus Boman, Paul Davidsson, and Håkan Younes. Artificial decision making under uncertainty in intelligent buildings. In *Proceedings of UAI'99*, 1999.
4. Susan Kalenka and Nicholas R. Jennings. Socially responsible decision making by autonomous agents. In *Proceedings of the 5th International Colloquium on Cognitive Science*, 1997.
5. Johan Kummeneje, David Lybäck, and Håkan Younes. Ubu - an object-oriented robocup team. Internal working note, Department of Computer and Systems Sciences, Stockholm Univerisity and the Royal Institute of Technology, Sweden, 1999.
6. Harko Verhagen. On the learning of norms. Maamaw Poster.
7. Håkan Younes. Current tools for assisting intelligent agents in realtime decision making. Master's thesis, Department of Computer and Systems Sciences, the Royal Institute of Technology, 1998. DSV Report 98-x-073.

YowAI

Takashi Suzuki

Department of Computer Science
The University of Electro-Communications, Tokyo, Japan

1 Introduction

I want to do various researches by using RoboCup Seccor Server in my laboratory. However, the history of my laboratory in RoboCup is shallow and only half a year has passed since the research began. The research theme at which we aimed at first was "Real time, Distribution, and Cooperation" and it was the approaches from three sides. The outline of the team is described in this paper around World Modeling which should be called the result of "Real time and Distribution" thought to have reached at a standard level by present.

We know the effect is thin in the gauge of winning by the game even if an advanced function is provided for the agent who has the operation of the uncertainty experiencing prototyping in short time of half a year. Even if this proposes a very excellent idea, this shows no appearance as a clear result too much. Then, we did the effort to decrease uncertainty of infomation other than unavoidable uncertainty of information which the server added as much as possible. As a result, the agents succeeded in the considerable, highly accurate construction of World Model[1, 2] (That is, it is various informations on agent's field). This result invented surprising result. When the algorithm only of the chase the ball and the kick to the goal was installed to evaluate World Model, the team were performed close games and Andhill98, one of strong teams in Japan. Afterwards, the agents has grown up to handle the game equally to the team at a top level of Japan by having introduced the system which the introduction of a dynamic role of goalie and which efficiently uses of stamina. The architecture of agent of the team and the composition method of World Model that the uncertainty are few is described in the following.

This agent came to leave a good result of the world rally best 8 and the semi-Japanese rally victory though operated by a very simple algorithm.

2 Team Development

Team Leader: Takashi Suzuki
Team Members:
  Takashi Suzuki
    – Japan
    – master
    – remote participation
Web page http://ne.cs.uec.ac.jp/ suu/RoboCup99

The agent executes continuously the cycle of acknowledgment, judgment, and action by synchronizing simulation cycle of SoccerServer because of the single process. The part of acknowledgment constructs World Model from infomation received from SoccerServer. The part of judgment decides the goal at present from World Model. Then the agent computes various conditions which is the optimal dynamic role, the offside line, and the intercepting location of the ball etc. to decide the goal. The part of action decides a short goal to achieve the decided goal and issues the command to SoccerServer actually. The behavior of the agent basically is decided by the judge by World Model at time and is put into practice.

3 World Model

Here, an important element for composing World Model is described. The synchronization which can be put on simulation cycle between Client/Server is necessary in the beginning. This is a condition to use all cycles effectively. It is that it is synchronization of step time of simulation cycle between Client/Server next. This is the most important element for composing World Model. Because, it is the only clue to know what when one information which has been sent from SoccerServer or I did when. World Model information is propagated by say message at the end. As a result, other agents can supplement information which a certain agent cannot understand and the thing to decrease the difference between each agent's World Model in addition becomes possible.

The synchronization of simulation cycle and the synchronization of step time exceed 99.9at the origin of the best environment. In addition, the error in the information of the position and the direction of the agent itself is about 0.3 meter and about 0.5 degrees respectively.

Table 1. The result of varius experiments

Experimant	Ratio of success
syncronization of simulation cycle	99.9%
syncronization of step time	99.9%
success of command execution	99.9%

4 Communication

My agent uses the communication between agents only to transmit own world model. Whether my agent communicates between agents of what time has been

Table 2. The error of value for player itself on World Model

Value	Error
x coordinate	0.2 meter
y coordinate	0.2 meter
body direction	0.5 degree

decided to each agent beforehand. Actually, only one agent uses the communication between agents once every two cycles.

5 Skills

My agent's main skill is to kick running after the ball, and the ball toward the goal of the enemy. Importance because of running after the ball is to forecast the position where the player comes in contact with the ball. Moreover, when the ball is kicked toward the goal of the enemy, my agent kicks the ball as twice as possible continuously.

6 Strategy

My agent is behaving by a simple algorithm of kicking the ball toward the goal of the enemy running after the ball. This algorithm is an algorithm originally used to debug the world model.

7 Special Team Features

7.1 Robust property

The team realized the robust property. The agent of the team has static role and dynamic role, for example goalie, defender etc. As Figure 1 when a agent with goalie of static role intercepted a ball, other agent with defender of static role defend goal. Then dynamic role is exchanged. This is judgment of agent itself.

7.2 Stamina system

The agent wastes the stamina, because the agent always chases the ball. But the quantity of movement of the agent is larger than that of the agent of the other team overall of a match. The agent dose not use the stamina over threshold decreased the effort or the recovery if it is unnecessary. When the agent can shot or the enemy agent will shot, the agent use the stamina over threshold. However, the quantity of the stamina decreased for the movement increased to be concernd with the version up to 5.xx of the SeccorServer. Therefore, I must try to design the new stamina system.

Zeng99 : RoboCup Simulation Team with Hierarchical Fuzzy Intelligent Control and Cooperative Development

Junji Nishino, Tomomi Kawarabayashi, Takuya Morishita, Takenori Kubo, Hiroki Shimora, Hironori Aoyagi, Kyoichi Hiroshima and Hisakazu Ogura

Department of Human and Artificial Intelligent Systems, Fukui University, JAPAN

Abstract. This paper discusses the design of the team Zeng99. The goal of team Zeng99 is to show a performance of Hierarchical Fuzzy Intelligent Control system in the field of multi agent problems. It worked well at RoboCup99 competition, even with little error in an invoking clients. It also allow independent/cooperative development client by client.

1 Introduction

The goal of robocup simulation team Zeng99 is to show a performance of Hierarchical Fuzzy Intelligent Control system (HiFIC) [1] in the field of multi agent problems, such as soccer simulation. The HiFIC is a scheme of controller for ill-defined/described objects.

These days, there are many studies on intelligent control systems to perform high level control such as human operators do. Human knowledge based controller model is an approach to realize an intelligent control system. In this paper, HiFIC is adopted to soccer agent cooperative behavior planning and reactive control.

The HiFIC controller is a derivative of three layered control model by Jens Rasmussen[2]. HiFIC also consist of three levels: lower layer to regulate primitive reactive control, middle layer to perform skill level behavior and highest layer to make decision on strategic and tactical playing plan. A main ability of this system is easy construction of hybrid controller which combines feedback loop regulator facility and feed-forward control facility.

2 Team Development

Zeng99 team development model was a independent and/or cooperative clients programming.

Team Leader: Tomomi Kawarabayashi (Offensive Mid-fielder)
  – Graduate student
  – attend the competition
Team Director: Junji Nishino

- Faculty: Assistant professor
- attend the competition

Team Members: – all members below did not attend the competition
Takuya Morishita (Defensive Mid-Fielder), Hiroki Shimora (Goalie)
- Graduate student

Takenori Kubo (Defender, Sweeper), Kyoichi Hiroshima (Goalie 2)
- Graduate student, PhD candidate

Hironori Aoyagi (Forward)
- Undergraduate student

Hisakazu Ogura
- Faculty: Professor

Web page http://bishop.fuis.fukui-u.ac.jp/~tomomi/index_e.html

We have six independent client programs that was created by different programmers respectively. The clients were preassigned to their own position in the 4-4-2 soccer formation that is shown in the figure 1.

3 World Model and Communications

Zeng99 use libsclient4.0 for offensive mid-fielder, the source cord of CMUnited-98[3] for defender and side mid-fielder, and special developed low level libraries for goalie. They are just ordinary models for time interpolation.

Zeng99 didn't use any on-line communication for emerging cooperative play, in another words the team didn't use say command. They used eye-contact communication, which is occurred by matching of behavior rules.

4 Skills

All clients except the Goal keeper don't have tuned skills such as a boll keeping faint move. CMUnited-98 based clients only have the skills of the CMUnited-98 source code.

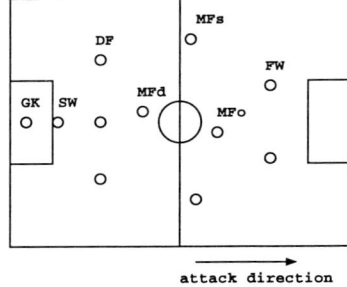

Fig. 1. Zeng99 4-4-2 Formation

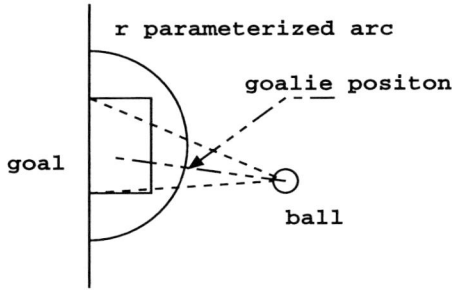

Fig. 2. Goalie positioning

The Goal keeper have two special skills. It have 1) middle term boll prediction and 2) trace moving on an arc. Here, middle term means several simulation steps among two and five. The goalie predicts two to five steps ahead, and then tests the boll position and his possible position to determine whether he can catch it or not at that time. If he decides that it can be caught, then he goes to appropriate position and catch the ball. If not, he goes to goalie position described in the figure 2 .

5 Special Team Features and Strategy

For the Zeng99 team's strategy, human soccer players' knowledge is implemented in the form of HiFIC. The HiFIC system allow us to write a knowledge in natural language like form of fuzzy inference model with ambiguous words by fuzzy logic facility.

The team clients were developed separately under the agreement of programmers, that is cooperative strategy with 4-4-2 formation system. In the process of development, it takes repeated three phases. The steps are as follows; 1) analyzing human players' knowledge, 2) brain storming and discussion on the knowledge by whole development members, 3) independent development and testing. This sequence was repeated once a week.

Programmers had a common workspace to store current version of one's client so that they were able to test his client with newest another position clients. They were brushed up by trial and error toward the agreements made at a discussion phase. This style of development could provide quickly actual prototypes of several clients which are pegged on particular rolls. In the knowledge analyzing phase, we interviewed real armature human soccer players and/or studied soccer books.

Zeng99 have neither learning mechanism nor modeling ability at the version of Stockholm. Though the Goalie was tuned very well. it needed a sweeper just in front of our goal. This style is not available very often in the usual soccer play. However, indeed, a combination play between goalie and back ground sweeper was worked very well. Because goalie moves strictly only ON THE ARC LINE shown in the figure 2, he couldn't catch balls behind him, even if the ball run very slow.

6 Results

We had errors in the startup script instead of the program, so the Zeng99 team couldn't gain a point at the regular competition. Therefor we like to show some result from friendly match results which has played in the simulation league site and on the same machines. At the regular round-robbin, because of startup script

Table 1. Results

Regular round-robbin group A		Friendly match results		
0-11	CMUnited99	2-0	Footux99	12-0	Polytech
0-1	UlmSparrows	1-5	Cyberoos99	9-0	Robolog
0-3	HCIII	2-0	UlmSparrows	0-7	BrainStormers
				4-0	NITStones

error, zeng99 clients couldn't read correct server.conf file in the game against CMUnited99, the champion team. Thus they couldn't kick the ball. Therefore CMUnited99 always keep the ball, however the goalie and the sweeper had saved their goal 30 times in 41 shoot trial. At the friendly matches, the goalie, the sweeper and defender also played well, repaired the error.

7 Conclusion

Human knowledge based cooperative soccer playing rules are implemented using Hierarchical Fuzzy Intelligent Control System architecture and they worked well on matches with other teams. This type developments allow us to separately developments. In the future work, make a common knowledge based strategic level controller for both simulator and real robots.

References

1. J. NISHINO, A. TAGAWA, H. SHIRAI, T. ODAKA, and H. OGURA. Hierarchical fuzzy intelligent controller for gymnastic bar actions. *Journal of Advanced Computational Intelligence*, Vol. 3, No. 2, pp. 106–113, 1999.
2. Jens Rasmussen. Skills, rules, and knowledge; signals, signs, and symbols, and other distinctions in human performance models. *IEEE Trans. SMC*, Vol. 13, No. 3, pp. 257–266, 1983.
3. Manuela Veloso Peter Stone and Patrick Riley. *RoboCup-98: Robot Soccer World Cup II*, Vol. 1604 of *Lecture Notes in Artificial Intelligence*, chapter Champion Teams: The CMUnited-98 Champion Simulator Team, pp. 61–76. Springer, 1999.

All Botz

Jacky Baltes, Nicholas Hildreth, and David Maplesden

Centre for Imaging Technology and Robotics
University of Auckland
Auckland, New Zealand
j.baltes@auckland.ac.nz

1 Introduction

This paper discusses some important features, which make the All Botz, the University of RoboCup team, a very unique team. In particular, the use of cheap hardware and the design of the video server.

Instead of custom built robots, the All Botz team consists of cheap remote controlled cars, which can be purchased at any toy store. The cars use coarse D-A converters to provide proportional steering and speed control. Cost is the main advantage of the All Botz; a standard team costs around $10,000 USD, compared to the $200 USD for the All Botz team.

Like most teams competing in the small sized league, the All Botz use a global vision system. However, there are a number of important differences between the All Botz and other teams. Firstly, it is not necessary that the camera is mounted overhead, but it can be mounted on any angle as long as the whole field can be viewed. Secondly, the toy cars are not modified, the video data is the *only* source of information for the agents.

Although disadvantaged by their cheap hardware, the All Botz achieved a respectable result.The All Botz had to play in a tough group, which included the two best teams in the world (Big Red from Cornell University, and Robotis from Korea).

In the first game, the All Botz lost 33:0 to Big Red. The biggest problem in this game was that our path planner took to long to formulate paths for our agents. Another problem was that the Big Red team, like all other teams, are much heavier and powerful than our robots, which meant that our robots were pushed around a lot.

In the next game, the All Botz faced Robotis. The Robotis team is a very exciting team to watch, since their robots move at a significantly faster than any other team. The All Botz lost 35:0 to Robotis. Our goalie made most of the first saves, but was not fast enough to clear the ball.

In our last game, we beat last year's third place finisher 5DPO 3:1. 5DPO is a difficult team to score against since their strategy is to cover the ball in their own half. Still, the All Botz dominated the game and territory. The only goal that 5DPO scored was due to an error of our goalie.

2 Team Development

Team Leader: Jacky Baltes
Team Members:
  Jacky Baltes
   - CITR, University of Auckland
   - New Zealand
   - Lecturer
   - did attend the competition
  Nicholas Hildreth
   - CITR, University of Auckland
   - New Zealand
   - Graduate Student
   - did attend the competition
  David Maplesden
   - CITR, University of Auckland
   - New Zealand
   - Graduate Student
   - did attend the competition
Web page http://www.citr.auckland.ac.nz/~jacky

3 Sensing

The most obvious difference between the All Botz and other teams is that the camera is looking at the playing field from the side rather than from directly overhead. This means that the All Botz are more flexible and can play under a wider range of conditions than other teams.

The side view introduces large perspective distortions in the image, which must be compensated for by the video server. This problem can be overcome by a sophisticated camera model and an accurate camera calibration. The All Botz use Tsai's camera calibration method that corrects for the radial distortion of the lens.

The calibration of this camera model requires a large set of calibration points, distributed across the playing field. Instead of using the points on the playing field itself (e.g., the corners and centre points), the All Botz use a calibration carpet (a duvet cover with a square pattern of white and blue square on it). Once a picture of the calibration carpet is taken, a program finds and sorts the squares in the image and assigns them real world coordinates. These mapped points are then used in the Tsai camera calibration to compute the parameters of the camera model.

This system has proven to be very robust. Camera calibration now takes less than 30 minutes and results are accurate to within one centimeter even at the far side of the image.

4 Communication

To improve robustness, the agents only use implicit communication to tell other agents about their goals and fall back on their autonomous behavior if communication fails.

The limitations of the car-like robot makes it very important that agents communicate. A lot of tasks that can easily be achieved by a single holonomic robot, such as clearing the ball by the goalie, can only be achieved efficiently by coordinating the activities of multiple agents using car-like robots.

Currently, the All Botz use two groups of players that communicate amongst each other: the two strikers, and the goalie/defender combo. The communication is similar in both groups and we will focus on the strikers in this paper.

Since it is impossible for our cars to move sideways, the strikers implement a "cycling" behavior. One striker moves in for a shot on goal, whereas the other striker moves into a position to wait for a rebound or to shoot at the goal next. Since we want to avoid having both strikers try to shoot at the goal at once, they need to communicate their intention. Both strikers evaluate their relative position to the ball and the goal. They then compute an estimate of the cost of a goal shot and the striker with the smaller cost will shoot at the goal whereas the other striker will move into the rebound position. The estimate of the goal shot cost is based on the holonomic path distance. After the first striker shoots at the goal, the world state will change and the striker waiting for the rebound will then start its attack run.

This scheme is augmented by: (a) implementing a hysteresis function for switching from goal shot to rebound to avoid oscillation, and (b) a time horizon scheme. The striker agent has only a limited time to show progress. If the goal shooter does not make progress towards the goal shot, because for example, it is being blocked by an opposing robot, the rebounder will attempt a goal shot. Once the rebounder is closer to the ball than the goal shooter, their roles will change and the striker will become the rebounder and it will attempt to move towards the rebound position.

5 Special Team Features

This section discusses the path planning problem and the problem of path following control for car-like robots and our solutions to these problems.

5.1 Car-like robot control

Controlling a car-like robot at high speeds with noisy vision data as the only source of information is a challenging and difficult problem.

Even though the non-holonomic control problem is more difficult than that of wheeled robots, the control of our agents is as good as that of other teams (with the exception of the Robotis team). The All Botz robots are controlled to a maximum speed of 1 m/s. However, the lack of local control means that the

All Botz system is much more robust. Schemes that use shaft encoders and stall detectors to augment the localization are unable to cope with rough terrain or slippery surfaces.

The state of the art in non-holonomic control was not sufficient to control car-like robots at high speeds. The All Botz developed a number of different control algorithms, the best ones being a look ahead controller and a controller that uses reinforcement learning to learn the control function. We believe that these are the best practical implementations of control for car-like robots in the world today.

5.2 Non-holonomic path planning

Efficient Path planning in highly dynamic environments is a difficult problem for holonomic vehicles. For car-like robot the problem is even worse.

Initially, the All Botz use a non-holonomic path planner, which is an extension of visibility graphs. This path planner performs well in a static environment, but its time complexity makes it unsuitable for dynamic environments.

Currently, the All Botz use an adaptive case-based path planning system. Instead of re-planning from scratch whenever an object moves in the domain, the path planner tries to adapt the current plan to match the new world state. The path planner uses the following set of adaptations for each path segment: translation, rotation, lengthening, shortening, change of turn radius, change from straight line to curve and vice versa, deletion, and insertion of a new path segment.

Another planner we are currently developing is a *any-time* path planner. The basic idea is that the agent can ask the planner for the best currently available plan. If given little time, the planner may return an incorrect or sub optimal plan. However, if given more time, the planner will eventually return the optimal plan.

6 Conclusion

For next year, the reinforcement learner will be enhanced to include look-ahead, which will allow us to control the cars at much higher speeds (2 - 3 m/s).

After emphasizing and finding scalable, practical solutions to the low level problems, such as control and path planning, we will focus more on the design of team strategies and agent coordination.

Currently, the agent architecture is being extended to include explicit communication. Also, the interaction between the controller and the path planner is being made more expressive. This allows the path planner to use short macro sequences (e.g., a quick kick followed by recovery) in the planning stage.

Also, currently, the strikers only attempt a direct shot on goal. The set of aims of the individual players will be extended to include passes, double plays, and give and go plays.

We hope that the addition of these strategic components will make the All Botz a team to be reckoned with in 2000. Our goal is to be among the best eight teams in the world.

Big Red: The Cornell Small League Robot Soccer Team

Raffaello D'Andrea, Jin-Woo Lee, Andrew Hoffman, Aris Samad-Yahaya, Lars B. Cremean, and Thomas Karpati

Mech. & Aero. Engr., Cornell University, Ithaca, NY 14853, USA
{rd28, jl206, aeh5, as103, lbc4, tck4}@cornell.edu
http://www.mae.cornell.edu/robocup

1 Introduction

In this paper we describe Big Red, the Cornell University Robot Soccer team. The success of our team at the 1999 competition can be mainly attributed to three points:

1. An integrated design approach; students from mechanical engineering, electrical engineering, operations research, and computer science were involved in the project, and a rigorous and systematic design process[2] was utilized.
2. A thorough understanding of the system dynamics, and ensuing control.
3. A high fidelity simulation environment that allowed us to quickly explore AI and control strategies well in advance of working prototypes.

The paper is organized as follows. In Section 2, we describe the overall aspects of the project, followed by the artificial intelligence and strategy in Section 3. We include some of the other features of our team in Section 4.

2 Characteristics of Overall System

The robots have a design mass of 1.5 kg and a maximum linear acceleration of 5.1 m/s^2 and a maximum linear velocity of 2.5m/s. The goalkeeper design is independent of the field player design, and thus the goalkeeper exhibits significantly different skills. The goalkeeper is equipped with a holding and kicking mechanism that can catch a front shot on goal, hold it for an indefinite amount of time required to find a clear pass to a teammate, and make this pass. Configuration constraints result in decreased angular but increased linear performance, as compared to the field players.

The main features of our electronics are wireless communication, motor control and the kicking control. Considering the speed, memory space, I/O capability, and the extension flexibility, we decided to use 16bit 50MHz microcontroller. An infrared system is used to detect the ball. It informs the microcontroller when the ball has come into contact with the front of the robot.

After careful considerations and trade-off analysis, the wireless communication was limited to one-way transmission from the global AI computer to each robot. The main justification for the decision is the lack of on-board local sensing information. The one-way transmission saves the communication time comparing to the two-way communication, and makes the AI strategy and the on-board program easier. The our wireless communication system takes 12.5ms to transmit the whole information from AI computer to the robots.

The vision system perceives the current state of the game and communicates this state to the artificial intelligence computer allowing decisions to be made in real-time in response to the current game play. The vision system captures frames at half resolution (320x240) at a rate of 35-40 Hz. Object detection and tracking is not determined by the previous locations of objects, and the full frame is searched for targets.

Listed below are the summary of main characteristics of our system:

Characteristic	Goal Keeper	Field Player
Weight	1.78 kg	1.65 kg
Max. Acceleration	5.90 m/s^2	5.10 m/s^2
Max. Velocity	1.68 m/s	2.53 m/s
Max. Kicking Speed	4.18m/s	2.6 m/s
Operating time	30 min per battery pack	
Special function	Ball Holding mechanism only for goalkeeper	
Wireless Transmission	40 kbit/s	
Other Sensor	Infrared systems to detect the ball	
Vision Speed	35 - 40 Hz	
Vision Resolution	320 by 240 pixels	

3 Strategy and Artificial Intelligence

The artificial intelligence subsystem is divided into two parts: high level AI and low level AI. The low-level AI computes the trajectory to the target point and computes the wheel velocities to transmit to a robot. The high-level AI takes the current game state (robot and ball positions, velocities, and game history) as input, and generates new targets and kicker commands for the robots as output. A target consists of a field location and a desired final state (time-to-target, robot orientation, and robot velocity).

3.1 Trajectory Control

Compared to reactive control strategies, such as those in [1] for example, we perform a global trajectory optimization for each robot and take advantage of the mechanical characteristics of the robots. It takes as inputs the current state of the robot and the desired ending state. The current state of a robot consists of the robot x and y coordinates, orientation, and left and right wheel velocities. A desired target state consists of the final x and y coordinates, final orientation, final velocity as well as the desired amount of time for the robot to reach the destination.

Our geometric path is represented by two polynomials in the x and y coordinates of the robots. The x coordinate polynomial is a fourth-degree polynomial and the y coordinate polynomial is third degree. The task is to solve for the nine polynomial coefficients for a particular path requirement.

Once a trajectory is generated, the velocity of both wheels are easily calculated. Even though each run of this algorithm generates a pre-planned path from beginning to end, it can be used to generate a new path after every few cycles to compensate for robot drift. The continuity of the paths generated is verified through testing.

3.2 Role Based Strategy

High-level AI maintains a list of roles that the four field players can assume. Every cycle, the high-level AI computes the priority and feasibility of each role as a function of the game state. To perform this, it first sorts the roles by priority. Then, it rearranges the list of roles based on their feasibilities. Certain role-robot pairings are ruled out if determined to be completely infeasible. For example, if the trajectory of the ball relative to a robot R is such that it is difficult to shoot it into the goal, not only will the feasibility value of this role for R be relatively low, it will reflect how difficult it is for R to shoot.

Finally, the high-level AI chooses the most suitable role assignments from the top of the list, and calls the respective role functions to give instructions to the robots.

4 Other Team Features

4.1 Vision Calibration

The vision calibration consists of 4 main parts. They are:

- barrel distortion correction
- scaling
- rotation
- parallax correction

Barrel distortion is a function of the lens of the camera and is radially symmetric from the center of the image. To invert the distortion, points are measured from the center of the image to the corner of the image and a look-up table is built to map a point in image-coordinates into a new coordinate equidistant coordinate system. The scaling is then computed such that the sides of the field are computed. Since the camera cannot be mounted perfectly, the rotations about the center axis of the camera also need to be taken out. The points are rotated so that the sides of the field are have constant x-coordinates along the widthwise walls and constant y-coordinates along the lengthwise walls. Finally the parallax error that results from differences in object height is removed by scaling the x- and y-coordinates proportionally to the distance that the object is from the center of the camera projected onto the field plane.

4.2 High Fidelity Simulation

To provide a realistic testing platform for our artificial intelligence system, we have constructed a simulation of the playing field that models the dynamics of our environment.

The dynamic modeling of our system is performed by a Working Model 2D[3] rendering of the complete playing field. The model includes two teams of five individual players, the game ball, and the playing field. Real world forces and constraints are modeled, including the modeling of the motion of the tires and the inertia of the robots and ball. Additionally, the physical interactions between the players and each other, the ball, and the playing environment are all modeled in Working Model's two dimensional environment.

5 Conclusion

Even though our team performed well at the competition last year, there are many subsystems and components that need to be improved. The main ones are outlined below:

- A more robust vision system. The current vision system performs well when operational, but does fail on occasion. In addition, it takes a very long time to calibrate the system. One of our objectives for next year is to construct a reliable vision system that can be set up in less than 30 minutes.
- Role coordination. This will allow us to implement set plays.
- More refined trajectory generation, obstacle avoidance, and trajectory control.
- Reduce the system latency.
- Innovative electro-mechanical designs.

References

1. RoboCup-98:Robot Soccer WorldCup II, Hiroaki Kitano(Ed.), Springer 1998
2. S. Blanchard and W.J. Fabrycky. System Engineering and Analysis. Prentice Hall, 3rd Edition, (1997)
3. Working Model. Knowledge Revolution, San Mateo, CA.

The CMUnited-99 Small-Size Robot Team

Manuela Veloso, Michael Bowling, and Sorin Achim

School of Computer Science
Carnegie Mellon University
Pittsburgh, PA 15213-3891
{mmv,mhb,sorin}@cs.cmu.edu

One of the necessary steps in entering a small-size RoboCup team is the actual construction of the robots. We have successfully built robots for RoboCup-97 and RoboCup-98, leading to two champion teams, namely CMUnited-97 [2] and CMUnited-98 [1].

Given that our research is focused on team work in multirobot systems and not particularly on the mechanical construction of robots, and given that we had already built two different teams of robots, we decided that we would use the CMUnited-98 robots for the RoboCup-99 competition. We still had to rebuild the goalie robot, as our RoboCup-98 goalie was slightly beyond the size of the allowed robots, according to the revised rules for RoboCup-99.

Our purpose therefore entering RoboCup-99 was to concentrate our effort in new research directions involving more elaborate robot motion and strategic team work, namely:

- Extension of the motion control algorithm to allow for a preferred side, so that attackers would make more and better use of the kicking device.
- Increase of the number of defenders to two from the single one used in CMUnited-98. The number of defenders would vary dynamically in response to the number of attackers.
- Improvement of the coordination between the goalie and the defenders, to maximize the coverage of the defense area and not endanger our own goal.
- Additional strategic planning to dynamically adjust the number of attackers and defenders based on the current situation (e.g. position of the ball, the "formation" of the opponent, or the current score).

We partially successfully pursued our research. We created a sophisticated simulator where we developed an interesting algorithm for the coordination of the goalkeeper and multiple defenders. We expect to use this algorithm in our future RoboCup teams and refine it then, when effectively tested on real robots.

We had however limited success carrying ahead our new approaches in the robot team, as we encountered several unexpected difficulties with the radio control of the five robots. We also realized earlier, but mainly at the competition, that our robots' hardware was indeed worn out, most probably from the several trips and its extensive usage for two years.

We look forward to participating in future RoboCup competitions necessarily with new robots!

We thank Kwun Han for the development of the vision algorithm of CMUnited-98, which we also used in CMUnited-99. We also thank Peter Stone, Rune Jensen, Jim Bruce, and Tucker Balch for their help during the competition.

References

1. Manuela Veloso, Michael Bowling, Sorin Achim, Kwun Han, and Peter Stone. The CMUnited-98 champion small robot team. In Minoru Asada and Hiroaki Kitano, editors, *RoboCup-98: Robot Soccer World Cup II*, pgs. 61-76, Springer, 1999.
2. Manuela Veloso, Peter Stone, Kwun Han, and Sorin Achim. The CMUnited-97 small robot team. In Hiroaki Kitano, editors, *RoboCup-97: Robot Soccer World Cup I*, pgs. 242-256, Springer, 1998.

5dpo Team Description

Paulo Costa[(*)(1)(2)], António Moreira[(3)], Armando Sousa[(1)(2)],
Paulo Marques[(2)], Pedro Costa[(1)], Aníbal Matos[(1)(2)]

paco@fe.up.pt, amoreira@fe.up.pt, asousa@fe.up.pt,
pamarques@riff.fe.up.pt, pedrogc@fe.up.pt, anibal@fe.up.pt
[(*)] Team Leader
[(1)] Lecturer at the Faculdade de Engenharia da Universidade do Porto (FEUP)
[(2)] PhD. Sutdent at the FEUP, [(3)] Assistant Professor at the FEUP

Abstract - This paper describes the 5dpo team. The paper will be divided into three main sections, corresponding to three main blocks: the Global Level, the Local Level and the Interface Level. These Levels, their subsystems and some implementation details will be described next.

1 Introduction

This is our second participation in the Robocup Competition. Our robots comply with the F-180 League Regulations that constrain their dimensions: the occupied floor area must not exceed 180 square centimeters and the height must be bellow 15 centimeters. That limits the processing power and the kind of sensory devices that can be fitted in. On the plus side it also limits the costs and eases the mechanical design.

The main options that shape the way the 5dpo team was designed were: the use of a global vision system with more than one camera, the relative autonomy (in a short time frame) expected from the robots and the unidirectional nature of the radio link.

The whole team can be seen as a system divided in three basic levels.

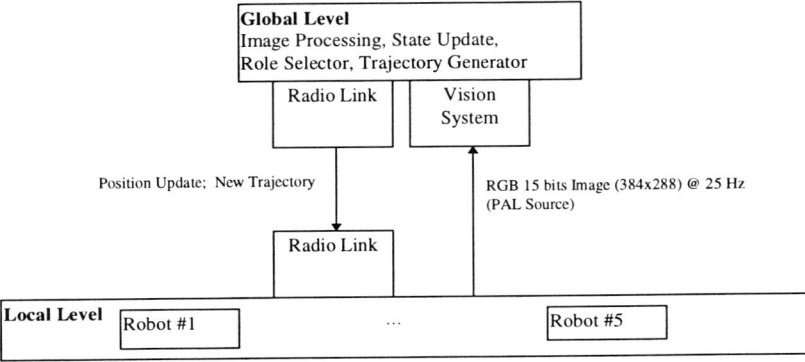

Fig. 1. The three levels with their interrelationships and the information flowing between them.

The global vision system saves us from building robots with onboard cameras and the corresponding processing power. It also gives us a global view which is independent from the robots' state. As the radio link cannot be completely reliable we tried to fit the robots with some autonomy so that they can survive a small starvation of orders from the global controller. That can ease the problem of lost packets over the air.

We will now describe each level and its subsystems.

2 The Global Level

This is the global Control Level. The global state of the system is updated based on the vision. Data fusion is attempted and adversary robot moves are tracked. A rule based engine is used to classify opponents intents. By observing the present system state as well as a global mid-term strategy, short term orders are generated and sent to the players.

This layer closes the global loop. It must be stressed that while the "sampling" frequency of this loop is 25 Hz, there is some intrinsic lag that degrades its optimal performance. The PAL signal takes 20 ms to deliver the frame, then some time is lost processing it, the reasoning unit must decide the new course of action and it is necessary to wait for the next time slot to send some orders to the robots. That is why the local loop, running at 50 Hz (only the double), can show a much better performance in some tasks than a globally closed loop.

Short term orders try to account for typical adversary moves. They are anticipated using a model for their behavior. Naturally, the quality of this model has a direct link with the quality of the tactic behavior achieved by the team.

Some team tactics are maintained at all times like some defense mechanisms that are enforced during the match. In any case, the defense robots stand in alignment in such a way that the robot that holds the ball can't easily shoot for goal. This implies the presence of a path planner not only for the robots but also for the ball and the opponents.

All these systems are implemented in C++ in DOS with a 32 bits extender. That was the only way to ensure the hard real-time nature of the tasks. Other operating systems could not guarantee hard real-time behavior or there were no drivers available for some of our hardware.

3 The Interface Level

3.1 The Radio Interface

The radio link allows sending and receiving short packets that carry messages from the Global Controller to the robots. We have the radio channel time slotted synchronously with the global Control Loop.

3.2 The Video Interface

Our team uses a global vision system as a primary sensorial source. This is our global positioning system for our robots, for the opponent robots and for the ball.

This system consists in one or more color video cameras, placed directly above the playing field. The TV signal from these cameras is feed to a video acquisition board placed in a PC. This board is capable of placing a digitalization of each image frame in the PC memory without CPU intervention, thus wasting almost no processor time. In the end of this process the board can signal the processor the conclusion of that task.

As we are using PAL cameras the image frequency is 50 Hz with alternating even and odd frames. For the single camera setup we are only using even frames therefore we have an image update frequency of 25 Hz. For the two camera setup, the synchronization procedure (even while using all frames), forces the update to 25 Hz too.

Based on the acquired image we must identify the ball position and also the robots' position and orientation. As the robots are fitted with colored Ping-Pong balls, the second problem is similar to the first one. It is easier because the background is stable and, in the case of our team's robots, chosen by us. Better, there is not any kind of occultation for these balls. The playing ball can be partially or totally hidden by the robot's body. It can have the green or the white of the lines as background. That makes its tracking more difficult. There is another problem: typically, the ball's speed can be grater than the robots speed.

Another problem is distinguishing each of our robots. That is being achieved by reading a bar code placed on the top of each robot. The bar code uses only three bars which give us 8 possible numbers, excluding the all whites and all blacks case we remain with 6 possible codes. That is enough for the five robots team.

4 The Local Level

In this layer we have the local control system that runs in each robot.

A robot is an autonomous unit considering a short time frame. The robots are capable of retaining a queue of tasks to be performed. These tasks may include following a specified trajectory, holding the ball, passing it along to another team member or maybe shooting for goal. The local control system tries to enforce those orders in the predefined sequence.

Next we describe the basic mechanical and electrical design of the robots.

The robots are fitted with two differential wheels. The wheels are driven independently by separated stepper motors. There is no third wheel and the robot is sustained by a pod. A castor would result in a more complex mechanical design as well as an increased uncertainty in its dynamical model.

The robots are presently powered by embedded Ni-Cd batteries. The motors are driven by two H-bridges that are directly powered from the batteries. The on board controller is a 8-bit RISC microcontroller (Atmel AVR90S4414). All digital circuitry gets its power from a low dropout linear regulator.

Two small single frequency RF modules (433 MHz and 418MHz) are used to communicate with the external global controller. Only one can be used at a time.

5 Future Work

A disclaimer must be made at this point: between now and the competition there is yet a lot of time to make changes in the described setup. As we test the performance of each subsystem and find better alternatives we will try to implement them. We want the overall system to show a more robust and efficient operation. There is the possibility of adding some kind kicking device to some of our robots.

The Decision System is, right now, very defensive and not very actively cooperative in the attack. That is being improved.

6 Conclusions

In this paper we described the 5dpo team and the solutions we found to this problem. Recognizing the overall system state (the ball position and speed, our team's robots' state and the adversarial robots' state) using vision is still a very difficult task. And the quality of the team behavior is very dependent from the accuracy of that system.

The decision of what to do, even with accurate knowledge of the system status, it is a major task on its own. The range of options, some discrete and some continuous has many dimensions and cannot be easily searched. A lot of heuristic rules must be used to trim the possibilities and the best framework to represent and find them is a matter that requires still a lot of research.

We hope to achieve a performance level that can leave us with the idea that our approach to this problem was justified and worthy of more development.

References

1. Arthur Gelb, Joseph Kasper Jr.,Raymond Nash Jr., Charles Price, Arthur Sutherland Jr.: Applied Optimal Estimation, The M.I.T. Press, (1989)
2. J. Borenstein, H. R. Everett, L. Feng, S. W. Lee and R. H. Byrne: Where am I? Sensors and Methods for Mobile Robot Positioning, (1996)
3. J. Carvalho: Dynamic Systems and Automatic Control, Prentice-Hall, (1993)
4. Huibert Kwakernaak, Raphael Sivan: Linear Optimal Control Systems, Willey-Interscience, (1972)
5. Jean-Claude Latombe: Robot Motion Planning, Kluwer Academic Publishers, (1991)
6. Lenart Ljung: System Identification: Theory For The User, Prentice-Hall, (1987)

FU-Fighters Team Description

Sven Behnke, Bernhard Frötschl, Raúl Rojas, Peter Ackers, Wolf Lindstrot, Manuel de Melo, Andreas Schebesch, Mark Simon, Martin Sprengel, and Oliver Tenchio

Free University of Berlin, Institute of Computer Science
Takustr. 9, 14195 Berlin, Germany
{behnke|froetsch|rojas|ackers|lind|melo |schebesc|simon|sprengel|tenchio}
@inf.fu-berlin.de, http://www.inf.fu-berlin.de/~robocup

Abstract. This paper describes the team FU-Fighters that won the second place in the RoboCup'99 F180-league competition.
The paper presents the mechanical and electrical design of our robots, including a kicking device. We also explain the hierarchical control architecture we used to generate the behavior of individual agents and the team. This reactive approach is mainly based on the Dual Dynamics concept developed by H. Jäger. In addition we describe, how the problems of vision and radio communication have been addressed.

Our group was a first-time participant in the RoboCup competition. Our main motivation was to use RoboCup as a platform for developing approaches for problems like vision and control using neural networks and learning [2]. Our main design goal for RoboCup'99 was to build a reliable team of robots that utilize innovative kicking devices and that are controlled by a new hierarchical reactive approach.

1 Mechanical and Electrical Design

Our robots were designed in compliance with the new F180 size regulations. We built four identical field players and a goal keeper. All robots have sturdy aluminum frames that protect the sensitive inner parts. They have a differential drive with two active wheels in the middle and are supported by a passive sphere that can rotate in any direction. Two Faulhaber DC-motors allow for a maximum speed of about 1 m/s. The motors have an integrated 19:1 gear and an impulse generator with 16 ticks per revolution. One distinctive feature of our robots is a kicking device (Fig. 1) that consists of a rotating plate that can accumulate the kinetic energy produced by a small motor and release it to the ball on contact.

For local control we use C-Control units from Conrad electronics. They include a microcontroller Motorola HC05 running at 4 MHz with 8 KB EEPROM for program storage, two pulse-length modulated outputs for motor control, a RS-232 serial interface, a free running counter with timer, analog inputs, and digital I/O. The units are attached to a custom board containing a stabilized

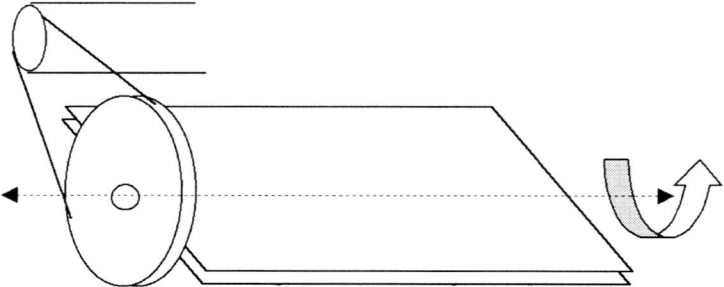

Fig. 1. Sketch of the kicking device.

power supply, a dual-H-bridge motor driver L298, and a radio transceiver SE200 working in the 433MHz band that can be tuned to 15 channels in 100kHz steps.

The robots receive commands via a wireless serial link with a speed of 9600 baud. The host sends 8-byte packets that include address, control bits, motor speeds, and checksum. The microcontroller decodes the packets, checks their integrity, and sets the target values for the control of the motor speeds. No attempt is made to correct transmission errors, since the packets are sent redundantly.

The robots are powered by 8 + 4 Ni-MH rechargeable mignon batteries. To be independent from the charging state of the batteries, we implemented a closed loop control of the motor speeds. The microcontroller counts the impulses from the motors 122 times per second, computes the differences to the target values, and adjusts the pulse length ratio for the motor drivers accordingly.

2 Tracking Colored Objects in the Video Input

The only physical sensor for our control software is a S-VHS camera that looks from above at the playground and outputs a video stream in NTSC format. Using a PCI-framegrabber we input the images into a PC. We capture RGB-images of size 640×480 at a rate of 30 fps and interpret them to extract the relevant information about the world. Since the ball as well as the robots are color-coded, we designed our vision software to find and track multiple colored objects. These objects are the orange ball and the robots marked with two colored dots in addition to the yellow or blue team ball.

To track the objects we predict their positions in the next frame and then inspect the video image first at a small window centered around the predicted position. We use an adaptive saturation threshold and intensity thresholds to separate the objects from the background. The window size is increased and larger portions of the image are investigated only if an object is not found.

The decision whether or not the object is present is made on the basis of a quality measure that takes into account the hue and size distances to the model and geometrical plausibility. When we find the desired objects, we adapt our model of the world using the estimates for position color, and size.

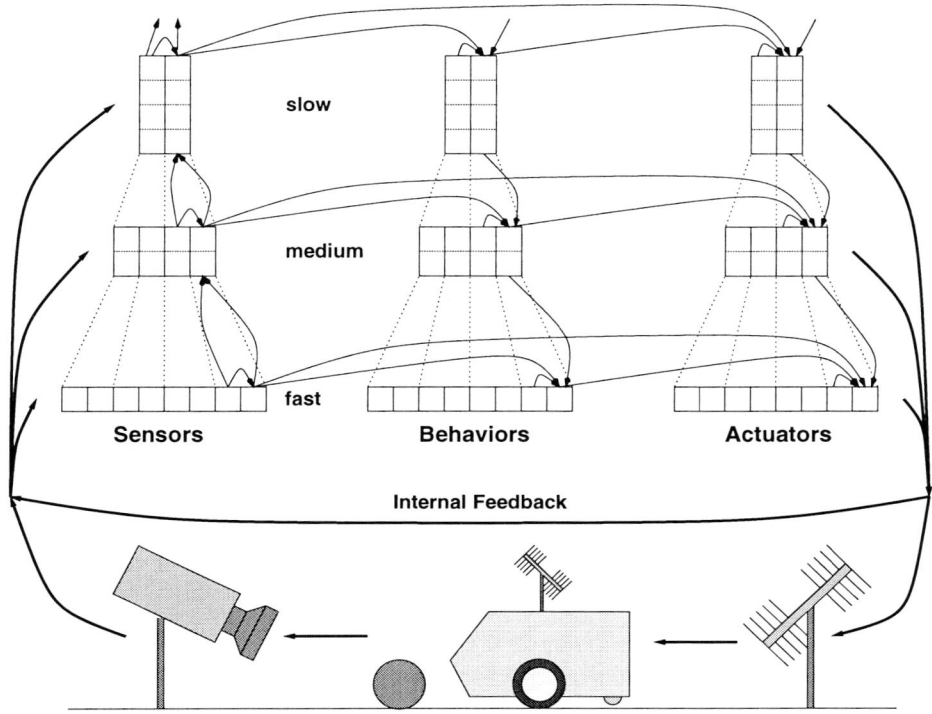

Fig. 2. Sketch of the control architecture.

3 Hierarchical Generation of Reactive Behavior

In 1992, the programming language PDL was developed by Steels and Vertommen for the stimulus driven control of autonomous agents [5]. This language has been used by a number of groups working in behavior oriented robotics [4]. It allows the description of parallel processes that react to sensor readings by influencing actuators. Many primitive behaviors, like taxis, are easily formulated in such a framework. On the other hand, it is difficult to implement more complex behaviors in PDL that need information about slow changes in the environment.

The Dual Dynamics control architecture, developed by Herbert Jäger [3], describes reactive behaviors in a hierarchy of control processes. Each layer of the system is partitioned into two modules: the activation dynamics that determines whether or not a behavior tries to influence actuators, and the target dynamics, that determines strength and direction of that influence. The different levels of the hierarchy correspond to different time scales. The higher level behaviors configure the lower level control loops via activation factors that determine the mode in which the primitive behaviors are. These can produce qualitatively different reactions if the agent encounters the same stimulus again, but has changed its mode due to stimuli that it saw in the meantime.

Our control architecture is based on these ideas, as shown in Figure 2. A more detailed description is given in [1]. The robots are controlled in closed loops that use different time scales. We extend the Dual Dynamics scheme by introducing a third dynamics, namely the perceptual dynamics shown on the left side. Here, either slow changing physical sensors are plugged in at higher levels, or readings of fast changing sensors, like the ball position, are aggregated to slower and longer lasting percepts. Since we use a subsampling in time, we can afford to implement an increasing number of sensors, behaviors and actuators in the higher layers without an explosion of computational costs.

The behaviors are constructed in a bottom up fashion: First, processes that should react quickly to fast changing stimuli are designed. Their critical parameters, e.g. a mode parameter or a target position, are determined. When the fast processes work reliably, the next level can be added to the system. This level can now influence the environment either directly by moving slow actuators or indirectly by changing the critical parameters of the processes in the lower level.

Each of our robots is controlled autonomously from the lower levels of the hierarchy using a local view to the world. For instance, we present the angle and the distance to the ball and the nearest obstacle to each agent. In the upper layers of the control system the focus changes. Now we regard the team as the individual. It has a slow changing global view to the playground and coordinates the robots as its extremities to reach strategic goals.

4 Summary

We designed robust and fast robots featuring a kicking device, reliable radio communication, and high speed vision. To generate actions, we implemented a reactive control architecture with interacting behaviors on different time scales. The system worked as designed at the RoboCup'99 F180 competition, where we finished second, next to Big Red from Cornell University.

We thank the companies Conrad ELECTRONICS GmbH, Dr. Fritz Faulhaber GmbH & Co KG, Siemens ElectroCom Postautomation GmbH, and Lufthansa Systems Berlin GmbH for their support that made this research possible.

References

1. Behnke, S., Frötschl, B., Rojas, R., Ackers, P., Lindstrot, W., de Melo, M., Preier, M., Schebesch, A., Simon, M., Sprengel, M, Tenchio, M.: Using hierarchical dynamical systems to control reactive behavior. RoboCup-Workshop IJCAI'99, 28–33.
2. Christaller, T.: Cognitive Robotics: A New Approach to Artificial Intelligence. In: Artificial Life and Robotics, Springer, 3/1999
3. Jäger, H., Christaller, T.: Dual Dynamics: Designing Behavior Systems for Autonomous Robots. In: Fujimura, S., Sugisaka, M. (eds:) Proceedings International Symposium on Artificial Life and Robotics (AROB '97), Beppu, Japan, (1997) 76–79
4. Schlottmann, E., Spenneberg, D., Pauer, M., Christaller, T., Dautenhahn, K.: A Modular Design Approach Towards Behaviour Oriented Robotics. GMD report 1088
5. Steels, L.: The PDL reference manual. VUB AI Lab memo 92-5, Brussels

Linked99

Junichi Akita[1], Jun Sese[2], Toshihide Saka[3], Masahiro Aono[4],
Tomomi Kawarabayashi[5] and Junji Nishino[5]

[1] Dept. of Elec. and Comp. Eng., Kanazawa University
[2] Dept. of Complexity Science and Eng., University of Tokyo
[3] Dept. of Architecture, University of Tokyo
[4] Dept. of Architecture, Tokyo Institute of Technology
[5] Dept. of Info. Science, Fukui University

1 Introduction

Figure 1 describes the overview of our team's system. The concept of our team is to employ simple, inexpensive robots and to control them by high-speed and actual vision feedback. The special hardware for color detection is developed and employed for our vision system to exract coordinates of ball and markers. Global strategy for the team is realized by rule based logic on the host computer.

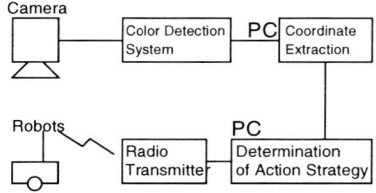

Fig. 1. Overview of "Linked99" system

2 Team Development

Team Leader: Junichi Akita
Team Members:
  Jun Sese
   – Dept. of Complexity Science and Eng., University of Tokyo, Japan
   – graduate student
   – did not attend the competition
  Toshihide Saka
   – Dept. of Architecture, University of Tokyo, Japan
   – undergraduate student
   – did not attend the competition
  Masahiro Aono

- Dept. of Architecture, Tokyo Institute of Technology, Japan
- undergraduate student
- did not attend the competition

Tomomi Kawarabayashi
- Dept. of Info. Science, Fukui University, Japan
- graduate student
- attended the competition

Junji Nishino
- Dept. of Info. Science, Fukui University, Japan
- research associate
- attended the competition

Web page http://www.ec.t.kanazawa-u.ac.jp/staffs/akita/yrobo99.html

3 Sensing

The vision system is composed of two parts; real-time color detection system implemented by special hardware, and coordinate extraction implemented by the software processing of PC.

Fig. 2. Structure of real-time color detection system

Figure 2 shows the structure of our real-time color detection system. The video signal from global vision CCD camera is converted to RGB signal at first and then converted to digial signal of 6bits respectively. The pixel's signal pair of RGB signal is then converted the pair of HSV, Hue and Saturation and Value, which is more invariable to lighting condition than RGB which is implemented by look-up table of 4Mbits ROM. The color information of each pixel expressed in HSV is then judged the following four color detectors, which each detects that all of hue, saturation, and value are put in the preset ranges, and converted to gray NTSC signal according to the color detected, to be captured by gray scale frame grabber of PC, which is three times faster than full-color frame grabber. The details of these system will be presented in [1].

The gray-scale image based on color detection generated by previous color detection system is captured by gray-scale frame grabber, PX610 by PC, and then they are labeled for each color independently, and then the coordinates of ball and each marker are calculated as their center of gravity. It is easy to process these sequences in video frame rate, since it is not needed to process for color information. The CPU, OS and programming language of the PC are PentiumII/450MHz, Windows95, and VisualC++4.0, respectively. These coordinate informations are sended to the other PC to determine the robots' action strategy through Ethernet.

4 Communication

The host PC for strategy issues motion commands for each robot. The commands are one byte short format; step go forward, step turn left or right, step go back. The employed radio module uses frequency of 318MHz, and its transfer bitrate is 4800bps.

5 Skills

(a)   (b)

Fig. 3. Photograph of player robot(a) and goalie robot(b)

We have developed very simple, inexpensive robot, by assuming the high-speed and actual control based on the information of global vision camera. The player robot, as shown in Fig.3(a), has a fixed arm to handle the ball when it goes forward.

Figure 3(b) shows the photograph of developed goalie robot. The goalie robot has special structure; three ultra-sonic sensors to detect coming ball, one DC motor and four wheels to move horizontally, and wall detection sensor to recognize its position aroud the goal.

6 Strategy

Control model of this system is a Hierarchical Behavior Controller with rule-based decision making processor. This hierarchy consists of three layer; local controller, behavior rules, strategy-base. The lowest control level: local control is mounted on robots. This sub system can accept and execute six radio-commands such as 'go forward', 'turn right'. This process may execute by local processor (PIC) on each robot.

To estimate the efficiency of RoboCup special image processor, strategy is constructed in a simple style. Global strategy for the team is realized by rule based logic on the host computer. Robots role is as follows.

1. goalie: 1 robot, have reactive motion control on its local processor, to make a ball clear just in front of his goal.
2. defense: 1 robot, just help the goalie. The goalie can not kick a ball from area near the goal to the middle of the pitch, thus the defense robot should kick it. This robot have ability to play role as forward play. It, however, places side form like as the goalie to assist him, and will get a position between goal and ball.
3. forward: 3 robots, make the game in middle of field through the opponent goal. They are in triangle position each other. These robots have each roles; ie. right wing, left wing, center top. Both side robots pass the ball simply to the point middle of between center top robot and the opponent's goal.

7 Special Team Features

We have newly developed the custom LSI for fast color detection as preprocessing for image processing PC, and we developed the very simple, inexpensive robot to be controled by visual feedback.

8 Conclusion

We employ very simple robot, with the high-speed and actual control based on the information of global vision camera. The special hardware for color detection is employed in order to high-speed and actual vision system.

In RoboCup99, we have learned that our motion commands for each robot does not work well, and we are planing to employ the robot control protorol for real-time control, such as radio-controlled car, with continuous transimission of motor states of each robot by host computer.

We are also improve our strategy algorithm, with developing basic library of robot action.

Acknowledgements

This team is aided by "Yugen Club Found" of Japanese Association for Mathematical Science.

References

[1] J.Akita, "Real-time Color Detection System using Custom LSI for High-Speed Machine Vision," *RoboCup Workshop in IJCAI99*, 1999. (to be presented)

This article was processed using the LaTeX macro package with LLNCS style

OWARI-BITO

Tadashi Naruse[1] Tomoichi Takahashi[2] Kazuhito Murakami[1] Yasunori Nagasaka[3] Katsutoshi Ishiwata[3] Masahiro Nagami[4] and Yasuo Mori[5]

[1] Aichi Prefectural University, Faculty of Information Science and Technology
Nagakute-cho Aichi pref. 480-1198 Japan
{naruse, murakami}@ist.aichi-pu.ac.jp
[2] Chubu University, College of Business Administration and Information Science
Matsumoto-cho Kasugai-shi 487-8501 Japan
ttaka@isc.chubu.ac.jp
[3] Chubu University, College of Engineering
Matsumoto-cho Kasugai-shi 487-8501 Japan
{zishi,any}@nn.solan.chubu.ac.jp
[4] Chukyo University, School of Computer and Cognitive Science
Kaizu-cho Toyota-shi 470-0393 Japan
nagami@grad.sccs.chukyo-u.ac.jp
[5] Mechatro Systems Co., LTD. Nakamura-ku Nagoya-shi 450-0003 Japan
info@robos.co.jp

1 Introduction

OWARI-BITO team consists of 5 small robots, each of which is sized in 10 cm wide, 10 cm deep and 17 cm high (except an antenna). The purposes of the research project are the study on the cooperation among robots, the advanced local vision system and the robust communication environment, in addition to be able to win the game.

The competition results in RoboCup 99 was 0 - 3 in the round robin group C.

2 Team Development

Team Leader: Tomoichi Takahashi
Team Members: Team members are listed in the above title field.
Web page http://kiyosu.isc.chubu.ac.jp/robocup

3 Hardware

3.1 Robot

The features of robot hardware are the followings.

- Two processors are in the robot, named a main processor and a vision processor, respectively. The main processor controls the robot and communicates with the host computer. The vision processor processes the images captured

by the camera on robot. These processors are connected through the parallel ports.
- Two motors drive the robot. Each motor is a DC brush motor with an encoder.
- There are twelve proximity sensors and two acceleration sensors. The proximity sensors are composed of an infrared LED and a photo diode. Sensing range is about 20 cm. Each robot has four sensors both in front and back side, and two both in left and right side. The acceleration sensors measure the accelerations in x-direction and y-direction.
- A radio system uses a spectrum spread method with 2.4 GHz band that is in accordance with the RCR STD-33A. The maximal communication speed is 38.4 kbps, however, we use 19.2 kbps communication for ensuring the reliability. Our radio system has four communication modes. We use one (host) to many (robots) packet communication mode.
- The video capturing hardware can capture 30 images (size 323 × 267 pixels/image) per second in maximum.

3.2 Host system

A host processor is a typical AT-compatible computer. The host processor computes the team strategy and processes the images captured by the global vision camera set at the ceiling. A radio communication system is connected to the processor through a serial line. The operating system for this processor is the LINUX.

4 Sensing

We employ a global vision system and a local vision system. The global vision system, which is implemented on the host processor, senses the field and calculates the positions of the current robots and ball. The local vision system, which is implemented on each robot, is employed to track the ball. These vision systems work exclusively. The timing that the local vision system is invoked is determined by the strategy process. In this section, we describe a summary of the global vision process. (See section 8 for local vision system.)

Initial positions of robots and ball are given by human. Once they are given, the global vision sistem tracks these positions as follows.

- The world image (size:640 × 480pixels) is captured.
- A local window (size: 71 × 71 pixels) is extracted for each robot and ball. Its position is determined by the previous image of the center positions of robots and ball.
- A new position of robot is determined in the local window. If it misses, the whole image has to be searched.

The position data are sent to the strategy process.

5 Communication

In the packet communication mode, each radio system has its own identification number(ID). We use the master-slave communication, where the host is a master and the robot is a slave. The host establishes the connection line to a robot by designating the ID of the robot. The broadcasting communication is also available, using special ID.

To reduce the communication overhead, five move-commands (one for each robot) are put together into one packet and broadcasted to all the robots. Each robot gets the packet and extracts his own command. When the robot gets his move-command, current executing move-command is canceled. Approximately, a half of the total communication time is reduced, compared with the case of one packet for each command.

A host processor makes an above packet and sends it to all robots at the time when the strategy process have determined the motion of each robot. In our system, this communication is able to occur 15 times per second.

6 Skills

All actions of the robots are planned by the predicted future ball position which is calculated from the current ball position, moving direction and speed. If none of our robots keep the ball, two robots which are close to the ball move to the place where is the ball to get it. In the attacking circumstance, attacking robots move to the place where they can kick to the opponent goal. In the defending circumstance, defending robots move to the position between the ball and our goal to disturb the shoot of the opponent robot. To kick the ball, the robot moves along the shortest path between the ball and the robot itself. We don't have special dribble skill now. When the dribble is necessary, the robot repeats a small kick.

7 Strategy

A strategy of OWARI-BITO is based on the simplified strategy of human soccer. Distinctive features are the following three points.

- Each robot is having a basic role such as Forward or Midfielder. However, it flexibly changes its role depending on the game situation.
- When our team is not having a ball, two robots which are close to the ball approach to the ball and try to get it. We call this action "Press".
- When our robot attacks the opponent goal, if the opponent robots surround it, it tries to put the ball out to the open space. It makes the other attackers to have the chance of free shoot.

The basic role of the five robots are a Forward (FW), two Midfielders (MF), a Defender (DF) and a Goalkeeper (GK). All robots other than GK can change their role dynamically in the game. For example, if three of our robots (other than GK) are in our side, two robots close to the ball change into MFs and try to press. The remaining robot changes into DF and defends the goal. The role of the robot is determined by the present position, not by the present role.

The basic strategy is as follows,

Attacking: FW and two MFs attend to the attack. DF and GK remain in our side to defend the opponent's counter attack. In the opponent side,if MF keeps the ball, it kicks the ball to the opponent goal direction. FW moves in front of the opponent goal and catches the ball which the MF passes. If FW can keep the ball, it shoots immediately. If all three attacking robots move to the ball, tere will be jammed. To avoid the jam, at least one attacking robot stands in an open space, away from the ball. The robots, in the jam area, try to pass the ball to the robot, in the open space. The robot, in the open space, tries the passing action, if it catches the ball.

Defending: Two MFs, DF and GK attempt the defense, and only FW remains in the opponent side to make counter attack. In the defense area, two MFs press the ball. DF moves to the place to disturb the opponent shooting.

8 Special Team Features

There are two kinds of local vision systems, a normal-vision and an omni-vision. The goal keeper robot and one of midfielder robots have the omni-vision systems and remaining robots have the normal vision systems. Each robot has the world model of the field. If he finds the ball but does not find the goal in his eye, he pushes the ball to the goal using the world model. In case of finding both the ball and the goal, he pushes the ball using real model. (Obstacle avoidance algorithm is not implemented yet.)

9 Conclusion

Main issues for future work are the followings.

- It is necessary to compensate the positional error caused by the slip. For example, a compensation using the acceleration sensors, or a compensation using a GPS-like positioning system should be considered.
- In a local vision mode, as the landmark such as goals and field corners are often occluded by the other robots, it may happen not to detect the landmark correctly. A robust algorithm to overcome such a case should be developed.
- To elaborate the strategy, a precise simulator of the robot motion should be developed.

Rogi 2 Team Description

Josep Lluís de la Rosa, Rafel García, Bianca Innocenti, Israel Muñoz,
Albert Figueras, and Josep Antoni Ramon

Institut d'Informàtica i Aplicacions
Universitat de Girona & LEA-SICA
C/Lluís Santaló s/n
E-17071 Girona, Catalonia
{peplluis, rafa, bianca, imunoz, figueras, jar} @ eia.udg.es

Abstract. This paper describes the main features of the new Rogi Team and some research applied focused on dynamics of physical agents. It explains the vision system, the control system and the robots, so that the research on dynamical physical agents could be performed. It presents part of the research done in physical agents, especially consensus of properly physical decisions among physical agents, and an example applied to passing.

1 Introduction

Our team started in 1996 at the first robot-soccer competition as the result of a doctorate course in multiagent systems. In 1997 and 1998 it took part at the international workshops held in Japan and Paris. The main goal has been always the implementation and experimentation in dynamical physical agents and autonomous systems. This year a step further towards the platform to develop this type of research is done

1.1 New Features in the 1999 Generation

The new team has evolved from past generations. This new generation has solved many important problems existing in previous versions and are more focused to deal with *dynamical* physical agents. Here, dynamical means related to dynamics of the robots, from the point of view of automatic control. This generation is designed to let study further the impact of dynamics of the body in the co-operative world.

- The robots have been improved and its structure has changed in order to have a better-fit dynamical behaviour for control.
- The new vision system is able to process up to 50 frames/sec, locating ten robots a and a ball, with a dedicated hardware result of our research.
- A rational physical agent's approach is operative for robots co-operation, for instance, applied in passing actions. This is also result of our research.

2 Team Description

The system that implements micro-robot soccer is made up of three parts: robots, vision system and control system. The vision and control systems are implemented in two different computers. The control system is also called the host system. The control system and the vision system are connected by means of a LAN, using TCP/IP protocol. This fact allows remote users to perform tests over the micro-robots platform and lets co-operative research. The vision system provides data to the control system that analyses the data and takes decisions. The decision is split up into individual tasks for each robot and is sent to them using a FM emitter at the host computer.

2.1 Robots' Description

The robots have on board 8 bits microprocessors 80C552 from Philips and RAM/EPROM memories of 32kBytes. The robots receive data from the host computer by means of a FM receiver. The FM receiver is prepared to work with two frequencies 418/433 MHz in half-duplex communication. The information sent by the host computer is converted to RS-232C protocol. The two motors have digital magnetic encoders with 265 counts per turn. They need 9V to work and consume 1.5W at a nominal speed of 12.300 rpm. 9 batteries of 1.2 V supply the energy. It is compensate to have clear dynamics, which is non-linear but linear piece-wise.

2.2 Vision System

A specific hardware has been designed to perform the vision tasks, merging specific components for image processing (video converters, analog filters, etc.) with multiple purpose programmable devices (FPGAs). A real time image-processing tool is obtained, which can be reconfigured to implement different algorithms.

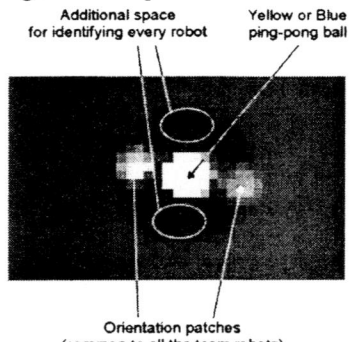

Fig. 1. *Top view with the team-colored Ping-Pong ball and orientation patches.*

According to RoboCup F-180 League Rules, each robot has to be marked using a yellow or blue Ping-Pong ball mounted at the center of their top surface. In order to provide angle orientation, additional color markings are allocated to the top of the robots. As shown in **Fig. 1**, purple and olive-green color patches have been added as the orientation patches in a basic configuration for our team robots. However, the robots may incorporate additional color patches to be distinguished from each other.

2.2.1 Algorithm

In order to locate the robots and the ball, the first step consists in their segmentation from the scene. The discriminatory properties of two color attributes, *hue* and *saturation*, are used so as to segment the objects. In this way, a pixel labeling of the image is obtained.

Since the size of the objects to track is rather small (5-6 pixels of diameter) mathematical morphology is applied in a 3×3 neighborhood of the processed pixel. Erosion and dilation operations are performed at video rate by using the tracking processor. In this way, particles smaller than the tracked patches are removed from the image and the remaining blobs are classified. The position and orientation of the 5 robots is computed using the blobs and the knowledge of the robot-patches geometry. A data association process solves the temporal matching problem at the highest abstraction level. However, the *identification* process is executed periodically to check that the data association has kept track properly of the robots' locations. As a last step, the measured locations of the objects are filtered in a sequence of images by means of an Extended Kalman Filter.

The low-level image processing operations are performed in hardware, while identification, data association, post-filtering and prediction are implemented in software. The cycle time is 20 ms, being limited by the PAL video standard.

2.3 Control System

The control system contains the strategy and the team decision making. The control system, using the positions of the robots and the ball provided by the vision system, has to determine which is the best decision to score a goal or to prevent the opponent team from scoring a goal. The system has been implemented using Lab-Windows. There are some advantages, but some problems have come up using it.

3 Research Challenges

The research is based on physical agents and focuses on the dynamics of the agents. We try to demonstrate that considering it when deciding prevents from undesirable situations. Knowing that a controller modifies the dynamics of the agents, we propose agents that are aware (introspection) of the set of controller their physical body has.

AGENT0 is used as an agent language. In this language an agent is described as an entity whose state consists of mental components such us *beliefs, capabilities, choices* and *commitments*. In our point of view, the capabilities are precisely the ones that let us represent the dynamics of the system. We believe that some of the capabilities are associated to the control of the system and we proposed them as a way for the agent to be aware of what it can or cannot do.

As a first approach our research is applied to passing the ball between two robots. This experiment consists on having to robots with crossing trajectories, that is, the trajectories have a common point and which will be the meeting point of one robot and the ball. We consider that there are no obstacles in the trajectories and the robots have several controllers to move forward in a one-dimensional linear movement. We know the transfer function of the robots and the ball.

The undesirable situations in this example could be that the robot 2 is too slow to be in the crossing point at the same time that the ball, or that the robot 1 does not give the necessary impulse to the ball to arrive to the crossing point, etc. Both robots have to agree on the time of doing the pass based on the knowledge they have of their dynamics. To do the pass, robot 1 will apply an impulsive kick to the ball and robot 2 will catch it in the crossing point. We want the ball arrives there at the same time that robot 2. Thus, the amplitude of the impulse will depend on distance to the meeting point, on the way in which the ball responds to this impulse (its transfer function) and on the time needed by robot 2 to arrive to the meeting point. But the latter condition depends on distance and on the transfer function of robot 2. The way to do it is modifying the parameters of the controllers. Once agreed, the robot 1 must kick the ball and robot 2 must be in the crossing point at the same time that the ball.

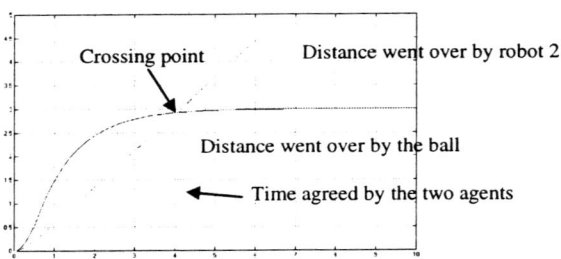

Fig. 2 *Distances went over by robot 2 and ball, the crossing point and the agreed time.*

Visit our web page at http://rogiteam.udg.es, especially the demos at http://rogiteam.udg.es/descrobots.html#sequencies

Temasek Polytechnic RoboCup Team-TPOTs

Nadir Ould Khessal

Temasek Engineering School, Temasek Polytechnic
21 Tampines Avenue 1 Singapore
nadirok@tp.edu.sg

1. Introduction

TPOTS is a Team of small size robots designed using a hybrid control architecture distributed among the robots and the host computer. The major characteristic of the RoboCup soccer competition is the dynamic nature of the environment in which robots operate. The only static object in the competition field is the field itself. Team and opponent robots as well as the ball can be placed anywhere in the field, be it a purposeful strategic positioning, a missed action or a forced displacement. This has led many researchers to shift from the traditional model-based top down control [1,2] to a reactive behavior based approach [3,4,5,6,7]. Robots need not waste a huge amount of resources building maps and generating paths that might prove useless at the time of action. Instead robots are supposed to react to the actual changes in the environment in a simple stimulus-response manner [8]. However due to the size limitations imposed by the RoboCup small robots league (15cm diameter circle) and rich visual input, on-board vision proved to be a complex and expensive task.
The 1999 RoboCup competition was the first world RoboCup experience for TPOTS. The team played three games in the round robin stage during which it scored 13 goals and conceded 10 (Table 1).

	CMUnited	Linked 99	Rogi2
TPOTS	0-4	9-0	4-6

Table 1. Temasek Polytechnic RoboCup team-tpots games results

In this Paper we will describe the overall architecture of TPOTS. Detailed descriprion of the Multi-Agent System (MAS) architecture can be found in [9].

2. TPOTS Development

Our approach in implementing the control architecture of the robots is based on dividing each robot controller into two parts: Embedded agent running on the on-board processor and situated in the environment (field) and Remote agent running in the off-board host computer and situated in an abstract model of the field. The embedded agent consists of several reactive behaviors competing with each other through the use of activation levels (inhibition and suppression). The main role of the embedded agent is to execute commands issued by the remote agent and navigate safely the soccer field while avoiding other robots and obstacles. The remote agent on the other hand implements strategies generated by the reasoning module.

Team Members:

Nadir Ould Khessal
- Affiliation: Temasek Eng. School
- Country: Singapore.
- Position: Lecturer.
- Attended the competition.

Sreedharan Gopalsamy
- Affiliation: Temasek Eng. School
- Country: Singapore.
- Position: Lecturer.
- Attended the competition.

Lim Hock Beng
- Affiliation: Temasek Eng. School
- Country: Singapore.
- Position: Lecturer.
- Did not attend the competition.

Kan Chi Ming, Dominic
- Affiliation: Temasek Eng. School
- Country: Singapore.
- Position: Lecturer.
- Attended the competition.

Chia Loong Suan, Alex
- Affiliation: Temasek Eng. School
- Country: Singapore.
- Position: Technical Staff.
- Attended the competition.

Hang ping
- Affiliation: Temasek Eng. School
- Country: Singapore.
- Position: Student.
- Attended the competition.

3. TPOTS System Architecture

The system hardware consists of a Pentium host computer, a vision system based on Newton labs Cognachrome vision card, RF transmission system and five robots (figure 1).

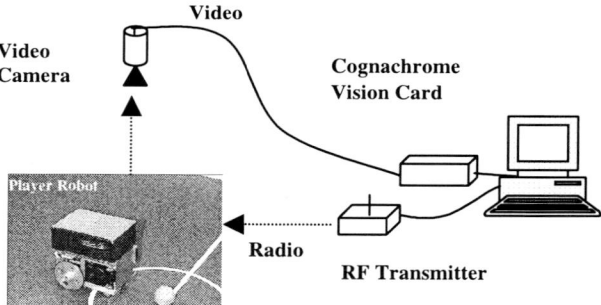

Fig. 1. System Overview

3.1 On-Board Perception

Each robot is equipped with three infrared sensors capable of detecting objects within the range of 20 cm. The on-board agent to avoid other robots placed in its path uses these sensors.

3.2 Off-Board Perception

Remote agents use the vision system as their perceptual module. A global vision system, which consists of color camcorders and a special image processor (MC68332), is used. The system is able to segment and track the robots and ball at a high frame rate. Each robot has two color pads. The image processor is trained to see the different colors and gives the locations of the center of gravity of the two color pads. Hence the orientation and robot position are known. Color pad areas are used to distinguish between different robots and minimize latency.

3.3 Agents Communication

The host computer transmits commands to the robot via radio transceivers utilizing UHF radio waves. Each robot has its own transceiver and a unique node address. The low-powered wireless system transmits less than 1mw of power and is effective over distances of 3 to 30 meters. Two-way communication rates of up to 38.4Kbps are possible. The command set is transmitted as text code piggybacking on the transmission protocol. Commands are sent and received from the transceiver using an RS-232 interface.

4. TPOTS Behaviors and Skills

4.1 Embedded Agents

Obstacle avoidance is the main autonomous task done by the robot. Using the three infrared sensors, the robot moves away from obstacles using the *Avoid Left* and *Avoid right* machines. The robot than moves a certain distance until a straight-line path to the target position is clear. The remote agent detects that the robot is out of the previously computed path and re-computes and transmits a new path.

4.2 Remote Agents

Intercept_ball: This machine enables the robot to move behind a predicted ball position before kicking it towards a target area. The target area could be the opponent goal keeper area (in an attempt to score a goal), a clear area in front of a team member (ball passing) or simply the opposite side of the field, in the case of a defending robot.

Follow: This machine is designed to keep the robot following a target object. The target can be the ball, a team robot or an opponent robot. This is done to keep the robot nearer to the ball and therefore in a better position to intercept the ball.

*Homing***:** Depending on the strategy being executed robots could be required to be placed at a certain position for the purpose of forming a defense wall for example.

5. Conclusion and Future Work

We will continue our research on developing further the distributed MAS architecture. However, work towards entering the Melbourne 2000 competition will focus on the development of a robust vision system, capable of adapting to the varying lighting conditions.

References

1. Albus, James S. *Hierarchical Control Of Intelligent Machines Applied to Space Station Telerobots.* IEEE Transactions on Aerospace and Electronic Systems, Vol. 24. NO. 5 Sept. 1988.
2. Housheng Hu and Michael Brady, *A Parallel Processing Architecture for Sensor Based Control of Intelligent Mobile Robots.* Robotics and Autonomous Systems, vol.17,1996, pp 235-257.
3. Brooks, Rodney A.. *Achieving Artificial Intelligence Through Building Robots* A.I.Memo 899, Artificial Intelligence Laboratory, MIT, May 1986.
4. Brooks , Rodney A. *Intelligence without Reason.* Computers and thought, IJCAI-91, Also A.I. Memo 1293 Artificial Intelligence Laboratory, MIT.
5. Maja J. Mataic. *Interaction and Intelligent Behavior.* Ph.D. Thesis, Department of Electrical Engineering And Computer Science, Artificial Intelligence Laboratory, MIT, 1994.
6. Arkin, Ronald C. *The Impact of Cybernetics on the Design of a Mobile Robot System: A Case Study.* IEEE Transactions on Systems, Man , and Cybernetics Vol. 20 No 6 Nov./DEC 1990.
7. Kenneth Moorman and Ashwin Ram.. *A Case-Based Approach to Reactive Control for Autonomous Robots.* AAAI Fall Symposium " AI for real time Mobile Robots", Cambridge MA, Cot 1992.
8. Brooks, Rodney A.. *A Layered Intelligent Control System for a Mobile Robot.* IEEE Journal Robotics And Automation . RA-2, March 1986, 14-23.
9. Nadir Ould Khessal. *Towards a Distributed Multi-Agent System for a Robotic Soccer Team.* In Ed. Manuela Veloso, Enrico Pagello and Hiroaki Kitano, *RoboCup-99: Robot Soccer World Cup III.* Springer Verlag, 2000, in this volume.

The VUB AI-lab RoboCup'99 Small League Team

Andreas Birk, Thomas Walle, Tony Belpaeme, and Holger Kenn

Vrije Universiteit Brussel, Artificial Intelligence Laboratory, Belgium
c/o cyrano@arti.vub.ac.be, http://arti.vub.ac.be/~cyrano

Abstract. The VUB AI-lab team is mainly interested in the two loosely linked aspects of on-board control and heterogeneity. One major effort for fostering both aspects within RoboCup's small robots league is our development of a so-to-say robot construction-kit, allowing to implement a wide range of players with on-board control. For the '99 competition, the existing RoboCube controller-hardware has been further improved. In addition, some solid and precise mechanical building-blocks were developed, which can easily be mounted on differently shaped bottom-plates. On top of these engineering efforts, we report here a computational inexpensive but efficient algorithm for motion-control, including obstacle avoidance. Furthermore, we shortly address the issue of increased difficulties of coordinating so-to-say multiple teams due to the possible variations based on heterogeneity. Operational semantics based on abstract data-types and patter matching capabilities can be a way out of this problem.

1 Introduction

As we already pointed out in a contribution to RoboCup'98 [BWB+98], RoboCup is not laid out as a single event, but as a long-term process where robots, concepts, and teams co-evolve through iterated competitions. Within this process, we belief that two loosely linked aspects are especially important, namely the exploitation of heterogeneous systems and on-board control.

Heterogeneity is an almost kind of "natural" aspect for soccer systems. Body aspects as well a behavioral aspects are typically linked to trade-offs like for example speed versus strength. Thus, there are no generally optimal players, but only suited players for certain situations. Heterogeneity of a team, including a rich set of players on the bench, allows to adapt the set of players on the field by substitutions much like in real soccer. Furthermore, diversity in the body features and behavioral aspects of the players plays an important role in the co-evolutionary process of iterated competitions, leading to constantly improving teams and scientific insights.

On-board control is in so far linked to heterogeneity as it is desirable to have a kind of construction-kit, which allows to design a wide range of different types of robot-players, including an easy implementation and change of motor and mechanical aspects as well as sensor systems. When using "string-puppets", i.e., radio-controlled toy-cars with off-board computation on a host, a support of a wide range of motor and sensor features is severely restricted due to bandwidth limitations. But there is an additional reason for the significance of on-board control within the RoboCup framework. Namely, RoboCup is an ideal testbed for the investigation of Autonomous Systems, i.e., networked embedded devices with physical interfaces in form of sensors and motors as well as stand-alone capabilities. This type of devices has in shortest time grown into a significant market and it will be one of the key technologies of the new millennium.

2 The body aspects

For RoboCup'98, the VUB AI-lab team focussed on the development of a suited hardware architecture, which allows to implement a wide range of different robots. The basic features of this so-called RoboCube-system are described in [BKW98]. For RoboCup'99, the system is further improved and extended.

In addition to improvements on the electronics and computational side, the mechanical approach for our robots has completely changed. Instead of using mechanical toy-kits like LEGO^{TM} as we did in the previous year, we developed a solid but still flexible solution based on metal components.

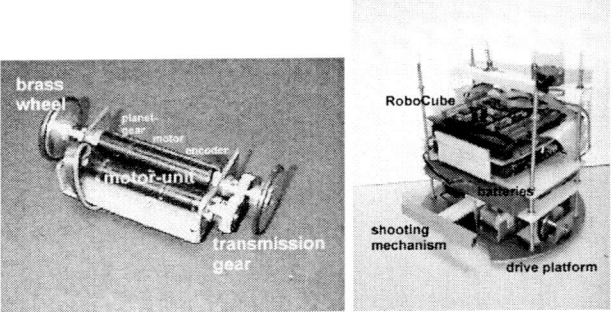

Fig. 1. The drive unit (left) as a mechanical building-block, which can be integrated into several different robots like e.g. the one shown on the right.

Keeping the basic philosophy of construction-kits, a "universal" building block is used for the drive (figure 1, left side) of the robots. The drive can be easily mounted onto differently shaped metal bottom-plates, forming the basis for different body-forms like e.g. the one shown in figure 1 on the right side. The motor-units in the drive exist with different ratios for the planetary gears, such that several trade-offs for speed versus torque are possible.

Other components, like e.g. shooting-mechanisms and the RoboCube, are added to the bottom-plate in a piled-stack-approach, i.e., four threaded rods allow to attach several layers of supporting plates.

3 On-board control

Based on these engineering efforts, it is now possible to implement quite some different types of robots with on-board control capabilities. But to actually use them, two major issues have to be solved, namely the software implementation of on-board control with the limited computational means of the RoboCube and the coordination of the so-to-say multiple teams due to the possible variations based on heterogeneity.

In this section, the on-board control is discussed. For a general discussion of this issue, the interested reader is referred to [BKW99].

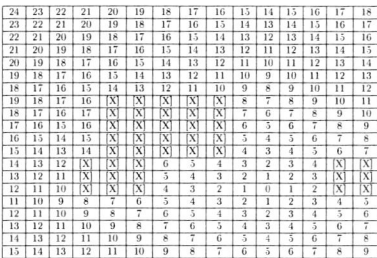

Fig. 2. A kind of potential field for motion-control based on Manhattan distances. Each cell in the grid shows the distance to a destination (marked accordingly with Zero) while avoiding obstacles.

For the RoboCup'99 team, we significantly increased the amount of processing, which is actually taking place on the robots. In addition to the basic control of driving and shooting, the complete motion-planning and simple strategies are computed on the robots themselves.

Especially the motion-planning, including obstacle-avoidance, is with most common approaches rather computationally expensive. We developed a kind of potential-field algorithm based on Manhattan-distances. Given a destination and a set of arbitrary obstacles, the algorithm computes for each cell of a grid the distance to the destination while avoiding the obstacles (figure 2). Thus, the cells can be used as gradients to guide the robot. The algorithm is very fast, namely linear in the number of cells.

4 Team-coordination and heterogeneity

As mentioned before, heterogeneity is an important feature for soccer with human as much as with robot players. It is the main basis for adaptation of a team, let it be to different opponent teams within a tournament, or to the general progress of a particular game, or to very momentary situations. Heterogeneity within soccer can range from high-level different roles of players in a team like forward or defender, down to different body features covering a wide-range of physical trade-offs like e.g. speed versus torque.

Straight-forward approaches to team coordination with the expressive power of finite state automata are doomed to fail under such a wide-ranges of heterogeneity due to the combinatorial explosion of states. Therefore, we investigate coordination schemes based on operational semantics, which allow an extremely compact and modular way of specifying team behaviors. One step in this direction is the *Protocol Operational Semantics (POS)*, an interaction protocol based on abstract data-types and patter matching capabilities. So far, it has only been tested in simulations, but the results are very encouraging. A detailed description can be found in [OBK99].

5 Conclusion and Acknowledgments

The paper describes the RoboCup'99 small robots league team of the VUB AI-lab. Our main interest is in on-board control and heterogeneous agents.

The VUB AI-Lab team thanks Sanders Birnie BV as supplier and Maxon Motors as manufacturer for sponsoring our motor-units. Andreas Birk is a research fellow of the Flemish Institute for Applied Science (IWT); research on RoboCup is partially financed within this framework (OZM980252). Research on the RoboCube is partially funded with a TMR-grant for Thomas Walle (ERB4001GT965154). Tony Belpaeme is a Research Assistant of the Fund for Scientific Research - Flanders (Belgium) (F.W.O.).

References

[BKW98] Andreas Birk, Holger Kenn, and Thomas Walle. Robocube: an "universal" "special-purpose" hardware for the robocup small robots league. In *4th International Symposium on Distributed Autonomous Robotic Systems*. Springer, 1998.

[BKW99] Andreas Birk, Holger Kenn, and Thomas Walle. From string-puppets to autonomous systems: On the role of on-board control in the small robots league. Technical Report under review, Vrije Universiteit Brussel, AI-Laboratory, 1999.

[BWB+98] Andreas Birk, Thomas Walle, Tony Belpaeme, Johan Parent, Tom De Vlaminck, and Holger Kenn. The small league robocup team of the vub ai-lab. In *Proc. of The Second International Workshop on RoboCup*. Springer, 1998.

[OBK99] Pierre-Yves Oudeyer, Andreas Birk, and Jean-Luc Koning. Interaction protocols with operational semantics and the coordination of heterogeneous soccer-robots. Technical Report under review, Vrije Universiteit Brussel, AI-Laboratory, 1999.

Agilo RoboCuppers: RoboCup Team Description

Thorsten Bandlow, Robert Hanek, Michael Klupsch, Thorsten Schmitt

Forschungsgruppe Bildverstehen (FG BV) – Informatik IX
Technische Universität München, Germany
{bandlow,hanek,klupsch,schmittt}@in.tum.de
http://www9.in.tum.de/research/mobile_robots/robocup/

Abstract. This paper describes the *Agilo RoboCuppers* [1] – the RoboCup team of the image understanding group (FG BV) at the Technische Universität München. With a team of five Pioneer 1 robots, equipped with CCD camera and a single board computer each and coordinated by a master PC outside the field we participate in the Middle Robot League of the Third International Workshop on RoboCup in Stockholm 1999. We use a multi-agent based approach to represent different robots and to encapsulate concurrent tasks within the robots. A fast feature extraction based on the image processing library HALCON provides the data necessary for the onboard scene interpretation. In addition, these features as well as the odometric data of the robots are sent over the net to the master PC, where they are verified with regard to consistency and plausibility and fusioned to one global view of the scene. The results are distributed to all robots supporting their local planning modules. This data is also used by the global planning module coordinating the team's behaviour.

1 Introduction

The aim of our activities on robot soccer is to develop software components, frameworks, and tools which can be used flexibly for several tasks within different scenarios under basic conditions, similar to robot soccer. This can be used for teaching students in vision, artificial intelligence, robotics, and, last but not least, in developing large dynamic software systems. For this reason, our basic development criterion is to use inexpensive, easy extendible standard components and a standard software environment.

2 Hardware Architecture

Our RoboCup team consists mainly of five Pioneer 1 robots [1] each equipped with a single board computer. They are supported by a master PC or coach, and one monitor PC for displaying the robot's data and states. Since the team size was reduced to four robots, the fifth robot is used as a substitute. The single board computers are mounted on the top of the robots, firmly fixed –

---

[1] The name is derived from the Agilolfinger, which were the first Bavarian ruling dynasty in the 8th century, with Tassilo as its most famous representative.

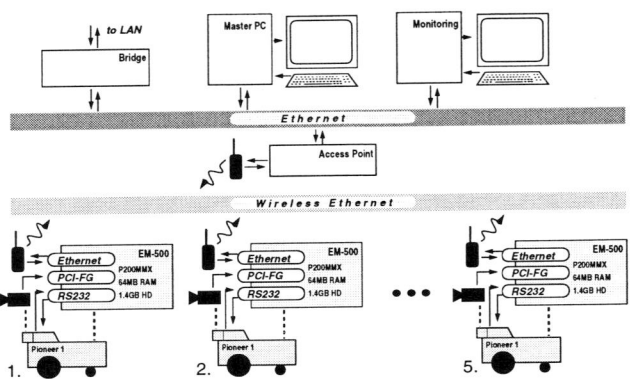

Fig. 1. Hardware architecture.

mechanically and electrically. All robot computers are linked via a 10 Mbps radio ethernet network [4, 5]. A master computer is located outside the soccer field and is linked to the radio ethernet, too. It can also be used for debugging purposes, monitoring the robots' planning states and feature extraction processes. The operating system for all computers is Linux. Figure 1 gives an overview of the hardware architecture.

Figure 2 (a) shows one of our Pioneer 1 robots. Each of them measures 45 cm × 36 cm × 56 cm in length, width, and height and weighs about 12 kg. Inside the robot a Motorola microprocessor is in charge for controlling the drive motors, reading the position encoders, for the seven ultrasonic sonars, and for communicating with the client. In our case this is a single board computer (EM-500 from [2]) which is mounted within a box on the topside of the robot. It is equipped with a Pentium 200 MHz processor, 64 MB RAM, 2.5" hard disk, onboard ethernet and VGA controller, and an inexpensive BT848-based [7] PCI video capture card [3]. PC and robot are connected via a standard RS232 serial port. A PAL color CCD camera is mounted on top of the robot console and linked to the S-VHS input of the video capture card. Gain, shutter time, and white balance of the camera are adjusted manually.

3 Fundamental Software Concepts

The software architecture of our system is based on several independent modules, each these performs a specific task. Software agents control the modules, they decide what to do next and are able to adapt the behavior of the modules they are in charge for according to their current goal. For this, several threads run in parallel. of our system. The modules are organized hierarchically, within the main modules basic or intermediate ones can be used. The main modules are image (sensor) analysis, robot control, local planning, information fusion, and global planning. The latter two run on the master PC outside the field, the others on the single board computers on the robots. Beside the main modules there are some auxiliary modules, one for monitoring the robots, extracted sensor data and

Fig. 2. (a) Odilo – one of our Pioneer 1 robots – and (b) what he percepts of the world around him.

planning decisions, one for interacting with the system or with particular robots, and one for supervising the running processes. A large number of basic functions define fundamental robot behaviors, provide robot data, and realize different methods for extracting particular sets of vision data. For the communication between different modules, we strictly distinguish between controlling and data flow. One module can control another by sending messages to the appropriate agent. Data accessed by various modules is handled in a different manner. For this, a special sequence object class was defined. This offers a consistent concept for exchanging dynamic data between arbitrary components [9].

4 Vision

The vision module is a key part of the whole system. Given a raw video stream, the module has to recognize relevant objects in the surrounding world and provide their positions on the field to other modules. This is done with the help of the image processing library HALCON (formerly known as HORUS [8, 6]). This tool provides efficient functions for accessing, processing and analysing iconic data, including framegrabber access and data management. The framegrabber interface was extended to features for capturing gray scale images and color regions at the same time. For this we use the YUV-image data provided by the video capture card. The color regions can be achieved very fast by a two-dimensional histogram-based classifier, which describes color classes as regions in the UV-plane and uses a brightness intervall as an additional restriction. Gray scale images, color regions and the extracted data are provided by sequence objects as described in section 3. As a compromise between accuracy and speed we capture the images with half the PAL resolution clipping the upper 40 percent. This results in a resolution of 384 × 172 with a frame rate of 7 to 10 images per second.

In general, the task of scene interpretation is a very difficult one. However, its complexity strongly depends on the context of a scene which has to be interpreted. In RoboCup, as it is defined in the present, the appearance of relevant objects is well known. For their recognition, the strictly defined constraints of

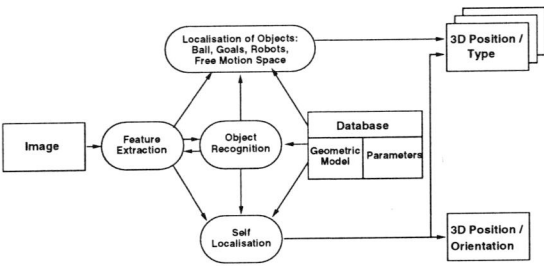

Fig. 3. Data flow diagram of the vision module.

color and shape are saved in the model database and can be used. These constraints are matched with the extracted image features such as color regions and line segments (see Fig. 2 (b) and 3).

Besides recognizing relevant objects with the help of the color regions, a second task of the image interpretation module is to localize the recognized objects and to perform a self-localization on the field if needed. To localize objects we use the lowest point of the appropriate color regions over the floor in conjunction with a known camera pose relative to the robot. From this we can determine their distance and position relative to the robot. Self-localization is performed by matching the 3D geometric field model to the extracted line segments of the border lines and – if visible – to a goal. A subpixel accurate edge filter performed on the gray-scale image (Y-channel) supplies contours from which, after removing radial distortions, straight line segments are extracted. Both, an absolute initial localization as well as a successive refinement, compensating the error of the odometric data have been implemented.

Building up and maintaining a RoboCup team is a great challenge and needs huge personal efforts and a lot of time. Thus we hope that we will still have enough resources in future to continue our interesting and promising work.

References

1. ActivMedia Robotics, *http://www.activmedia.com/robots/*
2. Lanner Electronics Inc., *http://www.lannerinc.com/*
3. Videologic Inc., *http://www.videologic.com/ProductInfo/capt_pci.htm*
4. RadioLAN Inc., *http://www.radiolan.com*
5. Delta Network Software GmbH, *http://www.delta-net.de*
6. MVTec Software GmbH, *http://www.mvtec.com*
7. Brooktree: BT848 Single-Chip Video Capture Processor, *http://www.brooktree.com/intercast/bt848.html*
8. Eckstein W., Steger C., *Architecture for Computer Vision Application Development within the HORUS System*, Electronic Imaging: 6(2), pp. 244–261, April 1997.
9. Klupsch, M., *Object-Oriented Representation of Time-Varying Data Sequences in Multiagent Systems*, 4th International Conference on Information Systems and Synthesis – ISAS '98, pp. 33–40, 1998.

ART99 - Azzurra Robot Team

Daniele Nardi[1], Giovanni Adorni[2], Andrea Bonarini[3], Antonio Chella[4], Giorgio Clemente[5], Enrico Pagello[6], and Maurizio Piaggio[7]

[1] Dipartimento di Informatica e Sistemistica, Università di Roma "La Sapienza"
[2] Dipartimento di Ingegneria dell'Informazione, Università degli Studi di Parma
[3] Dipartimento di Elettronica e Informazione, Politecnico di Milano
[4] Dipartimento di Ingegneria Elettrica, Università di Palermo and CERE - CNR
[5] Consorzio Padova Ricerche
[6] Dipartimento di Elettronica ed Informatica, Università di Padova
[7] Dipartimento di Informatica Sistemistica e Telematica, Università di Genova

1 Introduction

Azzurra Robot Team (ART) is the National Italian Team for F-2000 RoboCup league, developed within the RoboCup Italia project. ART99 is formed by six academic groups and Consorzio Padova Ricerche. ART started with RoboCup-98, and its goal is to exploit the expertise and ideas from all groups in order to build a team where players have different features (hw and sw), but retain the ability to coordinate their behaviour within the team. ART99 obtained the second place in RoboCup-99 F-2000 league, and coordination among players is, in our view, the most significant achievement of the team.

2 Team Development

In ART99 each group was responsible either of developing a robot and/or of developing specific hw/sw components to be used by the other team groups. Each group had a local coordinator, typically a Faculty, who is listed first, PhD students and undergraduate students. The project has been managed by scheduling regular team meetings, a one-week school (Rome, February 1999), to provide the necessary background to the students involved in the project, a one-week final preparation stage in Padova.

Team Leader: Daniele Nardi email:nardi@dis.uniroma1.it

Team Members: *Univ. Genova*: Maurizio Piaggio, Antonio Sgorbissa (Phd), Alessandro Scalzo; *Univ. Padova*: Enrico Pagello, Alessandro Vaglio, Walter Zanette, Robert Rosati, Nikita Scattolin, Alberto Speranzon, Roberto Polesel, Alessandro Modolo, Mattia Lorenzetti, Paolo De Pascalis, Massimo Ferraresso, Matteo Peluso; *Univ. Parma*: Giovanni Adorni, Stefano Cagnoni(Fac.), Monica Mordonini (Phd), Carlo Bernardi, Cristiano Rota; *Univ. Palermo*: Pietro

---

* The project has been supported by all the institutions the authors are affiliated with, Consiglio Nazionale delle Ricerche - Progetto "Robot Calciatori", Facoltà Ingegneria "La Sapienza" Roma, Politecnico Milano, Facoltà Ingegneria Padova, Sony Italia, Vesta Pneumatics, ImageS, Tekno.

Storniolo (PhD), Rosario Sorbello (PhD), Carmelo Amoroso, Vito Morreale; *Politecnico Milano*: Andrea Bonarini, Paolo Meriggi, Michele Lucioni, Paolo Aliverti, Andrea Marangon, Nicoletta Ghironi; *Univ. Roma "La Sapienza"*: Daniele Nardi, Luca Iocchi(PhD), GianPaolo Pucci, Claudio Castelpietra, Giorgio Grisetti, Alberto Gilio, Luca Luminari, Piero Tramontano, Domenico Mastrantuono.
Web page: http://www.dis.uniroma1.it/~ART

3 Robots

ART99 includes several types of players, that are built on top of two hw bases: BaseART and Mo^2Ro. Below we briefly describe them and present the main features of each type of player that we have developed on top of them.

BaseART was developed in preparation to the 1998 RoboCup by assembling several out-of-the-shelf, low-cost components, with the goal of keeping it very standard in terms of hw and, therefore, easily extensible with new devices. The mobile basis is the Pioneer 1, where we added a conventional PC, running LINUX, for onboard computing. We have reached a compromise between weight and power consumption, where the player has enough autonomy to play games. We also have a wireless high bandwidth connection that is used during development to obtain accurate information about the situation onboard, and supports the exchange of information among the players during the game, but it is not used to transfer raw data among the players. The vision system which is constituted by a low-cost frame grabber based on the BT848. At Robocup-99 we have used a Sony XC-999P color camera with about 100° aperture angle. The cameras are positioned differently on different types of players. Finally, BaseART provides a kicking device driven by air pressure, with two actuators differently arranged on the players, that enable different types of kicks (left, right or both).

Mo^2Ro is a *Mo*dular *Mo*bile *Ro*bot base designed and implemented at Politecnico di Milano Artificial Intelligence and Robotics Lab. as a general purpose robot matching the Robocup specifications, but also to support other needs. Mo^2Ro can run up to 60 cm/sec, and may have more then 40 kg as payload. The hw is functionally layered, and any module can be easily added or removed. At the first level, we have mounted, in the different implementations of Mo^2Ro: a sonar belt, bumpers, encoders, and different vision sensors; among these: two different types of omnidirectional sensors [3], and a camera mounted on top of a 5 DOF arm. Among the actuators that we have adopted up to now, we have two DC motors for movement, a kicker, and the arm. On the second layer, control and data acquisition can be done either by commercial or by home made cards, including one based on a Motorola 68HC12 fuzzy chip, for low level control.

TinoZoff is the goal keeper of ART99. The physical layout of the goalie is considerably different from the other players' structure. This goalie has a vision system based on two wide-angle cameras placed on top of it, having an aperture angle of about 70° vertically and 110° horizontally. This allows the robot to extend its field of view to over 200°, considering that the fields of view of the two cameras overlap by about 20° in the central region right in front of the goalie. As for its cinematics, the two driving wheels are located in the middle of the chassis, one on the front and the other on the rear. This makes translational movements more precise and accidental turns less likely. Balance is ensured by a pair of spheres, on which the robot leans, that are positioned along an axis at 90° to the wheel axis, and passing through its center. Turning is possible because the two wheels can be operated independently. Just ahead of the front wheel is a pneumatic kick device, whose air hangs just above. The vision system and the self-localization method developed for it are described in [1].

RonalTino and **TotTino** are middlefield players developed on BaseART. Their essential features from the hw viewpoint are: a specialized vision system with a camera rotating on 360° and infrared sensors to better control the kicking when the ball is close to the kickers. The control system of these players is designed on top of SAPHIRA, an environment developed to implement the robot control both in terms of actions, realized as programs in the Colbert language, and in terms of fuzzy behaviours, that are executed by a fuzzy controller. In [5] we discuss our experience in the design of the control system for RoboCup based on fuzzy rules. We have developed several tools to support the designer in the debugging and experimental activities. We also implemented several self-localization methods relying on the vision-based recognition of the goals, on the information coming from the compass and on the vision-based analysis of the lines in the field [6]. We have compared them trying to identify the conditions under which each source of information for localizing the player is reliable.

Relè, **Bart** and **Homer** are middlefield players, developed on baseART, provided with different settings for the kicking device and characterised by a novel sw planning and control architecture based on the ETHNOS real-time programming environment [8]. ETHNOS exploits the Linux RT multithreaded operating system and provides additional support from different points of view. From the communication perspective it supports and optimises transparent inter-robot information exchange and co-ordination across wireless media. From the runtime perspective it provides support for the real-time execution of periodic, sporadic and background tasks (called Experts), schedulability analysis, event handling, and resource allocation and synchronisation. From the sw engineering perspective it provides support for rapid development, platform independence and sw integration and re-use. The whole set of sw modules for controlling the players, managing communication, as well as the vision system, have been developed over the ETHNOS' Kernel. ETHNOS' Kernel has been selected because of the flexibility of its architecture, allowing the real time scheduling of both occasional Experts, that are conditionally activated, such as the arbitration module, and periodic Experts (i.e. Vision Experts, and Map Building Experts).

Rullit is implemented on a Mo^2Ro base. Its design is centered around the omnidirectional vision sensor we have implemented for Robocup. It consists of a mirror studied to exploit at best all the camera definition both in the neighborough of the robot and at a large distance: the ball corresponds to a reasonable number of pixels from 10 to 400 cm all around the robot. The vision system is implemented mimicking natural mechanisms for fast tracking and color interpretation: we have distances from all the visible, classified objects at a rate higher than 20 frames/sec. This makes it possible to implement behaviors and strategies that take advantage of the knowledge about most of what surrounds the robot. A self-localization module also has enough reliable information to provide an approximate, but satisfactory extimation of the position in most of the situations. Behavior modules are implemented in ETHNOS and a fuzzy low-level control system provides reliable actuation to the fuzzy behaviors.

4 Special Team Features and Conclusion

The most characterizing feature of ART99 has been the ability to coordinate the behaviour of players developed by different groups on different hw and on different sw platforms. This variety originates from the organization of the project at a national level, involving several research groups each one bringing its technical solutions. Coordination has been achieved through a communication layer based on broadcast TCP and on a coordination protocol based on the exchange of information concerning the state of the world and the robot's intentions. Such information was used both to adjust the robot's viewpoint of the world and to dynamically assign the role of each player (excluding the goalie). We have also set up the protocol to allow for different team strategies, but we have not actually adopted this feature during the competition.

Not withstanding the difficulties that we had to overcome to coordinate the activities of groups operating in different sites, we had two major outcomes from a nation-wide project. The first one is that the non-homogeneity of the players has been an advantage, especially considering that the Pioneer-1 based architectures have been pushed at the maximum of their capabilities. The second outcome is the focus on coordination issues that have been critical for the success of the overall team.

References

1. G. Adorni, S. Cagnoni, M. Mordonini, Landmark-Based Robot Self-Localization: a Case Study for the RoboCup Goal-Keeper. *Proc. ICIIS99.*
2. C. Amoroso, A. Chella, V. Morreale, P. Storniolo, A Segmentation System for Soccer Robot Based on Neural Networks, in this volume.
3. A. Bonarini, P. Aliverti, M. Lucioni, An omnidirectional vision sensor for fast tracking for mobile robots. *Proc. IMTC99*, IEEE Computer Press.
4. A. Bonarini, The body, the mind, or the eye first? in this volume.
5. L. Iocchi, D. Nardi. Design and Implementation of Robotic soccer Behaviors: A user viewpoint, *Proc. IIA-SOCO'99.*
6. L. Iocchi, D. Nardi. Self-Localization in the RoboCup Environment, in this volume.
7. E. Pagello, A. D'Angelo, F. Montesello, F. Garelli, C. Ferrari, Cooperative behaviors in multi-robot systems through implicit cooperation, *Robotics and Autonomous Systems.*29, 1999.
8. M.Piaggio, A. Sgorbissa, R. Zaccaria, Programming Real Time Distributed Multiple Robotic Systems, in this volume.

CoPS-Team Description

N.Oswald, M.Becht, T.Buchheim, G. Hetzel, G. Kindermann, R.Lafrenz,
P.Levi, M.Muscholl, M.Schanz, M.Schulé

Institute for Parallel and Distributed High Performance Systems (IPVR)
Applied Computer Science - Image Understanding
University of Stuttgart, Breitwiesenstr. 20-22, 70565 Stuttgart, Germany
robocup@informatik.uni-stuttgart.de

Abstract. This paper presents the hardware and software design principles of the medium size RoboCup Team *CoPS* which are developed by the image understanding group at the Institute for Parallel and Distributed High Performance Systems (IPVR) of the University of Stuttgart. By adapting already successfully tested multiagent software concepts by our group to the domain of robotic soccer we intend to improve those concepts at the field of realtime applications with uncertain sensory data.

1 Introduction

Multiagent theory has become a popular research area in the context of artificial intelligence. Although applied to many domains, the full potential of this paradigm developes especially in situations where decisions have to to be made upon uncertain data or partial information as e.g. in robotics. Deficiencies of a single agent in perceiving its environment and having only partial knowledge about its surroundings shall be compensated by its ability to exchange information with other agents.

Furthermore multiagent systems support the idea of the whole being more than the sum of its parts, meaning that the problem solving potential of a multiagent system exceeds by far the capabilities of the agents within the system. This can be achieved by providing the agents with proper communicational abilities to negotiate and coordinate their actions.

The research group **Image Understanding** of the **Institute for Parallel and Distributed High Performance Systems** has been working for several years in the field of multiagent-systems developing a multiagent architecture [1], a theoretical framework for cooperating agents [2] and applying cooperative concepts in computer vision [3]. With our medium size robotic soccer team *CoPS* (Cooperative Soccer Playing Robots) those developed concepts are tested and adapted to the real world application of a soccer match.

2 Hardware Equipment

Our RoboCup Team consists of 5 Nomad robots of type *Super Scout* equipped with additional sensorics and a kicking device. The onboard computer system of the Super Scouts consists of a Pentium 233 MMX with 64 MB and a hard disc of 4 GB capacity. The two wheel differential drive allows robot motion with a maximum speed of $1m/s$ and an acceleration of $2m/s^2$.

Each of these robots is already equipped with odometric sensors, a tactile bumper ring of 6 independent sensors and 16 ultrasonic sensors arranged around the outside perimeter of the robot. The 6 tactile sensors allow a coarse localization of physical contacts with the environment in the front-left/center/right or the rear-left/center/right of the robot.

The ultrasonic sensors are merely used for obstacle avoidance purposes. Nevertheless, there are situations where physical contacts with the environment are inevitable. In those cases the bumper sensor information is needed to react appropriately (e.g. stopping a backward movement, if the bumper sensor in the rear is activated).

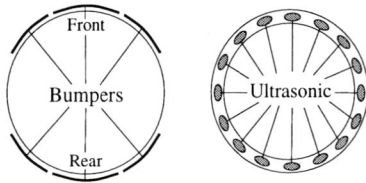

Fig. 1. Pre-installed Sensorics of the Super Scout

2.1 Modifications and additions

We installed a 1/3"-Chip CCD camera with a 752x582 resolution delivering a video signal to a MATROX frame grabber on the Pentium board. By image processing algorithms the ball is extracted from the grabbed pictures and its current position relative to the robot is estimated.

As the precision of the odometric sensors degrades constantly with every movement of the robot, we equipped each robot with a SICK laser range finder [1] for a more robust and exact self localization. Within an angle of 180° the range finder provides depth information with a resolution of 0.5° and an accuracy of 5cm. To derive the actual position from the laser data the measured laser image is iteratively rotated and shifted until it fits best a predefined model of the football field. For a faster computation of the robot's position by laser we make use of the odometric data which is taken as starting point for the iterative search. After accomplishing the self localization the odometric values are updated with the new values. However, because of this rather time consuming procedure, a localization by laser should be applied as rarely as possible.

For further computational power a Toshiba Portégé 7010 CT Laptop was mounted on top each robot. The Pentium board as well as the notebook are

---

[1] We'd like to thank the SICK AG,Germany for lending us three Laser Range Finders

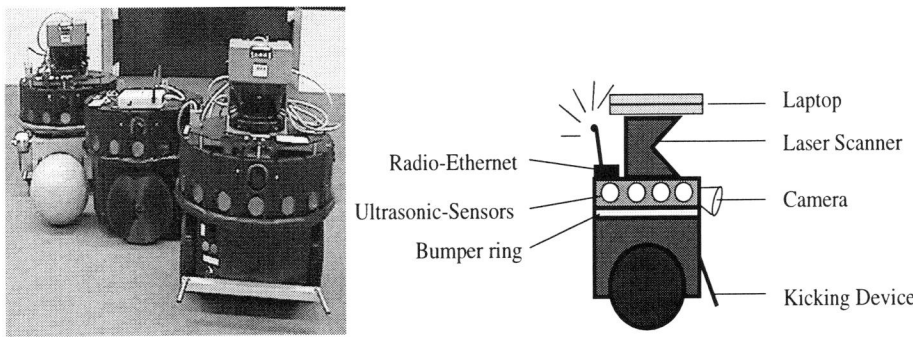

Fig. 2. Equipment of a soccer robot

connected to a radio-ethernet HUB enabling communication between the board and the laptop as well as communication with other robot agents.

For passing and kicking the ball we constructed a special kicking device which is driven by two solenoids. The camera, laser range finder and kicking device are mounted towards the front of the robot players. Only the goalkeeper's sensors are shifted by 90° as it only moves unidirectionally forwards or backwards to defend the goal.

3 Software Concepts

3.1 Agent Architecture

In our software model each robot consists of a set of concurrent software modules, so called elementary agents (EA). Each EA has special plans to perform tasks which can be requested by other elementary agents. We classify EAs into three categories according to the level of abstraction of the tasks they perform.

EAs concerned with simple control tasks of the sensors and actuators are situated at the lowest layer within the architecture, **the reflexive layer**. Reflexive tasks generally involve only simple planning or no planning at all, guaranteeing an immediate response on a request. The Pilot-EA e.g. performs tasks like driving a given route straight ahead and rotation by a given angle, while the Localizer-EA determines the current position by odometric and laser range finder data.

More complex tasks are performed by the EAs at **the tactical layer** of our software architecture. An EA concerned with scene-detection aggregates and evaluates information from the sensors of a robot and builds a model of the environment which can be requested by other agents. A navigator-EA, also located at this level and responsible for vehicle control, is concerned with tasks like *getting the ball* or *moving towards the goal*. Planning such tasks requires information about the environment which is provided by the scene-detection.

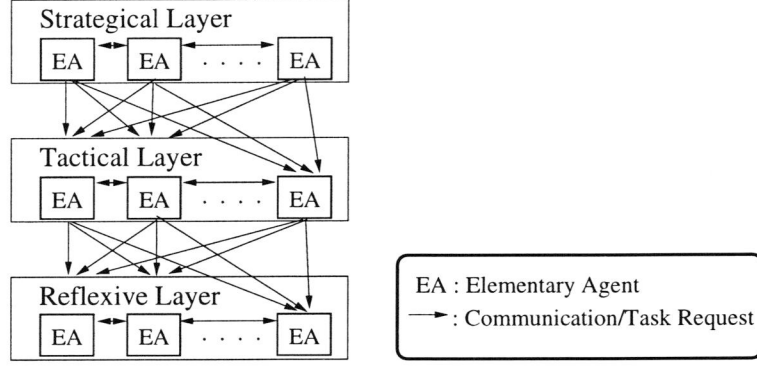

Fig. 3. Cooperative Agent Architecture

Finally, at the strategical layer, the Strategist-EA is concerned with long term goals and team coordination. It generally has to react only on external signals like e.g. *start of the game, goal* and *end of the game*. All EAs can send task-requests to each other within the same robot as well as to elementary agents of other robots. Thus, data-uncertainties concerning the environment can be compensated and a means for coordinating joint actions is provided. For a more detailed description of our software architecture we refer to our technical paper [4].

3.2 Implementation Issues

The elementary agents were all designed as separate multi threaded processes on a LINUX system, a communication thread waiting for task requests from other agents and further executional threads processing requested tasks. For interagent communication we used the freely available CORBA implementation MICO which facilitated the distributed agent modelling a lot.

References

1. P. Levi, M. Becht, R. Lafrenz, and M. Muscholl. COMROS - A Multi-Agent Robot Architecture. In *DARS 3*. Springer-Verlag, 1998.
2. M. Becht, M. Muscholl, and P. Levi. Transformable multi-agent systems: A specification language for cooperation processes. In *Proceedings of the World Automation Congress (WAC), Sixth International Symposium on Manufacturing with Applications (ISOMA)*, 1998.
3. N. Oswald and P. Levi. Cooperative vision in a multi-agent architecture. In *LNCS*, volume 1310, pages 709–716. Springer-Verlag, 1997.
4. N.Oswald, M.Schulé, and R.Lafrenz. A cooperative architecture to control multi-agent based robots. Submitted to : Proceedings of the IJCAI RoboCup Workshop, 1999.

CS Freiburg '99[*]

B. Nebel, J.-S. Gutmann, W. Hatzack

Universität Freiburg, Institut für Informatik
Am Flughafen 17, D-79110 Freiburg, Germany

1 Introduction

One of the interesting challenges in designing a successful robotic soccer team is the need to cover the entire loop from sensing over deliberation to acting. For example, successful ball passing needs good estimations of the position and velocity of the other players and the ball, projections into the future, planning ahead in order to create and exploit opportunities, and, finally, it requires to act accordingly.

One of our main goals in participating in RoboCup'99 was to enhance the design of our team *CS Freiburg* [5], which participated successfully in RoboCup'98 [1], in a way such that the robots can pass balls and are more flexible in their role assignment. For this purpose, we worked on enhancing the sensor data gathering and sensor data interpretation components, redesigned the deliberation components, and refined the behavior-based control module. The hardware design is basically the same. While we are aware of the fact that there are better alternatives for the basic platform and the kicker design, we decided to live with their limitations because they have proved to be reliable and robust enough for our purposes.

In RoboCup'99, our team lost the first game in its entire history, the semifinal against the Italian team. Nevertheless, we count the game as a success since this game was a pleasure to watch. In addition, we were able to demonstrate our ability to pass a ball (intentionally!). All in all, we came out as the 3rd in this competition. Counting our 1st place in RoboCup'98 (July 1998), the 1st place in the German open VISION RoboCup'98 (October 1998), and the 1st place in the German open VISION RoboCup'99 (October 1999), CS Freiburg is one of the most successful robotic soccer teams.

2 Team Development

Bernhard Nebel is head of the team, *Steffen Gutmann* is the main designer and coordinator of the development team, and *Wolfgang Hatzack* is responsible for the software development process, the global fusion component, and the user

---

[*] This work has been partially supported by *Deutsche Forschungsgemeinschaft* (DFG) as part of the graduate school on *Human and Machine Intelligence*, by *Medien- und Filmgesellschaft Baden-Württemberg mbH* (MFG), and by *SICK AG*, who donated a set of new generation laser range finders.

interface. In addition, the following graduate students (re-) designed and implemented components of the system. *Boris Bauer* and *Andreas Hill* (behavior redesign), *Markus Dietl* (global fusion, in particular ball position estimation), *Burkhard Dümmler* (integration of new laser scanners), *Immanuel Herrmann* (all mechanical components, in particular kicker design), *Kornel Marko* (simulator), *Christian Reetz* (tactical decision making and behaviors), *Augustinus Topor* (path planning), and *Maximilian Thiel* (vision and low-level interfaces to the Cognachrome board).

3 Robots

The robot hardware we use is described in detail [5]. As in last year's competition, we used Pioneer 1 robots enhanced by custom-built kickers, SICK laser scanners, and the Cognachrome vision system. For local information processing we used Librettos 110CT. For communication between the robots and the off-field computer, the WaveLan radio ethernet was employed.

4 Perception

The main sensors we use are *laser range finders* (now the new *LMS 200* range finders, which have an accuracy of 1cm) and the commercially available *Cognachrome* vision system [5]. In the '98 team design we used only 5 laser scans and 8 frames per second, although the devices could give us 35 laser scans and up to 60 frames per second. Furthermore, we only had very inaccurate time stamps for the measurements. In order to raise the data rate to the maximal possible rate, we modified the Cognachrome software and implemented new modules for gathering the data. Additionally, we started to use a real-time extension of Linux — *RTLinux* [2] — in order to cope with the high data rate from the laser range finder (500 KBaud) and to assign millisecond accurate time stamps to all measurements. Using the higher data rate, we got much better estimates for the velocity of moving objects on the field.

While self-localization based on the laser range finders give us very accurate and robust estimations of our own positions [6], the estimations of the ball is not very accurate. In the vision module, we now use the shape of the ball to exclude false positives and to increase the accuracy of the estimation of the ball position. Additionally, in order to compensate for the lack of stereo vision, we use the entire group of robots to estimate the ball position more reliably and precisely than any single robot with monocular vision can do using ideas from [4].

5 World Model

The world model is similar to the one we developed for RoboCup'98 [5]. Each robot builds a local world model about its own position on the field, the ball position, and the position of other players. This model is extended by the results of the global fusion component that runs on the off-field computer and combines all estimates from all other players. While this component gives much more accurate estimations, in particular of the ball position, and enables us to distinguish friends and foes, it is always a bit out-dated (100–200 msec).

6 Communication

Our robots communicate – using the WaveLan radio ethernet – in order to build up the global world model, to negotiate about which robot is going to the ball, and to initiate ball passing.

7 Skills

The basic ball handling skills are, from our experience, very important. However, it is very difficult to implement them in a robust way. For example, when arriving in Stockholm we noticed that our behaviors had to be tuned to the carpet which was significantly different from the one we used on our exercise field in Freiburg. While one robot can handle a ball usually adequately, we were not able to implement a reliable *ball intercepting* behavior because the responsiveness of the Pioneer is not adequate.

8 Strategy and Tactics

One of the main differences to our '98 team is that we use now a more principled way for choosing actions. We use an approach based on behavior networks as developed by Maes [7] and refined for the purpose of playing (simulated) robotic soccer by Dorer [3]. This approach enabled us to express our tactics in much more modular and extensible way so that we were able to modify our tactics in a significant way even during the competition in Stockholm. Furthermore, we extended our cooperative play approach. First of all, we do not have fixed areas of competence anymore, but roles that can be filled (and reassigned), such as defender, mid-fielder, and forward. In connection with that, the players negotiate which robot is going to the ball. Secondly, we have true cooperative play when the ball has to be passed. The player possessing the ball calls for a team mate to go to good position and plays the ball when the team mate signals that it has reached this position. Another significant difference to the '98 team is our new path planning component. Now we use a potential field approach that tries to stay away from obstacles, while in 1998 we used a geometric path planner that tried to compute the shortest path.

9 Special Team Features

Our special feature used to be that we use laser range finders in order to do self-localization and object recognition [5,6]. This year, the teams from Stuttgart and

Tübingen used laser range finders as well. As it turned out, laser range finders alone do not guarantee success.

Furthermore, other teams, such as the Italian and the Munich team demonstrated that reliable and accurate self-localization can be done solely based on vision. Although we demonstrated that our team is still competitive, the other teams proved to be very good (either at RoboCup'99 or the German open VISION RoboCup'99). In particular, we noticed that the factor of speed (who is first at the ball?) seems to become a crucial issue once the sensor interpretation and world modeling problem appears to be "solved."

10 Conclusion

Although we were satisfied (well, ...) with our performance at RoboCup'99, there are, of course, a number of points where our team can be improved. Some of these points are on an abstract level such as model-based object recognition, cooperative path planning, situation adapted placement of players, adaptable tactics, more adaptive vision, and so on. Other points are on the hardware level such as improving the responsiveness and speed of our players and improving the robustness of the vision (using other cameras and vision hardware), building better kickers, etc. Which of these points we are able address until the next competition is not clear. However, we intend to participate in the next RoboCup and hope to increase th level of play again.

References

1. M. Asada and H. Kitano, editors. *RoboCup-98: Robot Soccer World Cup II*. Springer-Verlag, Berlin, Heidelberg, New York, 1999.
2. M. Barabanov and V. Yodaiken. Introducing real-time Linux. *Linux Journal*, 34, 1997.
3. K. Dorer. Behavior networks for continuous domains using situation-dependent motivations. In *Proceedings of the 16th International Joint Conference on Artificial Intelligence (IJCAI-99)*, Stockholm, Sweden, Aug. 1999. Morgan Kaufmann.
4. J.-S. Gutmann, W. Burgard, D. Fox, and K. Konolige. An experimental comparison of localization methods. In *Proceedings of the International Conference on Intelligent Robots and Systems (IROS'98)*. IEEE/RSJ, 1998.
5. J.-S. Gutmann, W. Hatzack, I. Herrmann, B. Nebel, F. Rittinger, A. Topor, T. Weigel, and B. Welsch. The CS Freiburg robotic soccer team: Reliable self-localization, multirobot sensor integration, and basic soccer skills. In Asada and Kitano [1].
6. J.-S. Gutmann, T. Weigel, and B. Nebel. Fast, accurate, and robust self-localization in polygonal environments. In *Proceedings of the International Conference on Intelligent Robots and Systems (IROS'99)*. IEEE/RSJ, 1999.
7. P. Maes. Situated agents can have goals. In P. Maes, editor, *Designing Autonomous Agents: Theory and Practice from Biology to Engineering and Back*, pages 49–70. MIT Press, Cambridge, MA, 1990.

DREAMTEAM 99: Team Description Paper

Wei-Min Shen, Jafar Adibi, Rogelio Adobbati,
Jay Modi, Hadi Moradi, Behnam Salemi, Sheila Tejada
Computer Science Department / Information Sciences Institute
University of Southern California
4676 Admiralty Way, Marina del Rey, CA 90292-6695
email: {shen, dreamteam}@isi.edu

Abstract. The annual Robocup soccer competition is an excellent opportunity for our robotics and agent research. We view the competition as a rigorous testbed for our methods and a unique way of validating our ideas. After two years of competition, we have begun to understand what works (we won the competition in Tokyo 97) and what does not work (we failed to advance to the second round in Paris 98). This paper presents an overview of our goals in Robocup, our philosophy in building soccer playing robots and the methods we are employing in our efforts.

1 Introduction

The annual Robocup soccer competition is an excellent opportunity for our robotics and agent research. We view the competition as a rigorous testbed for our methods and a unique way of validating our ideas. As everyone knows, it is often easy to build robots and program them with algorithms that work well in controlled environments. However, in order to build robots that are robust and flexible, we feel it is essential to be able to test these ideas in more challenging arenas. Robocup provides us with such an environment. We have learned a great deal from two years of experience in the Robocup competition. We have begun to understand what works (we won the competition in Tokyo 97) and what does not work (we failed to advance to the second round in Paris 98). Using this experience, we plan to field a team of robots in Stockholm this year that will showcase our abilities in building autonomous physical agents and the state-of-the-art in robotic soccer.

2 Philosopy and Goals

Our primary goal in the Robocup project is to build autonomous physical robots that can function robustly in a challenging environment. Obviously, this implies two things about our robots:
  Requirement 1: They must be autonomous.
  Requirement 2: They must be robust.
  These requirements have significant implications on the methodology we use to build and program our robots. In particular, Requirement 1 implies that processing must be distributed and on-board. No remote computing or centralized

control is allowed. Requirement 2 implies that algorithms and hardware must be simple enough to guarantee reliability. Indeed, a guiding philosophy in building these robots is to favor robustness over sophistication. However this does not mean that we are satisfied with simple robots. Instead, we see our robots as evolving, becoming more advanced every year. (In fact our robots from two years ago already seem arcane to us.) This is an ongoing process and we continue to make significant improvements.

Adhering to the requirements outlined above, our efforts can be decomposed into three specific specialties; hardware, vision and learning. Section 2 below describes the hardware of our physical robots, section 3 describes our vision system and section 4 describes learning. Section 5 has a brief description of our soccer playing algorithm.

3 Hardware

Unlike many of the other teams at Robocup in previous years, our robots are entirely constructed from scratch and by our team. We feel this is a significant strong point of our robots and an area of expertise that our team brings to the competition. The flexibility to modify our custom-built robots gives us an added dimension for experimentation. As we learn more about what capabilities are needed by an autonomous physical agent interacting with its environment, we are able to easily adapt and extend our custom-built robots. (For example, in the past two years alone, we have been able to add dual cameras, replace motors, and redesign the base.) The next paragraph describes the hardware of our robots in detail.

The base of each robot is a modified 4-wheel, 2x4 drive DC model car. Specifically, we have lowered and widened the base for added stability. The wheels are independently controlled, allowing in-place turning and easy maneuverability. We have replaced the stock motors with stronger, heavy-duty motors to support the increased weight of the car. Mounted above the base is an on-board computer. It is an all-in-one 133MHz 586 CPU board extensible to connect various I/O devices. Attached to the top of the body are twin commercial digital color QuickCam cameras made by Connectix Corp. One faces forward, the other backward. Also, we have affixed fish-eye lenses to each camera to provide a wide-angle view of the environment. The two drive motors are independently controlled by the on-board computer through two serial ports. The hardware interface between the serial ports and the motor control circuits are custom built by our team. The images from the cameras are sent into the computer through a parallel port. On board are three batteries, one for each of the two motors and one for the CPU and cameras.

This year, we plan to incorporate additional hardware. In particular, we are going to extend the sensory capabilities of the robot by adding touch sensors to the body. This will allow the robot to avoid obstacles more effectively. Also, we are going to add shaft encoders. These devices allow the robot to measure the actual revolution of the wheels. We hope this will allow the robot to move about more accurately.

4 Vision

We view color-based vision as one of salient challenges in the Robocup initiative and one of the scientific issues on which we intend to focus. Building an accurate, reliable vision system that can work under a variety of conditions is one of our team's primary goals. We are continually improving this component of our robots. The following describes our current vision system.

Our vision system is entirely custom-built. It is a specialized software component developed specifically for detecting balls, goals and other robots. Visual information is extracted from an image of 658x496 RGB pixels, received from the on-board camera via a set of basic routines from a free package called CQ-CAM, provided by Patrick Reynolds from the University of Virginia. Since the on-board computing resources for an integrated robot are very limited, it is a challenge to design and implement a vision system that is fast and reliable. In order to make the recognition procedure fast, we have developed a sample-based method that can quickly focus attention on certain objects. Depending on the object that needs to be identified, this method will automatically select certain number of rows or columns in an area of the frame where the object is most likely to be located. For example, to search for a ball in a frame, this method will selectively search only a few horizontal rows in the lower part of the frame. If some of these rows contain segments that are red, then the program will report the existence of the ball (recall that the ball is painted red). Using this method, the speed to reliably detect and identify objects, including capture time, is greatly improved; we have reached frame rates of up to 6 images per second.

A significant drawback of our current vision system is its sensitivity to lighting conditions. Its parameters must be hand-tuned to a specific environment and this is a time consuming task. We are currently exploring more automated approaches that will reduce this burdensome task.

5 Learning

Any robot situated in a dynamic environment must be able to discover new things at run-time. We view autonomous learning as our holy grail and the dream that we are striving for. We also consider it the most difficult scientific issue. There are many well-known learning algorithms that work well in simulations or on a desktop, but what happens when you attempt to run these algorithms on a physical, situated robot? In our Robocup project, we have found that there is a large gap. In particular, we found that our efforts in learning have been limited by the previous two areas, hardware and vision. But we feel that we are finally beginning to achieve a critical mass in those areas to allow implementation of on-board learning algorithms. Some of our team members are actively involved in research in the field of multiagent learning and our team has significant expertise in the area of agent learning. We are trying to apply this work to our physical robots as a validation of the research.

6 Programming Approach

Our robotic soccer team consists of four identical robots. They all share the same general architecture and basic hardware. However, they differ in their programming. We have developed three specialized roles; the forward role, the defender role and the goalie role. Each role consists of a set of behaviors organized as a state machine. For example, the forward role contains a shoot_ball behavior, dribble_ball behavior, a search_for_ball behavior, etc. The state transitions occur in response to percepts from the environment. For example, the forward will transition from the search_for_ball behavior to the shoot_ball behavior if it detects the ball and the goal from its sensory input. At game time, each robot is loaded with the program for the role it has been assigned. Note that each robot has the integrated physical abilities to play any role (i.e. detect_ball, move_forward, turn, etc...). We feel this is a natural, flexible, efficient approach to programming the robots to play soccer.

7 Conclusion

In summary, we have stated that our primary goal is to advance the state-of-art in building autonomous, robust physical robots. We aim to accomplish this goal by focusing on three important areas; physical hardware, robot vision and agent learning. We view Robocup as an exciting, rigorous testbed for our project and hope to prove the viability of our ideas and approaches by success in the soccer tournaments.

Description of the GMD RoboCup-99 Team

Ansgar Bredenfeld, Wolf Göhring, Horst Günter, Herbert Jaeger, Hans-Ulrich Kobialka, Paul-Gerhard Plöger, Peter Schöll, Andrea Siegberg, Arend Streit, Christian Verbeek, and Jörg Wilberg

GMD, Sankt Augustin, Germany

Abstract. This article gives a brief sketch of the scientific and engineering approach taken at the GMD RoboCup Team. We sketch (i) the robot hardware, (ii) the "Dual Dyanmics" model of behavior control that we develop, and (iii) the integrated "Dual Dynamics Designer" environment that we use for programming, simulation, documentation, and code generation.

1 Introduction

Kicking a ball into the right direction is a very hard task for robots, and designing such robots is an equally hard task for human engineers. From a traditional engineering perspective, it cries out for a modularized, hybrid approach.

However, there are indications that the classical divide-and-conquer approach is not fully appropriate for football-playing robots [2]. Classical modular system consist of subsystems that communicate with each other over relatively narrow channels according to strict protocols, hiding from each other most of what is going on inside them. In our opinion, this kind of modularity is too inflexible to enable the kind of swift, "holistic", dynamic responses required from a football robot.

Thus, a fundamental challenge for mobile robotics is to reconcile, (i) the need for *some* sort of modular design, which results from the necessity of bringing together diverse techniques, with (ii) the "holistic" way in which the robot should respond to the situation.

At the Behavior Engineering (BE) research group at GMD we explicitly address this challenge. On the robot side, we develop a mathematical model of a behavior control system which aims at an integration of modularity and hierarchical structuring with the flexibility of self-organizing dynamical systems. This is the Dual Dynamics (DD) model [3]. On the other hand, we develop and utilize a design tool that fosters a close collaboration of engineers, by providing everyone with a unified access to the the entire robot control system under construction. This is the Dual Dynamics Designer (DDD) tool [1].

In this article, we first describe our robot hardware (Section 2), then give a quick overview of the DD model (Section 3), and finally describe the DDD tool (Section 4).

2 Our robots

We have built our robots from scratch[1]. They feature the following equipment:

PC and wavelan. Each robot carries a small laptop Pentium PC and a wavelan ethernet card.

Micro-controllers. Three C 167 micro-controllers host elementary sensor interfacing and motor control routines.

CAN communication. The computer components are linked to each other via a CAN-bus.

Vision system Our vision system is a Newton Lab "Cognachrome" system mounted on a pan unit.

Obstacle avoidance sensors. A bumper ring features custom-built force detectors with a wide measuring range. For detecting distant objects (up to 60 cm), an IR-based range detector (exploiting triangulation) is mounted on a "radar"-like sweeping servo.

Motor and chassis. We use two 40 Watt, high-quality Maxon motors that are mounted on a very solid, mill-cut aluminium frame.

Figure 1 shows one of our robots in a pose that is reminiscent of Rodin's *thinker*.

Fig. 1. One of our robots and one of our balls.

Morale. Building and maintaining one's own brand of robots, as opposed to buying an off-the-shelf system, does not make life easier, nor cheaper. Nobody knows so well how the perfect robot should be made as the ones that know all the loose wires by name, and have no time and money left to build the perfect one.

---

[1] We are indebted to Karl-Ludwig Paap and his group for their generous assistance

3 Our formalism

The Dual Dynamics scheme is a mathematical model of a behavior control system for autonomous mobile robots. It has grown from three roots: the behavior-based approach to robotics, the dynamical systems approach to cognition, and the mathematical theory of self-organizing dynamical systems [3].

A DD model specifies a collection of behaviors as a comprehensive dynamical system. It consists of many coupled subsystems, each of which is responsible for controlling a particular behavior. These behavior subsystems are specified through ordinary differential equations.

Behaviors are ordered in levels in the DD scheme. At the bottom level of a DD behavior hierarchy, one finds *elementary* behaviors. These are sensomotoric coordinations with direct access to external sensor data and actuators. Typical examples are `kick` and `bumpRetract`. Higher levels are constituted by increasingly comprehensive behaviors. They also have access to sensoric information but cannot directly activate actuators. Typical examples are `attack` and `defend`.

Elementary behaviors are made from two subsystems. This has given the approach its name, "dual dynamics".

The first of these subsystems is called the *target dynamics*. It calculates target trajectories for all relevant actuators.

The other subsystem of an elementary behavior is its *activation dynamics*. It regulates a single variable, the behavior's *activation*.

Only the activation dynamics is allowed to undergo bifurcations. The control parameters which induce these bifurcations are the activation variables of higher-level behaviors. This is the core idea behind DD. The ways of how, exactly, these bifurcations are induced, are highly constrained. These constraints warrant the transparency of the DD scheme.

Emphatically, an elementary behavior is not "called to execute" from higher levels. The level of elementary behaviors is fully operative on its own and would continue to work even if the higher levels were cut off. The effect of higher levels is not to "select actions", but to change the overall characteristics of the elementary level.

Morale. A good mathematical model of what one is doing is really satisfactory for the mathematically inclined. But it is also good for debugging.

4 Our development environment

In order to come closer to the ideal of a robot that behaves "holistically", the very design process must be maximally transparent. To this end, we have developed a unified software developing environment, the "Dual Dynamics Designer" (DDD).

The primary graphical user interface for designing a DD model is shown in Fig. 2. It includes icons of sensors, sensor filters and intermediate sensor representations, elementary and higher-level behaviors. Important global variables and constants (time constants, especially) appear highlighted besides the concerned icons. By clicking on the icons, context-sensitive editor windows pop up in which equations and/or ODEs can be specified in an intuitive syntax.

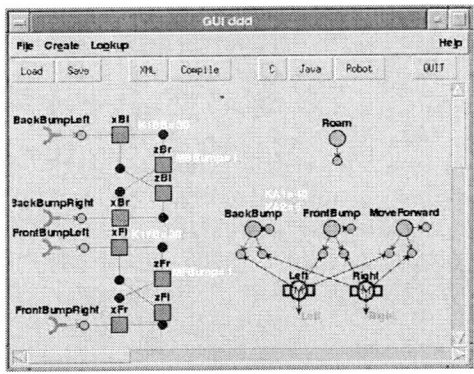

Fig. 2. The primary DDD user interface.

After designing the network of behaviors and preprocessing filters, a syntax check, global and local variable detection and checking for cyclic dependencies between equations is performed in a compilation step.

By hitting the C, Java, and Robot buttons, executable standard C code, Java code, and robot C++ code is generated. At the same time, a HTML documentation of the entire current system is generated automatically. The Java code can be fed into a simulation engine.

The DDD tool itself is constructed with the Rapid Prototyping Environment APICES. Readers interested in software engineering aspects can find more details on the software architecture and development process of the DDD tool in [1].

Morale. The DDD tool makes editing, code generation, simulation, and documentation so easy that we just don't understand why it still is so difficult to make robots kick a ball into a goal.

5 Conclusion

Science can be fun and still be science, even the more so.

References

1. A. Bredenfeld. Co-design tool construction using APICES. In *Proc. of the 7th Int. Workshop on Hardware/Software Co-Design (CODES'99)*, 1999. To appear.
2. A. Bredenfeld, W. Göhring, H. Günter, H. Jaeger, H.-U. Kobialka, P.-G. Plöger, P. Schöll, A. Siegberg, A. Streit, C. Verbeek, and J. Wilberg. Behavior engineering with "dual dynamics" models and design tools. 1999. Submitted to the IJCAI-99 RoboCup Workshop.
3. H. Jaeger and Th. Christaller. Dual dynamics: Designing behavior systems for autonomous robots. *Artificial Life and Robotics*, 2:108–112, 1998. http://www.gmd.de/People/ Herbert.Jaeger/Publications.html.

ISocRob — Intelligent Society of Robots*

Rodrigo Ventura, Pedro Aparício, Carlos Marques, Pedro Lima, and Luís Custódio

Instituto de Sistemas e Robótica/Instituto Superior Técnico (ISR/IST),
Av. Rovisco Pais – 1, 1049-001 Lisboa
PORTUGAL
{yoda,aparicio,pal,lmmc}@isr.ist.utl.pt

Abstract. The SocRob project was born as a challenge for multidisciplinary research on broad and generic approaches for the design of a cooperative robot society, involving Control, Robotics and Artificial Intelligence researchers. In this paper the basic aspects of last year implementation as well as the improvements made meanwhile are briefly recalled and presented. Naturally, a special emphasis is given here to the novel solutions proposed for this year implementation, the results obtained and the expected future developments.

1 Introduction

The Artificial Intelligence and the Intelligent Control groups of the ISR/IST have started almost two years ago a joint project on Cooperative Robotics, denominated SocRob, to foster research on methodologies for the definition of functional, hardware and software architectures to support intelligent autonomous behavior and evaluate performance of a group of *real* robots, either as a society and as individuals.

The utilization of real robotic agents to perform on real environments, for instance, a robotic soccer game, raises several new questions and perspectives that turn the development of a multi-agent system a much more difficult and challenging problem [5].

The robots used by the ISocRob team were developed from scratch, so that both conceptual and implementation issues were considered [1]. For the 1999 competition, some adjustments and improvements were made both on hardware and software components: new robot wheels, a kicker device, development of a self-localization system, a friendly man-machine interface, and a new software framework based on the multi-agent system paradigm.

2 Hardware and Software Description

Each robot hardware is divided in four main blocks: sensors, main processing unit, actuators and communications. Currently, from the hardware architecture standpoint, the population is composed of homogeneous mobile robots.

* This work has been supported by a grant from the Fundação Calouste Gulbenkian and the Portuguese Foundation for Science and Technology (ISR/IST programmatic funding).

2.1 Hardware

The sensors and actuators available in each robot are those mentioned in last year's team report [1]. In terms of sensors, the novelty for this year was the development of a self-localization system (a pose sensor), but unfortunately this sensor was not available at the time of RoboCup'99. In terms of actuators, it has been developed a new kicking device similar for all robots.

Pose sensor Depending on the type of application involved, each robot of the society needs to regularly update its current pose (position and orientation) with respect to a reference frame (e.g., located in the field center). This may be accomplished based on the triangulation principle using for instance a *convex mirror for full scene image* system based on a vision camera and a mirror with a special geometry. Since the RoboCup environment has available a sufficient number of visual landmarks, the SocRob project team decided to experiment the "mirror" solution. The idea is to allow robots, using only one vision camera, to acquire images from the mirror, appropriately positioned above the robot, in order to obtain a global view of the environment. If images are sufficiently broad to include three different and static beacons (e.g., goal plus two field corners), robots may apply the triangulation principle to determine their position.

Kicker The kicking ability enables soccer players to move the ball into places that otherwise would not be accessible. The kicker device is divided in two main parts: electronic and mechanical structures. The kicker electronics is composed of a micro-controller, an IR beam circuit and a power actuator. The micro-controller runs the control program and generates a signal modulation to be used in the IR beam. This signal consists of a square wave, rated 40kHz that is fed to the amplifier powering the IR led's. The 40KHz detector output is also directly connected to the controller. If an object is obstructing the beam the demodulator delivers a 0V constant signal, otherwise should a 40kHz IR beam be received, a 5V constant signal is obtained. The controller output is connected to the circuitry that drives the servo-motor. This solution can be seen as an instinctive reaction when the robot senses the ball. However this behavior can be disabled by the processor unit in order the robot to perform different type of actions. The kicker mechanics is based on a automobile door opening servo-motor, which when powered with opposite polarities moves a piston in opposite directions. The piston course is approximately 3cm.

2.2 Software

In what concerns software components, the two main improvements made this year were the development of a man-machine interface and a multi-agent software framework for programming each one of the robots. Instead of having a set of procedures to implement each of tasks needed (*e.g.*, motors control, image processing, communications, behaviour control, etc.), it was implemented a new software framework based on the multi-agent paradigm. For each task it is created a specific (micro-)agent[1] prepared to perform the task. All micro-agents are implemented using the concept of *thread* for multitasking programming and communicate with each other through a common, but distributed, repository of information (blackboard).

The man-machine interface allows to observe the behaviour of all robots using telemetric information (encoder data), images acquired from the frontal camera and some relevant parametric information. The interface permits also to manually control the robots through a *space mouse* device or directly from the keyboard, change some robot internal variables and specify certain game parameters.

3 Functional Architecture

From a functional standpoint, the whole robot society is composed of functionally heterogeneous robots. The functional architecture is *scalable* regarding the number of robots (or *agents*) involved. This means that, when a new robot joins the society, no changes have to be made to the overall system. The functional architecture establishes three levels: an organizational level dealing with the issues common to the whole society, a relational level where groups of agents cooperate/negotiate in order to establish a mutual agreement(commitment) concerning the execution of a particular action or the achievement of some objective, and an individual level encompassing all the available behaviors of each robot. A behavior is a set of purposive primitive tasks sequentially and/or concurrently executed. These primitive tasks consist of *sense-think-act loops*, a generalization of a closed loop control system which may include motor control, ball tracking, ball following, etc. For this year participation in RoboCup, some modifications were made on the relational and individual levels, especially in what concerns cooperation among robots [2].

Individual behaviors can be temporarily modified to allow cooperative relations between teammates. The negotiation implemented concerns two Forward players who actively try to get the ball. If two or more Forward players see the ball, a communication protocol is used by all players involved in order to determine which player is closest the ball. So, all players broadcast the estimation of its distance to the ball, and after that negotiate which player will follow the ball and which should return to a pre-defined location near one of the goals.

4 Agent-based Programming Language

The idea beyond the development of a programming language specially adequate for implementing multi-agent systems follows the work previously done by some members of our team — an agent-based programming language called RUBA [4]. The goal is to have a way for defining agents' architecture, creating agents, establishing communication links among agents, specifying cooperation mechanisms (based on a particular teamwork model), creating and deleting temporary sub-groups, and removing agents.

---

[1] Since this agent implements a primitive task running on a robot, it is called here a micro-agent.

The initial version of the computational model for the language consists of two classes of objects: *agents* and *blackboards*. A blackboard is the basic communication medium among agents, either to communicate among themselves, or between them and the external world. In what concerns RUBA, the current language specifications propose several improvements: extension of the blackboard for a distributed system, efficient blackboard indexing using a hierarchical namespace, and event-driven programming.

Conceptually, a blackboard is a centralized repository of data. The idea of a distributed blackboard is to distribute the information (data) among the agents. Practically, a blackboard is a mapping of symbols (hierarchically organized in nested name-spaces, *e.g.* `robot0.sensors.collision.2`) to variables. This scheme is supposed to uniformly implement several and different processes, such as message passing, shared memory, distributed data and local variables. A blackboard is implemented with an hash table of names to variables. Each variable has a set of attributes, such as scope, location, policy, type and lock. Also, there are a set of primitives to access the variable: `read`, `write`, `hook` and `lambda` [3].

5 Conclusions and Future Work

Currently, our robots are capable of simple but essential behaviors, composed of primitive tasks, such as following a ball, kicking a ball, scoring goals and defending the goal, using vision-based sensors and the other available sensors. Our current and future work is centered on concluding the development of the self-localization system based on a vision camera and a mirror, updating and tuning of the low-level software, design and implementation of an agent-based programming language suitable for multi-agent systems, study and development of a teamwork model and its integration with our functional architecture.

References

1. P. Aparicio, R. Ventura, P. Lima, and C. Pinto-Ferreira. *ISocRob - Team Description*, chapter In Minoru Asada and Hiroaki Kitano, editors, RoboCup-98: Robot Soccer World Cup II. Springer-Verlag, Berlin, 1999.
2. P. Lima, R. Ventura, P. Aparício, and L. Custódio. *A Functional Architecture for a Team of Fully Autonomous Cooperative Robots*, chapter in this book. Springer-Verlag, Berlin, 1999.
3. R. Ventura, P. Aparício, and P. Lima. Agent-based programming language for multi-agent teams. Technical Report RT-701-99, RT-401-99, Instituto de Sistemas e Robótica, IST-Torre Norte, March 1999.
4. Rodrigo M. M. Ventura and Carlos A. Pinto-Ferreira. Problem solving without search. In Robert Trappl, editor, *Cybernetics and Systems '98*, pages 743–748. Austrian Society for Cybernetic Studies, 1998. Proceedings of EMCSR-98, Vienna, Austria.
5. Alex S. Fukunaga Y. Uny Cao, Andrew B. Kahng, and Frank Meng. Cooperative Mobile Robotics: Antecedents and Directions. In *http://www.cs.ucla.edu:8001/-Dienst/UI/2.0/Describe/ncstrl.ucla_cs%2f950049?abstract=Cooperation*, December 1995.

KIRC: Kyutech Intelligent Robot Club

Takeshi OHASHI, Masato FUKUDA, Shuichi ENOKIDA, Takaichi YOSHIDA
and Toshiaki EJIMA

Kyushu Institute of Technology

1 Introduction

Autonomous soccer robots should recognize the environment from the captured image from a video camera and plan to proper behavior. Furthermore, when some robots play cooperatively, communication system between robots is important inputs. We choose simple vision and actuator system, then the gap between the real world and simulation environment are small. Our research target is to accomplish multiagent system using reinforcement learning.

From the competition result, we had two wins, two losses and one draw. In this paper, our robots' hardware and software are described. Then current problems and future works are discussed.

2 Team Development

Team Leader: Toshiaki Ejima, Prof., attended the competition
Team Members:
- Takaichi YOSHIDA, Associate Prof., attended
- Takeshi OHASHI, Research associate, attended
- Shuichi ENOKIDA, Graduate student, attended
- Masato FUKUDA, Graduate student, attended
- Yuudai KARASUYAMA, Graduate student, attended
- Tatsuya ASAZU, Graduate student, attended
- Four support students did not attend the competition

Web page http://www.mickey.ai.kyutech.ac.jp/KIRC/kirc.html

3 Hardware configuration

Our robots were designed simple and low cost. Main parts are not special devices. Then the robot could be used for not only soccer play but also a general platform of an autonomous robot. A field player and a goal keeper are shown as Fig.1. The robots use a radio control tank chassis, a note type personal computer and a video camera. Main parts are listed in Fig.2. Its chassis is used a M4 Sherman base that has independent left and right motors and gearboxes and tracks. It can pivot on the spot and run over rough field.

The field players mainly keep attention to their front view. Because their tasks are finding a ball and goals, avoiding other players and walls. Then they

Fig. 1. Our field player and goal keeper

have a wide lens video camera toward front which tilt slightly down to remove the noise from the outside the field.

The goal keeper should watch not only front but also left and right side. There are some solutions of this problem. They are using panoramic movable camera, using two or three cameras and using a fish-eye lens or a wide mirror. If it has a panoramic movable camera, it should keep consistency of its body direction and the camera direction. It will increase the computational cost in the action planning. If it has more than one camera system, image capturing and processing cost are in proportion to the number of cameras. If it has a fish-eye lens or a wide mirror, viewable seen area is wider, but valuable image area in the image plain is reduced. We considered these merit and demerit. We choose the goal keeper have an omunidirectional vision system, because increasing the computational processing cost is more serious than reducing the image resolution. Our omunidirectional vision system uses a video camera and a half sphere mirror.

3.1 Motor controller interface

This radio control tank kit has independent left and right motors and gearboxes. The digital control unit has two control mode. Mode 1 is simulating steering wheel system. Mode 2 is supporting direct control each motors separately. Our robot uses the mode 2, then each motor is controlled separately. This controller is connected in parallel port of the computer through a CPLD(Complicate Programmable Logic Device) interface unit.

The digital speed controller received digital pulse signals. The duty ratio is variable from 6.8% to 11.4%. If the duty ratio is higher than 9.1%, the motor rotates forward, else if the ratio is lower, it rotates back. The interface logic is written in Verilog-HDL and down load to Xilinx CPLDX95108 board. The interface board has a parallel port adapter and DMD controller connecters and a serial port for rewrite the internal logic.

Fig. 2. Main components

Main components	Product name
Chassis	Tamiya M4 Sherman
Note PC	IBM ThinkPad-235
Video camera	Sony CCD-MC100
Capture card	IBM Smart Capture Card
Control interface	CPLD board
Wireless LAN	NCR WaveLAN/IEEE
Omni vision	half sphere mirror

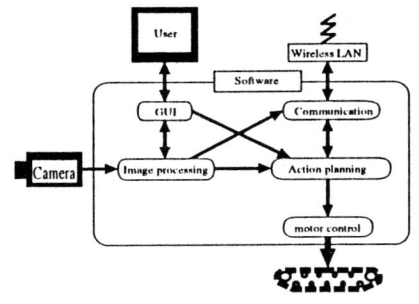

Fig. 3. Software modules

4 Software configuration

Operating system of the robots are FreeBSD 3.2-RELEASE with the PAO3 mobile package. An application program is written in C++ language. Base robot class has fundamental common API for primitive behavior and communication. Each robot program is written as a derived class. It is easy to prepare many kind of player, for example, the variation is depending on its positions and policy.

Software modules diagram is shown in Fig.3. Most modules are written as POSIX threads and the motor control module is written as another process, because the motor controller use timer alarm to archive a fail safe.

Image processing: The image processing module captures an image from a video camera and extract a ball, goals, walls and robots by their template colors. First, the captured image is covert from RGB color space to YUV color space. Next, each pixel is calculated the distance from the target objects' color. If the distance is lower than a threshold value, the pixel is labeled as the target object. Last, the labeled image is scanned same labeled area, if the area size is larger than a threshold value then the target is found out. Same processes are applied for the other targets.

Motor control: FreeBSD 3.2-RELEASE has a special device, for example /dev/ppi0, which is used for direct access from user land process to a parallel port. The robot has ten primitive actions, slow forward, fast forward, forward left, forward right, pivots left, pivot right, back left, back right and back. A primitive action selected by the action planning module is sent to the motor control module with a valid interval. The motor control module sets an alarm for valid interval of the action. The alarm is override for each control message. If the alarm interrupt is invoked, the motor is stopped. This mechanism continues the same or similar actions without any stop, but it avoids going out of control when new actions are lost or late.

Communications: The communication module is consisted by two threads, one is to send the processed result as a broad cast packet to other robots over wireless LAN, another is to receive any data from other robots or a start/stop controller. Communication protocol uses UDP packet and does not wait ACK, because sometime during the competition the wireless LAN would down, it avoids the dead lock state. This protocol is similar to human players shouting protocol.

Behavior planning: Each robot selects same primitive actions, but it should choice a proper action that depends on the situation.

5 Future works

From the result of our experience to attend RoboCup-99, fundamental components, which are recognize the environment from the image processing, agent communication over wireless LAN and prior knowledge based action planning, worked well. However, the following problems were realized.

1. Under the high lighting field, YUV color space has larger white and black area. It is hard to separate the color objects. It causes the radius of the sensable area is reduced.
2. Our robots do not have a kicking device. It makes ambiguity of dribble and shoots, it is difficult learn by reinforcement learning.

To solve the first problem, we will use TSL color space[1] which is more robust than YUV color space. In the future, RoboCup will remove the walls. We consider making a line detection module and use them for the action planning.

We want applies the reinforcement leaning method to the real and complicated tasks. Our current research interest is how too approximate to the actual action value function using continual base functions[2] and layered reinforcement learning[3]. In addition, we will study about multiagent cooperative work. We want applies these approaches to real robot competitions.

Acknowledgment

Special thanks to Mr. Kouichiro Tanaka who is at Center for Microelectronic Systems, Kyushu Institute of Technology. This project was also supported by many laboratory members.

References

1. Jean-Christophe Terrillon and Shigeru Akamatsu: Comparative Performance of Different Chrominance Spaces for Color Segmentation and Detection of Human Faces in Complex Scene Images Vision Interface '99, Trois-Rivieres, Canada, pp.18-21, 1999.
2. Shuichi Enokida, Takeshi Ohashi, Takaichi Yoshida, Toshiaki Ejima: Stochastic Field Model for autonomous robot learning, 1999 IEEE International Conference on Systems, Man and Cybernetics, 1999.
3. Masato Fukuda, Shuich Enokida, Takeshi Ohashi, Takaichi Yoshida and Toshiaki Ejima Behavior Acquisition of Soccer Robot Based on Vision Sensor Information, Technical report of IEICE, PRMU99-82, pp. 31–38, (in Japanese), 1999-09.

The Concept of Matto

Kosei Demura, Kenji Miwa, Hiroki Igarashi, and Daitoshi Ishihara

Matto Laboratories for Human Information Systems
Kanazawa Institute of Technology
3-1 Yatsukaho Matto Ishikawa 924-0838, JAPAN
demura@his.kanazawa-it.ac.jp

Abstract. This paper describes our research interests and technical information of our team for RoboCup-99. Our robots have been developed to have advantages for playing soccer. That is, the capability of kicking a ball and high mobility. We developed pneumatic kickers and omnidirectional bases.

1. Concept

We study and analyze robots of the RoboCup-97, 98 to design our robots, and decide the design concept as follows
- Kicking device
- High mobility

Considering the kicking device, only a few teams such as the CS Freiburg team [1] and the UTTORI United team [2] equipped it. Kicking device will change the tactics of the RoboCup dramatically in the middle size league. It will be a pass-based tactics like the modern soccer.

To accomplish the tactics, there are a lot of hard problems to be solved. Mobility is a key point. The matches at RoboCup-97 and 98 show that most robots were like tortoises. Thus, we developed robots that have a fast and an omnidirectional mobile capability.

We learned from RoboCup Japan Open 99 that the complicated system is no use for the real robot soccer games. To make the robot system simple and reliable, standard PC/AT notebook computers were adopted as the processing system of our robots. Because the notebooks have been designed to deal with the nocks and shocks, moreover have excellent batteries such as the lithium ion batteries.

Fig.1 shows one of our robots. It measures 390mm × 360mm × 330mm in length, width, and height. The weight is about 8kg with a laptop PC and two 12VDC, 2.2Ah sealed lead-acid batteries.

The vision system is the most important of all sensorial systems. Commercial video capture PCMCIA cards (IBM Smart Capture Card and Ratoc System REX-9590) are used for the vision system. These capture cards can capture 320x240 images at a frame-rate of 30 per second and have device drivers for Linux. Capturing performance is based on CPU power. Therefore we used powerful CPUs such as Mobile Cerelons 300MHz and Mobile Pentium IIs 333MHz.

Fig. 1. One of our robot: HIKARU

2. Architecture

We developed the actuator system and the interface system for RoboCup-99.

2.1 Actuator System

Kicking Device: The kicking device is composed of an air tank, an electric valve and an air cylinder as shown in Fig.2 (a) and installation as shown in Fig.2 (b).

Omnidirectional Mobile System: The omnidirectional mobile system has been developed for two of our robots (the rest of robots are conventional mobile system). This type of system has also adopted by several teams, e.g. RMIT[3], Uttori United [2]. Those teams have developed a new system. Our system is conventional, however the reliability is very high and the max speed 2.0m/s is expected. There are 4 pairs of omniwheels as shown in Fig.3 (b) and 4 DC gearmotors as shown in Fig.3 (a). Each pair of omniwheels is simply driven by the DC motor.

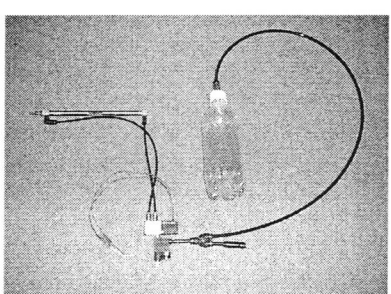
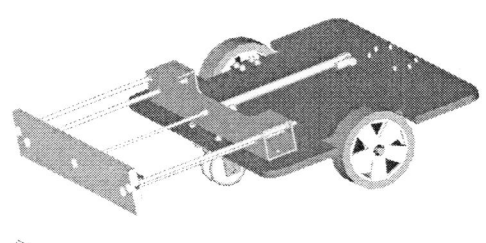

(a) Kicking Device   (b) Kicking Device and Base

Fig. 2. Kicking device

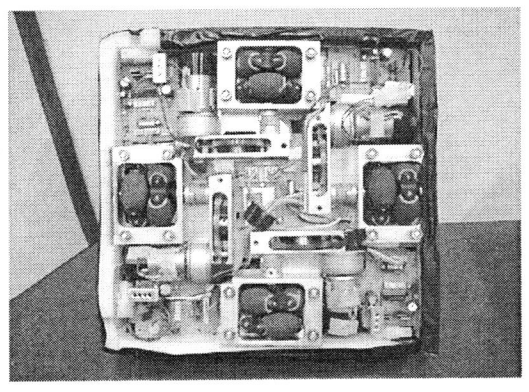
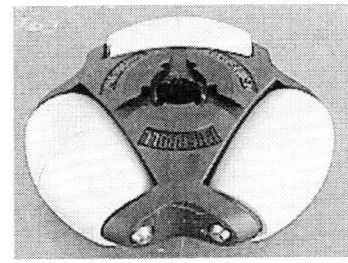

(a) Omnidirectional Base (b) Omniwheel

Fig. 3. Omniwheel

2. 2 Interface System

Interface System: A small notebook computer is suitable for a processing system of the mobile robot. Because it is reliable and self-contained.

However, the interface system between the PC and other devices (sensors, motors) is few and expensive on the market. Therefore we developed the interface system composed of a hub and motor drivers.

Fig.4 shows the hardware architecture of our interface system. The interface system can be two-way communication. The character of the interface system is that each device is connected by a serial bus with the SPI (Serial Peripheral Interface) protocol [4]. It provides support for a high bandwidth (1 Mbps) network connection amongst CPUs and other devices.

Hub: The hub unit transforms a signal between PC and the device (motor drivers, sensors). Hub unit works as the FIFO buffer, too. Power of the devices is supplied from the hub unit. Therefore it is need only one cable, when we increase a new device. The communication speed between the hub unit and the device is about 10 Kbps.

Motor Driver: It is the H-bridge PWM motor driver. The control unit of the motor driver is composed of the PIC micro controller (PIC16F84)[5]. Therefore, it gives intelligent PWM control for a DC motor. The frequency and the resolution of PWM are about 1 kHz and 5Bit. The frequency of PWM is low, but it is sufficient for the middle size league and easy to clock up by replacing the micro controller. The cost of the motor driver is less than $50.

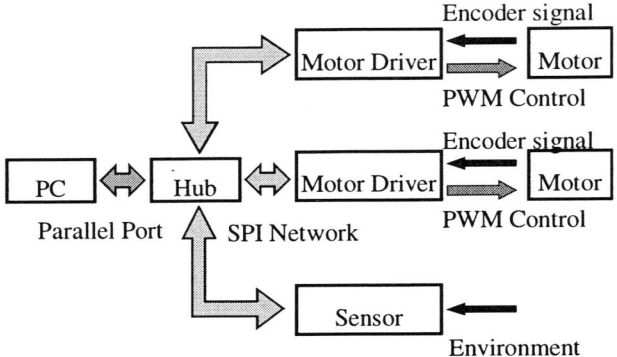

Fig. 4. Interface System

3. Conclusions

This article presents the details of our team. RoboCup-99 is our first challenge. We have been spent a lot of time to build our robots. Developing a reliable and suited robot platform for the soccer game is very important not only to win the competition, but also to make the study.

Therefore, we have developed the interface system, the kicking device and the omnidirectional base. They are indispensable to accomplish the pass-based tactics.

Acknowledgements

We are grateful for a Let's Note CF-S51VXJ8 notebook computer provided by Panasonic Computer for a processing system of our robot.

References

[1] J. Gutmann, W. Hatzack, I. Herrmann, B. Nebel, F. Rittinger, A. Topper, T. Weigel, and B. Welsch: Proceedings of the 2nd RoboCup Workshop, pp.451-457, 1998.
[2] K. Yokota, K. Ozaki, A. Matsumoto, K. Kawabata, H. Kaetsu and H. Asama: Omnidirectional Autonomous Robots Cooperating for Team Play, pp. 333-347, RoboCup-97: Robot Soccer World Cup I, Lecture Notes in Artificial Intelligence, Springer, 1997.
[3] A. Price, A. Jennings and J. Kneen: RoboCup97: An Omnidirectional Perspective, pp. 320-332, RoboCup-97: Robot Soccer World Cup I, Lecture Notes in Artificial Intelligence, Springer, 1997.
[4] The serial peripheral interface (SPI): http://mot-sps.com/
[5] PIC micro controller: http://www.microchip.com/

The RoboCup-NAIST

T. Nakamura K. Terada H. Takeda
A. Ebina H. Fujiwara

Graduate School of Information Science, Nara Institute of Science and Technology
8916-5, Takayama-cho, Ikoma, Nara 630-0101, Japan

1 Introduction

Through robotic soccer issue, we focus on **"perception"** and **"situation and behavior"** problem among RoboCup physical agent challenges [1]. So far, we have implemented some behaviors for playing soccer by combining four primitve processes (motor control, camera control, vision, and behavior generation processes)[2]. Such behaviors were not sophisticated very much because they were fully implemented by the human programmer. In order to improve the performance of such behaviors, a kind of learning algorithm would be useful during off/on-line skill development phase.

To acquire purposive behavior for a goalie, we have developed a robot learning method based on system identification approach. We also have developed the vision system with on-line visual learning function [3]. This vision system can adapt to the change of lighting condition in realtime. This year, we refined some behaviors using such learning algorithms. Furthermore, we constructed a new robot equipped with an omnidirectional camera in addition to an active vision system so as to enlarge view of our soccer robot. Using this omnidirectional camera, such robot can recognize its location in the soccer field.

In the RoboCup99 competition, our team had 9 games including roundrobin, wild card match, and finals. We won 5 games and lost 4 games. Finally, our team reached the quaterfinals. But, our team lost the game against the CS Freiburg.

2 Team Development

Team Leader: Takayuki Nakamura
Team Members:

- Hideaki Takeda, Aossicate Professor, who attended the RoboCup since 1998.
- Kazunori Terada, Ph.D candidate student, who attended it since 1998.
- Akihiro Ebina, Master course student, who attended it for the first time.
- Hiromitsu Fujiwara, Master course student, who attended it for the first time.

Web page http://robotics.aist-nara.ac.jp

3 Robots

We have developed a compact multi-sensor based mobile robot for robotic soccer as shown in **Fig.**1. As a controller of the robot, we have chosen to use a Libretto 100 (Toshiba) which is small and light-weight PC. We utilize a wireless LAN PC card (WaveLAN(AT&T)) for communication on our soccer robot.

Motor control system

A motor control system is used for driving two DC motors and is actually an interface board between a portable PC and motors on the chassis of our soccer robot. This control board is plugged into a parallel port on the portable PC. The motor speed is controlled by PWM. To generate PWM pulses, we use a PIC16C87 microcontroller. The motor control command is actually 8 bits binary commands for one motor. This board can receive control commands from the portable PC and generate PWM signals to right and left motors.

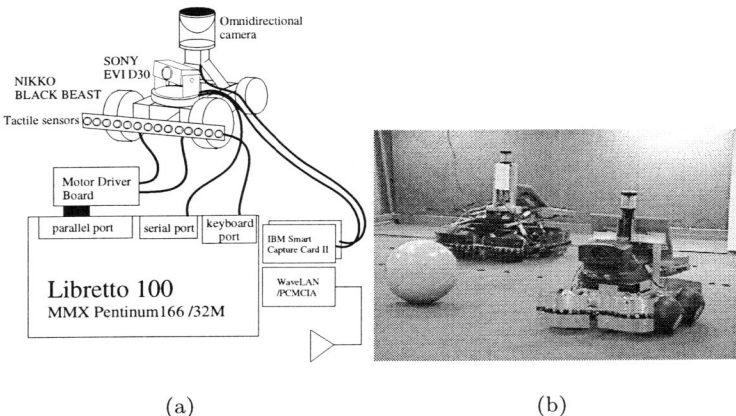

(a)   (b)

Fig. 1. Our soccer robot.

4 Perception

Tactile sensing system

We constructed a cheap tactile sensing system [2] by remodeling a keyboard which is usually used as an input device for PC. A keyboard consists of a set of tactile sensors each of which is a ON/OFF switch called a key. If a key is pressed, the switch is ON. If not, the switch is OFF. By using these sensors which are set around the body of soccer robot., a tactile sensing system can detect contact with the other objects such as a ball, teammates, opponents and a wall.

Visual sensing system

We use an active vision system and an omnidirectional camera system. As an active vision system, we have chosen a color CCD camera (SONY EVI D30, hereafter EVI-D30) which has a motorized pan-tilt unit. An omnidirectional camera system consists of a hyperbolic mirror and a color CCD camera of which optical axis is aligned with the vertical axis of the mirror. In order to capture two images from both vision systems, we use two video capture PCMCIA cards (IBM Smart Capture Card II, hereafter SCCII) which can be easily plugged into a portable PC.

Vision Module

The vision module provides some information about the ball, goal and teammates in the image. The vision module provides the area of the targets(ball, goal and a colored marker), the coordinates of their center and the both maximum and minimum horizontal coordinates of the goal and so on.

We actually implemented a color image segmentation and object tracking processing in the vision module. Even if surroundings such as lighting condition changes, our vision module can adapt to the change since such vision module with on-line visual learning capability based on fuzzy ART model [3].

5 World Model

In order to control our hardware systems, we use a shared memory and 5 software components which are the motor controller, camera controller, tactile sensor module, vision module and behavior generator. All software components read and write the same shared memory in order to acquire and give the states of our hardware systems. Using this shared memory, they can communicates each other asynchronously. For example, the behavior generator takes the state of camera, vision, tactile and motor in the shared memory as input vectors. Then, it combines these information with programmer's knowledge and decides the robot's action at next time step. Finally, it writes the motor command for the motor controller on the shared memory.

6 Communication

There is no communication between our robots in the competition.

7 Strategy

The behavior generator decides the robot's behavior such as avoiding a wall (called avoiding behavior), shooting a ball into a goal (called shooting behavior) and defending own goal (called goalie behavior). We make a simple strategy for shooting the ball into the goal. To shoot the ball to the goal, it is important that the robot can see both ball and goal. Therefore, the robot must round the ball until the robot can see both ball and goal with the camera toward the ball. Finally, the robot kicks the ball strongly. **Fig. 2 (a)** shows the shooting behavior. Avoiding behavior is implemented in a reflex way based on the tactile information.

8 Special Team Features

A word is enough to the wise. We are pursuing research issues focused on the realization of such learning capability. We're going to develop a robot learning method based on system identification approach. Our method utilizes GMDH algorithm[4] which is a kind of system identification method and expands it so that multiinput-multioutput type system can be applied to. Suppose that a set of visual information is input data and a set of motor commands to the robot is output data, identifying a mapping function between input and output data is equivalent to resolving skill acquisition problem.

Now, our robots succeeds in acquiring a simple strategy for preventing a ball from entering a goal. When a ball is approaching to our goal, our robot can move left/right with the center of robot body toward a ball. The home position of the goalie is the center of a line close to our goal. The goalie only moves along that line. **Fig. 2 (b)** shows the goalie behavior.

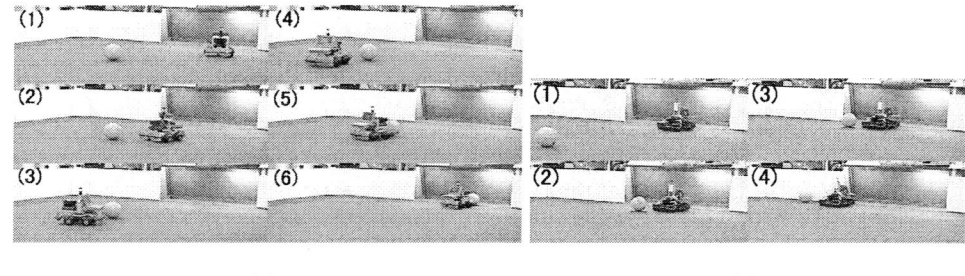

(a) (b)
Fig. 2. Sequences of shooting and goalie behaviors

9 Conclusion

An accurate localization method is a key technology for successful accomplishment of tasks in cooperative way. Next year, we will develop a method for estimating position and orientation of multiple robots using multiple omnidirectional images based on geometrical constraints.

References

1. M. Asada, Y. Kuniyoshi, A. Drogoul, and et.al. "The RoboCup Physical Agent Challenge:Phase I(Draft)". In *Proc. of The First International Workshop on RoboCup*, pages 51–56, 1997.
2. T. Nakamura, K. Terada, and et al. "Development of a Cheap On-board Vision Mobile Robot for Robotic Soccer Research". In *Proc. of IEEE/RSJ International Conference on Intelligent Robots and Systems 1998 (IROS '98)*, pages 431–436, 1998.
3. T. Nakamura and T. Ogasawara. "On-Line Visual Learning Method for Color Image Segmentation and Object Tracking". In *Proc. of IEEE/RSJ International Conference on Intelligent Robots and Systems 1999 (to appear)*, 1999.
4. Hema R. Madala and Alexy G. Ivakhnenko. *Inductive Learning Algorithm for Complex System Modeling*. CRC Press, 1993.

Robot Football Team from Minho University

Carlos Machado, Ilídio Costa, Sérgio Sampaio, Fernando Ribeiro

Grupo de Automação e Robótica, Industrial Electronics Department, University of Minho
Campus de Azurém, 4000 Guimarães, PORTUGAL
fernando@dei.uminho.pt
http://www.robotica.dei.uminho.pt/

Abstract. This paper describes an Autonomous Mobile Robot team which plays football, developed by the Group of Automation and Robotics at the Industrial Electronics department of the University of Minho, in Guimarães (Portugal). In this competition each team is free to use and/or build all the different electronics, sensory systems, playing algorithms, etc. as far as they cope with the rules imposed by the organisation. Instead of using several different sensors increasing electronics complexity, this team decided to use only one major sensor: a vision system with a small colour camera. All the image processing algorithms were developed from scratch and they consist on the heart of the whole project. This vision system uses an innovative approach: in order to see the whole field, a convex mirror was placed at the top of the robot looking downwards with the video camera looking upwards towards the mirror. This way, the robot can see all around itself with a top view, which means continuous vision of the ball, goals and other robots.

1 Introduction

Autonomous mobile robots are ever increasing their number of different applications, even in ludic applications or in sports. In the last few years, several robotic football competitions have been organised with participating teams from all over the world. The University of Minho (Portugal) decided to accept the challenge of participating in this competition and found it to be a quite interesting experience and a lot was learned. The final classification of this team at RoboCup'99 was not as good as expected mainly due to problems that have nothing to do with the RoboCup'99 competition. The robots arrived at Stockholm only after the competition had started; equipment had been damaged in the transport; the postponing of some games forced this team to play 4 games in one day with the robots not completely assembled. Besides all these problems, some other teams did not take into consideration the collision avoidance rule, crashing and destroying some of our robots. Our robots were properly programmed to avoid collisions but not to run away from other dangerous robots. However, participating was an extremely good experience, and this team expects to participate again next year hopefully without all these problems.

2 Robots

The robot base used was one from another competition and is made up of very light wood. It consists of a two levels platform with the two wheel voids. On the bottom level was placed the DC/DC converter (between the two wheels) and the two 12V 7Ah batteries (one at the front side and the other at the rear side). At the top level, it was placed the computer mother board and respective boards (video and graphics boards). The hardware consisted on a personal computer mother board with a Pentium processor running at 200 MHz (MMX), with 32 Mbytes of memory (although the DOS operating system was used and therefore only 1Mbyte was used). The hard disk had 2 Gbytes. A colour video camera is used with a frame grabber type Bt848. The communication hardware and software was not ready in time for the RoboCup99 competition and therefore each robot played on his own.

Fig. 1. Photograph of a team's Robot

3 World Model

At any given time, these robots are aware of the ball (knowing its direction), both goals direction and their own approximate position on the field. Robots on the field are also seen but not distinguished whether their belong to own or opponent team. This information is kept in the form of a direction variable, and when required they move towards it, updating that direction variable at every frame captured. These robots do not memorise anything else.

4 Perception

The only sensor to perceive all the items needed to play a game, is a simple colour camera with its frame grabber plugged on a computer slot. To perceive all those items, these robots grab one image every 20 ms, and the software finds the peak of a certain colour (after removing noise). For example, to track the ball, the software searches a peak of red. Since colours depend very much on the light conditions, a calibration is made prior to a game, in order to inform the software what is the minimum value for a red to be a ball. The software tracks down the following items: the ball (by its red colour); the two goals (also by their yellow and blue colours); all other robots (mainly black coloured); the field surrounding walls (mainly white coloured).

These robots avoid collisions, by perceiving as uncollidable items all the black and/or white items. This is the way they avoid walls (mainly white) and other robots (mainly black). The white lines on the green field are ignored because what the robot sees is not "mainly white" due to the slim thickness of the lines.

5 Communication

It was this team intention to implement communication, but due to lack of time this was not ready in time for the competition. However, the system is hereby described. It consists of a radio frequency module containing an emitter and a receiver, plugged in each robot, plus one more module in a remote computer allowing to "see" what is going on, on the field, through a graphical animation.

Each of these modules are able to send a message to a certain robot of to all robots. The message is not longer than 255 bytes and contains instructions about actual information (a position, or a decision) or what to do next, and these messages have different levels of importance. Each robot can communicate with one other particular robot or with all of them at the same time. Being so, a complete confusion could be generated and therefore different levels of importance are used. This level depends on the owner position, or distance to the ball, goals, opponent robots, etc. The robots communicate only when it is needed (not all the time) in order to keep the radio environment free for urgent messages to pass through.

6 Strategy

In order to drive the ball, these robots use an arch with a re-entrance of 7cm (allowed by the rules). This way, ball control is achieved just by pushing it, although a sudden change of direction might mean loosing the ball. These sudden changes of direction are avoided by the robot's software by following longer and wider trajectories.

These robots intercept the ball very easy. When they see the ball, they just go towards it, avoiding collisions with the opponent robot, but insisting and never giving up, until the opponent robot looses the ball. Once they have the ball, they move towards the opponent goal dribbling the opponents (and avoiding collisions). In case they loose the ball, instantaneously start the procedure "following ball" again.

When owning the ball near the opponent goal, these robots do not kick the ball. They run into the goal pushing the ball with their body. This means a disadvantage since most goalies are very good and attack the ball sufficiently fast to avoid scoring.

This team's goalie is different from the other players only what concerns the direction of the wheels. These are rotated 90 degrees in order to be fast defending the goal rather than moving towards the front. The goalie software is very simple and consists of looking and observing the ball all the time. That is possible with that convex mirror. It then moves sideways in order to keep its body always in the ball's direction no matter how distant this is. When the ball approaches, the goalie kicks the ball with its arch rotating its body, doing a movement like a tennis player with its racket. This movement is very beautiful and improves the quality of the game. This technique not only avoids a goal but also kicks the ball far away from its goal.

Since these robots always have an eye on the ball, their reaction is very simple and efficient. When they don't have the ball, they go towards it and don't give up until they get the ball. Once they have the ball they go towards the opponent goal in order to score, and avoiding obstacles. If, for some rare reason they don't see the ball, they start moving in a spiral until they see the ball (avoiding the walls, of course).

7 Special Team Features

The image processing was the most important aspect of this team. It proved to be very consistent, fast and original. All the video drivers were re-written in assembly language in order to take the most out of the video board. This way, 50 frames per second were achieved making the rest of the control program an easy task. All the image processing routines were written in assembly language in order to increase speed and it proved to be very necessary. The general control program and strategy were written in C language since it did not need to be extremely fast. Sometimes, in two consecutive cycles the same image was analysed (giving the same result as the previous cycle), proving that it was not necessary to have a so fast processor. The mirror technique also proved to be very efficient since everything can be seen at all time. With everything on sight, it is much easier to make the flux control program.

8 Conclusions

As main conclusions it can be said that the image processing developed and used by this team is the most important characteristic. It is very reliable, consistent, fast.

The robot movements are very smooth and acceptably unpredictable unlike a typical algorithm with known steps used by many teams.

All the hardware and software of the robots was designed, developed and built by graduate students at the University of Minho. Only four robots were built due to lack of budget. These 4 robots were designed, built, programmed and tested by three industrial electronics students only, plus the team leader making a team of 4.

Even though this team had had many problems during its participation on RoboCup, the team learned a lot and gained experience. The rules are now clear, new ideas came up by looking at other teams playing, and new improvements will be implemented. This team intends to participate next year, after improving the robots.

Real MagiCol 99: Team Description

C. Moreno[1], A. Suárez[1], Y. Amirat[2], E. González[1], H. Loaiza[1,3]

Université d'Evry Val d'Essonne, LAMI[1] – CEMIF[3],
Cours Monseigneur Roméro, 91025 Evry Courcouronnes, France.
{ moreno, suarez, gonzalez }@lami.univ-evry.fr, amirat@univ-paris12.fr,
hloaiza@eeie.univalle.edu.co,
Université Paris XII – LIIA
122, Rue Paul Armangot 94 Vitry S/Seine
Universidad del Valle - PAyRA. Escuela de Ingeniería Eléctrica y Electrónica,
Ciudad Universitaria, Cali, Colombia

Abstract. The hardware and software architectures of the Real MagiCol robots are presented. The hardware remains the same as the 1998 Robot Cup one, our research bearing on the cooperative architecture based on agent concept. The Real Magicol soccer players team for Robot Cup 99 is based on a *Behavior Oriented Commands* (BOC) architecture extension which combines reactive and deliberative reasoning by the distribution of the knowledge system into modules called behaviors.

1. Introduction

The middle size team "Real MagiCol" (Realismo Mágico [1] Colombiano) is a joint effort of institutions in France and Colombia. In addition to participating in RoboCup99, the robots will be used in the future for research and educational activities in Colombian universities and will be employed in a permanent exhibit of the interactive science museum "Maloka". In fact, we are the first Latin-American middle size team in RoboCup.

We decided to build our own robots for ROBOCUP 98. This allows a greater insight and a complete mastering of the robot's technology. We designed an open, easily reconfigurable PC based architecture in order to allow for future evolutions.

The Real MagiCol team features our hybrid software architecture called Behavior Oriented Commands (BOC) [2] [3]. It allows a soccer robot to plan complex deliberative actions while offering good reactivity in a very dynamical environment. BOCs provide a high level distributed intelligence model which is directly translated into a real time application.

This article presents the main aspects of the robots hardware and vision system, as well as their control architecture and the team strategy.

2. Robot Description

The Real MagiCol team consists of five robots sharing the same hardware design. Each robot has an external diameter of 44 cm and a height of 18 cm (37 with the optional turret and vision system). Lineal speeds of almost 2m/s with accelerations of 1 m/s^2 are possible. The hardware architecture of the robots is shown in Figure 1. Figure 2 shows our goalkeeper in action.

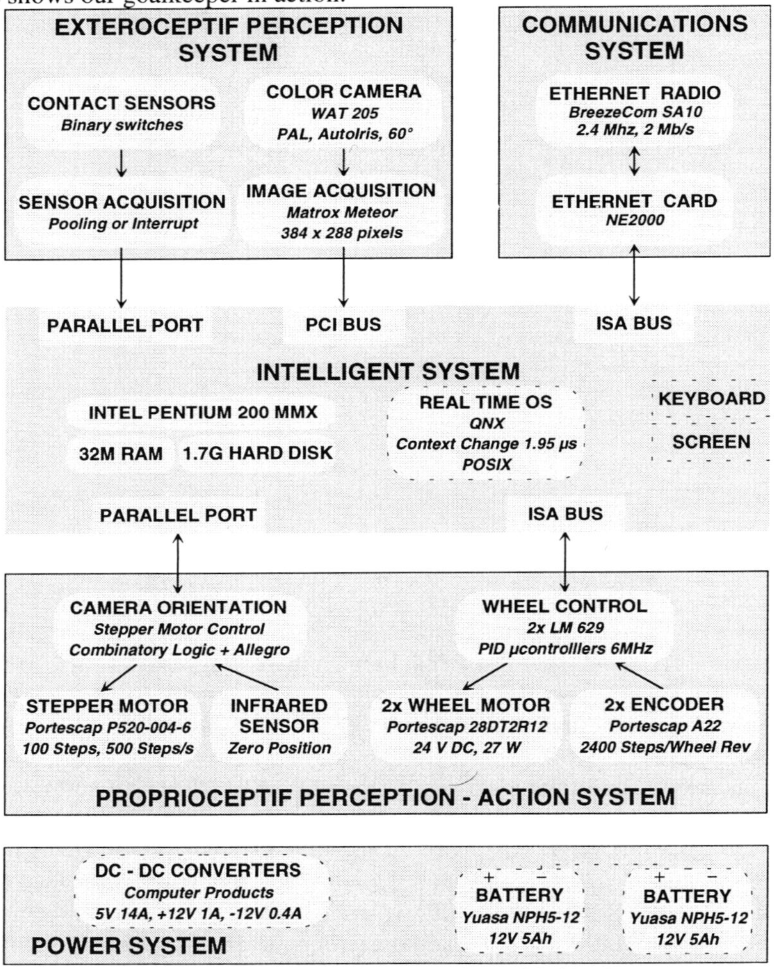

Fig. 1. Hardware Architecture of MagiCol Robots.

Fig. 2. General View of Robot and Ball

3. Local Vision System

Each robot possesses a color mini-camera with a 3.8 mm focal distance and a 51°(h) x 40°(v) vision field. The images are acquired in the RGB color model [4]. The detection algorithm of the elements (color-objects) of the terrain (ball, robot markers, goals, etc.) uses thresholded LUTs applied to the three color fields. A logical function between these images results in eight binary images, one for each color-object. A composed 8-bit image is obtained after bit-shifts and combination of the binary images. While performing the last step a LUT is applied to discriminate and label ambiguous pixels. A 9-level image is finally constructed.

The attributes of each color-object are obtained after segmentation of the 9-level image. For each detected color-object, the center of gravity, the surface and the enclosing window are calculated in image coordinates. Color-objects presenting a small size are rejected. The vision system was calibrated to carry out a 3D position reconstitution of the objects, taking advantage of the fact that the height of the ball, goals and robots are known to reduce the unknown variables in the camera model.

The *local map* is the set of objects recently seen by the robot. An object is characterized by its relative position, speed and uncertainty in local coordinates. The color-object information provided by the image treatment module and the odometer allows to estimate the position and speed of objects. Newly detected objects are incorporated into the map with an initial uncertainty value based on their distance and image surface. Objects re-detected in new images are updated. The uncertainty of undetected objects is increased until they reach the forget threshold.

The robot localization is carried out using objets known to be static. We plan to integrate the information from the local modules in the different robots into a *global map*, in order to improve the precision of the position and speed estimates of objects of interest.

We expect to improve the robustness by implementing an HSI (Hue, Saturation, Intensity) model based algorithm for the color-objets detection. These model could allow the color camera calibration under unstable illumination conditions.

4. Roles and Formations

In a soccer game robots need to be organised to play coherently [9]. Our robots incorporate the rules to execute individual actions depending on its role, team formation and game context.

A role assigns responsibilities and actions to a robot. The generic roles of goalie, defender and attacker are defined. The goalie stays parallel and close to the goal, trying always to be directly between its goal and the ball. The attacker moves to the ball trying to kick towards the opponent's goal; when the ball is in the attacking side it attempts to have a good non interfering attacking position. The defender maintains a good defending position between its goal and the ball/opponent, to move near the ball and pass it to the attackers; it also tries to place itself between the ball and its own goal when the ball is far [5][6]. We also define a new role, the *coach*. It performs global localisation, role and formation distribution, supervision activities, and manages external information.

The generic roles are specialised in sub-roles according to the robot playing region in the field and its attitude towards the game. The field is divided into three regions: left, right and central allowing to decline the roles as left-handed, right-handed and central. Robots can also play a role having different attitude towards the game, for example, a defender may be *prudent* (always staying in a defensive position) or *aggressive* (always trying to kick the ball). This specialisation by attitude allows to easily built teams playing different tactics without modifying its formation.

A formation is a team structure that defines a set of roles in a particular game [6]. A formation assigns a specialised role for each robot. The selection of the team formation depends on the game situation, particularly on the score and opponent's strategy. At start information concerning global team formation affects the way individual sub-role rules are interpreted which allows to have collective conscience.

5. BOC Implementation

A real time control architecture should be used to implement a soccer mobile robot that deals with a dynamic environment. Our robot control system is implemented using the hybrid architecture BOC, which combines reactivity and deliberative reasoning by the distribution of the knowledge system into modules called *behaviours*. A *BOC* is a service carried out by a set of cooperative associated behaviours (*ABs*) executing in parallel. The co-ordination of the *AB*s is performed by a control unit (CU) using synchronisation signals.

A general description of MagiCol robots using the *BOC* architecture was presented in our previous paper [3]. Figure 3 presents relevant aspects of the actual BOC model implementation of our goalie; ellipses and rectangles represent *behaviours* and *control units* respectively.

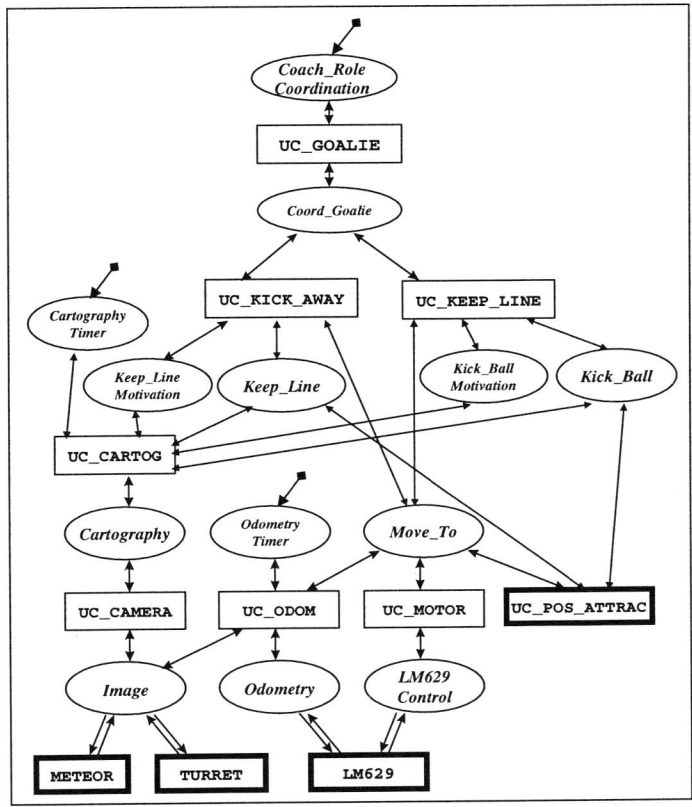

Fig. 3. BOC Model of our Goalie for RoboCup98.

The coach assigns roles by requesting a high level *BOC* to each robot. The command parameters include zone and attitude specialization attributes of the role. In our goalie, UC_GOALIE activates the *behavior Coord_Goalie*, which decides the next action to execute. Two possible high level goalie actions are defined: *Keep_Line* and *Kick_Away*. A request to the concerned *UC* is performed. When the action is finished an acknowledge is received, the state is modified and the next action to perform is selected using the control rules embedded in the *behavior*.

The execution control of the command Keep_Line is performed by UC_KEEP_LINE that activates three *ABs*. The behaviour *Position_Line* modifies the attraction point to make robot move to the best defensive position over its line (parallel to the goal entry). The *behavior Keep_Line_Motivation* monitors if the motivations for doing this action are still valid, when motivation falls under a threshold (specified by *Coord_Goalie*) an end signal is generated thus stopping all *ABs* and finishing the *BOC*. The third *AB Move_To* drive the robot to arrive to the attraction point specified by other cooperative *behaviors*. High level behaviors have access to sensors and actuators through low level commands.

6. Conclusions

The hardware and software architectures of the Real Magicol Robots were presented. These robots were built for the RoboCup competition, but future research in other subjects has also been considered.

The BOC architecture was used with few modifications and proofs to be well adapted to this kind of challenge, allowing a straightforward well structured real-time implementation of the proposed concepts.

In our current implementation, collective behaviours emerge as a result of role attribution and team formation. We plan to extend the task parallelism to the team as a whole by adding explicit communication between the players. The increased information exchange, should also allow our coach to detect specific strategy patterns from the opponent team in order to adapt our own strategy.

Acknowledges. This work was financed by the Fundación de la Universidad del Valle and Colciencias-Colombia. We also wish to thank the SIN.OVIA, Sinfor, and Renault Automation.

7. References

1. Realismo Magico. http://artcon.rutgers.edu/artists/magicrealism/magic.html
2. Cuervo J., González E.,Suárez A., Moreno , «Behavior-Oriented Commands: From Distributed Kwoledge Representation to Real Time Implementation », *Proc.Euromicro Workshop on Real-Time Systems,* Jun. 1996. pp. 151-156
3. Loaiza H., Suarez A., González E., Lelandais S., Moreno C., «Real MagiCol: Complex Behavior through simpler Bahavior Oriented Commands », *Proc.of the second RoboCup Workshop,* July. 1998. pp. 475-482
4. Marszalec E., Pietikäinen M., «Some Aspects of RGB Vision and its Applications in Industry», *Machine Vision for Advanced Production,* Series in Machine Perception and Artificial Intelligence, vol. 22, 1996, pp 55-72.
5. Shen Wei-Min, Adibi Jafar, Adobbati Rogelio, Cho Bonghan « Building Integrated Mobile Robots for Soccer Competition », *http://www.isi.edu*
6. Veloso Manuela, Stone Peter, Han Kwun « The CMUnited-97 Robotic Soccer Team : Perception and Multiagent Control, *http://www.cs.cmu.edu/{~mmv,~pstone,~kwunh}*

RMIT Raiders

James Brusey, Andrew Jennings, Mark Makies, Chris Keen, Anthony Kendall, Lin Padgham*, and Dhirendra Singh

RMIT University, Melbourne, Australia
linpa@cs.rmit.edu.au

1 Introduction

The RMIT Raiders team is composed of three custom-made robots and one Pioneer[1] robot. The most significant feature of the custom platform is a powerful kicking device, which proved itself in the first game by kicking a goal from the centre of the field. Compared with previous competitions, our custom robots were much more reliable and the batteries lasted much longer.

The strategic component of the system is based on a commercial agent development system, JACK[2] [4]. Due to the difficulties of testing with physical robots, we developed a simulator for testing and debugging of plans.

We kept our existing vision mechanism but made some improvements in terms of robustness of recognition under varying lighting and improved distance estimation.

Our competition results were: 1-0, 0-1, 0-3, 0-6, 1-2.

2 Team Development

Team Leader: Lin Padgham, Associate Professor
Team Members:

- Daniel Bradby, Undergraduate student, did attend competition
- James Brusey, Graduate student, did attend competition
- Andrew Jennings, Professor, did attend competition
- Mark Makies, Design and Development Engineer, did attend competition
- Chris Keen, Senior Technical Officer, did not attend competition
- Anthony Kendall, Honours student, did not attend competition
- Dhirendra Singh, Undergraduate student, did attend competition

Web page: http://www.robocup.rmit.edu.au

---

* Lin Padgham is currently on secondment to CSIRO, though still fully involved in the RMIT RoboCup project.
[1] Pioneer robots are produced by ActivMedia Inc., http://www.activrobots.com
[2] JACK is a java based agent programming system developed by Agent Oriented Software, http://www.agent-software.com.au

3 Robots

We use two types of robot bases: a Pioneer Model 1 for the goalie, and a custom-made robot (called a Socbot) for the three other players. All robots use the same vision sensor, a Logitech QuickCam VC camera, and all have wheel encoder based odometry [1, 2]. The Socbots also have an electronic compass. The Socbots have a special kicker described below, while the Pioneer uses a less powerful, solenoid actuated kicker. We use Pentium II laptops for the majority of the computation, including vision processing. Each robot has a Lucent WaveLAN PCMCIA card for communication with other robots. Figure 1 summarizes the system architecture.

The Socbot uses two 6W MAXON motors controlled by a PID (Proportional Integral Derivative) controller. The controller makes use of wheel encoders that give 1800 counts per wheel revolution. It runs on a Motorola M68HC11 microprocessor and makes use of a Xilinx XC3090 Field Programmable Gate Array for some functions.

The Socbot's kicking device is spring loaded (using a bungee cord). It is wound up using a worm drive motor, and then held in a loaded position by a custom-made clutch. The clutch is released by a solenoid. A microswitch is used as a trigger to ensure that the kicker is not fired until the ball is close to the optimum position. The kicker is quite powerful. In testing we found that, at full setting, it could kick the ball over 20 metres, and then take only about 5 seconds before being ready to kick again.

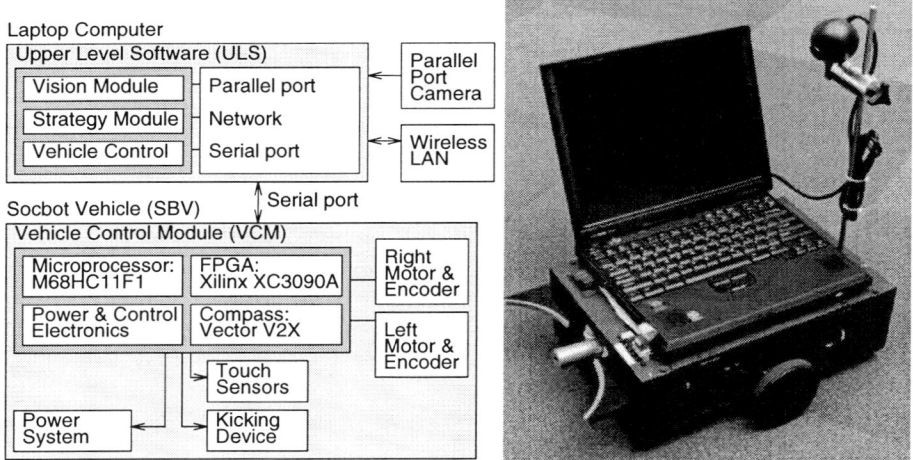

Fig. 1. Architecture of soccer robot and a picture of the Socbot. For the goalie, a Pioneer substitutes for the Socbot but uses the same software

4 Perception

The robots attempt to perceive the ball position, the position of other robots, and the position of the walls, through the vision system. Also, some information is broadcast to other robots to assist them building a world model.

The vision system is based on an inexpensive, parallel port camera: Logitech's QuickCam VC. The vision system outputs distance and angle information for each object that it finds. Although we had used the QuickCam in the previous tournament, we made some significant improvements in colour recognition using techniques described elsewhere [3]. Also, we changed from using the size of the image segment as the determinant of distance and used the vertical angle to the base of the object instead. We still check the size of the image segment, however, to see if it roughly matches the distance obtained from the vertical angle.

We use information about wall segments, together with information from the compass, in order to correct for errors in odometry.

5 World Model

We used an allocentric (i.e. non-egocentric) view of the field, so that objects were represented in terms of field coordinates rather than relative to the robot. The robots keep information about own position and heading, the ball position and velocity, and other robots' position and velocity. Own position and heading is determined through wheel encoders with correction from the compass and wall sightings. The vision gives relative position to other objects and this is translated into an absolute position. The world model is updated from this using a simplified Kalman filtering approach.

In addition, there are interpreted state variables, such as flags for whether the robot is lined up for a shot at goal.

6 Communication

The robots communicate via a radio ethernet LAN using multi-cast packets. All robots transmit where they think they are and where they think the ball is. This information is sent from each robot about ten times per second. We noticed that the communication was specifically helpful to the goalie in tracking the ball when the ball was far away, however we had some problems with wrong information being transmitted back and forth between robots.

7 Skills

To get to the ball, our robots used a series of waypoints to avoid bumping into the ball when trying to get to the other side of it. Due to the explicit modeling of the ball in the world model, they were able to do this without seeing the ball for much of the manœuver. When the ball is close to the front of the robot, the kicking mechanism is "armed", which means that the kicker will fire automatically when a microswitch triggers on the front of the robot. If the kicker doesn't fire for some reason, we will still try to dribble the ball by constantly trying to move towards it, as long as it is still roughly in line with the goal.

The goalie robot has special code for blocking the goal. Since our goalie is quite slow, we use some of the prediction mechanism available from the world model to try to predict the path of the ball.

8 Strategy

The robot's basic strategy is straightforward: Find the ball by spinning; line up the ball and the goal; approach the ball and kick it. While doing all of this, it avoids collision with other robots and uses a modified approach for when the ball is near the wall. Much of the sophistication of the software is in building a model of the field.

9 Conclusions

Although the game scores do not reveal it, we believe that the team has greatly improved over previous years and is beginning to look competitive. Our main failing was to rely heavily on the compass, which behaved unpredictably in the electromagnetically noisy environment. The robots turned out to be slower than expected, largely due to the software limitation of only being able to turn when stopped. The defence was poor mainly because no defensive plans had been written. Our kicker worked excellently however, and in general the systems were robust and reliable.

Next year's competition will be held in Melbourne, Australia and so we want to field a much more competitive team. The main areas that we will attempt to improve are the vision system, where we plan to use a digital signal processor (DSP) to perform much of the computation, and the strategy component, which will be rewritten to make more use of the JACK agent architecture. Part of this rewrite will include improvements in cooperation.

References

1. Ronald Arkin. *Behavior-Based Robotics*. MIT Press, 1998.
2. Johann Borenstein. Where am I? Sensors and methods for autonomous mobile robot localization. Technical Report UM-MEAM 94-21, The University of Michigan, December 1994.
3. James Brusey and Lin Padgham. Techniques for obtaining robust, real-time, colour-based vision for robotics. In Manuela Veloso, editor, *Robocup-99 : Proceedings of the Third RoboCup Workshop*, July 1999.
4. P. Busetta, R. Rönnquist, A. Hodgson, and A. Lucas. JACK intelligent agents – components for intelligent agents in Java. *AgentLink Newsletter*, January 1999.

Design and Construction of a Soccer Player Robot ARVAND

M. Jamzad[1], A. Foroughnassiraei[2], E. Chiniforooshan[1], R. Ghorbani[2],
M. Kazemi[1], H. Chitsaz[1], F. Mobasser[1], and S.B. Sadjad[1]

[1] Computer Engineering Deptartment
Sharif University of Technology
Tehran, Iran.
jamzad@sina.sharif.ac.ir, {chinif, chitsaz,
sajjad}@linux.ce.sharif.ac.ir, {mobasser, kazemi}@ce.sharif.ac.ir
http://www.sharif.ac.ir/~ceinfo

[2] Mechanical Engineering Deptartment
Sharif University of Technology
Tehran, Iran.
{ghorbani, forough}@linux.ce.sharif.ac.ir
http://www.sharif.ac.ir/~mechinfo

Abstract. Arvand is a robot specially designed and constructed for playing soccer according to RoboCup rules and regulations for the medium size robots. This robot consists of three main parts: mechanics (motion mechanism and kicker), hardware (image acquisition, processing unit and control unit) and software (image processing, wireless communication, motion control and decision making). The motion mechanism is based on a drive unit, a steer unit and a castor wheel. We designed a special control board which uses two microcontrollers to carry out the software system decisions and transfers them to the robot mechanics. The software system written in C++ performs real time image processing and object recognition. Playing algorithms are based on deterministic methods. We have constructed 4 such robots and successfully tested them in a soccer field according to RoboCup regulations for middle size robots.

1 Introduction

In order to prepare a suitable ground for research in many aspects involved in Robocup, we designed and constructed all parts of the robots by our group members. These robots have a controllable speed of maximum 0.53 m/sec. In addition to the basic movements of robot, the special design of its mechanics, allows it to rotate around any point in the field. In practice, the distance between ball center and robot geometrical center is calculated and the robot can be commanded to rotate around the ball center until seeing the opponent team goal.

The machine vision system uses a widely available video camera and a frame grabber. Our fast image processing algorithm can process up to 16 frames per second and recognize objects in this speed. Objects are recognized according to their color and size. Communications between robots are done using wireless network under TCP protocols. The non deterministic software algorithms are designed based on object oriented methods and are written in C++ using DJGPP compiler in MS/DOS.

2 Mechanical Architecture

According to the motion complexity of a soccer player robot, proper design of its mechanics can play a unique role in simplifying the playing algorithms. In this regard, a specific mechanism was designed and implemented that together with the sensors and control feedbacks, to a good extent, verified our expectancy.

2.1 Motion Mechanism

Arvand consists of two motion units in front of the robot and one castor wheel in the rear. Each motion unit has a drive unit and a steer unit. The functionality of drive unit is moving the robot and that of steer unit is rotating the drive unit round the vertical axis of its wheel. The drive unit consists of a wheel which is moved by a DC motor and a gearbox of 1:15 ratio [1]. The steer unit uses a DC motor and a gearbox of 1:80 ratio. For controlling the steer unit, the optical encoders are mounted on the respective motor shafts and their resolutions are such that one pulse represents 0.14 degrees of drive unit rotation.

This mechanism has the following capabilities:

1. By rotating the drive unit round its vertical axis the rotation center of the robot changes accordingly and this allows the robot to turn around any point in the plane. This point can be selected inside or outside the robot. It is necessary to adjust the speed of two drive units according to the following formula [2]:

$$v_1.r_2 = v_2.r_1 \tag{1}$$

   where v_1 and v_2 are speeds of the left and right motors respectively, r_1 is the distance of the left drive unit from the rotation center and r_2 is the distance of the right drive unit from the rotation center. Therefore, the robot rotation center will not depend on the robot gravity center and on the position of drive units in the robot.
2. In our software system we can set the drive units to be parallel to each other and also have a specific angle related to robot front. This mechanism is useful for taking out the ball when stuck in a wall corner and also dribbling other robots.
3. The kicker consists of simple crowbar that connects the solenoid to a kicking arm. The kicking power is controlled by duration of 24 DC voltage applied to it.

3 Hardware Architecture

The goal of our hardware architecture is to provide a control unit independent from software system as much as possible and also reduce the robots mechanical errors.

Arvand hardware system consists of three main parts: Image acquisition unit, processing unit and control unit.

For all robots including the goal keeper we used a PixelView CL-GD544XP+ capture card which has an image resolution of 704x510 with the frame rate of 30 frames per second. The image acquisition system of goal keeper consists of a Topica PAL color CCD camera with 4.5 mm lens in front and two digital Connectix Color QuickCam2 for the rear view. For other robots we used a Handycam in front which could record the robot vision too.

The processing unit consists of an Intel Pentium 233 MMX together with a main board and 32MB RAM. Two serial ports onboard are used as communication means with the control unit. A floppy disk drive is installed on the robot from which the system boots and runs the programs.

The control unit senses the robot and informs the processing unit of its status. It also fulfills the processing unit commands. Communication between the control unit and the processing unit is done via two serial ports with RS-232 standard[3]. Two microcontrollers AT8952 and AT8951 [4] are used in control unit. They control the drive units, steer units, kicker and limit switches. Two limit switches are mounted on each steer unit. Microcontroller counts the number of pulses generated by the encoders mounted on a motor shaft to control the drive unit rotation. Each pulse represents 0.14 degrees of the drive unit rotation. The motors speed are controlled by PWM pulse frequency of about 70kHz

4 Software Architecture

Software architecture of **Arvand** consists of four main parts: Real time object recognition, Motion control, Communication and Decision making module. Due to the object oriented design of the software, we have defined 5 classes such as: Camera class (all related functions for working with frame grabber), Image class (machine vision functions), Motion class (motion functions which is the interface between software and hardware), Communication class (all TCP related wireless networking) and Motion class (all robot playing methods and algorithms).

4.1 Real time object recognition

Object recognition is based on detecting its color. We used HSI color model [5]. In this model a color can be detected by determining its domain in HSI space. To find all objects in a scene the image matrix is processed from top to bottom only once. In order to speed up this routine, instead of examining each single pixel in the image matrix, only one point from subwindows of size $m_w \times m_h$ (that can be the size of the smallest object) is tested. If this point has the desired color, then move one pixel upward until hitting a border point.

At this point a clockwise contour tracing algorithm is performed and border points of the object are marked. If the object size is larger than a predefined threshold then it is recognized as an object, otherwise it is marked to be a noise.

To find the next object the search is continued from the start point from which the previous object was found. In our search for the next object the marked points are not checked again. At the end of this step, the objects marked as noise are deleted and for the remaining objects their size, distance from camera and angle are calculated.

4.2 Motion Control

This module is responsible for receiving the motion commands from the "Decision Making Module" and putting the hardware to work. As it is mentioned in the hardware architecture section, the communication between the processing unit and the control unit is via two onboard PC serial ports using RS-232. So, just some basic computations are done in this module and commands are sent via serial ports to the microcontroller where they are executed.

For example, some commands are kick, go(forward), go(backward), rotate(left), rotate(right), rotate_round(left, 10) (this stands for rotation around a point 10 centimeters straight from the robot geometrical center) and etc.

4.3 Communication

Communication between robots is done by wireless LAN under TCP protocol. The main kernel of communication class can be downloaded from [6]. Each robot has a wireless network card, and there is a server machine outside the field which coordinates messages between robots. The server also provides a useful user interface to command robots manually. Server's main responsibility is to receive the robots messages and inform them about each robot status. For example, if one robot knows that another robot is holding the ball it will not go for the ball.

4.4 Decision making

Principally, the Decision making module is referred to that part of **Arvand** software that processes the results of Real time object recognition, decides accordingly and finally commands the Motion control software. We have taken deterministic approach in these routines. This module is a finite state machine (FSM) whose inputs come from changing state are machine vision results, motion control hardware feedbacks and server messages. Each robot playing algorithm kernel is finding the ball, catching it, finding the opponent goal and finally carrying the ball toward the goal and kicking. But there are a large number of parameters that affect this main kernel and cause interruppts in its sequence. For example, the main method for finding the ball is rotating. When our robot is moving inside the field it tries not to collide with other robots. This is done by calculating the distance and angle of other robots and change the speed of its motors such that object collision is avoided.

In addition, robot ability to measure the motors current feed back, allows it to determine the stuck situations and thus making appropriate move to come out of that state.

5 Conclusion

Arvand is the 2nd generation of robots constructed by our team. One advantage of **Arvand** is its mechanics capability to rotate around any point in the plane. This makes it possible for the robot to rotate around ball center while finding the goal position. In practice, this capability enabled us to implement special individual playing techniques in dribbling, coming out when stuck and taking out the ball from a wall corner. Another advantage of our robot is its use of MS/DOS operating system, because it can be executed on a floppy disk which is a cheep and reliable device on mobile robot. Our robots showed a good performance in real games and we are going to improve our software algorithms based on individual techniques and also team play. The wireless LAN system used in our robots enabled the communication between robots which is the key to the success of team play algorithms and also many individual techniques.

References

1. Shigley, J.E., *Mechanical Engineering Design*, McGraw-Hill, 1986.
2. Meriam, J.L., *Dynamics*, John Wiley, 1993.
3. Mazidi, M.A., and Mazidi, J.G., *The 80x86 IBM PC and Compatible Computers, Volume II*, Prentice Hall, 1993.
4. MacKenzie, I.S., *The 8051 Microcontroller*, Prentice Hall, 1995.
5. Gonzalez, R.C., and Woods, R.E., *Digital Image Processing*, Addison-Wesley, 1993.
6. WATTCP http://www.geocities.com/SiliconValley/Vista/6552/l6.html

The Team Description of Osaka University "Trackies-99"

Sho'ji Suzuki[1], Tatsunori Kato[1], Hiroshi Ishizuka[1],
Hiroyoshi Kawanishi[1], Takashi Tamura[1], Masakazu Yanase[1],
Yasutake Takahashi[1], Eiji Uchibe[1], and Minoru Asada[1]

Dept. of Adaptive Machine Systems, Graduate School of Engineering,
Osaka University, Suita, Osaka 565-0871, Japan

Abstract. This is the team description of Osaka University "Trackies" for RoboCup-99. We have worked two issues for our new team. First, we have changed our robot system from a remote controlled vehicle to a self-contained robot. The other, we have proposed a new learning method based on a Q-learning method so that a real robot can aquire a bhevior by reinforcement learning.

1 Introduction

We are interesting in how a robot acquires a behavior in dynamic environments and how robots cooperate without explicit communication. in the context of cooperative distributed vision [1]. We have applied a Q-learning method, one of major method of reinforcement learning, to real robots and tested in RoboCup competitions[2][3].

However, the performance of the behavior is not enough because;

1. the controll of the robot is not reliable.
2. the applied method is not enough to adapt to the real robot.

In RoboCup-99, we will improve these problems by building a new platform and propsing a new learning method. In the rest of this paper, we describe our new robot and propse a new learning method.

2 The Robot of Osaka University "Trackies-99"

In RoboCup-97 and 98, we used a radio controlled model car as a robot body and equiped a CCD camera and a video transmitter on it. The image captured by the camera was transmitted to a host computer and it sent control signals to the robot. We could test various image processor and software development tools, however, the control of the robot was not reliable due to noises on radio links[2][3] and it brought poor performance of the robot's behavior.

To escape from noise problem, we have build a self-contained robot where the host computer is equiped on the robot. Figure 1(a) shows the robot of the team of Osaka University "Trackies-99". Figure 1(b) shows configuration of its controller including following devices:

base of the vehicle is a product of Mechatro Systems. Left and right wheel is driven by a DC motor and other two wheels are caster.

CCD Camera is Sony EVI-D30. Pan and tilt angle of the camera is controllable from CPU via serial communication.

CPU and Operating System Linux is running on Pentium MMX 233MHz.

image processor is Hitachi IP-5005. It is a fast color image processing board where basic operations are installed.

LAN WaveLAN is used for monitoring and debug.

video transmitter is used for monitoring a processed image

The size of the robot is 400[mm] of length, 360[mm] of width, and 450[mm] of height. The weight is 10[Kg].

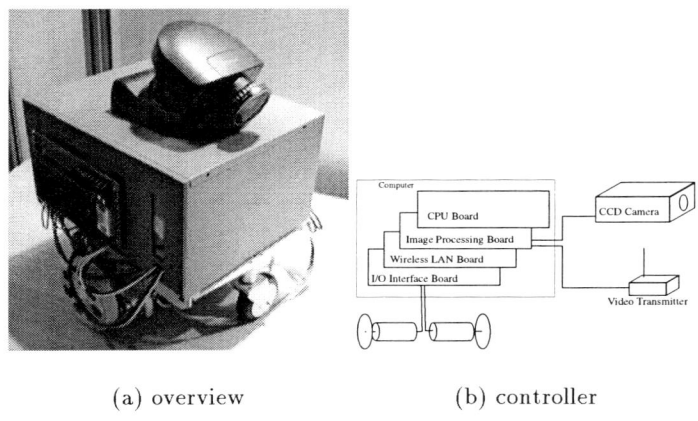

(a) overview (b) controller

Fig. 1. Robots of Osaka University "Trackies-99"

3 A Learning method for a real robot

We propose a continuous valued Q-learning for real robot applications. Unlike the conventional real-valued Q-learning methods, the proposed method dose not need well-defined quantized state and action spaces to converge. The basic idea for continuous value representation of state, action, and reward in $Q-learning$ is to describe them as contribution vectors of representative states, actions, and rewards.

First, we tessellate the state space into n-dimensional hyper cubes[1]. The vertices of all hyper cubes can be the representative state vectors $\bm{x}^i = (x_1^i, x_2^i, \cdots x_n^i)$

---
[1] the unit length is determined by normalizing the length of each axis appropriately

$i = 1, \cdots, N$ (here, N denotes the number of the vertices), and we call each vertex the representative state s_i. The contribution value w_i^x for each representative state s_i when the robot perceives the input $x = (x_1, x_2, \cdots x_n)$ is defined as follows:

1. Specify a hyper cube including the input $x = (x_1, x_2, \cdots x_n)$.
2. Tessellate the cube into 2^n hyper boxes based on the input x (see Figure 2 for the two dimensional case)
3. Calculate the volume of each hyper box.
4. Assign the volume w_i^x of the box diagonal to the state s_i.
5. If the input x is on the surface of the hyper cube, the volume can be reduced to the area or the length.
6. Any other contribution values for the states which do not compose the above cube are all zeros.

Mathematical formulation of the above process is given by

$$w_i^x = \prod_{k=1}^{n} l_i(x_k), \qquad (1)$$

where

$$l_i(x_k) = \begin{cases} 1 - |x_k^i - x_k| & \text{(if } |x_k^i - x_k| \leq 1) \\ 0 & \text{(else)} \end{cases} \qquad (2)$$

Figure 2 shows the case of two-dimensional sensor space. The area w_i^x is assigned as a contribution value for state s_i. The summation of contribution values w_i^x for the input x is one, that is,

$$\sum_{i=1}^{N} w_i^x = 1 \qquad (3)$$

Thus, the state representation corresponding to the input x is given by a state contribution vector $\boldsymbol{w}^x = (w_1^x, \cdots, w_N^x)$. Similarly, the action representation corresponding to the output $\boldsymbol{u}$ is given by an action contribution vector $\boldsymbol{w}^u = (w_1^u, \cdots, w_M^u)$, where M denotes the number of the representative actions a_j in the tessellated action space.

To show the validity of the method, we applied the method to a vision-guided mobile robot of which task is to chase the ball. Figure 3 shows a sequence of the aquired behavior. Although the task was simple, the performance was quite impressive.

4 conclusions

This work was supported by the Cooperative Distributed Vision project in the Research for the Future Program of the Japan Society for the Promotion of Science (JSPS-RFTF96P00501).

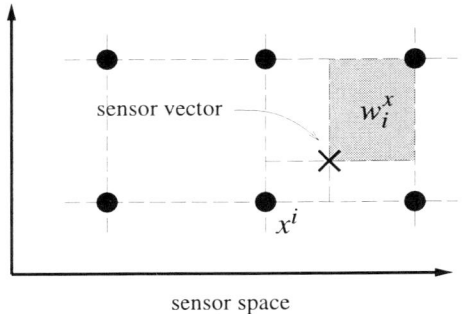

Fig. 2. Calculation of contribution value w_i for the representative state s_i in the case of two-dimensional state space

Fig. 3. A part of video image sequence

References

1. T. Matsuyama. Cooperative Distributed Vision – Dynamic Integration of Visual Perception, Action, and Communication –. In *Proc. of Image Understanding Workshop*, 1998.
2. S. Suzuki, Y. Takahashi, E. Uchibe, M. Nakamura, C. Mishima, and M. Asada. "Vision-Based Learning Towards RoboCup: Osaka University 'Trackies' ". *RoboCup-97: Robot Soccer World Cup I*, Springer, pp.305–319, 1997.
3. S. Suzuki, T. Kato, H. Ishizuka, Y. Takahashi, E. Uchibe, and M. Asada. "An Application of Vision-Based Learning in RoboCup for a Real Robot with an Omnidirectional Vision System and the Team Description of Osaka University "Trackies" ", *RoboCup-98: Robot Soccer World Cup II*, Springer, 1999(to be published).

5dpo-2000 Team Description

Paulo Costa[(*)(1)(2)], António Moreira[(3)], Armando Sousa[(1)(2)],
Paulo Marques[(2)], Pedro Costa[(1)], Aníbal Matos[(1)(2)]

paco@fe.up.pt, amoreira@fe.up.pt, asousa@fe.up.pt,
pamarques@riff.fe.up.pt, pedrogc@fe.up.pt, anibal@fe.up.pt

[(*)]Team Leader
[(1)]Lecturer at the Faculdade de Engenharia da Universidade do Porto (FEUP)
[(2)] PhD. Sutdent at the FEUP, [(3)] Assistant Professor at the FEUP

Abstract - This paper describes the 5dpo-2000 team. The paper will be divided into three main sections, corresponding to three main blocks: the Global Level, the Local Level and the Interface Level. These Levels, their subsystems and some implementation details will be described next.

1   Introduction

This is our first participation in the Robocup Competition in the F-2000 League. We have already played in the F-180 League and we tried to incorporate our experience in the design of this team. We think the first and most important issue to be dealt with is the sensorial problem. Without knowing their position and the ball position, the robots are not able to deploy any kind of coherent action. So we are trying to tackle the problem of extracting this information from multiple cameras and other sensors. All the information is acquired by each robot and so we have distributed system. The data must be fused, taking in account their reliability and the communication influence: bandwidth, delays and possible interference.

Each robot has a video acquisition system and a radio link to communicate with the other robots and with the coach. In addition to the video there are other sensors like, infrared and acoustic range finders, contact sensors, etc.

As the radio link cannot be completely reliable we tried to fit the robots with some autonomy so that they can survive a small starvation of orders from the Coach. That can ease the problem of lost packets over the air.

The whole team can be seen as a system divided in two basic levels.

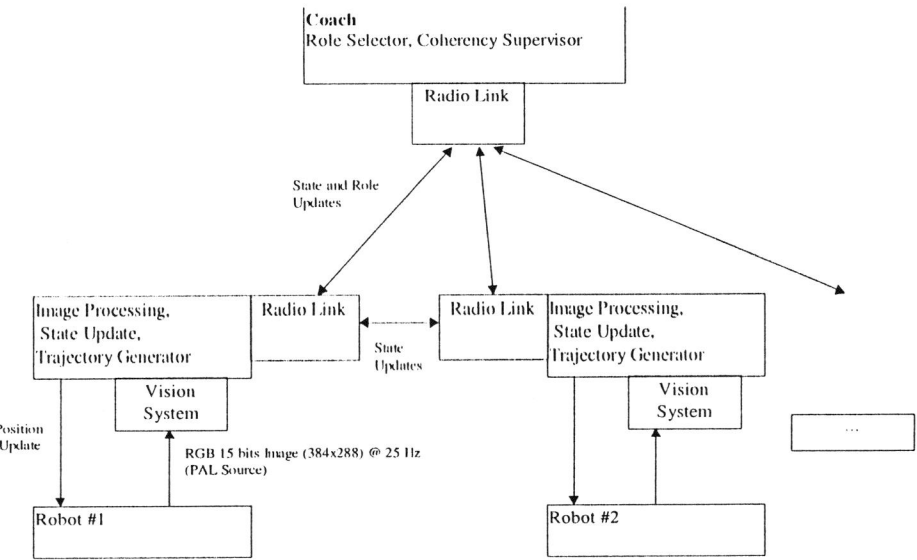

Fig. 1. Robots and Coach and the information flowing between them.

We will now describe the team and its subsystems.

2 The Coach

This is the global Control Level. The global state of the system is updated based on the information flowing from the robots. Data fusion is attempted and adversary robot moves are tracked. A rule based engine is used to classify opponents intents. By observing the present system state as well as a global mid-term strategy, new roles are assigned and sent to the players.

This level closes the global loop but there is some intrinsic lag that degrades its optimal performance. Each robot acquires the image, then some time is lost processing it and more time is lost to transmit the data to the coach. There, the reasoning unit must decide the new course of action and it is necessary to wait for the next time slot to send information to the robots. That is why the local loop, running in each robot, can show a much better performance in some tasks than the globally closed loop.

Other role of the coach is to maintain a "official" global state that can be used to ensure the coherency of the global system state viewed by each robot.

3 The Communications

The information flowing is used by all the robots to update their view of the global state. In that way we can have each robot gaining extra information from the other robots' sensors.

3.1 The Radio Link

The radio link allows sending and receiving short packets that carry messages from the Coach to the robots and between robots. We have the radio channel time slotted synchronously with a global clock. This global clock is kept by each robot using the time of arrival of the radio signals to compensate any possible drift.

3.2 The Alternative Link

The quality of the communication system is paramount to the team performance. So we are considering adding an alternative communication link, probably an infrared link, to improve the reliability of the overall communication system.

4 The Robots

A robot is an autonomous unit considering a short time frame. The robots are capable of generating a queue of tasks to be performed. These tasks may include following a specified trajectory, holding the ball, passing it along to another team member or maybe shooting for goal. The local control system tries to enforce those orders in the predefined sequence.

Some team tactics are maintained at all times like some defense mechanisms that are enforced during the match. In any case, the defense robots stand in alignment in such a way that the robot that holds the ball can't easily shoot for goal. This implies the presence of a path planner present in each robot.

Next we describe the basic mechanical and electrical design of the robots.

The robots are fitted with two differential wheels. The wheels are driven independently by separated stepper motors. Two extra free wheels ensure the static stability.

The robots are presently powered by embedded Lead-Acid batteries. The motors are driven by two H-bridges that are directly powered from the batteries. The on board controller is a 8-bit RISC microcontroller (Atmel AVR90S4414). A PC deals with the higher level functions.

Two small single frequency RF modules (433 MHz and 418MHz) are used to communicate with the coach and other robots. Only one can be used at a time.

The PC code was done in C++ in DOS with a 32 bits extender. That was the only way to ensure the hard real-time nature of the tasks. Other operating systems could not guarantee hard real-time behavior.

Our team uses a local vision system as a primary sensorial source for each member. This is our positioning system for the robots, for the opponent robots and for the ball.

This system consists in one or more color video cameras, placed on the robot. The TV signal from these cameras is feed to a video acquisition board placed in the local PC. This board is capable of placing a digitalization of each image frame in the PC memory without CPU intervention, thus wasting almost no processor time. In the end of this process the board can signal the processor the conclusion of that task.

As we are using PAL cameras the image frequency is 50 Hz with alternating even and odd frames. We are only using even frames therefore we have an image update frequency of 25 Hz. Based on the acquired image we try to identify the ball position and also the robots' position and orientation.

5   Future Work

A disclaimer must be made at this point: between now and the competition there is yet a lot of time to make changes in the described setup. As we test the performance of each subsystem and find better alternatives we will try to implement them. We want the overall system to show a more robust and efficient operation. There is the possibility of adding some kind kicking device to some of our robots.

6   Conclusions

In this paper we described the 5dpo-2000 team and the solutions we found to this problem. Recognizing the overall system state (the ball position and speed, our team's robots' state and the adversarial robots' state) using vision in distributed environment is still a very difficult task. And the quality of the team behavior is very dependent from the accuracy of that system.

The decision of what to do, even with accurate knowledge of the system status, it is a major task on its own. The range of options, some discrete and some continuous has many dimensions and cannot be easily searched. A lot of heuristic rules must be used to trim the possibilities and the best framework to represent and find them is a matter that requires still a lot of research.

References

1. Arthur Gelb, Joseph Kasper Jr.,Raymond Nash Jr., Charles Price, Arthur Sutherland Jr.: Applied Optimal Estimation, The M.I.T. Press, (1989)
2. J. Borenstein, H. R. Everett, L. Feng, S. W. Lee and R. H. Byrne: Where am I? Sensors and Methods for Mobile Robot Positioning, (1996)
3. J. Carvalho: Dynamic Systems and Automatic Control, Prentice-Hall, (1993)
4. Huibert Kwakernaak, Raphael Sivan: Linear Optimal Control Systems, Willey-Interscience, (1972)
5. Jean-Claude Latombe: Robot Motion Planning, Kluwer Academic Publishers, (1991)
6. Lenart Ljung: System Identification: Theory For The User, Prentice-Hall, (1987)

Team ARAIBO

Yuichi KOBAYASHI and Hideo YUASA

The University of Tokyo

1 Introduction

Our team focused on manipulating the ball. We developed the kicking and heading motion. Two kinds of neural networks are utilized in order to recognize the ball and kick it. The recognition and manipulation did not work sufficiently, but intdicates our interest of playing soccer with legged robots. We confronted the team LRP and the team McGill in the round robin. We defeated McGill by PK, and lost the game with LRP 0-2.

2 Team Development

Team Leader: Tamio ARAI
Team Members: Tamio ARAI
  − Dept. of Precision Machinery Engineering, The University of Tokyo
  − Japan
  − Professor
  − did not attend the competition
Hideo YUASA
  − Dept. of Precision Machinery Engineering, The University of Tokyo
  − Japan
  − Professor
  − attended the competition
Yuichi KOBAYASHI
  − Dept. of Precision Machinery Engineering, The University of Tokyo
  − Japan
  − Student in the doctor course
  − attended the competition
Masaki Fukuchi
  − Dept. of Precision Machinery Engineering, The University of Tokyo
  − Japan
  − Student in the master course
  − attended the competition
Jun'ichi IMANISHI
  − Dept. of Precision Machinery Engineering, The University of Tokyo
  − Japan
  − Student in the master course
  − attended the competition
Toru IWATA

- Dept. of Precision Machinery Engineering, The University of Tokyo
- Japan
- Student in the master course
- attended the competition

Web page http://www.arai.pe.u-tokyo.ac.jp/robocup

3 Complete Robot Architecture

We did not integrate reflex hehaviors such as avoiding obstacles or other exceptional behaviors. So the robots decided actions based on the recognition of the ball and the information of self-localization.

When the robot does not recognize the ball, it continues to turn at the same place with its head swinging. In order to avoid dead-rock state, it keeps the time since the current walking command has been decided. If the walking action does not change over a threshold time value, it changes its walking action by a random value unconditionally.

4 Vision

The 8 channels of the color detector are set, orange, green, white, skyblue, yellow, pink, darkblue and red. Darkblue and red are the colors of robots. They are recognized by the color detector but are not used for action dicision. When the color pink is detected, upper and lower regions of the center of pink are referenced. The type of the landmark is decided by comparing the number of pixels of each color. The goals are recognized when skyblue and yellow are detected without pink.

We made the color table as follows:

- Take images and save in memory sticks.
- Create image file in the 'ppm' format.
- Pick up the region where we want to recognize.
- Make a ppm image file that consists of black and white. Black corresponds to the color we want to recognize, and white other colors.
- Pick up YUV values of each pixel which correspond to a black pixel.
- Decide max and minimum YUV values of the color by collecting data.

In order to measure the ball position more precisely, we used Back Propagation algorithm. The ordinal way of recognizing the ball position is to utilize the pan and tilt joint values of the head while the head is tracking the ball. This way is not so much reliable especially when the ball comes near to the body. We developed the way of refining the measurement by utilizing the four parameters of the ball image.

First, we collect image data of the ball at a fixed position changing the pan and tilt joint angles of the head. The pan and tilt joint angles are not always the same even when the position of the ball is the same. The collected data consist of pan and tilt joint angle values and four parameters mentioned above. These six variables are measured mostly around ten times with different pan and tilt angles.

The first neural network is used to calculate the ideal pan and tilt joint. Here the word 'ideal joint' means that the CCD camera captures the ball exactly in the center of the visual field with the pan and tilt joint angles. In such a state, these angles provides enough information to specify the position of the ball. The stored data are used to solve the inverse problem. The training data of the network consist of x and y value of the center of the ball in the visual field as inputs and the pan and tilt joint values as outputs. After training, the ideal pan and tilt joint value is calculated by giving this network the center of the visual field.

Then the position of the ball is changed and fixed. The same process is iterated by changing the ball position. The second neural network is used to express the relation between six parameters and ideal pan and tilt joint values. Input data consists of actual pan and tilt joint values and four image parameters. Output data consists of ideal pan and tilt joint values which are calculated by the first neural network. By training with the data set of them, ideal pan and tilt joint values can be calculated from the actual joint values and image parameters.

5 Control

The manipulation actions for the ball are kicking and heading. When the robot kicks the ball keeping the posture of standing, the ball does not go so far because the reach of the kicking leg is limited. So, we gave up preserving the standing posture during the kicking motion. First, the robot bends its right (or left) front leg. The robot begins to fall down, and then the robot extends its bent leg. The ball gets farthest when the extending action sinchronizes with the fall of the body.

When the goal is the side of the robot, it hits the ball by its head. Two motions are combined to realize this heading action. One is the swinging motion of the head, and the other is the translation of the body by the legs. shows the motion of legs to translate the body. The heading action is achieved by the conbination of this action and the swinging action of the head.

6 Localization

We used roughly estimating algorithm from only one landmark. When one landmark is recognized, the current direction of the robot can be estimated very roughly.

When the position and posture of the robot is estimated in some way, these variables are transformed to the relative relation between the robot and the goal. The robot can update these relative relations by dead reckoning.

7 Strategy

The main strategy of our robots is to 'find the ball, and get close to the ball', and to 'manipulate the ball according to the direction of the goal'. The two modes ball-searching and ball-tracking correspond to the most basic condition.

When the mode is ball-tracking, the robot approaches to the ball and decides how to manipulate the ball. Here we mean kicking and heading action for the word 'manipualte'. The robot changes walking command so as to adjust the pan angle of the head to zero.

If the pixel number of the orange ball exceeds a threshold value, it decides the manipulation action according to the roughly estimated direction of the goal.

When the mode is ball-searching, the robot turns its body while swinging its head. In order to avoid dead-rock state, it changes the walking action when the same walking action continues over a threshold time.

8 Special Team Features

We developed learning system to realize 'Kick the ball where the robot wants to'. This architecture is based on Kohonen's Self-Organizing Maps(SOMs). Here we used SOM as associative memory. Action and its evaluation data is stored beforehand by the experiment. The weight vector of each node consist of as state inputs, as action outputs and as evaluation signals. Here, the state inputs are the ideal pan and tilt joint angle values calculated by the neural network. The action outputs are the direction of the kicking leg. The evaluation signals are the direction and distance of the ball after kicking. The distance is estimated by the number of pixels in the visual field.

The best matching node is calculated with the state inputs and desired evaluation signals. Action outputs and estimation of evaluation signals are decided by the best matching node. Actually the difference of the action output is not so influential with the direction of the ball. At least, this stored data can be utilized to estimate the direction of the ball before kicking. When the robot knows that the ball doesn't go to the disired direction with any action output, it can change the position by walking.

But this learning architecture did not work sufficiently in Stockholm. This is partly because the calculation of the ball position mentioned above is not precise enough.

9 Conclusion

We are going to participate the league next year. Development of the ability of ball operation is indispensable in order to realize more complicated strategy. We should focus on recognition and manipulation of the ball in front of the body, continuously. Learning for acquisition of the relation between manipulation and observation will be the main problem of our approach.

BabyTigers-99: Osaka Legged Robot Team

Noriaki Mitsunaga and Minoru Asada

Dept. of Adaptive Machine Systems, Osaka University, Suita, Osaka, 565-0871, Japan

1 Introduction

Our interests are learning issues such as action selection, emergence of walking, and self localization without 3D-reconstruction. We implemented teaching, self localization without 3D-reconstruction and embodied trot walking.

Our embodied walking showed the fastest movement in the all nine teams. We got second place in the RoboCup Challenge. Although due to the fatal bug revealed in the game, we lost in the semi-final game and we got the fourth place in the competition.

2 Team Development

Team Leader: Minoru Asada
Team Members:
   Minoru Asada
      − Osaka University
      − Japan
      − Professor
      − Attended the competition
   Noriaki Mistunaga
      − Osaka University
      − Japan
      − Ph.D candidate
      − Attended the competition
   Tatsuro Nohara
      − Osaka University
      − Japan
      − Master course student
      − Attended the competition
Web page http://www.er.ams.eng.osaka-u.ac.jp/

3 Complete Robot Architecture

Since everything changes rapidly in the real world, we consider it important for the robot to make fast decisions. Then we adopted to execute the learned movements rather than do planning during the games.

We wrote a program which consists of three objects. One is for the vision, one for the walking and head movements, and one for cognition and decision making.

4 Vision

In the RoboCup Legged Robot League field [2], seven colors (aqua-blue, yellow, orange, blue, red, pink, green) are used and robots need to detect and discriminate them. The Sony's legged robot has the color detection facility in hardware that can handle up to eight colors at frame rate. To detect colors with this facility, we need to specify each color in terms of subspace in YUV color space. YUV subspace is expressed in a table called Color Detection Table(CDT). In this table, Y are equally separated into 32 levels and in each Y level we specify one rectangle $(u_{\min i}, v_{\min i}), (u_{\max i}, v_{\max i})$ $(i = 1, ..., 32)$.

In order to make CDTs, we used the same method used in last year. Briefly, 1) take an image which includes the object to detect with the robot, 2) specify pixels to detect with GUI program and make a list of YUV to detect, 3) order the program to classify each pixel according to the Y level as they are classified in CDT and make a bounding box of UV in each level, 4) check if detection satisfies the need and if not do 1) again. Iterate these procedure for each color with several images.

If some parts of an object are detected by a CDT, but still other parts of it remain not detected, one can extend the CDT with continuousness of color in YUV space. We also made this kind of CDT expansion tool and used it (Fig.1).

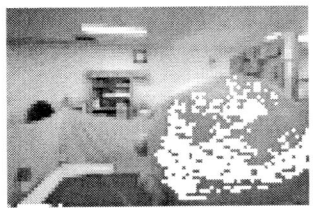
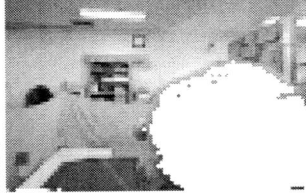

(a) A partially detected image by CDT

(b) Mostly detected image by expanded CDT

Fig. 1. An example of expanding CDT

Objects recognition was done by extraction of connected areas with 8-neighbor method in the color detected images. After connected areas are extracted, object recognition, including landmarks (all landmarks are consisted of two colors) recognition are done by concatenating the areas. To overcome noises in the image, the order was determined empirically. Details of 1)-4) and objects recognition are described in [1].

5 Control

Since it is important to move fast and efficiently, we developed an embodied trot walking, which enables simple controlling program and fast movement. Details will appear elsewhere.

For the posture control we used getting up program designed by SONY. We wrote down the program to sense of falling down, and called the SONY's recovering program when needed.

Head movement for searching and tracking the ball and observing landmarks about 180 degrees have been implemented beforehand. Also we prepared six discrete movements, forward, left forward, right forward, left turn, right turn, and chase the ball for 2.4 seconds (four walking periods). Since we decided to use teaching for behavior acquisition, we did not implement behavior such as avoiding obstacles or keeping its own goal.

6 Localization

In order to determine the behavior to take, it is important to localize itself in the soccer field. Location estimation with occupancy grids is time and memory consuming and it is difficult to do dead recogning with trot walking due to slippage. Although the 180 degree view is unique in most locations of the RoboCup Legged Robot League field, the robot has to pan or memorize how the landmarks were seen for 3-D reconstruction, due to the limited view angle of the camera.

To overcome such difficulties, we used the direction of the landmarks from the robot for its localization. The robot sees landmarks panning the camera and calculates in which direction they are and assigns the quantized direction to each landmark. Depending on the directions of the landmarks and the ball, the robot behaves as the human trainer taught. To reduce the time for panning, we also implemented estimation of landmarks. The robot only observes if estimation and current view of landmarks are not sufficient to determine an action to take.

7 Strategy

We used two kinds of programs. One is to demonstrate the fast trot walking, which only chases the ball. The other is to play back the taught behavior. Therefore our strategy mostly depended on the human trainer. Basic teaching policy was, 1)if the ball can be seen, 1-a) chase the ball if the target goal can be seen, 1-b) otherwise, turn around the ball as to see the target goal, 2)if the ball can not be seen, 2-a) if near the own goal, turn outside not to make own goal, 2-b) otherwise, turn to center of the field expecting to see the ball. Unfortunately, we have not considered team works in the teaching yet.

8 Special Team Features

We have proposed and implemented a new method for decision making based on information criterion. The basic idea is to construct a decision tree and prediction trees of the landmarks, which enable a robot with a limited visual angle to localize itself in the environment with fewer observations. The number of pannings (observation) are reduced by predicting the locations of landmarks and judging if the current observation and prediction of landmarks are sufficient for the decision making. We used a teaching method to collect example data for making the decision and prediction trees. Details will appear elsewhere.

9 Conclusion

We implemented an embodied trot walking and showed fast movements but had difficulty in making small or steady movements. Embodied steady static walking which can easily combined with dynamic walking should be developed.

We employed the direct teaching method for learning. For the self localization, we used quantized direction of landmarks from the camera rather than 3-D reconstruction. We also implemented a decision making method to see or to take an action dependent on estimation. Team work and self learning are future issues.

References

1. N. Mitsunaga, M. Asada, and C. Mishima. Babytigers-98: Osaka legged robot team. In M. Asada and H. Kitano, editors, *RoboCup-98: Robot Soccer World Cup II*, pages 498–506. Springer, Lecture Note in Artificail Intelligence (1604), 1999.
2. M. Veloso, W. Uther, M. Fujita, M. Asada, and H. Kitano. Playing soccer with legged robot. In *Proceedings of the 1997 IEEE/RSJ International Conference on Intelligent Robots and Systems*, volume 1, pages 437–442, 1998.

CM-Trio-99

Manuela Veloso, Scott Lenser, Elly Winner, and James Bruce

School of Computer Science
Carnegie Mellon University
Pittsburgh, PA 15213-3891
{mmv,slenser,elly,jbruce}@cs.cmu.edu

1 Overview

The robots used in this competition were generously provided by Sony [3]. The robots are the same as the commercial AIBO robots except for slight hardware changes and programming capabilities. These autonomous robots are about 30cm long and have 18 degrees of freedom. The neck pans $\pm 90°$ allowing the robot to scan the field with its on board camera. Six uniquely colored landmarks are placed around the field (at the corners and center-line) to help the robots localize. Each team consists of three robots. Like our team last year, CMTrio-98 [5], we divided our team between two identical attackers and one goalie.

We divided our system into three main components: vision processing, localization, and behaviors. The vision system is responsible for calculating distance, angle, and confidence measures for all objects visible by the robot. The localization is responsible for calculating the position of the robot on the field given the movements executed and the landmarks seen (we did not implement goals for localization this year). The behaviors are responsible for taking this information and winning the game.

Results from our matches in RoboCup-99 at Stockholm show our algorithms to be effective. Our team won all but one of its games, and the one it lost was lost by only one goal. Our team was the only one in this year's league to score goals against opposing teams and never to score a goal against itself. Our goaltender was the only one in this year's league to score a goal itself.

2 Vision

The vision system processes images captured by the robot's camera to report the locations of the ball, the 6 unique location markers, the two goals, and the robots. The main steps in vision processing are: 1) capture an image and classify each pixel's color in hardware using predetermined color thresholds, 2) find connected regions of the same color, 3) merge close regions of the same color, 4) use geometric filters to remove false positives, 5) calculate distance and angle of objects in ego-centric coordinates.

The on-board camera provides 88x60 images in the YUV space at about 15Hz. Hardware Color classification is then performed on these images. The thresholds for color segmentation are created by a supervised learning method based upon hand labelled images captured from the dog's camera. This results in a new image indicating color class membership rather than raw camera colors. This image is run length encoded (RLE) for further processing.

The region finding [1] method employs a tree-based *union find* with path compression. The algorithm outputs a forest of disjoint stubby trees corresponding to a connected region in the image. We next extract region information. The bounding box, centroid, and size of each region is calculated incrementally in a single pass over the forest data structure. The regions are separated by color and sorted by size (putting larger, more important blobs first).

A problem with connected components is that a single row of pixels incorrectly classified can separate an object into two components. We used a density based merging scheme to try to overcome this problem. We merge close regions of the same color if the resulting new region has a sufficiently high density.

The next step is to calculate the location of the various objects given the colored regions. Various top down and geometric object filters are applied to limit the occurrence of false positives and serve as the basis for confidence values. The largest orange blob below the horizon is labelled as the ball. The confidence value and distance is calculated based on image area and a circular object model. The field markers are detected as pink regions with green, cyan, or yellow regions nearby. The confidence is the ratio between the squared distance between the centers of the regions and the area of each region. The distance is calculated from the distance between the centers of the two regions. The goals are the largest yellow or cyan regions below the horizon. The very course distance approximation is based on the angular height of the goal in the camera image. The confidence is based on the aspect ratio of the goal in the image. The final objects detected are opponents and teammates. Due to the multiple complicated markers present on each robot, no distance or confidence was estimated and the marker regions are returned in raw form.

The system performed well in practice; it had a good detection rate and was robust to the unmodeled noise experienced in a competition due to competitors and crowds. The distance metrics and confidence values were also useful in this noisy environment.

3 Localization

Our localization algorithm is based upon a classical Bayesian approach which updates the location of the robot in two stages, one for incorporating robot movements and one for incorporating sensor readings. This approach represents the location of the robot as a probability density over possible positions of the robot. Our localization algorithm, called Sensor Resetting Localization (SRL) [4], is based upon a popular approach called Monte Carlo Localization (MCL) which represents the probability density using a sampling approach.

MCL [2] represents the probability density for the location of the robot as a set of discrete samples. The density of samples within an area is proportional to the probability that the robot is in that area. We calculated the robot's position from these samples by taking their mean. We estimated the uncertainty by calculating the standard deviation of the samples. We encountered some problems implementing MCL for the robot dogs. MCL took more samples to do global localization than we could actually run on the hardware. MCL also has problems dealing with our large modelling errors.

SRL is motivated by the desire to use fewer samples, handle larger errors in modelling, and handle unmodeled movements. SRL adds a new step to the sensor update phase of the MCL algorithm. If the sensor reading and locale belief state disagree, we replace some of our samples with samples consistent with the current sensor readings. In this way, we effectively through out our history and reset. Note that when tracking is working well no resetting is done and SRL behaves exactly the same as MCL.

This resetting step allows SRL to adapt to large systematic errors in movement by occasionally resetting itself. SRL is also able to recover from large unmodeled movements easily by using this same resetting methodology. Unexpected movements happen frequently in the robotic soccer domain we are working in due to collisions with the walls and other robots. Collisions are difficult to detect on our robots and thus cannot be modelled. We also incur teleportation due to application of the rules by the referee.

Robot movement was modelled as three Gaussians with hand measured parameters; one Gaussian for distance travelled, one for direction travelled, and one for heading change. Sensor readings were modelled as two Gaussians, one for distance and one for angle. Standard deviations were estimated by testing.

We used 400 samples in actual competition. We weren't quite capable of keeping up with real time this way if we saw a lot of markers, so we through out some sensor readings from time to time to catch up. The localization is accurate to about 10cm and $15°$ while the robot is looking around for markers and moving. Performance drops somewhat when the robot goes long periods of time without looking around for markers as often happens during play. We observed that the localization algorithm quickly resets itself when unmodeled errors such as being picked up occur.

4 Behaviors

Choosing behaviors for the robot is a difficult challenge. The robot must act under uncertainty and varying amounts of localization information. The robot can affect the amount of information available to it by actively localizing. Actively localizing involves stopping the robot and scanning for markers, a process taking 15–20 seconds. Stopping the robot reduces the computational load on the localization system allowing it to operate in real time. The behavior system has to balance the time spent localizing with time spent acting. Every moment spent looking around provides an opportunity to the opponent robots.

Our solution is structured as a finite state machine. Each state corresponds to a set of behaviors that all accomplish the same goal. Each behavior expects a different amount of information to be available. The best behavior that has all its required information available is chosen to be executed. In this way, the robot takes advantage of all the information that is available to it. We call this approach multi-fidelity behaviors [6].

Switching between action and localization is controlled by timeouts that switch the robot between states of the finite state machine. These timeouts make sure that the robot localizes occasionally and that the robot spends enough time acting. The robot needs to localize occasionally to prevent the localization output from drifting from reality without being detected. The timeout for action ensures that the robot doesn't spend all of its time localizing. The localization timeout is turned off during behaviors that do not require localization information.

Our behavior state machine has 5 main states: score, recover ball, search for ball, approach ball, and localize. Search for ball and approach ball do not require localization. Search for ball employs a random search method that alternates between walking forward/backward random distances and rotating in place a random number of degrees. Approaching the ball uses the visual input to approach the ball. If localization information is available, the robot attempts to approach a point behind the ball to save time. The scoring behavior circles the ball (while facing it) to get behind it and then pushes the ball towards the goal. If sufficient localization information is available, the robot: circles in the quickest direction, avoids circling into the wall, and does not bother visually acquiring the goal. The recover ball behavior backs up when the robot looses track of the ball. This optimizes for the common case of the robot loosing the ball by walking past it. The localization mode stops the robot and scans for markers.

We have specialized behaviors for kickoff and goal protection. At kickoff, we charge the ball to ensure the best ball position possible. The outcome of the kickoff often decided who would score next. Our goalie runs a specialized set of behaviors. The goalie scans for the ball from its home position in front of its goal. When the goalie sees that the ball is close enough, the goalie clears the ball and then returns to the home position using the goals as landmarks for navigation. When the goalie is clearing the ball, the goalie uses localization information to hit the ball off-center such that the ball heads towards the opponents half of the field. This tactic was sufficient to allow our goalie to clear a ball all the way into the opponents goal.

Acknowledgments We thank Sony for providing the robots for our research. We thank Tucker Balch for his many insights on this work.

References

1. J. Bruce, T. Balch, and M. Veloso. Fast and cheap color image segmentation for interactive robots. 2000. *Proceedings of WIRE-2000*.
2. D. Fox, W. Burgard, F. Dellaert, and S. Thrun. Monte Carlo localization: Efficient position estimation for mobile robots. In *Proceedings of AAAI-99*, 1999.
3. M. Fujita, M. Veloso, W. Uther, M. Asada, H. Kitano, V. Hugel, P. Bonnin, J.-C. Bouramoue, and P. Blazevic. Vision, strategy, and localization using the Sony legged robots at RoboCup-98. *AI Magazine*, 1999.
4. S. Lenser and M. Veloso. Sensor resetting localization for poorly modelled mobile robots. In *Proceedings of ICRA-2000*, 2000.
5. M. Veloso and W. Uther. The CMTrio-98 Sony legged robot team. In M. Asada and H. Kitano, editors, *RoboCup-98: Robot Soccer World Cup II*, pages 491–497. Springer Verlag, Berlin, 1999.
6. E. Winner and M. Veloso. Multi-fidelity robotic behaviors: Acting with variable state information. 2000. *Proceedings of AAAI-2000*.

Humboldt Hereos in RoboCup-99
(Team description)*

Hans-Dieter Burkhard, Matthias Werner, Michael Ritzschke, Frank Winkler,
Jan Wendler**, Andrej Georgi, Uwe Düffert, and Helmut Myritz

Humboldt University Berlin, Department of Computer Science,
D-10099 Berlin, Germany
email: hdb/mwerner/ritzschk/fwinkler/wendler/georgi/dueffert/
myritz@informatik.hu-berlin.de,
WWW: http://www.ki.informatik.hu-berlin.de

1 Introduction

The team members include students as well as members of the teaching stuff from the Department of Computer Science at the Humboldt University. They represent the groups of Artificial Intelligence, Responsive Computing, and Signal Processing, respectively. It was the aim of the project to combine the skills of these disciplines to program soccer playing legged robots.

An underlying idea was to use the experiences from the simulation league for the general structure of the robot software. We still think that this concept is realistic. But the restricted time forced us to use a very simple reactive approach for the RoboCup 1999 world championship.

Our general research interests can be described as follows: We are interested in the development of skills on higher level decision protocols using methods from Machine Learning, especially from Case Based Reasoning. For knowledge processing and deliberation we are using mental models from Artificial Intelligence. We are specifically interested in developing normal consensus protocols, collision avoidance protocols and would like to develop new models of faults, e.g., the opposing soccer team would be considered as a new type of a fault. Furthermore we are interested in novel algorithms for image processing and their implementation in embedded systems. We would like to apply parallel computing structures for image processing using the pixel-bit parallelism principles of distributed arithmetic. Scalable resolution allows simultaneous suppression of noise, sharpening of discontinuities and labelling of important data.

2 Team Development

Team Leader: Prof. Hans-Dieter Burkhard

---

* This work was partly sponsored by TecInno GmbH Kaiserslautern, Daimler Chrysler AG Research & Technology Berlin and PSI AG Berlin
** This work has been supported by the German Research Society, Berlin-Brandenburg Graduate School in Distributed Information Systems (DFG grant no. GRK 316).

Team Members:
Prof. Hans-Dieter Burkhard
- leader of the AI group
- did lead the design and did consulting
- did attend the competition

Dr. Matthias Werner and Dr. Michael Ritzschke
- research assistant at the responsive computing / signal processing group
- did lead Aperios and OPEN-R integration / did lead design of vision and did consulting
- did attend the competition

Dr. Frank Winkler and Peter Tröger
- research assistant at the signal processing group / undergraduate student
- did lead development of vision and acoustic tools and did consulting / did implementation of acoustic communication
- did not attend the competition

Jan Wendler
- PhD student at the AI group
- did design and implementation of the world model
- did attend the competition

Helmut Myritz, Uwe Düffert and Andrej Georgi
- undergraduate students
- did the implementation and debugging
- did attend the competition

Web page http:// www.ki.informatik.hu-berlin.de/RoboCup/RoboCup99/index_e.html

3 Complete Robot Architecture

We have distinguished four main parts which we call Cortex, Brain, Body, and Communication. Messages are passed between these modules according to the underlying control structure.

The general idea is to transmit the plans computed by the Brain to the Body and perform it by the available skills. The Body controls the movement of the legs in order to turn, move, kick etc. Additionally, there exists a direct information flow between Cortex and Body for immediate actions, e.g., for keeping track of the ball. This imposes some rudimentary layered architecture.

4 Vision

The Cortex uses the Color Detection Engine (CDT) to identify the objects in the image by common procedures of image-processing to find the object parameters, e.g. position, width, center-point.

It is well known, there are some possibilities to describe the colour in a picture. Common colour spaces are for instance RGB, YUV (the PAL/European standard for colour television broadcasting) and HLS, where H is for the Hue

(the H value is a degree value through colour families), L for Lightness (1 = white, 0 = black) and S fore Saturation (that is the degree of strength of a colour - greater is S, the purest is the colour).

The robot-eye use the YUV space and so we can analyze the YUV-values in a robot-image. But we want to create CDTs for a wide range of lightness/darkness and of saturation, because the robot sees different colours in the pictures if he looks from different viewpoint to the same objects. On the other side we found in our testing-period, that every robot had from the same viewpoint under the same light conditions little different YUV-values. Therefore we realize a way to develop our CDT's with four steps:

- Shooting session to get some images from all relevant objects (ball, goals, playerdress, landmarks), we use different viewpoints and - if enough time, all our robots.
- Analyzing of the YUV values and transformation via RGB in the HLS space. Our tool allows us to use the mouse for moving a reticule over the object. With the help of the statistical componente we get for H,L,S the mean value and the standard deviation.
- Now we use a second tool to simulate possible combinations of HLS around the mean values of H,L,S - with the help of random numbers of the constant distribution like radio noise. Every generated HLS-point (we use more then 1000 points) will be transformated in the YUV space and the resulting borderlines of YU and YV plane built our CDTs, which we can save in a file.
- We check the quality of our CDTs with a further program using originally robot-images. This tool allows - if it is necessary - manually corrections of the CDTs.

5 Control

We did not develop our own skills. This was a real drawback since the usage of the available skills caused several problems, e.g. the main disadvantage was a missed possibility to interrupt a movement in process. Another disadvantage was a lack of movements' precision.
We tried to overcome these disadvantages by several means:

- To interrupt a movement, we insert an interception function between the skill module and the robot module. Its task was to catch outgoing commands and to report success to the skill module.
- To improve the real-time behavior, we used a priority queue to transmit commands to the body. Sent but not yet executed commands that became superfluous are thrown away.
- Critical parts of a movement are executed in the step-wise mode. That shall increase the movement's accuracy.

However, the creation of a real-time walking and posture control is a mayor objective to become an appropriate competitor in the next competition.

6 Localization

The robot has to stop each time he tries to localize himself on the field. Then he turns his head to scan the full area around him all seen flags and goals are stored. The dog can now calculate his own position and body direction with these informations. The calculations are most of the times correct, only if the number of seen flags is very low or the distance to a flag or a goal is quite wrong, than the calculation can went wrong very hard. Usually we have to wait a small amount of time until the next localization is of advantage.

7 Strategy

The software architecture of the Brain is oriented on our simulation league agents, the AT Humboldt team, which is using a BDI architecture for the mental modelling. According to the BDI architecture the Brain transforms the received data into an internal world representation ("belief"). It identifies possible options ("desire") and commits for useful plans ("intention").

Actually, we did not finish the work on this concept for RoboCup. Instead, we used a simple reactive approach:

- Look for the ball
- Run to the ball
- Search and positioning for the opponent goal
- Go with the ball to the opponent goal
- Kick if you are near the opponent goal

8 Special Team Features

Because of the hearing and speaking possibilites of the sony legged robot, we decided to implement an acoustic communication. The speaking ability was a big advantage for debugging, the robot could tell us what he had seen by playing a predefined wave-file.

Unfortunately SONY did not supported the hearing capacity this year, therefore an acoustic communication between the robots was not possible.

9 Conclusion

We are very thankful to SONY for giving us the opportunity to work with such an exciting device. We are full of plans for the next year.

In future we will continue to develop in a close cooperation of all groups (Artificial Intelligence, Responsive Computing and Signal Processing). We will try to optimize coordination of all movements of the robot. Additionaly the problem of localization is another major point of interest in next years work.

During this year all three groups together are organizing one lecture "Intelligent Robotics", where motivated students are able to work with the Humboldt Heroes team. Certainly we don't forget the simulation league, the AT Humboldt 2000 is already in preparation.

McGill RedDogs

Richard Unger

Center for Intelligent Machines, McGill University

1 Introduction

After a period of experimentation and investigation into the dog's possibilities we defined six core areas in which we would concentrate our development efforts. In concert, we felt, these areas of functionality would give a strong system capable of succeeding at the proposed challenges. The six major areas of functionality can be roughly divided into two groups, informally labeled input and output. Input consists of the tasks of Vision, odometry and localization/map-building. Output embodies the tasks of moving, path planning and AI (decision making). These areas are treated in greater detail in the remainder of this document, after a brief overview of our system's infrastructure.

2 Complete Robot Architecture

In this section we will briefly outline the underlying architecture of our system. It is important to describe this in order to help make clear how the other areas of our system (outlined above) were implemented. After receiving the hardware we experimented extensively with different models for control, sharing and transferring of data and synchronization of separate modules of code. We implemented several different versions of a core software engine, using different approaches, and finally settled on the version we called 'monolithic', for reasons which will become clear in the next few paragraphs. First a couple of definitions:

- Core software engine, or engine will be used to refer to the code that is the bottom layer of our system. While this code does not actually solve any robotics problems in terms of the areas described in the rest of this article, it nevertheless offers important functionality. It interacts with Aperios for the other functional units of our system, and provides important scheduling, synchronization and information management functionality for them.
- Object will be used to mean a C++ object.
- Ooobject will be used to mean an Aperios Ooblet object.
- Process will be used to refer to one higher level functional unit of our system. A Process is something like our Path Planner, or our Vision System.

During our experimentation it became clear that what we really wanted was to have separate threads of execution in our code so that we might have independent and easily separable functional units, whose code could be kept specialized, focused and clean. We envisioned these functional units (e.g.: Vision System,

Map, Path Planner, etc...) to be implemented as Objects, with simple APIs for interacting with each other. Although this seems to be exactly what the mCOOP layer of Aperios offers, we found synchronization and sharing of information to be very difficult to achieve using the Observer/Subject communication channels. Also, we found the architecture was conducive to designs that were fairly serial in the way information and control was distributed among the Oobjects, or at best pipelines. It was hard to see how really parallel software models were to be implemented.

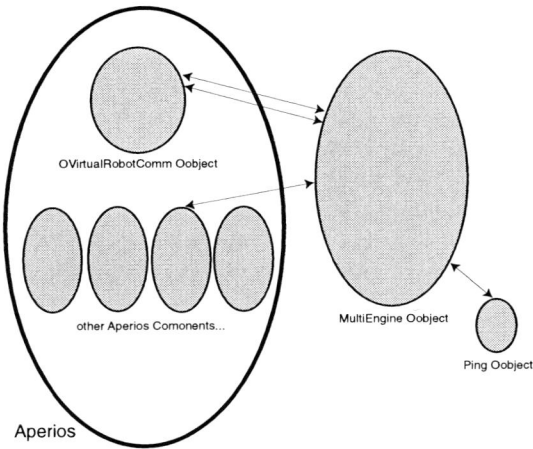

Fig. 1. The Architecture of our system within Aperios

All these considerations led us to the monolithic engine approach. We decided to implement one very large Oobject, composed of a huge number of Objects which would be our entire program, from vision to AI. Within this Oobject is a layer which accepts control from Aperios, quickly processes any data the message from Aperios may carry, and then calls an entry function in one of the higher layers of code according to a well defined scheduling scheme. The higher layers of code are also designed according to a well defined API to inherit from a base class called 'Process' and offer a simple interface consisting of a process ID number and an entry point. The higher layers are written with the constraint that any entry point which is called must be sure to return to its caller within a fairly short time, certainly less than 8ms. This constraint is necessary because there is no preemptive scheduling that we were able to access in the mCOOP layer of Aperios. However, this also makes our synchronization tasks somewhat easier, since we can guarantee that while our Oobject has control from Aperios, no other code in this Oobject is executing at the same time, and we can change variables or set flags without risk of being in the same critical section as another functional unit. To transfer control back to Aperios we must return from our entry function. All this is implemented by having a master Object that contains all the Aperios Ready/Notify functions for all the channels to which our system needs access. The same object also maintains a list of Processes, basically an array of references to Process objects and some associated information. Each process is associated with a state flag, a process number, entry point, and a delay value. During initialization each process is registered with the master object (placed in the list). Whenever an entry from Aperios occurs, the master object checks how much time has passed and decrements all process delays by this amount. Any process whose delay drops below zero is placed in a round robin queue for execution. Finally the process at the front of the queue is dequeued and its entry function is invoked. When this function returns, control is returned to Aperios, to give time to other parts of the OS to do their job and process our requests. While testing this core engine we found we were not getting the execution speed and frequency of entries from Aperios that we needed for reliable operation. For this reason we added one more, extremely small Ooblet to our system. Simply named 'Ping', all this Ooblet is capable of is to receive a message through its notify channel and upon receiving it, immediately send out a ready. This Oobject is used to have a continuous exchange of messages from and to our master object in the principle Oobject, so that we have more transfers of control to and from Aperios. This resulted in the desired performance. We found this engine to work extremely well, and to be fairly easy to write code for. Each process was able to focus on its task, and it was easy to pass information between and synchronize the processes. Other supporting code could simply be written as single objects, and once compiled into the Oobject were available to any process. No process had to deal directly with the Subject/Observer channels, and a good API was implemented to enable to the processes to suspend, resume or wait on each other, as well as set their delays or entry points. The shortcomings of this architecture were the fact that there was no real preemption available to us, so all processes had to be implemented to return control very quickly. This meant that long operations had to be distributed over several entry points or iterations of the same entry point. Great care had to be taken not to misuse the shared memory or the scheduling and synchronization APIs.

3 Vision

During our experimentation with the dog's camera and vision hardware, we took many pictures of the playing field and relevant objects in order to accumulate training data and find the appropriate setup for the hardware segmentation. In doing this analysis we found it difficult to come up with good bounding boxes that would adequately separate all the colours we were interested in detecting and we decided that it might be worthwhile pursuing our own strategy for segmentation. This work was begun by creating a representation of the colour space for the robot that would allow us to detect all the classes of objects we wanted. Since bounding boxes were not sufficient to capture the complexity of the arrangement of the colour classes, we decided to represent the YUV space explicitly, as a 64 by 64 by 64 colour cube. Each element in this three dimensional array was a single byte encoding two colour classes to which that point might belong, the 'best' colour class in the high order nibble, and the 'second best' colour class in the low order nibble. This allowed for 32 different colour classes, and the resulting data structure occupied 256K on the memory stick. In order to determine the best and second best colour class for each point in YUV space, colour training is necessary. To do this, several utility programs were written in C and Java which are also included in our source package. All the pictures for a class are preprocessed manually, masking or cutting out those areas unimportant to the colour class in question. The entire vision system is implemented as a process in our overall control engine. The process begins by requesting a new picture from the camera hardware. When Aperios delivers the new picture, processing begins in a series of small steps. This is necessary to ensure that Aperios and the other processes in the system receive enough time to do their jobs. The vision process achieves the segmentation task by building blobs of like colour classes in the image, initially creating the first set of blobs in a single pass over the image, and then processing the list of blobs thus created until the segmentation is stable. A blob is simply a contiguous area of pixels, associated with some information, for example giving the area's size and its best/second best colour class. The first pass over the image scans the image row by row, assembling horizontal segments of like colour classes. A segment is a straight line segment consisting of pixels with the same two colour classes. This could be compared to run length encoding the image. Each segment thus found is added to the list of discovered blobs according to the following rules:

- If the segment touches no other blobs with the same two colour classes, create a new blob with this segment.
- If the segment touches one other blob with the same two colour classes, add the segment to the blob, updating the blob information.
- If the segment touches more than one blob of the same two colour classes, add it to one blob, but record the connections to the other blobs it touches.

After this operation has completed (in a single pass for the entire image) we are left with a list of blobs and connections between them. We now want to simplify this representation, until we are left with only the blobs belonging to relevant objects in the game. To do this, we define 4 rules:

- Merge any two touching blobs with the same two colour classes.
- Define a minimum size, and a maximum size. Do not merge blobs larger than the maximum size.
- Merge blobs smaller than the minimum size to any touching larger blob.
- Merge blobs between the maximum and minimum size if they have a neighbour with the same best colour class.

Using these rules we iterate over the blob list until it converges (no more blobs merge). During the iteration, we use only the first rule in the first pass, adding the second rule on the second pass, and so on. The fourth and subsequent passes use all the rules. When merging, we merge a blob with the first matching blob it touches. While this is clearly less good than merging with the best-matching blob, it is faster, and it worked well in practice. The final step in our vision system is to use the contextual information we have about the game to improve our segmentation results. After our blob merging and the contextual analysis are complete, we have a fairly reliable segmentation of the field, as well as information about possible occlusion relationships. We were very pleased with the results of our vision system. Test we conducted showed that it was able to see the ball at larger distances than when using the hardware vision, and was also less prone to error when segmenting the difficult colours. The final version of the system was delivering frame rates of 8-30fps, depending on the complexity of the scene to be segmented.

4 Odometry

We considered good knowledge of the dogs position and pose important to solve higher level tasks. Initial experimentation showed that computing the dogs position after some amount of motion using only the information from its internal sensors would be a difficult task. Testing with the accelerometer and gyroscope showed that these would be difficult to use for this purpose, the error on their measurements too high to make them useful. Instead we considered the approach of using the forward kinematics of the dogs joints to compute its change pose, and to perform an integration on this operation over time to estimate the dogs global position from a known origin. The goal was to obtain a level of accuracy sufficient for playing the soccer game in the short term, with the position being corrected by localization when possible. In addition, the odometry level of our system was to include a complete API providing access to details of the robots pose, allowing us to retrieve the head direction vector and head position, for example, if needed by other parts of the code. The odometry is computed from two consecutive robot poses. We assume in making odometry calculations that the floor is level, and the robot's roll, pan and 'up vector' are the same in the two poses. We obtain from this equations which are solved according to the forward kinematics of the robot to recover the new x, y and orientation components of the robot's pose.

5 Mapping and Localization

Using the vision system described above as a base we thought it possible to build a simple mapping and localization module for our system. Having a map would allow the higher level AI layer of the system to deal with objects in the dogs world in a practical high level way. The goal was to allow the AI layer to query the map for information such as the ball's location, etc, and the map would provide these in the same coordinate system as used by the path planner. At the same time the map would provide a good abstraction, freeing the AI from having to deal with the vision system, converting observations into coordinates, and keeping information updated as events in the game (the robot moving, falling over, relocalizing) change it. The mapping layer would also incorporate the localization step. The reason for this design was that the map, taking as input all the raw information from the vision, would be best able to tell when there are sufficient markers or other visual cues in the input to permit a robust attempt at localizing. In addition, the map is the principle consumer of relocalization information, since relocalization changes foremost and most importantly the information in the map. Our map was kept fairly simple, essentially it can be thought of as a list of objects the robot has seen in the world. The object's locations are remembered in unlocalized coordinates. The map can be queried for the absolute or relative position of an object. When absolute positions are requested, the localization (see below) is taken into account. The necessary localization information is stored as a matrix, so to recover the absolute position of an observed object is a simple coordinate transformation. The objects on the map are remembered for varying amounts of time after they disappear from the robot's field of view. The exact time depends on the role of the object in the game, but is on the order of a few seconds. When an object is seen many times in succession, its observed locations are combined in an average location estimate, to make the map more stable. The localization operation was not deemed critical to playing the game, but important to making some higher level decisions. We decided to put the localization step into the mapping logic of our engine, since the markers used for localization are objects on the map, and it is this part of the controlling software that most easily permits localizing at the right times. To recover our robot's x y position and orientation we proceed in two steps, first recovering the x,y position and then the orientation. While both mapping and localization appeared to work fairly well, these are hard things to test, and we were not really able to fully test either functionality. Initial tests confirmed the map and localization to be functioning, while not to perfect accuracy, to an accuracy that was very useable for the purposes of playing soccer. We believe some work is needed in this are, particularly with regard to establishing a good policy for the management of map information over time. When an object disappears from the dog's vision, should it be deleted from the map? What should the update policy for the map be, particularly with regard to destructive updates? Can much be gained by keeping a history of object positions? (Would this allow, for example, some recovery of velocity of an object?) All these questions deserve some attention in a future iteration of our system.

6 Ambulation and Actuation

Clearly moving is a very important part of the robocup problem. Initially, we had little knowledge of the complications and problems to expect when attempting to obtain locomotion for a legged robot, although working on it has certainly given us new respect for the problem. We considered writing our walking methods important since better walking than that provided by Sony would probably give a significant advantage to our dogs, while not using the Sony walking APIs was also attractive from the point of view of controllability of the robot. Our own walking would permit us to work in our own kicks, have greater control over how and when the robots stops and starts and performs transitions between different modes of walking. Our system included three major layers of code for ambulation, implemented in several process objects and utility objects. These layers consisted of an interface layer for talking to the Sony APIs for walking in the OMLE2 Ooblet, a actuation toolkit layer to allow flexible control of all or some of the dog's joints over time, and a layer that used our actuation toolkit to achieve walking. The interface to the Sony walking APIs was simple to implement, and worked well from the outset. Essentially this layer was simply a wrapper for the Subject/Observer channel to the OMLE2 object, so that we could treat walking using Sony's API and walking using our own system in the same fashion from the higher layered functional units of our engine. The actuation toolkit consists of several objects designed to be used by other processes in the system to move the robots joints. These objects permit programming motions for the dog in a fast and flexible manner. An object called 'Effector' allows the buffered, automated dispatch of vectors of joint commands to the dogs joints or a subset of its joints. Lowest level joint commands (essentially lists of angles) are written into the buffers in advance, and dispatched to the Aperios actuation drivers when necessary. Another object, called 'MotionTable' allows the creation of more sophisticated motion sequences by allowing the programmer to set up a 'motion program' which is executed to generate the lists of low level joint commands. We found ambulation a very difficult problem to solve for our system. We believe that while our actuation toolkit worked extremely well for other types of motion, its architecture of creating and executing larger predefined motion sequences was not well suited to walking. An architecture allowing instantaneous control of the joints based on frequently and rapidly computed decisions would be better suited, and we feel a system that actively took into account sensors and situation when computing a gait could have made for better walking. Another thing that made developing ambulation very difficult was the lack of precise timing and process control in the mCOOP layer of Aperios. To effectively develop walking we would like to be able to use interrupt driven routines and timer driven routines, to allow us to control the robot with greater precision and respond more quickly to problems. Improved walking would definitely be an area worth investigating for next year's robocup, although the problem seems so difficult, and other areas also need attention, that using Sony's walking seems like an attractive option.

7 Path Planning

Our path planner allowed the robot to move to a desired x-y point on the field with a desired final orientation. The path planner retrieved the robot current position information from the Localization process, the other objects position from the Map and the robots destination from the Brain process. All the calculations were made in the field reference frame.

The path planner used 3 types of curves to build the paths. Family of curves:

- Straight Line
- 200 mm radius left turn
- 200 mm radius right turn

The basic path was built in the following way. First the path planner computed the beginning of the path and the end of the path by placing 2 circles at the start point and 2 others at the destination point. As shown in figure 2. These circles were tangential to the start and end directions. The next step was to draw the 16 tangents to the four circles. Only 4 of these tangents result in a continuous path. the resulting shortest path was chosen.

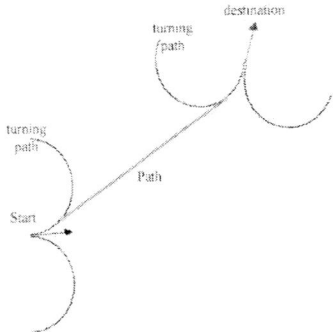

Fig. 2. Simple Path

The more complex case of path planning was also the basis of our object avoidance scheme. Instead of having the Brain worry about the collision avoidance, the path planner took objects on the field into account automatically. If an object intersected a basic path the path planner recomputed the path after inserting that obstacle in its calculation. This was done by including an intermediate path which was an 200 mm arc centered about the obstacle. Figure 3 shows the two resulting paths. Here the top path would have been chosen since it is the shortest one.

8 The AI Layer

The AI Layer was implemented as two different objects, called AttackerBehaviour and GoalieBehaviour. Although they are intended to be mutually exclusive, and only one was used on any one dog in the competition, there is no reason why both could be run on the same dog, with the two processes suspending and resuming each other in response to changing conditions in the game. In this way

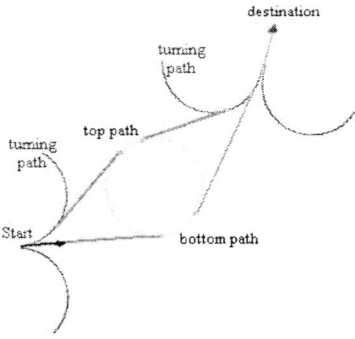

Fig. 3. Compliacted Path

dynamic role switching behaviour could be quickly implemented if the necessary trigger conditions could be defined and detected. Unfortunately time constraints prevented us from implementing a sufficiently complicated AI system. We were forced to settle for a very simple approach in order to be able to implement and test it in time for the competition. However, as mentioned above, we spent some effort in making clean, easy to use APIs for our lower layers of functionality, so that the AI component would be able to deal with the dog in a high level way. Issuing commands like go to point or asking questions like where is ball and being able to work with a map and well defined coordinate systems make programming higher level tasks much simpler. The Attacker really covers both attacking and defending, and is implemented as a simple state machine. The dog is initially in Defending state. In this state he attempts to get between his goal and the ball, facing the ball, and if close enough, will attempt to kick the ball away from his goal. If the ball is observed to be far enough away from the dog and towards the opponent goal, the dog will enter attacking mode. In attacking mode the dog attempts to position itself behind the ball on the line joining the ball and the opponent goal. It then moves close enough to the ball to kick it, and attempts to kick it towards the opponent goal. If the ball is observed to be more towards the dogs own goal, in other words if the dog is beyond the ball with respect to his own goal by a certain amount, he enters defending mode again. As mentioned we were not able to give this component of our system nearly the amount of attention it needed. However, we were able to write and modify what behaviour we did implement very quickly and easily thanks to the clean separation between the functional units and the high level APIs they provided. Unfortunately lack of testing and tuning, as well as general insufficiency was very obvious in the performance of this module of the system, and this part will clearly be receiving the most attention in the next iteration of our software. In particular a more complicated model is needed for the soccer game, and a more sophisticated way to make decisions about the percepts and game situation, as well as taking a richer set of percepts into account in the first place. Our goals for next year include: detection and dealing with tricky situations (the corners) and more reaction to teammates and enemies (avoiding ganging up on the ball), as well as dynamic switching of roles.

9 Conclusion

It has been a great opportunity for us to work with the Sony dogs on the RoboCup problem, and we have certainly learnt a lot from our first year's efforts. We are most happy with the core architecture of our system and the vision component. The other areas still need some work, and in particularly AI and path planning will be our primary focus for the next year. It is very much our hope that next year all the teams will exhibit behaviour that looks much more like soccer playing.

10 Team Development

Team Leader: Richard Unger
Team Members:
Faculty

- Pr. Gregory Dudek (MRL)
- Pr. Martin Buehler (ARL)
- Pr. Jeremy Cooperstock (SRL)

Undergraduate Students

- Andrew Ladd (MRL)
- Giulliaume Marceau (MRL)
- Didier Papadopoulos (ARL)
- Haig Hugo Vrej Djambazian (SRL)

Graduate Students

- Scott Burlington (MRL)
- Francois Belair (MRL)
- Sami Obaid (ARL)
- Shawn Arseneau (SRL)
- Richard Unger (MRL)

MRL: Mobile Robotics Laboratory
ARL: Ambulatory Robotics Laboratory
SRL: Shared Reality Laboratory

Web page: http://www.cim.mcgill.ca/~mumeteam

Team Sweden

M. Boman[2], K. LeBlanc[1], C. Guttmann[2], and A. Saffiotti[1]

[1] AASS, Center for Applied Autonomous Sensor Systems
Dept. of Technology and Science, Örebro University
Fakultetsgatan 1, S-701 82 Örebro, Sweden
[2] DSV, Department of Computer and Systems Sciences
Stockholm University and The Royal Institute of Technology
Electrum 203, S-164 40 Kista, Sweden

1 Introduction

"Team Sweden" is the Swedish national team that entered the Sony legged robot league at the RoboCup 99 competition. We had two main requirements in mind when preparing our entry to the competition:

1. The entry should effectively address the specific challenges present in this domain; in particular, it should be able to tolerate errors and imprecision in perception and execution; and
2. it should illustrate our research in autonomous robotics, by incorporating general techniques that can be reused in different robots and environments.

While the first requirement could have been met by writing some *ad hoc* competition software, the second one led us to develop principled solutions that drew upon our current research, and that pushed it further ahead.

2 Team Development

The work has been distributed over several universities in Sweden, which has made the project organization especially demanding. The three main cities of activity were Stockholm, Örebro, and Ronneby, which are separated by a geographical distance of up to 600 Km.

Team Leader: Magnus Boman (Stockholm)
Team Members: in addition to the authors of this paper, the team included:
- from Stockholm: M. Ericmats (MSc student), J. Kummeneje (PhD student), and A. Tollet (MSc student);
- from Örebro: M. Karlström (undergrad), D. Petersson (undergrad), and Z. Wasik (PhD student);
- from University of Karlskrona/Ronneby: P. Davidsson (ass. professor), M. Fredriksson (PhD student), and S. Johansson (PhD student).

Many other people contributed to the team in several ways. Their names are listed in the team web page.
Web page: http://www.dsv.su.se/~robocup/teamsweden/. Information on the current progress of the team: http://aass.oru.se/Living/RoboCup/.

3 Robot Architecture

Each robot was endowed with a layered architecture for autonomy inspired by the Thinking Cap,[1] sketched in Fig. 1. The lower layer provides an abstract interface to the physical functionalities of the robot. The middle layer is responsible for maintaining a local representation of the space around the robot (PAM), and for implementing a set of robust tactical behaviors (HBM). The higher layer maintains a global map of the field (GM) and makes real-time strategic decisions (RP). More detailed information can be found at the Team web sites.

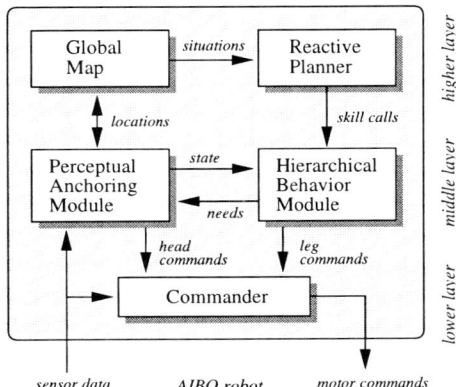

Fig. 1. The variant of the Thinking Cap architecture used at RoboCup '99.

4 Perception

The main locus of perception is the PAM, which acts as a short term memory of the location of the objects around the robot. The position of each object is updated by three mechanisms: by *perceptual anchoring*, whenever the object is detected by the camera; by *global information* —for the fixed objects only— whenever the robot registers its own location in the global map; and by *odometric clamping*, whenever the robot moves.

Each object is associated with two fuzzy predicates, that take truth values in the $[0, 1]$ interval: the *anchored* predicate measures how much the object's position is supported by recent perception; the *needed* predicate measures how much that object is needed by the currently running behaviors. These are used to measure how much it is *important* to re-acquire a given object o by

$$\text{important}(o) = \text{needed}(o) \cdot (1 - \text{anchored}(o)). \tag{1}$$

We use importance to decide the perceptual focus by selecting the object f with the highest importance. (We need focus control because we have only one

---

[1] The agent architecture based on fuzzy logic in use at Örebro University. This is a successor of the architecture originally developed for the robot Flakey [1].

camera, and many potential objects of interest.) The camera is pointed to the expected position of f, and a head scan is done if the object is not found. The use of (1) results in selecting all of the *needed* objects in a round-robin fashion.

Object recognition relies on the color detection hardware in the Sony robot, to which we provide the intended color signatures (produced off-line from color samples). We combine color blobs into features by a model-based approach: for instance, a green blob and a pink blob in a given geometric relation are fused into a landmark feature. These features are checked against the three criteria illustrated in Fig. 2. First, we reject features which are "astray": e.g., the pink band of a landmark cannot be a ball feature since it does not lay on the ground. Second, we select features which are in the "anchoring zone," a small fovea in the image where we can reliably measure the object's position. The remaining features are consider to be in "tracking zone," meaning that they should be brought into the anchoring zone by moving the camera.

 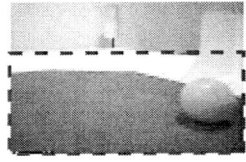 

Fig. 2. A ball feature in the astray (left), tracking (center), and anchoring zone (right).

5 Control

The HBM implements a set of motion behaviors realized using fuzzy logic techniques and organized in a hierarchical way. The use of fuzzy logic brings two main advantages: firstly, fuzzy behaviors can tolerate some amount of uncertainty in the position estimates; secondly, it is easy to write complex control strategies using a simple rule based language [2]. As an illustration, the following set of fuzzy rules implement the "GoToPosition" behavior.

```
IF (AND(NOT(PositionHere), PositionLeft))    TURN (LEFT);
IF (AND(NOT(PositionHere), PositionRight))   TURN (RIGHT);
IF (OR(PositionHere, PositionAhead))         TURN (AHEAD);
IF (AND(NOT(PositionHere), PositionAhead))   GO (FAST);
IF (OR(PositionHere, NOT(PositionAhead)))    GO (STAY);
```

The fuzzy predicates in the antecedents rely on the information in the PAM; 'GO' and 'TURN' are fuzzy set-points for linear and rotational velocity, respectively. These are passed to the Commander module, which translates them to an appropriate walking style. The translation simplifies the writing of motion behaviors, and makes them more portable between different platforms. It also allows us to easily modify the walking style used to implement each velocity pair.

Simple behaviors can be composed into increasingly more complex behaviors by using fuzzy rules that activate concurrent sub-behaviors. For instance, the following rules implement the "GoAndScore" behavior.

```
IF (NOT(NearBall))              USE (GoToPosition,Ball);
IF (AND(NearBall, NOT(Aligned))) USE (Align,Net1);
IF (AND(NearBall, Aligned))     USE (Kick,Net1);
IF TRUE                         NEED (Ball);
IF (NearBall)                   NEED (Net1);
```

These rules also show how behaviors communicate the current perceptual needs to the PMA via the 'NEED' keyword. For instance, when the ball is at 400 mm from the robot the truth value of "NearBall" is 0.7, and the above rules assert a value of needed of 1.0 for the object Ball and of 0.7 for Net1. These values are used by the gaze control mechanism to track the relevant features.

6 Localization

Spatial information is represented at two levels: locally, in the PAM; and globally, in the GM. The GM incorporates prior knowledge of the relative position of the fixed objects in the field. The information in the PAM and in the GM is registered by estimating the robot's posture in the field using the observed position of the landmarks. Unfortunately, we could not test this mechanism to a sufficient extent by time of the competition, so we only used local information.

7 Strategy

We used hierarchical behavior composition to write simple fixed strategies like "GoAndKick." More complex strategies are dynamically generated by the RP, which implements a decision making scheme inspired by cognitive psychology [3]. The RP decides the top-level behavior to activate by a voting mechanism, where votes correspond to motivations for using a given behavior in a given situation. The RP was not fully developed by the time of the competition, so we only used very basic strategies for the players and for the goal-keeper.

8 Conclusion

Our experience at RoboCup '99 has shown that the general principles and techniques developed in our research could be successfully applied to a radically different domain. Fuzzy logic proved beneficial in writing robust behaviors, and in developing an effective gaze control strategy. We believe that fuzzy logic could also bring substantial advantages in self localization and in flexible reactive planning. We plan to explore these issues in preparing our entry to RoboCup '2000.

Acknowledgements We thank Aline Dahlke at Sony France and the Sony staff in Japan for being so understanding of the difficulties of our distributed work situation. Founding was provided by the Universities of Örebro, of Stockholm, and of Ronneby.

References

1. A. Saffiotti, K. Konolige, and E.H. Ruspini. A multivalued-logic approach to integrating planning and control. *Artificial Intelligence*, 76:481–526, 1995.
2. A. Saffiotti. Using fuzzy logic in autonomous robot navigation. *Soft Computing*, 1(4):180–197, 1997. Online at http://aass.oru.se/Living/FLAR/.
3. C. Guttmann *A software architecture for four-legged robots*. M.Sc. Thesis, Dept. of Computer and System Science, Stockholm University, 1999.

UNSW United

Mike Lawther and John Dalgliesh

University of New South Wales

1 Introduction

The system we developed uses a hierarchial approach to software design, and consists of three subsystems : vision, acting, and planning. We also developed an offline classification system that uses concepts derived from Machine Learning.

2 Team Development

Our team is from the University of New South Wales in Sydney, Australia. All members were present at RoboCup-99 in Stockholm.

Team Leader: Claude Sammut
Team Members:
    Claude Sammut
      – Supervisor
      – Professor
    Mike Lawther
      – Core Developer
      – Undergraduate
    John Dalgliesh
      – Core Developer
      – Undergraduate
    Philip Preston
      – Technical Manager
      – Engineer
Web page http://www.cse.unsw.edu.au/~robocup/

3 Complete Robot Architecture

Fig 1 shows how the three subsystems of our architecture are connected. The system is driven by input from the camera. With each new frame, the Vision subsystem recognises objects in the robot's vision, and uses these to update a world model. A localisation algorithm is executed as part of this process.

    The Planning subsystem then suggests an action for the robot to perform. It makes this decision based on the current state of the world as represented in the world model, as well as its own internal state.

    The Acting subsystem monitors the action that the robot is currently performing. If Acting subsystem determines that the action has completed, the next action from the Planning subsystem is accepted, and begins execution.

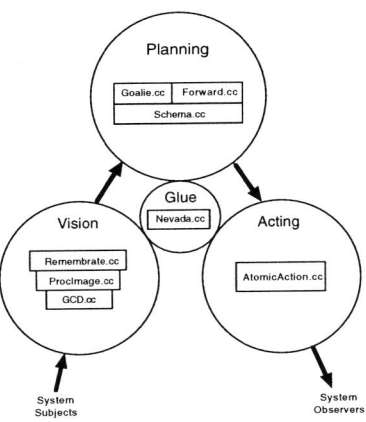

Fig. 1. System architecture

4 Vision

The goal of the Vision subsystem is the accurate detection of colours, allowing it to classify each pixel as one of up to eight colours. In order to calibrate this classification process, sample images are taken with the robot's camera and these are classified by hand.

The definition of a 'colour' for classification purposes is a single polygon in UV space — irrespective of a pixel's Y value. These polygons are learnt offline by a program which tries to evolve a polygon that minimises classification error within the sample images, for each colour, as shown in Fig 2.

Fig. 2. Polygons learnt from sample images of the main field on 30th July 1999

The onboard hardware colour detection is not used. Instead, each camera frame is classified using the polygon scheme described above. This classification results in a bitmap for each colour. Four-connected blobs are formed from these bitmaps, and these blobs are used by the object recognition algorithms. Our system recognises the ball, the goals, and the six beacons. Although red and blue colours can be detected, these are not used to recognise the robots.

After the objects have been detected, they are integrated into the world model. Every object has a number of properties, including a heading, elevation and distance to the object, relative to the base of the robot's neck. Objects also have a confidence factor associated with them. In the world model, the confidence factor of an object indicates the certainty of its location.

These confidence factors are used in a weighted average calculation when updating the world model with new information. Mathematically, a confidence factor is the factor c in the probability distribution of the object, $p(d) = c * e^{-\left(\frac{d}{1.3-c}\right)^2}$. The argument d is the distance of the object from its nominal position. The updated confidence factor is calculated to minimise the error in the addition of the distributions corresponding to the existing and input confidence factors.

5 Acting

The basic unit of robot control is an *atomic action*. This is an uninterruptible sequence of steps, which may control leg motors directly or send commands to the OPENR MoNet.

We have two styles of walks as actions, a fast walk and a precise walk. These use the Trot and Fast OPENR walks respectively. A kick was also developed, but was not used due to the time required to position the robot accurately.

In parallel with general leg-based actions, the Acting subsystem also controls the movement of the head, which tracks objects as instructed by the Planning subsystem. If the object is not in the world model, then the head searches for it in a circle. There are two distinct search 'circles', one for the ball, and one for all other objects. The ball circle takes care to look down so the robot does not miss the ball if it is at its feet.

6 Localisation

We use a passive localisation scheme. For each frame, the system attempts to derive its position on the field, based on the objects in the world model.

The algorithm is based on trilateration. Because goals are large, it is difficult to accurately calculate distances and angles to them.

Therefore goals are excluded from consideration for this calculation. The x, y and θ (orientation) coordinates of the robot relative to the field are determined solely by sightings of beacons.

The algorithm maintains a confidence factor for the robot's position, similar to that in the world model described in Section 4. When high-level control wants to locate a fixed object, such as a goal, if the robot is confident of its position on the field, then the object's position can be determined even if it cannot be seen.

7 Planning

Our Planning subsystem consists of *plans* and *roles*. Roles are top-level behaviours such as *Goalie* and *Forward*. A plan performs some simple function that may be common to several roles.

One important plan is 'PositionForKick'. A robot following this plan attempts to determine the angle at which it must kick the ball so as to move it towards a goal. If necessary, the robot must turn itself around the ball.

When playing the game, our robots use a dribbling technique to move the ball. This was implemented in the 'DribbleBall' plan. A robot following this plan would dribble the ball until it loses sight of the ball. At this point, the 'DribbleBall' plan completes. The high-level controller must then decide what to do next.

The Goalie uses simple heuristics to localise itself. It simply walks straight into its own goal. Once in its goal area, it monitors the ball's position. If the ball becomes too close, it follows the 'DribbleBall' plan to clear the ball.

The other role used in our team is the Forward. The Forward tries to move the ball into the opponent's goal. If the robot cannot see the ball, the role instructs the Acting subsystem to look for the ball. Once the ball is acquired, the Forward will follow the 'PositionForKick' plan, until the robot is aligned correctly with the ball and the goal. 'DribbleBall' is used to move the ball until the robot loses sight of it.

None of the roles recognises any other robot on the field.

For the RoboCup Challenge, the Forward role was used without modification.

8 Special Team Features

Our team employs machine learning for the colour classification. This is performed offline. When the robot is on the field, no learning takes place.

One of the advantages of the modular design of our software is that it is possible to add and change plans and roles quickly. This proved valuable during the competition.

9 Conclusion

UNSW plans to enter another team in RoboCup2000.

The main focus of our development will be to investigate further use of machine learning in vision and control.

For next year's competition, we would like to see a slightly smaller, and softer ball This will allow the robots a higher degree of control when kicking or dribbling. We would also like to have the floor of the goal carpeted to avoid slippage. Another improvement would be lessening the slope of the walls surrounding the field, allowing the robots to walk on them. This would reduce the 'huddles' and 'ball-stuck' situations that occured frequently in RoboCup_99.

UPennalizers: The University of Pennsylvania RoboCup Legged Soccer Team

James P. Ostrowski

University of Pennsylvania

1 Introduction

The main areas of focus for our team were the development of solid algorithms for merging vision data with other inputs and basic strategies for moving the ball towards the goal and for defending the goal. Our team came in fifth place overall in the competition. We defeated Team Sweden 2-0, though both goals were "own-goals" (scored by Team Sweden). We lost 2-0 to Osaka Univ., in a game where the first half we spent 90% of the time in the corner of the field near their goal, unable to score, while in the second half they scored one quick goal, just making it past our (slow) goalkeeper, and then a second goal later on in the game. We also had two scrimmage games– a 2-0 victory over Team Sweden, and a 0-0 tie against the team from Humboldt.

2 Team Development

Team Leader: Jim Ostrowski
- Department of Mechanical Engineering and Applied Mechanics, University of Pennsylvania
- United States
- Assistant Professor
- Attended the competition

Team Members:
Aveek Das and Kenneth A. McIsaac
- Department of Mechanical Engineering and Applied Mechanics, University of Pennsylvania
- United States
- Graduate Students
- Attended the competition

Thomas J. J. Ferguson, V and Mike Portnoy
- Department of Computer Science and Engineering, University of Pennsylvania
- United States
- Sophomores
- Attended the competition

Austin J. Parker
- Haverford College
- United States
- Sophomore
- Attended the competition

Tim Ledlie
- Harvard University
- United States
- Sophomore
- Did not attend the competition

Max Mintz
- Department of Computer and Information Science, University of Pennsylvania
- United States
- Professor
- Did not attend the competition

The team was also advised by in the area of vision processing by Prof. C.J. Taylor and in the area of simulation and strategies by Prof. Vijay Kumar.

Web page ftp://ftp.cis.upenn.edu/pub/extra/RoboCup99/public_html/index.html

3 Complete Robot Architecture

The robot architecture we developed is a mode-based architecture in which higher level modes, such as going to the ball or kicking, are chosen deterministically based on the current and predicted state. This architecture is described in more detail in [1]. We have divided the system into four basic components, as shown in Figure 1, and described individually below:

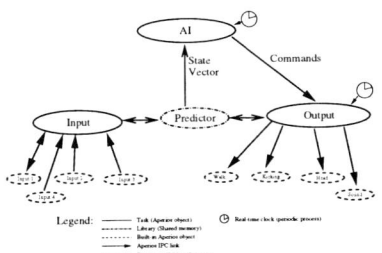

Fig. 1. Software architecture for our team

Input task: The input task is made up of several objects with responsibilities for reading the various sensors. Each of these objects sends messages to the *Predictor* object when they have received (and processed) their data. For example, the object handling image input processes the image to locate the position of items of interest in the image (ball, goals, markers ...) and sends only the condensed information (ball position, etc.) to the *Predictor*.

Predictor task: The *Predictor* task collects and fuses the information coming from the sensors. Confidence values are also maintained for each piece of information, so as to allow for decisions to be made based on how "fresh" the data is. The *Predictor* also performs any necessary filtering, especially Kalman filtering for prediction of object positions at future times (hence the name). All objects in the system have access to a pointer to the shared memory *Predictor* that gives them (read-only) access to the state of the system.

AI task: The decision-making task (or *AI* task for short) performs all high-level decision-making for the robot. Using the state information in the *Predictor*, and based on past history implemented as a finite state machine, the AI is able to determine what its current obligations and goals (approaching ball, attacking goal, etc.) should be. Based on its decisions, the AI sends high-level *Commands* to the *Output* task for processing. The AI is also responsible for determining if the robot is "lost"– a state that means no useful information about ball or self position is known. At this time, the AI moves into a search and localize mode.

Output task: The *Output* task carries out operations contained in the high-level *Commands* sent by the AI. These *Commands* are at the level of "move to the ball", or "orbit the ball until facing goal". They require an "inner control loop" to be carried out. The *Output* task, therefore, receives timer messages and carries out the appropriate actions at each time step for the current *Command* being processed.

4 Vision

There are three primary concepts that were involved in our use of the vision sensor for RoboCup99: making color tables, blob and marker detection, and distance and orientation computations.

Making color tables: The development of the color tables for pre-processing of color images was quite involved. Our basic algorithms relied on making masks (by hand) of multiple images which included different lighting conditions and views of the colored markers and goals. From these masks, a weighted index was formed which was then used to generate by hand the actual color boundaries for each color.

Blob and marker detection: In order to perform a variety of tasks, including finding the ball or the goal and performing self-localization, the ability to locate, isolate, and even track colored blobs was essential. To this end, we developed a set of blob tracker routines. The blob detection routines (called the "Multi-blob Tracker") use the images extracted using the CDT hardware, and so work with color segmented images.

The algorithm that we used sought a compromise between computational efficiency and a robust and thorough search. To detect the blobs, we first scan the picture horizontally and sum each of the columns in the image to determine whether there are objects in the horizontal projection of the image. In other words, we project the blobs in the image onto a single horizontal row, and then detect blobs of point on this 1-D set (see Figure 2 below). Then the same thing is done with the vertical pixels, so that we get a set of clumps of pixels (representing objects projected horizontally from the image) in a single column. We then do some simple matching to determine where the actual objects lie, and to compute centroid location and horizontal and vertical sizes of the object. In the scenarios we encountered, it was generally the case that there were not multiple objects of the same color aligned vertically, so this algorithm worked quite well.

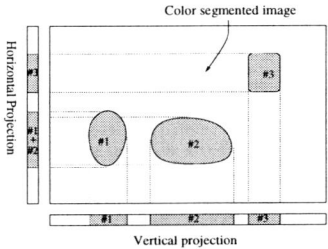

Fig. 2. Projections for multiblob detection

On top of the simple blob detection, we also do feature matching to determine which blobs correspond to markers, and which are other objects, such as the goal. This is done by stepping through the different blob location data structures and performing matching based on known marker geometry and colors. Thus, the Multi-blob Tracker routines return both the locations and sizes of all the blobs that are seen, as well as the locations of each of the markers.

Distance and orientation computations: For determining distance to the ball and the marker, we used a simple calibration lookup table, which we found to be reasonably reliable. We independently calibrated for both the width and height of the object, to provide additional robustness. For orientation, the angular heading to the target was computed using the pixel location of the centroid of the object in the image plane and the current configuration of the head/neck motors. Based on some simple geometry, this gave a good calculation of the actual bearing of the object to the robot.

5 Control

For low-level control, we utilized the walking and posture control algorithms given to us by Sony, with some minor modifications in how steps were executed. One of our innovations was in generalized walking, where we implemented a routine that would allows us to walk around the ball. This was done by side-stepping until the ball was at a large enough angle relative to the head of the robot, and then rotating until the ball again was closer to straight ahead. Combining these two behaviors generated a net effect of rotating around the ball, which worked quite well in most circumstances.

6 Localization

In the attacking strategy, localization was limited to determining the heading to the goal we were attacking. We also developed localization routines that weren't used in the tournament, which were based on a well-known surveyor's algorithm for determining location from the bearing to three known markers. We use this algorithm because it relies on the minimal needed information (scanning for markers can be time-consuming, and so we try to get away with as few as possible), and it relies only on angular heading and not distance. We have tried to use angular information whenever possible, since it is well-known that estimating distance using vision can be very error prone and particularly sensitive to occlusions.

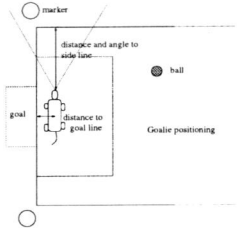

Fig. 3. Geometry for goalie stance

The goalie localization, on the other hand, was based on detecting the goal line and observing the distance to the corner marker (see Figure 3). For the goal line, the goalie would periodically look down towards its front feet. If it was the correct distance from the goal line, then the color detection algorithm should pick up the goal line a certain distance from the center of the image. Furthermore, a least-squares fit was used to calculate the slope (and offset) of the line that matched the goal-line. This fit would tell us at what angle (and distance) the robot was standing with respect to the goal line. Corrections could then be made if the orientation or offset error was significant enough. For keeping its position along the goal mouth, the robot generally used dead reckoning (step counting) to determine whether it was to the left, middle, or right of the goal. Periodically, the goalie would look up to find the corner marker. Using the pixel size of the marker in the image, along with a calibration lookup table, the robot would determine the distance to the corner, and hence its own position in the goal mouth.

7 Strategies: Attacker and Goalkeeper

Our overall strategy focused on two areas: attacking with the ball towards the goal and defending the goal.

Algorithm: Attacker (code-name Predictor-Writer) The strategy of the attacker was kept quite simple– find the ball, approach it, walk around it until the ball was between the robot and the goal (preferably an opening to one side of the goal keeper), and then charge towards the goal. For this, we used a simple scan routine to locate the ball, and Sony's walking routines to move us towards the ball. Once in close enough range, we utilized our own walking modification to allow us to move around the ball. As we walked around the ball, we continuously looked for the goal that we were to be attacking. Once this was found, the robot would try to dribble forward towards the goal, primarily relying on bumping the ball forward with its body.

One important lesson learned during the games was that in most situations it was not possible to get fully around the ball, so that the ball would be between the robot and the goal. This was due to the fact that there was very often some object blocking the robot from completing this maneuver, whether it was the side wall (particularly in the corners), or another robot from either team. For this reason, we implemented a "time-out" condition that would have the robot drive the ball forward if it had been trying to move around the ball for an extended period of time.

Algorithm: Goalie (code-name Stonewall) Our goalie got its nickname (Stonewall) because it was very good at getting in the way of the ball and staying in the way. It also tended to move a little slower than some goalies. The basic strategy was to keep the body of the robot faced parallel to the goal line, so as to present the maximum cross-sectional area for defense. Also, the goalie always stayed in a range having the middle of its body aligned with limits at the right goal post and the left goal post. The main (and very simple) idea was to keep the robot in front of the ball as long as the ball was in view. When the ball went out of view for any significant amount of time, the goalie would then retreat to the back side (right side) of the goal, so as to guarantee that the ball couldn't be slipped behind the goalkeeper.

8 Conclusion

We plan on entering a team for next year's competition (RoboCup2000) in Australia. There are many components that we will focus on for next year, and we are extremely excited to have enough time to work on developing many different areas. Some of the focal points will be: inter-robot communication using audio, team strategies, dynamic tracking of markers used for estimating position, and generalized walking gaits for moving in arbitrary directions.

References

1. K. A. McIsaac, A. K. Das, J. M. Esposito, and J. P. Ostrowski. A hierarchical structure for hybrid control applied to AIBO robotic soccer. In *Proc. IEEE/RSJ Intl. Conf. on Intelligent Robots and Systems (IROS)*, Japan, 2000. (Submitted).

Author Index

Achim, S. 661
Ackers, P. 186, 667
Adibi, J. 707
Adobbati, R. 460, 707
Adorni, G. 695
Akita, J. 128, 671
Amirat, Y. 390, 735
Amoroso, C. 136
Aono, M. 671
Aoyagi, H. 649
Aparício, P. 378, 715
Asada, M. 1, 519, 750, 762
Asai, N. 608

Balch, T. 1
Baltes, J. 148, 162, 653
Bandlow, T. 174, 691
Becht, M. 699
Behnke, S. 186, 667
Belpaeme, T. 687
Birk, A. 196, 687
Bittencourt, G. 274
Blazevic, P. 74
Boß, T. 638
Boman, M. 642, 784
Bonarini, A. 210, 695
Bonnin, P. 74
Borrajo, D. 292
Bowling, M. 222, 661
Braxmeier, H. 638
Bredenfeld, A. 231, 711
Bremer, B. 481
Bruce, J. 766
Brusey, J. 243, 741
Buchheim, T. 699
Buck, S. 588
Burkhard, H.-D. 531, 542, 770
Butler, M. 546

Cassinis, R. 254
Chang, C. F. 572
Chella, A. 136, 695
Chiniforooshan, E. 61, 745
Chitsaz, H. 61, 745
Christaller, T. 231

Cisternino, A. 263, 580
Clemente, G. 695
Coradeschi, S. 1
da Costa, A. L. 274
Costa, I. 731
Costa, Paulo 286, 663, 754
Costa, Pedro 286, 663, 754
Cremean, L. B. 49, 657
da Cruz, J. J. 439
Custódio, L. 378, 715

D'Andrea, R. 49, 657
Dalgliesh, J. 788
Demura, K. 723
Dettinger, M. 638
Dietz, M. 638
Dilger, S. 588
Dorer, K. 600
Du, X. 608
Düffert, U. 770
Duhaut, D. 74

Ebina, A. 509, 727
Ehrmann, R. 588
Ejima, T. 719
Enderle, S. 638
Enokida, S. 719

Fernández, F. 292
Ferrari, C. 618
Figueras, A. 434, 679
Folkerts, H. 638
Foroughnassiraei, A. 61, 745
Frank, I. 114
Frommberger, L. 588
Frötschl, B. 186, 667
Fujita, M. 1
Fujiwara, H. 509, 727
Fukuda, M. 719

García, R. 679
Garelli, F. 618
Georgi, A. 770
Ghorbani, R. 61, 745
Ghose, A. 572
Giebel, J. 638

Girault, F. 567
Göhring, W. 231, 711
González, E. 390, 735
Günther, H. 231, 711
Günther, R. 424
Gutmann, J.-S. 304, 703
Guttmann, C. 784

Hanek, R. 174, 691
Hannebauer, M. 531
Harvey, P. 572
Hatzack, W. 703
Hecht, H. 331
Heintz, F. 563
Hetzel, G. 699
Hildreth, N. 653
Hiroshima, K. 649
Hoffman, A. 49, 657
Hofmann, A. 588
Howard, T. 546
Hu, H. 366, 558
Hugel, V. 74
Hunter, M. 558

Igarashi, Harukazu 596
Igarashi, Hiroki 723
Ihlenburg, J. 424
Innocenti, B. 434, 632, 679
Iocchi, L. 318
Ishihara, D. 723
Ishii, N. 608
Ishiwata, K. 675
Ishizuka, H. 750
Ito, N. 608

Jaeger, H. 231, 711
Jamzad, M. 61, 745
Jennings, A. 741
Jung, B. 331
Jung, D. 424

Kaiser, J. 638
Kaminka, G. A. 345
Karlsson, L. 1
Karpati, T. 49, 657
Kato, T. 750
Kawanishi, H. 750
Kawarabayashi, T. 649, 671
Kazemi, M. 61, 745
Keen, C. 741

Kemppainen, J. 611
Kendall, A. 741
Kenn, H. 196, 687
Khessal, N. O. 357, 683
Kindermann, G. 699
Kinoshita, S. 550
Kitano, H. 1
Klupsch, M. 174, 691
Kobayashi, Y. 758
Kobialka, H.-U. 231, 711
Kostiadis, K. 366, 558
Kosue, S. 596
Kozhushkin, A. N. 623
Kraetzschmar, G. 1, 638
Kubo, T. 649
Kummeneje, J. 642
Kylmäoja, J. 611

Lafrenz, R. 699
Lawther, M. 788
LeBlanc, K. 784
Lee, J.-W. 49, 657
Lenser, S. 766
Levi, P. 699
Lima, P. 378, 715
Lin, Y. 162
Lindstrot, W. 186, 667
Lipman, J. 572
Livani, M. 638
Loaiza, H. 390, 735
Lönneberg, M. 576
Lybäck, D. 642

Machado, C. 731
Makies, M. 741
Maplesden, D. 653
Marques, C. 715
Marques, P. 286, 663, 754
Marsella, S. 85
Maschke, D. 638
Matos, A. 663, 754
Matsubara, H. 114
Matsumura, T. 554
Meinert, T. 531, 542
Meis, U. 638
de Melo, M. 186, 667
Merke, A. 588
Mitsunaga, N. 762
Miwa, K. 723
Mobasser, F. 61, 745

Mochizuki, K. 509
Modi, J. 460, 707
Montaner, M. 434, 632
Moradi, H. 707
Moreira, A. 286, 663, 754
Moreno, C. 390, 735
Mori, Y. 675
Morishita, T. 649
Morreale, V. 136
Muñoz, I. 434, 632, 679
Murakami, K. 675
Murray, J. 481, 628
Muscholl, M. 699
Myritz, H. 531, 542, 770

Nagami, M. 675
Nagasaka, Y. 675
Nakagawa, K. 608
Nakamura, T. 509, 727
Nardi, D. 318, 695
Naruse, T. 675
Nebel, B. 304, 703
Neubeck, A. 638
Nilsson, P. 576
Nishida, T. 509
Nishino, J. 649, 671
Noda, I. 114, 400, 584

Obst, O. 481, 628
Oesker, M. 331
Ogura, H. 649
Ohashi, T. 719
Ostrowski, J. P. 792
Oswald, N. 699

Padgham, L. 243, 741
Pagello, E. 1, 618, 695
Palm, G. 638
Piaggio, M. 412, 695
Plagge, M. 424
Plöger, P.-G. 231, 711
Polani, D. 604
Pourazin, S. 614
Prokopenko, M. 546

Raines, T. 85
Räsänen, J. 611
Ramon, J. A. 434, 679
Raulet, L. 74
Reali Costa, A. H. 439

Ribeiro, F. 731
Riedmiller, M. 588
Riley, P. 35
Ritter, M. 638
Ritzschke, M. 770
Rizzi, A. 254
Rojas, R. 186, 667
de la Rosa, J. L. 434, 632, 679

Sablatnög, S. B. 638
Sadjad, S. B. 61, 745
Saffiotti, A. 784
Saka, T. 671
Sakurai, T. 596
Salemi, B. 460, 707
Samad-Yahaya, A. 49, 657
Sampaio, S. 731
Sander, G. 531, 542
Scardua, L. A. 439
Scerri, P. 450, 576
Schaeffer, P. 638
Schanz, M. 699
Schebesch, A. 186, 667
Schmitt, T. 174, 691
Schöll, P. 231, 711
Schulé, M. 699
Seabrook, M. 558
Sese, J. 671
Sgorbissa, A. 412
Shen, W.-M. 460, 707
Shimora, H. 649
Shinjoh, A. 469
Siegberg, A. 231, 711
Simi, M. 263, 580
Simon, M. 186, 667
Singh, D. 741
Sinner, A. 588
Sousa, A. 286, 663, 754
Sprengel, M. 186, 667
Stinckwich, S. 567
Stolzenburg, F. 481, 628
Stone, P. 1, 35, 495
Storniolo, P. 136
Streit, A. 231, 711
Suárez, A. 390, 735
Suzuki, S. 750
Suzuki, T. 646

Takahashi, T. 103, 592, 675
Takahashi, Y. 750

Takeda, H. 509, 727
Tambe, M. 85
Tamura, T. 750
Tanaka-Ishii, K. 114
Tejada, S. 707
Tenchio, O. 186, 667
Terada, K. 509, 727
Thate, O. 588

Uchibe, E. 519, 750
Ueno, A. 509
Unger, R. 774
Uthmann, T. 604

Veloso, M. 1, 35, 222, 495, 661, 766
Ventura, R. 378, 715
Verbeek, C. 231, 711
Voutilainen, V. 611

Walle, T. 687

Weigel, T. 304
Wendler, J. 531, 542, 770
Werner, M. 770
Wilberg, J. 231, 711
Winkler, F. 770
Winner, E. 766
Wiren, T. 576
Wong, W. Y. 546

Yamamoto, Y. 550
Yanase, M. 750
Ydrén, J. 450, 576
Yoshida, S. 469
Yoshida, T. 719
Younes, H. 642
Yuasa, H. 758

Zaccaria, R. 412
Zell, A. 424

Lecture Notes in Artificial Intelligence (LNAI)

Vol. 1739: A. Braffort, R. Gherbi, S. Gibet, J. Richardson, D. Teil (Eds.), Gesture-Based Communication in Human-Computer Interaction. Proceedings, 1999. XI, 333 pages. 1999.

Vol. 1744: S. Staab, Grading Knowledge: Extracting Degree Information from Texts. X, 187 pages. 1999.

Vol. 1747: N. Foo (Ed.), Adavanced Topics in Artificial Intelligence. Proceedings, 1999. XV, 500 pages. 1999.

Vol. 1757: N.R. Jennings, Y. Lespérance (Eds.), Intelligent Agents VI. Proceedings, 1999. XII, 380 pages. 2000.

Vol. 1759: M.J. Zaki, C.-T. Ho (Eds.), Large-Scale Parallel Data Mining. VIII, 261 pages. 2000.

Vol. 1760: J.-J. Ch. Meyer, P.-Y. Schobbens (Eds.), Formal Models of Agents. Poceedings. VIII, 253 pages. 1999.

Vol. 1761: R. Caferra, G. Salzer (Eds.), Automated Deduction in Classical and Non-Classical Logics. Proceedings. VIII, 299 pages. 2000.

Vol. 1771: P. Lambrix, Part-Whole Reasoning in an Object-Centered Framework. XII, 195 pages. 2000.

Vol. 1772: M. Beetz, Concurrent Reactive Plans. XVI, 213 pages. 2000.

Vol. 1775: M. Thielscher, Challenges for Action Theories. XIII, 138 pages. 2000.

Vol. 1778: S. Wermter, R. Sun (Eds.), Hybrid Neural Systems. IX, 403 pages. 2000.

Vol. 1788: A. Moukas, C. Sierra, F. Ygge (Eds.), Agent Mediated Electronic Commerce II. IX, 239 pages. 2000.

Vol. 1792: E. Lamma, P. Mello (Eds.), AI*IA 99: Advances in Artificial Intelligence. Proceedings, 1999. XI, 392 pages. 2000.

Vol. 1793: O. Cairo, L.E. Sucar, F.J. Cantu (Eds.), MICAI 2000: Advances in Artificial Intelligence. Proceedings, 2000. XIV, 750 pages. 2000.

Vol. 1794: H. Kirchner, C. Ringeissen (Eds.), Frontiers of Combining Systems. Proceedings, 2000. X, 291 pages. 2000.

Vol. 1804: B. Azvine, N. Azarmi, D.D. Nauck (Eds.), Intelligent Systems and Soft Computing. XVII, 359 pages. 2000.

Vol. 1805: T. Terano, H. Liu, A.L.P. Chen (Eds.), Knowledge Discovery and Data Mining. Proceedings, 2000. XIV, 460 pages. 2000.

Vol. 1809: S. Biundo, M. Fox (Eds.), Recent Advances in AI Planning. Proceedings, 1999. VIII, 373 pages. 2000.

Vol. 1810: R. López de Mántaras, E. Plaza (Eds.), Machine Learning: ECML 2000. Proceedings, 2000. XII, 460 pages. 2000.

Vol. 1813: P.L. Lanzi, W. Stolzmann, S.W. Wilson (Eds.), Learning Classifier Systems. X, 349 pages. 2000.

Vol. 1821: R. Loganantharaj, G. Palm, M. Ali (Eds.), Intelligent Problem Solving. Proceedings, 2000. XVII, 751 pages. 2000.

Vol. 1822: H.H. Hamilton, Advances in Artificial Intelligence. Proceedings, 2000. XII, 450 pages. 2000.

Vol. 1831: D. McAllester (Ed.), Automated Deduction – CADE-17. Proceedings, 2000. XIII, 519 pages. 2000.

Vol. 1834: J.-C. Heudin (Ed.), Virtual Worlds. Proceedings, 2000. XI, 314 pages. 2000.

Vol. 1835: D. N. Christodoulakis (Ed.), Natural Language Processing – NLP 2000. Proceedings, 2000. XII, 438 pages. 2000.

Vol. 1836: B. Masand, M. Spiliopoulou (Eds.), Web Usage Analysis and User Profiling. Proceedings, 2000. V, 183 pages. 2000.

Vol. 1847: R. Dyckhoff (Ed.), Automated Reasoning with Analytic Tableaux and Related Methods. Proceedings, 2000. X, 441 pages. 2000.

Vol. 1849: C. Freksa, W. Brauer, C. Habel, K.F. Wender (Eds.), Spatial Cognition II. XI, 420 pages. 2000.

Vol. 1856: M. Veloso, E. Pagello, H. Kitano (Eds.), RoboCup-99: Robot Soccer World Cup III. XIV, 802 pages. 2000.

Vol. 1860: M. Klusch, L. Kerschberg (Eds.), Cooperative Information Agents IV. Proceedings, 2000. XI, 285 pages. 2000.

Vol. 1861: J. Lloyd, V. Dahl, U. Furbach, M. Kerber, K.-K. Lau, C. Palamidessi, L. Moniz Pereira, Y. Sagiv, P.J. Stuckey (Eds.), Computational Logic – CL 2000. Proceedings, 2000. XIX, 1379 pages.

Vol. 1864: B. Y. Choueiry, T. Walsh (Eds.), Abstraction, Reformulation, and Approximation. Proceedings, 2000. XI, 333 pages. 2000.

Vol. 1865: K.R. Apt, A.C. Kakas, E. Monfroy, F. Rossi (Eds.), New Trends Constraints. Proceedings, 1999. X, 339 pages. 2000.

Vol. 1866: J. Cussens, A. Frisch (Eds.), Inductive Logic Programming. Proceedings, 2000. X, 265 pages. 2000.

Vol. 1867: B. Ganter, G.W. Mineau (Eds.), Conceptual Structures: Logical, Linguistic, and Computational Issues. Proceedings, 2000. XI, 569 pages. 2000.

Vol. 1881: C. Zhang, V.-W. Soo (Eds.), Design and Applications of Intelligent Agents. Proceedings, 2000. X, 183 pages. 2000.

Vol. 1886: R. Mizoguchi, J. Slaney (Eds.), PRICAI 2000: Topics in Artificial Intelligence. Proceedings, 2000. XX, 835 pages. 2000.

Vol. 1889: M. Anderson, P. Cheng, V. Haarslev (Eds.), Theory and Application of Diagrams. Proceedings, 2000. XII, 504 pages. 2000.

Lecture Notes in Computer Science

Vol. 1852: T. Thierauf: The Computational Complexity of Equivalence and Isomorphism Problems. VIII, 135 pages. 2000.

Vol. 1853: U. Montanari, J.D.P. Rolim, E. Welzl (Eds.), Automata, Languages and Programming. Proceedings, 2000. XVI, 941 pages. 2000.

Vol. 1854: G. Lacoste, B. Pfitzmann, M. Steiner, M. Waidner (Eds.), SEMPER — Secure Electronic Marketplace for Europe. XVIII, 350 pages. 2000.

Vol. 1855: E.A. Emerson, A.P. Sistla (Eds.), Computer Aided Verification. Proceedings, 2000. X, 582 pages. 2000.

Vol. 1856: M. Veloso, E. Pagello, H. Kitano (Eds.), RoboCup-99: Robot Soccer World Cup III. XIV, 802 pages. 2000. (Subseries LNAI).

Vol. 1857: J. Kittler, F. Roli (Eds.), Multiple Classifier Systems. Proceedings, 2000. XII, 404 pages. 2000.

Vol. 1858: D.-Z. Du, P. Eades, V. Estivill-Castro, X. Lin, A. Sharma (Eds.), Computing and Combinatorics. Proceedings, 2000. XII, 478 pages. 2000.

Vol. 1860: M. Klusch, L. Kerschberg (Eds.), Cooperative Information Agents IV. Proceedings, 2000. XI, 285 pages. 2000. (Subseries LNAI).

Vol. 1861: J. Lloyd, V. Dahl, U. Furbach, M. Kerber, K.-K. Lau, C. Palamidessi, L. Moniz Pereira, Y. Sagiv, P.J. Stuckey (Eds.), Computational Logic – CL 2000. Proceedings, 2000. XIX, 1379 pages. 2000. (Subseries LNAI).

Vol. 1862: P.G. Clote, H. Schwichtenberg (Eds.), Computer Science Logic. Proceedings, 2000. XIII, 543 pages. 2000.

Vol. 1863: L. Carter, J. Ferrante (Eds.), Languages and Compilers for Parallel Computing. Proceedings, 1999. XII, 500 pages. 2000.

Vol. 1864: B. Y. Choueiry, T. Walsh (Eds.), Abstraction, Reformulation, and Approximation. Proceedings, 2000. XI, 333 pages. 2000. (Subseries LNAI).

Vol. 1865: K.R. Apt, A.C. Kakas, E. Monfroy, F. Rossi (Eds.), New Trends Constraints. Proceedings, 1999. X, 339 pages. 2000. (Subseries LNAI).

Vol. 1866: J. Cussens, A. Frisch (Eds.), Inductive Logic Programming. Proceedings, 2000. X, 265 pages. 2000. (Subseries LNAI).

Vol. 1867: B. Ganter, G.W. Mineau (Eds.), Conceptual Structures: Logical, Linguistic, and Computational Issues. Proceedings, 2000. XI, 569 pages. 2000. (Subseries LNAI).

Vol. 1868: P. Koopman, C. Clack (Eds.), Implementations of Functional Languages. Proceedings, 1999. IX, 199 pages. 2000.

Vol. 1869: M. Aagaard, J. Harrison (Eds.), Theorem Proving in Higher Order Logics. Proceedings, 2000. IX, 535 pages. 2000.

Vol. 1872: J. van Leeuwen, O. Watanabe, M. Hagiya, P.D. Mosses, T. Ito (Eds.), Theoretical Computer Science. Proceedings, 2000. XV, 630 pages. 2000.

Vol. 1876: F. J. Ferri, J.M. Iñesta, A. Amin, P. Pudil (Eds.), Advances in Pattern Recognition. Proceedings, 2000. XVIII, 901 pages. 2000.

Vol. 1877: C. Palamidessi (Ed.), CONCUR 2000 – Concurrency Theory. Proceedings, 2000. XI, 612 pages. 2000.

Vol. 1878: J.P. Bowen, S. Dunne, A. Galloway, S. King (Eds.), ZB 2000: Formal Specification and Development in Z and B. Proceedings, 2000. XIV, 511 pages. 2000.

Vol. 1879: M. Paterson (Ed.), Algorithms – ESA 2000. Proceedings, 2000. IX, 450 pages. 2000.

Vol. 1880: M. Bellare (Ed.), Advances in Cryptology – CRYPTO 2000. Proceedings, 2000. XI, 545 pages. 2000.

Vol. 1881: C. Zhang, V.-W. Soo (Eds.), Design and Applications of Intelligent Agents. Proceedings, 2000. X, 183 pages. 2000. (Subseries LNAI).

Vol. 1883: B. Triggs, A. Zisserman, R. Szeliski (Eds.), Vision Algorithms: Theory and Practice. Proceedings, 1999. X, 383 pages. 2000.

Vol. 1886: R. Mizoguchi, J. Slaney (Eds.), PRICAI 2000: Topics in Artificial Intelligence. Proceedings, 2000. XX, 835 pages. 2000. (Subseries LNAI).

Vol. 1889: M. Anderson, P. Cheng, V. Haarslev (Eds.), Theory and Application of Diagrams. Proceedings, 2000. XII, 504 pages. 2000. (Subseries LNAI).

Vol. 1892: P. Brusilovsky, O. Stock, C. Strapparava (Eds.), Adaptive Hypermedia and Adaptive Web-Based Systems. Proceedings, 2000. XIII, 422 pages. 2000.

Vol. 1893: M. Nielsen, B. Rovan (Eds.), Mathematical Foundations of Computer Science 2000. Proceedings, 2000. XIII, 710 pages. 2000.

Vol. 1896: R. W. Hartenstein, H. Grünbacher (Eds.), Field-Programmable Logic and Applications. Proceedings, 2000. XVII, 856 pages. 2000.

Vol. 1897: J. Gutknecht, W. Weck (Eds.), Modular Programming Languages. Proceedings, 2000. XII, 299 pages. 2000.

Vol. 1899: H.-H. Nagel, F.J. Perales López (Eds.), Articulated Motion and Deformable Objects. Proceedings, 2000. X, 183 pages. 2000.

Vol. 1900: A. Bode, T. Ludwig, W. Karl, R. Wismüller (Eds.), Euro-Par 2000 Parallel Processing. Proceedings, 2000. XXXV, 1368 pages. 2000.

Vol. 1912: Y. Gurevich, P.W. Kutter, M. Odersky, L. Thiele (Eds.), Abstract State Machines. Proceedings, 2000. X, 381 pages. 2000.

Vol. 1913: K. Jansen, S. Khuller (Eds.), Approximation Algorithms for Combinatorial Optimization. Proceedings, 2000. IX, 275 pages. 2000.